Lecture Notes in Computer Science   9301

Commenced Publication in 1973
Founding and Former Series Editors:
Gerhard Goos, Juris Hartmanis, and Jan van Leeuwen

More information about this series at http://www.springer.com/series/7407

Vladimir P. Gerdt · Wolfram Koepf
Werner M. Seiler · Evgenii V. Vorozhtsov (Eds.)

Computer Algebra in Scientific Computing

17th International Workshop, CASC 2015
Aachen, Germany, September 14–18, 2015
Proceedings

 Springer

Editors
Vladimir P. Gerdt
Laboratory of Information Technologies
Joint Institute of Nuclear Research
Dubna
Russia

Wolfram Koepf
Institute for Mathematics
Universität Kassel
Kassel
Germany

Werner M. Seiler
Institute for Mathematics
Universität Kassel
Kassel
Germany

Evgenii V. Vorozhtsov
Russian Academy of Sciences
Institute of Theoretical
  and Applied Mechanics
Novosibirsk
Russia

ISSN 0302-9743   ISSN 1611-3349   (electronic)
Lecture Notes in Computer Science
ISBN 978-3-319-24020-6   ISBN 978-3-319-24021-3   (eBook)
DOI 10.1007/978-3-319-24021-3

Library of Congress Control Number: 2015949455

LNCS Sublibrary: SL1 – Theoretical Computer Science and General Issues

Springer Cham Heidelberg New York Dordrecht London
© Springer International Publishing Switzerland 2015

Printed on acid-free paper

Springer International Publishing AG Switzerland is part of Springer Science+Business Media
(www.springer.com)

Preface

The ongoing progress both in theoretical computer algebra and in its expanding applications has led to a need for forums bringing together both the scientists working in the area of computer algebra methods and systems and the researchers who apply the tools of computer algebra for the solution of problems in scientific computing, in order to foster new and closer interactions. This need has led to the series of CASC (Computer Algebra in Scientific Computing) workshops, which started in 1998 and since then has been held annually.

This year the seventeenth CASC conference takes place in Aachen (Germany). Computer algebra in Aachen has a long and a fruitful history.

RWTH Aachen University is the birthplace of the system GAP: Groups, Algorithms and Programming (www.gap-system.org). GAP was initiated in 1986 by J. Neubüser and was managed by himself until his retirement in 1997. It is one of the most influential computer algebra systems oriented towards computational group theory. The big number of GAP citations from applied sciences alone confirms its widely recognized interdisciplinary importance.

It is interesting to know that several predecessors of GAP and – to some extent – also the system MAGMA, including the system CAYLEY, were developed with very active participation of people from Aachen. Moreover, developments "made in Aachen" include the popular package MEATAXE by M. Ringe for working with matrix representations over finite fields (www.math.rwth-aachen.de/homes/MTX/), the package CHEVIE, providing symbolic calculations with generic character tables of groups of Lie type, Coxeter groups, Iwahori-Hecke algebras and other related structures, and the Modular Atlas project (www.math.rwth-aachen.de/~MOC/) by G. Hiss, F. Lübeck, and others.

Over the years, a number of packages for MAGMA (magma.maths.usyd.edu.au/magma/) have been written in Aachen as well.

In the group led by W. Plesken, both GAP and MAPLE (www.maplesoft.com/) have been used and developed further from very early times on. Numerous computer algebra packages evolved, notably the MAPLE suite OREMODULES (wwwb.math.rwth-aachen.de/OreModules/) by D. Robertz (RWTH) and A. Quadrat (INRIA), which includes several subpackages written not only by the authors. OREMODULES provides rich functionality for linear control systems over noncommutative Ore algebras.

In addition, a notable implementation of the involutive algorithms, designed by V.P. Gerdt (Dubna) and Yu.A. Blinkov (Saratov) and based on Janet's monomial division, has been done in this group. The MAPLE package INVOLUTIVE (wwwb.math.rwth-aachen.de/Janet/involutive.html) computes Janet bases and Janet-like Gröbner bases for submodules of free modules over a polynomial ring, while JANET (wwwb.math.rwth-aachen.de/Janet/janet.html) can do these computations for linear systems of partial differential equations and LDA

(LINEAR DIFFERENCE ALGEBRA) for linear difference systems. A further MAPLE package called JANETORE (wwwb.math.rwth-aachen.de/Janet/janetore.html) can carry out these computations for submodules of free left modules over certain noncommutative Ore algebras. More recently, two algorithms for the triangular Thomas decomposition of both algebraic (ALGEBRAICTHOMAS) and differential (DIFFERENTIALTHOMAS) systems of equations and inequations, respectively, were designed and implemented in MAPLE by T. Bächler, M. Lange-Hegermann, and D. Robertz in cooperation with V.P. Gerdt (wwwb.math.rwth-aachen.de/thomasdecomposition/).

The open source project GINV (invo.jinr.ru/ginv/index.html), which implements in C++ the involutive algorithm by V.P. Gerdt and Y.A. Blinkov for polynomial systems, was also partially developed in Aachen.

M. Barakat initiated the homalg project (homalg.math.rwth-aachen.de/) for constructive homological algebra, which has evolved into a multi-author multi-package open source software project. The project exploits many capabilities of the GAP4 programming language (the object model, the method selection mechanism, and the deduction system). It implements on one hand high-level algorithms for constructive Abelian categories, e.g., spectral sequences of bicomplexes, and on the other hand concrete realizations of such constructive categories, e.g., categories of finitely presented modules over the wide class of so-called computable rings. The ring arithmetic and all matrix operations over such rings are delegated to other dedicated computer algebra systems.

The group of E. Zerz and V. Levandovskyy works with SINGULAR (www.singular.uni-kl.de) and has been implementing various algebraic tools towards, among other things, system and control theory. V. Levandovskyy has furthermore been developing the noncommutative extensions PLURAL (for a broad class of Noetherian domains) and LETTERPLACE (for free associative algebras) of SINGULAR, which provide extended functionality, based on Gröbner bases, for various noncommutative rings. In particular, several libraries for SINGULAR:PLURAL contain sophisticated implementations of fundamental algorithms for algebraic D-module theory.

The above-listed impressive activities in several wide areas of computer algebra have predetermined, to a large extent, the choice of Aachen as a venue for the CASC 2015 workshop.

This volume contains 33 full papers submitted to the workshop by the participants and accepted by the Program Committee after a thorough reviewing process. Additionally, the volume includes two invited talks.

Polynomial algebra, which is at the core of computer algebra, is represented by contributions devoted to the computation of resolutions and Betti numbers with the aid of a combination of Janet bases and algebraic discrete Morse theory, automatic reasoning in reduction rings using the THEOREMA software system, estimation of the complexity of a new algorithm for recognizing a tropical linear variety, conversion of a zero-dimensional standard basis into a standard basis with respect to any other local ordering, simplification of cylindrical algebraic decomposition formulas with the aid of a new multi-level heuristic algorithm, solving polynomial systems with polynomial homotopy continuation, implementation of solvable polynomial rings in the object oriented computer algebra system JAS (Java Algebra System), explicit construction of the tangent cone of a variety using the theory of regular chains, computation of the limit points of the regular chain quasi-component by using linear changes of coordinates, obtaining new bounds for the largest positive root of a univariate polynomial with real coefficients, real root isolation by means of root radii approximation, and application of the algebra of resultants for distance evaluation between an ellipse and an ellipsoid.

Among the many existing algorithms for solving polynomial systems, perhaps the most successful numerical ones are the homotopy methods. The number of operations that these algorithms perform depends on the condition number of the roots of the polynomial system. Roughly speaking the condition number expresses the sensitivity of the roots with respect to small perturbations of the input coefficients. The invited talk by E. Tsigaridas deals with the problem of obtaining effective bounds for the condition number of polynomial systems with integer coefficients. The provided bounds depend on the number of variables, the degree, and the maximum coefficient bitsize of the input polynomials. Such bounds allow one to estimate the bit complexity of algorithms like the homotopy algorithms that depend on the condition number for solving polynomial systems.

Two papers deal with the solution of difference systems: hypergeometric solutions of first-order linear difference systems with rational-function coefficients; computation of regular solutions of linear difference systems with the aid of factorial series.

The topics of two further papers are related to problems in linear algebra and the theory of matrices: in one of them, it is proposed to perform a randomized preprocessing of Gaussian elimination without pivoting with the aid of circulants; the other paper proposes a new form of triangular decomposition of a matrix, which is termed the LDU-decomposition.

Several papers are devoted to using computer algebra for the investigation of various mathematical and applied topics related to ordinary differential equations (ODEs): obtaining the algebraic general solutions of first order algebraic ODEs, obtaining the analytic-form solutions of two-point boundary problems with one mild singularity with the aid of the Green's function and the *Theorema* software system, investigation of quasi-steady state phenomena arising in the solutions of parameter-dependent ODE systems, in particular, for reaction networks, symbolic computation of polynomial first integrals of systems of ODEs, and homotopy analysis of stochastic differential equations with maxima.

Two papers deal with applications of symbolic and symbolic-numeric computations for investigating and solving partial differential equations (PDEs) in mathematical physics: partial analytical solution of a PDE system from Kirchhoff's rod theory, symbolic-numeric solution of boundary-value problems for the Schrödinger equation.

Among the numerical methods to approximately solve partial differential equations on complicated domains, finite element methods are often the preferred tool. The invited talk by V. Pillwein shows how symbolic computations can be used in the construction of higher-order finite element methods. The analysis and construction methods range from Gröbner basis computations and cylindrical algebraic decomposition to algorithms for symbolic summation and integration.

Applications of symbolic and symbolic-numeric algorithms in mechanics and physics are represented by the following themes: investigation of the stability of relative equilibria of oblate axisymmetric gyrostat by means of symbolic-numerical modeling, investigation of the influence of constant torque on stationary motions of satellite, investigation of invariant manifolds and their stability in the problem of motion of a rigid body under the influence of two force fields, development of a symbolic algorithm for generating irreducible bases of point groups in the space of $SO(3)$ group for its application in molecular and nuclear physics, analysis of reaction network systems using tropical geometry, and approximate quantum Fourier transform and quantum algorithm for phase estimation.

The remaining topics include the use of resultants and cylindrical algebraic decomposition for the symbolic determination of the topology of a plane algebraic curve, the solution of the problem of interpolating the reduced data in cases when the interpolation knots are unknown, and safety verification of hybrid systems using certified multiple Lyapunov-like functions.

The CASC 2015 workshop was supported financially by a generous grant from Deutsche Forschungsgemeinschaft (DFG). Further financial support was obtained from our sponsors Additive, Dr. Hornecker Software-Entwicklung and IT-Dienstleistungen, and Maplesoft.

Our particular thanks are due to the members of the CASC 2015 local organizing committee at RWTH Aachen University, i.e., Eva Zerz and Viktor Levandovskyy, who provided us with the history of the computer algebra activities at RWTH and handled all the local arrangements in Aachen. Furthermore, we would like to thank all the members of the Program Committee for their thorough work. We are grateful to Matthias Seiß (Kassel University) for his technical help in the preparation of the camera-ready manuscript for this volume. Finally, we are grateful to the CASC publicity chair Andreas Weber (Rheinische Friedrich-Wilhelms-Universität Bonn) together with Hassan Errami for the design of the conference poster and the management of the conference web page.

July 2015   Vladimir P. Gerdt
                                                              Wolfram Koepf
                                                              Werner M. Seiler
                                                           Evgenii V. Vorozhtsov

Organization

CASC 2015 was organized jointly by the Institute of Mathematics at Kassel University and the Department of Mathematics at RWTH Aachen University.

Workshop General Chairs

Vladimir P. Gerdt, Dubna   Werner M. Seiler, Kassel

Program Committee Chairs

Wolfram Koepf, Kassel   Evgenii V. Vorozhtsov, Novosibirsk

Program Committee

Moulay Barkatou, Limoges
François Boulier, Lille
Hans-Joachim Bungartz, München
Jin-San Cheng, Beijing
Victor F. Edneral, Moscow
Dima Grigoriev, Lille
Jaime Gutierrez, Santander
Sergey A. Gutnik, Moscow
Jeremy Johnson, Philadelphia
Victor Levandovskyy, Aachen
Marc Moreno Maza, London, Canada

Alexander Prokopenya, Warsaw
Georg Regensburger, Linz
Eugenio Roanes-Lozano, Madrid
Valery Romanovski, Maribor
Markus Rosenkranz, Canterbury
Doru Stefanescu, Bucharest
Thomas Sturm, Saarbrücken
Jan Verschelde, Chicago
Stephen M. Watt,
    W. Ontario, Canada
Kazuhiro Yokoyama, Tokyo

Additional Reviewers

Parisa Alvandi
Serge Andrianov
Alexander Batkhin
Paola Boito
Charles Bouillaguet
Changbo Chen

Thomas Cluzeau
Xavier Dahan
Alicia Dickenstein
Xiaojie Dou
Stephane Gaubert
Domingo Gomez

Valentin D. Irtegov
Hidenao Iwane
Maximilian Jaroschek
Xiaohong Jia
Kai Jin
Marek Kosta
Ryszard Kozera
Franois Lemaire
Songxin Liang
Diane Maclagan
Gennadi Malaschonok
Bernard Mourrain

Ralf Mundani
Tobias Neckel
Masayuki Noro
Franois Ollivier
Franz Pauer
Marko Petkovsek
Clemens G. Raab
Daniel Robertz
Yosuke Sato
Takafumi Shibuta
Jacques-Arthur Weil
Tobias Weinzierl

Local Organization

Eva Zerz, Aachen

Viktor Levandovskyy, Aachen

Publicity Chair

Andreas Weber, Bonn

Website

http://www.casc.cs.uni-bonn.de/2015
(Webmaster: Hassan Errami)

Contents

Hypergeometric Solutions of First-Order Linear Difference Systems with Rational-Function Coefficients

S.A. Abramov[1,*], M. Petkovšek[2,**], and A.A. Ryabenko[1,*]

[1] Computing Centre of the Russian Academy of Sciences,
Vavilova, 40, Moscow 119991, Russia
sergeyabramov@mail.ru, anna.ryabenko@gmail.com
[2] University of Ljubljana, Faculty of Mathematics and Physics,
Jadranska 19, SI-1000 Ljubljana, Slovenia
Marko.Petkovsek@fmf.uni-lj.si

Abstract. Algorithms for finding hypergeometric solutions of scalar linear difference equations with rational-function coefficients are known in computer algebra. We propose an algorithm for the case of a first-order system of such equations. The algorithm is based on the resolving procedure which is proposed as a suitable auxiliary tool, and on the search for hypergeometric solutions of scalar equations as well as on the search for rational solutions of systems with rational-function coefficients. We report some experiments with our implementation of the algorithm.

1 Introduction

As a rule, both in scientific literature and in practice, algorithms for finding solutions of a certain kind for scalar differential or difference equations appear earlier than for systems of such equations. It may also be that a direct algorithm for systems is known in theory but does not yet have an available computer implementation (e.g., there is no such implementation in commonly used software packages). In this case, one makes an effort to find solutions of a system through some auxiliary scalar equations which are constructed for the system.

In [20, 18, 15], algorithms for finding hypergeometric solutions of scalar linear difference equations with rational-function coefficients were described. Using those algorithms, we propose below an algorithm to find hypergeometric solutions of linear normal first-order systems of the form $y(x+1) = A(x)y(x)$, where $A(x)$ is a square matrix whose entries are rational functions. Our algorithm differs somewhat from the algorithms based on the cyclic vector approach and is faster, as our experiments show. It is also worthy to note that even if $A(x)$ is singular, this is not an obstacle for our algorithm.

* Supported in part by the Russian Foundation for Basic Research, project no. 13-01-00182-a.
** Supported in part by the Ministry of Education, Science and Sport of Slovenia research programme P1-0294.

© Springer International Publishing Switzerland 2015
V.P. Gerdt et al. (Eds.): CASC 2015, LNCS 9301, pp. 1–14, 2015.
DOI: 10.1007/978-3-319-24021-3_1

Generally, direct algorithms work faster than algorithms that first uncouple the system. Thus, very likely, also the search for hypergeometric solutions of systems will become faster with the advent of full direct algorithms (it is known, e.g., that work is under way on such an algorithm for normal first-order systems with rational-function coefficients [11]). Until then, our algorithm can be useful for solving systems – all the more so since our experiments show that it works in reasonable time.

As an example of a computational problem which requires finding hypergeometric solutions of a first-order linear difference systems with rational-function coefficients we mention the important OPERATOR FACTORIZATION PROBLEM: Given a linear difference operator L of order n with rational-function coefficients and a positive integer $r < n$, find a linear difference operator R of order r with rational-function coefficients which divides L from the right, or prove that no such R exists. This problem can be solved by noticing that the $N = \binom{n}{r}$ maximal minors of the nth generalized Casoratian of a fundamental set of solutions of R must satisfy a system of first-order linear difference equations with rational-function coefficients easily obtainable from the coefficients of L, and that the coefficients of R must be proportional to certain $r + 1$ of these minors (cf. [13]).

2   The Problem

Let K be an algebraically closed field of characteristic 0. Denote by H_K the K-linear space of finite linear combinations of hypergeometric terms over K (i.e., $\frac{h(x+1)}{h(x)} \in K(x)$ for each hypergeometric term $h(x)$ under consideration) with coefficients in K.

Let E be the shift operator: $Ev(x) = v(x+1)$, and let $A(x)$ be an $m \times m$-matrix whose entries are in $K(x)$. We consider systems of the form

$$Ey = A(x)y, \quad y = (y_1(x), \ldots, y_m(x))^T, \tag{1}$$

and propose an algorithm which for a given system of the form (1) constructs a basis for the space of its solutions belonging to H_K^m. The basis consists of elements of the form

$$h(x)R(x), \tag{2}$$

where $h(x)$ is a hypergeometric term and $R(x) \in K(x)^m$.

We will say that an element of H_K^m is *related* to a hypergeometric term $h(x)$ if it can be represented in the form (2) (i.e., if each of its nonzero components is similar to $h(x)$).

For a system of the form (1), we will use the short notation $[A(x)]$.

3   The Reasoning Behind the Algorithm

3.1   The Resolving Equation and Matrix

Let y be any solution of (1). It follows by induction on j that

$$E^j y = \left(\prod_{i=1}^{j} E^{j-i} A(x) \right) y, \quad j = 0, 1, 2, \dots$$

Let $c \in K(x)^m$ be an arbitrary row vector, and $t = cy$ a scalar function. Then $E^j t = (E^j c)(E^j y) = c^{[j]} y$ where

$$c^{[j]} = (E^j c) \left(\prod_{i=1}^{j} E^{j-i} A(x) \right) \in K(x)^m, \quad j = 0, 1, 2, \dots,$$

are row vectors. We can construct the sequence $c^{[0]}, c^{[1]}, \dots$ step by step, using the recurrence relation

$$c^{[0]} = c, \quad c^{[i]} = (Ec^{[i-1]})A, \quad i = 1, 2, \dots \tag{3}$$

As $c^{[0]}, c^{[1]}, \dots, c^{[m]}$ are $m+1$ vectors of length m, they are linearly dependent over $K(x)$. Let $k \in \{0, 1, \dots, m\}$ be the least integer such that $c^{[0]}, c^{[1]}, \dots, c^{[k]}$ are linearly dependent over $K(x)$. Then there are $u_0(x), u_1(x), \dots, u_k(x) \in K(x)$, with $u_k(x) \neq 0$, such that $\sum_{j=0}^{k} u_j(x) c^{[j]} = 0$. So $\sum_{j=0}^{k} u_j(x) E^j t = \sum_{j=0}^{k} u_j(x) c^{[j]} y = 0$ as well. In particular, for

$$c = (\underbrace{0, \dots, 0}_{i-1}, 1, \underbrace{0, \dots, 0}_{m-i}) \tag{4}$$

we have $t = cy = y_i$, hence

$$\sum_{j=0}^{k} u_j(x) E^j y_i = 0 \tag{5}$$

is a scalar equation of order k satisfied by y_i for any solution y of $[A(x)]$.

Definition 1. *Let row vectors* $c^{[0]}, c^{[1]}, \dots, c^{[k]}$ *and* $u_0(x), u_1(x), \dots, u_k(x) \in K(x)$ *be constructed as it is described above. Then we call (5) the* y_i-*resolving equation, and the full rank* $k \times m$-*matrix* $B(x)$ *whose* jth *row, for* $j = 1, \dots, k$, *is* $c^{[j-1]}$, *the* y_i-*resolving matrix of* $[A(x)]$.

With the y_i-resolving matrix $B(x)$ we have

$$B(x)y = (y_i, Ey_i, \dots, E^{k-1} y_i)^T \tag{6}$$

for any solution y of $[A(x)]$.

3.2   The Minimal Subspace Containing All Solutions with $y_i \neq 0$

Fix i and pick one solution from each set of similar hypergeometric terms satisfying (5). Let the selected hypergeometric terms be

$$h_1(x), \dots, h_l(x). \tag{7}$$

For each $h_j(x)$ substitute $y(x) = h_j(x)z(x)$ into $[A(x)]$, where $z(x) = (z_1(x), \dots, z_m(x))^T \in K(x)^m$ is a new unknown vector. If $\frac{h_j(x+1)}{h_j(x)} = r_j(x) \in K(x)$ then we get the system

$$Ez(x) = \frac{1}{r_j(x)} A(x)z(x). \tag{8}$$

If $R_{j,1}(x), \dots, R_{j,s_j}(x) \in K(x)^m$ is a basis for rational solutions of system (8) then we obtain K-linearly independent hypergeometric solutions

$$h_j(x)R_{j,1}(x), \dots, h_j(x)R_{j,s_j}(x) \tag{9}$$

of $[A(x)]$. Consider all such $h_j(x)$ for $j = 1, \dots, l$. The solutions

$$h_1(x)R_{1,1}(x), \dots, h_1(x)R_{1,s_1}(x), \dots, h_l(x)R_{l,1}(x), \dots, h_l(x)R_{l,s_l}(x) \tag{10}$$

of the system $[A(x)]$ generate over K all the solutions of $[A(x)]$ which have the form $h(x)R(x)$, where $R(x) \in K(x)^m$ and $h(x)$ is a hypergeometric term similar to one of (7). In particular, they generate all the solutions with $y_i(x) \neq 0$. In this sense, (10) is a basis of the minimal subspace containing all solutions with $y_i \neq 0$.

3.3   The Use of RNF

If $h_j(x) \in K(x)$ for some j then the system transformation leading to (8) is not needed since if a solution is related to a rational function then it is related to 1. More generally, if $h_j(x)$ is a hypergeometric term and $\frac{h_j(x+1)}{h_j(x)} = r_j(x)$ then we can construct the rational normal form (RNF) of $r_j(x)$, i.e., represent $r_j(x)$ in the form $U_j(x)\frac{V_j(x+1)}{V_j(x)}$ with $U_j(x), V_j(x) \in K(x)$ where $U_j(x)$ has the numerator and the denominator of minimal possible degrees [7]. We can use $U_j(x)$ instead of $r_j(x)$ in (8). In this case we have to use in (9) the hypergeometric term $\frac{1}{V_j(x)}h_j(x)$ instead of $h_j(x)$.

3.4   The Space of Solutions with $y_i = 0$

Here we are interested in the solutions of $[A(x)]$ belonging to H_K^m with $y_i(x) = 0$.

Proposition 1. *Let equation (5) with $k < m$ be the y_i-resolving equation for $[A(x)]$, $1 \leq i \leq m$. Then there are $m - k$ indices $1 \leq i_1 < i_2 < \cdots < i_{m-k} \leq m$, and an $(m - k) \times (m - k)$-matrix $\tilde{A}(x)$ with entries in $K(x)$ such that if in some space Λ over $K(x)$, the system $[A(x)]$ has a solution $y(x)$ with $y_i(x) = 0$, then*

1. the vector $\tilde{y}(x) = (y_{i_1}(x), \ldots, y_{i_{m-k}}(x))^T$ satisfies $E\tilde{y} = \tilde{A}(x)\tilde{y}$,
2. each $y_j(x)$ with $j \notin \{i_1, \ldots, i_{m-k}\}$ can be expressed as a linear form in $y_{i_1}, \ldots, y_{i_{m-k}}$ having coefficients from $K(x)$.

If $k = m$ in (5) then $y_i(x) = 0$ implies $y_j(x) = 0$ for all $j = 1, \ldots, m$.

Proof. Note that if $y_i(x)$ is zero then $Ey_i(x), E^2 y_i(x), \ldots, E^{k-1} y_i(x)$ are zero as well. Since $c^{[j]}y = E^j(cy) = E^j y_i = 0$, this yields a system of k independent linear algebraic equations

$$B(x)y = 0 \qquad (11)$$

for the unknown $y(x)$, where $B(x)$ is the y_i-resolving matrix of $[A(x)]$ (see Section 3.1). The matrix $B(x)$ has full rank, and hence there exist $m - k$ entries $y_{i_1}, \ldots, y_{i_{m-k}}$ of y such that by means of this system, the other k entries of y can be expressed as linear forms in $y_{i_1}, \ldots, y_{i_{m-k}}$ having coefficients from $K(x)$. Now we can transform $[A(x)]$ as follows:

For each $1 \leq j \leq m$ such that $j \notin \{i_1, \ldots, i_{m-k}\}$ we

(a) remove the equation

$$Ey_j = a_{j1}y_1 + \cdots + a_{jm}y_m$$

from $[A(x)]$,
(b) in all other equations, replace y_j by the corresponding linear form in $y_{i_1}, \ldots, y_{i_{m-k}}$ (in particular, y_i will be replaced by 0, since the first row of $B(x)$ is $c^{[0]} = c$ as given in (4), hence system (11) contains the equation $y_i = 0$, and $i \notin \{i_1, \ldots, i_{m-k}\}$).

Denote the matrix of the resulting system by $\tilde{A}(x)$. If $[A(x)]$ has a solution $y(x) \in \Lambda$ such that $y_i(x) = 0$, then the vector $\tilde{y}(x) = (y_{i_1}(x), \ldots, y_{i_{m-k}}(x))^T$ satisfies $E\tilde{y} = \tilde{A}(x)\tilde{y}$, and each $y_j(x)$ with $j \notin \{i_1, \ldots, i_{m-k}\}$ can be expressed as a linear form in $y_{i_1}, \ldots, y_{i_{m-k}}$, having coefficients from $K(x)$.

If $k = m$ then $B(x)$ is an invertible $m \times m$-matrix of the linear algebraic system (11). Thus, if in addition $y_i = 0$ then $y(x)$ satisfies (11), and $y(x) = 0$.

The proof of Proposition 1 contains an algorithm for constructing the matrix $\tilde{A}$. This matrix is independent of the space Λ. We will use it in Section 4 for the case $\Lambda = H_K^m$.

3.5 When $k = m$

Suppose that $k = m$ in (5). Then $B(x)$ is an invertible $m \times m$-matrix, and c in (4) is a *cyclic vector* (see, e.g., [16, 8, 12, 14] and Section 6.3 below) for the original system $[A(x)]$.

Let $h_1(x), \ldots, h_l(x)$ be a basis for solutions of (5) that belong to H_K. Then by solving the inhomogeneous linear algebraic systems

$$B(x)y = (h_i(x), Eh_i(x), \ldots, E^{m-1}h_i(x))^T, \quad i = 1, \ldots, l, \qquad (12)$$

we obtain a basis for solutions of $[A(x)]$ that belong to H_K^m.

3.6   Selection of y_i

The simplest way to select y_i is just to pick the first unknown from those under consideration. However, it is probably more reasonable to find such a row of the matrix of the difference system which is the least "cumbersome" of all the rows which contain the largest number of zero entries (the "cumbersome" criterion should be clarified). Then we select the unknown y_i so that Ey_i corresponds to the selected row in the matrix of the difference system.

4   The Algorithm

Input: A system of the form (1) (or, equivalently, $[A(x)]$).
Output: A basis for the space of all solutions of $[A(x)]$ belonging to H_K^m, with basis elements in the form (2).

1. $\ell := \emptyset$; $M(x) := A(x)$.

2. Select y_i (Section 3.6). Construct the y_i-resolving equation and the y_i-resolving matrix $B(x)$ of the system $[M(x)]$ (Section 3.1). Let k be the order of the y_i-equation. Compute a basis b for solutions of the constructed y_i-equation that belong to H_K; the elements of b are hypergeometric terms.

3. Include into ℓ those elements of b that are not similar to any of the elements already in ℓ; in each moment the elements of ℓ are pairwise non-similar hypergeometric terms.

4. If $m = k$ then compute a basis $h_1(x), \ldots, h_l(x)$ for the solutions of (5) that belong to H_K. Then by solving inhomogeneous linear algebraic systems (12) find a basis for solutions of $[A(x)]$ that belong to H_K^m. (All the systems (12) can be considered as one system with the left hand side $B(x)y$ and a finite collection of right hand sides.) STOP.

Comment: The equality $k = m$ may be satisfied only at the first execution of Step 4. In all the subsequent executions, k will be less than m.

5. If the order k of equation (5) is less than the order of the matrix $M(x)$ then construct $[\tilde{M}(x)]$ using $B(x)$ (Section 3.4), set $M(x) := \tilde{M}(x)$, and go to 2.

6. For each $h_j(x)$ belonging to ℓ, apply the RNF transformation (Section 3.3) to the rational function $\frac{h_j(x+1)}{h_j(x)}$. If the result is $U_j(x)\frac{V_j(x+1)}{V_j(x)}$ then set $r_j(x) := U_j(x)$ and use it in (8) to construct a basis for the space of those solutions of the system $[A(x)]$ which are related to $h_j(x)$ (Section 3.2). (If $r_j(x) = 1$ the original system does not change.) The union of all such bases gives a basis for the space of solutions of $[A(x)]$ that belong to H_K^m.

In Examples 1 and 2, the unknown y_i is always selected as the first unknown from all the unknowns of the current system.

Example 1. Let

$$A(x) = \begin{pmatrix} \dfrac{x-1}{x} & 0 & -\dfrac{x-1}{x+1} & 0 \\[2mm] 1 & 0 & \dfrac{2}{x+1} & -x \\[2mm] -1 & 1 & x-1 & 1 \\[2mm] -\dfrac{x+2}{x} & \dfrac{x+1}{x} & \dfrac{x^2-x-1}{x(x+1)} & \dfrac{x^2+x+1}{x} \end{pmatrix}$$

With this matrix as input the algorithm proceeds as follows:

1. $\ell := \emptyset$; $M(x) := A(x)$.

2. Set $y_i = y_1$. The y_i-resolving matrix and equation of $[M(x)]$ are

$$B(x) = \begin{pmatrix} 1 & 0 & 0 & 0 \\[2mm] \dfrac{x-1}{x} & 0 & -\dfrac{x-1}{x+1} & 0 \\[2mm] \dfrac{2(x^2+x-1)}{(x+1)(x+2)} & -\dfrac{x}{x+2} & -\dfrac{x(x-1)(x^2+3x+3)}{(x+1)^2(x+2)} & -\dfrac{x}{x+2} \end{pmatrix},$$

and

$$\begin{aligned}
& -x(x-1)(x+2)(x+1)(x^2-x-1)y_1(x) + \\
& 2x(x+2)(x^4+x^3-x^2-x-1)y_1(x+1) - \\
& (x-1)(x+1)(x^4+6x^3+12x^2+8x+4)y_1(x+2) + \\
& x^2(x-1)(x+3)(x+2)y_1(x+3) = 0.
\end{aligned} \tag{13}$$

A basis b for the solutions of the resolving equation that belong to H_K consists of only one element which happens to be a rational function:

$$h_1(x) = \frac{1}{x-1}.$$

3. In accordance with Section 3.3 we set

$$\ell = \left\{ \frac{1}{x-1} \right\}.$$

4. Since $k = 3$ and $m = 4$, we go to Step 5.

5. Using $B(x)$, the matrix $M(x)$ is transformed into $\tilde{M}(x)$ which is a 1×1-matrix, and the system $[\tilde{M}(x)]$ is

$$y_4(x+1) = xy_4(x).$$

The set ℓ is extended by the hypergeometric term $\Gamma(x)$:

$$\ell = \left\{ \frac{1}{x-1}, \Gamma(x) \right\}.$$

6. For the first element of ℓ, we get $r_1(x) = 1$ since the RNF of $\frac{x-1}{x}$ is $1 \cdot \frac{1/x}{1/(x-1)}$. The system (8) with $r_1(x) = 1$ has no rational solutions, thus, there is no solution of the original system which is related to $\frac{1}{x-1}$.

Since $\frac{\Gamma(x+1)}{\Gamma(x)} = x$ and the RNF of x is $x \cdot \frac{1}{1}$, we use $r_2(x) = x$ in (8). This system has a one-dimensional space of rational solutions, generated by

$$R(x) = (0, -1, 0, 1)^T.$$

Finally, we obtain the basis of the (one-dimensional) space of all solutions of $[A(x)]$ belonging to H_K^4. It contains the single element

$$\Gamma(x)R(x) = (0, -\Gamma(x), 0, \Gamma(x))^T.$$

Remark 1. *Example 1 shows that the proposed resolving approach is not a modification of the block-diagonal form algorithm [8]: if the constructed y_1-resolving equation (13) corresponds to a diagonal block of the original system then the system would have a rational solution. However, this is not the case.*

Example 2. Let

$$A(x) = \begin{pmatrix} \dfrac{x^3 + 4x^2 + 4x - 2}{(x+4)(x+2)(x+1)} & \dfrac{x^2 + 3x + 1}{(x+2)(x+1)} & \dfrac{x+1}{x+4} & \dfrac{2(x+2)}{x+4} \\[3mm] -\dfrac{x^3 + 4x^2 + 4x - 2}{(x+4)(x+2)(x+1)} & \dfrac{1}{(x+2)(x+1)} & -\dfrac{x+1}{x+4} & -\dfrac{x}{x+4} \\[3mm] -\dfrac{x(2x^2 + 8x + 9)}{(x+4)(x+2)(x+1)} & -\dfrac{x^2 + 3x + 1}{(x+2)(x+1)} & -\dfrac{2(x+1)}{x+4} & -\dfrac{2x}{x+4} \\[3mm] \dfrac{x+1}{x+4} & 0 & \dfrac{x+1}{x+4} & \dfrac{x}{x+4} \end{pmatrix}.$$

With this matrix as input the algorithm proceeds as follows:

1. $\ell := \emptyset$; $M(x) := A(x)$.

2. Set $y_i = y_1$. The y_i-resolving matrix is a 4×4-matrix and the y_i-resolving equation is

$$(x^2 + 9x + 21)(x^5 + 17x^4 + 111x^3 + 339x^2 + 453x + 167)(x+2)^2 y_1(x) +$$

$$(x^8 + 30x^7 + 383x^6 + 2727x^5 + 11919x^4 + 33035x^3 +$$

$$57308x^2 + 57507x + 25746)y_1(x+1) -$$

$$(x^2 + 5x + 5)(x^9 + 28x^8 + 341x^7 + 2360x^6 + 10158x^5 + 27884x^4 +$$

$$47833x^3 + 47264x^2 + 21093x + 462)y_1(x + 2) +$$

$$(x + 4)(x^9 + 25x^8 + 267x^7 + 1564x^6 + 5268x^5 + 9116x^4 +$$

$$1933x^3 - 21905x^2 - 36519x - 19110)y_1(x + 3) +$$

$$(x + 5)(x + 4)(x + 7)(x + 1)(x^2 + 7x + 13)(x^5 + 12x^4 + 53x^3 + 98x^2 +$$

$$45x - 42)y_1(x + 4) = 0$$

A basis b for the solutions of the y_i-equation belonging to H_K consists of four elements:

$$h_1(x) = \frac{(-1)^x(2x + 5)}{(x + 2)(x + 3)}, \quad h_2(x) = \frac{1}{(x + 3)(x + 2)},$$

$$h_3(x) = \frac{(-1)^x(x^3 + 7x^2 + 16x + 12)}{\Gamma(x + 4)}, \quad h_4(x) = \frac{x^3 + 5x^2 + 6x}{\Gamma(x + 4)}.$$

3. ℓ is the same as b.

4. We decide to solve (6) and get the following basis for solutions of $[A(x)]$:

$$\left(\frac{(-1)^x(2x + 5)}{(x + 2)(x + 3)}, -\frac{(-1)^x(2x + 5)}{(x + 2)(x + 3)}, -\frac{(-1)^x(6x^2 + 23x + 19)}{(x + 3)(x + 2)(x + 1)}, \frac{(-1)^x(2x + 5)}{(x + 2)(x + 3)} \right)^T,$$

$$\left(\frac{1}{(x + 3)(x + 2)}, -\frac{1}{(x + 3)(x + 2)}, -\frac{x - 1}{(x + 3)(x + 2)(x + 1)}, \frac{1}{(x + 3)(x + 2)} \right)^T,$$

$$\left(\frac{(-1)^x(x^3 + 7x^2 + 16x + 12)}{\Gamma(x + 4)}, \frac{(-1)^x(x^2 + 5x + 6)}{-\Gamma(x + 4)}, \frac{(-1)^x(x^3 + 7x^2 + 16x + 12)}{-\Gamma(x + 4)}, 0 \right)^T,$$

$$\left(\frac{x(x^2 + 5x + 6)}{\Gamma(x + 4)}, \frac{x^2 + 5x + 6}{\Gamma(x + 4)}, -\frac{x(x^2 + 5x + 6)}{\Gamma(x + 4)}, 0 \right)^T.$$

5   On the Resolving Procedure

Now we investigate the complexity of the *resolving procedure* that was presented in Sections 3.1, 3.4. The procedure transforms a linear difference system $[A(x)]$, where $A(x)$ is an $m \times m$-matrix ($m \geq 2$) whose entries belong to $K(x)$, into one or several scalar difference equations with coefficients in $K(x)$. The sum of the orders of those scalar equations does not exceed m. If $[A(x)]$ has a solution of the form (2) then at least one of the obtained equations has a solution of the form $h(x)r(x)$ with $r(x) \in K(x)$.

Proposition 2. *The complexity (the number of field operations in $K(x)$ in the worst case) of the resolving procedure is $O(m^3)$.*

Proof. Let the size of the input matrix resp. the resolving matrix be $n \times n$ resp. $l \times n$ where $l \leq n$. Then there is a constant C such that the number of field operations in $K(x)$ needed for constructing the resolving equation, the resolving matrix and the matrix $\tilde{A}(x)$ (see the proof of Proposition 1) does not exceed Cln^2. Using this, it is easy to prove by induction on the number of steps needed to construct the y_i-resolving equations and matrices that the number of field operations needed to perform the resolving procedure does not exceed Cm^3. Indeed, if the number of steps is 1 the claim is evident. Otherwise, let the size of the y_i-resolving matrix constructed on the first step be $k \times m$ where $k < m$. In this case, the general number of operations in $K(x)$ does not exceed $C(m - k)^3 + Ckm^2$. It remains to note that $m^3 - (m - k)^3 - km^2 = k^2(m - k) + 2k(m - k)^2 \geq 0$.

Algorithms for constructing a cyclic vector and related uncoupling algorithms for normal first-order systems are well known in computer algebra. The paper [12] contains a review of such algorithms and corresponding references. In that paper, a new algorithm for constructing a cyclic vector based on fast linear algebra algorithms ([21]) is also proposed. The existence of cyclic vector algorithms suggests that there is no need for resolving procedures – all the more so since cyclic vector procedures and the uncoupling are multi-purpose procedures which may be useful not only for finding hypergeometric or exponential-logarithmic solutions of systems. Besides, the resolving procedure only solves part of the problem. Solutions of the operators belonging to the resolving sequence are a "half-finished product", since we have in addition to find rational solutions of other difference systems. However, the resolving equations approach can have some advantage over the cyclic vector approach. First, a sequence of resolving equations is constructed in a single pass, while in practical cyclic vector algorithms numerous random candidates are considered (if a candidate is not appropriate then another one is generated, and all calculations are resumed from the beginning).

Second, asymptotically fast linear algebra algorithms have advantages over classical algorithms only for large inputs. The orientation on asymptotic complexity estimates does not seem to be very productive here. The use of classical linear algebra algorithms as auxiliary tools for searching for solutions of difference systems can be advantageous in many cases.

Third, the resolving system constructed by our algorithm is such that the sum of the orders of its operators does not exceed the order of the operator obtained by a cyclic vector algorithm. As a rule, it is easier to solve a few equations of small orders than a single equation of a large order (which provides motivation to develop factorization algorithms). Even when we obtain a unique resolving operator (in the case $k = m$, Section 3.5), we can profit since such a cyclic vector is of a very simple form, and the scalar equation to solve will not have "too cumbersome" coefficients. In general, we have an opportunity to use various heuristics to obtain operators with possibly less cumbersome coefficients.

As for the second stage when some solutions of additional difference systems have to be found, this can quite often be done in reasonable time. The search for rational solutions of difference systems having rational-function coefficients is not time consuming ([2, 4, 5, 6, 9, 10, 17]).

Our experimental comparison demonstrates a definite advantage of a resolving procedure over the cyclic vector approach.

6   Implementation and Experiments

6.1   Implementation

We have implemented the algorithm in Maple 18 ([22]). The implemented procedures are put together in the package LRS (Linear Recurrence Systems). The main procedure of the package is HypergeometricSolution.

To find a basis of hypergeometric solutions of the y_i-resolving equation (5), the procedure hypergeomsols from the package LREtools is used. It implements the algorithm from [18]. To find a basis of rational solutions of the system (8) we use the procedure RationalSolution from the package LinearFunctionalSystems. This procedure implements the algorithms from [2, 19, 1, 3]. To perform RNF transformation, the procedure RationalCanonicalForm from the package RationalNormalForms is used. It implements the algorithms from [7].

To select y_i, the procedure SelectIndicator from LRS finds all the rows of $A(x)$ with the greatest number of zero entries. In this set of rows, it finds the rows having the least sum of degrees in x of all the numerators and denominators of nonzero entries. If the resulting set has several rows, one of them is selected by the standard Maple procedure rand. The procedure returns the number i of the selected row. The arguments of the procedure HypergeometricSolution are a square matrix with rational-function entries and a name of an independent variable. The output is a list of vectors whose entries are hypergeometric terms.

6.2   Some Experiments

Example 3. Applying HypergeometricSolution to the matrix $A(x)$ from Example 1 we get the result in 0.303 CPU seconds[1]:

```
> A1 := Matrix([[(x-1)/x, 0, -(x-1)/(x+1), 0],
                [1, 0, 2/(x+1), -x],
                [ -1, 1, x-1, 1],
                [-(x+2)/x, (x+1)/x, (x^2-x-1)/((x+1)*x),
                                        (x^2+x+1)/x]]):

> st := time():
  LRS:-HypergeometricSolution(A1, x);
  time()-st;
```

---

[1] By Maple 18, Ubuntu 8.04.4 LTS, AMD Athlon(tm) 64 Processor 3700+, 3GB RAM.

$$\left[\left[\begin{array}{c} 0 \\ -\Gamma(x) \\ 0 \\ \Gamma(x) \end{array}\right]\right]$$

0.303

Example 4. For $A(x)$ from Example 2 we get the result:

```
> A2 := Matrix([
  [(x^3+4*x^2+4*x-2)/((x+4)*(x+2)*(x+1)),
      (x^2+3*x+1)/((x+2)*(x+1)), (x+1)/(x+4), (2*x+4)/(x+4)],
  [ -(x^3+4*x^2+4*x-2)/((x+4)*(x+2)*(x+1)),
          1/((x+2)*(x+1)), -(x+1)/(x+4),  -x/(x+4)],
  [-x*(2*x^2+8*x+9)/((x+4)*(x+2)*(x+1)),
      -(x^2+3*x+1)/((x+2)*(x+1)), -(2*x+2)/(x+4), -2*x/(x+4)],
  [(x+1)/(x+4), 0, (x+1)/(x+4), x/(x+4)]]):


> st := time():
  LRS:-HypergeometricSolution(A2, x);
  time()-st;
```

$$\left[\left[\left[\begin{array}{c} \frac{(-1)^x(2x+5)}{(x+2)(x+3)} \\[2mm] -\frac{(-1)^x(2x+5)}{(x+2)(x+3)} \\[2mm] -\frac{(-1)^x(6x^2+23x+19)}{(x+3)(x+2)(x+1)} \\[2mm] \frac{(-1)^x(2x+5)}{(x+2)(x+3)} \end{array}\right], \left[\begin{array}{c} \frac{1}{(x+3)(x+2)} \\[2mm] -\frac{1}{(x+3)(x+2)} \\[2mm] -\frac{x-1}{(x+3)(x+2)(x+1)} \\[2mm] \frac{1}{(x+3)(x+2)} \end{array}\right], \left[\begin{array}{c} -\frac{(-1)^x(x+2)}{\Gamma(x+2)} \\[2mm] \frac{(-1)^x}{\Gamma(x+2)} \\[2mm] \frac{(-1)^x(x+2)}{\Gamma(x+2)} \\[2mm] 0 \end{array}\right], \left[\begin{array}{c} -\frac{x}{\Gamma(x+2)} \\[2mm] -\frac{1}{\Gamma(x+2)} \\[2mm] \frac{x}{\Gamma(x+2)} \\[2mm] 0 \end{array}\right]\right]\right]$$

0.410

Here, `SelectIndicator` selects $i = 4$, because the fourth matrix row has one zero. The corresponding resolving equation is of order $k = 2$ and has a two-dimensional hypergeometric solutions space. The reduced system $[\tilde{M}(x)]$ has a second-order resolving equation with a two-dimensional hypergeometric solutions space.

Example 5. We tested `HypergeometricSolution` for systems with a 16×16-matrix. We cannot present here this matrix and a basis of the hypergeometric solutions space since they are too large. The matrix is such that 80% of its entries are zeros. The maximum degree of the numerators of its entries is 13. The maximum degree of their denominators is 11. The procedure finds a two-dimensional hypergeometric solutions space in 346.046 CPU seconds.

The code and examples of applications of `HypergeometricSolution` are available from http://www.ccas.ru/ca/doku.php/lrs.

6.3   Comparison with the Cyclic Vector Approach

Besides the resolving procedure, we implemented also the search for hypergeometric solutions based on the cyclic vector approach (procedure `CyclicVector`). In our experiments, we used a traditional randomized version of a cyclic vector algorithm which allows to obtain

- a scalar difference equation of order m with rational-function coefficients, and
- a matrix $B(x)$ with rational-function entries.

These objects are such that if $h_1(x), \ldots, h_l(x)$ form a basis for solutions of the scalar equation that belong to H_K then by solving the inhomogeneous linear algebraic systems $B(x)y = (h_i(x), Eh_i(x), \ldots, E^{m-1}h_i(x))^T$, $i = 1, \ldots, l$, one obtains a basis for solutions of $[A(x)]$ that belong to H_K^m:

1. Randomly choose a row vector $c^{[0]}$ containing polynomials of degree 0.

2. Create an $m \times m$-matrix $B(x)$:
   for i from 1 to m do
      the i-th row of $B(x)$ is $c^{[i-1]}$;
      $c^{[i]} := (Ec^{[i-1]})A(x)$.

3. If $B(x)$ is not invertible, go to step 2 with a new random row vector $c^{[0]}$ of polynomials of degree $m - 1$. Otherwise, solve the linear system $u(x)B(x) = c^{[m]}$ to obtain a vector $u(x) = (u_0(x), \ldots, u_{m-1}(x))$.

4. Return the scalar equation for a new unknown $t(x)$

$$E^m t = u_{m-1}(x)E^{m-1}t + \cdots + u_0(x)t \qquad (14)$$

and the matrix $B(x)$.

Instead of the CPU time demonstrated in Examples 3, 4, 5, i.e., 0.303, 0.410, and 346.046 seconds, the computation using the cyclic vector procedure takes, resp., 0.410, 0.474, 1063.747 seconds.

Acknowledgments. The authors are thankful to M. Barkatou for numerous discussions about the problem of the search for hypergeometric solutions of difference systems, to A. Bostan for consultations on cyclic vector algorithms, and to anonymous referees for useful comments.

References

1. Abramov, S.: EG-Eliminations. J. Difference Equations Appl. 5, 393–433 (1999)
2. Abramov, S., Barkatou M.A.: Rational solutions of first order linear difference systems. In: ISSAC 1998 Proceedings, pp. 124–131 (1998)
3. Abramov, S., Bronstein, M.: On solutions of linear functional systems. In: ISSAC 2001 Proceedings, pp. 1–6 (2001)

4. Abramov, S., Gheffar, A., Khmelnov, D.: Factorization of polynomials and gcd computations for finding universal denominators. In: Gerdt, V.P., et al. (eds.) CASC 2010. LNCS, vol. 6244, pp. 4–18. Springer, Heidelberg (2010)

5. Abramov, S., Gheffar, A., Khmelnov, D.: Rational solutions of linear difference equations: universal denominators and denominator bounds. Programming and Comput. Software 37 (2), 78–86 (2011); Transl. from Programmirovanie 2, 28–39 (2011)

6. Abramov, S., Khmelnov, D.: Denominators of rational solutions of linear difference systems of arbitrary order. Programming and Comput. Software 38 (2), 84–91 (2012); Transl. from Programmirovanie 2, 45–54 (2012)

7. Abramov, S., Petkovšek, M.: Rational normal forms and minimal representation of hypergeometric terms. J. Symb. Comp. 33, 521–543 (2002)

8. Barkatou, M.: An algorithm for computing a companion block diagonal form for a system of linear differential equations. AAECC 4, 185–195 (1993)

9. Barkatou, M.: A fast algorithm to compute the rational solutions of systems of linear differential equations. RR 973–M– Mars 1997, IMAG–LMC, Grenoble (1997).

10. Barkatou, M.: Rational solutions of matrix difference equations: problem of equivalence and factorization. In: ISSAC 1999 Proceedings, pp. 277–282 (1999)

11. Barkatou, M.: Hypergeometric solutions of systems of linear difference equations and applications. http://www-sop.inria.fr/cafe/SA08/SA08talks/#barkatou

12. Bostan, A., Chyzak, F., de Panafieu, E.: Complexity estimates for two uncoupling algorithms. In: ISSAC 2013 Proceedings, pp. 85–92 (2013)

13. Bronstein, M., Petkovšek, M.: An introduction to pseudo-linear algebra. Theoret. Comput. Sci. 157, 3–33 (1996)

14. Churchill, R.C., Kovacic, J.: Cyclic vectors. In: Differential algebra and related topics, pp. 191–218. World Sci. Publ., River (2002)

15. Cluzeau, T., van Hoeij, M.: Computing hypergeometric solutions of linear difference equations. AAECC 17, 83–115 (2006)

16. Cope, F.T.: Formal solutions of irregular linear differential equations. Part II. Amer. J. Math. 58(1), 130–140 (1936)

17. van Hoeij, M.: Rational solutions of linear difference equations. In: ISSAC 1998 Proceedings, pp. 120–123 (1998)

18. van Hoeij, M.: Finite singularities and hypergeometric solutions of linear recurrence equations. J. Pure Appl. Algebra 139, 109–131 (1999)

19. Khmelnov, D.: Search for polynomial solutions of linear functional systems by means of induced recurrences. Programming Comput. Software 30(2), 61–67 (2004)

20. Petkovšek, M.: Hypergeometric solutions of linear recurrences with polynomial coefficients. J. Symb. Comp. 14, 243–264 (1992)

21. Storjohann, A.: High-order lifting and integrality certification. J. Symb. Comp. 36, 613–648 (2003)

22. Maple online help. http://www.maplesoft.com/support/help/

Janet Bases and Resolutions in CoCoALib

Mario Albert, Matthias Fetzer, and Werner M. Seiler

Institut für Mathematik, Universität Kassel, 34132 Kassel, Germany
{albert,fetzer,seiler}@mathematik.uni-kassel.de

Abstract. Recently, the authors presented a novel approach to computing resolutions and Betti numbers using Pommaret bases. For Betti numbers, this algorithm is for most examples much faster than the classical methods (typically by orders of magnitude). As the problem of δ-regularity often makes the determination of a Pommaret basis rather expensive, we extend here our algorithm to Janet bases. Although in δ-singular coordinates, Janet bases may induce larger resolutions than the corresponding Pommaret bases, our benchmarks demonstrate that this happens rarely and has no significant effect on the computation costs.

1 Introduction

Computing resolutions represents a fundamental task in algebraic geometry and commutative algebra. For some problems like computing derived functors one needs indeed the full resolution, i. e. including the differential. For other applications, the Betti numbers measuring the size of the resolution are sufficient, as they contain already important geometric and topological information.

Determining a minimal resolution is generally rather expensive. For a module of projective dimension p, the costs correspond roughly to those of p Gröbner bases computation. Theoretically, computing only the Betti numbers should be considerably cheaper, as one does not need the differential. However, all implementations we are aware of read off the Betti numbers of the minimal resolutions and thus in practice their costs are the same as for the whole resolution.

In the recent article [2], we presented a novel approach to computing resolutions and Betti numbers based on a combination of the theory of Pommaret bases [12], a special form of involutive bases, and algebraic discrete Morse theory [15]. Within this approach, it is possible to compute Betti numbers without first determining a whole resolution. In fact, it is even possible to determine individual Betti numbers without the remaining ones.

While Pommaret bases are theoretically very nice, as they provide simple access to many invariants [12], they face from a practical point of view the problem of δ-regularity, i. e. for positive-dimensional ideals they generally exist only after a sufficiently generic coordinate transformation. There are deterministic approaches to the construction of (hopefully rather sparse) δ-regular coordinates [8], but nevertheless the computation of a Pommaret basis is usually significantly more expensive than that of a Janet basis (a more detailed analysis of this topic will appear in the forthcoming work [3]).

© Springer International Publishing Switzerland 2015
V.P. Gerdt et al. (Eds.): CASC 2015, LNCS 9301, pp. 15–29, 2015.
DOI: 10.1007/978-3-319-24021-3_2

Here, we show that the ideas of [2] also work Janet instead of Pommaret bases and thus remove an important bottleneck in their application. As a by-product, we show that the degree of a Janet basis can never be smaller than that of a Pommaret basis. For the resolution induced by the Janet basis this implies that in general it is longer than the minimal one and can extend to higher degrees.

2 Involutive Bases and Free Resolutions

Involutive bases are Gröbner bases with additional combinatorial properties. They were introduced by Gerdt and Blinkov [5,6] who combined Gröbner bases with ideas from the algebraic theory of partial differential equations. Surveys over their basic theory and further references can be found in [11] or [13, Chapts. 3/4].

Throughout this work, $\Bbbk$ denotes an arbitrary field and $\mathcal{P} = \Bbbk[x_1, \ldots, x_n] = \Bbbk[\mathcal{X}]$ the polynomial ring in n variables over $\Bbbk$ together with the standard grading. The standard basis of the free module $\mathcal{P}^t$ is denoted by $\{\mathbf{e}_1, \ldots, \mathbf{e}_t\}$. Our conventions for the Janet division require that we define term orders always "reverse" to the usual conventions, i.e. we revert the ordering of the variables. In the sequel, $0 \neq \mathcal{U} \subseteq \mathcal{P}^t$ will always be a graded submodule and all appearing elements $\mathbf{f} \in \mathcal{P}^t$ are homogeneous.

The basic idea underlying involutive bases is that each generator $\mathbf{f}$ in a basis may only be multiplied by polynomials in a restricted set of variables, its *multiplicative variables* $\mathcal{X}(\mathbf{f}) \subseteq \mathcal{X}$. The remaining variables are called the *non-multiplicative* ones $\overline{\mathcal{X}}(\mathbf{f}) = \mathcal{X} \setminus \mathcal{X}(\mathbf{f})$. Different involutive bases differ in the way the multiplicative variables are chosen. We will use here only Janet bases.

For them, the assignment of the multiplicative variables depends not only on the generator $\mathbf{f}$, but on the whole basis. Given a finite set $\mathcal{F}$ of terms and a term $x^\mu \mathbf{e}_\alpha \in \mathcal{F}$, we introduce for each $1 \leq k \leq n$ and each $1 \leq \alpha \leq t$ the subsets

$$(\mu_{k+1}, \ldots, \mu_n)_\alpha = \{x^\nu \mathbf{e}_\alpha \in \mathcal{F} \mid \nu_{k+1} = \mu_{k+1}, \ldots, \nu_n = \mu_n\} \subseteq \mathcal{F}$$

and put $x_k \in \mathcal{X}_{J,\mathcal{F}}(x^\mu \mathbf{e}_\alpha)$, if $\mu_k = \max\{\nu_k \mid x^\nu \mathbf{e}_\alpha \in (\mu_{k+1}, \ldots, \mu_n)_\alpha\}$. For finite sets $\mathcal{F}$ of polynomial vectors, we reduce to the monomial case via a term order $\prec$ by setting $\mathcal{X}_{J,\mathcal{F},\prec}(\mathbf{f}) = \mathcal{X}_{J,\mathrm{lt}\,\mathcal{F}}(\mathrm{lt}\,\mathbf{f})$.

Definition 1. *A finite set of terms $\mathcal{H} \subset \mathcal{P}^t$ is a Janet basis of the monomial module $\mathcal{U} = \langle \mathcal{H} \rangle$, if as a $\Bbbk$-linear space $\mathcal{U} = \bigoplus_{\mathbf{h} \in \mathcal{H}} \Bbbk[\mathcal{X}_{J,\mathcal{H}}(\mathbf{h})] \cdot \mathbf{h}$. For every term contained in the involutive cone $\Bbbk[\mathcal{X}_{J,\mathcal{H}}(\mathbf{h})] \cdot \mathbf{h}$, we call $\mathbf{h}$ an involutive divisor. A finite polynomial set $\mathcal{H} \subset \mathcal{P}^t$ is a Janet basis of the polynomial submodule $\mathcal{U} = \langle \mathcal{H} \rangle$ for the term order $\prec$, if all elements of $\mathcal{H}$ possess distinct leading terms and these terms form a Janet basis of the leading module $\mathrm{lt}\,\mathcal{U}$.*

Every submodule admits a Janet basis for any term order [5,11]. Arbitrary involutive bases can be characterised similarly to Gröbner bases [5,11]. The S-polynomials in the theory of Gröbner bases are replaced by products of the generators with one of their *non*-multiplicative variables. A key difference is the uniqueness of involutive standard representations.

Proposition 2. *[11, Thm. 5.4] The finite set $\mathcal{H} \subset \mathcal{U}$ is a Janet basis of the submodule $\mathcal{U} \subseteq \mathcal{P}^t$ for the term order $\prec$, if and only if every element $0 \neq \mathbf{f} \in \mathcal{U}$ possesses a unique involutive standard representation $\mathbf{f} = \sum_{\mathbf{h} \in \mathcal{H}} P_{\mathbf{h}} \mathbf{h}$ where each non-zero coefficient satisfies $P_{\mathbf{h}} \in \Bbbk[\mathcal{X}_{J,\mathcal{H},\prec}(\mathbf{h})]$ and $\mathrm{lt}\,(P_{\mathbf{h}} \mathbf{h}) \preceq \mathrm{lt}\,(\mathbf{f})$.*

Proposition 3. *[11, Cor. 7.3] Let $\mathcal{H} \subset \mathcal{P}^t$ be a finite set and $\prec$ a term order such that no leading term in $\mathrm{lt}\,\mathcal{H}$ is an involutive divisor of another one. The set $\mathcal{H}$ is a Janet basis of the submodule $\langle \mathcal{H} \rangle$ with respect to $\prec$, if and only if for every $\mathbf{h} \in \mathcal{H}$ and every non-multiplicative variable $x_j \in \overline{\mathcal{X}}_{J,\mathcal{H},\prec}(\mathbf{h})$ the product $x_j \mathbf{h}$ possesses an involutive standard representation with respect to $\mathcal{H}$.*

The classical Schreyer Theorem [10] describes how every Gröbner basis induces a Gröbner basis of its first syzygy module for a suitable chosen term order. If $\mathcal{H} = \{\mathbf{h}_1, \ldots, \mathbf{h}_s\} \subset \mathcal{P}^t$ is a finite set with s elements and $\prec$ an arbitrary term order on $\mathcal{P}^t$, then the *Schreyer order* $\prec_{\mathcal{H}}$ is the term order induced on the free module $\mathcal{P}^s$ by setting $x^\mu \mathbf{e}_\alpha \prec_{\mathcal{H}} x^\nu \mathbf{e}_\beta$, if $\mathrm{lt}\,(x^\mu \mathbf{h}_\alpha) \prec \mathrm{lt}\,(x^\nu \mathbf{h}_\beta)$ or if these leading terms are equal and $\beta < \alpha$.

The Schreyer order $\prec_{\mathcal{H}}$ depends on the ordering of $\mathcal{H}$. For the involutive version of the Schreyer Theorem, we assume that $\mathcal{H}$ is a Janet basis and order its elements in a suitable manner. We associate a directed graph with $\mathcal{H}$. Its vertices are given by the elements in $\mathcal{H}$. If $x_j \in \overline{\mathcal{X}}_{J,\mathcal{H},\prec}(\mathbf{h})$ for some generator $\mathbf{h} \in \mathcal{H}$, then $\mathcal{H}$ contains a unique generator $\bar{\mathbf{h}}$ such that $\mathrm{lt}\,\bar{\mathbf{h}}$ is an involutive divisor of $\mathrm{lt}\,(x_j \mathbf{h})$. In this case we include a directed edge from $\mathbf{h}$ to $\bar{\mathbf{h}}$. The graph thus defined is called the *J-graph* of the Janet basis $\mathcal{H}$. We require that the ordering of $\mathcal{H}$ satisfies the following condition: whenever the *J*-graph of $\mathcal{H}$ contains a path from $\mathbf{h}_\alpha$ to $\mathbf{h}_\beta$, then we must have $\alpha < \beta$. We then speak of a *J-ordering*. One can show that such orderings always exist [12]. In fact, one easily verifies that an explicit *J*-ordering is provided by any module order (applied to the leading terms of $\mathcal{H}$) that restricts to the lexicographic order when applied to terms living in the same component of $\mathcal{P}^t$.

Assume that $\mathcal{H} = \{\mathbf{h}_1, \ldots, \mathbf{h}_s\}$ is a Janet basis of the polynomial submodule $\mathcal{U} \subseteq \mathcal{P}^t$. According to Proposition 3, we have for every non-multiplicative variable x_k of a generator $\mathbf{h}_\alpha$ an involutive standard representation $x_k \mathbf{h}_\alpha = \sum_{\beta=1}^s P_\beta^{(\alpha;k)} \mathbf{h}_\beta$ and thus a syzygy $\mathbf{S}_{\alpha;k} = x_k \mathbf{e}_\alpha - \sum_{\beta=1}^s P_\beta^{(\alpha;k)} \mathbf{e}_\beta$. Let $\mathcal{H}_{\mathrm{Syz}}$ be the set of all these syzygies.

Lemma 4. *[12, Lemma 5.7] If the finite set $\mathcal{H} \subset \mathcal{P}$ is a J-ordered Janet basis, then we find for all admissible values of α and k with respect to the Schreyer order $\prec_{\mathcal{H}}$ that $\mathrm{lt}\,\mathbf{S}_{\alpha;k} = x_k \mathbf{e}_\alpha$.*

Theorem 5. *[12, Thm. 5.10] Let $\mathcal{H}$ be a J-ordered Janet basis of the submodule $\mathcal{U} \subseteq \mathcal{P}^t$. Then $\mathcal{H}_{\mathrm{Syz}}$ is a Janet basis of the syzygy module $\mathrm{Syz}(\mathcal{H})$ for the Schreyer order $\prec_{\mathcal{H}}$.*

Like the classical Schreyer Theorem, we can iterate Theorem 5 and obtain then a free resolution of the submodule $\mathcal{U}$. However, in contrast to the classical situation, the involutive version yields the full shape of the arising resolution without any further computations. We present here a bigraded version of this result which is obtained by a trivial extension of the arguments in [12]. It provides sharp upper bounds for the Betti numbers of $\mathcal{U}$.

Theorem 6. *[12, Thm. 6.1, Rem. 6.2] Let the finite set $\mathcal{H} \subset \mathcal{P}^t$ be a Janet basis of the polynomial submodule $\mathcal{U} \subseteq \mathcal{P}^t$. Furthermore, let $\beta_{0,j}^{(k)}$ be the number of generators $\mathbf{h} \in \mathcal{H}$ of degree j having k multiplicative variables and set $d = \min\{k \mid \exists j : \beta_{0,j}^{(k)} > 0\}$. Then $\mathcal{U}$ possesses a finite free graded resolution[1]*

$$0 \longrightarrow \bigoplus \mathcal{P}[-j]^{r_{n-d,j}} \longrightarrow \cdots \longrightarrow \bigoplus \mathcal{P}[-j]^{r_{1,j}} \longrightarrow \bigoplus \mathcal{P}[-j]^{r_{0,j}} \longrightarrow \mathcal{U} \longrightarrow 0 \tag{1}$$

of length $n - d$ where the ranks of the free modules are given by

$$r_{i,j} = \sum_{k=1}^{n-i} \binom{n-k}{i} \beta_{0,j-i}^{(k)}. \tag{2}$$

For the proof, one shows that the Janet basis $\mathcal{H}_j$ of the jth syzygy module $\mathrm{Syz}^j(\mathcal{H})$ with respect to the Schreyer order $\prec_{\mathcal{H}_{j-1}}$ consists of the syzygies $\mathbf{S}_{\alpha;\mathbf{k}}$ with an ordered integer sequence $\mathbf{k} = (k_1, \ldots, k_j)$ where $1 \leq k_1 < \cdots < k_j \leq n$ and all variables x_{k_i} are non-multiplicative for the generator $\mathbf{h}_\alpha \in \mathcal{H}$. These syzygies are defined recursively. We denote for any $1 \leq i \leq j$ by $\mathbf{k}_i$ the sequence obtained by eliminating k_i from $\mathbf{k}$. Now $\mathbf{S}_{\alpha;\mathbf{k}}$ arises from the involutive standard representation of $x_{k_j} \mathbf{S}_{\alpha;\mathbf{k}_j}$: $x_{k_j} \mathbf{S}_{\alpha;\mathbf{k}_j} = \sum_{\beta=1}^s \sum_\ell P_{\beta;\ell}^{(\alpha;\mathbf{k})} \mathbf{S}_{\beta;\ell}$. Here the second sum is over all ordered integer sequences ℓ of length $j - 1$ such that for all entries ℓ_i the variables x_{ℓ_i} is non-multiplicative for the generator $\mathbf{h}_\beta \in \mathcal{H}$. Lemma 4 implies that $\mathrm{lt}\, \mathbf{S}_{\alpha;\mathbf{k}} = x_{k_j} \mathbf{e}_{\alpha;\mathbf{k}_j}$ and that the coefficient $P_{\beta;\ell}^{(\alpha;\mathbf{k})}$ depends only on those variables which are multiplicative for the syzygy $\mathbf{S}_{\beta;\ell}$. If $\overline{\mathcal{X}}_{J,\mathcal{H},\prec}(h_\alpha) = \{x_{i_1}, x_{i_2}, \ldots, x_{i_{n-k}}\}$, then we find for the first syzygies that $\mathcal{X}_{J,\mathcal{H}_{\mathrm{Syz}},\prec_\mathcal{H}}(\mathbf{S}_{\alpha;i_j}) = \mathcal{X} \setminus \{x_{i_{j+1}}, \ldots, x_{i_{n-k}}\}$. Iteration yields the multiplicative variables for the higher syzygies. The simple form of the leading terms yields via a simple combinatorial computation the ranks $r_{i,j}$ of the modules in the resolution (1).

Corollary 7. *Let $\mathcal{H}$ be a Janet basis of the submodule $\mathcal{U} \subseteq \mathcal{P}^t$. If we set again $d = \min\{k \mid \exists j : \beta_{0,j}^{(k)} > 0\}$ and $q = \deg(\mathcal{H}) = \max\{\deg \mathbf{h} \mid \mathbf{h} \in \mathcal{H}\}$, then the projective dimension and the Castelnuovo-Mumford regularity of $\mathcal{U}$ are bounded by $\mathrm{pd}(\mathcal{U}) \leq n - d$ and $\mathrm{reg}(\mathcal{U}) \leq q$.*

Proof. The first estimate follows immediately from the resolution (1) induced by the Janet basis $\mathcal{H}$. Furthermore, the ith module of this resolution is obviously generated by elements of degree less than or equal to $q + i$. This observation implies that $\mathcal{U}$ is q-regular and thus the second estimate.   □

Starting from an arbitrary Janet basis, more information cannot be obtained in a simple manner. If, however, δ-regular coordinates are used, i.e. if the Janet

---

[1] We use here the usual shift notation: $(\mathcal{P}[j])_d = \mathcal{P}_{d+j}$.

basis is simultaneously a Pommaret basis (which is generically the case), then stronger results hold: the resolution (1) is then always of minimal length, i.e. $\mathrm{pd}\,(\mathcal{U}) = n - d$ [12, Thm. 8.11] (which also implies by the Auslander-Buchsbaum formula that d is nothing but the depth of the submodule $\mathcal{U}$), and also the second estimate becomes an equality: $\mathrm{reg}\,(\mathcal{U}) = \deg\,(\mathcal{H})$ [12, Thm. 9.2]. By contrast, for a Janet basis the difference between the regularity and the degree of the basis can become arbitrarily big, as the next example demonstrates.

Example 8. Consider the polynomial ring $\mathcal{P} = \Bbbk[x_1, x_2, x_3]$. For a moment, we switch to the standard conventions $x_1 \succ x_2 \succ x_3$ and define $\mathcal{I}^{(d)}$ as the ideal generated by the lexsegment terminating at x_2^d for an arbitrary degree $d \in \mathbb{N}$. Thus $d = 3$ yields for example

$$\mathcal{I}^{(3)} = \langle x_1^3, x_1^2 x_2, x_1^2 x_3, x_1 x_2^2, x_1 x_2 x_3, x_1 x_3^2, x_2^3 \rangle\,.$$

The regularity of $\mathcal{I}^{(d)}$ is of course independent of the used ordering of the variables. For the ordering $x_1 \succ x_2 \succ x_3$, the above mentioned generating set is both a Janet and a Pommaret basis. Thus we find $\mathrm{reg}\,\mathcal{I}^{(d)} = d$.

For the ordering $x_1 \prec x_2 \prec x_3$, the ideal $\mathcal{I}^{(d)}$ possesses no Pommaret basis, as it contains no term of the form x_3^e for some $e \in \mathbb{N}$ which was necessary for the existence of a Pommaret basis. We can write the lexsegment as the union of the following two sets:

$$\mathcal{L}_1 = \left\{ x_1^e x_2^f \mid e \geq 0,\ f \geq 0,\ e + f = d \right\}\,,$$
$$\mathcal{L}_2 = \left\{ x_1^e x_2^f x_3^g \mid e > 0,\ f \geq 0,\ 1 \leq g \leq d - 1,\ e + f + g = d \right\}\,.$$

Then the Janet basis of $\mathcal{I}^{(d)}$ is given by

$$\mathcal{H}^{(d)} = \mathcal{L}_1 \cup \mathcal{L}_2 \cup \left\{ x_2^d x_3^g \mid 1 \leq g \leq d - 1 \right\} \cup$$
$$\left\{ x_1^e x_2^{f+f'} x_3^g \mid x_1^e x_2^f x_3^g \in \mathcal{L}_2,\ f' > 0,\ f + f' < d \right\}\,.$$

The degree of this basis is $2d - 1$ and hence $\deg \mathcal{H}^{(d)} - \mathrm{reg}\,\mathcal{I}^{(d)} = d - 1$ can become arbitrarily large.

3 Free Resolutions with Janet Bases

Theorem 6 describes only the shape of the induced resolution; it provides no information about the higher syzygies. In [2], it is shown how algebraic discrete Morse theory allows an explicit determination of all differentials in the case of a Pommaret basis. Now we will show here that this is also possible with Janet bases. First we recall some of the material presented in [2] and then point out where the crucial differences occur.

Definition 9. *A graded polynomial module $\mathcal{M} \subseteq \mathcal{P}^t$ has initially linear syzygies, if $\mathcal{M}$ possesses a finite presentation*

$$0 \longrightarrow \ker \eta \longrightarrow \mathcal{W} = \bigoplus_{\alpha=1}^{s} \mathcal{P}\mathbf{w}_\alpha \xrightarrow{\eta} \mathcal{M} \longrightarrow 0 \qquad (3)$$

such that with respect to some term order $\prec$ on the free module $\mathcal{W}$ the leading module lt $\ker \eta$ *is generated by terms of the form $x_j \mathbf{w}_\alpha$. We say that $\mathcal{M}$ has initially linear minimal syzygies, if the presentation is minimal in the sense that $\ker \eta \subseteq \mathfrak{m}^s$ with $\mathfrak{m} = \langle x_1, \ldots, x_n \rangle$ the homogeneous maximal ideal.*

The construction begins with the following two-sided Koszul complex $(\mathcal{F}, d_\mathcal{F})$ defining a free resolution of $\mathcal{M}$. Let $\mathcal{V}$ be a $\Bbbk$-linear space with basis $\{\mathbf{v}_1, \ldots, \mathbf{v}_n\}$ (with n still the number of variables in $\mathcal{P}$) and set $\mathcal{F}_j = \mathcal{P} \otimes_\Bbbk \Lambda_j \mathcal{V} \otimes_\Bbbk \mathcal{M}$ which obviously yields a free $\mathcal{P}$-module. Choosing a $\Bbbk$-linear basis $\{m_a \mid a \in A\}$ of $\mathcal{M}$, a $\mathcal{P}$-linear basis of $\mathcal{F}_j$ is given by the elements $1 \otimes v_\mathbf{k} \otimes m_a$ with ordered sequences $\mathbf{k}$ of length j. The differential is now defined by

$$d_\mathcal{F}(1 \otimes \mathbf{v}_\mathbf{k} \otimes m_a) = \sum_{i=1}^{j} (-1)^{i+1} \left(x_{k_i} \otimes \mathbf{v}_{\mathbf{k}_i} \otimes m_a - 1 \otimes \mathbf{v}_{\mathbf{k}_i} \otimes x_{k_i} m_a \right) \qquad (4)$$

where $\mathbf{k}_i$ denotes the sequence $\mathbf{k}$ without the element k_i. Here it should be noted that the second term on the right hand side is not yet expressed in the chosen $\Bbbk$-linear basis of $\mathcal{M}$. For notational simplicity, we will drop in the sequel the tensor sign $\otimes$ and leading factors 1 when writing elements of $\mathcal{F}_\bullet$.

Under the assumption that the module $\mathcal{M}$ has initially linear syzygies via a presentation (3), Sköldberg [15] constructs a Morse matching leading to a smaller resolution $(\mathcal{G}, d_\mathcal{G})$. He calls the variables

$$\text{crit} (\mathbf{w}_\alpha) = \{x_j \mid x_j \mathbf{w}_\alpha \in \text{lt} \ker \eta\} ; \qquad (5)$$

critical for the generator $\mathbf{w}_\alpha$; the remaining *non-critical* ones are contained in the set ncrit $(\mathbf{w}_\alpha)$. Then a $\Bbbk$-linear basis of $\mathcal{M}$ is given by all elements $x^\mu \mathbf{h}_\alpha$ with $\mathbf{h}_\alpha = \eta(\mathbf{w}_\alpha)$ and $x^\mu \in \Bbbk[\text{ncrit} (\mathbf{w}_\alpha)]$. We define $\mathcal{G}_j \subseteq \mathcal{F}_j$ as the free submodule generated by those vertices $\mathbf{v}_\mathbf{k} \mathbf{h}_\alpha$ where the ordered sequences $\mathbf{k}$ are of length j and such that every entry k_i is critical for $\mathbf{w}_\alpha$. In particular $\mathcal{W} \cong \mathcal{G}_0$ with an isomorphism induced by $\mathbf{w}_\alpha \mapsto \mathbf{v}_\emptyset \mathbf{h}_\alpha$.

The description of the differential $d_\mathcal{G}$ is based on reduction paths in the associated Morse graph (for a detailed treatment of these notions, see [2,9,14]) and expresses the differential as a triple sum. If we assume that, after expanding the right hand side of (4) in the chosen $\Bbbk$-linear basis of $\mathcal{M}$, the differential of the complex $\mathcal{F}_\bullet$ can be expressed as

$$d_\mathcal{F}(\mathbf{v}_\mathbf{k} \mathbf{h}_\alpha) = \sum_{\mathbf{m},\mu,\gamma} Q_{\mathbf{m},\mu,\gamma}^{\mathbf{k},\alpha} \mathbf{v}_\mathbf{m}(x^\mu \mathbf{h}_\gamma) , \qquad (6)$$

then $d_\mathcal{G}$ is defined by

$$d_\mathcal{G}(\mathbf{v}_\mathbf{k} \mathbf{h}_\alpha) = \sum_{\ell,\beta} \sum_{\mathbf{m},\mu,\gamma} \sum_p \rho_p(Q_{\mathbf{m},\mu,\gamma}^{\mathbf{k},\alpha} \mathbf{v}_\mathbf{m}(x^\mu \mathbf{h}_\gamma)) \qquad (7)$$

where the first sum ranges over all ordered sequences ℓ which consists entirely of critical indices for $\mathbf{w}_\beta$. Moreover the second sum may be restricted to all values such that a polynomial multiple of $\mathbf{v_m}(x^\mu \mathbf{h}_\gamma)$ effectively appears in $d_{\mathcal{F}}(\mathbf{v_k h}_\alpha)$ and the third sum ranges over all reduction paths p going from $\mathbf{v_m}(x^\mu \mathbf{h}_\gamma)$ to $\mathbf{v_\ell h}_\beta$. Finally ρ_p is the reduction associated with the reduction path p satisfying

$$\rho_p\big(\mathbf{v_m}(x^\mu \mathbf{h}_\gamma)\big) = q_p \mathbf{v_\ell h}_\beta \tag{8}$$

for some polynomial $q_p \in \mathcal{P}$.

A key point for applying this construction in the context of involutive bases is that any Janet basis has initially linear syzygies. Thus given a Janet basis we have two resolutions available: (1) and the one obtained by Sköldberg's construction. The main result of this section will be that the two are isomorphic.

Lemma 10. *Let* $\mathcal{H} = \{\mathbf{h}_1, \ldots, \mathbf{h}_s\}$ *be the Janet basis of the polynomial submodule* $\mathcal{U} \subseteq \mathcal{P}^t$. *Then* $\mathcal{U}$ *has initially linear syzygies[2] for the Schreyer order* $\prec_{\mathcal{H}}$ *and* $\mathrm{crit}(\mathbf{w}_\alpha) = \overline{\mathcal{X}}_{J,\mathcal{H},\prec}(\mathbf{h}_\alpha)$, *i.e. the critical variables of the generator* $\mathbf{w}_\alpha$ *are the non-multiplicative variables of* $\mathbf{h}_\alpha = \eta(\mathbf{w}_\alpha)$.

The reduction paths can be divided into elementary ones of length two. There are essentially three types of reductions paths [2, Section 4]. The elementary reductions of *type 0* are not of interest [2, Lemma 4.5]. All other elementary reductions paths are of the form

$$\mathbf{v_k}(x^\mu \mathbf{h}_\alpha) \longrightarrow \mathbf{v}_{\mathbf{k}\cup i}\big(\frac{x^\mu}{x_i}\mathbf{h}_\alpha\big) \longrightarrow \mathbf{v}_\ell(x^\nu \mathbf{h}_\beta).$$

Here $\mathbf{k}\cup i$ is the ordered sequence which arises when i is inserted into $\mathbf{k}$; likewise $\mathbf{k}\setminus i$ stands for the removal of an index $i \in \mathbf{k}$.

Type 1: Here $\ell = (\mathbf{k} \cup i)\setminus j$, $x^\nu = \frac{x^\mu}{x_i}$ and $\beta = \alpha$. Note that $i = j$ is allowed. We define $\epsilon(i; \mathbf{k}) = (-1)^{|\{j \in \mathbf{k} | j > i\}|}$. Then the corresponding reduction is

$$\rho(\mathbf{v_k} x^\mu \mathbf{h}_\alpha) = \epsilon(i; \mathbf{k} \cup i)\epsilon(j; \mathbf{k} \cup i)x_j \mathbf{v}_{(\mathbf{k}\cup i)\setminus j}\big(\frac{x^\mu}{x_i}\mathbf{h}_\alpha\big).$$

Type 2: Now $\ell = (\mathbf{k} \cup i) \setminus j$ and $x^\nu \mathbf{h}_\beta$ appears in the involutive standard representation of $\frac{x^\mu x_j}{x_i}\mathbf{h}_\alpha$ with the coefficient $\lambda_{j,i,\alpha,\mu,\nu,\beta} \in \Bbbk$. In this case, by construction of the Morse matching, we have $i \neq j$. The reduction is

$$\rho(\mathbf{v_k} x^\mu \mathbf{h}_\alpha) = -\epsilon(i; \mathbf{k} \cup i)\epsilon(j; \mathbf{k} \cup i)\lambda_{j,i,\alpha,\mu,\nu,\beta}\mathbf{v}_{(\mathbf{k}\cup i)\setminus j}(x^\nu \mathbf{h}_\beta).$$

These reductions follow from the differential (4): The summands appearing there are either of the form $x_{k_i}\mathbf{v}_{k_i} m_a$ or of the form $\mathbf{v}_{k_i}(x_{k_i} m_a)$. For each of these summands, we have a directed edge in the Morse graph $\Gamma_{\mathcal{F}_\bullet}^A$. Thus for an elementary reduction path

$$\mathbf{v_k}(x^\mu \mathbf{h}_\alpha) \longrightarrow \mathbf{v}_{\mathbf{k}\cup i}\big(\frac{x^\mu}{x_i}\mathbf{h}_\alpha\big) \longrightarrow \mathbf{v}_\ell(x^\nu \mathbf{h}_\beta),$$

---

[2] Note that we apply here Definition 9 directly to $\mathcal{U}$ and not to $\mathcal{M} = \mathcal{P}^t/\mathcal{U}$, i.e. in (3) one must replace $\mathcal{M}$ by $\mathcal{U}$.

the second edge can originate from summands of either form. For the first form we then have an elementary reduction path of type 1 and for the second form we have type 2.

For completeness, we repeat some simple results from [2] which we need to show that the free resolution $\mathcal{G}$ is isomorphic to the resultion induced by a Janet basis $\mathcal{H}$. Some of the proofs in [2] use the class of a generator in $\mathcal{H}$, a notion arising in the context of Pommaret bases. When working with Janet bases, one has to replace it by the index of the maximal multiplicative variable.

Lemma 11. [2, Lemma 4.3] For a non-multiplicative index[3] $i \in \operatorname{crit}(\mathbf{h}_\alpha)$ let $x_i \mathbf{h}_\alpha = \sum_{\beta=1}^{s} P_\beta^{(\alpha;i)} \mathbf{h}_\beta$ be the involutive standard representation. Then we have $d_{\mathcal{G}}(\mathbf{v}_i \mathbf{h}_\alpha) = x_i \mathbf{v}_\emptyset \mathbf{h}_\alpha - \sum_{\beta=1}^{s} P_\beta^{(\alpha;i)} \mathbf{v}_\emptyset \mathbf{h}_\beta$.

The next result states that if one starts at a vertex $\mathbf{v}_i(x^\mu \mathbf{h}_\alpha)$ with certain properties and follows through all possible reduction paths in the graph, one will never get to a point where one must calculate an involutive standard representation. If there are no critical (i. e. non-multiplicative) variables present at the starting point, then this will not change throughout any reduction path. In order to generalise this lemma to higher homological degrees, one must simply replace the conditions $i \in \operatorname{ncrit}(\mathbf{h}_\alpha)$ and $j \in \operatorname{ncrit}(\mathbf{h}_\beta)$ by ordered sequences $\mathbf{k}, \boldsymbol{\ell}$ with $\mathbf{k} \subseteq \operatorname{ncrit}(\mathbf{h}_\alpha)$ and $\boldsymbol{\ell} \subseteq \operatorname{ncrit}(\mathbf{h}_\beta)$.

Lemma 12. [2, Lemma 4.4] Assume that $i \cup \operatorname{supp}(\mu) \subseteq \operatorname{ncrit}(\mathbf{h}_\alpha)$. Then for any reduction path $p = \mathbf{v}_i(x^\mu \mathbf{h}_\alpha) \to \cdots \to \mathbf{v}_j(x^\nu \mathbf{h}_\beta)$ we have $j \in \operatorname{ncrit}(\mathbf{h}_\beta)$. In particular, in this situation there is no reduction path $p = \mathbf{v}_i(x^\mu \mathbf{h}_\alpha) \to \cdots \to \mathbf{v}_k \mathbf{h}_\beta$ with $k \in \operatorname{crit}(\mathbf{h}_\beta)$.

In the sequel, we use Schreyer orders on the components of the complex $\mathcal{G}$. We define $\mathcal{H}_0$ as the Janet basis of $d_{\mathcal{G}}(\mathcal{G}_1) \subseteq \mathcal{G}_0$ with respect to the Schreyer order $\prec_\mathcal{H}$ induced by the term order $\prec$ on $\mathcal{P}^t$ and $\mathcal{H}_i$ as the Janet basis of $d_{\mathcal{G}}(\mathcal{G}_{i+1}) \subseteq \mathcal{G}_i$ for the Schreyer order $\prec_{\mathcal{H}_{i-1}}$.

Lemma 13. Let $\mathcal{H}$ be a Janet basis. $\mathbf{h}_\beta$ is greater or equal than $\mathbf{h}_\alpha$ according to the J-order if $\operatorname{lt}(\mathbf{h}_\beta)$ is an involutive divisor of $x^\mu \operatorname{lt}(\mathbf{h}_\alpha)$.

Proof. There is another way to compute the involutive divisor of $x^\mu \operatorname{lt}(\mathbf{h}_\alpha)$: We choose x_i, such that $\deg_i(x^\mu) \neq 0$ and compute the involutive divisor of $x_i \operatorname{lt}(\mathbf{h}_\alpha)$. Then $x^\mu \operatorname{lt}(\mathbf{h}_\alpha) = \frac{x^\mu}{x_i} x^\nu \operatorname{lt}(\mathbf{h}_\gamma)$. Then we check if $\frac{x^\mu}{x_i} x^\nu$ contains a non-multiplicative variable for $\operatorname{lt}(\mathbf{h}_\gamma)$. If not we are finished and $\operatorname{lt}(\mathbf{h}_\gamma)$ is an involutive divisor of $x^\mu \operatorname{lt}(\mathbf{h}_\alpha)$. In the other case we repeat the procedure above. This procedure must end after a finite number of steps. If this were not the case we have found a cycle in the J-graph. But this is not possible [12].

To compute the involutive divisor of $x^\mu \operatorname{lt}(\mathbf{h}_\alpha)$, we have constructed a chain $\mathbf{h}_\alpha = \mathbf{h}_{\gamma_1}, \cdots, \mathbf{h}_{\gamma_m} = \mathbf{h}_\beta$ above. Due to the procedure we see that $\mathbf{h}_{\gamma_i}$ must be smaller than $\mathbf{h}_{\gamma_{i+1}}$ according to the J-order. Hence $\mathbf{h}_\alpha$ is smaller or equal

---

[3] For notational simplicity, we will often identify sets X of variables with sets of the corresponding indices and thus simply write $i \in X$ instead of $x_i \in X$.

than $\mathbf{h}_\beta$ according to the J-order (in fact equality only happens when x^μ is multiplicative for $\mathbf{h}_\alpha$). $\qquad\square$

Lemma 14. *Let $p = \mathbf{v}_i(x^\mu \mathbf{h}_\alpha) \to \cdots \to \mathbf{v}_j(x^\nu \mathbf{h}_\beta)$ be a reduction path that appears in the differential (7) (possibly as part of a longer path). If $\rho_p\big(\mathbf{v}_i(x^\mu \mathbf{h}_\alpha)\big) = x^\kappa \mathbf{v}_j(x^\nu \mathbf{h}_\beta)$, then $\mathrm{lt}_{\prec_{\mathcal{H}_1}}(x^{\kappa+\nu}\mathbf{v}_j \mathbf{h}_\beta) \preceq_{\mathcal{H}_1} \mathrm{lt}_{\prec_{\mathcal{H}_1}}(x^\mu \mathbf{v}_i \mathbf{h}_\alpha).$*

Proof. We prove the assertion only for an elementary reduction path p and the general case follows by induction over the path length. If p is of type 1 we can easily prove the assertion by using the same arguments as for the corresponding lemma in the Pommaret case [2, Lemma 4.6].

If p is of type 2, there exists an index $j \in \mathrm{supp}(\mu)$ (implying $j \in \mathrm{ncrit}(\mathbf{h}_\alpha)$) and thus $j \in \mathcal{X}_{J,\mathcal{H},\prec}(\mathbf{h}_\alpha)$, a multi index ν and a scalar $\lambda \in \mathbb{k}$ such that $\rho_p(\mathbf{v}_i(x^\mu \mathbf{h}_\alpha)) = \lambda \mathbf{v}_j(x^\nu \mathbf{h}_\gamma)$ where $x^\nu \mathbf{h}_\gamma$ appears in the involutive standard representation of $\frac{x^\mu x_i}{x_j} \mathbf{h}_\alpha$ with a non-vanishing coefficient. Lemma 12 implies now $j \in \mathrm{crit}(\mathbf{h}_\gamma)$. By construction, $\mathrm{lt}_\prec(\frac{x_i x^\mu}{x_j}\mathbf{h}_\alpha) \succeq \mathrm{lt}_\prec(x^\nu \mathbf{h}_\gamma)$.

Here we have to distinguish between equality and strict inequality. If strict inequality holds, then also $\mathrm{lt}_\prec(x_i x^\mu) \succ \mathrm{lt}_\prec(x_j x^\nu \mathbf{h}_\gamma)$. Hence by definition of the Schreyer order we get $\mathrm{lt}_{\prec_{\mathcal{H}_1}} x^\mu \mathbf{v}_i \mathbf{h}_\alpha \succ_{\mathcal{H}_1} \mathrm{lt}_{\prec_{\mathcal{H}_1}}(x^{\kappa+\nu}\mathbf{v}_j \mathbf{h}_\beta)$. In the case of equality, we note that $x^\nu \mathrm{lt}_\prec(\mathbf{h}_\gamma)$ must be an involutive divisor of $\frac{x_i x^\mu}{x_j} \mathrm{lt}_\prec(\mathbf{h}_\alpha)$. Hence Lemma 13 guarantees that $\mathbf{h}_\alpha$ is smaller than $\mathbf{h}_\gamma$ according to the J-order and hence the claim follows for this special case. $\qquad\square$

For notational simplicity, we formulate the two decisive corollaries only for the special case of second syzygies, but they remain valid in any homological degree. They assert that there is a one-to-one correspondence between the leading terms of the syzygies contained in the free resolution (1) and of the syzygies in Sköldberg's resolution, respectively.

Corollary 15. *If $i < j$, then $\mathrm{lt}_{\prec_{\mathcal{H}_1}}\big(d_{\mathcal{G}}(\mathbf{v}_{(i,j)}\mathbf{h}_\alpha)\big) = x_j \mathbf{v}_i \mathbf{h}_\alpha$.*

Proof. We assume that the elements of the given Janet basis are numbered according to a J-order. Consider now the differential $d_{\mathcal{G}}$. We first compare the terms $x_i \mathbf{v}_j \mathbf{h}_\alpha$ and $x_j \mathbf{v}_i \mathbf{h}_\alpha$. Lemma 12 (or the minimality of these terms with respect to any order respecting the used Morse matching) entails that there are no reduction paths $[\mathbf{v}_j \mathbf{h}_\alpha \rightsquigarrow \mathbf{v}_k \mathbf{h}_\delta]$ with $k \in \mathrm{crit}(\mathbf{h}_\delta)$ (except trivial reduction paths of length 0). By definition of the Schreyer order, we have $x_i \mathbf{v}_j \mathbf{h}_\alpha \prec_{\mathcal{H}_1} x_j \mathbf{v}_i \mathbf{h}_\alpha$.

Now consider any other term in the sum. We will prove $x_j \mathbf{v}_i \mathbf{h}_\alpha \succ_{\mathcal{H}_1} x^\kappa \mathbf{v}_i \mathbf{h}_\beta$, where $x^\kappa \mathbf{h}_\beta$ effectively appears in the involutive standard representation of $x_j \mathbf{h}_\alpha$. Then the claim follows from applying Lemma 14 with $x_j \mathbf{v}_i \mathbf{h}_\alpha \succ_{\mathcal{H}_1} x^\kappa \mathbf{v}_i \mathbf{h}_\beta \succeq_{\mathcal{H}_1} \mathrm{lt}_{\prec_{\mathcal{H}_1}}\big(\rho_p(\mathbf{v}_i x^\kappa \mathbf{h}_\beta)\big)$.

We always have $\mathrm{lt}_\prec(x_j x_i \mathbf{h}_\alpha) \succeq \mathrm{lt}_\prec(x^\kappa x_i \mathbf{h}_\beta)$. If this is a strict inequality, then $x_j \mathbf{v}_i \mathbf{h}_\alpha \succ_{\mathcal{H}_1} x^\kappa \mathbf{v}_i \mathbf{h}_\beta$ follows at once by definition of the Schreyer order. So now assume $\mathrm{lt}_\prec(x_j x_i \mathbf{h}_\alpha) = \mathrm{lt}_\prec(x^\kappa x_i \mathbf{h}_\beta)$. By construction, $x^\kappa \in \mathbb{k}[\mathcal{X}_{J,\mathcal{H}}(\mathbf{h}_\beta)]$. Again by definition of the Schreyer order, the claim follows, if we can prove

$\text{lt}_{\prec_{\mathcal{H}_0}}(x_j x_i \mathbf{v}_{\emptyset} \mathbf{h}_\alpha) \succ_{\mathcal{H}_0} \text{lt}_{\prec_{\mathcal{H}_0}}(x^\kappa x_i \mathbf{v}_{\emptyset} \mathbf{h}_\beta)$. Since $j \in \text{crit}(\mathbf{h}_\alpha)$ and $\text{lt}(x_j \mathbf{h}_\alpha)$ is involutively divisible by $\text{lt}(\mathbf{h}_\beta)$, we have $\alpha < \beta$, by definition of a J-ordering. As we have $\text{lt}_\prec(x_j \mathbf{h}_\alpha) = \text{lt}_\prec(x^\kappa \mathbf{h}_\beta)$, this implies $\text{lt}_{\prec_{\mathcal{H}_0}}(x_j x_i \mathbf{v}_{\emptyset} \mathbf{h}_\alpha) \succ_{\mathcal{H}_0} \text{lt}_{\prec_{\mathcal{H}_0}}(x^\kappa x_i \mathbf{v}_{\emptyset} \mathbf{h}_\beta)$ and therefore $\text{lt}_{\prec_{\mathcal{H}_1}}(x_j \mathbf{v}_i \mathbf{h}_\alpha) \succ_{\mathcal{H}_1} \text{lt}_{\prec_{\mathcal{H}_1}}(x^\kappa \mathbf{v}_i \mathbf{h}_\beta)$. □

Corollary 16. *The set* $\{d_{\mathcal{G}}(v_\mathbf{k} \mathbf{h}_\alpha) \mid |\mathbf{k}| = 2; \mathbf{k} \subseteq \text{crit}(\mathbf{w}_\alpha)\}$ *is a Janet basis with respect to the term order* $\prec_{\mathcal{H}_0}$.

With Lemma 10 and these two corollaries, we are able to prove that Sköldberg's resolution is isomorphic to the resolution (1). The proof is essentially the same as for a Pommaret basis, only the mentioned lemmata and corollaries must be replaced by their Janet version.

Theorem 17. *Let* $\mathcal{H} = \{\mathbf{h}_1, \ldots, \mathbf{h}_s\}$ *be the Janet basis of the polynomial submodule* $\mathcal{U} \subseteq \mathcal{P}^t$, *where* $\mathcal{H}$ *is a J-ordered Janet basis of* $\mathcal{U}$. *Then the resolution* $(\mathcal{G}, d_{\mathcal{G}})$ *is isomorphic to the resolution induced by* $\mathcal{H}$.

4 Benchmarks

We describe now the results from a large set of benchmarks comparing our approach with standard methods. As already discussed in [2], we focus for this comparison on the determination of Betti numbers, as currently our implementation is not yet competitive for computing minimal resolution because of the rather naive minimisation strategy used. Indeed, an important aspect for the performance of our approach is the size of the generally non-minimal resolution (1). For a given ideal $\mathcal{I}$, we call its length the *projective pseudo-dimension* $\text{ppd}\,\mathcal{I}$ and the maximal degree of a generator appearing in it the *pseudo-regularity* $\text{preg}\,\mathcal{I}$. It follows from the results above that for the resolution induced by a Janet basis $\text{ppd}\,\mathcal{I}$ is just the maximal number of non-multiplicative variables and $\text{preg}\,\mathcal{I}$ the maximal degree of a generator. As the rough measure for the size of the whole minimal solution, we define the *Betti rank* $\text{brk}\,\mathcal{I}$ as the sum of all Betti numbers and similarly the *Betti pseudo-rank* $\text{bprk}\,\mathcal{I}$.

We have implemented the algorithms explained above in the computer algebra system COCOALIB [1]. The implementation is very similar to the one based on Pommaret bases which we described in [2]. The main difference is that we can no longer guarantee that $\text{preg}\,\mathcal{I} = \text{reg}\,\mathcal{I}$ and $\text{ppd}\,\mathcal{I} = \text{pd}\,\mathcal{I}$. But it is straightforward to accomodate for this effect. For comparison purposes, we used as benchmark the implementations of the standard algorithms in SINGULAR [4] and MACAULAY2 [7].

For many geometrical and topological applications, it is sufficient to know only the Betti numbers; the differential of the minimal resolution is not required. To our knowledge, all current implementations read off the Betti numbers from a free resolution. By contrast, our method can determine Betti numbers without computing a complete resolution. We briefly sketch our method described in more details in [2].

Firstly, we compute only the constant part of the complex $\mathcal{G}_\bullet$. If we perform an elementary reduction of type 2 the degree of the map does not change. For an elementary reduction of type 1, the degree increases by one. Thus we obtain the constant part of $\mathcal{G}_\bullet$ by only applying reductions of type 2 on the constant part of the complex $\mathcal{F}_\bullet$. It follows from the explicit form (4) of the differential $d_\mathcal{F}$ that the left summand yields always elements of degree one and the right summand elements of degree zero. Hence, by simply skipping the left summands, we directly obtain the constant part of $\mathcal{F}_\bullet$.

Because of the above proven isomorphy between the complex $\mathcal{G}_\bullet$ and the resolution induced by the Janet basis, the bigraded ranks $r_{i,j}$ of the components of the complex $\mathcal{G}_\bullet$ can be directly determined with (2). Then, as described above, we construct (degreewise) the constant part of the matrices of $d_\mathcal{G}$. Subtracting their ranks from the corresponding $r_{i,j}$ yields the Betti numbers $b_{i,j}$. It should be noted that this approach also allows to compute directly individual Betti numbers, as the explicit expressions for the differentials in the complexes $\mathcal{F}_\bullet$ and $\mathcal{G}_\bullet$, resp., show that the submatrices relevant for the different Betti numbers are independent of each other.

Our testing environment consists of an Intel i5-4570 processor with 8GB DDR3 main memory. As operating system we used Fedora 20 and as compiler for the CoCoALib gcc 4.8.3. The running times are given in seconds and we limited the maximal time usage to two hours and the maximal memory consumption to 7.5 GB. A * marks when we run out of time and ** marks when we run out of memory. A bold line indicates that the given example is δ-singular, i. e. that no Pommaret basis exists for it in the used coordinates. As benchmarks, we took a number of standard examples given in [16]. As most of these ideals are not homogeneous, we homogenised them by adding a new smallest variable. Furthermore, we always chose $\mathbb{k} = \mathbb{Z}/101\mathbb{Z}$ as base field.

Singular and Macaulay2 apply the command `res` for computing a free resolution at first. In a second step both systems extract the Betti numbers from the resolution. Singular uses the classical Schreyer Theorem to compute a free resolution, which is possibly not minimal, and then determines the graded Betti numbers from it. Macaulay2 uses La Scalas method to compute a minimal free resolution and read off the graded Betti numbers. For Singular and Macaulay2 we took as input the reduced Gröbner basis of the ideal; for our algorithm the Janet basis. Because of our choice of a small coefficient field, the time needed for the determination of these input bases is neglectable (for almost all examples less than two seconds).

The benchmarks presented in Table 1 show that our approach is generally much faster than the standard methods requiring a complete resolution (often by orders of magnitude!). In particular, it scales much better when examples are getting larger. Even for δ-singular ideals, we are in general much faster than the standard methods. In fact, there is no obvious difference between δ-singular and δ-regular examples.

In Table 2 we collect some data about the examples in Table 1. The following list describes the columns:

Table 1. Various examples for computing Betti diagramms

Example	Time Macaulay2	Time Singular	Time CoCoALib
butcher8	**126.25**	**19.92**	**1.20**
camera1s	**0.09**	**6.00**	**0.13**
chandra6	0.64	8.00	0.13
cohn2	**0.03**	**1.00**	**0.03**
cohn3	**1.47**	**5.90**	**0.32**
cpdm5	14.71	5.05	0.64
cyclic6	0.99	1.26	0.37
cyclic7	1 093.66	*	37.42
cyclic8	*	*	**1 663.00**
des18_3	433.45	20.84	3.15
des22_24	*	**	52.19
dessin1	428.13	20.89	3.10
dessin2	*	*	32.90
f633	**591.08**	**7.70**	**49.06**
hcyclic5	**0.03**	**2.00**	**0.09**
hcyclic6	**11.00**	**47.12**	**7.41**
hcyclic7	*	*	**3 688.01**
hemmecke	**0.00**	**0.00**	**2.69**
hietarinta1	**443.15**	**170.29**	**4.12**
katsura6	51.41	13.90	1.22
katsura7	**	1 373.70	15.87
katsura8	*	**	412.90
kotsireas	**51.89**	**17.84**	**0.83**
mckay	**0.84**	**3.20**	**0.38**
noon5	0.13	6.00	0.27
noon6	15.14	5.07	5.25
noon7	6 979.40	821.64	122.61
rbpl	**58.81**	**22.69**	**57.91**
redcyc5	**0.02**	**2.00**	**0.01**
redcyc6	**6.79**	**1.95**	**0.13**
redcyc7	*	*	**8.26**
redcyc8	*	**	**207.02**
redeco7	2.72	2.20	0.42
redeco8	355.30	11.83	5.01
redeco9	**	312.49	84.89
redeco10	**	**	2 694.05
reimer4	0.01	1.00	0.01
reimer5	1.39	5.00	0.35
reimer6	1 025.89	176.08	19.01
speer	**0.20**	**3.00**	**0.13**

Example: name of the example
#JB: number of elements in the minimal Janet basis
#GB: number of elements in the reduced Gröbner basis
$\frac{\#JB}{\#GB}$: the quotient of #JB and #GB
ppd: the projective pseudo-dimension
pd: the projective dimension
preg: the pseudo-regularity
reg: the regularity
bprk: the Betti pseudo-rank
brk: the Betti rank
$\frac{bprk}{brk}$: the quotient of bprk and brk.

In our test set there is only a very small subset of examples where the standard algorithms perform better than our new algorithm. For example, for *hemmecke* our method needs 2.69 seconds to compute the Betti diagram, whereas SINGULAR and MACAULAY2 do not even need a measurable amount of time. If we take a look at the size of the minimal Janet basis and the reduced Gröbner basis in Table 2, we see immediately why this example is bad for our algorithm. The reduced Gröbner basis consists of 9 elements, but the minimal Janet basis contains 983 elements.[4] Therefore the Betti pseudo-rank is with 6 242 much larger than the real Betti rank of 38. As a consequence, we must spend first much time to compute the constant part of a large resolution and then even more time for reducing it. In comparatively small examples like *cyclic6* one notices in our approach overhead effects because of the need to set up complex data structures. In general, one observes that the larger the example (in particular, the larger its projective dimension) the better our algorithm fares in comparison to the standard methods.

It seems that the quotient $\frac{\#JB}{\#GB}$ provides a good indication whether or not our algorithm is fast relative to the standard methods. One could think that the quotient $\frac{bprk}{brk}$ is also a good indicator for efficiency. But in our test set we cannot identify such a correlation. In fact, even if the factor is greater than 100, somewhat surprisingly our algorithm can be faster (see *redeco10*).

There are two aspects which may explain this observation for *redeco10*. The first one is that we only perform matrix operations over the base field, which are not only much more efficient than polynomial computations but also consume much less memory. The second one could be the relatively large projective dimension 10 of *redeco10*. Classical methods have to compute roughly pd$\mathcal{I}$ Gröbner bases to determine the Betti numbers. Our approach requires always only one Janet basis and some normal form computations.

Another interesting observation in Table 2 concerns the difference of the projective pseudo-dimension and the pseudo-regularity to the true values pd$\mathcal{I}$ and reg$\mathcal{I}$ for δ-singular ideals. In our test set only for two examples (*hcylic5* and *mckay*) the values differ and in only one of them (*mckay*) a significant difference occurs. Thus it seems that in typical benchmark examples no big differences in

---

[4] Although this example is not a toric ideal, it shares certain characteristic features of toric ideals. It is well-known that for such ideals special techniques must be employed.

Table 2. Statistics for examples from Table 1

Example	#JB	#GB	$\frac{\text{\#JB}}{\text{\#GB}}$	ppd	pd	preg	reg	bprk	brk	$\frac{\text{bprk}}{\text{brk}}$
butcher8	64	54	1.19	8	8	3	3	3 732	2 631	1.42
camera1s	59	29	2.03	6	6	4	4	863	337	2.56
chandra6	32	32	1.00	6	6	5	5	684	64	10.69
cohn2	33	23	1.43	4	4	7	7	179	67	2.67
cohn3	106	92	1.15	4	4	7	7	696	370	1.88
cpdm5	83	77	1.08	5	5	9	9	1 020	100	10.20
cyclic6	46	45	1.02	6	6	9	9	1 060	320	3.31
cyclic7	210	209	1.00	7	7	11	11	10 356	1 688	6.14
cyclic8	384	372	1.03	8	8	12	12	34 136	6 400	5.33
des18_3	104	39	2.67	8	8	4	4	8 132	2 048	3.97
des22_24	129	45	2.87	10	10	4	4	32 632	6 192	5.27
dessin1	104	39	2.67	8	8	4	4	8 132	2 048	3.97
dessin2	122	46	2.65	10	10	4	4	22 760	6 192	3.68
f633	153	47	3.26	10	10	3	3	17 390	4 987	3.49
hcyclic5	52	38	1.37	6	5	11	10	932	32	29.13
hcyclic6	221	99	2.23	7	7	14	14	9 834	146	67.36
hcyclic7	1 182	443	2.67	8	8	17	17	105 957	1 271	83.37
hemmecke	983	9	109.22	4	4	61	61	6 242	38	164.26
hietarinta1	52	51	1.02	10	10	2	2	6 402	3 615	1.77
katsura6	43	41	1.05	7	7	6	6	1 812	128	14.16
katsura7	79	74	1.07	8	8	7	7	6 900	256	26.95
katsura8	151	143	1.06	9	9	8	8	27 252	512	53.23
kotsireas	78	70	1.11	6	6	5	5	1 810	1 022	1.77
mckay	126	51	2.47	4	4	15	9	840	248	3.39
noon5	137	72	1.90	5	5	8	8	1 618	130	12.45
noon6	399	187	2.13	6	6	10	10	9 558	322	29.68
noon7	1 157	495	2.34	7	7	12	12	56 666	770	73.59
rbpl	309	126	2.45	7	7	14	14	13 834	1 341	10.32
redcyc5	23	10	2.30	5	5	7	7	276	88	3.14
redcyc6	46	21	2.19	6	6	9	9	1 060	320	3.31
redcyc7	210	78	2.69	7	7	11	11	10 356	1 688	6.14
redcyc8	371	193	1.92	8	8	12	12	32 459	6 973	4.65
redeco7	48	33	1.45	7	7	5	5	1 708	128	13.34
redeco8	96	65	1.48	8	8	6	6	6 828	256	26.67
redeco9	192	129	1.49	9	9	7	7	27 308	512	53.34
redeco10	384	257	1.49	10	10	8	8	109 228	1 024	106.67
reimer4	19	17	1.12	4	4	6	6	118	16	7.38
reimer5	55	38	1.45	5	5	9	9	694	32	21.69
reimer6	199	95	2.09	6	6	12	12	5 302	64	82.84
speer	49	44	1.11	5	5	7	7	359	133	2.70

the sizes of the induced resolutions for Janet and Pommaret bases, respectively, occur, although we showed in Example 8 that theoretically the difference may become arbitrarily large.

References

1. Abbott, J., Bigatti, A.: CoCoALIB 0.99535: a C++ library for doing computations in commutative algebra (2014). http://cocoa.dima.unige.it/cocoalib
2. Albert, M., Fetzer, M., Sáenz-de Cabezón, E., Seiler, W.: On the free resolution induced by a Pommaret basis. J. Symb. Comp. **68**, 4–26 (2015)
3. Albert, M., Hashemi, A., Pytlik, P., Schweinfurter, M., Seiler, W.: Deterministic genericity for polynomial ideals. Preprint in preparation (2015)
4. Decker, W., Greuel, G.M., Pfister, G., Schönemann, H.: Singular 4-0-2 – a computer algebra system for polynomial computations (2012). http://www.singular.uni-kl.de
5. Gerdt, V., Blinkov, Y.: Involutive bases of polynomial ideals. Math. Comp. Simul. **45**, 519–542 (1998)
6. Gerdt, V., Blinkov, Y.: Minimal involutive bases. Math. Comp. Simul. **45**, 543–560 (1998)
7. Grayson, D., Stillman, M.: Macaulay2 1.6, a software system for research in algebraic geometry (2013). http://www.math.uiuc.edu/Macaulay2/
8. Hausdorf, M., Sahbi, M., Seiler, W.: δ- and quasi-regularity for polynomial ideals. In: Calmet, J., Seiler, W., Tucker, R. (eds.) Global Integrability of Field Theories, pp. 179–200. Universitätsverlag Karlsruhe, Karlsruhe (2006)
9. Jöllenbeck, M., Welker, V.: Minimal Resolutions via Algebraic Discrete Morse Theory. Memoirs Amer. Math. Soc. 197, Amer. Math. Soc. (2009)
10. Schreyer, F.: Die Berechnung von Syzygien mit dem verallgemeinerten Weierstraßschen Divisionssatz. Master's thesis, Fakultät für Mathematik, Universität Hamburg (1980)
11. Seiler, W.: A combinatorial approach to involution and δ-regularity I: Involutive bases in polynomial algebras of solvable type. Appl. Alg. Eng. Comm. Comp. **20**, 207–259 (2009)
12. Seiler, W.: A combinatorial approach to involution and δ-regularity II: Structure analysis of polynomial modules with Pommaret bases. Appl. Alg. Eng. Comm. Comp. **20**, 261–338 (2009)
13. Seiler, W.: Involution – The Formal Theory of Differential Equations and its Applications in Computer Algebra. Algorithms and Computation in Mathematics, vol. 24. Springer-Verlag, Berlin (2010)
14. Sköldberg, E.: Morse theory from an algebraic viewpoint. Trans. Amer. Math. Soc. **358**, 115–129 (2006)
15. Sköldberg, E.: Resolutions of modules with initially linear syzygies. Preprint arXiv:1106.1913 (2011)
16. Yanovich, D., Blinkov, Y., Gerdt, V.: Benchmarking (2001). http://invo.jinr.ru/ginv/benchmark.html

Regular Chains under Linear Changes of Coordinates and Applications

Parisa Alvandi[2], Changbo Chen[1], Amir Hashemi[3,4], and Marc Moreno Maza[2]

[1] Chongqing Key Laboratory of Automated Reasoning and Cognition, Chongqing Institute of Green and Intelligent Technology, Chinese Academy of Sciences
[2] ORCCA, University of Western Ontario
[3] Department of Mathematical Sciences, Isfahan University of Technology Isfahan, 84156-83111, Iran
[4] School of Mathematics, Institute for Research in Fundamental Sciences (IPM), Tehran, 19395-5746, Iran
palvandi@uwo.ca, changbo.chen@hotmail.com, amir.hashemi@cc.iut.ac.ir, moreno@csd.uwo.ca

Abstract. Given a regular chain, we are interested in questions like computing the limit points of its quasi-component, or equivalently, computing the variety of its saturated ideal. We propose techniques relying on linear changes of coordinates and we consider strategies where these changes can be either generic or guided by the input.

1 Introduction

Applying a change of coordinates to the algebraic or differential representation of a geometrical object is a fundamental technique to obtain a more convenient representation and reveal properties. For instance, random linear change of coordinates are performed in algorithms for solving systems of polynomial equations and inequations in the algorithms of Krick and Logar [11], Rouillier [15], Verschelde [17] and Lecerf [12].

For polynomial ideals, one desirable representation is Noether normalization, which was introduced by Emmy Noether in 1926. We refer to the books [9,8] for an account on this notion and, to the articles [16,10] for deterministic approaches to compute Noether normalization, when the input ideal is given by a Gröbner basis.

In regular chain theory, one desirable and challenging objective is, given a regular chain T, to obtain the (non-trivial) limit points of its quasi-component $W(T)$, or equivalently, computing the variety of its saturated ideal $\mathrm{sat}(T)$. The set $\lim(W(T))$ of the non-trivial limit points of $W(T)$ satisfies $V(\mathrm{sat}(T)) = \overline{W(T)} = W(T) \cup \lim(W(T))$. Hence, $\lim(W(T))$ is the set-theoretic difference $V(\mathrm{sat}(T)) \backslash W(T)$. Deducing $\lim(W(T))$ or $V(\mathrm{sat}(T))$ from T is a central question which has theoretical applications (like the so-called *Ritt Problem*) and practical ones (like removing redundant components in triangular decomposition).

Of course $V(\mathrm{sat}(T))$ can be computed from T via Gröbner basis techniques. But this approach is of limited practical interest. In fact, considering the case

© Springer International Publishing Switzerland 2015
V.P. Gerdt et al. (Eds.): CASC 2015, LNCS 9301, pp. 30–44, 2015.
DOI: 10.1007/978-3-319-24021-3_3

where the base field is $\mathbb{Q}$, we are looking for approaches that would run in polynomial time w.r.t. the degrees and coefficient heights of T. Thanks to the work of [7], algorithms for change of variable orders (and more generally, algorithms for linear changes of coordinates) are good candidates.

Returning to Noether normalization[1], we ask in Section 4 how "simple" can T be if we assume that sat(T) is in Noether position. Unfortunately, an additional hypothesis is needed in order to obtain a satisfactory answer like "all initials of T are constant", see Theorem 1 and Remark 1.

In Section 5 and 6, we develop a few criteria for computing $\lim(W(T))$ or $V(\text{sat}(T))$. Our techniques (see Proposition 2, Theorem 2, Theorem 3, Theorem 4 and Algorithm 1) rely on linear changes of coordinates and allow us to relax the "dimension one" hypothesis in our previous paper [1], where $\lim(W(T))$ was computed via Puiseux series.

Therefore, the techniques proposed in this paper can be used to compute $\lim(W(T))$ or $V(\text{sat}(T))$ without Gröbner basis or Puiseux series calculations. Moreover, these new techniques can handle cases where the results of our previous paper [1] could not apply. One of the main ideas of our new results (see for instance Theorem 2) is to use a linear change of coordinates so as to replace the description of $\overline{W(T)}$ by one for which $\overline{W(T)} \cap V(h_T)$ can be computed by means of standard operations on regular chains. Nevertheless, our proposed techniques do not cover all possible cases and the problem of finding a "Gröbner-basis-free" general algorithm for $\lim(W(T))$ or $V(\text{sat}(T))$ remains unsolved.

2   Preliminaries

Throughout this paper, polynomials have coefficients in a field $\mathbf{k}$ and variables in a set $\mathbf{x}$ of n ordered variables $x_1 < \cdots < x_n$. The corresponding polynomial ring is denoted by $\mathbf{k}[\mathbf{x}]$. Let F be a subset of $\mathbf{k}[\mathbf{x}]$. We denote by $\langle F \rangle$ the ideal generated by F in $\mathbf{k}[\mathbf{x}]$. Recall that a polynomial $f \in \mathbf{k}[\mathbf{x}]$ is *regular* modulo the ideal $\langle F \rangle$ whenever f does not belong to any prime ideals associated with $\langle F \rangle$, thus, whenever f is neither null nor a zero-divisor modulo $\langle F \rangle$. Further, $\overline{\mathbf{k}}$ stands for the algebraic closure of $\mathbf{k}$ and $V(F) \subset \overline{\mathbf{k}}^n$ for the algebraic set consisting of all common zeros of all $f \in F$. For a set $W \subset \overline{\mathbf{k}}^n$, we denote by $\overline{W}$ the *Zariski closure* of W, that is, the intersection of all algebraic sets containing W.

We briefly review standard notions and concepts related to regular chains and we refer to [2,6] for details. For a non-constant $f \in \mathbf{k}[\mathbf{x}]$, we denote by $\text{mvar}(f)$, $\text{mdeg}(f)$ and $\text{init}(f)$, the variable of greatest rank appearing in f, the degree of f w.r.t. that variable and the leading coefficient of f w.r.t. that same variable. The quantities $\text{mvar}(f)$, $\text{mdeg}(f)$ and $\text{init}(f)$ are called respectively the *main variable*, *main degree* and *initial* of f. A set T of non-constant polynomials from $\mathbf{k}[\mathbf{x}]$ is called *triangular* if no two polynomials from T have the same main variable. Let $T \subset \mathbf{k}[\mathbf{x}]$ be a triangular set. Observe that T is necessarily finite and that every subset of T is itself triangular. For a variable $v \in \mathbf{x}$, if there exists

---

[1] Section 4 contains a brief review of Noether normalization which makes our paper self-contained.

$f \in T$ such that $\mathrm{mvar}(f) = v$, we denote this polynomial by T_v and say that v is *algebraic* w.r.t. T, otherwise we say that v is *free* w.r.t. T; in all cases, we define $T_{<v} := \{g \in T \mid \mathrm{mvar}(g) < v\}$ and denote by $\mathrm{free}(T)$ the set of the variables from $\mathbf{x}$ which are free w.r.t. T. We denote by h_T the product of the polynomials $\mathrm{init}(f)$, for $f \in T$. We say that T is *strongly normalized* if all variables occurring in h_T are in $\mathrm{free}(T)$; when this holds, it is easy to check that T is a Gröbner basis of the ideal that T generates in $\mathbf{k}(\mathbf{u})[\mathbf{x} \setminus \mathbf{u}]$ where $\mathbf{u} := \mathrm{free}(T)$ and $\mathbf{k}(\mathbf{u})$ is the field of rational functions over $\mathbf{k}$ and with variables in $\mathbf{u}$. Moreover, we say that T is *monic* whenever $h_T \in \mathbf{k}$ holds. The *saturated ideal* of T, written $\mathrm{sat}(T)$, is defined as the column ideal $\mathrm{sat}(T) = \langle T \rangle : h_T^{\infty}$. The *quasi-component* of T is the basic constructible set given by $W(T) := V(T) \setminus V(h_T)$. The following two properties are easy to prove:

$$\overline{W(T)} \;=\; V(\mathrm{sat}(T)) \quad \text{and} \quad \overline{W(T)} \;=\; W(T) \;\cup\; \lim(W(T)), \qquad (1)$$

where $\lim(W(T)) := \overline{W(T)} \cap V(h_T)$ holds and the points of that latter set are called the *(non-trivial) limit points* of $W(T)$, for reasons explained in [1]. We say that T is a *regular chain* whenever T is empty or $T_{<w}$ is a regular chain and the initial of T_w is regular modulo $\mathrm{sat}(T_{<w})$, where w is the largest main variable of a polynomial in T. If T consists of $n - d$ polynomials, for $0 \leq d < n$, then $\mathrm{sat}(T)$ has dimension d and either $\lim(W(T))$ is empty or has dimension $d - 1$; moreover, we have $\mathbf{k}[\mathbf{u}] \cap \mathrm{sat}(T) = \langle 0 \rangle$, where $\mathbf{u} := \mathrm{free}(T)$.

Let $F \subset \mathbf{k}[\mathbf{x}]$ be finite. Let $T_1, \ldots, T_e$ be finitely many regular chains of $\mathbf{k}[\mathbf{x}]$. We say that $\{T_1, \ldots, T_e\}$ is a *Kalkbrener triangular decomposition* of $V(F)$ if we have $V(F) = \cup_{i=1}^{e} \overline{W(T_i)}$. We say that $\{T_1, \ldots, T_e\}$ is a *Lazard-Wu triangular decomposition* of $V(F)$ if we have $V(F) = \cup_{i=1}^{e} W(T_i)$.

We call *linear change of coordinates in* $\overline{\mathbf{k}}^n$ any bijective map A of the form

$$A : \; \overline{\mathbf{k}}^n \; \to \; \overline{\mathbf{k}}^n \\ \mathbf{x} \; \longmapsto \; (A_1(\mathbf{x}), \ldots, A_n(\mathbf{x})) \qquad (2)$$

where $A_1, \ldots, A_n$ are linear forms over $\overline{\mathbf{k}}$. Hence $A(\mathbf{x})$ can be written as $M\mathbf{x}$ where M is an invertible matrix over $\overline{\mathbf{k}}$. For the algebraic set $V(F)$, we denote $V^A(F) := V(\{f^A \mid f \in F\})$, where $f^A(\mathbf{x}) := f(A_1(\mathbf{x}), \ldots, A_n(\mathbf{x}))$. Observe that if $V(F)$ is irreducible, then so is $V^A(F)$. Similarly, the image of $W(T)$ under A is $W^A(T) = V^A(T) \setminus V^A(h_T)$.

3   Algorithm for Linear Change of Coordinates

The goal of this section is to explain how to obtain a practically efficient algorithmic solution to the following problem.

Problem 1. *Given a regular chain $T \subset \mathbf{k}[\mathbf{x}]$ and given a linear change of coordinates A in $\overline{\mathbf{k}}^n$, compute finitely many regular chains $C_1, \ldots, C_e$ such that*

$$\overline{W^A(T)} \;=\; \overline{W(C_1)} \cup \cdots \cup \overline{W(C_e)}.$$

In the literature, see [3,4,7], the following related problem has been addressed.

Problem 2. *Given two total orderings $\mathcal{R}$ and $\overline{\mathcal{R}}$ on $\{x_1, \ldots, x_n\}$, given $T \subset \mathbf{k}[x_1, \ldots, x_n]$, assuming that*

1. *T is a regular chain for the ordering $\mathcal{R}$ on $\{x_1, \ldots, x_n\}$ and,*
2. *the saturated ideal $\mathrm{sat}(T, \mathcal{R})$ (which is an alias of $\mathrm{sat}(T)$ with a second argument recalling the ordering) of T of $\mathbf{k}[x_1, \ldots, x_n]$ is prime,*

compute $C \subset \mathbf{k}[x_1, \ldots, x_n]$ such that

3. *C is a regular chain for the ordering $\overline{\mathcal{R}}$ on $\{x_1, \ldots, x_n\}$ and,*
4. *the saturated ideal $\mathrm{sat}(C, \overline{\mathcal{R}})$ of C in $\mathbf{k}[x_1, \ldots, x_n]$ is equal to $\mathrm{sat}(T, \mathcal{R})$.*

We call this second problem *change of variable order*. The articles [3,4] are actually dedicated to the case of differential regular chains, where a differential counterpart of Problem 2 is termed *ranking conversion*. However, these articles suggest that, from the differential case, a solution to Problem 2 could be derived and they call it `PALGIE`, which is an acronym for Prime ALGebraic IdEal.

Next, towards Problem 1, we consider the following extension of Problem 2 where the primality assumption is relaxed.

Problem 3. *Given two total orderings $\mathcal{R}$ and $\overline{\mathcal{R}}$ on $\{x_1, \ldots, x_n\}$, given $T \subset \mathbf{k}[x_1, \ldots, x_n]$, assuming that T is a regular chain for the ordering $\mathcal{R}$ on $\{x_1, \ldots, x_n\}$, compute finitely many regular chains $C_1, \ldots, C_e$ such that the radical of the saturated ideal $\mathrm{sat}(T, \mathcal{R})$ of T in $\mathbf{k}[x_1, \ldots, x_n]$ is equal to the intersection of the radicals of the saturated ideals $\mathrm{sat}(C_i, \overline{\mathcal{R}})$ of C_i in $\mathbf{k}[x_1, \ldots, x_n]$, for $1 \le i \le e$.*

Extending the `PALGIE` algorithm (as suggested in [3]) to a solution of Problem 3 can be achieved by standard techniques from regular chain theory, see [6].

Before further extending the `PALGIE` algorithm to a solution of Problem 1, we argue that Problem 2 deals with a special case of Problem 1, that is, ranking conversions are, indeed, a special case of linear change of coordinates.

As in the statement of Problem 2, consider two total orderings $\mathcal{R}$ and $\overline{\mathcal{R}}$ on $\{x_1, \ldots, x_n\}$ as well as a regular chain $T \subset \mathbf{k}[x_1, \ldots, x_n]$ for the order $\mathcal{R}$ such that its saturated ideal $\mathrm{sat}(T, \mathcal{R})$ is prime. W.l.o.g. we can assume that the order $\mathcal{R}$ on $\{x_1, \ldots, x_n\}$ is given by $x_1 < \cdots < x_n$. Then, the change of variable order from $\mathcal{R}$ to $\overline{\mathcal{R}}$ can be interpreted as a permutation σ of the sequence $(x_1, \ldots, x_n)$. Let A be the linear change of coordinates replacing the column vector $(x_1, \ldots, x_n)^t$ with $M_\sigma (y_1, ..., y_n)^t$ where $(y_1, \ldots, y_n)$ stand for the new coordinates and M_σ is the matrix of σ w.r.t. the canonical basis of $\overline{\mathbf{k}}^n$ as a vector space over $\overline{\mathbf{k}}$. Running the extended version of the `PALGIE` algorithm solving Problem 1 we obtain a regular chain C such that we have

$$\mathrm{sat}(C) = \mathrm{sat}(T)^A.$$

Then simply renaming y_i with $x_{\sigma(i)}$, for $1 \le i \le n$, in C produces a regular chain D satisfying the output specifications of the original version of the `PALGIE` algorithm whose purpose is to perform change of variable order. To make the proof strict, requiring that T and D be strongly normalized (and reduced Gröbner bases over the field of rational functions $\mathbf{k}(\mathsf{free}(T))$) make them unique which completes the proof.

We turn our attention back to Problem 1 and suggest how a solution of Problem 3 can lead to a solution of Problem 1. Let $T \subset \mathbf{k}[\mathbf{x}]$ be a regular chain and let A be a linear change of coordinates in $\overline{\mathbf{k}}^n$. We denote by d the dimension of $\mathsf{sat}(T)$. W.l.o.g. we assume that the variables $x_1 < \cdots < x_d$ are algebraically independent modulo $\mathsf{sat}(T)$, that is, $\mathsf{free}(T) = \{x_1, \ldots, x_d\}$. Let us write $T = \{t_{d+1}, \ldots, t_n\}$ such that t_i has main variable x_i and initial h_i. We apply the extended version of the PALGIE algorithm (that is, the one solving Problem 3) to the solving of the polynomial system S below

$$
\begin{cases}
t_n^A(\mathbf{x}) = 0 \\
\quad \vdots \;\; \vdots \;\; \vdots \\
t_{d+1}^A(\mathbf{x}) = 0 \\
h_{d+1}^A(\mathbf{x}) \cdots h_n^A(\mathbf{x}) \neq 0
\end{cases}
\tag{3}
$$

We denote by $Z(S) \subset \overline{\mathbf{k}}^n$ the zero set of S. Observe that for all polynomials $f \in \mathbf{k}[\mathbf{x}]$, we have

$$
f \in \langle Z(S) \rangle \quad \Longleftrightarrow \quad f^{A^{-1}} \in \sqrt{\mathsf{sat}(T)}.
\tag{4}
$$

where $\langle Z(S) \rangle$ is the ideal of $\mathbf{k}[\mathbf{x}]$ consisting of all polynomials vanishing on $Z(S)$. Relation (4) allows one to easily adapt the *master - student relationship* described in Section 3.2 of [4] and thus to adapt the (extended version of the) PALGIE algorithm so as to solve Problem 1.

4   Noether Normalization and Regular Chains

In this section, we study the relation between Noether normalization and regular chains. Our initial quest was to determine whether, for a prime ideal $\mathcal{P} \subset \mathbf{k}[\mathbf{x}]$ in Noether position, one could find a monic regular chain T whose saturated ideal is precisely $\mathcal{P}$. For this purpose, we start by reviewing basic properties of Noether normalization, following Logar's paper [14].

Let $\mathcal{P} \subset \mathbf{k}[\mathbf{x}]$ be a (proper) prime ideal and G the reduced lexicographical Gröbner basis of $\mathcal{P}$. Recall that $\mathbf{x}$ counts n variables ordered as $x_1 < \cdots < x_n$. We assume that $\mathbf{k}$ is an infinite field. We denote by $T_{\mathcal{P}}$ the set defined by $T_{\mathcal{P}} = \{v \in \mathbf{x} \mid (\forall g \in G) \; \mathsf{mvar}(g) \neq v\}$. This set satisfies two important properties:

- $T_{\mathcal{P}}$ is algebraically independent modulo $\mathcal{P}$ that is, $\mathcal{P} \cap \mathbf{k}[T_{\mathcal{P}}] = \langle 0 \rangle$,
- the number of elements in $T_{\mathcal{P}}$ gives the dimension of $\mathcal{P}$, that is, $\dim(\mathcal{P}) = \mathsf{card}(T_{\mathcal{P}})$.

A variable $x_s \in \mathbf{x}$ is said *integral* over $\mathbf{k}[x_1, \ldots, x_{s-1}]$ modulo $\mathcal{P}$ if there exists $f \in \mathcal{P} \cap \mathbf{k}[x_1, \ldots, x_{s-1}, x_s]$ such that $\mathsf{mvar}(f) = x_s$ and $\mathsf{init}(f) \in \mathbf{k}$. Integral variables satisfy two important properties:

– A variable $x_s \in \mathbf{x}$ is integral over $\mathbf{k}[x_1, \dots, x_{s-1}]$ modulo $\mathcal{P}$ if and only if there exists $g \in G$ such that $\mathrm{lm}(g) = x_s^{d_s}$ for some positive integer d_s,

– if a variable $x_s \in \mathbf{x}$ is integral over $\mathbf{k}[x_1, \dots, x_{s-1}, \mathbf{u}]$ modulo $\mathcal{P}$, with $\mathbf{u} \subseteq T_{\mathcal{P}}$ disjoint from $\{x_1, \dots, x_s\}$, then x_s is also integral over $\mathbf{k}[x_1, \dots, x_{s-1}]$ modulo $\mathcal{P}$.

Thanks to the above properties, we may assume w.l.o.g. that if $d = \dim(\mathcal{P})$ then we have $T_{\mathcal{P}} = \{x_1, \dots, x_d\}$. Consider a linear change of coordinates A in $\overline{\mathbf{k}}^n$ defined by a matrix M of the following form:

$$
M \;=\; \left(\begin{array}{c|ccc}
 & a_{1,d+1} & \cdots & a_{1,n} \\
\mathrm{I}_{d\times d} & \vdots & \vdots & \vdots \\
 & a_{d,d+1} & \cdots & a_{d,n} \\
\hline
\mathbf{0} & \multicolumn{3}{c}{\mathrm{I}_{(n-d)\times(n-d)}}
\end{array} \right)
\tag{5}
$$

where $a_{i,j} \in \mathbf{k}$. We denote by $\mathcal{P}^A$ the ideal generated by f^A for all $f \in \mathcal{P}$. Then, by Noether normalization lemma, for a generic choice of $a_{1,d+1}, \dots, a_{d,n}$ the following properties hold:

1. $x_1, \dots, x_d$ are algebraically independent modulo $\mathcal{P}^A$,
2. x_{d+i} is integral over $\mathbf{k}[x_1, \dots, x_d]$ modulo $\mathcal{P}^A$ for all $i = 1, \dots, n - d$.

In this case, we say that $\mathcal{P}^A$ is in *Noether position*.

We turn our attention to the regular chain representation of the prime ideal $\mathcal{P}$. To this end, using Theorem 3.3 of [2], one can extract, in an algorithmic fashion, a subset T of G such that T is a regular chain whose saturated ideal is precisely $\mathcal{P}$. Let H be the reduced lexicographical Gröbner basis of $\mathcal{P}^A$ and C be the regular chain extracted from H using the same theorem from [2].

Theorem 1. *If T generates its saturated ideal, then the regular chain C is monic, that is, for each polynomial $f \in C$ we have $\mathrm{init}(f) \in \mathbf{k}$.*

PROOF. Assume by contradiction that there exists $f \in C$ such that $\mathrm{init}(f) \notin \mathbf{k}$ and let us choose such an f with minimum main variable. Since $x_1, \dots, x_d$ are algebraically independent modulo $\mathcal{P}^A$ and since C is a regular chain, one can compute a polynomial f' such that $\mathrm{init}(f') \in \mathbf{k}[x_1, \dots, x_d]$ and $\mathrm{sat}(C') = \mathrm{sat}(C)$ holds with $C' = C \setminus \{f\} \cup \{f'\}$.

Let $\mathrm{mvar}(f) = x_r$. Since $\mathcal{P}^A$ is in Noether position, it follows from [14] that there exists a polynomial $H_{x_r} \in H$ whose leading monomial is of the form $x_r^{d_r}$. Since $\mathrm{init}(H_{x_r}) \in \mathbf{k}$, we have $\deg(f', x_r) = \deg(f, x_r) < d_r = \deg(H_{x_r}, x_r)$. Indeed, otherwise the polynomial H_{x_r} would have been selected as an element of the regular chain C.

From the choice of f and the assumption on T, the regular chain $C' \cap \mathbf{k}[x_1, \dots, x_r]$ is a basis of $\mathcal{P}^A \cap \mathbf{k}[x_1, \dots, x_r]$. Therefore, the polynomial H_{x_r} reduces to zero through multivariate division by $C' \cap \mathbf{k}[x_1, \dots, x_r]$ and thus by $C \cap \mathbf{k}[x_1, \dots, x_r]$. This contradicts the fact that H is a reduced Gröbner basis. $\qquad\square$

Remark 1. *Theorem 1 states that if T generates $\mathrm{sat}(T)$ and $\mathcal{P}^A$ is in Noether position, then C is monic. Unfortunately, if T does not generate $\mathrm{sat}(T)$, then the previous conclusion may not hold as shown by the following example.*

Example 1. *Consider the regular chain* $T := \{x_2^5 - x_1^4, x_1 x_3 - x_2^2\} \subset \mathbb{Q}[x_1 < x_2 < x_3]$ *which does not generate its saturated ideal. Consider also the linear change of coordinates* A *defined by the matrix below*

$$M = \begin{pmatrix} 1 & 0 & -1 \\ 0 & 1 & 0 \\ 0 & 0 & 1 \end{pmatrix}.$$

Then $\langle T \rangle^A$ *is in Noether position and under this new change of coordinates we can compute the regular chain* $C = \{c_1, c_2\}$ *such that* $\sqrt{\mathrm{sat}(C)} = \sqrt{\mathrm{sat}(T)^A}$ *where* $c_1 = x_2^5 - 2x_2^4 + x_2^3 + 4x_1^2 x_2^2 - x_1^4$ *and* $c_2 = \left(-x_1^3 + 2x_2^2 x_1\right) x_3 + x_1^2 x_2^2 - x_2^4 + x_2^3$. *As you can see* $\mathrm{init}(c_2) \notin \mathbb{Q}$.

5   Applications of Random Linear Changes of Coordinates

Let $T \subset \mathbf{k}[\mathbf{x}]$ be a regular chain whose saturated ideal has dimension d. Let $\mathbf{u}$ be the free variables of T. Recall that h_T stands for the product of the $\mathrm{init}(f)$ for $f \in T$. Let A be a linear change of coordinates in $\overline{\mathbf{k}}^n$. Assume that the extended version of the PALGIE algorithm (see Problem 3 in Section 3) applied to T and A produces a single regular chain $C \subset \mathbf{k}[\mathbf{x}]$, thus satisfying $\overline{W^A(T)} = \overline{W(C)}$. Let h_T and h_C be the products of the initials of T and C, respectively. Let r_T^A and r_C be the iterated resultants (see [6] for this term) of h_T^A and h_C w.r.t. C.

Proposition 1 gathers elementary properties of r_T^A and r_C. Proposition 2 provides conditions for deriving a basis of $\mathrm{sat}(T)$ from the calculation of C while Theorem 2 provides a condition for deriving $\lim(W(T)) = \overline{W(T)} \cap V(h_T)$ from the calculation of C. The basic idea of Theorem 2 is to use a linear change of coordinates so as to replace the description of $\overline{W(T)}$ by one for which $\overline{W(T)} \cap V(h_T)$ can be computed by set-theoretic operations on constructible sets (represented by regular chain as in [5]). Moreover, Corollary 1 shows that, if T generates $\mathrm{sat}(T)$, then the computation of $\lim(W(T))$ can always be achieved by the techniques of [5].

Proposition 1. *The following properties hold:*

(i) *the polynomial* h_T^A *is regular w.r.t.* $\mathrm{sat}(C)$,
(ii) *the polynomials* r_T^A *and* r_C *belong to* $\mathbf{k}[\mathbf{u}]$ *and are non-zero.*

PROOF. Property (i) is by construction, that is, following the extended PALGIE algorithm applied to T and A. Property (ii) follows from (i) and the relations between regular chains and iterated resultants, see [6].   □

Proposition 2. *The following properties hold:*

(i) *if* $\mathrm{sat}(T)$ *is radical and if the ideal* $\langle h_T, (h_C^{A^{-1}}) \rangle$ *equals the whole ring* $\mathbf{k}[\mathbf{x}]$, *then* $T \cup C^{A^{-1}}$ *generates* $\mathrm{sat}(T)$,
(ii) *if the regular chain* C *is monic, then* $C^{A^{-1}}$ *generates* $\mathrm{sat}(T)$.

PROOF. We prove Property (i). Since sat(T) is radical, the relations $\overline{W^A(T)} = \overline{W(C)}$ implies $C^{A^{-1}} \subset$ sat(T). Hence, we "only" need to prove that if a polynomial f belongs to sat(T), then f is generated by $T \cup C^{A^{-1}}$. So let $f \in$ sat(T). On one hand, there exists a non-negative integer e such that $h_T^e f \in \langle T \rangle$. On the other, there exists a non-negative integer d such that $(h_C^{A^{-1}})^d f \in \langle C^{A^{-1}} \rangle$. Since the ideal $\langle h_T^e, (h_C^{A^{-1}})^d \rangle$ is the whole ring $\mathbf{k}[\mathbf{x}]$, then we can write f as an element of $\langle T, C^{A^{-1}} \rangle$. Now we prove (ii). Since C is monic, it is a Gröbner basis of sat(C), and, from the specifications of the PALGIE algorithm, a basis of sat$(T)^A$ as well. Thus $C^{A^{-1}} := \{ f^{A^{-1}} \mid f \in C \}$ is a basis of sat(T). □

From now on, we assume that the coefficients of the matrix $M = (m_{ij})$ are pairwise different variables. We view the coefficients of M, as well as the coefficients of all polynomials, as elements of the field of rational functions $\mathbf{k}(m_{ij})$. Moreover, the base field $\mathbf{k}$ is either $\mathbb{R}$ or $\mathbb{C}$ so that the affine space $\overline{\mathbf{k}}^n$ is endowed with the Euclidean topology. In this context, we recall from [1] that the quasi-component $W(T)$ has the same closure in both the Euclidean and the Zariski topologies.

Theorem 2. *For all values of (m_{ij}) such that $V(r_T^A, r_C)$ is empty, we have*

$$\lim(W(T)) = \{ A^{-1}(\mathbf{y}) \mid \mathbf{y} \in V(h_T^A) \cap W(C) \}. \tag{6}$$

PROOF. Observe first that $V(r_T^A, r_C)$ is empty if and only if $V(r_T, r_c^{A^{-1}})$ is empty. Observe next that any zero $\zeta \in \overline{\mathbf{k}}^n$ of h_T extends a zero $\zeta' \in \overline{\mathbf{k}}^d$ of r_T, see [5]. Therefore, for any choice of the parameters (m_{ij}) such that $V(r_T^A, r_C)$ is empty, one can let $(x_1, \ldots, x_n)$ approach a given root of h_T while staying within a bounded open set of $W^{A^{-1}}(C)$ leading to finitely many (possibly zero) finite limits for $(x_1, \ldots, x_n)$. Since, by construction, the constructible sets $W^{A^{-1}}(C)$ and $W(T)$ have the same Zariski closure, it follows that the points of $V(h_T^A) \cap W(C)$ are the images by A of the desired limit points of $W(T)$. □

Example 2. *Consider the regular chain $T := \{ x_4, x_2 x_3 + x_1^2 \} \subset \mathbb{Q}[x_1 < x_2 < x_3 < x_4]$ and the linear change of coordinates A corresponding to the matrix*

$$M = \begin{pmatrix} 0 & 0 & 0 & 1 \\ 0 & 1 & 1 & 0 \\ 0 & 1 & 0 & 0 \\ 1 & 0 & 0 & 0 \end{pmatrix}.$$

Using the extended of PALGIE, we can compute $C := \{ x_4, x_3^2 + x_2 x_3 + x_1^2 \}$ and consequently, $r_T^A = x_1^2$ and $r_C = 1$. Then $\langle r_T^A, r_C \rangle = \langle 1 \rangle$ holds. Using Triangularize *command of Maple, one can get*

$$\langle C, h_T^A \rangle^{A^{-1}} = \langle x_4, x_2, x_1 \rangle = \lim(W(T)).$$

Corollary 1. *Assume that T generates sat(T). Then we have*

$$\lim(W(T)) = V(T) \setminus W(T) \tag{7}$$

Hence, $\lim(W(T))$ *can be obtained by set-theoretic operations on constructible sets. Moreover, generically, the set* $\lim(W(T))$ *is determined by* $V(h_T^A) \cap W(C)$.

PROOF.   We prove the first claim. The hypothesis implies $V(T) = \overline{W(T)}$. Since $V(T) = W(T) \cup (V(T) \cap V(h_T))$, the conclusion follows. The second claim follows immediately from Theorems 2 and 1.   □

6   On the Computation of $\lim(W(T))$ and $\mathrm{sat}(T)$

Let T be a regular chain whose saturated ideal has dimension d. A driving application of this paper is the computation of $\lim(W(T))$. Section 5 was primarily dedicated to the case where T is a basis of its saturated ideal, while in the present section we replace this assumption by others. Recall that we have the follow equalities:

$$V(\mathrm{sat}(T)) = \overline{W(T)} = \left(\overline{W(T)} \cap V(h_T) \right) \cup W(T) = \lim(W(T)) \cup W(T).$$

Therefore, computing $\lim(W(T))$ and computing $V(\mathrm{sat}(T))$ are equivalent problems. Theorems 3, 4 and Lemma 2 below deal with the latter problem while Proposition 3 is concerned with the former. All these results make some assumption on T and we do not know a general procedure for computing either $\lim(W(T))$ or $V(\mathrm{sat}(T))$ that would avoid Gröbner basis calculation.

Lemma 1. *Let* $\mathcal{I}$ *be a radical ideal of* $\mathbf{k}[\mathbf{x}]$. *Let* $h \in \mathbf{k}[\mathbf{x}]$. *Assume that the dimension of any associated prime* $\mathfrak{p}$ *of* $\mathcal{I}$ *is at least* d. *Then* $\dim(V(\mathcal{I}, h)) < d$ *implies that* h *is regular modulo* $\mathcal{I}$. *If the dimension of any associated prime* $\mathfrak{p}$ *of* $\mathcal{I}$ *is* d, *that is, if* $\mathcal{I}$ *is an unmixed ideal of dimension* d, *then* $\dim(V(\mathcal{I}, h)) < d$ *holds if and only if* h *is regular modulo* $\mathcal{I}$.

PROOF.   Let $\mathcal{I} = \cap_{i=1}^{s} \mathfrak{p}_i$, where $\mathfrak{p}_i$ are the associated prime of $\mathcal{I}$. Assume that $\dim(V(\mathcal{I}, h)) < d$, it is enough to show that h does not belong to any $\mathfrak{p}_i$. On the other hand, we have $V(\mathcal{I}, h) = \cup_{i=1}^{s} V(\mathfrak{p}_i, h)$. If h belongs to some $\mathfrak{p}_i$, then $V(\mathfrak{p}_i, h) = V(\mathfrak{p}_i)$. Since $\dim(\mathfrak{p}_i) \geq d$, we know that $\dim(V(\mathcal{I}, h)) \geq d$, which is a contradiction to the assumption that $\dim(\mathcal{I}, h) < d$.

If $\mathcal{I}$ is an unmixed ideal of dimension d, by the above argument, $\dim(V(\mathcal{I}, h)) < d$ implies that h is regular modulo $\mathcal{I}$. On the other hand, if h is regular modulo $\mathcal{I}$, then h does not belong to any $\mathfrak{p}_i$. Thus $\dim(V(\mathcal{I}, h)) = \max(\dim(V(\mathfrak{p}_i, h))) < \max(\dim(\mathfrak{p}_i)) = d$.

□

Theorem 3. *Let* $T \subset \mathbf{k}[\mathbf{x}]$ *be a regular chain with free variables* $x_1, \ldots, x_d$. *Let* h_T *be the product of the initials of the polynomials in* T. *Then, we have* $\sqrt{\langle T \rangle} = \sqrt{\mathrm{sat}(T)}$ *if and only if* $\dim(V(T, h_T)) < d$ *holds.*

PROOF.   First we claim that for any associated prime $\mathfrak{p}$ of $\sqrt{\langle T \rangle}$, we have $\dim(\mathfrak{p}) \geq n - d$. To prove this, we first notice that the associated primes $\mathfrak{p}$ of

$\sqrt{\langle T \rangle}$ are exactly the minimal associated primes $\mathfrak{p}$ of $\langle T \rangle$. On the other hand, since $\langle T \rangle \subseteq \mathrm{sat}(T)$ and $V(\mathrm{sat}(T)) \neq \emptyset$ hold, we know that $\langle T \rangle$ generates a proper ideal. By Krull's principle ideal theorem, for any minimal associated prime $\mathfrak{p}$ of $\langle T \rangle$, the height of $\mathfrak{p}$ is less than or equal to $|T|$. Since $|T| = n - d$, we have $\dim(\mathfrak{p}) \geq n - d$. The claim is proved.

Now we prove that we have $\sqrt{\langle T \rangle} = \sqrt{\mathrm{sat}(T)}$ if and only if $\dim(V(T, h_T)) < d$ holds. First, we show that the condition is sufficient. If $\dim(V(T, h_T)) < d$ holds, with the previous claim and Lemma 1, we deduce that h_T is regular modulo $\sqrt{\langle T \rangle}$. Thus, we have $\sqrt{\mathrm{sat}(T)} = \sqrt{\langle T \rangle} : h_T^\infty = \sqrt{\langle T \rangle} : h_T^\infty = \sqrt{\langle T \rangle}$. Next, we show that the condition is necessary. If $\sqrt{\langle T \rangle} = \sqrt{\mathrm{sat}(T)}$, then $\sqrt{\langle T \rangle}$ is an unmixed ideal and h_T is regular modulo $\sqrt{\langle T \rangle}$. Thus, $\dim(V(T, h_T)) < d$ holds by Lemma 1. □

Remark 2. *As an immediate corollary, we have $V(T) = \overline{W(T)}$ if and only if $\dim(V(T, h_T)) < d$. There are many ways to compute the dimension of an algebraic set. In particular, this dimension can be determined by computing a Kalkbrener triangular decomposition. We denote by* IsClosure *a procedure to test $V(T) = \overline{W(T)}$, by applying Theorem 3.*

Example 3. *Consider the regular chain $T := \{x_1 x_2 + x_1, x_1 x_3 + 1\}$ of $\mathbb{Q}[x_1 < x_2 < x_3]$. Since the first polynomial is not primitive w.r.t. x_2, T is not a primitive regular chain in the sense of [13]. Since $V(T, x_1) = \emptyset$ holds, applying Theorem 3, we have $\sqrt{\langle T \rangle} = \sqrt{\mathrm{sat}(T)}$. Actually $\langle T \rangle = \mathrm{sat}(T)$ also holds.*

Theorem 4. *Let T be a regular chain of $\mathbf{k}[\mathbf{x}]$ with free variables $x_1, \ldots, x_d$. Let $C_1, \ldots, C_s \subset \mathbf{k}[\mathbf{x}]$. Assume that $\langle C_i \rangle \subseteq \sqrt{\mathrm{sat}(T)}$ holds, for all $i = 1, \ldots, s$. Let $\mathcal{I} = \langle T, C_1, \ldots, C_s \rangle$. Then $\sqrt{\mathrm{sat}(T)} = \sqrt{\mathcal{I}}$ if and only if there exist regular chains T_i, $i = 1, \ldots, t$, such that each of the following properties hold:*

(i) $\sqrt{\mathcal{I}} = \cap_{i=1}^t \sqrt{\mathrm{sat}(T_i)}$,
(ii) $|T_1| = \cdots = |T_t| = n - d$,
(iii) h_T is regular modulo all $\sqrt{\mathrm{sat}(T_i)}$.

Proof. The direction "⇒" obviously holds. Next we prove the direction "⇐". By (i) and (ii), we know that $\sqrt{\mathcal{I}}$ is an unmixed ideal of dimension d. Since h_T is regular modulo all $\sqrt{\mathrm{sat}(T_i)}$, by Lemma 1, we have $\dim(V(h_T, \mathrm{sat}(T_i))) < d$. Thus $\dim(V(\mathcal{I}, h_T)) < d$ holds. Applying Lemma 1 again, we know that h_T is regular modulo $\sqrt{\mathcal{I}}$. Thus $\sqrt{\mathcal{I}} = \sqrt{\mathcal{I}} : h_T^\infty = \sqrt{\mathcal{I} : h_T^\infty}$ holds. On the other hand, we have $\langle T \rangle \subseteq \mathcal{I}$, thus we deduce that $\sqrt{\mathrm{sat}(T)} \subseteq \sqrt{\mathcal{I}}$. Since $\mathcal{I} = \langle T, C_1, \ldots, C_s \rangle$ and $\langle C_i \rangle \subseteq \sqrt{\mathrm{sat}(T)}$, we also have $\mathcal{I} \subseteq \sqrt{\mathrm{sat}(T)}$. The theorem is proved. □

Remark 3. *In Theorem 4, if $s = 0$, then the theorem trivially holds for $t = 1$ and $T_1 = T$. In practice, for example in Algorithm 1, the polynomial sets C_i, for all $i = 1, \ldots, s$, are regular chains for different orderings such that $\sqrt{\mathrm{sat}(C_i)} = \sqrt{\mathrm{sat}(T)}$ holds. Let $T_1, \ldots, T_t$ be regular chains in the output of* Triangularize$(\mathcal{I})$. *Then (i) automatically holds. If condition (ii) is satisfied, then $\overline{W(T)} = V(\mathcal{I})$ holds if and only if (iii) holds, which is easy to check by computing iterated*

Algorithm 1. Closure(T)

Input: A non-empty regular chain T of $\mathbf{k}[x_1 < \cdots < x_n]$.
Output: Return $\emptyset$ or a polynomial set G such that $\overline{W(T)} = V(G)$. If $\emptyset$ is returned, this means that the algorithm fails to compute $\overline{W(T)}$.

```
1  begin
2  │  G := ∅;
3  │  for i from 1 to n do
4  │  │  if i = 1 then
5  │  │  │  C := T;
6  │  │  else
7  │  │  │  let R be the ordering x_i < x_{i+1} < ⋯ < x_n < x_1 ⋯ < x_{i-1};
8  │  │  │  D := PALGIE(T, R);
9  │  │  │  if |D| ≠ 1 then
10 │  │  │  │  return ∅
11 │  │  │  else
12 │  │  │  │  let C be the only regular chain in D;
13 │  │  if IsClosure(C) then
14 │  │  │  return C;
15 │  │  else
16 │  │  │  G := G ∪ C;
17 │  │  │  D := Triangularize(G, mode = K)// compute a Kalkbrener
   │  │  │      triangular decomposition of V(G)
18 │  │  │  if all regular chains in D have dimension d and h_T is regular w.r.t.
   │  │  │  each of them then
19 │  │  │  │  return G
20 │  return ∅;
```

---

resultants of h_T w.r.t. the regular chains T_i. Thus, this theorem provides an algorithmic recipe which may compute $\overline{W(T)}$ in somes cases, see Algorithm 1.

Example 4. We illustrate Algorithm 1 on one example. Consider the regular chain $T := \{x_2^5 - x_1^2, x_1 x_3 - x_2^2(x_2 + 1)\}$ of $\mathbb{Q}[x_1 < x_2 < x_3]$. Then $V(T, x_1) := \{(x_1, x_2, x_3) \mid x_1 = x_2 = 0\}$, whose dimension is 1. By Theorem 3, we know that $V(T) \neq V(sat(T))$. Let $C := \{x_2 x_3^2 - x_2^2 - 2x_2 - 1, x_3 x_1 - x_2^3 - x_2^2\}$ be another regular chain of $\mathbb{Q}[x_2 < x_3 < x_1]$. One can verify that $sat(C) = sat(T)$ holds. Let $\mathcal{I} := \langle C, T \rangle$. A Kalkbrener triangular decomposition of $\mathcal{I}$ w.r.t. the order $x_1 < x_2 < x_3$ consists only of one regular chain, which is T itself. Thus by Theorem 4, we have $V(sat(T)) = V(\mathcal{I})$.

Remark 4. We selected 22 one-dimensional non-primitive regular chains to test Algorithm 1. For 10 of them, the algorithm could successfully compute $\overline{W(T)}$. We also tested some random examples. The random regular chains are generated as follows. We choose a pair of random polynomials with 4 variables and of total degree 2. Then we apply Triangularize to this pair, thus obtaining 2-dimensional regular chains. In this way, we generated 20 regular chains, out of which 16 turned out to be non-primitive regular chains. Algorithm 1 successfully computed $\overline{W(T)}$ for 10 of those 16 examples.

Lemma 2. Let $T = \{t_2(x_1, x_2), t_3(x_1, x_3), \ldots, t_s(x_1, x_s)\}$ be a regular chain of $\mathbf{k}[x_1 < \cdots < x_s]$. Assume that for all $i = 2, \ldots, s$, the polynomial t_i is a primitive polynomial w.r.t. its main variable x_i. Then, the regular chain T generates its saturated ideal.

PROOF. To prove this lemma, it is enough to prove by induction that $\mathrm{sat}(T_i) = \langle T_i \rangle$, for $i = 2 \ldots, s$, where $T_i := \{t_2, \ldots, t_i\}$. The lemma clearly holds for $i = 2$. Assume that the regular chain T_{i-1} is generating its saturated ideal. If $\mathrm{tail}(t_i)$ is invertible modulo $\langle \mathrm{init}(t_i)\} \cup T_{i-1} \rangle$, then $\langle T_i \rangle = \mathrm{sat}(T_i)$ holds (see [13]). Suppose that $\mathrm{tail}(t_i)$ is not invertible modulo $\langle \{\mathrm{init}(t_i)\} \cup T_{i-1} \rangle$, then $\langle \{\mathrm{init}(t_i)\} \cup T_{i-1} \rangle$ generates a proper zero-dimensional ideal, since $\mathrm{init}(t_i)$ is regular modulo $\langle T_{i-1} \rangle$. Let $\mathfrak{p}$ be an associated prime of this ideal. If $\mathrm{tail}(t_i)$ is not regular modulo $\mathfrak{p}$, then all the coefficients of t_i belong to $\mathfrak{p}$. On the other hand, since $t_s(x_1, x_i)$ is primitive, the ideal formed by the coefficients of t_i is the field $\mathbf{k}$, a contradiction. $\qquad\square$

Remark 5. If a regular chain T has the same shape as in Lemma 2, except that the polynomials t_i are not necessarily primitive, for $i = 2, \ldots, s$, then by making all the polynomials t_i primitive, we obtain a new regular chain T' such that we have $\langle T' \rangle = \mathrm{sat}(T') = \mathrm{sat}(T)$.

Example 5. Let $T := \{x_3^2 - 2x_1, 3x_2^3 + 4x_1^2\} \subset \mathbb{Q}[x_1 < x_2 < x_3]$ be a 1-dimensional regular chain. As you can see both elements of T are primitive bivariate polynomials. Then Lemma 2 implies that T generates its saturated ideal.

Example 6. The above lemma clearly does not hold for regular chains with more than one free variable. Consider for example the regular chain $T := \{x_1 x_3 + x_2, x_1 x_4 + x_2\}$, where $x_1 < x_2 < x_3 < x_4$. It is clear that $x_4 - x_3 \notin \langle T \rangle$. However, one can prove that $x_4 - x_3 \in \mathrm{sat}(T)$ because $x_1 x_4 + x_2 = x_1(x_4 - x_3)$ modulo $\langle x_1 x_3 + x_2 \rangle$.

Lemma 3. Let $T \subset \mathbf{k}[\mathbf{x}]$ be a regular chain with free variable x_1. Let $C_2 = T$ and let C_i, for $3 \leq i \leq n$, be regular chains w.r.t. the order $x_1 < x_i < \mathbf{x} \setminus \{x_1, x_i\}$ such that $\sqrt{\mathrm{sat}(C_i)} = \sqrt{\mathrm{sat}(T)}$. Assume that all the polynomials of C_i are primitive w.r.t. their main variables for $i = 2, \ldots, n$. Then $\dim(V(C_2, \ldots, C_n)) = 1$ holds.

PROOF. By the fact that $\overline{W(T)} = \overline{W(C_i)}$, we know that $\overline{W(T)} \subseteq V(C_2, \ldots, C_n)$, which implies that $\dim(V(C_2, \ldots, C_n)) \geq 1$. Let c_i be the polynomial in C_i with the main variable x_i. Then the set $C := \{c_2, \ldots, c_n\}$ is clearly a regular chain since $\mathrm{init}(c_i) \in \mathbf{k}[x_1]$ holds for each $i = 2, \ldots, s$. Moreover C generates its saturated ideal by Lemma 2. Thus $\dim(V(C)) = 1$. Since $V(C_2, \ldots, C_n) \subseteq V(C)$, we know that $\dim(V(C_2, \ldots, C_n)) \leq 1$. Thus the lemma holds. $\qquad\square$

Example 7. *Let* $T := \{x_2^5 - x_1^4, x_1 x_3 - x_2^2\}$ *be a regular chain of* $\mathbb{Q}[x_1 < x_2 < x_3]$. *Let also* $C := \{x_3^5 - x_1^3, x_3^2 x_2 - x_1^2\}$ *be a regular chain of* $\mathbb{Q}[x_1 < x_3 < x_2]$ *for which we have* $\mathrm{sat}(C) = \mathrm{sat}(T)$. *One can verify that* $\dim(V(T, C)) = 1$. *Indeed a Kalkbrener triangular decomposition of* $T \cup C$ *computed by the* Triangularize *command of* RegularChains *library w.r.t. the order* $x_1 < x_2 < x_3$ *is* $\{T, D\}$, *where* $D := \{x_1, x_2, x_3\}$.

It is easy to observe that the decomposition computed by Triangularize *is redundant, that is we have* $\mathrm{sat}(T) \subseteq \mathrm{sat}(D)$ *holds. By Theorem 4, we conclude that* $\sqrt{\langle T, C \rangle} = \sqrt{\mathrm{sat}(T)}$. *However, for this example, Algorithm 1 fails to compute the set* G *such that* $\overline{W(T)} = V(G)$, *since* T *and* D *do not have the same height.*

Lemma 3, Example 7 and Theorem 4 show that it is possible to compute $\mathrm{sat}(T)$ by a change of order of the variables. One might wonder if this is always true. In particular, we ask the following two questions.

Question 1. *Let* $C_1, \ldots, C_n$ *be regular chains of* $\mathbf{k}[\mathbf{x}]$ *w.r.t. the order* $x_i < x_{i+1} < \cdots < x_n < x_1 \cdots < x_{i-1}$, *for* $i = 1, \ldots, n$. *Assume that* $\sqrt{\mathrm{sat}(C_1)} = \cdots = \sqrt{\mathrm{sat}(C_n)}$. *Does* $\sqrt{\mathrm{sat}(C_1)} = \sqrt{\langle \cup_{i=1}^n C_i \rangle}$ *always hold?*

Question 2. *Let* $C_1, \ldots, C_n$ *be polynomial sets of* $\mathbf{k}[\mathbf{x}]$ *such that* C_i *is a regular chain for the order* $x_i < x_{i+1} < \cdots < x_n < x_1 \cdots < x_{i-1}$, *for* $i = 1, \ldots, n$. *Assume that* $\sqrt{\mathrm{sat}(C_i)} = \sqrt{\mathrm{sat}(C_j)}$ *for all* $1 \leq i < j \leq n$. *Let* $P_i \in C_i$ *be the polynomial of least rank. Let* H_1 *be the product of the initials of* C_1. *Does the relation*

$$\lim(W(C_1)) = V(C_1 \cup \{P_1, \ldots, P_n, H_1\})$$

always hold?

To answer the two questions, we investigated over 35 different polynomial systems, and all of them succeeded but two of which failed. Here is one of them.

Example 8. *Suppose* $T := \{t_1, t_2\} \subset \mathbb{Q}[x_1 < x_2 < x_3 < x_4]$ *is a regular chain of dimension two, where* $t_1 = -93 x_1 x_2^2 + (53 x_1 - 35) x_2 + 93 x_1^3 - 26 x_1^2 - 57 x_1$ *and* $t_2 = 93 x_1 x_4 + ((3233 x_1 - 2135) x_2 + 5673 x_1^3 + 213 x_1^2 - 3477 x_1) x_3 + (-530 x_1^2 - 3091 x_1) x_2 - 930 x_1^4 + 6119 x_1^3 + 570 x_1^2 - 1767 x_1$. *One can verify that* T *does not generate its saturated ideal.*

Following the notations if Question 1, using PALGIE, *we will be able to compute regular chains* C_i *for* $i = 1, \ldots, 4$ *w.r.t the orders mentioned in Question 1. To see whether the statement of Question 1 is true or not, on one hand, we can find the Kalkbrener triangular decomposition* $\{C_1, R_1, R_2\}$ *for* $V(\cup_{i=1}^4 C_i)$ *where* $C_1 = T$, $R_1 := \{x_4 - 19, x_2, x_1\}$, *and* $R_2 := \{961 x_4^2 + 42428 x_4 + 279756, x_3, x_2, x_1\}$.

On the other hand, using methods based on Gröbner bases computations to find a generator for $\mathrm{sat}(C_1)$, *one can find the Kalkbrener triangular decomposition* $\{C_1, R_1\}$ *for* $V(\mathrm{sat}(C_1))$.

Therefore, we have

$$V(\mathrm{sat}(C_1)) = W(C_1) \cup W(R_1) \neq V(\cup_{i=1}^4 C_i) = W(C_1) \cup W(R_1) \cup W(R_2).$$

This shows that the statement of Question 1 is not true. Furthermore,

$$V(C_1 \cup \{P_1, \ldots, P_4, H_1\}) \;=\; W(R_1) \;\cup\; W(R_2)$$

where H_1 is the product of the initials of C_1 and P_i is the polynomial in C_i with least rank for $i = 1, \ldots, 4$. But the correct limit points are only represented by R_1 which means $\lim(W(C_1)) \neq V(C_1 \cup \{P_1, \ldots, P_4, H_1\})$. Cosequently, for this example, the answer to both Questions 1 and 2 is negative.

In Example 8, as one can see, we computed the limit points plus some extra points. The extra component R_2 in this example is of dimension 0 while the limit points we are expecting are of dimension 1.

Proposition 3. *Let T be a regular chain such that $\langle \mathrm{sat}(T) \rangle$ has dimension d and let $F \subset \langle \mathrm{sat}(T) \rangle$ such that either $V(T \cup F \cup \{h_T\})$ has dimension $d-1$ and is irreducible. Suppose also that $\lim(W(T))$ is not empty. Then, we have $\lim(W(T)) = V(T \cup F \cup \{h_T\})$.*

PROOF. The proof is straightforward. □

Example 9. *Consider the regular chain $T := \{x_1\,x_3 + x_2, x_2\,x_4 + x_1\} \subset \mathbb{Q}[x_1 < x_2 < x_3 < x_4]$. One can consider F to be the regular chain computed by applying* PALGIE *to T w.r.t. the variable order $x_3 < x_4 < x_1 < x_2$ and consequently, "fish" the polynomial $x_3\,x_4 - 1 \in \mathrm{sat}(T)$. Then*

$$
\begin{aligned}
V(T \cup F \cup \{h_T\}) &= V(x_1\,x_3 + x_2, x_2\,x_4 + x_1, x_3\,x_4 - 1, x_1\,x_2) \\
&= V(x_1, x_2, x_3\,x_4 - 1) \\
&= \lim(W(T)).
\end{aligned}
$$

7 Conclusion

Among all the methods we have considered for computing $\lim(W(T))$ and $\mathrm{sat}(T)$, those based on linear changes of coordinates seem very promising. They are a good trick for finding a subset $F \subset \mathrm{sat}(T)$ such that $F \cup T$ is a basis of $\mathrm{sat}(T)$, see Proposition 3. To develop that direction further, we are currently investigating the following related questions:

- decide whether $\lim(W(T))$ is empty
- decide whether $W(R) \subseteq \lim(W(T))$ for a given regular chain.

Acknowledgements. The authors would like to thank the referees for careful reading of the manuscript and for their helpful suggestions. The research of the second author was partially supported by NSFC (11301524,11471307,61202131). The research of the third author was in part supported by a grant from IPM (No. 93550420).

References

1. Alvandi, P., Chen, C., Maza, M.M.: Computing the limit points of the quasi-component of a regular chain in dimension one. In: Gerdt, V.P., Koepf, W., Mayr, E.W., Vorozhtsov, E.V. (eds.) CASC 2013. LNCS, vol. 8136, pp. 30–45. Springer, Heidelberg (2013)

2. Aubry, P., Lazard, D., Maza, M.M.: On the theories of triangular Sets. J. Symb. Comput. **28**(1–2), 105–124 (1999)

3. Boulier, F., Lemaire, F., Moreno Maza, M.: Pardi! In: Proceedings of International Symposium on Symbolic and Algebraic Computation, ISSAC 2001, pp. 38–47 (2001)

4. Boulier, F., Lemaire, F., Maza, M.M.: Computing differential characteristic sets by change of ordering. J. Symb. Comput. **45**(1), 124–149 (2010)

5. Chen, C., Golubitsky, O., Lemaire, F., Maza, M.M., Pan, W.: Comprehensive triangular decomposition. In: Ganzha, V.G., Mayr, E.W., Vorozhtsov, E.V. (eds.) CASC 2007. LNCS, vol. 4770, pp. 73–101. Springer, Heidelberg (2007)

6. Chen, C., Maza, M.M.: Algorithms for computing triangular decomposition of polynomial systems. J. Symb. Comput. **47**(6), 610–642 (2012)

7. Dahan, X., Jin, X., Maza, M.M., Schost, É.: Change of order for regular chains in positive dimension. Theor. Comput. Sci. **392**(1–3), 37–65 (2008)

8. Eisenbud, D.: Commutative Algebra with a View toward Algebraic Geometry. Springer, New York (1995)

9. Greuel, G.M., Pfister, G.: A Singular Introduction to Commutative Algebra. Springer, Berlin (2002)

10. Hashemi, A.: Effective computation of radical of ideals and its application to invariant theory. In: Hong, H., Yap, C. (eds.) ICMS 2014. LNCS, vol. 8592, pp. 382–389. Springer, Heidelberg (2014)

11. Krick, T., Logar, A.: An algorithm for the computation of the radical of an ideal in the ring of polynomials. In: Mattson, H.F., Mora, T., Rao, T.R.N. (eds.) Applied Algebra, Algebraic Algorithms and Error-Correcting Codes, AAECC 1991. LNCS, vol. 539, pp. 195–205. Springer, Heidelberg (1991)

12. Lecerf, G.: Computing the equidimensional decomposition of an algebraic closed set by means of lifting fibers. J. of Complexity **19**(4), 564–596 (2003)

13. Lemaire, F., Maza, M.M., Pan, W., Xie, Y.: When does $\langle T \rangle$ equal sat(T)? J. Symb. Comput. **46**(12), 1291–1305 (2011)

14. Logar, A.: A computational proof of the noether normalization lemma. In: Mora, T. (ed.) Applied Algebra, Algebraic Algorithms and Error-Correcting Codes, AAECC 1988. LNCS, vol. 357, pp. 259–273. Springer, Heidelberg (1989)

15. Rouillier, F.: Solving zero-dimensional systems through the rational univariate representation. Appl. Algebra Eng. Commun. Comput. **9**(5), 433–461 (1999)

16. Seiler, W.M.: A combinatorial approach to involution and δ-regularity II: Structure analysis of polynomial modules with Pommaret bases. Appl. Alg. Eng. Comm. Comp. **20**, 261–338 (2009)

17. Sommese, A.J., Verschelde, J.: Numerical homotopies to compute generic points on positive dimensional algebraic sets. J. Complexity **16**(3), 572–602 (2000)

A Standard Basis Free Algorithm for Computing the Tangent Cones of a Space Curve

Parisa Alvandi[1], Marc Moreno Maza[1], Éric Schost[1], and Paul Vrbik[2]

[1] Department of Computer Science, University of Western Ontario
[2] School of Mathematical and Physical Sciences,
The University of Newcastle Australia

Abstract. We outline a method for computing the tangent cone of a space curve at any of its points. We rely on the theory of regular chains and Puiseux series expansions. Our approach is novel in that it explicitly constructs the tangent cone at arbitrary *and possibly irrational* points without using a standard basis.

Keywords: Computational algebraic geometry, tangent cone, regular chain, Puiseux series.

1 Introduction

Traditionally, standard bases, Gröbner bases and cylindrical algebraic decomposition are the fundamental tools of computational algebraic geometry. The computer algebra systems CoCoA, Macaulay 2, Magma, Reduce, Singular have well-developed packages for computing standard bases or Gröbner bases, on which they rely in order to provide powerful toolkits to algebraic geometers.

Recent progress in the theory of regular chains has exhibited efficient algorithms for doing local analysis on algebraic sets. One of the algorithmic strengths of the theory of regular chains is its *regularity test* procedure. In algebraic terms, this procedure decides whether a hypersurface contains at least one irreducible component of the zero set of the saturated ideal of a regular chain. Broadly speaking, this procedure separates the zeros of a regular chain that belong to a given hypersurface from those which do not. This regularity test permits to extend an algorithm working over a field into an algorithm working over a direct product of fields. Or, to phrase it in another way, it allows one to extend an algorithm working at a point into an algorithm working at a group of points.

Following that strategy, the authors of [8] have proposed an extension of Fulton's algorithm for computing the intersection multiplicity of two plane curves at the origin. To be precise, this paper extends Fulton's algorithm in two ways. First, thanks to the regularity test for regular chains, the construction is adapted such that it can work correctly at any point in the intersection of two plane curves, whether this point has rational coordinates or not. Secondly, an algorithmic criterion, see Theorem 1, is proposed for reducing intersection multiplicity computation in arbitrary dimension to the case of two plane curves. This

© Springer International Publishing Switzerland 2015
V.P. Gerdt et al. (Eds.): CASC 2015, LNCS 9301, pp. 45–60, 2015.
DOI: 10.1007/978-3-319-24021-3_4

algorithmic criterion requires to compute the tangent cone $TC_p(\mathscr{C})$ of a space curve $\mathscr{C}$ at one of its points p. In principle, this latter problem can be handled by means of standard basis (or Gröbner basis) computation. Available implementations (like those in MAGMA or SINGULAR) require that the point p is uniquely determined by the values of its coordinates. However, when decomposing a polynomial system, a point may be defined as one of the roots of a particular sub-system (typically a regular chain h). Therefore, being able to compute the tangent cones of $\mathscr{C}$ at all ts points defined by a given regular chain h, becomes a desirable operation it is desirable operation. Similarly, and as discussed in [8], another desirable operation is the computation of the intersection multiplicity of a zero-dimensional algebraic set V at all its points defined by a given regular chain h. This type of tangent cone computation is addressed in the present paper.

Tangent cone computations can be approached at least in two ways. First, one can consider the formulation based on homogeneous components of least degree, see Definition 1. The original algorithm of Mora [9] follows this point of view. Secondly, one can consider the more "intuitive" characterization based on limits of secants, see Lemma 1. This second approach, that we follow in this paper, requires to compute limits of algebraic functions. For this task, we take advantage of [2] where the authors show how to compute the limit points of the quasi-component of a regular chain. This type of calculation can be used for computing the Zariski closure of a constructible set. In the present paper, it is used for computing tangent cones of space curves, thus providing an alternative to the standard approaches based on Gröbner bases and standard bases.

The contributions of the present paper are as follows

1. In Section 3, we present a proof of our algorithmic criterion for reducing intersection multiplicity computation in arbitrary dimension to the plane case; this criterion was stated with no justification in [8].
2. In Section 4.1, with Lemma 2, under a smoothness assumption, we establish a natural method for computing $TC_p(\mathscr{C})$; as limit of intersection of tangent spaces.
3. In Section 4.2, we relax the assumption of Section 4.1 and exhibit an algorithm for computing $TC_p(\mathscr{C})$.

This latter algorithm is implemented, in the `AlgebraicGeometryTools` subpackage [1] of the `RegularChains` library which is available at `www.regularchains.org`. Section 4.4 offers examples. However, an issue with MAPLE's `algcurves[puiseux]` command that we have no control over prohibits us from providing meaningful experimental results at this time. For those test cases which do not encounter error from the `algcurves[puiseux]` command we indeed calculate the correct tangent cone. We are currently re-implementing MAPLE's `algcurves[puiseux]` command and we will provide experimental results in a future report.

2   Preliminaries

Throughout this article, we denote by $\mathbb{K}$ a field with algebraic closure $\overline{\mathbb{K}}$, and by $\mathbb{A}^{n+1}(\overline{\mathbb{K}})$ the $(n + 1)$-dimensional affine space over $\overline{\mathbb{K}}$, for some positive integer n. Let $\boldsymbol{x} := x_0, \ldots, x_n$ be $n + 1$ variables ordered as $x_0 \succ \cdots \succ x_n$. We denote by $\mathbb{K}[\boldsymbol{x}]$ the corresponding polynomial ring. Let $\boldsymbol{h} \subset \mathbb{K}[\boldsymbol{x}]$ be a subset and $h \in \mathbb{K}[\boldsymbol{x}]$ be a polynomial. We say that h is *regular* modulo the ideal $\langle h \rangle$ of $\mathbb{K}[\boldsymbol{x}]$ whenever h does not belong to any prime ideals associated with $\langle h \rangle$, thus, whenever h is neither null nor a zero-divisor modulo $\langle h \rangle$. The *algebraic set* of $\mathbb{A}^{n+1}(\overline{\mathbb{K}})$ consisting of the common zeros of the polynomials in $\boldsymbol{h}$ is written as $\mathbf{V}(\boldsymbol{h})$. For $\mathbf{W} \subset \mathbb{A}^{n+1}(\overline{\mathbb{K}})$, we denote by $\mathbf{I}(\mathbf{W})$ the ideal of $\mathbb{K}[\boldsymbol{x}]$ generated by the polynomials vanishing at every point of $\mathbf{W}$. The ideal $\mathbf{I}(\mathbf{W})$ is radical and when $\overline{\mathbb{K}} = \mathbb{K}$ holds, Hilbert's Nullstellensatz states that $\sqrt{\langle \boldsymbol{h} \rangle} = \mathbf{I}(\mathbf{V}(\boldsymbol{h}))$.

In the next two sections, we review the main concepts used in this paper, namely tangent cones and regular chains. For the former, we restrict ourselves to tangent cones of a space curve and refer to [4] for details and the general[1] case. For the latter concept, we refer to [3], in particular for the specifications of the basic operations on regular chains.

2.1   Tangent Cone of a Space Curve

As above, let $\boldsymbol{h} \subset \mathbb{K}[\boldsymbol{x}]$. Define $\mathbf{V} := \mathbf{V}(\boldsymbol{h})$ and let $p := (p_0, \ldots, p_n) \in \mathbf{V}$ be a point. We denote by $\dim_p(\mathbf{V})$ the maximum dimension of an irreducible component C of $\mathbf{V}$ such that we have $p \in C$. Recall that the *tangent space* of $\mathbf{V}(\boldsymbol{h})$ at p is the algebraic set given by

$$T_p(\boldsymbol{h}) := \mathbf{V}(\; \mathbf{d}_p(f) : f \in \mathbf{I}(\mathbf{V}))$$

where $\mathbf{d}_p(f)$ is the *linear part* of f at p, that is, the affine form $\frac{\partial f}{\partial x_0}(p)(x_0 - p_0) + \cdots + \frac{\partial f}{\partial x_n}(p)(x_n - p_n)$. Note that $T_p(\boldsymbol{h})$ is a linear space. We say that $\mathbf{V}(\boldsymbol{h})$ is *smooth* at p whenever the dimension of $T_p(\boldsymbol{h})$ is $\dim_p(\mathbf{V})$ and *singular* otherwise. The *singular locus* of $\mathbf{V}(\boldsymbol{h})$, denoted by $\text{sing}(\boldsymbol{h})$, is the set of the points $p \in \mathbf{V}(\boldsymbol{h})$ at which $\mathbf{V}(\boldsymbol{h})$ is singular.

Let $f \in \mathbb{K}[\boldsymbol{x}]$ be a polynomial of total degree d and $p := (p_0, \ldots, p_n) \in \mathbb{A}^{n+1}(\overline{\mathbb{K}})$ be a point such that $f(p) = 0$ holds. Let $\alpha = (\alpha_0, \ldots, \alpha_n) \in \mathbb{N}^{n+1}$ be a $(n + 1)$-tuple of non-negative integers. Denote: $(\boldsymbol{x} - p)^\alpha := (x_0 - p_0)^{\alpha_0} \cdots (x_n - p_n)^{\alpha_n}$, where $|\alpha| = \alpha_0 + \cdots + \alpha_n$ is the total degree of $\boldsymbol{x} - p$. Since the polynomial $f \in \mathbb{K}[\boldsymbol{x}]$ has total degree d, it writes as a $\mathbb{K}$-linear combination of the form:

$$f = \sum_{|\alpha|=0} c_\alpha (\boldsymbol{x} - p)^\alpha + \cdots + \sum_{|\alpha|=d} c_\alpha (\boldsymbol{x} - p)^\alpha$$

with all coefficients c_α belonging to $\mathbb{K}$. Each summand

---

[1] Note that in the book [3], and other classical algebraic geometry textbooks like [12], the tangent cone of an algebraic set at one of its points, is also an algebraic set. Two equivalent definitions appear in [3] and are recalled in Definition 1 and Lemma 1.

$$\mathrm{HC}_p\left(f;\,j\right):=\sum_{|\alpha|=j}c_\alpha(\boldsymbol{x}-p)^\alpha$$

is called the *homogeneous component in $\boldsymbol{x}-p$ of f in degree j*. Moreover, the *homogeneous component of least degree* of f in $\boldsymbol{x}-p$ is given by $\mathrm{HC}_p\left(f;\,\min\right):=\mathrm{HC}_p\left(f;\,j_{\min}\right)$ where $j_{\min}=\min(j\in\mathbb{N}:\mathrm{HC}_p\left(f;\,j\right)\neq0)$.

Definition 1 (Tangent Cone of a Curve). *Let $\mathscr{C}\subset\mathbb{A}^{n+1}(\overline{\mathbb{K}})$ be a curve and $p\in\mathscr{C}$ be a point. The* tangent cone *of $\mathscr{C}$ at a point p is the algebraic set denoted by $TC_p(\mathscr{C})$ and defined by $TC_p(\mathscr{C})=\mathbf{V}\left(\mathrm{HC}_p\left(f;\,\min\right):f\in\mathbf{I}(\mathscr{C})\right)$.*

One can show that $TC_p(\mathscr{C})$ consists of finitely many lines, all intersecting at p.

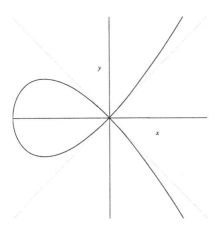

Fig. 1. This figure displays the typical "fish" curve, which is a planar curve given by $h=y^2-x^2(x+1)\in\mathbb{Q}[x,y]$. Clearly, two tangent lines are needed to form a "linear approximation" of the curve at the origin. Elementary calculations show these two lines actually form the tangent cone of the fish curve at the origin.

If $\mathbf{I}(\mathscr{C})$ is generated by a single polynomial then computing $TC_p(\mathscr{C})$ is easy. Otherwise, this is a much harder computation. Let $\boldsymbol{h}\subset\mathbb{K}[\boldsymbol{x}]$ be such that $\mathbf{V}(\boldsymbol{h})=\mathscr{C}$. As pointed out by Mora et al. in [10], one can compute $\langle\,\mathrm{HC}_p\left(f;\,\min\right):f\in\mathbf{I}(\mathscr{C})\,\rangle$ via a graded Gröbner basis, say $\boldsymbol{G}$, of the *homogenization* of $\boldsymbol{h}$ (a process where an additional variable x_{n+1} is used to make every $h\in\boldsymbol{h}$ a homogeneous polynomial in $\mathbb{K}[\boldsymbol{x}][x_{n+1}]$). *Dehomogenizing* $\boldsymbol{G}$ by letting $x_{n+1}=1$ produces the tangent cone of $\boldsymbol{h}$ [4, Chapter 9.7, Proposition 4].

Tangent cones are intimately related to the notion of intersection multiplicity that we review below. As mentioned in the introduction, computing intersection multiplicities is the main motivation of the algorithm presented in this paper.

Definition 2. *Let $\boldsymbol{h}\subset\mathbb{K}[\boldsymbol{x}]$. The* intersection multiplicity *of p in $\mathbf{V}(\boldsymbol{h})$ is defined by $\mathrm{im}(p;\,\boldsymbol{h}):=\dim_{vec}(\mathscr{O}/\langle\boldsymbol{h}\rangle)$ where $\mathscr{O}:=\{f/g:f,g\in\overline{\mathbb{K}}[\boldsymbol{x}],\,g(p)\neq0\}$ is the* localization ring *of $\mathbb{K}[\boldsymbol{x}]$ at p and $\dim_{vec}(\mathscr{O}/\langle\boldsymbol{h}\rangle)$ is the dimension of $\mathscr{O}/\langle\boldsymbol{h}\rangle$ as a vector space over $\mathbb{K}$. Note by [5, Chapter 4.2, Proposition 11] we may substitute the power series ring $\mathbb{K}[[\boldsymbol{x}-p]]$ for $\mathscr{O}$.*

Example 1. Let $\boldsymbol{x} = [x, y, z]$ and $\boldsymbol{h} = \{x, y - z^3, z^2 (z^4 + 1)\}$. We have:

$$\mathbb{K}[[\boldsymbol{x}]] / \langle \boldsymbol{h} \rangle = \mathbb{K}[[\boldsymbol{x}]] / \langle x, y - z^3, z^2 \rangle = \mathbb{K}[[\boldsymbol{x}]] / \langle x, y, z^2 \rangle = \{a + bz : a, b \in \mathbb{K}\}$$

implying $\mathrm{im}\,(\boldsymbol{0}; \boldsymbol{h}) = 2$.

2.2 Regular Chains

Broadly speaking, a *regular chain* of $\mathbb{K}[\boldsymbol{x}]$ is a system of equations and inequations defined by polynomials in $\mathbb{K}[\boldsymbol{x}]$ such that each equation specifies, in an implicit manner, the possible values of one of the variables x_i as a function of the variables of least rank, namely $x_{i+1}, \ldots, x_n$. Regular chains are a convenient way to describe the solution set of a polynomial system. More precise statements follow.

Let $h \in \mathbb{K}[\boldsymbol{x}]$ be a non-constant polynomial. The *main variable* of h is the largest variable $x \in \boldsymbol{x}$ (for the ordering $x_0 \succ \cdots \succ x_n$) such that h has a positive degree in x. The *initial* of h, denoted $\mathrm{init}(h)$, is the *leading coefficient* of h w.r.t. its main variable. For instance the initial of $zx + t$ is x in $\mathbb{Q}[x \succ y \succ z \succ t]$ and 1 in $\mathbb{Q}[t \succ z \succ y \succ x]$.

Let $\boldsymbol{t} \subset \mathbb{K}[\boldsymbol{x}]$ consist of non-constant polynomials. Then, the set $\boldsymbol{t}$ is said *triangular* if any two polynomials in $\boldsymbol{t}$ have different main variables. When $\boldsymbol{t}$ is a triangular set, denoting by $I_{\boldsymbol{t}}$ the product of the initials $\mathrm{init}(f)$ for $f \in \boldsymbol{t}$, we call *saturated ideal* of $\boldsymbol{t}$, written $\mathrm{sat}(\boldsymbol{t})$, the column ideal $\mathrm{sat}(\boldsymbol{t}) = \langle \boldsymbol{t} \rangle : I_{\boldsymbol{t}}^{\infty}$ and we call *quasi-component* of $\boldsymbol{t}$ the basic constructible set $\mathbf{W}(\boldsymbol{t}) := \mathbf{V}(\boldsymbol{t}) \setminus \mathbf{V}(I_{\boldsymbol{t}})$.

Definition 3 (Regular Chain). *The triangular set $\boldsymbol{t} \subset \mathbb{K}[\boldsymbol{x}]$ is a regular chain if either $\boldsymbol{t}$ is empty or the initial of f is regular modulo $\mathrm{sat}(\boldsymbol{t} \setminus \{f\})$, where f is the polynomial in $\boldsymbol{t}$ with largest main variable.*

Regular chains are used to decompose both algebraic sets and radical ideals, leading to two types of decompositions called respectively *Wu-Lazard* and *Kalkbrener* decompositions. More precisely, we have the following definition.

Finitely many regular chains $\boldsymbol{t}_0, \ldots, \boldsymbol{t}_e \subset k[\boldsymbol{x}]$ form a Kalkbrener decomposition of $\sqrt{\langle \boldsymbol{h} \rangle}$ (resp. a Wu-Lazard decomposition of $\mathbf{V}(\boldsymbol{h})$) whenever we have $\sqrt{\langle \boldsymbol{h} \rangle} = \sqrt{\mathrm{sat}(\boldsymbol{t}_0)} \cap \cdots \cap \sqrt{\mathrm{sat}(\boldsymbol{t}_e)}$ (resp. $\mathbf{V}(\boldsymbol{h}) = \mathbf{W}(\boldsymbol{t}_0) \cup \cdots \cup \mathbf{W}(\boldsymbol{t}_e)$). These two types of decompositions are different since the quasi-component of a regular chain $\boldsymbol{t}$ may not be an algebraic set. One should note that the Zariski closure of $\mathbf{W}(\boldsymbol{t})$ (that is, the intersection of all algebraic sets containing $\mathbf{W}(\boldsymbol{t})$) is the zero set (i.e. algebraic set) of $\mathrm{sat}(\boldsymbol{t})$. One should observe, however, that if $\mathrm{sat}(\boldsymbol{t})$ is zero-dimensional then the quasi-component $\mathbf{W}(\boldsymbol{t})$ and the algebraic set $\mathbf{V}(\boldsymbol{t})$ coincide. Practically efficient algorithms computing both types of decompositions appear in [3].

Regular chains enjoy important algorithmic properties. One of them is the ability to test whether a given polynomial $f \in \mathbb{K}[\boldsymbol{x}]$ is regular or not modulo the saturated ideal of a regular chain $\boldsymbol{t} \subset \mathbb{K}[\boldsymbol{x}]$. This allows us to specify an operation, called Regularize, as follows. The function call Regularize $(f, \boldsymbol{t})$ computes regular chains $\boldsymbol{t}_0, \ldots, \boldsymbol{t}_e \subset \mathbb{K}[\boldsymbol{x}]$ such that $\sqrt{\mathrm{sat}(\boldsymbol{t})} = \sqrt{\mathrm{sat}(\boldsymbol{t}_0)} \cap \cdots \cap \sqrt{\mathrm{sat}(\boldsymbol{t}_e)}$ holds and for $i = 0, \ldots, e$, either f is zero modulo $\mathrm{sat}(\boldsymbol{t}_i)$ or f is regular modulo $\mathrm{sat}(\boldsymbol{t}_i)$.

When sat(t) is zero-dimensional, one can give a simple geometrical interpretation to Regularize: this operation separates the points of $\mathbf{V}(t)$ belonging to $\mathbf{V}(f)$ from those which do not lie on $\mathbf{V}(f)$.

3   Computing Intersection Multiplicities in Higher Dimension

Our interest in a standard-basis free algorithm for computing tangent cones comes by way of an overall goal to compute intersection multiplicities in arbitrary dimension. As mentioned in the introduction, in a previous paper [8], relying on the book of Fulton [7] and the theory of regular chains, we derived an algorithm for computing intersection multiplicities of planar curves. We also sketched an algorithm criterion, see Theorem 1 below, for reducing the computation of intersection multiplicities in arbitrary dimension to computing intersection multiplicities in lower dimension. When applicable, successive uses of this criterion reduces intersection multiplicity computation in arbitrary dimension to the bivariate case.

Theorem 1. *For $\boldsymbol{h} = h_0, \ldots, h_{n-1}, h_n \in \mathbb{K}[\boldsymbol{x}]$ such that $\mathbf{V}(h_0, \ldots, h_{n-1}, h_n)$ is zero-dimensional, for $p \in \mathbf{V}(h_n)$, if the hyper-surface $\mathbf{V}(h_n)$ is not singular at p and if that the tangent space π of $\mathbf{V}(h_n)$ at p intersects transversally[2] the tangent cone of the curve $\mathbf{V}(h_0, \ldots, h_{n-1})$ at p, then we have*

$$\operatorname{im}(p; h_0, \ldots, h_{n-1}, h_n) = \operatorname{im}(p; h_0, \ldots, h_{n-1}, \pi),$$

hence, there is a polynomial map which takes $\boldsymbol{h}$ to a lower dimensional subspace while leaving the intersection multiplicity of $\mathbf{V}(\boldsymbol{h})$ at p invariant.

Checking whether this criterion is applicable, requires to compute the tangent cone of the curve $\mathbf{V}(h_0, \ldots, h_{n-1})$ at p, which motivates the present paper. This algorithmic criterion was stated in [8] without justification, although the authors had a long and technical proof available in a technical report extending [8]. In the PhD thesis of the fourth author [13], a simpler proof was obtained and we present it below.

Proof. The theorem follows from results of [12, Chapter IV]; we reuse the same notation as in that reference when feasible.

Since p is an isolated point of $\mathbf{V}(\boldsymbol{h})$, any irreducible component of $\mathbf{V}(h_0, \ldots, h_{n-1})$ through p must have dimension one. By Lemma 2 in [12, Chapter IV.1.3] it follows $\overline{\mathscr{O}}$ is a one-dimensional local ring, where

$$\overline{\mathscr{O}} := \mathscr{O} / \langle h_0, \ldots, h_{n-1} \rangle.$$

---

[2] Two algebraic sets V_0 and V_1 in $\mathbb{A}^{n+1}(\overline{\mathbb{K}})$ *transversally intersect* at a point $p \in V_0 \cap V_1$ whenever their tangent cones intersect at $\{p\}$ *only* once or not at all. Note that if one of V_0 is a linear space, then it is its own tangent cone at p. Note also that, for a sake of clarity, we have restricted Definition 1 to tangent cones of curves, although tangent cones of algebraic sets of higher dimension are defined similarly, see [4].

Let $\mathscr{C}_0, \ldots, \mathscr{C}_r$ be the irreducible components of $\mathbf{V}(h_0, \ldots, h_{n-1})$ passing through p and let $\mathfrak{p}_0, \ldots, \mathfrak{p}_r$ be their respective defining (prime) ideals in $\mathscr{O}$. Our transversality assumption ensures h_n and π are both nonzero divisors in $\overline{\mathscr{O}}$ and consequently, since $\overline{\mathscr{O}}$ is a one-dimensional local ring, we use Equation (6) from [12, Chapter IV.1.3] to deduce

$$\operatorname{im}(p; h_0, \ldots, h_{n-1}, h_n) = \sum_{i=0}^{r} m_i \dim_{\mathrm{vec}}(\overline{\mathscr{O}}/\langle \mathfrak{p}_i, h_n \rangle) \tag{1}$$

and

$$\operatorname{im}(p; h_0, \ldots, h_{n-1}, \pi) = \sum_{i=0}^{r} m_i \dim_{\mathrm{vec}}(\overline{\mathscr{O}}/\langle \mathfrak{p}_i, \pi \rangle) \tag{2}$$

for some constants $m_1, \ldots, m_r$ that we need not define more precisely.

Remark 1. In the original reference the dimensions above are written as lengths but [6, Example A.1.1] permits us to use the vector space dimension instead. This holds for all the dimensions written below as well.

Because $\langle h_0, \ldots, h_{n-1} \rangle \subset \mathfrak{p}_i$ for all i, we can rewrite (1) and (2) as (resp.) $\dim_{\mathrm{vec}}(\mathscr{O}/\langle \mathfrak{p}_i, h_n \rangle)$ and $\dim_{\mathrm{vec}}(\mathscr{O}/\langle \mathfrak{p}_i, \pi \rangle)$. Hence it suffices to prove, exploiting that $\langle h_0, \ldots, h_{n-1} \rangle$ has been replaced by a dimension one prime ideal, that

$$\dim_{\mathrm{vec}}(\mathscr{O}/\langle \mathfrak{p}_i, h_n \rangle) = \dim_{\mathrm{vec}}(\mathscr{O}/\langle \mathfrak{p}_i, \pi \rangle)$$

for all $i = 1, \ldots, r$ to conclude.

Fix i for the remainder of this proof. The prime ideal $\mathfrak{p}_i$ defines a curve $\mathscr{C} \subset \overline{\mathbb{K}}^{n+1}$. Let $\mathscr{C}' \subset \overline{\mathbb{K}}^{n'+1}$ be a normalization of $\mathscr{C}$ given by $\nu : \mathscr{C}' \to \mathscr{C}$; thus $\mathscr{C}'$ is non-singular. It follows from [12, Chapter IV.1.3.(9)] that

$$\dim_{\mathrm{vec}}(\mathscr{O}/\langle \mathfrak{p}_i, h_n \rangle) = \sum_{\nu(p')=p} \dim_{\mathrm{vec}}(\mathscr{O}_{\mathscr{C}',p'}/h_n^*),$$

when $\mathscr{O}_{\mathscr{C}',p'}$ is the local ring of $\mathscr{C}'$ at p' and h_n^* is the *pull-back* of h_n by ν. A similar expression holds for π.

Now fix p' in the fiber $\nu^{-1}(p)$. We prove

$$\dim_{\mathrm{vec}}(\mathscr{O}_{\mathscr{C}',p'}/h_n^*) = \dim_{\mathrm{vec}}(\mathscr{O}_{\mathscr{C}',p'}/\pi^*).$$

Without loss of generality shift to the origin, that is, assume $p = 0 \in \overline{\mathbb{K}}^{n+1}$ and $p' = 0 \in \overline{\mathbb{K}}^{n'+1}$ and also let t be a *uniformizer* for $\mathscr{C}'$ at p' (remember that $\mathscr{C}'$ is non-singular). Finally, write $\nu = (\nu_0, \ldots, \nu_n)$, with all ν_i in $\overline{\mathbb{K}}[\mathscr{C}']$.

Expanding $\nu = (\nu_0, \ldots, \nu_n)$ in power series at the origin permits us to view them as in $\overline{\mathbb{K}}[[t]]^{n+1}$. With this in mind, and without loss of generality, assume ν_0 has the smallest valuation among $\nu_0, \ldots, \nu_n$ (otherwise, do a change of coordinates in $\overline{\mathbb{K}}^{n+1}$). Call this valuation r, so that we can write, for all i:

$$\nu_i(t) = \nu_{i,r} t^r + \nu_{i,r+1} t^{r+1} + \cdots$$

It follows the component of the $TC_0(\mathscr{C})$ corresponding to the image $\nu(\mathscr{C}')$ around p' is the limit of secants having directions

$$\left(\frac{\nu_0(t)}{\nu_0(t)}, \frac{\nu_1(t)}{\nu_0(t)}, \ldots, \frac{\nu_n(t)}{\nu_0(t)}\right).$$

This limit is a line with direction

$$\left(1, \frac{\nu_{1,r}}{\nu_{0,r}} \ldots, \frac{\nu_{n,r}}{\nu_{n,r}}\right),$$

or equivalently $(\nu_{1,r}, \ldots, \nu_{n,r})$. Because we assumed p is the origin, h_n has a writing

$$h_n(x_0, \ldots, x_n) = \pi + \text{higher order terms}$$

with $\pi = h_{n,0}\, x_0 + \cdots + h_{n,n}\, x_n$; the transversality assumption implies

$$h_{n,0}\, \nu_{0,r} + \cdots + h_{n,n}\, \nu_{n,r} \neq 0.$$

Using the local parameter t, the multiplicities

$$\dim_{\text{vec}}(\mathscr{O}_{\mathscr{C}',p'}/h_n^*) \quad \text{and} \quad \dim_{\text{vec}}(\mathscr{O}_{\mathscr{C}',p'}/\pi^*)$$

can be rewritten as the respective valuations in t of h_n^* and π^*, that is, of

$$h_n(\nu_0(t), \ldots, \nu_n(t)) \quad \text{and} \quad \pi(\nu_0(t), \ldots, \nu_n(t)).$$

The latter is easy to find; it reads

$$\pi(\nu_0(t), \ldots, \nu_n(t)) =$$
$$(h_{n,0}\, \nu_{0,r} + \cdots + h_{n,n}\, \nu_{n,r})t^r + (h_{n,0}\, \nu_{0,r+1} + \cdots + h_{n,n}\, \nu_{n,r+1})t^{r+1} + \cdots.$$

Due to the shape of h_n, the former is

$$h_n(\nu_0(t), \ldots, \nu_n(t)) = (h_{n,0}\, \nu_{0,r} + \cdots + h_{n,n}\, \nu_{n,r})t^r + \text{higher order terms}.$$

Since we know $h_{n,0}\, \nu_{0,r} + \cdots + h_{n,n}\, \nu_{n,r} \neq 0$, both expressions must have the same valuation r, so we are done.   □

4   Computing Tangent Lines as Limits of Secants

From now on, the coefficient field $\mathbb{K}$ is the field $\mathbb{C}$ of complex numbers and the affine space $\mathbb{A}^{n+1}(\mathbb{C})$ is endowed with both Zariski topology and the Euclidean topology. While Zariski topology is coarser than the Euclidean topology, we have the following key result (Corollary 1 in Section I.10 of Mumford's book [11]): For an irreducible algebraic set $\mathbf{V}$ and a subset $U \subseteq \mathbf{V}$ open in the Zariski topology induced on $\mathbf{V}$, the closure of U in Zariski topology and the closure of U in the Euclidean topology are both equal to $\mathbf{V}$. It follows that, for a regular chain $t \subset \mathbb{C}[x]$ the closure of $\mathbf{W}(t)$ in Zariski topology and the closure of $\mathbf{W}(t)$ in the Euclidean topology are equal, thus both equal to $\mathbf{V}(\text{sat}(t))$. This result provides a bridge between techniques from algebra and techniques from analysis. The authors of [2] take advantage of Mumford's result to tackle the following problem: given a regular chain $t \subset \mathbb{C}[x]$, compute the (non-trivial) limit points of the quasi-component of t, that is, the set $\lim(\mathbf{W}(t)) := \overline{\mathbf{W}(t)} \setminus \mathbf{W}(t)$.

In the present paper, we shall obtain the lines forming the tangent cone of a space curve at a point by means of a limit computation process. And in fact, this limit computation will reduce to computing $\lim(\mathbf{W}(t))$ for some regular chain t. To this end, we start by stating the principle of our method in Section 4.1. Then, we turn this principle into an actual algorithm in Section 4.2 via an alternative characterization of a tangent cone, based on *secants*.

4.1 An Algorithmic Principle

Let $h = \{h_0, \ldots, h_{n-1}\} \subset \mathbb{C}[x]$ be n polynomials such that $\mathscr{C} = \mathbf{V}(h)$ is a curve, that is, a one-dimensional algebraic set. Let $p \in \mathscr{C}$ be a point. The following proposition is well-known, see Theorem 6 in Chapter 9 of [4].

Lemma 1. *A line L through p lies in the tangent cone $TC_p(\mathscr{C})$ if and only if there exists a sequence $\{q_k : k \in \mathbb{N}\}$ of points on $\mathscr{C} \setminus \{p\}$ converging to p and such that the secant line L_k containing p and q_k becomes L when q_k approaches p.*

Under some mild assumption, we derive from Lemma 1 a method for computing $TC_p(\mathscr{C})$. We assume that for each $h \in h$, the hyper-surface $\mathbf{V}(h)$ is non-singular at p. This assumption allows us to approach the lines of $TC_p(\mathscr{C})$ with the intersection of the tangent spaces $T_q(h_0), \ldots, T_q(h_{n-1})$ when $q \in \mathscr{C}$ is an sufficiently small neighborhood of p. A more precise description follows.

For each branch of a connected component $\mathscr{D}$ through p of $\mathscr{C} = \mathbf{V}(h)$ there exists a neighborhood B about p (in the Euclidean topology) such that $\mathbf{V}(h_0), \ldots, \mathbf{V}(h_{n-1})$ are all non-singular at each $q \in (B \cap \mathscr{D}) \setminus \{p\}$. Observe also that the singular locus $\text{sing}(\mathscr{D})$ contains a *finite* number of points. It follows that we can take B small enough so that $B \cap \text{sing}(\mathscr{D})$ is either empty or $\{p\}$. Define

$$v(q) := T_q(h_0) \cap \cdots \cap T_q(h_{n-1}),$$

where $T_q(h_i)$ is the tangent space of $\mathbf{V}(h_i)$ at p.

Lemma 2 states that we can obtain $TC_p(\mathscr{C})$ by finding the limits of $v(q)$ as q approaches p. Since $TC_p(\mathscr{C})$ is the union of all the $TC_p(\mathscr{D})$, this yields a method for computing $TC_p(\mathscr{C})$.

Lemma 2. *The collection of limits of lines $v(q)$ as q approaches p in $(B \cap \mathscr{D}) \setminus \{p\}$ gives the tangent cone of $\mathscr{D}$ at q. That is to say*

$$TC_p(\mathscr{D}) = \lim_{q \to p} v(q) = \lim_{q \to p} T_q(h_0) \cap \cdots \cap T_q(h_{n-1}).$$

Proof. There are two cases, either

1. $\mathscr{D}$ is *smooth* at p and $B \cap \text{sing}(\mathscr{D}) = \emptyset$, or
2. $\mathscr{D}$ is *singular* at p and $B \cap \text{sing}(\mathscr{D}) = \{p\}$.

Case 1. Assume $q \in B \cap \mathscr{D}$ is arbitrary and observe $\mathscr{D}$ is smooth within B and thereby the tangent cone of $\mathscr{D}$ is simply the tangent space (i.e. $TC_q(\mathscr{D}) = T_q(\mathscr{D})$).

Notice $T_q(\mathscr{D})$ is a sub-vector space of $v(q)$. Indeed, let $w \in T_q(\mathscr{D})$ be any tangent vector to $\mathscr{D}$ at q. As $\mathscr{D}$ is a curve in each $\mathbf{V}(h)$ for $h \in \boldsymbol{h}$ it follows w is a vector tangent to each $\mathbf{V}(h)$ as well. Correspondingly $w \in T_q(h)$ for any $h \in \boldsymbol{h}$ and thus $w \in v(q)$.

Finally, since $h_0, \ldots, h_{n-1}$ form a local complete intersection in B, we know $v(q)$ is a one-dimensional subspace of each $T_q(h_0)$. Since $w \in T_q(h)$ for each $h \in \boldsymbol{h}$, the vector w must span this subspace. Thus, for each $q \in B \cap \mathscr{D}$, we have

$$T_q(\mathscr{D}) = T_q(h_0) \cap \cdots \cap T_q(h_{n-1}).$$

Taking the limit of each side of the above equality, when q approaches p and using again the fact that $\mathscr{D}$ is smooth at $q = p$, we obtain the desired result, that is, $TC_p(\mathscr{D}) = \lim_{q \to p} v(q)$.

Case 2. Assume $\mathscr{D} \cap B - \{p\}$ is a finite union of smooth curves $\mathscr{D}_0, \ldots, \mathscr{D}_j$. These are the smooth branches of $\mathscr{D} \cap B$ meeting at the singular point p. Each j corresponds to a unique line

$$L_j = \lim_{q \to p} v(q) \subset T_p(\mathscr{D})$$

as q approaches p *along* $\mathscr{D}_j$.

By Lemma 1 the tangent cone $TC_p(\mathscr{D})$ is the collection of limits to p of secant lines through p in $\mathscr{D}$. Such lines given by secants along $\mathscr{D}_j$ must coincide with L_j. More precisely

$$L_0 \cup \cdots \cup L_j \subset TC_p(\mathscr{D}).$$

Because each $\mathscr{D}_j$ is smooth there is only one secant line for each j and thereby

$$L_0 \cup \cdots \cup L_j = TC_p(\mathscr{D})$$

as desired.

4.2   Algorithm

Under a smoothness assumption, Lemma 2 states a principle for computing $TC_p(\mathscr{C})$. Let us now turn this principle into a precise algorithm and relax this smoothness assumption as well. To this end, we make use of Lemma 1.

Let q be a point on the curve $\mathscr{C} = \mathbf{V}(h)$ with coordinates $\boldsymbol{x}$. Further let $\widehat{pq}$ be a unit vector in the direction of $\overline{pq}$ (i.e. the line through p and q). To exploit Lemma 1 we must calculate the set

$$\left\{ \lim_{\substack{q \to p \\ q \neq p}} \widehat{pq} \right\},$$

which is indeed a set because $\mathscr{C}$ may have several branches through p yielding several lines in the tangent cone $TC_p(\mathscr{C})$.

Let $\boldsymbol{t} \subset \mathbb{C}[\boldsymbol{y}][\boldsymbol{x}]$ be a zero-dimensional regular chain encoding[3] the point p, that is, such that we have $\mathbf{V}(\boldsymbol{t}) = \{p\}$. Note that the introduction of $\boldsymbol{y}$ for the coordinates of p is necessary because the "moving point" q is already using $\boldsymbol{x}$ for its own coordinates. Consider the polynomial set

$$\boldsymbol{s} = \boldsymbol{t} \cup \boldsymbol{h}.$$

and observe that the ideal $\langle \boldsymbol{s} \rangle$ is one-dimensional in the polynomial ring $\mathbb{C}[x_{n-1} \succ \cdots \succ x_0 \succ y_{n-1} \succ \cdots \succ y_0]$. Let $\boldsymbol{t}_0, \ldots, \boldsymbol{t}_e \subset \mathbb{C}[\boldsymbol{y}][\boldsymbol{x}]$ be one-dimensional regular chains forming a Kalkbrener decomposition of $\sqrt{\langle \boldsymbol{s} \rangle}$. Thus we have

$$\mathbf{V}(\boldsymbol{s}) = \overline{\mathbf{W}(\boldsymbol{t}_0)} \cup \cdots \cup \overline{\mathbf{W}(\boldsymbol{t}_e)}.$$

Computing with the normal vector $\widehat{pq}$ is unnecessary and instead we divide the vector $\overrightarrow{pq}$ by $x_n - y_n$. Since the n-th coordinate of $\frac{\overrightarrow{pq}}{x_n - y_n}$ is 1, this vector remains non-zero when q approaches p. However, this trick leads to a valid limit computation provided that $x_n - y_n$ vanishes finitely many times in $\mathbf{V}(\boldsymbol{s})$. When this is the case, the lines of the tangent cone, that are not contained in the hyperplane $y_n = x_n$, can be obtained via limits of meromorphic functions (namely Puiseux series expansions) by letting x_n approach y_n and using the techniques of [2]. As we shall argue below, an ordering of $\boldsymbol{x}$, for which $x_n - y_n$ is regular, always exists. Hence, up to variable re-ordering, this tricks applies.

Since the tangent cone may have lines contained in the hyperplane $y_n = x_n$, additional computations are needed to capture them. There are essentially two options:

1. Perform a random linear change of the coordinates so as to assume that, generically, $y_n = x_n$ contains no lines of $TC_p(\mathscr{C})$.
2. Compute in turn the lines not contained in the hyperplane $y_i = x_i$ for all $i = 0, \ldots, n$ and remove the duplicates; indeed no lines of the tangent cone can simultaneously satisfy $y_i = x_i$ for all $i = 0, \ldots, n$.

Our experiments with theses two approaches suggest that, although the second one seems computationally more expensive, it avoids the expression swell of the first one and is practically more efficient.

---

[3] In practice, we may use a zero-dimensional regular chain $\boldsymbol{t} \subset \mathbb{C}[\boldsymbol{y}][\boldsymbol{x}]$ such that $\{p\} \subseteq \mathbf{V}(\boldsymbol{t}) \subseteq \mathscr{C}$ holds. Then, the following discussion will bring the tangent cone at several points of $\mathscr{C}$ instead of p only.

From now on, we focus on computing the lines of the tangent cone *not* contained in the hyperplane $y_n = x_n$. We note that, deciding whether $x_n - y_n$ vanishes finitely many times in $\mathbf{V}(s)$ can be done algorithmically by testing whether $x_n - y_n$ is regular modulo the saturated ideal of each regular chain $t_0, \ldots, t_e$. The operation Regularize described in Section 2 performs this task.

Consider now t_j, that is, one of the regular chains $t_0, \ldots, t_e$. Thanks to the specifications of Regularize, we may assume w.l.o.g. that either $x_n - y_n$ is regular modulo $\mathrm{sat}(t_j)$ or that $x_n - y_n \equiv 0 \bmod \mathrm{sat}(t_j)$ holds.

Consider the latter case first. If $x_n - y_n \equiv 0 \bmod \mathrm{sat}(t_j)$ then $\overline{\mathbf{W}(t_j)} \subseteq \mathbf{V}(x_n - y_n)$ holds and we try to divide each component of $\overline{pq}$ by $x_{n-1} - y_{n-1}$ instead of $x_n - y_n$. A key observation is that there is $d \in [0, n]$ such that $x_d - y_d \not\equiv 0 \bmod \mathrm{sat}(t_j)$ necessarily holds. Indeed, if $x_i - y_i \equiv 0 \bmod \mathrm{sat}(t_j)$ would hold for all $i \in [0, n]$ then $\overline{\mathbf{W}(t_j)} \subset \mathbf{V}(x_0 - y_0) \cap \cdots \cap \mathbf{V}(x_n - y_n)$ would hold as well. Since the $\boldsymbol{y}$ coordinates are fixed by t, the algebraic set $\overline{\mathbf{W}(t_j)}$ would be zero-dimensional—a contradiction.

Hence, up to a variable renaming, we can assume that $x_n - y_n$ is regular modulo $\mathrm{sat}(t_j)$. Therefore, the algebraic set $\mathbf{V}(x_n - y_n) \cap \overline{\mathbf{W}(t_j)}$ is zero-dimensional, thus, each component of $\overline{pq}$ is divisible by $x_n - y_n$, when q is close enough to p, with $q \neq p$. Define

$$m_0 = \frac{x_0 - y_0}{x_n - y_n}, \ldots, m_n = \frac{x_n - y_n}{x_n - y_n}.$$

and regard $\boldsymbol{m} = m_0, \ldots, m_n$ as new variables, that we call *slopes*, for clear reasons. Observe that the vector of coordinates $(m_0, \ldots, m_n, 1)$ is a normal vector of the secant line $\overline{pq}$. Thus, our goal is to "solve for" $\boldsymbol{m}$ when x_n approaches y_n with $(y_0, \ldots, y_n, x_0, \ldots, x_n) \in \mathbf{W}(t_j)$.

We turn this question into one computing the limit points of a one-dimensional regular chain, so as to use the algorithm of [2]. To this end, we extend the regular chain t_j to the regular chain $M_j \subset \mathbb{C}[\boldsymbol{m}][\boldsymbol{y}][\boldsymbol{x}]$ given by

$$M_j = t_j \cup \begin{cases} m_0(x_0 - y_0) - (x_n - y_n) \\ \vdots \\ m_n(x_n - y_n) - (x_n - y_n) \end{cases}.$$

Note that M_j is one-dimensional in this extended space and computing $\lim(\mathbf{W}(M_j))$, using the algorithm of [2], solves for $\boldsymbol{m}$ when $x_n \to y_n$ with $(\boldsymbol{x}, \boldsymbol{y}) \in \mathbf{W}(t_0)$. Therefore and finally, the desired set $\{\lim_{q \to p, q \neq p} \widehat{pq}\}$ is obtained as the limit points of the quasi-components of $M_0, \ldots, M_n$.

Remark 2. Observe that the above process determines the slopes $m_0, \ldots, m_n$ as roots of the top n polynomials of zero-dimensional regular chains in the variables $m_n \succ \cdots \succ m_0 \succ x_n \succ \cdots \succ x_0 \succ y_n \cdots \succ y_0$. Performing a change of variable ordering to $\boldsymbol{x} \succ \boldsymbol{m} \succ \boldsymbol{y}$ expresses $m_0, \ldots, m_{n-1}$ as functions of the coordinates of the point p only. We consider this a more desirable output.

4.3 Equations of Tangent Cones

In the previous section, we saw how to compute the tangent cone $TC_p(\mathscr{C})$ in the form of the slopes of vectors defining the lines of $TC_p(\mathscr{C})$. Instead, one may prefer to obtain $TC_p(\mathscr{C})$ in the form of the equations of the lines of $TC_p(\mathscr{C})$. We explain below how to achieve this. Let S be an arbitrary point with coordinates $(X_0, \ldots, X_n)$. This point belongs to one of the lines of the tangent cone (corresponding to the branches of the curve defined by $\overline{W}(t_j)$) if and only if the vectors

$$\frac{\vec{pq}}{x_n - y_n} = \begin{pmatrix} 1 \\ m_{n-1} \\ \vdots \\ m_0 \end{pmatrix} \quad \text{and} \quad \overline{pS} = \begin{pmatrix} X_n - y_n \\ X_{n-1} - y_{n-1} \\ \vdots \\ X_0 - y_0 \end{pmatrix}$$

are collinear. That is, if and only if we have the following relations

$$\begin{cases} X_n = m_n(x_n - y_n) + y_n \\ \quad \vdots \\ X_0 = m_0(x_n - y_n) + y_0. \end{cases} \tag{3}$$

Consider a regular chain (obtained with the process described in Remark 2) thus expressing the slopes $m_0, \ldots, m_{n-1}$ as functions of the coordinates $y_0, \ldots, y_n$ of p. Let us extend this regular chain with the relations from Equation (3), so as to obtain a one-dimensional regular chain in the variables $X_n \succ \cdots \succ X_0 \succ m_{n-1} \succ \cdots \succ m_0 \succ y_n \cdots \succ y_0$. Next, we eliminate the variables $m_0, \ldots, m_{n-1}$, with the above equations. This is, indeed, legal since the only point of a line of the tangent cone where the equation $x_n = y_n$ holds is p itself. Finally, this elimination process consists simply of substituting $\frac{X_i - y_i}{x_n - y_n}$ for m_i into the equations defining $m_0, \ldots, m_n$.

4.4 Examples

The following examples illustrates our technique for computing tangent cones as limits. We write tangent cones using unions to save vertical space and to separate slope from point.

Example 2. Consider calculating the tangent cone of the fish $h = y^2 - x^2(x+1)$ at the origin. The Puiseux expansions of h at $x = 0$ in T are given by

$$\begin{cases} y = -T - \frac{1}{2}T^2 + O(T^3) \\ x = T \end{cases} \quad \text{and} \quad \begin{cases} y = T + \frac{1}{2}T^2 + O(T^3) \\ x = T \end{cases}$$

and substituting these values into $ym - x$ produces

$$\left(-\tfrac{1}{2}T^2 - T\right)m - T \quad \text{and} \quad \left(\tfrac{1}{2}T^2 + T\right)m - T.$$

Call these expressions M_0 and M_1 respectively.

To find the value of m at $T = 0$ we find the Puiseux series expansions for M_0 and M_1 at $T = 0$ in U; these are respectively.

$$\begin{cases} m = -1 + \frac{1}{2}U - \frac{1}{4}U^2 + O(U^3) \\ T = U \end{cases} \quad \text{and} \quad \begin{cases} m = 1 - \frac{1}{2}U + \frac{1}{4}U^2 + O(U^3) \\ T = U \end{cases}.$$

Taking $U \to 0$ in the above produces the (expected) slopes of 1 and -1.

Fig. 2. Limiting secants along $\mathbf{V}\left(x^2 + y^2 + z^2 - 1, \ x^2 - y^2 - z\right)$.

Example 3. Consider Figure 2, i.e. secants along the curve $\boldsymbol{h} = \{x^2 + y^2 + z^2 - 1, \ x^2 - y^2 - z\} \subset \mathbb{K}[x, y, z]$ limiting to a point given by a zero dimensional regular chain $\boldsymbol{t} = \langle\, x + y, \ 2y^2 - 1, \ z \,\rangle$.

$$TC_t(\boldsymbol{h}) = \begin{cases} m_1 - 1 \\ m_2 \\ m_3 \end{cases} \quad \cup \quad \begin{cases} 2x^2 - 1 \\ 2y^2 - 1 \\ z \end{cases}$$

or alternatively (using equations of lines instead)

$$TC_t(\boldsymbol{h}) = \left\{ z \pm \frac{4x}{\sqrt{2}} + 2, \ y - x \pm \frac{2}{\sqrt{2}} \right\}.$$

Notice the slope for *four* points are encoded here. In particular the points

$$\left\{ \left(\frac{1}{\pm\sqrt{2}}, \ \frac{1}{\pm\sqrt{2}}, \ 0 \right), \ \left(-\frac{1}{\pm\sqrt{2}}, \ \frac{1}{\mp\sqrt{2}}, \ 0 \right) \right\}$$

have slope given by the vector $\langle 1, 0, 0 \rangle$.

Fig. 3. Secants along $\mathbf{V}\left(x^2 + y^2 + z^2 - 1\right) \cap \mathbf{V}\left(x^2 - y^2 - z(z-1)\right)$ limiting to $(0, 0, 1)$.

Example 4. Consider Figure 3, i.e. secants along the curve $\boldsymbol{h} = \{x^2 + y^2 + z^2 - 1, x^2 - y^2 - z(z-1)\} \subset k[x, y, z]$ limiting to $(0, 0, 1)$

$$TC_{(0,0,1)}(\boldsymbol{h}) = \begin{cases} m_1 + m_2 \\ 2m_2^2 - 6m_2 + 3 \\ m_3 \end{cases} \cup \begin{cases} x \\ y \\ z - 1 \end{cases}$$

or alternatively (using equations of lines instead)

$$TC_{(0,0,1)}(\boldsymbol{h}) = \left\{z - 1, \, y^2 - 3x^2\right\}.$$

Notice the values of the slopes here are in the algebraic closure of the coefficient ring. In particular, they are

$$\left\{\left(\tfrac{3}{2} + \sqrt{6}, \tfrac{3}{2} + \sqrt{6}, 0\right), \left(\tfrac{3}{2} - \sqrt{6}, \tfrac{3}{2} - \sqrt{6}, 0\right)\right\}.$$

5   Conclusion

We presented an alternative and Gröbner-free method for calculating the tangent cone of a space curve at any of its points. In essence, this is done by simulating a limit calculation along a curve using variable elimination. From this limit we can construct each line of the tangent cone by solving for the vector of instantaneous slope along each tangents corresponding secant lines. Finally, this slope vector can be converted into equations of lines.

References

1. Alvandi, P., Chen, C., Marcus, S., Maza, M.M., Schost, É., Vrbik, P.: Doing algebraic geometry with the regularchains library. In: Hong, H., Yap, C. (eds.) ICMS 2014. LNCS, vol. 8592, pp. 472–479. Springer, Heidelberg (2014)
2. Alvandi, P., Chen, C., Maza, M.M.: Computing the limit points of the quasi-component of a regular chain in dimension one. In: Gerdt, V.P., Koepf, W., Mayr, E.W., Vorozhtsov, E.V. (eds.) CASC 2013. LNCS, vol. 8136, pp. 30–45. Springer, Heidelberg (2013)

3. Chen, C., Maza, M.M.: Algorithms for computing triangular decomposition of polynomial systems. J. Symb. Comput. **47**(6), 610–642 (2012)
4. Cox, D., Little, J., O'Shea, D.: Ideals, Varieties, and Algorithms, 1st edn. Spinger (1992)
5. Cox, D., Little, J., O'Shea, D.: Using Algebraic Geometry. Graduate Text in Mathematics, vol. 185. Springer, New York (1998)
6. Fulton, W.: Introduction to intersection theory in algebraic geometry. CBMS Regional Conference Series in Mathematics, vol. 54. Published for the Conference Board of the Mathematical Sciences, Washington, DC (1984)
7. Fulton, W.: Algebraic curves. Advanced Book Classics. Addison-Wesley (1989)
8. Marcus, S., Maza, M.M., Vrbik, P.: On fulton's algorithm for computing intersection multiplicities. In: Gerdt, V.P., Koepf, W., Mayr, E.W., Vorozhtsov, E.V. (eds.) CASC 2012. LNCS, vol. 7442, pp. 198–211. Springer, Heidelberg (2012)
9. Mora, F.: An algorithm to compute the equations of tangent cones. In: Calmet, J. (ed.) Computer Algebra. LNCS, vol. 144, pp. 158–165. Springer, Berlin Heidelberg (1982)
10. Mora, T., Pfister, G., Traverso, C.: An introduction to the tangent cone algorithm issues in robotics and non-linear geometry. Advances in Computing Research **6**, 199–270 (1992)
11. Mumford, D.: The Red Book of Varieties and Schemes, 2nd edn., Springer-Verlag (1999)
12. Shafarevich, I.R.: Basic algebraic geometry. 1, 2nd edn. Springer, Berlin (1994)
13. Vrbik, P.: Computing Intersection Multiplicity via Triangular Decomposition. PhD thesis, The University of Western Ontario (2014)

Research on the Stability of Relative Equilibria of Oblate Axisymmetric Gyrostat by Means of Symbolic-Numerical Modelling

Andrei V. Banshchikov

Matrosov Institute for System Dynamics and Control Theory
of Siberian Branch of Russian Academy of Sciences,
P.O. Box 292, 134, Lermontov str., Irkutsk, 664033, Russia

Abstract. The conditions on parameters of the system ensuring stability or instability of relative equilibria of the orbital girostat mentioned in the title were found. The parametrical analysis of conditions of gyroscopic stabilization of the unstable equilibria was carried out. Propositions about the solution of corresponding system of inequalities in the form of intervals of values of the parameter defining one of two nonzero components of a constant vector of gyrostatic moment were formulated. The research was conducted with "Mathematica" built-in tools for symbolic-numerical modelling.

1 Introduction

The rigid body with the fixed axis of a flywheel that rotates with constant relative angular velocity and is counterbalanced statically and dynamically is a gyrostat. The system circles the Kepler orbit in a central Newtonian field of forces around the attracting center. The restricted formulation of the problem is considered. This means that the motion around the center of mass does not affect the orbit of the satellite.

Sufficient conditions of stability of the relative equilibria of an orbital gyrostat were studied for different positioning of a flywheel's axis of rotation by many authors (see, for example, [1] – [3]). However, the analysis of necessary conditions of stability of the relative equilibria is made only for positioning of the vector of gyrostatic moment along with any principal central axis of inertia of the system.

A more general case in which the constant vector of gyrostatic moment is on one of the principal central plane of inertia is considered in [4]. The necessary conditions of stability of the relative equilibria are obtained. The question on an opportunity of gyroscopic stabilization of unstable equilibria is also considered. This paper partly develops some results from [4] and shall present a more detailed parametrical analysis of the conditions of gyroscopic stabilization.

Constructing the mathematical model (the differential equations of motion) for complex mechanical objects, searching for its solutions and conducting the qualitative analysis of their properties, it is necessary to operate with bulky analytical expressions. Thus, it is relevant to apply computer algebra systems

© Springer International Publishing Switzerland 2015
V.P. Gerdt et al. (Eds.): CASC 2015, LNCS 9301, pp. 61–71, 2015.
DOI: 10.1007/978-3-319-24021-3_5

and develop specialized software on the basis of these systems. The software (see, for example, [5] – [7]) used in this paper is designed to model the systems of interconnected absolutely rigid bodies, and also to research the questions of stability and stabilization of solutions of linearized models on the basis of classical theorems of stability of motion.

2   Relative Equilibria

For the description of a motion of the system, two right rectangular Cartesian coordinate systems with the poles in the system's mass center O are introduced. $OZ_1 Z_2 Z_3$ is an orbital coordinate system ($\mathbf{a}_k$ is a unit vector of the corresponding axis). The OZ_3 axis is directed by the radius-vector drawn from the attracting center into the mass center of a gyrostat. The OZ_2 axis is perpendicular to the plane of the orbit, with $\boldsymbol{\omega} = \omega \mathbf{a}_2$, where $\omega = |\boldsymbol{\omega}|$. The coordinate system $Oz_1 z_2 z_3$ ($\mathbf{i}_j$ is the unit vector of the Oz_j axis) rigidly connected to a body has the axes directed on the principal central axes of inertia of a gyrostat. Let us choose the numbering of axes so that $B > A = C$ (A, B, C are the moments of inertia concerning axes Oz_1, Oz_2, Oz_3) and the components of a vector $\mathbf{h}$ of gyrostatic moment divided by ω accordingly, are $h_1 > 0, h_2 > 0, h_3 = 0$. For definition of a relative positioning of the OZ_k and Oz_j axes, the directional cosines $a_{kj} = \mathbf{a}_k \cdot \mathbf{i}_j$, defined by aircraft angles α, β, γ, are used (see, for example, [8]).

According to the angles of turns introduced and the classification of the relative equilibria taken (see, for example, [9]), the equations describing the equilibrium positions in regard to the orbital system of the coordinates of axisymmetric ($A = C$) gyrostat define the equilibria ($\dot{\alpha} = 0, \dot{\beta} = 0, \dot{\gamma} = 0$) of, accordingly, the third and second classes:

$$\begin{cases} \alpha = \alpha_0 = n\pi, \ (n = 0, 1); \quad \gamma = \gamma_0 = 0; \\ \beta = \beta_0 = const: \quad h_2 \sin\beta_0 - \cos\beta_0(h_1 + (A - B)\sin\beta_0) = 0. \end{cases} \tag{1}$$

$$\begin{cases} \alpha = \alpha_0 = \frac{\pi}{2} + n\pi, \ (n = 0, 1); \quad \gamma = \gamma_0 = 0; \\ \beta = \beta_0 = const: \quad h_2 \sin\beta_0 - \cos\beta_0(h_1 + 4(A - B)\sin\beta_0) = 0. \end{cases} \tag{2}$$

3   Construction of Symbolical Model and Parametrization of a Problem

The equations of motion of a gyrostat in a circular orbit with a constant vector of gyrostatic moment are quite well-known (see, for example, [8]). With the help of developed software, the following results in a symbolic form on PC are obtained:
- kinetic and potential energy of a system (as given in [8]);
- nonlinear equations in the form of Lagrange of the 2nd kind describing the motion of a gyrostat in the central Newtonian field of forces;
- existence conditions of equilibrium positions (1) and (2);

- matrices of equations of perturbed motion in the 1st approximation in the vicinity of unperturbed motion (1) and (2);
- coefficients of the system's characteristic equation.

In the vicinity of (1) or (2), the linear equations of perturbed motion look this way:

$$M\ddot{q} + G\dot{q} + Kq = 0, \tag{3}$$

where $q = (\alpha, \beta, \gamma)^T$ is a vector of deviations of generalized coordinates from unperturbed motion; M is a positive definite matrix of kinetic energy; G is a skew-symmetric matrix of gyroscopic forces; K is a symmetric matrix of potential forces; all derivatives are calculated on already dimensionless time $\tau = \omega t$.

So, for example, for equilibrium positions (1), these matrices have the following form:

$$M = \begin{pmatrix} m_{11} & 0 & A\sin\beta_0 \\ 0 & A & 0 \\ A\sin\beta_0 & 0 & A \end{pmatrix}; \quad K = \begin{pmatrix} k_{13}\sin\beta_0 & 0 & k_{13} \\ 0 & k_{22} & 0 \\ k_{13} & 0 & k_{33} \end{pmatrix}; \quad G = \begin{pmatrix} 0 & g_{12} & 0 \\ -g_{12} & 0 & g_{23} \\ 0 & -g_{23} & 0 \end{pmatrix},$$

where $m_{11} = B\cos^2\beta_0 + A\sin^2\beta_0$, $k_{22} = (B-A)\cos2\beta_0 + h_1\sin\beta_0 + h_2\cos\beta_0$,

$k_{13} = 3(B-A)\sin\beta_0$, $k_{33} = (B-A)(3 + \cos^2\beta_0) + h_2\cos\beta_0$,

$g_{12} = h_1\cos\beta_0 - h_2\sin\beta_0 + (A-B)\sin2\beta_0$, $g_{23} = h_2 + (B-2A)\cos\beta_0$.

Let us introduce dimensionless parameters:

$$H_1 \equiv \frac{h_1}{B}; \quad H_2 \equiv \frac{h_2}{B}; \quad J \equiv \frac{A}{B}; \quad p_c \equiv \cos\beta_0; \quad p_s \equiv \sin\beta_0. \tag{4}$$

The values of the parameters belong to the intervals:

$$H_1 > 0; \quad H_2 > 0; \quad \tfrac{1}{2} < J < 1; \quad -1 < p_c < 1, \quad \left(p_c \neq 0, \; p_s = \pm\sqrt{1 - p_c^2}\right). \tag{5}$$

The restrictions on the parameter J come from the conditions $B < A+C$, $C = A$. With the values of $p_c = 0$ (or $p_s = 0$), from equation in (1), it then follows that $h_2 = 0$ (or $h_1 = 0$), what contradicts the conditions $h_1 > 0$, $h_2 > 0$.

4   Stability of the Third Class Equilibria

4.1   Necessary Conditions of Stability

Using (4), let us resolve the equation from (1) with respect to the parameter H_1:

$$H_1 = \frac{p_s (H_2 + (1 - J)p_c)}{p_c}. \tag{6}$$

Taking into account notations (4) and expression (6), the equations of the first approximation (3) have matrices:

$$M = \begin{pmatrix} Jp_s^2 + p_c^2 & 0 & p_s J \\ 0 & J & 0 \\ p_s J & 0 & J \end{pmatrix}; \quad G = \begin{pmatrix} 0 & p_c p_s(J-1) & 0 \\ p_c p_s(1-J) & 0 & H_2 + p_c(1-2J) \\ 0 & p_c(2J-1) - H_2 & 0 \end{pmatrix};$$

$$K = \begin{pmatrix} 3(1-J)\,p_s^2 & 0 & 3p_s(1-J) \\ 0 & (1-J)\,p_c^2 + \frac{H_2}{p_c} & 0 \\ 3p_s(1-J) & 0 & H_2 p_c + (3+p_c^2)(1-J) \end{pmatrix} \qquad (7)$$

The parameter p_s enters the coefficients of the system's characteristic equation:

$$\det\left(M\lambda^2 + G\lambda + K\right) = v_3\lambda^6 + v_2\lambda^4 + v_1\lambda^2 + v_0 = 0\,, \qquad (8)$$

only in even degrees. Let us eliminate it, considering $p_c^2 + p_s^2 = 1$. Let us write down these coefficients depending on three parameters J, p_c, H_2, in an explicit form:

$$v_3 = J^2 p_c^2\,; \quad v_0 = 3(1-J)\left(1-p_c^2\right)(H_2 + (1-J)\,p_c)\left(H_2 + (1-J)p_c^3\right);$$

$$v_1 = H_2^2\left(1+2\,(1-J)\left(1-p_c^2\right)\right) + H_2\,p_c(1-J)\left(3 + 2p_c^2 - 2J\left(1-p_c^2\right)\right)$$
$$+\, p_c^2(1-J)\left(4p_c^2 - J\left(7p_c^2 - 3\right)\right); \qquad (9)$$

$$v_2 = H_2^2\left(J+(1-J)p_c^2\right) + H_2\,p_c(1-J)\left(J+(2-J)p_c^2\right) +$$
$$p_c^2\left((J-1)^2 p_c^2 + (3-2J)J\right).$$

Last two coefficients are presented in (9) as quadratic polynomials with respect to H_2 (i.e., $v_i = a_i(J,p_c)H_2^2 + b_i(J,p_c)H_2 + c_i(J,p_c)$, $i = 1,2$).

The equation (8) contains λ only in even degrees. The stability of a trivial solution of equations (3) takes place when all roots of equation (8) with respect to λ^2, being simple, will be real negative numbers. The algebraic conditions providing specified properties of roots (necessary conditions of stability), represent the system of inequalities [10]:

$$\begin{cases} v_3 > 0\,, \quad v_2 > 0\,, \quad v_1 > 0\,, \quad v_0 > 0\,, \\ Dis \equiv v_2^2 v_1^2 - 4v_1^3 v_3 - 4v_2^3 v_0 + 18v_3 v_2 v_1 v_0 - 27v_0^2 v_3^2 > 0\,. \end{cases} \qquad (10)$$

The discriminant Dis of a cubic equation is an 8th degree polynomial in regard to H_2 with the coefficients depending in a complicated manner on the parameters J and p_c. This polynomial in an explicit analytical form, due to being immense, is not presented.

4.2   Stable and Unstable Equilibria

According to Kelvin–Chetaev's theorems [11], studying the questions on stability of equilibria begins with the analysis of a matrix of potential forces. Proceeding from the analysis of matrix of potential forces, in [4], propositions in the form of conditions on parameters of the system, providing stability or instability of relative equilibria are formulated and proved.

Let us note that $v_0 \equiv \det K$ and write the conditions of definite positiveness of a matrix K from (7):

$$(1-J)\left(1-p_c^2\right) > 0\,, \quad (1-J)\,p_c^2 + \frac{H_2}{p_c} > 0\,, \quad v_0 > 0\,. \qquad (11)$$

By means of "Mathematica" function *Reduce*, the solution for the system of inequalities (11) on intervals (5) is obtained:

$$\left(\frac{1}{2}<J<1\right)\wedge\left((-1<p_c<0\wedge 0<H_2<(J-1)\,p_c^3)\vee(0<p_c<1\wedge H_2>0)\right). \quad (12)$$

This function is assigned to find a symbolical (analytical) solution of systems of inequalities.

We shall formulate two propositions which are proved by [4].

Proposition 1. *At values of the parameters from (12), the relative equilibria (1) will be steady considering the nonlinear equations of motion.*

Proposition 2. *At values of the parameters from the range* $1/2<J<1 \wedge -1< p_c<0 \wedge (J-1)p_c^3 <H_2<(J-1)p_c$ *, the relative equilibria (1) will also be unstable for a nonlinear system of equations of motion.*

4.3   Gyroscopic Stabilization: Symbolic-Numerical Modelling

The domains of gyroscopic stabilization in space of three parameters and their two-parameter cuts are constructed in [4]. This paper shall present a more detailed parametrical analysis of system of inequalities (10) with respect to "the principal or foremost" parameter H_2 in the domain where a gyroscopic stabilization of the system is possible.

It is known that if the position of an equilibrium is unstable at potential forces, Kelvin–Chetaev's theorem [11] of influence of gyroscopic forces tells that gyroscopic stabilization is possible only for systems with an even degree of instability. According to Poincaré, evenness (or oddness) of the degree of instability is determined by positiveness (or negativity) of the determinant of a matrix of potential forces. Considering that the main diagonal minor of the 2nd order of a matrix K from (7) is negative, i.e. $minor2 = 3(1-J)(1-p_c^2)\left(\frac{H_2}{p_c} + (1-J)p_c^2\right) < 0$, the trivial solution of equations (3) will be unstable. Then, it is possible to raise the question on an opportunity of gyroscopic stabilization of the solution at $\det K > 0$.

The symbolical solution of a system of the inequalities defining the domain with an even degree of instability is found by means of the function

$$Reduce\,[\,1/2 <J<1 \wedge -1<p_c<1 \wedge p_c\neq 0 \wedge H_2>0 \wedge$$
$$\wedge\, minor2 <0 \wedge \det K > 0, \{\,J\,,p_c\,,H_2\,\}\,].$$

and looks the following way

$$1/2 < J < 1 \wedge -1 < p_c < 0 \wedge H_2 > (J-1)\,p_c. \quad (13)$$

Let us note that the first condition in (10) is satisfied ($v_3 \equiv \det M > 0$), and the condition $v_0 \equiv \det K > 0$ is satisfied in range (13) by definition. For the detection of a property of gyroscopic stabilization, it is necessary to find in what part of the domain with an even degree of instability the remaining inequalities from (10) are fulfilled (except for $v_3 > 0$, $v_0 > 0$). By means of function *Reduce* mentioned above, it is not difficult to prove and make the following remarks.

Remark 1. The following conditions $a_i(J, p_c) > 0$, $b_i(J, p_c) < 0$, $c_i(J, p_c) > 0$, $(i = \overline{0,2})$ on coefficients of quadratic polynomials v_i from (9) for the values of the parameters J, p_c from the range (13) are satisfied.

Remark 2. There can be any roots of a polynomial v_1 in range (13), but the roots of polynomial v_2 can only be complex conjugate (negativity of a discriminant of a quadratic equation in this range). For the complex conjugate roots, considering remark 1, corresponding conditions $v_1 > 0$, $v_2 > 0$, are certainly fulfilled.

Remark 3. Both of the real roots $H_2^{(1)}$, $H_2^{(2)}$, of the polynomial v_1:

$$H_2^{(1)}, H_2^{(2)} = \frac{p_c \left(\left(2(J^2 - 1)\, p_c^2 + (5 - 2J)J - 3 \right) \mp \sqrt{D_1} \right)}{2\left(3 - 2J + 2(J-1)p_c^2 \right)}, \quad \text{where}$$

$$D_1 = (J-1)\left(4p_c^2 \left(9 + J((23 - 2J)J - 33) + (J-1)(9 + (J-12)J)p_c^2 \right) \right.$$
$$\left. + (2J - 3)(3 + J(2J - 17)) \right).$$

have only positive values and lie more to the right of the second root $(J-1)p_c$ of polynomial v_0 (for the values of the parameters J, p_c from the range (13)).

Considering the remarks made above, the graphs of coefficients $y_i = v_i(H_2)$, $(i = \overline{0,2})$ of a characteristic equation from (9) and the location of their roots in the domain with an even degree of instability are presented in Fig. 1. The intervals $(J-1)p_c < H_2 < H_2^{(1)} \lor H_2 > H_2^{(2)}$ of the values H_2 in which all coefficients are positive, are highlighted with bold black lines on abscissa axis.

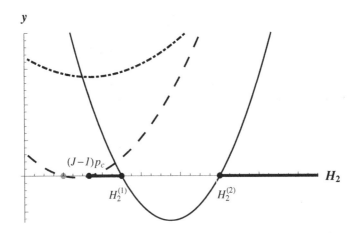

Fig. 1. The range of positivity of coefficients (9), where parabolas —— $v_1(H_2)$, $- - - v_0(H_2)$, $- \cdot - \cdot$ $v_2(H_2)$.

With the help of the functions of symbolic-numerical modelling and the means of the programming language of "Mathematica" system where the values of the parameters J, p_c varied in a grid with a step 10^{-3}, the following facts, drawn as remark 4 and proposition 3, are identified.

Remark 4. In the domain with an even degree of instability, the polynomial *Dis* has two or four real roots with respect to H_2.

Proposition 3. *The solution of the system of inequalities (10), defining the domain of gyroscopic stabilization of equilibria (1), is presented as a union of two intervals $(J-1)p_c < H_2 < x_1 \vee H_2 > x_{2,4}$ of the values for the parameter H_2, where x_1, $x_{2,4}$ are accordingly the first and last (the second or the fourth) real roots of the polynomial Dis.*

For example, Fig. 2 shows the graph of function $y = Dis(H_2)$ (at values $J = 501/1000$, $p_c = -479/500$) and a location of its two real roots x_1 and x_2, while the graphic interpretation of proposition 3 (i.e., the domain of a solution of the entire system of inequalities (10)) is highlighted on the axis of abscissas with bold black lines. Thus, the well-known fact that gyroscopic stabilization is possible only in some part of the domain with an even degree of instability is confirmed.

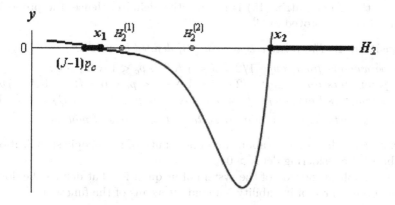

Fig. 2. The domain of gyroscopic stabilization.

5   Parametrical Analysis of the Conditions of Gyroscopic Stabilization for Second Class Equilibria

The algorithm of the investigation of gyroscopic stabilization of the relative equilibria (2) is reduced to repetition of the steps described in paragraphs 3, 4.1 – 4.3 with respective alterations.

Let us find the solution for the equation from (2) in parameters (4) in regard to H_1:

$$H_1 = \frac{p_s\,(H_2 + 4\,(1 - J)\,p_c)}{p_c}. \tag{14}$$

For equilibria (2), matrixes G and K of equation (3) and, accordingly, the coefficients v_i, $(i = \overline{0,2})$ of the characteristic equation (8) change. Taking into account (14), the matrix of potential forces becomes:

$$K = \begin{pmatrix} 3(J-1)p_s^2 & 0 & 3(J-1)p_s \\ 0 & 4(1-J)p_c^2 + \frac{H_2}{p_c} & 0 \\ 3(J-1)p_s & 0 & H_2 p_c + (1-J)(2(p_c^2 - p_s^2) - 1) \end{pmatrix}, \quad (15)$$

and the coefficients (after the elimination of parameter p_s)

$$v_3 = J^2 p_c^2 > 0; \quad v_2 = H_2^2\big((1-J)\,p_c^2 + J\big) + H_2\,p_c(1-J)\big(2p_c^2 + 7J(1-p_c^2)\big)$$

$$+ \ p_c^2\big(p_c^2 + (12J^3 - 16J^2 + 5J)(1-p_c^2) + (1-J)\,J\big);$$

$$v_1 = H_2^2\big(4\,(1-J)\,p_c^2 + 4J - 3\big) + H_2\,p_c(1-J)\big(5p_c^2 - (21-25J)(1-p_c^2)\big) \quad (16)$$

$$+ \ p_c^2(1-J)\big(4(1-J)\,p_c^2 + (60J - 36J^2 - 27)(1-p_c^2)\big);$$

$$v_0 \equiv \det K = 3\,(1-J)(H_2 + 4\,(1-J)\,p_c)\big(p_c^2 - 1\big)\big(H_2 + 4\,(1-J)\,p_c^3\big).$$

The main diagonal minor of the 1st order of the matrix K on intervals (5) is negative, therefore, matrix (15) is not positive-definite. Hence, the proposition is formulated and proved in [4]:

Proposition 4. *The relative equilibria (2), defined by values:*
a) for parameters from range $1/2 < J < 1 \wedge 0 < p_c < 1 \wedge H_2 > 0$;
b) for parameters from range $1/2 < J < 1 \wedge -1 < p_c < 0 \wedge H_2 > 4(J-1)\,p_c$;
c) for parameters from range $1/2 < J < 1 \wedge -1 < p_c < 0 \wedge 0 < H_2 < 4(J-1)\,p_c^3$
will be also unstable for a nonlinear system of equations of motion.

Let us raise the question about an opportunity of gyroscopic stabilization of equilibria (2) considering $\det K > 0$.

The symbolical solution of the system of inequalities that defines the domain with an even degree of instability is found by means of the function

$$Reduce\left[\frac{1}{2} < J < 1 \wedge -1 < p_c < 1 \wedge p_c \neq 0 \wedge H_2 > 0 \wedge \det K > 0, \{J, p_c, H_2\}\right]$$

and has the form

$$\frac{1}{2} < J < 1 \wedge -1 < p_c < 0 \wedge 4(J-1)\,p_c^3 < H_2 < 4(J-1)\,p_c. \quad (17)$$

Remark 5. It is clear to notice that the conditions $a_0(J, p_c) < 0$, $b_0(J, p_c) > 0$, $c_0(J, p_c) < 0$ on the coefficients of a quadratic polynomial v_0 from (16) are satisfied at any values of the parameters J, p_c from (17).

Applying the *Reduce* function, let us analytically find several facts on the location of the roots of quadratic polynomials v_1, v_2 from (16) and the signs of the coefficients of these polynomials. As a result, several remarks are made. Let us cite the following:

Remark 6. In range (17), the roots $H_2^{(1)}$, $H_2^{(2)}$ of the polynomial v_1 have only real values (i.e., the discriminant of the quadratic equation is positive).

Remark 7. In range (17), the roots $H_2^{(3)}$, $H_2^{(4)}$ of the polynomial v_2:

$$H_2^{(3)}, H_2^{(4)} = \frac{p_c\left((J-1)\left((7J-2)p_c^2 - 7J\right) \mp \sqrt{J}\sqrt{D_2}\right)}{2\left((J-1)p_c^2 - J\right)}, \quad \text{where}$$

$$D_2 = J(25 + (J-30)J) - 2J(J-21)(J-1)p_c^2 + (J-12)(J-1)^2 p_c^4$$

can be both real (in this case $H_2^{(4)} > H_2^{(3)}$) or complex conjugate.

Remark 8. The positivity of the coefficients v_i, $(i = \overline{0,3})$ from (16) (as part of the system of inequalities from (10)) is possible only inside a parallelepiped:

$$\frac{1}{2} < J < \frac{3}{4} \wedge -1 < p_c < -\frac{1}{2} \wedge \frac{1}{2} < H_2 < 2. \tag{18}$$

¿From remark 8, it is evident that if the values of the parameters J, p_c, H_2 do not belong to intervals from (18), the investigated system of inequalities (10) will be inconsistent.

The various types of graphs of behavior of functions $y_i = v_i(H_2)$, $(i = \overline{0,2})$ are drawn, with the parameters J, p_c of (18) changing. For example, for the values of parameters from the intervals $\frac{1}{2} < J < \frac{3}{4} \wedge -1 < p_c < -\frac{1}{2}\sqrt{\frac{4J-3}{J-1}}$, the range $H_2^{(4)} < H_2 < 4(J-1)p_c$ of the positivity of the coefficients v_i is highlighted in Fig. 3 with a bold black line on the axis of abscissas.

With the help of the functions of symbolic-numerical modelling and the means of the programming language of "Mathematica" system where the values of the parameters J, p_c varied in a grid with a step 10^{-3} within the limits of the intervals from (18) already, the following facts, drawn as remark 9 and proposition 5, are formulated.

Remark 9. The polynomial *Dis* from (10) has from one to four real roots with respect to H_2, belonging to the interval $1/2 < H_2 < 2$.

Proposition 5. *The solution of the system of inequalities (10), defining the domain of gyroscopic stabilization of equilibria (2), has one of four forms:*
1) $x_2 < H_2 < 4(J-1)p_c$ (in case of existence in the range (18) of two real roots of the polynomial Dis), where x_2 is the second (right) root of the polynomial;
2) $x_2 < H_2 < x_3$ (in case of existence in the range (18) of three real roots of the polynomial Dis), where x_2 and x_3 are the second and third roots of the polynomial;

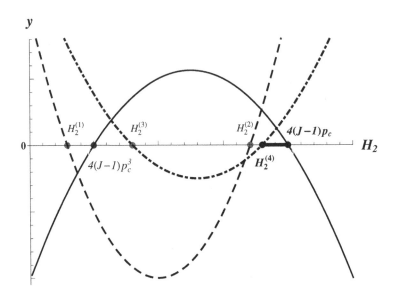

Fig. 3. The range of positivity of coefficients (16), where parabolas —— $v_0(H_2)$ — — $v_1(H_2)$, – - – $v_2(H_2)$.

3) $x_2 < H_2 < x_3 \vee x_4 < H_2 < 4(J-1)p_c$ *(in case of existence in the range (18) of four real roots of a polynomial Dis), where x_2, x_3, and x_4 are the second, third, and fourth roots of a polynomial;*
4) *FALSE (in case of existence in the range (18) of one real root of the polynomial Dis). It means that for some values of the parameters J, p_c of the domain with an even degree of instability, the system of inequalities (10) is inconsistent.*

Let us note that numerical values of the real roots of the polynomial Dis in range (17) are quite close to each other. Thus, the gyroscopic stabilization of the equilibria (2) is possible, but in the very narrow limited intervals of H_2 values.

6   Conclusion

We have presented a parametric analysis of the conditions for gyroscopic stabilization (10), where, as the "principal" parameter, H_2 has been selected. The remarks and propositions about the location of roots of the quadratic polynomials v_i $(i = \overline{0,2})$, about the quantity of real roots of the polynomial Dis of the eighth degree (with respect to H_2) in the domain with an even degree of instability, and also about the intervals of values H_2 which are a solution of the system of inequalities (10) for equilibria (1) and (2) have been formulated. Results have been obtained either in analytical form or by carrying out a numerical experiment with graphic interpretation. To conduct symbolic-numerical computations, means of the programming language and the built-in functions of a computer algebra system "Mathematica" have been used.

References

1. Sarychev, V.A., Mirer, S.A., Degtyarev, A.A.: Dynamics of a gyrostat satellite with the vector of gyrostatic moment in the principal plane of inertia. Cosmic Research **46**(1), 60–73 (2008)
2. Gutnik, S.A., Sarychev, V.A.: Symbolic-Numerical investigation of gyrostat satellite dynamics. In: Gerdt, V.P., Koepf, W., Mayr, E.W., Vorozhtsov, E.V. (eds.) CASC 2013. LNCS, vol. 8136, pp. 169–178. Springer, Heidelberg (2013)
3. Gutnik, S.A., Sarychev, V.A.: Dynamics of an axisymmetric gyrostat satellite. Equilibrium positions and their stability. J. Appl. Math. Mech. **78**(3), 249–257 (2014)
4. Chaikin, S.V., Banshchikov, A.V.: On gyroscopic stabilization of the relative equilibriums of oblate axisymmetric gyrostat. Matematicheskoe Modelirovanie **25**(5), 109–122 (2013) (in Russian)
5. Banshchikov, A.V., Bourlakova, L.A.: Information and Research System "Stability". Journal of Computer and Systems Sciences International **35**(2), 177–184 (1996)
6. Irtegov, V.D., Titorenko, T.N.: Using the system "Mathematica" in problems of mechanics. Mathematics and Computers in Simulation **57**(3–5), 227–237 (2001)
7. Irtegov, V.D., Titorenko, T.N.: Computer Algebra Methods in the Study of Nonlinear Differential Systems. Comp. Math. Math. Phys. **53**(6), 845–857 (2013)
8. Sarychev, V.A.: Problems of orientation of satellites. In: Itogi Nauki i Tekhniki. Series "Space Research". 11, VINITI Publ. Moscow (1978) (in Russian)
9. Anchev, A.: Equilibrium orientations of a satellite with rotors. Izd-vo Bulgarska Akademiia na Naukite, Sofia (1982)
10. Kozlov, V.V.: Stabilization of the unstable equilibria of charges by intense magnetic fields. J. Appl. Math. Mech. **61**(3), 377–384 (1997)
11. Chetaev, N.G.: Stability of Motion. Works on Analytical Mechanics, AS USSR, Moscow (1962) (in Russian)

A New Approach for Computing Regular Solutions of Linear Difference Systems

Moulay Barkatou[1], Thomas Cluzeau[1], and Carole El Bacha[2]

[1] University of Limoges, CNRS, XLIM UMR 7252
123 avenue Albert Thomas, 87060 Limoges cedex, France
{moulay.barkatou,thomas.cluzeau}@unilim.fr
[2] Lebanese University, Faculty of Science II, Department of Mathematics
P.O. Box 90656 Fanar-Matn, Lebanon
carole.bacha@ul.edu.lb

Abstract. In this paper, we provide a new approach for computing regular solutions of first-order linear difference systems. We use the setting of factorial series known to be very well suited for dealing with difference equations and we introduce a sequence of functions which play the same role as the powers of the logarithm in the differential case. This allows us to adapt the approach of [5] where we have developed an algorithm for computing regular solutions of linear differential systems.

Introduction

Let z be a complex variable and Δ the difference operator whose action on a function f is defined by

$$\Delta(f(z)) = (z-1)\,(f(z) - f(z-1)).$$

In the present paper, we consider first-order systems of linear difference equations of the form

$$D(z)\,\Delta(\mathbf{y}(z)) + A(z)\,\mathbf{y}(z) = 0, \tag{1}$$

where $D(z)$ and $A(z)$ are given $n \times n$ matrices with factorial series entries of the form $\sum_{i \geq 0} a_i\,z^{-[i]}$ with $a_i \in \mathbb{C}$ and

$$z^{-[0]} = 1, \quad \forall\, i \geq 1,\ z^{-[i]} = \frac{1}{z(z+1)\cdots(z+i-1)},$$

and $\mathbf{y}(z)$ is an n-dimensional vector of unknown functions of the complex variable z. We further assume that the matrix $D(z)$ is invertible so that System (1) can be written as

$$\Delta(\mathbf{y}(z)) = z^q\,B(z)\,\mathbf{y}(z), \tag{2}$$

where $q \in \mathbb{Z}$ and $B(z)$ is an $n \times n$ matrix whose entries are factorial series and $B(\infty) \neq 0$. When $q < 0$, System (2) has a fundamental matrix of factorial series solutions. In the particular case $q = 0$, System (2) is said to be *of the first kind*

© Springer International Publishing Switzerland 2015
V.P. Gerdt et al. (Eds.): CASC 2015, LNCS 9301, pp. 72–86, 2015.
DOI: 10.1007/978-3-319-24021-3_6

and it has been shown in [15] that it admits a basis of n formal solutions of the form

$$\mathbf{y}(z) = z^{-\rho} \left(\mathbf{y}_0(z) + \mathbf{y}_1(z) \log(z) + \cdots + \mathbf{y}_s(z) \log^s(z) \right), \tag{3}$$

where $s < n$, ρ is a complex number and the $\mathbf{y}_i(z)$ are vectors of factorial series. Solutions of the form (3) are called *(formal) regular solutions*. A method (close to that of Frobenius in the differential case) for computing a basis of regular solutions of System (2) with $q = 0$ has been developed in [15, 14, 3]. It generalizes the method of [21] restricted to scalar equations in order to handle systems directly. In [11], the authors consider systems of the first kind and give another approach to establish the existence of regular solutions of the form (3) where the powers of the log function are replaced by other functions. When $q \leq 0$, the factorial series involved in the solutions are convergent whenever those involved in the entries of the system are so (see [14, 12, 11]). Finally, when $q > 0$, in addition to potential regular solutions, System (2) may also have non-regular solutions.

In a previous work [5], we have provided an efficient algorithm for computing regular solutions of systems of linear differential equations with formal power series coefficients. The goal of the present paper is to adapt the method developed in [5] in order to give a new approach for computing regular solutions of systems of linear difference equations of the form (1). Note that although the theory of difference equations have close similarities with the theory of differential equations, the generalization of results and algorithms from the differential case to the difference case are in general not totally straightforward (see, e.g., [21, 14–16, 10, 3, 4, 12, 6, 7]).

The method developed in [5] consists in writing a regular solution of a linear differential system $D(z)\,\vartheta(\mathbf{y}(z)) + A(z)\,\mathbf{y}(z) = 0$, where $\vartheta = z\frac{d}{dz}$ denotes the Euler operator and $D(z)$ and $A(z)$ are $n \times n$ matrices with power series entries, as a series $\mathbf{y}(z)$ of the form

$$\mathbf{y}(z) = \sum_{m \geq 0} z^{\rho+m}\, \mathbf{w}_m(z), \tag{4}$$

where $\mathbf{w}_m(z)$ is a polynomial in $\log(z)$ with constant vector coefficients. Roughly speaking, it proceeds by plugging the ansatz $\mathbf{y}(z)$ defined by (4) into the system and identifying the coefficients of the resulting series to zero. This reduces constructively the problem to the simpler problem of computing appropriate solutions of linear differential systems with constant coefficients for which we have results on the existence of solutions of the desired form and algorithms for computing them. This approach mainly relies on the good behavior of the powers of z and $\log(z)$ with respect to the Euler operator ϑ, namely, we have

$$\forall n \in \mathbb{N}, \quad \vartheta(z^n) = n\,z^n, \quad \vartheta(\log^n(z)) = n\,\log^{n-1}(z).$$

The latter relations are not preserved when we replace the Euler operator ϑ by the difference operator Δ. Therefore, in order to apply the strategy of [5] in the setting of difference equations, we must replace power series by factorial series (i.e., z^n is replaced by $z^{-[n]}$), and we introduce a sequence of functions $(\phi_n)_{n \in \mathbb{N}}$ which plays a role analogous to $(\log^n)_{n \in \mathbb{N}}$ so that we have

$$\forall n \in \mathbb{N}, \quad \Delta(z^{-[n]}) = -n\, z^{-[n]}, \quad \Delta(\phi_n(z)) = n\, \phi_{n-1}(z).$$

Note that factorial series have been proved to be very well suited for studying linear difference equations: see [21, 14–16, 10, 3, 4, 12, 6, 7]. They are also useful for solving linear differential equations (see, e.g., [22]). Then, instead of using the form (3), we search for regular solutions of System (1) written as

$$\mathbf{y}(z) = \sum_{m \geq 0} z^{-[\rho+m]}\, \mathbf{y}_m(z), \quad z^{-[\rho]} = \frac{\Gamma(z)}{\Gamma(z + \rho)},$$

where Γ stands for the usual Gamma function and $\mathbf{y}_m(z)$ is now a finite linear combination of the functions ϕ_i with constant vector coefficients. For example, within this setting, the system of two linear difference equations given by

$$\Delta(\mathbf{y}(z)) + \begin{pmatrix} \rho & -1 \\ 0 & \rho \end{pmatrix} \mathbf{y}(z) = 0,$$

where ρ is a given complex number, admits two linearly independent regular solutions written as

$$\mathbf{y}_1(z) = z^{-[\rho]} \begin{pmatrix} 1 \\ 0 \end{pmatrix} \phi_0,$$

and

$$\mathbf{y}_2(z) = z^{-[\rho]} \left(\begin{pmatrix} 0 \\ 1 \end{pmatrix} \phi_0 + \begin{pmatrix} 1 \\ 0 \end{pmatrix} \phi_1 \right) + \sum_{m \geq 0} z^{-[\rho+m+1]} \begin{pmatrix} \frac{\rho\,(\rho+1)\,\cdots\,(\rho+m)}{m+1} \\ 0 \end{pmatrix} \phi_0.$$

The rest of the paper is organized as follows. In Section 1, we recall some background information on factorial series that are needed in the sequel. Then, in Section 2, we introduce the functions ϕ_n and give some of their properties. Finally, in Section 3, we develop the main contribution of the paper, i.e., our new approach for computing regular solutions of first-order linear difference systems. An appendix provides the detailed proofs of two results of the paper.

1 Factorial Series

In the theory of linear difference equations, *factorial series* are often used instead of the power series classically used in the theory of linear differential equations: see [21, 14–16, 10, 3, 4, 12, 6, 7]. Indeed, some interesting results of the theory of linear differential equations in connection with power series remain valid in the theory of linear difference equations when factorial series are involved.

We refer to [18, Chap. 10] or [19–21, 14, 17, 10, 3, 12, 7] for more details on the basic facts concerning factorial series that are recalled in this section.

1.1   Definition

Before defining factorial series, let us introduce a useful notation that we shall use in the rest of the paper. For a complex variable z and $\rho \in \mathbb{C}$, we denote

$$z^{-[\rho]} = \frac{\Gamma(z)}{\Gamma(z + \rho)}.$$

Note that $z^{-[\rho]}$ is not defined when $z \in \mathbb{Z}_{\leq 0}$ and if $\rho = n \in \mathbb{N}$, then we have

$$z^{-[0]} = 1, \quad \forall n \geq 1, \; z^{-[n]} = \frac{1}{z(z+1)\cdots(z+n-1)}.$$

A straightforward calculation shows that, for all $n \in \mathbb{N}$, $\Delta(z^{-[n]}) = -n\, z^{-[n]}$ so that the behavior of $z^{-[n]}$ with respect to Δ is the same as the one of z^n with respect to the Euler operator $\vartheta = z\frac{d}{dz}$. This is one of the reasons for replacing power series by factorial series when we deal with linear difference equations.

Definition 1. *A factorial series is a series of the form* $\sum_{n\geq 0} a_n\, z^{-[n]}$ *where* $a_n \in \mathbb{C}$.

Note that in the literature, factorial series defined as in Definition 1 are sometimes called *inverse factorial series*.

The domain of convergence of a factorial series is a half-plane $\Re(z) > \mu$, where μ is the abscissa of convergence whose formula is given in [18, Section 10.09, Theorem V] and $\Re(z)$ denotes the real part of the complex z. The convergence is uniform in the half-plane $\Re(z) > \mu' + \epsilon$ where $\mu' = \max(\mu, 0)$ and $\epsilon > 0$.

1.2   Ring Structure

The set of factorial series is endowed with a ring structure for the addition and the multiplication defined below. The addition is naturally defined by

$$\sum_{n\geq 0} a_n\, z^{-[n]} + \sum_{n\geq 0} b_n\, z^{-[n]} = \sum_{n\geq 0} (a_n + b_n)\, z^{-[n]}.$$

The multiplication of factorial series is not as simple as the multiplication of power series but a solution has been found by Nielsen [19] (see also [18, Chap. 10]). We first investigate the product of two *monomials* $z^{-[n]}$ and $z^{-[p]}$ for which we have

$$z^{-[n]}\, z^{-[p]} = \sum_{k\geq 0} C_{n,p}^k\, z^{-[n+p+k]}, \quad n, p \in \mathbb{N},$$

where the constants $C_{x,y}^k$ are defined by

$$C_{x,y}^k = \frac{x^{[k]} y^{[k]}}{k!}, \tag{5}$$

for all $x, y \in \mathbb{C}$ and $k \in \mathbb{N}$ with $z^{[k]}$ denoting the usual rising factorial given by

$$z^{[k]} = \prod_{j=0}^{k-1} (z + j).$$

A *multiplication formula* for factorial series can then be deduced as follows: the product of two factorial series $\sum_{n\geq 0} a_n z^{-[n]}$ and $\sum_{p\geq 0} b_p z^{-[p]}$ is defined as the factorial series $\sum_{s\geq 0} d_s z^{-[s]}$ given by

$$d_s = \sum_{\substack{n,p,k\geq 0 \\ n+p+k=s}} C_{n,p}^k a_n b_p.$$

Note that if both $\sum_{n\geq 0} a_n z^{-[n]}$ and $\sum_{p\geq 0} b_p z^{-[p]}$ are convergent series, then their product converges at least in the smallest of their convergence domains. The ring $\mathcal{R}$ of factorial series endowed with the above addition and multiplication is isomorphic to the ring $\mathbb{C}[[z^{-1}]]$ (see [10]). The isomorphism $\varphi : \mathbb{C}[[z^{-1}]] \to \mathcal{R}$ and its inverse φ^{-1} are given by: $\varphi(1) = 1$, $\varphi^{-1}(1) = 1$, and

$$\varphi\left(\frac{1}{z}\right) = z^{-[1]}, \ \forall n \geq 1, \ \varphi\left(\frac{1}{z^{n+1}}\right) = (-1)^n (n-1)! \sum_{k\geq 0} (-1)^k s(n+k, n) z^{-[n+k+1]},$$

$$\forall n \geq 1, \ \varphi^{-1}(z^{-[n]}) = \sum_{k\geq 0} (-1)^k S(n+k-1, n-1) \frac{1}{z^{n+k}},$$

where the constants $s(n, k)$ (resp. $S(n, k)$) are the Stirling numbers of the first (resp. second) kind (see [2, Sec. 24]).

1.3   Translation $z \mapsto z + \beta$

With the notation (5), we have the formula

$$\forall \beta \in \mathbb{C}, \quad z^{-[n]} = \sum_{k\geq 0} C_{\beta,n}^k (z + \beta)^{-[n+k]},$$

which can be used to express any series in $z^{-[n]}$ as a series in $(z + \beta)^{-[n]}$. We deduce the following *translation formula* for factorial series

$$\forall \beta \in \mathbb{C}, \quad \sum_{n\geq 0} a_n z^{-[n]} = a_0 + \sum_{p\geq 1}\left(\sum_{k=1}^{p} a_k C_{\beta,k}^{p-k}\right) (z + \beta)^{-[p]}.$$

Moreover, for all $m \in \mathbb{N}^*$, we have

$$\forall \beta \in \mathbb{C}, \quad \sum_{n\geq 0} a_n z^{-[n+m]} = \sum_{p\geq 0}\left(\sum_{k=0}^{p} a_k C_{\beta,k+m}^{p-k}\right) (z + \beta)^{-[p+m]}.$$

2   The Functions ϕ_n

In order to adapt the method of [5] to the setting of difference equations we shall replace the triple $(\vartheta = z\frac{d}{dz}, z^n, \log^n)$ used in the differential case and satisfying

$$\forall n \in \mathbb{N}, \quad \vartheta(z^n) = n\, z^n, \quad \vartheta(\log^n(z)) = n\, \log^{n-1}(z),$$

by a triple $(\Delta, z^{-[n]}, \phi_n)$ satisfying

$$\forall n \in \mathbb{N}, \quad \Delta(z^{-[n]}) = -n\, z^{-[n]}, \quad \Delta(\phi_n(z)) = n\, \phi_{n-1}(z). \tag{6}$$

The aim of this section is to introduce the sequence of functions $(\phi_n)_{n \in \mathbb{N}}$ which will play a role analogous to the one played by $(\log^n)_{n \in \mathbb{N}}$ in the differential case, i.e., which satisfy the last equality of (6).

2.1   Definition

For $z \in \mathbb{C} \setminus \mathbb{Z}_{\leq 0}$, let us consider the function $\rho \in \mathbb{C} \mapsto \frac{\Gamma(z)}{\Gamma(z+\rho)}$, where Γ denotes the well-known *Gamma function*. As the reciprocal $1/\Gamma$ of the Gamma function is an entire function (see [2, Sec. 6] or [18, Ch. 9]), the function $\rho \mapsto \frac{\Gamma(z)}{\Gamma(z+\rho)}$ is holomorphic at $\rho = 0$ and we define the functions ϕ_n from the coefficients of the Taylor series expansion of $\rho \mapsto \frac{\Gamma(z)}{\Gamma(z+\rho)}$ at $\rho = 0$:

Definition 2. *For $n \in \mathbb{N}$ and $z \in \mathbb{C} \setminus \mathbb{Z}_{\leq 0}$, we define*

$$\phi_n(z) = (-1)^n \frac{\partial^n}{\partial \rho^n} \left(\frac{\Gamma(z)}{\Gamma(z+\rho)} \right) \bigg|_{\rho=0}, \tag{7}$$

and the sequence of functions $(\phi_n)_{n \in \mathbb{N}}$ where, for $n \in \mathbb{N}$, ϕ_n is the function of the complex variable z defined by (7) for $z \in \mathbb{C} \setminus \mathbb{Z}_{\leq 0}$.

The Taylor series expansion of the function $\rho \mapsto \frac{\Gamma(z)}{\Gamma(z+\rho)}$ at $\rho = 0$ can thus be written as

$$\frac{\Gamma(z)}{\Gamma(z+\rho)} = \sum_{n \geq 0} \frac{(-1)^n}{n!} \phi_n(z)\, \rho^n.$$

From (7), we have $\phi_0(z) = 1$ and $\phi_1(z) = \Psi(z)$, where Ψ is the *Digamma or Psi function* defined as the logarithmic derivative of the Gamma function Γ, i.e., $\Psi(z) = \frac{\Gamma'(z)}{\Gamma(z)}$ (see [2, Sec. 6] or [18, Ch. 9]).

Remark 1. The function $z \mapsto \frac{\Gamma(z)}{\Gamma(z+\rho)}$ is asymptotically equivalent to $z^{-\rho}$ which has the following Taylor series expansion at $\rho = 0$

$$z^{-\rho} = e^{-\rho \log(z)} = \sum_{n \geq 0} \frac{(-1)^n}{n!} \log^n(z)\, \rho^n.$$

This illustrates the close relations between the functions ϕ_n and $\log^n$ (see also Lemma 2 below).

2.2 Properties

As we have seen above, the function ϕ_1 coincides with the Digamma function Ψ which satisfies the following:

Lemma 1. *The Digamma function Ψ satisfies the linear difference equation*

$$\Delta(\Psi(z)) = 1,$$

and has the asymptotic expansion

$$\Psi(z) \sim \log(z) - \frac{1}{2\,z} - \sum_{n \geq 1} \frac{B_{2n}}{2\,n\,z^{2n}}, \quad z \to \infty, \ |\arg(z)| < \pi, \tag{8}$$

where B_k represents the kth Bernoulli number ([2, Sec. 23]).

Proof. The first equality can be checked by a straightforward calculation. We refer to [2, Sec. 6.3] or [18, Ch. 9] for the asymptotic expansion (8).

We now give some properties of the functions ϕ_n, namely, the action of Δ on each function ϕ_n which generalizes the first statement of Lemma 1 and three distinct formulas for ϕ_n:

Proposition 1. *For $n \geq 1$, the functions ϕ_n satisfy*

$$\Delta(\phi_n(z)) = n\,\phi_{n-1}(z), \tag{9}$$

$$\phi_n(z) = \sum_{k=0}^{n-1} (-1)^k \binom{n-1}{k} \Psi^{(k)}(z)\,\phi_{n-k-1}(z), \tag{10}$$

$$\phi_n(z) = \Psi(z)\,\phi_{n-1}(z) - \phi'_{n-1}(z), \tag{11}$$

$$\phi_n(z) = (-1)^n\,\Gamma(z)\,\frac{d^n}{dz^n}\left(\frac{1}{\Gamma(z)}\right), \tag{12}$$

where $f^{(k)}$ (resp. f') denotes the kth order (resp. 1st order) derivative of a function f with respect to the complex variable z.

Proof. For completeness, a proof of Proposition 1 is provided in the appendix (Section 4.1) at the end of the paper.

Note that the formula (10) yields an expression of ϕ_n in terms of the *Polygamma functions* $\Psi^{(k)}$ (see [2, Sec. 6] or [18, Ch. 9]) and the previous ϕ_i, for $i < n$. One can also define the functions ϕ_n using the formula (12) (as it has been done in [11]) from which one can easily recover (11), (10) and (9). However, getting the expression (7) doesn't seem to be straightforward.

We now show that ϕ_n is asymptotically equivalent to the nth power of the log function.

Lemma 2. *For $n \in \mathbb{N}^*$, there exists $P_{n,i}(z^{-1}) \in \mathbb{C}[[z^{-1}]]$, $i = 0, \ldots, n-1$, such that the function ϕ_n has the asymptotic expansion*

$$\phi_n(z) \sim \log^n(z) + \sum_{i=0}^{n-1} P_{n,i}(z^{-1}) \log^i(z), \quad z \to \infty, \; |\arg(z)| < \pi.$$

Proof. Using the properties of the asymptotic expansions of holomorphic functions (see [22, Chap. 3]), the result can be proved by induction from the asymptotic expansion of the Digamma function Ψ given by (8) and Formula (11). For $n = 1$, we have $\phi_1 = \Psi$ so that $P_{1,0}(z^{-1}) = -\frac{1}{2z} - \sum_{n \geq 1} \frac{B_{2n}}{2n \, z^{2n}}$. For $n > 1$, $P_{n,i}(z^{-1})$ satisfies the recurrence $P_{n,i} = P_{1,0} \, P_{n-1,i} - P'_{n-1,i} + P_{n-1,i-1} - \frac{i+1}{z} P_{n-1,i+1}$, where for all $n \in \mathbb{N}$, $P_{n,n} = 1$, and $P_{n,i} = 0$, for $i > n$ or $i < 0$.

3 Regular Solutions of Linear Difference Systems

In this section, we develop our new approach for computing regular solutions of first-order linear difference systems of the form (1). This method is a generalization to the difference case of the method described in [5] for computing regular solutions of linear differential systems. It differs from the existing Frobenius-like methods given in [15, 14, 3] for computing regular solutions of linear difference systems of the form (2). Indeed, instead of looking for regular solutions of the form (3), we shall search for regular solutions written as

$$\mathbf{y}(z) = \sum_{m \geq 0} z^{-[\rho+m]} \mathbf{y}_m(z), \quad \mathbf{y}_0(z) \neq 0, \tag{13}$$

where $\rho \in \mathbb{C}$ is a complex number to be determined and the $\mathbf{y}_m(z)$, $m \geq 0$, are unknown finite linear combinations with constant vector coefficients of the functions ϕ_i defined by (7), i.e., for all $m \in \mathbb{N}$:

$$\mathbf{y}_m(z) = \mathbf{u}_{m,0} \, \phi_0(z) + \cdots + \mathbf{u}_{m,l_m} \, \phi_{l_m}(z), \; l_m \in \mathbb{N}, \; \mathbf{u}_{m,i} \in \mathbb{C}^n, \; i = 0, \ldots, l_m.$$

The matrix coefficients $D(z)$ and $A(z)$ of System (1) can be written as

$$D(z) = \sum_{i \geq 0} D_i \, z^{-[i]}, \quad A(z) = \sum_{i \geq 0} A_i \, z^{-[i]},$$

where, for $i \geq 0$, $D_i, A_i \in \mathbb{C}^{n \times n}$ are constant square matrices of size n.
We shall furthermore suppose that the matrix pencil $D_0 \lambda - A_0$ is regular, that is, its determinant does not vanish identically, i.e., $\det(D_0 \lambda - A_0) \not\equiv 0$ (see [13, 5]). This will be crucial in order to determine the possible values for ρ in (13): see Proposition 2 below.

Remark 2. The condition $\det(D_0 \lambda - A_0) \not\equiv 0$ is not restrictive since it can always be ensured by applying, if necessary, the algorithms developed in [1, 9, 3]. In a future work, we shall also adapt to the difference case the algorithm developed in [8] for the differential case.

Theorem 1. *With the above notation, $\mathbf{y}(z)$ defined by (13) is a regular solution of the linear difference system (1) if and only if the complex number ρ and the vector $\mathbf{y}_0$ are such that*

$$D_0\,\Delta(\mathbf{y}_0) - (D_0\,\rho - A_0)\,\mathbf{y}_0 = 0, \tag{14}$$

and for $m \geq 1$, $\mathbf{y}_m$ satisfies

$$D_0\,\Delta(\mathbf{y}_m) - (D_0\,(\rho + m) - A_0)\,\mathbf{y}_m = \mathbf{q}_m, \tag{15}$$

where, for $m \geq 1$, $\mathbf{q}_m$ is a linear combination with constant matrix coefficients of the $\mathbf{y}_i$ and $\Delta(\mathbf{y}_i)$, $i = 0, \ldots, m-1$, that can be effectively computed.

Proof. A detailed proof of Theorem 1 with explicit formulas for the $\mathbf{q}_m$, $m \geq 1$, is provided in the appendix (Section 4.2) at the end of the paper. The proof can be sketched as follows: we plug the ansatz $\mathbf{y}(z)$ defined by (13) into the linear difference system (1). Then, using the product and translation formulas for factorial series recalled in Section 1, we manage to write each term of the equation $D(z)\,\Delta(\mathbf{y}(z)) + A(z)\,\mathbf{y}(z) = 0$ in such a way that we can identify coefficients to get the equations (14) and (15) for the coefficients of the ansatz $\mathbf{y}(z)$.

Theorem 1 reduces the problem of computing regular solutions of linear difference systems (1) to the resolution of the linear difference systems with constant matrix coefficients given by (14) and (15).

We shall now state two results showing that System (14), resp. Systems (15) for $m \geq 1$, can always be solved for $\mathbf{y}_0$, resp. $\mathbf{y}_m$ for $m \geq 1$, of the desired form.

Proposition 2 gives a necessary and sufficient condition for System (14) to have a non trivial solution $\mathbf{y}_0$ as a finite linear combination of the ϕ_i defined by (7) with constant vector coefficients.

Proposition 2. *With the above notation and assumptions, the linear difference system with constant matrix coefficients (14) has a solution of the form*

$$\mathbf{y}_0(z) = \sum_{i=0}^{k} \frac{(-1)^i}{i!}\,\mathbf{v}_i\,\phi_i(z), \tag{16}$$

where $\mathbf{v}_0, \ldots, \mathbf{v}_k \in \mathbb{C}^n$ are constant vectors such that $\mathbf{v}_k \neq 0$, if and only if ρ is an eigenvalue of the matrix pencil $D_0\,\lambda - A_0$, i.e., $\det(D_0\,\rho - A_0) = 0$ and $\mathbf{v}_k, \mathbf{v}_{k-1}, \ldots, \mathbf{v}_0$ form a Jordan chain associated with ρ (see [13, 5]).

Proof. Plugging the ansatz (16) for $\mathbf{y}_0(z)$ into (14), we obtain

$$\frac{(-1)^{k+1}}{k!}(D_0\,\rho - A_0)\,\mathbf{v}_k\,\phi_k(z) + \sum_{i=0}^{k-1}\frac{(-1)^i}{i!}\left(-D_0\,\mathbf{v}_{i+1} - (D_0\,\rho - A_0)\,\mathbf{v}_i\right)\phi_i(z) = 0,$$

which is equivalent to the $k + 1$ linear algebraic systems

$$\begin{cases} (D_0\,\rho - A_0)\,\mathbf{v}_k = 0, \\ (D_0\,\rho - A_0)\,\mathbf{v}_i = -D_0\,\mathbf{v}_{i+1}, \quad \forall i = k-1, \ldots, 0. \end{cases} \tag{17}$$

The first system of (17) admits a non-trivial solution $\mathbf{v}_k$ if and only if ρ is chosen as an eigenvalue of $D_0 \lambda - A_0$. In this case, a Jordan chain $\mathbf{v}_k, \mathbf{v}_{k-1}, \ldots, \mathbf{v}_0$ associated with the eigenvalue ρ forms a solution of (17). See [13, 5].

Assuming now that ρ is an eigenvalue of $D_0 \lambda - A_0$ and $\mathbf{y}_0$ given by (16) is a solution of (14), then, for $m \geq 1$, $\mathbf{y}_m$ satisfies a non-homogeneous linear difference system with constant matrix coefficients whose right-hand side is a finite linear combination of the $\mathbf{y}_i$ and $\Delta(\mathbf{y}_i)$, $i = 0, \ldots, m-1$, with constant matrix coefficients. The following proposition shows that such a system always admits a solution of the desired form.

Proposition 3. *With the above notation and assumptions, let us further assume that the right-hand side $\mathbf{q}_m$ of (15) is a linear combination of the functions $\phi_0, \phi_1, \ldots, \phi_d$ with constant vector coefficients. Then System (15) has a solution $\mathbf{y}_m$ expressed as a linear combination of $\phi_0, \phi_1, \ldots, \phi_p$ with constant vector coefficients with $p \in \mathbb{N}$ such that*

$$\begin{cases} d \leq p \leq d + \max\{\kappa_i, i = 1, \ldots, m_g(\rho + m)\} & \text{if } \det(D_0\,(\rho + m) - A_0) = 0, \\ p = d & \text{otherwise,} \end{cases}$$

where $m_g(\rho+m)$ denotes the dimension of the kernel of the matrix $D_0\,(\rho+m)-A_0$ and the κ_i, $i = 1, \ldots, m_g(\rho + m)$, are the partial multiplicities of the eigenvalue $\rho + m$ of the matrix pencil $D_0 \lambda - A_0$ (see [13, 5]).

Proof. The proof of this proposition is similar to the one of [5, Proposition 1]. ∎

Theorem 1, Propositions 2 and 3 and their proofs provide an algorithm for computing regular solutions of first-order linear difference systems of the form (1). In particular, it relies on the computation of eigenvalues and Jordan chains of the matrix pencil $D_0 \lambda - A_0$.

4   Appendix

4.1   Proof of Proposition 1

For completeness, we provide here the proof of Proposition 1. We recall that the functions ϕ_n are defined from the Taylor series expansion of the holomorphic function $\rho \mapsto \frac{\Gamma(z)}{\Gamma(z+\rho)}$ at $\rho = 0$ given by

$$\frac{\Gamma(z)}{\Gamma(z + \rho)} = \sum_{n \geq 0} \frac{(-1)^n}{n!} \phi_n(z)\, \rho^n. \tag{18}$$

Proof. 1. Applying Δ on both sides of (18), we get

$$-\rho \frac{\Gamma(z)}{\Gamma(z + \rho)} = \sum_{n \geq 0} \frac{(-1)^n}{n!} \Delta(\phi_n(z))\, \rho^n.$$

Using again (18) to rewrite the left-hand side of the above equation as a power series in ρ, we find

$$\sum_{n\geq 1} \frac{(-1)^n}{(n-1)!} \phi_{n-1}(z)\,\rho^n = \sum_{n\geq 0} \frac{(-1)^n}{n!}\, \Delta(\phi_n(z))\,\rho^n.$$

Identifying the coefficients of ρ^n in the above equality yields (9).

2. By differentiating both sides of (18) with respect to ρ, we obtain

$$-\Psi(z+\rho)\frac{\Gamma(z)}{\Gamma(z+\rho)} = \sum_{n\geq 1}\frac{(-1)^n}{(n-1)!}\phi_n(z)\,\rho^{n-1}. \qquad (19)$$

Using the Taylor series expansion of $\rho \mapsto \Psi(z+\rho)$ at $\rho = 0$ and performing the Cauchy product with the expansion of $\frac{\Gamma(z)}{\Gamma(z+\rho)}$ given by (18), we get

$$\Psi(z+\rho)\frac{\Gamma(z)}{\Gamma(z+\rho)} = \sum_{n\geq 0}\left(\sum_{k=0}^{n}\frac{\Psi^{(k)}(z)}{k!}\frac{(-1)^{n-k}}{(n-k)!}\phi_{n-k}(z)\right)\rho^n.$$

Now identifying the coefficients of ρ^{n-1} in both sides of (19) leads to

$$\sum_{k=0}^{n-1}\frac{\Psi^{(k)}(z)}{k!}\frac{(-1)^{n-k}}{(n-k-1)!}\phi_{n-k-1}(z) = \frac{(-1)^n}{(n-1)!}\phi_n(z),$$

which yields (10).

3. We now proceed by differentiating both sides of (18) with respect to z and we write the result as a power series in ρ. For the left-hand side, we have

$$\begin{aligned}
\frac{\partial}{\partial z}\left(\frac{\Gamma(z)}{\Gamma(z+\rho)}\right) &= \frac{\Gamma'(z)}{\Gamma(z+\rho)} - \frac{\Gamma(z)\Gamma'(z+\rho)}{\Gamma(z+\rho)^2}, \\
&= \frac{\Gamma(z)}{\Gamma(z+\rho)}\left(\Psi(z) - \Psi(z+\rho)\right), \\
&= \left(\sum_{n\geq 0}\frac{(-1)^n}{n!}\phi_n(z)\rho^n\right)\left(-\sum_{n\geq 1}\frac{\Psi^{(n)}(z)}{n!}\rho^n\right), \\
&= -\rho\left(\sum_{n\geq 0}\frac{(-1)^n}{n!}\phi_n(z)\rho^n\right)\left(\sum_{n\geq 0}\frac{\Psi^{(n+1)}(z)}{(n+1)!}\rho^n\right), \\
&= \sum_{n\geq 0}\left(\sum_{k=0}^{n}\frac{\Psi^{(k+1)}(z)}{(k+1)!}\frac{(-1)^{n-k+1}}{(n-k)!}\phi_{n-k}(z)\right)\rho^{n+1},
\end{aligned}$$

and for the right-hand side we obtain

$$\frac{\partial}{\partial z}\left(\sum_{n\geq 0}\frac{(-1)^n}{n!}\phi_n(z)\,\rho^n\right) = \sum_{n\geq 0}\frac{(-1)^n}{n!}\phi'_n(z)\,\rho^n.$$

Thus, identifying the coefficients of ρ^{n-1} in both series, we get

$$\frac{(-1)^{n-1}}{(n-1)!} \phi'_{n-1}(z) = \sum_{k=0}^{n-2} \frac{\Psi^{(k+1)}(z)}{(k+1)!} \frac{(-1)^{n-k-1}}{(n-k-2)!} \phi_{n-k-2}(z),$$

and then,

$$-\phi'_{n-1}(z) = \sum_{k=0}^{n-2} (-1)^{k-1} \binom{n-1}{k+1} \Psi^{(k+1)}(z) \phi_{n-k-2}(z),$$

$$= \sum_{p=1}^{n-1} (-1)^p \binom{n-1}{p} \Psi^{(p)}(z) \phi_{n-p-1}(z).$$

Finally, we notice that $\Psi(z) \phi_{n-1}(z)$ coincides with the term of the latter sum for the index $p = 0$ so that

$$\Psi(z) \phi_{n-1}(z) - \phi'_{n-1}(z) = \sum_{p=0}^{n-1} (-1)^p \binom{n-1}{p} \Psi^{(p)}(z) \phi_{n-p-1}(z).$$

Using (10), we can thus conclude that (11) holds.

4. Dividing (11) by $\Gamma(z)$ and using $\Psi(z) = \frac{\Gamma'(z)}{\Gamma(z)}$, we get that, for $n \geq 1$, we have:

$$\frac{\phi_n(z)}{\Gamma(z)} = -\frac{d}{dz} \left(\frac{\phi_{n-1}(z)}{\Gamma(z)} \right).$$

By iteration, we then obtain

$$\frac{\phi_n(z)}{\Gamma(z)} = (-1)^n \frac{d^n}{dz^n} \left(\frac{\phi_0(z)}{\Gamma(z)} \right),$$

which yields (12) since we have $\phi_0(z) = 1$.

4.2   Proof of Theorem 1

We provide here a detailed proof of Theorem 1 which yields explicit formulas for the right-hand sides $\mathbf{q}_m$ of (15).

Proof. We first investigate the action of the difference operator Δ on one term of the series involved in the ansatz $\mathbf{y}(z)$ defined by (13), i.e., $z^{-[\rho+m]} \mathbf{y}_m$. Using the formula $\Delta(f(z) g(z)) = f(z-1) \Delta(g(z)) + \Delta(f(z)) g(z)$, we obtain

$$\Delta \left(z^{-[\rho+m]} \mathbf{y}_m \right) = (z-1)^{-[\rho+m]} \Delta(\mathbf{y}_m) - (\rho+m) z^{-[\rho+m]} \mathbf{y}_m. \qquad (20)$$

A straightforward computation shows that

$$(z-1)^{-[\rho+m]} = \frac{1}{z-1} z^{-[\rho-1]} (z+\rho-1)^{-[m]}, \quad z^{-[\rho+m]} = z^{-[\rho-1]} (z+\rho-1)^{-[m+1]}.$$

Thus, we can factor out the term $z^{-[\rho-1]}$ in the right-hand side of (20) to get

$$\Delta\left(z^{-[\rho+m]}\,\mathbf{y}_m\right) = z^{-[\rho-1]}\left(\frac{(z+\rho-1)^{-[m]}}{z-1}\,\Delta(\mathbf{y}_m) - (\rho+m)\,(z+\rho-1)^{-[m+1]}\,\mathbf{y}_m\right).$$

We shall now express $\Delta\left(z^{-[\rho+m]}\,\mathbf{y}_m\right)$ as a factorial series in z. Applying twice the translation formula for factorial series (see Section 1), we have

$$\frac{(z+\rho-1)^{-[m]}}{z-1} = \sum_{k\geq0} C^k_{-\rho+1,m}\,\frac{z^{-[m+k]}}{z-1},$$

$$= \sum_{k\geq0} C^k_{-\rho+1,m}\,(z-1)^{-[m+k+1]},$$

$$= \sum_{k\geq0}\left(\sum_{p=0}^{k} C^p_{-\rho+1,m}\,C^{k-p}_{1,m+1+p}\right) z^{-[m+1+k]},$$

$$= \sum_{k\geq0} E^k_{-\rho+1,m}\,z^{-[m+1+k]},$$

where, in order to simplify the expressions below, we define the new constant

$$E^k_{-\rho+1,m} = \sum_{p=0}^{k} C^p_{-\rho+1,m}\,C^{k-p}_{1,m+1+p} = (m+k)!\,\sum_{p=0}^{k} \frac{C^p_{-\rho+1,m}}{(m+p)!} \in \mathbb{C}.$$

Note that $E^0_{-\rho+1,m} = 1$. Using again the translation formula, we get

$$(z+\rho-1)^{-[m+1]} = \sum_{k\geq0} C^k_{-\rho+1,m+1}\,z^{-[m+1+k]}.$$

Hence, the action of Δ on a term $z^{-[\rho+m]}\,\mathbf{y}_m$ can be written as

$$\Delta\left(z^{-[\rho+m]}\,\mathbf{y}_m\right) = z^{-[\rho-1]}\sum_{k\geq0}\left(E^k_{-\rho+1,m}\,\Delta(\mathbf{y}_m) - (\rho+m)\,C^k_{-\rho+1,m+1}\,\mathbf{y}_m\right) z^{-[m+1+k]}.$$

Applying the latter equality on each term of the sum in the ansatz $\mathbf{y}(z)$ defined by (13) yields

$$\Delta(\mathbf{y}(z)) = z^{-[\rho-1]}\sum_{m\geq0}\sum_{k\geq0}\left(E^k_{-\rho+1,m}\,\Delta(\mathbf{y}_m) - (\rho+m)\,C^k_{-\rho+1,m+1}\,\mathbf{y}_m\right) z^{-[m+1+k]},$$

which can be rewritten as

$$\Delta(\mathbf{y}(z)) = z^{-[\rho-1]}\sum_{m\geq0} \mathcal{F}_{-\rho+1,m}\,z^{-[m+1]},$$

where we set

$$\mathcal{F}_{-\rho+1,m} = \sum_{k=0}^{m} \left(\mathrm{E}_{-\rho+1,m-k}^{k} \, \Delta(\mathbf{y}_{m-k}) - (\rho+m-k) \, \mathrm{C}_{-\rho+1,m-k+1}^{k} \, \mathbf{y}_{m-k} \right). \quad (21)$$

It is useful to notice that $\mathcal{F}_{-\rho+1,m} = (\Delta(\mathbf{y}_m) - (\rho+m)\,\mathbf{y}_m) + \tilde{\mathcal{F}}_{-\rho+1,m}$, where $\tilde{\mathcal{F}}_{-\rho+1,m}$ depends only on the $\Delta(\mathbf{y}_i)$ and $\mathbf{y}_i$ for $i = 0, \ldots, m-1$.
We now compute the product $D(z)\,\Delta(\mathbf{y}(z))$. Using the multiplication formula for factorial series recalled in Section 1, we find that

$$D(z)\,\Delta(\mathbf{y}(z)) = z^{-[\rho-1]} \left(\sum_{i\geq 0} D_i \, z^{-[i]} \right) \left(\sum_{m\geq 0} \mathcal{F}_{-\rho+1,m} \, z^{-[m+1]} \right),$$

$$= z^{-[\rho-1]} \sum_{m\geq 0} \left(D_0 \, \mathcal{F}_{-\rho+1,m} + \sum_{(s,p,k)\in J_{m+1}} \mathrm{C}_{s,p}^{k} \, D_s \, \mathcal{F}_{-\rho+1,p-1} \right) z^{-[m+1]},$$

$$(22)$$

where

$$J_{m+1} = \{(s,p,k) \in \mathbb{N}^3 \mid s,p \geq 1 \text{ and } s+p+k = m+1\}.$$

By convention, the corresponding sum is zero when J_{m+1} is the empty set, i.e., $m = 0$. In a similar way, we show that the product $A(z)\,\mathbf{y}(z)$ can be expressed as

$$A(z)\,\mathbf{y}(z) = z^{-[\rho-1]} \sum_{m\geq 0} \left(A_0 \, \mathcal{G}_{-\rho+1,m} + \sum_{(s,p,k)\in J_{m+1}} \mathrm{C}_{s,p}^{k} \, A_s \, \mathcal{G}_{-\rho+1,p-1} \right) z^{-[m+1]},$$

$$(23)$$

where we set

$$\mathcal{G}_{-\rho+1,m} = \sum_{k=0}^{m} \mathrm{C}_{-\rho+1,m-k+1}^{k} \, \mathbf{y}_{m-k}. \quad (24)$$

Notice that $\mathcal{G}_{-\rho+1,m}$ is a linear combination of $\mathbf{y}_0, \ldots, \mathbf{y}_m$ with constant coefficients in $\mathbb{C}$ and the coefficient of $\mathbf{y}_m$ equals 1.
Now, combining Equations (22) and (23) and equating the coefficients of $z^{-[m+1]}$ to zero, for all $m \in \mathbb{N}$, we find that $\mathbf{y}(z)$ defined by (13) is a solution of System (1) if and only if, for all $m \in \mathbb{N}$, we have:

$$D_0 \, \mathcal{F}_{-\rho+1,m} + A_0 \, \mathcal{G}_{-\rho+1,m} + \sum_{(s,p,k)\in J_{m+1}} \mathrm{C}_{s,p}^{k} \, (D_s \, \mathcal{F}_{-\rho+1,p-1} + A_s \, \mathcal{G}_{-\rho+1,p-1}) = 0, \quad (25)$$

where $\mathcal{F}_{-\rho+1,m}$ and $\mathcal{G}_{-\rho+1,m}$ are respectively defined by (21) and (24).
  Note that for a triple $(s,p,k) \in J_{m+1}$, since $s \geq 1$ and $s+p+k = m+1$, we necessarily have $p \leq m$ so that $\mathcal{F}_{-\rho+1,p-1}$ and $\mathcal{G}_{-\rho+1,p-1}$ contain terms in $\mathbf{y}_i$ and $\Delta(\mathbf{y}_i)$ only for $i = 0, \ldots, m-1$. This implies that for $m \in \mathbb{N}$, $\mathbf{y}_m$ and $\Delta(\mathbf{y}_m)$ only appear in first two terms of the left-hand side of (25). Finally, using the two observations made after the definitions of the terms $\mathcal{F}_{-\rho+1,m}$ and $\mathcal{G}_{-\rho+1,m}$ above, we can conclude that (25) yields (14) and (15) which ends the proof.

References

1. Abramov, S.A., Barkatou, M.A.: Rational solutions of first order linear difference systems. In: Proc. of ISSAC 1998, Rostock, pp. 124–131. ACM, New York (1998)
2. Abramowitz, M., Stegun, I.A.: Handbook of Mathematical Functions with Formulas, Graphs, and Mathematical Tables. Dover Publications Inc., New York (1992); reprint of the 1972 edition. First edition 1964
3. Barkatou, M.A.: Contribution à l'étude des équations différentielles et aux différences dans le champ complexe. PhD thesis, Institut National Polytech. de Grenoble (1989)
4. Barkatou, M.A.: Characterization of regular singular linear systems of difference equations. Numerical Algorithms **1**, 139–154 (1991)
5. Barkatou, M.A., Cluzeau, T., El Bacha, C.: Simple forms of higher-order linear differential systems and their applications in computing regular solutions. Journal of Symbolic Computation **46**, 633–658 (2011)
6. Barkatou, M.A., Duval, A.: Sur les séries formelles solutions d'équations aux différences polynomiales. Ann. Inst. Fourier **44**(2), 495–524 (1994)
7. Barkatou, M.A., Duval, A.: Sur la somme de certaines séries de factorielles. Annales de la Faculté des Sciences de Toulouse **6**(1), 7–58 (1997)
8. Barkatou, M.A., El Bacha, C.: On k-simple forms of first-order linear differential systems and their computation. Journal of Symbolic Computation **54**, 36–58 (2013)
9. Barkatou, M.A., Broughton, G., Pflügel, E.: A monomial-by-monomial method for computing regular solutions of systems of pseudo-linear equations. Mathematics in Computer Science **4**(2–3), 267–288 (2010)
10. Duval, A.: Équations aux différences dans le champ complexe. PhD thesis, Université de Strasbourg (1984)
11. Fitzpatrick, W.J., Grimm, L.J.: Convergent factorial series solutions of linear difference equations. Journal of Differential Equations **29**, 345–361 (1978)
12. Gerard, R., Lutz, D.A.: Convergent factorial series solutions of singular operator equations. Analysis **10**, 99–145 (1990)
13. Gohberg, I., Lancaster, P., Rodman, L.: Matrix Polynomials. Academic Press, New York (1982)
14. Harris Jr., W.A.: Linear systems of difference equations. Contributions to Differential Equations **1**, 489–518 (1963)
15. Harris Jr., W.A.: Equivalent classes of difference equations. Contributions to Differential Equations **1**, 253–264 (1963)
16. Harris Jr., W.A.: Analytic theory of difference equations. Lecture Notes in Mathematics, vol. 183. Springer, Berlin (1971)
17. Harris Jr., W.A., Turrittin, H.L.: Reciprocals of inverse factorial series. Funkcialaj Ekvacioj **6**, 37–46 (1964)
18. Milne-Thomson, L.M.: The Calculus of Finite Differences. Chelsea Publishing Company, New York (1981)
19. Nielsen, N.: Sur la multiplication de deux séries de factorielles. Rendiconti della R. Acc. dei Lincei, (5), 13, 517–524 (1904)
20. Nörlund, N.E.: Leçons sur les Séries d'Interpolation, pp. 170–227. Gauthiers Villars et Cie, Paris (1926)
21. Nörlund, N.E.: Leçons sur les Équations Linéaires aux Différences Finies. Gauthiers Villars et Cie, Paris (1929)
22. Wasow, W.: Asymptotic Expansions For Ordinary Differential Equations. Dover Publi., New York (1965)

Solving Polynomial Systems in the Cloud with Polynomial Homotopy Continuation*

Nathan Bliss, Jeff Sommars, Jan Verschelde, and Xiangcheng Yu

University of Illinois at Chicago, Department of Mathematics, Statistics,
and Computer Science, 851 S. Morgan Street (m/c 249),
Chicago, IL 60607-7045, USA
{nbliss2,sommars1,janv}@uic.edu, xiangchengyu@outlook.com

Abstract. Polynomial systems occur in many fields of science and engineering. Polynomial homotopy continuation methods apply symbolic-numeric algorithms to solve polynomial systems. We describe the design and implementation of our web interface and reflect on the application of polynomial homotopy continuation methods to solve polynomial systems in the cloud. Via the graph isomorphism problem we organize and classify the polynomial systems we solved. The classification with the canonical form of a graph identifies newly submitted systems with systems that have already been solved.

Keywords: Blackbox solver, classifying polynomial systems, cloud computing, graph isomorphism, internet accessible symbolic and numeric computation, homotopy continuation, mathematical software, polynomial system, web interface.

1   Introduction

The widespread availability and use of high speed internet connections combined with relatively inexpensive hardware enabled cloud computing. In cloud computing, users of software no longer download and install software, but connect via a browser to a web site, and interact with the software through a web interface. Computations happen at some remote server and the data (input as well as output) are stored and maintained remotely. Quoting [27], "Large, virtualized pools of computational resources raise the possibility of a new, advantageous computing paradigm for scientific research."

This model of computing offers several advantages to the user; we briefly mention three benefits. First, installing software can be complicated and a waste of time, especially if one wants to perform only one single experiment to check whether the software will do what is desired — the first author of [28] has an account on our web server. In cloud computing, the software installation is replaced with a simple sign up, as common as logging into a web store interface. One should not have to worry about upgrading installed software to newer versions. The second advantage is that for computationally intensive tasks, the web

---

* This material is based upon work supported by the National Science Foundation under Grant No. 1440534.

V.P. Gerdt et al. (Eds.): CASC 2015, LNCS 9301, pp. 87–100, 2015.
DOI: 10.1007/978-3-319-24021-3_7

server can be aided by a farm of compute servers. Thirdly, the input and output files are managed at the server. The user should not worry about storage, as the web server could be aided by file servers. A good web interface helps to formulate the input problems and manage the computed results.

In this paper, we describe a prototype of a first web interface to the blackbox solver of PHCpack [31]. This solver applies homotopy continuation methods to polynomial systems. Its blackbox solver, available as `phc -b` at the command line, seems the most widely used and popular feature of the software. Our web interface is currently running at `https://kepler.math.uic.edu`.

In addition to the technical aspects of designing a web interface, we investigate what it means to run a blackbox solver in the cloud. Because the computations happen remotely, the blackbox solver not only hides the complexity of the algorithms, but also the actual cost of the computations. In principle, the solver could use just one single core of a traditional processor, or a distributed cluster of computers, accelerated with graphics processing units [34], [35]. One should consider classifying the hardness of an input problem and allocating the proper resources to solve the problem. This classification problem could be aided by mining a database of solved problems.

To solve the classification problem we show that the problem of deciding whether two sets of support sets are isomorphic can be reduced to the graph isomorphism problem [21], [22]. Our classification problem is related to the isomorphism of polynomials problem [24] in multivariate cryptology [5], [10]. Support sets of polynomials span Newton polytopes and a related problem is the polytope isomorphism problem, see [18] for its complexity, which is as hard as the graph isomorphism problem.

2 Related Work and Alternative Approaches

The Sage notebook interface and the newer SageMathCloud can be alternative solutions to setting up a standalone cloud service. PHCpack is an optional package in Sage [30], available through the `phc.py` interface, developed by Marshall Hampton and Alex Jokela, based on the earlier efforts of Kathy Piret [26] and William Stein. We plan to upgrade the existing `phc.py` in Sage with `phcpy` [32].

The computational algebraic geometry software Macaulay2 [13] distributes `PHCpack.m2` [14], which is a package that interfaces to PHCpack. Macaulay2 runs in the cloud as well. Below are the input commands to the version of Macaulay2 that runs online. The output is omitted.

```
Macaulay2, version 1.7

i1 : loadPackage "PHCpack";
i2 : help(PHCpack);
i3 : help solveSystem;
i4 : R = CC[x,y,z];
i5 : S = {x+y+z-1, x^2+y^2, x+y-z-3};
i6 : solveSystem(S)
```

3   Design and Implementation

All software in this project is free and open source, as an application of the LAMP stack, where LAMP stands for Linux, Apache, MySQL, and Python. Our web server runs Red Hat Linux, Apache [6] as the web server, MySQL [9] as the database, and Python as the scripting language. Our interest in Python originates in its growing ecosystem for scientific computing [25]. In our current implementation we do not take advantage of any web framework. Our web interface consists mainly of a collection of Python CGI scripts.

We distinguish three components in the development of our web interface: the definition of the database, the sign up process, and the collection of Python scripts that are invoked as the user presses buttons. In the next three paragraphs we briefly describe these three components.

MySQL is called in Python through the module MySQLdb. The database manages two tables. One table stores data about the users, which includes their encrypted passwords and generated random names that define the locations of the folders with their data on the server. In the other table, each row holds the information about a polynomial system that is solved. In this original setup, mathematical data are not stored in the database. Every user has a folder which is a generated random 40-character string. With every system there is another generated 40-character string. The Python scripts do queries to the database to locate the data that is needed.

When connecting to the server, the index.html leads directly to the Python script that prints the first login screen. The registration script sends an email to the first time user and an activation script runs when that first time user clicks on the link received in the registration email.

The third component of the web interface consists of the Python scripts that interact with the main executable program, the phc built with the code in PHCpack. Small systems are solved directly. Larger systems are placed in a queue that is served by compute servers.

4   Solving by Polynomial Homotopy Continuation

When applying polynomial homotopy continuation methods to solve polynomial systems, we distinguish two different approaches. The first is the application of a blackbox solver, and the second is a scripting interface.

4.1   Running a Blackbox Solver

For the blackbox solver, the polynomials are the *only* input. The parameters that control the execution options are set to work well on a large class of benchmark examples; and/or tuned automatically during the solving. While the input is purposely minimal, the output should contain various diagnostics and checks. In particular, the user must be warned in case of ill conditioning and nearby singularities. The form of the output should enable the user to verify (or falsify) the computed results.

The current blackbox solver `phc -b` was designed in [31] for *square* problems, that is, systems with as many equations as unknowns. Polyhedral homotopies [15] [33] are optimal for sparse polynomial systems. This means that the mixed volume is a sharp root count for generic problems (`phc -b` calls MixedVol [12] for a fast mixed volume computation). Every path in a polyhedral homotopy ends at an isolated root, except for systems that have special initial forms [16]. For a survey, see e.g. [20].

Special cases are polynomials in one variable, linear systems, and binomial systems. A binomial system has exactly two monomials in every polynomial. Its isolated solutions are determined by a Hermite normal form of the exponent vectors. Its positive dimensional solution sets are monomial maps [1].

A more general blackbox solver should operate without any assumptions on the dimension of the solution sets. Inspired by tropical algebraic geometry, and in particular its fundamental theorem [17], we can generalize polyhedral homotopies for positive dimensional solution sets, as was done for the cyclic n-roots problems in [2], [3]. This general polyhedral method computes tropisms based on the initial forms and then develops Puiseux series starting at the solutions of the initial forms.

4.2   The Scripting Interface `phcpy`

The Python package `phcpy` replaces the input and output files by persistent objects. Instead of the command line interface of `phc` with its interactive menus, the user of `phcpy` runs Python scripts, or calls functions from the `phcpy` modules in an interactive Python shell. Version 0.1.4 of `phcpy` is described in [32].

The current version 0.2.5 exports most of the tools needed to compute a numerical irreducible decomposition [29]. With a numerical irreducible decomposition one gets all solutions, the isolated solutions as well as the positive dimensional sets. The latter come in the form of as many generic points as the degree of each irreducible component, satisfying as many random hyperplanes as the dimension of the component.

5   Pattern Matching with a Database

Solving a system of polynomial equations for the first time with polynomial homotopy continuation happens in two stages. In the first stage, we construct and solve a simpler system than the original problem. This simpler system serves as a start system to solve the original system in the second stage. Polynomial homotopy continuation methods deform systems with known solutions into systems that need to be solved. Numerical predictor-corrector methods track solution paths from one system to another.

In many applications polynomial systems have natural parameters and often users will present systems with the same input patterns.

5.1   The Classification Problem

If we solve polynomial systems by homotopy continuation, we first solve a similar system, a system with the same monomial structure, but with generic coefficients. If we could recognize the structure of a new polynomial system we have to solve, then we could save on the total solving time of the new polynomial system, because we skip the first stage of solving the start system. Furthermore, we could give a prediction of the time it will take to solve the new system with the specific coefficients.

Giving a name such as `cyclic n-roots` as a query to a search engine is likely to come up with useful results because this problem has been widely used to benchmark polynomial system solvers. But if one has only a particular formulation of a polynomial system, then one would like to relate the particular polynomials to the collection of polynomial systems that have already been solved. Instead of named polynomial systems, we work with anonymous mathematical data where even the naming of the variables is not canonical.

For example, the systems

$$\begin{cases} x^2 + xy^2 - 3 = 0 \\ 2x^2y + 5 = 0 \end{cases} \text{ and } \begin{cases} 3 + 2ab^2 = 0 \\ b^2 - 5 + 2a^2b = 0 \end{cases} \tag{1}$$

must be considered isomorphic to each other.

The support set of a polynomial is the set of all exponent tuples of the monomials that appear in the polynomial with nonzero coefficient. For the systems in the specific example above, the sets of support sets are

$$\{\{(2,0), (1,2), (0,0)\}, \{(2,1), (0,0)\}\} \tag{2}$$
$$\text{and } \{\{(0,0), (1,2)\}, \{(0,2), (0,0), (2,1)\}\}. \tag{3}$$

Definition 1. We say that two sets of support sets are *isomorphic* if there exists a permutation of their equations and variables so that the sets of support sets are identical.

The problem is then to determine whether the sets of support sets of two polynomial systems are isomorphic to each other. This problem is related to the isomorphism of polynomials problem [24]. Algorithms in multivariate cryptology apply Gröbner basis algorithms [10] and graph-theoretic algorithms [5].

5.2   The Graph Isomorphism Problem

If we encode the sets of support sets of a polynomial system into a graph, then our problem reduces to the graph isomorphism problem, for which practical solutions are available [22] and accessible in software [21]. The problem of determining whether two sets of support sets are isomorphic is surprisingly nontrivial. We begin with some theoretical considerations before moving on to implementation.

Definition 2. The *graph isomorphism problem* asks whether for two undirected graphs F, G there is a bijection ϕ between their vertices that preserves incidence– i.e. if a and b are vertices connected by an edge in F, then $\phi(a)$ and $\phi(b)$ are connected by an edge in G.

Proposition 1. *The problem of determining whether two sets of support sets are isomorphic is equivalent to the graph isomorphism problem.*

Proof. We will give a constructive embedding in both directions.

($\supseteq$) We start by showing that graph isomorphism can be embedded in checking isomorphism of sets of support sets. Recall that the incidence matrix of a graph $G = (V, E)$ is a matrix with $\#V$ rows and $\#E$ columns where the $(i, j)^{th}$ entry is 1 if the ith vertex and j^{th} edge are incident and 0 otherwise. It is straightforward to show that two graphs are isomorphic if and only if their incidence matrices can be made equal by rearranging their rows and columns.

Now suppose we have a graph with incidence matrix $A = (a_{ij})$. Construct a polynomial by considering the rows as variables and the columns as monomials. To be precise, set

$$p_A = \sum_j \prod_i x_i^{a_{ij}}. \tag{4}$$

For example, if a graph has incidence matrix

$$A = \begin{pmatrix} 1 & 1 \\ 1 & 0 \\ 0 & 1 \end{pmatrix}, \quad \text{then} \quad p_A = x_1 x_2 + x_1 x_3. \tag{5}$$

Switching two columns of the matrix corresponds to reordering the monomials in the sum; switching two rows corresponds to a permutation of variables. Therefore, to determine if two graphs are isomorphic, one may find their incidence matrices, form polynomials from them, and check if the polynomials (thought of as single-polynomial systems) are the same up to permutation of variables. Finally, it is important to note that this is a polynomial time reduction. To create the polynomial, start at the first column and iterate through all of the columns. When done with the first column, we have created the first monomial. Repeat this process for every column, touching every entry in the matrix exactly once. Given an incidence matrix of a graph $G = (V, E)$ with $\#V = n, \#E = m$, converting to a polynomial requires $\mathcal{O}(nm)$ operations.

($\subseteq$) We now show the other direction: checking isomorphism of sets of support sets can be embedded in graph isomorphism. To do so, we give a way of setting up a graph from a set of support sets, which is most easily seen via an example. The graph for the system $\{x^2 + xy, x^3 y^2 + x + 1\}$ is shown in Figure 1. As can be seen from the diagram, we build a graph beginning with a root node to ensure the graph is connected. We attach one node for each equation, then attach nodes to each equation for its monomials. Finally we put a row of variable nodes at the bottom and connect each monomial to the variables it contains via as many segments as the degree of the variable in the monomial.

At this point our graph setup is inadequate, as two different systems could have isomorphic graphs. Consider the graphs for $\{x, y\}$ and $\{x^4\}$, seen in Figure 2. Even though the systems are different, the graphs are clearly isomorphic. To remedy this, self-loops are added in order to partition the graph into root, equations, monomials, and variables by putting one self-loop at the

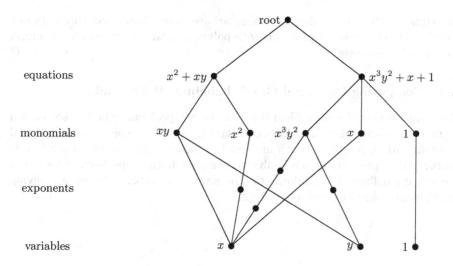

Fig. 1. Basic graph from system

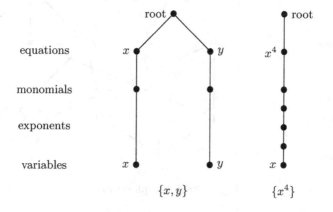

Fig. 2. Isomorphic graphs

root node, two on each equation node, etc. This graph will uniquely represent our system since any automorphism will be a permutation of nodes within their partitions. Partitioning by self-loops is possible because there are no self-loops in the initial setup. The graph in Figure 1 is drawn without the self-loops for the sake of readability.

The time to complete this process is easily seen to be polynomial in the number of monomials and the sum of their total degrees.

Because we have shown that two sets of support sets are isomorphic if and only if two uniquely dependent graphs are isomorphic, and had previously shown that two graphs are isomorphic if and only if two uniquely dependent polynomials are equal up to rearranging variables, we are done. Note that this puts our problem in **GI**, the set of problems with polynomial time reduction to the graph isomorphism problem.   □

5.3   Computing Canonical Graph Labelings With Nauty

The graph isomorphism problem is one of the few problems which is not known to be P or NP-complete [22]. Although there is no known worst-case polynomial time algorithm, solutions which are fast in practice exist. One such solution is **nauty** [21], a program which is able to compute both graph isomorphism and canonical labelings. More information on **nauty** and other software for solving graph isomorphism can be found in [22].

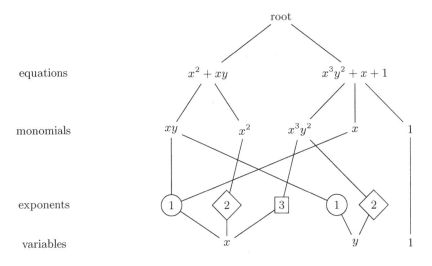

Fig. 3. Nauty graph setup

Because **nauty** can incorporate the added information of ordered partitions, we chose to revise our setup to take advantage of this and minimize the number of nodes. Instead of using self-loops to partition the graph, we specify equations, monomials, and variables to be three of the partitions. We then check for which exponents occur in the system, and attach nodes representing these exponents to the variables where appropriate. Instead of using a sequence of nodes and edges to record the degree as we did in the proof, we instead attach monomials to these exponent nodes. We group these exponent nodes so that all the nodes representing the lowest exponent are in the first partition, all representing the second lowest are in the second partition, etc. The setup is shown in Figure 3, where the different shapes of the exponent nodes represent the distinct partitions.

If we couple this graph with an ordered list of the exponents that occur in the system, we once again have an object that uniquely represents the isomorphism class of our system. In addition, nauty works by computing the automorphisms of a graph and determining one of them to be canonical; the automorphisms must respect the partition, hence partitioning reduces the number of possibilities to check. So not only does this use fewer vertices and edges than our previous setup, but the use of partitions speeds up nauty's calculations.

Another advantage to using nauty is that it computes generators of the automorphism group of the graph. Some of these generators permute the equations and monomials. These are unimportant and may be discarded, but the generators that permute the variables are quite useful, as PHCpack has the ability to take advantage of symmetry in polynomial systems. If a system remains the same under certain changes of variables, runtime may be significantly decreased if this symmetry is passed to phc.

It is worth noting that by Frucht's theorem [11] and our Proposition 1 we immediately obtain that for any group G there is a polynomial system with G as its symmetry structure. If we want to actually find systems with particular structures this is fairly impractical, however, as the proof of Frucht's theorem uses the Cayley graph which has a node for every group element, meaning that a system built with this method would have a variable for every group element.

We are primarily interested in using nauty in the context of storing tuples of support sets in a database. Because of this, the canonical labeling feature is much more useful than the ability to check graph isomorphism–looking up a system in a database ought not be done by going through a list of systems and querying nauty as to whether they are isomorphic to the one we are looking up, since this would be highly inefficient. Instead we simply parse the system into its graph form (including partition data), pass this information to nauty and compute the canonical labeling, and attach the exponent data, as a string, to nauty's output. This process gives us a string that uniquely corresponds to the isomorphism class of the system, which we can then store in a database.

5.4 Benchmarking the Canonization

Timings reported in this section were done on a 3.5 GHz Intel Core i5 processor in a iMac Retina 5K running version 10.10.2 of Mac OS X, with 16 GB RAM. Scripts are available at https://github.com/sommars/PolyGraph.

In order to design an intelligent storage system, it is necessary to know an upper bound of the length of the string. As polynomial systems can be arbitrarily large, leading to arbitrarily long strings from nauty, we chose to analyze a number of well known systems to act as benchmarks, giving us an idea of an upper bound. First, consider the cyclic n-root polynomial systems. These systems consist of n equations: $\sum_{i=1}^{n} \prod_{j=i}^{i+k} x_{j \bmod n} = 0$ for k from 0 to $n - 2$, and $x_1 \ldots x_n - 1 = 0$. For example, the cyclic-3 system is

$$\begin{cases} x_1 + x_2 + x_3 = 0 \\ x_1 x_2 + x_2 x_3 + x_3 x_1 = 0 \\ x_1 x_2 x_3 - 1 = 0. \end{cases} \tag{6}$$

These systems have a lot of symmetry, so they are an interesting case for us to benchmark this process. We tested against both small and large values of n to gain a thorough understanding of how this system will be canonized. Experiments in small dimensions are summarized in Table 1. The computation time, an average of three trials, is quite fast, though the length of the string grows quickly. In comparison, note that the calculation of the root counts in the blackbox solver of PHCpack (which includes the mixed volume computation) for the cyclic 10-roots problems takes 48.8 seconds.

Table 1. Cyclic n-root benchmark for small n. For each n, its takes only milliseconds. We list the number of nodes in the graph and the length of the canonical form.

n	time	#nodes	#characters
4	0.006	29	526
6	0.006	53	1,256
8	0.006	85	2,545
10	0.007	125	5,121
12	0.007	173	8,761

For large instances of cyclic n-roots, the exponential growth of the number of solutions increases so much that we may no longer hope to be capable of computing all solutions. Nevertheless, with GPU acceleration [34], we can manage to track a limited number of paths. Table 2 illustrates the relationship between the dimension n, the time, and the sizes of the data.

Table 2. For larger values of the dimension of the cyclic n-root problem, times and sizes of the data start to grow exponentially.

n	time	#nodes	#characters
16	0.010	293	20,029
32	0.045	1,093	168,622
48	0.265	2,405	601,702
64	1.200	4,229	1,427,890
80	4.316	6,565	2,778,546
96	15.274	9,413	4,784,390
112	38.747	12,773	8,595,408
128	80.700	16,645	13,094,752

Another interesting class of benchmark polynomial systems that we can formulate for any dimension is the computation of Nash equilibria [8]. We use the formulation of this problem as in [23]. The number of isolated solutions also grows exponentially in n. Table 3 summarizes our computational experiments. As before, the times represent an average of three trials.

Compared to the cyclic n-roots problem, the dimensions in Table 3 are just as small as in Table 1, but the time and sizes grow much faster for the Nash equilibria than for the cyclic n-roots. We suspect two factors. First, while structured,

Table 3. The cost of the canonization for the Nash equilibria polynomial systems for increasing dimension n, with the running time expressed in seconds, the number of nodes, and the size of the canonical form.

n	time	#nodes	#characters
4	0.006	47	977
5	0.006	98	2,325
6	0.007	213	7,084
7	0.013	472	18,398
8	0.054	1,051	51,180
9	0.460	2,334	134,568
10	4.832	5,153	331,456
11	73.587	11,300	872,893
12	740.846	24,615	2,150,512

the Nash equilibria systems are not as sparse as the cyclic n-roots problems. Second, unlike the cyclic n-roots problem every equation in the Nash equilibria system has the same structure, so the full permutation group leaves the sets of support sets invariant.

We end with a system formulated by S. Katsura, see [4] and [19]. Table 4 shows the cost of the canonization of this system. Because there is no symmetry in the support sets, the cost of the canonization increases not as fast in the dimension n as with the other two examples.

The actual cost of the computation of the canonical form may serve as an initial estimate on the cost of solving the system.

5.5   Storing Labelings in a Database

There are many different ways to design a database to store a given set of information. We do not contend that this is necessarily the best way, but it is certainly an effective way of storing our uniquely identifying information that will lead to fast lookup. Consider the explicitly described schema in Figure 4.

Table 4. The cost of the canonization of the Katsura system for increasing dimension n. The number of solutions for the n-dimensional version of the system equals 2^n.

n	time	#nodes	#characters
25	0.020	929	24,906
50	0.090	3,411	112,654
75	0.546	7,454	254,770
100	1.806	13,061	495,612
125	4.641	20,229	793,662
150	10.860	28,961	1,157,498
175	21.194	39,254	1,587,115
200	52.814	51,111	2,082,562
225	98.118	64,529	2,643,891

Section	Name	Datatype	Description
Graph Nodes	*n_node_variable*	INT	Number of variables nodes
	n_node_monomial	INT	Number of monomial nodes
	n_node_equation	INT	Number of equation nodes
	n_node_degree	INT	Number of degree nodes
Degree Set	*n_degree*	INT	Sum of all degrees
	degrees	VARCHAR	Set of all degrees
Graph	*graph_length*	INT	Length of complete graph
	graph_filename	VARCHAR	Complete graph in file
Polynomial Info	*poly_filename*	VARCHAR	Information and Reference file for polynomial system

Fig. 4. Database structure for polynomial system graph

Each of the data elements in it are used to partition the database from the previous elements, so it can be seen as a B-tree structure. For identifying whether or not a set of support sets is already in our database, this would lead to a search time of $O(\log n)$ [7].

6   Conclusions

The practical considerations of offering a cloud service to solve polynomial systems with polynomial homotopy continuation led to the classification problem of polynomial systems. To solve this problem, we linked the isomorphism problem for sets of support sets to the graph isomorphism problem and applied the software nauty.

Although no polynomial time algorithm are known to solve the graph isomorphism problem, we presented empirical results from benchmark polynomial systems that the computation of a canonical form costs much less than solving a polynomial system.

Acknowledgements. The computer that hosts our cloud service was purchased through a UIC LAS Science award. We thank the reviewers for their comments and suggestions.

References

1. Adrovic, D., Verschelde, J.: A polyhedral method to compute all affine solution sets of sparse polynomial systems. arXiv:1310.4128
2. Adrovic, D., Verschelde, J.: Computing Puiseux series for algebraic surfaces. In: van der Hoeven, J., van Hoeij, M. (eds.) Proceedings of the 37th International Symposium on Symbolic and Algebraic Computation (ISSAC 2012), pp. 20–27. ACM (2012)

3. Adrovic, D., Verschelde, J.: Polyhedral methods for space curves exploiting symmetry applied to the cyclic n-roots problem. In: Gerdt, V.P., Koepf, W., Mayr, E.W., Vorozhtsov, E.V. (eds.) CASC 2013. LNCS, vol. 8136, pp. 10–29. Springer, Heidelberg (2013)
4. Boege, W., Gebauer, R., Kredel, H.: Some examples for solving systems of algebraic equations by calculating groebner bases. J. Symbolic Computation **2**, 83–98 (1986)
5. Bouillaguet, C., Fouque, P.-A., Véber, A.: Graph-theoretic algorithms for the "isomorphism of polynomials" problem. In: Johansson, T., Nguyen, P.Q. (eds.) EUROCRYPT 2013. LNCS, vol. 7881, pp. 211–227. Springer, Heidelberg (2013)
6. Coar, K., Bower, R.: Apache Cookbook. 1st edn. O'Reilly Media, Inc. (2004)
7. Comer, D.: Ubiquitous b-tree. ACM Comput. Surv. **11**(2), 121–137 (1979)
8. Datta, R.: Finding all nash equilibria of a finite game using polynomial algebra. Economic Theory **42**(1), 55–96 (2009)
9. DuBois, P.: MySQL Cookbook. 2nd edn. O'Reilly Media, Inc. (2006)
10. Faugère, J.-C., Perret, L.: Polynomial equivalence problems: algorithmic and theoretical aspects. In: Vaudenay, S. (ed.) EUROCRYPT 2006. LNCS, vol. 4004, pp. 30–47. Springer, Heidelberg (2006)
11. Frucht, R.: Herstellung von Graphen mit vorgegebener abstrakter Gruppe. Compositio Mathematica **6**, 239–250 (1939)
12. Gao, T., Li, T., Wu, M.: Algorithm 846: MixedVol: a software package for mixed-volume computation. ACM Trans. Math. Softw. **31**(4), 555–560 (2005)
13. Grayson, D., Stillman, M.: Macaulay2, a software system for research in algebraic geometry. http://www.math.uiuc.edu/Macaulay2/
14. Gross, E., Petrović, S., Verschelde, J.: PHCpack in Macaulay2. The Journal of Software for Algebra and Geometry: Macaulay 2(5), 20–25 (2013)
15. Huber, B., Sturmfels, B.: A polyhedral method for solving sparse polynomial systems. Math. Comp. **64**(212), 1541–1555 (1995)
16. Huber, B., Verschelde, J.: Polyhedral end games for polynomial continuation. Numerical Algorithms **18**(1), 91–108 (1998)
17. Jensen, A., Markwig, H., Markwig, T.: An algorithm for lifting points in a tropical variety. Collectanea Mathematica **59**(2), 129–165 (2008)
18. Kaibel, V., Schwartz, A.: On the complexity of polytope isomorphism problems. Graphs and Combinatorics **19**(2), 215–230 (2003)
19. Katsura, S.: Spin glass problem by the method of integral equation of the effective field. In: Coutinho-Filho, M., Resende, S. (eds.) New Trends in Magnetism, pp. 110–121. World Scientific (1990)
20. Li, T.: Numerical solution of polynomial systems by homotopy continuation methods. In: Cucker, F. (ed.) Handbook of Numerical Analysis, vol. 11. Special Volume: Foundations of Computational Mathematics, pp. 209–304. North-Holland (2003)
21. McKay, B., Piperno, A.: nautyTraces, software distribution web page. http://cs.anu.edu.au/~bdm/nauty/ and http://pallini.di.uniroma1.it/
22. McKay, B., Piperno, A.: Practical graph isomorphism II. Journal of Symbolic Computation **60**, 94–112 (2014)
23. McKelvey, R., McLennan, A.: The maximal number of regular totally mixed Nash equilibria. Journal of Economic Theory **72**, 411–425 (1997)
24. Patarin, J.: Hidden fields equations (HFE) and isomorphisms of polynomials (IP): two new families of asymmetric algorithms. In: Maurer, U.M. (ed.) EUROCRYPT 1996. LNCS, vol. 1070, pp. 33–48. Springer, Heidelberg (1996)
25. Pérez, F., Granger, B., Hunter, J.: Python: An ecosystem for scientific computing. Computing in Science & Engineering **13**(2), 12–21 (2011)

26. Piret, K.: Computing Critical Points of Polynomial Systems using PHCpack and Python. PhD thesis, University of Illinois at Chicago (2008)

27. Rehr, J., Vila, F., Gardner, J., Svec, L., Prange, M.: Scientific computing in the cloud. Computing in Science & Engineering **12**(3), 34–43 (2010)

28. Shirt-Ediss, B., Sole, R., Ruiz-Mirazo, K.: Emergent chemical behavior in variable-volume protocells. Life **5**, 181–121 (2015)

29. Sommese, A., Verschelde, J., Wampler, C.: Numerical irreducible decomposition using PHCpack. In: Joswig, M., Takayama, N. (eds.) Algebra, Geometry, and Software Systems, pp. 109–130. Springer (2003)

30. Stein, W., et al.: Sage Mathematics Software (Version 6.5). The Sage Development Team. (2015). http://www.sagemath.org

31. Verschelde, J.: Algorithm 795: PHCpack: A general-purpose solver for polynomial systems by homotopy continuation. ACM Trans. Math. Softw. **25**(2), 251–276 (1999)

32. Verschelde, J.: Modernizing PHCpack through phcpy. In: de Buyl, P., Varoquaux, N. (eds.) Proceedings of the 6th European Conference on Python in Science (EuroSciPy 2013), pp. 71–76 (2014)

33. Verschelde, J., Verlinden, P., Cools, R.: Homotopies exploiting Newton polytopes for solving sparse polynomial systems. SIAM J. Numer. Anal. **31**(3), 915–930 (1994)

34. Verschelde, J., Yu, X.: Accelerating polynomial homotopy continuation on a graphics processing unit with double double and quad double arithmetic. arXiv:1501.06625, accepted for publication in the Proceedings of the 7th International Workshop on Parallel Symbolic Computation (PASCO 2015)

35. Verschelde, J., Yu, X.: Tracking many solution paths of a polynomial homotopy on a graphics processing unit. arXiv:1505.00383, accepted for publication in the Proceedings of the 17th IEEE International Conference on High Performance Computing and Communications (HPCC 2015)

Finding First Integrals Using Normal Forms Modulo Differential Regular Chains

François Boulier and François Lemaire

Univ. Lille, CRIStAL, UMR 9189, 59650 Villeneuve d'Ascq, France
{francois.boulier,francois.lemaire}@univ-lille1.fr

Abstract. This paper introduces a definition of polynomial first integrals in the differential algebra context and an algorithm for computing them. The method has been coded in the Maple computer algebra system and is illustrated on the pendulum and the Lotka-Volterra equations. Our algorithm amounts to finding linear dependences of rational fractions, which is solved by evaluation techniques.

Keywords: First integral, linear algebra, differential algebra, nonlinear system.

1 Introduction

This paper deals with the computation of polynomial first integrals of systems of ODEs, where the independent variable is t (for time). A first integral is a function whose value is constant over time along every solution of a system of ODEs. First integrals are useful to understand the structure of systems of ODEs. A well known example of first integral is the energy of a mechanical conservative system, as shown by Example 1.

Example 1. Using a Lagrange multiplier $\lambda(t)$, a pendulum of fixed length l, with a centered mass m submitted to gravity g can be coded by:

$$\Sigma \begin{cases} m\,\ddot{x}(t) & = \lambda(t)\,x(t) \\ m\,\ddot{y}(t) & = \lambda(t)\,y(t) + mg \\ x(t)^2 + y(t)^2 & = l^2. \end{cases} \quad (1)$$

A trivial first integral is $x(t)^2 + y(t)^2$ since $x(t)^2 + y(t)^2 = l^2$ on any solution. A less trivial first integral is $\frac{m}{2}\left(\dot{x}(t)^2 + \dot{y}(t)^2\right) - mg\,y(t)$ which corresponds to the total energy of the system (kinetic energy + potential energy).

When no physical considerations can be used, one needs alternative methods. Assume we want to compute a polynomial first integral of a system of ODEs. Our algorithm findAllFirstIntegrals, which has been coded in Maple, relies on the following strategy. We choose a certain number of monomials $\mu_1, \mu_2, \ldots, \mu_e$ built

© Springer International Publishing Switzerland 2015
V.P. Gerdt et al. (Eds.): CASC 2015, LNCS 9301, pp. 101–118, 2015.
DOI: 10.1007/978-3-319-24021-3_8

over t, the unknown functions and their derivatives (namely t, $x(t)$, $y(t)$, $\dot{x}(t)$, $\dot{y}(t)$, ... on Example 1), and look for a candidate of the form $q = \alpha_1\mu_1 + \cdots + \alpha_e\mu_e$ satisfying $\frac{dq(t)}{dt} = 0$ for all solutions, where the α_i are in some field $\mathbb{K}$. If the α_i are constant (i.e. $\frac{d\alpha_i}{dt} = 0$ for each α_i), then our problem amounts to finding α_i such that $\frac{dq(t)}{dt} = \alpha_1\frac{d\mu_1}{dt} + \cdots + \alpha_e\frac{d\mu_e}{dt}$ is zero for all solutions. On Example 1, μ_1, μ_2 and μ_3 could be the monomials $\dot{x}(t)^2$, $\dot{y}(t)^2$ and $y(t)$ and α_1, α_2 and α_3 could be $m/2$, $m/2$ and $-mg$.

Anticipating Section 4, differential algebra techniques combined with the Nullstellensatz Theorem permit to check that a polynomial p vanishes on all solutions of a radical ideal $\mathfrak{A}$ by checking that p belongs to $\mathfrak{A}$. Moreover, in the case where the ideal $\mathfrak{A}$ is represented[1] by a list of differential regular chains $M = [C_1, \ldots, C_f]$, checking that $p \in \mathfrak{A}$ can be done by checking that the normal form (see Section 4) of p modulo the list of differential regular chains M, denoted $\mathsf{NF}(p, M)$, is zero. Since the normal form is $\mathbb{K}$-linear (see Section 4), finding first integrals can be solved by finding a linear dependence between $\mathsf{NF}(\frac{d\mu_1}{dt}, M)$, $\mathsf{NF}(\frac{d\mu_2}{dt}, M)$, ..., $\mathsf{NF}(\frac{d\mu_e}{dt}, M)$.

This process, which is in the spirit of [3,8], encounters a difficulty: the normal forms usually involve fractions. As a consequence, we need a method for finding linear dependences between rational functions. A possible approch consists in reducing all fractions to the same denominator, and solving a linear system based on the monomial structure of the numerators. However, this has disadvantages. First, the sizes of the numerators might grow if the denominators are large. Second, the linear system built above is completely modified if one considers a new fraction. Finding linear dependences between rational functions is addressed in Sections 2 and 3 by evaluating the variables occurring on the denominators, thus transforming the fractions into polynomials. The main difficulty with this process is to choose adequate evaluations to obtain deterministic (i.e. non probabilistic) and terminating algorithms.

This paper is structured as follows. Section 2 presents some lemmas around evaluation and Algorithm findKernelBasis (and its variant findKernelBasisMatrix) which computes a basis of the linear dependences of some given rational functions. Section 3 presents Algorithm incrementalFindDependence (and its variant incrementalFindDependenceLU) which sequentially treats the fractions and stops when a linear dependence is detected. Section 4 recalls some differential algebra techniques and presents Algorithm findAllFirstIntegrals which computes a basis of first integrals which are $\mathbb{K}$-linear combinations of a predefined set of monomials.

In Sections 2 and 3, $\mathbf{k}$ is a commutative field of characteristic zero containing $\mathbb{R}$, $\mathbb{A}$ is an integral domain with unit, a commutative $\mathbf{k}$-algebra and also a $\mathbf{k}$-vector space equipped with a norm $\|\cdot\|_{\mathbb{A}}$. Moreover, $\mathbb{K}$ is the total field of fractions of $\mathbb{A}$. In Section 4, we will require, moreover, that $\mathbb{K}$ is a field of constants. In applications, $\mathbf{k}$ is usually $\mathbb{R}$, $\mathbb{A}$ is usually a commutative ring of multivariate polynomials over $\mathbf{k}$ (i.e. $\mathbb{A} = \mathbf{k}[z_1, \ldots, z_s]$) and $\mathbb{K}$ is the field of fractions of $\mathbb{A}$ (i.e. the field of rational functions $\mathbf{k}(z_1, \ldots, z_s)$).

---

[1] More precisely $\mathfrak{A} = [C_1] : H_{C_1}^{\infty} \cap \cdots \cap [C_f] : H_{C_f}^{\infty}$.

2   Basis of Linear Dependences of Rational Functions

This section presents Algorithms findKernelBasis and findKernelBasisMatrix which compute a basis of the linear dependences over $\mathbb{K}$ of e multivariate rational functions denoted by $q_1, \ldots, q_e$. More precisely, they compute a $\mathbb{K}$-basis of the vectors $(\alpha_1, \ldots, \alpha_e) \in \mathbb{K}^e$ satisfying $\sum_{i=1}^{e} \alpha_i q_i = 0$.

Those algorithms are mainly given for pedagogical reasons, in order to focus on the difficulties related to the evaluations (especially Lemma 2). The rational functions are taken in the ring $\mathbb{K}(Y)[X]$ where $Y = \{y_1, \ldots, y_m\}$ is a finite set of indeterminates and X is another (possibly infinite) set of indeterminates such that $Y \cap X = \emptyset$. The idea consists in evaluating the variables Y, thus reducing our problem to finding the linear dependences over $\mathbb{K}$ of polynomials in $\mathbb{K}[X]$, which can be easily solved for instance with linear algebra techniques. If enough evaluations are made, the linear dependences of the evaluated fractions coincide with the linear dependences of the non evaluated fractions.

Even if it is based on evaluations, our algorithm is not probabilistic. A possible alternative is to use [12] by writing each rational function into a (infinite) basis of multivariate rational functions. However, we chose not to use that technique because it relies on multivariate polynomial factorization into irreducible factors, and because of a possible expression swell.

2.1   Preliminary Results

This section introduces two basic definitions as well as three lemmas needed for proving Algorithm findKernelBasis.

Definition 1 (Evaluation). *Let $D = \{g_1, \ldots, g_e\}$ be a set of polynomials of $\mathbb{K}[Y]$ where $Y = \{y_1, \ldots, y_m\}$. Let $y^0 = (y_1^0, y_2^0, \ldots, y_m^0)$ be an element of $\mathbb{K}^m$, such that none of the polynomials of D vanish on y^0. One denotes σ_{y^0} the ring homomorphism from $\mathbb{K}[X, Y, g_1^{-1}, \ldots, g_e^{-1}]$ to $\mathbb{K}[X]$ defined by $\sigma_{y^0}(y_j) = y_j^0$ for $1 \leq j \leq m$ and $\sigma_{y^0}(x) = x$ for $x \in X$. Roughly speaking, the ring homomorphism σ_{y^0} evaluates at $y = y_0$ the rational functions whose denominator divides a product of powers of g_i.*

Definition 2 (Linear combination application). *Let E be a $\mathbb{K}$-vector space. For any vector $v = (v_1, v_2, \ldots, v_e)$ of E^e, Φ_v denotes the linear application from $\mathbb{K}^e$ to E defined by $\alpha = (\alpha_1, \alpha_2, \ldots, \alpha_e) \to \Phi_v(\alpha) = \sum_{i=1}^{e} \alpha_i v_i$.*

The notation Φ_v defined above is handy: if $q = (q_1, \ldots, q_e)$ is (for example) a vector of rational functions, then the set of the vectors $\alpha = (\alpha_1, \ldots, \alpha_n)$ in $\mathbb{K}^e$ satisfying $\sum_{i=1}^{e} \alpha_i q_i = 0$ is simply $\ker \Phi_q$.

The following Lemma 1 is basically a generalization to several variables of the classical following result: *a polynomial of degree d over an integral domain admitting $d + 1$ roots is necessarily the zero polynomial.*

Lemma 1. *Let $p \in \mathbb{K}[Y]$ where $Y = \{y_1, \ldots, y_m\}$. Assume $\deg_{y_i} p \leq d$ for all $1 \leq i \leq m$. Let $S_1, S_2, \ldots, S_m$ be m sets of $d + 1$ points in $\mathbb{Q}$. If $p(y^0) = 0$ for all $y^0 \in S_1 \times S_2 \times \ldots \times S_m$, then p is the zero polynomial.*

Proof. By induction on m. When $m = 1$, p has $d+1$ distinct roots and a degree at most d. Since p is a polynomial over a field $\mathbb{K}$, p is the zero polynomial.

Suppose the lemma is true for m. One shows it is true for $m + 1$. Seeing p as an element of $\mathbb{K}(y_1, \ldots, y_m)[y_{m+1}]$ and using the Lagrange polynomial interpolation formula [9, page 101] over the ring $\mathbb{K}(y_1, \ldots, y_m)[y_{m+1}]$, one has $p = \sum_{i=1}^{d+1} \left(p(y_1, \ldots, y_m, s_i) \prod_{j=1, j\neq i}^{d+1} \frac{y_{m+1} - s_j}{s_i - s_j} \right)$ where $S_{m+1} = \{s_1, \ldots, s_{d+1}\}$. For each $1 \leq i \leq d+1$, $p(y_1, \ldots, y_m, s_i)$ vanishes on all points of $S_1 \times S_2 \times \ldots \times S_m$. By induction, all $p(y_1, \ldots, y_m, s_i)$ are thus zero, and p is the zero polynomial. $\square$

The following lemma is quite intuitive. If one takes a nonzero polynomial g in $\mathbb{K}[Y]$, then it is possible to find an arbitrary large "grid" of points in $\mathbb{N}^m$ where g does not vanish. This lemma will be applied later when g is the product of some denominators that we do not want to cancel.

Lemma 2. *Let g be a nonzero polynomial in $\mathbb{K}[Y]$ and d be a positive integer. There exist m sets $S_1, S_2, \ldots, S_m$, each one containing $d + 1$ consecutive nonnegative integers, such that $g(y^0) \neq 0$ for all y^0 in $S = S_1 \times S_2 \times \cdots \times S_m$.*

The proof of Lemma 2 is the only one explicitly using the norm $\|\cdot\|_{\mathbb{A}}$. The proof is a bit technical but the underlying idea is simple and deserves a rough explanation. If g is nonzero, then its homogeneous part of highest degree, denoted by h, must be nonzero at some point $\bar{y}$, hence on a neighborhood of $\bar{y}$. Since g "behaves like" h at infinity (the use of $\|\cdot\|_{\mathbb{A}}$ will make precise this statement), one can scale the neighborhood to prove the lemma.

Proof. Since $\mathbb{K}$ is the field of fraction of $\mathbb{A}$, one can assume with no loss of generality that g is in $\mathbb{A}[Y]$.

Denote $g = h + p$ where h is the homogeneous part of g of (total) degree $e = \deg g$. Since, g is nonzero, so is h. By Lemma 1, there exists a point $\bar{y} \in \mathbb{R}_{>0}^m$, such that $h(\bar{y}) \neq 0$, where $\mathbb{R}_{>0}$ denotes the positive reals. Without loss of generality, one can assume that $\|\bar{y}\|_\infty = 1$ since h is homogeneous. There exists an open neighborhood V of $\bar{y}$ such that $V \subset \mathbb{R}_{>0}^m$ and $0 \notin h(V)$. Moreover, this open neighborhood V also contains a closed ball B centered at $\bar{y}$ for some positive real ϵ i.e. $B = \{y \in \mathbb{R}_{>0}^m \mid \|y - \bar{y}\|_\infty \leq \epsilon\} \subset V$.

Since B is a compact set and the functions h and t are continuous, the two following numbers are well-defined and finite: $m = \min_{y \in B} \|h(y)\|_{\mathbb{A}}$ and $M = \max_{y \in B}(\sum_{i \in I} \|a_i\|_{\mathbb{A}} m_i(y))$ (where $p = \sum_{i \in I} a_i m_i$ with $a_i \in \mathbb{A}$ and m_i is a monomial in Y). Moreover, $m > 0$ (since by a compactness argument there exists $y \in B$ such that $h(y) = m$).

Take $y \in B$ and $s > 1 \in \mathbb{R}$. Then $g(sy) = h(sy) + p(sy) = s^e h(y) + p(sy)$. By the reverse triangular inequality and the homogeneity of h, one has $\|g(sy)\|_{\mathbb{A}} \geq s^e \|h(y)\|_{\mathbb{A}} - \|p(sy)\|_{\mathbb{A}}$. Moreover

$$\|p(sy)\|_{\mathbb{A}} \leq \sum_{i \in I} \|a_i\|_{\mathbb{A}} m_i(sy) \leq \sum_{i \in I} \|a_i\|_{\mathbb{A}} s^{e-1} m_i(y) \leq s^{e-1} M.$$

Consequently, $\|g(sy)\|_{\mathbb{A}} \geq s^e m - s^{e-1} M$. If one takes s sufficiently large to ensure $s^e m - s^{e-1} M > 0$ and $s\epsilon \geq (d+1)/2$, the ball $\bar{B}$ obtained by uniformly scaling B

by a factor s contains no root of g. Since the width of the box $\bar{B}$ is at least $d+1$, the existence of the expected S_i is guaranteed. □

Roughly speaking, the following lemma ensures that if a fraction q in $\mathbb{K}(Y)[X]$ evaluates to zero for well chosen evaluations of Y, then q is necessarily zero.

Lemma 3. *Take an element q in $\mathbb{K}(Y)[X]$. Then, there exist an integer d, and m sets $S_1, S_2, \ldots, S_m$, each one containing $d+1$ nonnegative consecutive integers, such that $\sigma_{y^0}(q)$ is well-defined for all $y^0 \in S = S_1 \times S_2 \times \cdots \times S_m$. Moreover, if $\sigma_{y^0}(q) = 0$ for all y^0 in S, then $q = 0$.*

Proof. Let us write $q = p/g$ where $p \in \mathbb{K}[X, Y]$ and $g \in \mathbb{K}[Y]$. Consider $d = \max_{1 \leq i \leq m} \deg_{y_i}(p)$. By Lemma 2, there exist m sets $S_1, \ldots, S_m$ of $d+1$ consecutive nonnegative integers such that $g(y^0) \neq 0$ for all y^0 in $S = S_1 \times S_2 \times \cdots \times S_m$. Consequently, $\sigma_{y^0}(q)$ is well-defined for all y^0 in S. Let us assume that $\sigma_{y^0}(q) = 0$ for all y^0 in S. As a consequence, one has $\sigma_{y^0}(p) = 0$ for any y^0 in S. By Lemma 1, one has $p = 0$, hence $q = 0$. □

2.2   Algorithm findKernelBasis

We rely on the notion of *iterator*, which is a tool in computer programming that enables to enumerate all values of a structure (such as a list, a vector, ...). An iterator can be *finite* (if it becomes empty after enumerating a finite number of values) or *infinite* (if it never gets empty).

In order to enumerate evaluation points (that we take in $\mathbb{N}^m$ for simplicity), we use two basic functions to generate one by one all the tuples, which do not cancel any element of D, where D is a list of nonzero polynomials of $\mathbb{K}[Y]$. The first one is called newTupleIteratorNotCancelling(Y, D). It builds an iterator I to be used with the second function getNewTuple(I). Each time one calls the function getNewTuple(I), one retrieves a new tuple which does not cancel any element of D. The only constraint we require for getNewTuple is that any tuple of $\mathbb{N}^m$ (which does not cancel any element of D) should be output after a finite number of calls to getNewTuple. To ensure that, one can for example enumerate all integer tuples by increasing order (where the order is the sum of the entries of the tuple) refined by the lexicographic order for tuples of the same order. Without this constraint on getNewTuple, Algorithm findKernelBasis may not terminate.

Example 2. Take $\mathbb{K} = \mathbb{Q}(z)$, $Y = \{a\}$ and $X = \{x\}$. Consider $q = (q_1, q_2, q_3) = \left(\frac{ax+2}{a+1}, \frac{z(a-1-ax)}{1+a}, 1 \right)$. One can show that the only (up to a constant) $\mathbb{K}$-linear dependence between the q_i is $-q_1 - (1/z)q_2 + q_3 = 0$.

Apply Algorithm findKernelBasis. One has $D = [a + 1]$ at Line 3, so one can evaluate the indeterminate a on any nonnegative integer. At Line 7, $\bar{S}$ contains the tuple $s_1 = (0)$, corresponding to the evaluation $a = 0$. One builds the vectors v_1, v_2 and v_3 at Line 7 by evaluating q_1, q_2 and q_3 on $a = 0$. One obtains polynomials in $\mathbb{K}[x]$. Thus $v_1 = (2)$, $v_2 = (-z)$ and $v_3 = (1)$. A basis of $\ker \Phi_v$ computed at Line 10 could be $B = \{(-1/2, 0, 1), (z/2, 1, 0)\}$. Note that it is normal that both $v = (v_1, v_2, v_3)$ and B involve z since one performs linear

---

Input: Two lists of variables Y and X
Input: A vector $q = (q_1, \ldots, q_e)$ of e rational fractions in $\mathbb{K}(Y)[X]$
Output: A basis of $\ker \Phi_q$

1 **begin**
2   For each $1 \leq i \leq e$, denote the reduced fraction q_i as f_i/g_i
         with $f_i \in \mathbb{K}[Y, X]$, $g_i \in \mathbb{K}[Y]$;
3      $D \leftarrow [g_1, \ldots, g_e]$;
4      $I \leftarrow$ newTupleIteratorNotCancelling(Y, D);
5      $\bar{S} \leftarrow [\text{getNewTuple}(I)]$;
6   **while** *true* **do**
7   For each $1 \leq i \leq e$, denote v_i the vector $(\sigma_{s_1}(q_i), \sigma_{s_2}(q_i), \ldots, \sigma_{s_r}(q_i))$
             where $\bar{S} = [s_1, s_2, \ldots, s_r]$;
8   // each v_i is a vector of $r = |\bar{S}|$ elements of $\mathbb{K}[X]$, obtained by
             evaluating q_i on all points of $\bar{S}$
9   Denote $v = (v_1, \ldots, v_e)$;
10   Compute a basis B of the kernel of Φ_v using linear algebra;
11   // if $\ker \Phi_q \supset \ker \Phi_v$, one returns B
12   **if** $\sum_{i=1}^{e} b_i q_i = 0$ for all $b = (b_1, \ldots, b_e) \in B$ **then** **return** B;
13   Append to $\bar{S}$ the new evaluation getNewTuple(I);

Algorithm 1. findKernelBasis

algebra over $\mathbb{K}$, which contains z. The test at Line 12 fails because the vector $b_1 = (-1/2, 0, 1)$ of B does not yield a linear dependence over the q_i. Indeed, $-(1/2)q_1 + 0q_2 + q_3 \neq 0$.

Consequently, $\bar{S}$ is enriched with the new tuple $s_1 = (1)$ at Line 13 and one performs a second iteration. One builds the vectors v_1, v_2 and v_3 at Line 7 by evaluating q_1, q_2 and q_3 on $a = 0$ and $a = 1$. Thus $v_1 = (2, 1 + x/2)$, $v_2 = (-z, -xz/2)$ and $v_3 = (1, 1)$. A basis B computed at Line 10 could be $(-1, -1/z, 1)$. This time, the test at Line 12 succeeds since $-q_1 - (1/z)q_2 + q_3 = 0$. The algorithm stops by returning the basis $(-1, -1/z, 1)$.

Proof (Algorithm findKernelBasis*).* **Correctness.** If the algorithm returns, then $\ker \Phi_q \supset \ker \Phi_v$. Moreover, at each loop, one has $\ker \Phi_q \subset \ker \Phi_v$ since $\sum \alpha_i q_i = 0$ implies $\sum \alpha_i \sigma(q_i) = 0$ for any evaluation which does not cancel any denominator. Consequently, if the algorithm stops, $\ker \Phi_q = \ker \Phi_v$, and B is a basis of $\ker \Phi_q$. **Termination.** Assume the algorithm does not terminate. The vector space $\ker \Phi_v$ is smaller at each step since the set $\bar{S}$ grows. By a classical dimension argument, the vector space $\ker \Phi_v$ admits a limit denoted by E, and reaches it after a finite number of iterations. From $\ker \Phi_q \subset \ker \Phi_v$, one has $\ker \Phi_q \subset E$. Since the test at Line 12 always fails (the algorithm does not terminate), $\ker \Phi_q \subsetneq E$.

Take $\alpha \in E \setminus \ker \Phi_q$ and consider the set S obtained by applying Lemma 3 with $\bar{q} = \sum_{i=1}^{e} \alpha_i q_i$. Since the algorithm does not terminate, $\bar{S}$ will eventually contain S. By construction of v, one has $\sum_{i=1}^{e} \alpha_i \sigma_{y^0}(q_i) = 0 = \sigma_{y^0}(\bar{q})$ for all y^0 in S. By Lemma 3, one has $\bar{q} = \sum_{i=1}^{e} \alpha_i q_i = 0$ which proves that $\alpha \in \ker \Phi_q$. Contradiction since $\alpha \notin \ker \Phi_q$. □

Input: Two lists of variables Y and X
Input: A vector $q = (q_1, \ldots, q_e)$ of e rational fractions in $\mathbb{K}(Y)[X]$
Output: A basis of $\ker \Phi_q$

1 **begin**
2   For each $1 \le i \le e$, denote the reduced fraction q_i as f_i/g_i
       with $f_i \in \mathbb{K}[Y, X]$, $g_i \in \mathbb{K}[Y]$;
3    $D \leftarrow [g_1, \ldots, g_e]$;
4    $I \leftarrow \mathsf{newTupleIteratorNotCancelling}(Y, D)$;
5    $M \leftarrow$ the $0 \times e$ matrix;
6   **while** *true* **do**
7       $y^0 \leftarrow \mathsf{getNewTuple}(I)$;
8   // evaluate the vector q at y^0
9       $\bar{q} \leftarrow \sigma_{y^0}(q)$;
10   build $L = [\omega_1, \ldots, \omega_l]$ the list of monomials involved in $\bar{q}$;
11   build the $l \times e$ matrix $N = (N_{ij})$ where N_{ij} is the coefficient of $\bar{q}_j$ in the monomial ω_i ;
12      $M \leftarrow \begin{pmatrix} M \\ N \end{pmatrix}$;
13   Compute a basis B of the right kernel of M;
14   **if** $\sum_{i=1}^{e} b_i q_i = 0$ for all $b = (b_1, \ldots, b_e) \in B$ **then return** B;

Algorithm 2. findKernelBasisMatrix

2.3 A Variant of findKernelBasis

A variant of algorithm findKernelBasis consists in incrementally building a matrix M with values in $\mathbb{K}$, encoding all the evaluations. Each column of M corresponds to a certain q_j, and each line corresponds to a couple (s, m), where s is an evaluation point, and m is a monomial of $\mathbb{K}[X]$. This yields Algorithm findKernelBasisMatrix.

Example 3. Apply Algorithm findKernelBasisMatrix on Example 2. The only difference with Algorithm findKernelBasis is that the vectors v_i are stored vertically in a matrix M that grows at each iteration. At the end of the first iteration of the while loop, one has $y^0 = (0)$, $\bar{q} = (2, -z, 1)$, $L = [1]$, and $M = N = \begin{pmatrix} 2 & -z & 1 \end{pmatrix}$, and $B = \{(-1/2, 0, 1), (z/2, 1, 0)\}$. Another iteration is needed since $-(1/2)q_1 + 0q_2 + q_3 \ne 0$.

At the end of the second iteration, one has $y^0 = (1)$, $\bar{q} = (1 + x/2, -xz/2, 1)$,
$L = [1, x]$, $N = \begin{pmatrix} 1 & 0 & 1 \\ 1/2 & -z/2 & 0 \end{pmatrix}$, and $M = \begin{pmatrix} 2 & -z & 1 \\ 1 & 0 & 1 \\ 1/2 & -z/2 & 0 \end{pmatrix}$. A basis B is
$(-1, -1/z, 1)$ and the algorithm stops since $-q_1 - (1/z)q_2 + q_3 = 0$.

Proof (Algorithm findKernelBasisMatrix*).* The proof is identical to the proof of findKernelBasis. Indeed, the vector v_i in findKernelBasis is encoded vertically in the ith column of the M matrix in findKernelBasisMatrix. Thus, the kernel of Φ_v in findKernelBasis and the kernel of M in findKernelBasisMatrix coincide. $\qquad \square$

In terms of efficiency, Algorithm findKernelBasisMatrix is far from optimal for many reasons. First, the number of lines of the matrix M grows excessively. Second, a basis B has to be computed at each step of the loop. Third, the algorithm needs to know all the rational functions in advance, which forbids an incremental approach. Next section addresses those issues.

3   Incremental Computation of Linear Dependences

The main idea for building incrementally the linear dependences is presented in Algorithm incrementalFindDependence. It is a sub-algorithm of Algorithms findFirstDependence and findAllFirstIntegrals. Assume we have e linearly independent rational functions $q_1, \ldots, q_e$ and a new rational function q_{e+1}: either the $q_1, \ldots, q_{e+1}$ are also linearly independent, or there exists $(\alpha_1, \ldots, \alpha_{e+1}) \in \mathbb{K}^{e+1}$ such that $\sum_{i=1}^{e+1} \alpha_i q_i = 0$ with $\alpha_{e+1} \neq 0$. It suffices to iterate this idea by increasing the index e until a linear dependence is found. When such a dependence has been found, one can either stop, or continue to get further dependences.

3.1   Algorithm incrementalFindDependence

The main point is to detect and store the property that the $q_1, \ldots, q_e$ are linearly independent, hence the following definition. Moreover, for efficiency reasons, one should be able to update this property at least cost when a new polynomial q_{e+1} is considered.

Definition 3 (Evaluation matrix). *Consider e rational functions $q_1, \ldots, q_e$ in $\mathbb{K}(Y)[X]$, and e couples $(s_1, \omega_1), \ldots, (s_e, \omega_e)$ where each s_i is taken in $\mathbb{N}^m$ and each ω_i is a monomial in X. Assume that none of the s_i cancels any denominator of the q_i. Consider the $e \times e$ matrix M with coefficients in $\mathbb{K}$, defined by $M_{ij} = \mathsf{coeff}(\sigma_{s_i}(q_j), \omega_i)$ where $\mathsf{coeff}(p, m)$ is the coefficient of the monomial m in the polynomial p. The matrix M is called the* evaluation matrix *of $q_1, \ldots, q_e$ w.r.t $(s_1, \omega_1), \ldots, (s_e, \omega_e)$.*

Example 4. Recall $q_1 = \frac{ax+2}{a+1}$ and $q_2 = \frac{z(a-1-ax)}{1+a}$ from Example 2. Consider $(s_1, \omega_1) = (0, 1)$ and $(s_2, \omega_2) = (1, 1)$. Thus, $\sigma_{s_1}(q_1) = 2$, $\sigma_{s_1}(q_2) = -z$, $\sigma_{s_2}(q_1) = x/2 + 1$ and $\sigma_{s_2}(q_2) = -zx/2$. Then, the evaluation matrix for $(s_1, \omega_1), (s_2, \omega_2)$ is $M = \begin{pmatrix} 2 & -z \\ 1 & 0 \end{pmatrix}$. If w_2 were the monomial x instead of 1, the evaluation matrix would be $M = \begin{pmatrix} 2 & -z \\ 1/2 & -z/2 \end{pmatrix}$. In both cases, the matrix M is invertible.

The evaluation matrices computed in Algorithms incrementalFindDependence, findFirstDependence and findAllFirstIntegrals will be kept invertible, to ensure that some fractions are linearly independent (see Proposition 1 below).

Proposition 1. *Keeping the same notations as in Definition 3, if the matrix M is invertible, then the rational functions $q_1, \ldots, q_e$ are linearly independent.*

Input: Two lists of variables Y and X, and a list D of elements of $\mathbb{K}[Y]$
Input: A list $Q = [q_1, \ldots, q_e]$ of e rational functions in $\mathbb{K}(Y)[X]$
Input: A list $E = [(s_1, \omega_1), \ldots, (s_e, \omega_e)]$, with $s_i \in \mathbb{N}^m$ and ω_i a monomial in X
Input: M an invertible eval. matrix of $q_1, \ldots, q_e$ w.r.t. $(s_1, \omega_1), \ldots, (s_e, \omega_e)$
Input: q_{e+1} a rational function
Assumption: denote $q_i = f_i / g_i$ with $f_i \in \mathbb{K}[X,Y]$ and $g_i \in \mathbb{K}[Y]$ for
$\qquad\qquad 1 \leq i \leq e+1$. Each g_i divides a power of elements of D.
$\qquad\qquad$ Moreover, $\sigma_{s_i}(d_j) \neq 0$ for any s_i and $d_j \in D$.
Output: Case 1 : $(\alpha_1, \ldots, \alpha_e, 1)$ s.t. $q_{e+1} + \sum_{i=1}^{e} \alpha_i q_i = 0$
Output: Case 2 : $M', (s_{e+1}, \omega_{e+1})$ such that M' is the evaluation matrix of
$\qquad\qquad q_1, \ldots, q_{e+1}$ w.r.t. $(s_1, \omega_1), \ldots, (s_{e+1}, \omega_{e+1})$, with M' invertible

1 **begin**
2 $\quad b \leftarrow (\ldots, \mathsf{coeff}(\sigma_{s_j}(q_{e+1}), \omega_j), \ldots)_{1 \leq j \leq e}$;
3 $\quad$ solve $M\alpha = -b$ in the unknown $\alpha = (\alpha_1, \ldots, \alpha_e)$;
4 $\quad h \leftarrow q_{e+1} + \sum_{i=1}^{e} \alpha_i q_i$;
5 $\quad$ **if** $h = 0$ **then**
6 $\quad\quad \llcorner$ **return** $(\alpha_1, \ldots, \alpha_e, 1)$;   // Case 1: a linear dependence has been found
7 $\quad$ **else**
8 $\quad\quad I \leftarrow \mathsf{newTupleIteratorNotCancelling}(Y, D)$;
9 $\quad\quad$ **repeat** $s_{e+1} \leftarrow \mathsf{getNewTuple}(I)$ **until** $\sigma_{s_{e+1}}(h) \neq 0$;
10 $\quad\quad$ choose a monomial ω_{e+1} such that $\mathsf{coeff}(\sigma_{s_{e+1}}(h), \omega_{e+1}) \neq 0$;
11 $\quad\quad l \leftarrow (\mathsf{coeff}(\sigma_{s_{e+1}}(q_1), \omega_{e+1}) \quad \cdots \quad \mathsf{coeff}(\sigma_{s_{e+1}}(q_{e+1}), \omega_{e+1}))$;
12 $\quad\quad M' \leftarrow \left(\dfrac{M \mid b}{l} \right)$;
13 $\quad\quad$ **return** $M', (s_{e+1}, \omega_{e+1})$ // Case 2: $q_1, \ldots, q_{e+1}$ are linearly independent

Algorithm 3. incrementalFindDependence

Proof. Consider $\alpha = (\alpha_1, \ldots, \alpha_e)$ such that $\sum_{i=1}^{e} \alpha_i q_i = 0$. One proves that $\alpha = 0$. For each $1 \leq j \leq e$, one has $\sum_{i=1}^{e} \alpha_i \sigma_{s_j}(q_i) = 0$, and consequently $\sum_{i=1}^{e} \alpha_i \mathsf{coeff}(\sigma_{s_j}(q_i), \omega_j) = 0$. This can be rewritten as $M\alpha = 0$, which implies $\alpha = 0$ since M is invertible. $\qquad\square$

By Proposition 1, each evaluation matrix of Example 4 proves that q_1 and q_2 are linearly independent. In some sense, an invertible evaluation matrix can be viewed as a certificate (in the the computational complexity theory terminology) that some fractions are linearly independent.

Proposition 2. *Take the same notations as in Definition 3 and assume the evaluation matrix M is invertible. Consider a new rational function q_{e+1}. If the rational functions $q_1, \ldots, q_{e+1}$ are linearly independent then one necessarily has $q_{e+1} + \sum_{i=1}^{e} \alpha_i q_i = 0$ where α is the unique solution of $M\alpha = -b$, with $b = (\ldots, \mathsf{coeff}(\sigma_{s_j}(q_{e+1}), \omega_j), \ldots)_{1 \leq j \leq e}$.*

Proof. Since M is invertible and by Proposition 1, any linear dependence involves q_{e+1} with a nonzero coefficient, assumed to be 1. Assume $q_{e+1} + \sum_{i=1}^{e} \alpha_i q_i = 0$. Then for each j, one has $\sum_{i=1}^{e} \alpha_i \mathsf{coeff}(\sigma_{s_j}(q_i), \omega_j) = -\mathsf{coeff}(\sigma_{s_j}(q_{e+1}), \omega_j)$ which can be rewritten as $M\alpha = -b$. $\qquad\square$

Example 5. Consider the q_i of Example 2. Take $D = [a+1]$, and $(s_1, \omega_1) = (0, 1)$, and the 1×1 matrix $M = (2)$ which is the evaluation matrix of q_1 w.r.t. (s_1, ω_1). Apply algorithm incrementalFindDependence on Y, X, D, $[q_1]$, $[(s_1, \omega_1)]$, M and q_2. The vector b at Line 2 equals $(-z)$ since $q_2 = z(a - ax - 1)/(a+1)$ evaluates to $-z$ when $a = 0$. Solving $M\alpha = -b$ yields $\alpha = (z/2)$. Then $h = q_2 + (z/2)q_1 = \frac{az(2-x)}{2(a+1)} \neq 0$. One then iterates the *repeat until* loop until h evaluates to a non zero polynomial. The value $a = 0$ is skipped, and the *repeat until* loop stops with $s_2 = 1$. Choosing the monomial $w_2 = 1$ yields the matrix $M' = \begin{pmatrix} 2 & -z \\ 1 & 0 \end{pmatrix}$.

Proof (Algorithm incrementalFindDependence). **Correctness.** Assume the fractions $q_1, \ldots, q_{e+1}$ are not linearly independent. Then Proposition 2 ensures that h must be zero and Case 1 is correct. It the $q_1, \ldots, q_{e+1}$ are linearly independent, then necessarily h is non zero. Assume the *repeat until* loop terminates (see proof below). Since $\sigma_{s_{e+1}}(h) \neq 0$, a monomial ω_{e+1} such that $\mathsf{coeff}(\sigma_{s_{e+1}}(h), \omega_{e+1}) \neq 0$ can be chosen. By construction, the matrix M' is the evaluation matrix of $q_1, \ldots, q_{e+1}$ w.r.t $(s_1, \omega_1), \ldots, (s_{e+1}, \omega_{e+1})$. One just needs to prove that M' is invertible to end the correctness proof. Assume $M'v = 0$ with a non zero vector $v = (\alpha_1, \ldots, \alpha_e, \beta)$. If $\beta = 0$, then $M\alpha = 0$ where $\alpha = (\alpha_1, \ldots, \alpha_e)$ and $\alpha \neq 0$ since $v \neq 0$ and $\beta = 0$. Since M is invertible, $\alpha = 0$ hence a contradiction. If $\beta \neq 0$, then one can assume $\beta = 1$. The e first lines of $M'v = 0$ imply $M\alpha = -b$ where α is the vector computed in the algorithm. The last line of $M'v = 0$ implies that $l(\alpha_1, \ldots, \alpha_e, 1) = 0$, which implies $\mathsf{coeff}(\sigma_{s_{e+1}}(h, \omega_{e+1})) = 0$ and contradicts the choice of ω_{e+1}.

Termination. One only needs to show that the *repeat until* loop terminates. This follows from Lemma 3 in the case of the single fraction q_{e+1}.   □

3.2 Improvement Using a LU-decomposition

In order to optimize the solving of $M\alpha = -b$ in incrementalFindDependence, one can require a LU-decomposition of the evaluation matrix M. The specification of Algorithm incrementalFindDependence can be slightly adapted by passing a LU-decomposition of $M = LU$ (with L lower triangular with a diagonal of 1, and U upper triangular), and by returning a LU-decomposition of $M' = L'U'$ in Case 2. Note that a PLU-decomposition (where P is a permutation matrix) is not needed in our case as shown by Proposition 4 below.

Proposition 3 (Solving α). *Solving $M\alpha = -b$ is equivalent to solving the two triangular systems $Ly = -b$, then $U\alpha = y$.*

Proposition 4 (The LU-decomposition of M'). *The LU-decomposition of M' can be obtained by:*

- *solve $\gamma U = l_{1:e}$ (in the variable γ), where $l_{1:e}$ denotes the e first components of the line vector l computed in findFirstDependence*
- *solve $Ly = -b$ (in the variable y)*
- $L' \leftarrow \begin{pmatrix} L & 0_{e \times 1} \\ \gamma & 1 \end{pmatrix}$ *and* $U' \leftarrow \begin{pmatrix} U & -y \\ 0_{1 \times e} & l_{e+1} + \gamma y \end{pmatrix}$

Input: Two lists of variables Y and X
Input: A list D of elements of $\mathbb{K}[Y]$
Input: A finite iterator J which outputs rational functions $q_1, q_2, \ldots$ in $\mathbb{K}(Y)[X]$
Assumption: Denote the reduced fraction $q_i = f_i/g_i$ with $f_i \in \mathbb{K}[X, Y]$ and
$\qquad\qquad$ $g_i \in \mathbb{K}[Y]$. Each g_i divides a power of elements of D.
Output: Case 1: a shortest linear dependence i.e. a vector $(\alpha_1, \ldots, \alpha_e)$ and a
$\qquad\qquad$ list $[q_1, \ldots, q_e]$ with $\sum_{i=1}^{e} \alpha_i q_i = 0$ and e the smallest possible.
Output: Case 2: FAIL if no linear dependence exists
1 **begin**
2 $\quad$ $M \leftarrow$ the 0×0 matrix ; $\qquad\qquad\qquad\qquad$ // M is an evaluation matrix
3 $\quad$ $Q \leftarrow$ the empty list $[]$; $\qquad\qquad\qquad\qquad\quad$ // Q is the list of the q_i
4 $\quad$ // E is a list of (s, ω), where s is an evaluation and w is a monomial
5 $\quad$ $E \leftarrow$ the empty list $[]$;
6 $\quad$ $bool \leftarrow$ false;
7 $\quad$ **while** ($bool$=false) **and** (the iterator J is not empty) **do**
8 $\quad\quad$ $q \leftarrow$ getNewRationalFunction(J);
9 $\quad\quad$ $r \leftarrow$ incrementalFindDependence(Y, X, D, Q, E, M, q);
10 $\quad\quad$ append q at the end of Q;
11 $\quad\quad$ **if** r is a linear dependence **then** $\alpha \leftarrow r$; $\quad bool \leftarrow$ true ;
12 $\quad\quad$ **else** $M', (s, \omega) \leftarrow r$; $\quad M \leftarrow M'$; $\quad$ append (s, ω) at the end of E;
13 $\quad$ **if** $bool$=true **then** **return** (α, Q) **else** **return** FAIL ;

Algorithm 4. findFirstDependence

3.3 Finding the First Linear Dependence

The main advantage of Algorithm incrementalFindDependence is to limit the number of computations if one does not know in advance the number of rational fractions needed for having a linear dependence. Imagine the rational functions are provided by a finite iterator (i.e. a iterator that outputs a finite number of fractions), one can build the algorithm findFirstDependence which terminates on the first linear dependence it finds, or fails if no linear dependence exists.

Example 6. Let us apply Algorithm findFirstDependence on Example 2. The first fraction q to be considered is $\frac{ax+2}{a+1}$. The call to incrementalFindDependence returns the 1×1 matrix (2) and the couple $(s_1, \omega_1) = (0, 1)$. One can check that q evaluated at $a = 0$ yields 2 and that (2) is indeed the evaluation matrix for the couple $(s_1, \omega_1) = (0, 1)$.

Continue with the second iteration. One now considers $q = \frac{z(a - ax - 1)}{a+1}$. Example 5 shows that the call to incrementalFindDependence returns the 2×2 invertible matrix $\begin{pmatrix} 2 & -z \\ 1 & 0 \end{pmatrix}$ and the couple $(s_2, \omega_2) = (1, 1)$.

Finally, the third iteration makes a call to incrementalFindDependence. Line 2 builds the vector $b = (1, 1)$. Line 3 solves $M\alpha = -b$, yielding $\alpha = (-1, -1/z)$. Line 4 builds $h = q_3 - q_1 - (1/z)q_2$ which is zero, so incrementalFindDependence returns at Line 6. As a consequence, findFirstDependence detects that a linear dependence has been found at Line 11 and returns $((-1, -1/z), [q_1, q_2, q_3])$ at Line 13.

Proof (Algorithm findFirstDependence*).* **Termination.** The algorithm obviously terminates since the number of loops is bounded by the number of elements output by the iterator J.

Correctness. One first proves the following loop invariant: M is invertible, and M is the evaluation matrix of Q w.r.t. E. The invariant is true when first entering the loop (even if the case is a bit degenerated since M, Q and E are all empty). Assume the invariant is true at some point. The call to incrementalFindDependence either detects a linear dependence α, or returns an evaluation matrix M' and a couple (s, ω). In the first case, M, Q and E are left unmodified so the invariant remains true. In the second case, M, Q and E are modified to incorporate the new fraction q, and the invariant is still true thanks to the specification of incrementalFindDependence. The invariant is thus proven.

When the algorithm terminates, it returns either (α, Q) or FAIL. If it returns (α, Q), this means that the variable *bool* has changed to true at some point. Consequently a linear dependence α has been found, and the algorithm returns (α, Q). The dependence is necessarily the one with the smallest e because of the invariant (ensuring M is invertible) and Proposition 1. This proves the Case 1 of the specification.

If the algorithms returns FAIL, the iterator has been emptied, and the variable *bool* is still false. Consequently, all elements of the iterator have been stored in Q, and because of the invariant and Proposition 1, the elements of Q are linearly independent. This proves the Case 2 of the specification.   □

Remark 1. Please note that Algorithm findFirstDependence can be used with an infinite iterator (i.e. an iterator that never gets empty). However, Algorithm findFirstDependence becomes a semi-algorithm since it will find the first linear dependence if it exists, but will never terminates if no such dependence exists.

3.4   Complexity of the Linear Algebra

When findFirstDependence terminates, it has solved at most e square systems with increasing sizes from 1 to e. Assuming the solving of each system $M\alpha = b$ has a complexity of $O(n^\omega)$ [7] arithmetic operations, where n is the size of the matrix and ω is the exponent of linear algebra, the total number of arithmetic operations for the linear algebra is $O(e^{\omega+1})$. If the LU-decomposition variant is used in Algorithm incrementalFindDependence, then the complexity of drops to $O(e^3)$, since solving a triangular system of size n can be made in $O(n^2)$ arithmetic operations. As for the space complexity of algorithm findFirstDependence, it is $O(e^2)$, whether using the LU-decomposition variant or not.

4   Application to Finding First Integrals

In this section, we look for first integrals for ODE systems. Roughly speaking, a first integral is an expression which is constant over time along any solution of the ODE system. This is a difficult problem and we will make several simplifying hypotheses. First, we work in the context of differential algebra, and assume that the ODE system is given by polynomial differential equations. Second, we will only look for polynomial first integrals.

4.1 Basic Differential Algebra

This section is mostly borrowed from [6] and [2]. It has been simplified in the case of a single derivative. The reference books are [11] and [10]. A *differential ring* R is a ring endowed with an[2] abstract *derivation* δ i.e. a unary operation which satisfies the axioms $\delta(a + b) = \delta(a) + \delta(b)$ and $\delta(a\,b) = \delta(a)\,b + a\delta(b)$ for all $a, b \in R$. This paper considers a differential polynomial ring R in n *differential indeterminates* $u_1, \ldots, u_n$ with coefficients in the field $\mathbb{K}$. Moreover, we assume that $\mathbb{K}$ is a field of constants (i.e. $\delta k = 0$ for any $k \in \mathbb{K}$). Letting $U = \{u_1, \ldots, u_n\}$, one denotes $R = \mathbb{K}\{U\}$, following Ritt and Kolchin. The derivation δ generates a monoid w.r.t. the composition operation. It is denoted: $\Theta = \{\delta^i, i \in \mathbb{N}\}$ where $\mathbb{N}$ stands for the set of the nonnegative integers. The elements of Θ are the *derivation operators*. The monoid Θ acts multiplicatively on U, giving the infinite set ΘU of the *derivatives*.

If A is a finite subset of R, one denotes (A) the smallest ideal containing A w.r.t. the inclusion relation and $[A]$ the smallest differential ideal containing A. Let $\mathfrak{A}$ be an ideal and $S = \{s_1, \ldots, s_t\}$ be a finite subset of R, not containing zero. Then $\mathfrak{A} : S^\infty = \{p \in R \mid \exists\, a_1, \ldots, a_t \in \mathbb{N},\ s_1^{a_1} \cdots s_t^{a_t}\, p \in \mathfrak{A}\}$ is called the *saturation* of $\mathfrak{A}$ by the multiplicative family generated by S. The saturation of a (differential) ideal is a (differential) ideal [10, chapter I, Corollary to Lemma 1].

Fix a *ranking*, i.e. a total ordering over ΘU satisfying some properties [10, chapter I, section 8]. Consider some differential polynomial $p \notin \mathbb{K}$. The highest derivative v w.r.t. the ranking such that $\deg(p, v) > 0$ is called the *leading derivative* of p. It is denoted $\mathsf{ld}\, p$. The leading coefficient of p w.r.t. v is called the *initial* of p. The differential polynomial $\partial p / \partial v$ is called the *separant* of p. If C is a finite subset of $R \setminus \mathbb{K}$ then I_C denotes its set of initials, S_C denotes its set of separants and $H_C = I_C \cup S_C$.

A differential polynomial q is said to be *partially reduced* w.r.t. p if it does not depend on any proper derivative of the leading derivative v of p. It is said to be *reduced* w.r.t. p if it is partially reduced w.r.t. p and $\deg(q, v) < \deg(p, v)$. A set of differential polynomials of $R \setminus \mathbb{K}$ is said to be *autoreduced* if its elements are pairwise reduced. Autoreduced sets are necessarily finite [10, chapter I, section 9]. To each autoreduced set C, one may associate the set $L = \mathsf{ld}\, C$ of the leading derivatives of C and the set $N = \Theta U \setminus \Theta L$ of the derivatives which are not derivatives of any element of L (the derivatives "under the stairs" defined by C).

In this paper, we need not recall the (rather technical) definition of differential regular chains (see [2, Definition 3.1]). We only need to know that a differential regular chain C is a particular case of an autoreduced set and that membership to the ideal $[C] : H_C^\infty$ can be decided by means of normal form computations, as explained below.

---

[2] In the general setting, differential ring are endowed with finitely many derivations.

4.2 Normal Form Modulo a Differential Regular Chain

All the results of this section are borrowed from [2] and [6]. Let C be a regular differential chain of R, defining a differential ideal $\mathfrak{A} = [C]\!:\!H_C^\infty$. Let $L = \mathsf{ld}\, C$ and $N = \Theta U \setminus \Theta L$. The normal form of a rational differential fraction is introduced in [2, Definition 5.1 and Proposition 5.2], recalled below.

Definition 4. *Let a/b be a rational differential fraction with b regular modulo $\mathfrak{A}$. A normal form of a/b modulo C is any rational differential fraction f/g such that*

1 *f is reduced with respect to C,*
2 *g belongs to $\mathbb{K}[N]$ (and is thus regular modulo $\mathfrak{A}$),*
3 *a/b and f/g are equivalent modulo $\mathfrak{A}$ (in the sense that $a\,g - b\,f \in \mathfrak{A}$).*

Proposition 5. *Let a/b be a rational differential fraction, with b regular modulo $\mathfrak{A}$. The normal form f/g of a/b exists and is unique. The normal form is a $\mathbb{K}$-linear operation. Moreover*

4 *a belongs to $\mathfrak{A}$ if and only if its normal form is zero,*
5 *f/g is a canonical representative of the residue class of a/b in the total fraction ring of $R/\mathfrak{A}$,*
6 *each irreducible factor of g divides the denominator of an inverse of b, or of some initial or separant of C.*

The interest of [6] is that it provides a normal form algorithm which always succeeds (while [2] provides an algorithm which fails when splittings occur).

Recall that the normal form algorithm relies on the computation of inverses of differential polynomials, defined below.

Definition 5. *Let f be a nonzero differential polynomial of R. An* inverse *of f modulo C is any fraction p/q of nonzero differential polynomials such that $p \in \mathbb{K}[N \cup L]$ and $q \in \mathbb{K}[N]$ and $f\,p \equiv q \mod \mathfrak{A}$.*

4.3 Normal Form Modulo a Decomposition

This subsection introduces a new definition which in practice is useful for performing computations modulo a radical ideal $\sqrt{[\Sigma]}$ expressed as an intersection of differential regular chains (i.e. $\sqrt{[\Sigma]} = [C_1]\!:\!H_{C_1}^\infty \cap \cdots \cap [C_f]\!:\!H_{C_f}^\infty$). Such a decomposition can be computed with the RosenfeldGroebner algorithm [1,4].

Definition 6 (Normal form modulo a decomposition). *Let Σ be a set of differential polynomials, such that $\sqrt{[\Sigma]}$ is a proper ideal. Consider a decomposition of $\sqrt{[\Sigma]}$ into differential regular chains $C_1, \ldots, C_f$ for some ranking, that is differential regular chains satisfying $\sqrt{[\Sigma]} = [C_1]\!:\!H_{C_1}^\infty \cap \ldots \cap [C_f]\!:\!H_{C_f}^\infty$. For any differential fraction a/b with b regular modulo each $[C_i]\!:\!H_{C_i}^\infty$, one defines the normal form of a/b w.r.t. to the list $[C_1, \ldots, C_f]$ by the list*

$$[\mathsf{NF}(a/b, C_1), \ldots, \mathsf{NF}(a/b, C_f)]\,.$$

It is simply denoted by $\mathsf{NF}(a/b, [C_1, \ldots, C_f])$.

Since it is an open problem to compute a canonical (e.g. minimal) decomposition of the radical of a differential ideal, the normal form of Definition 6 depends on the decomposition and not on the ideal.

Proposition 6. *With the same notations as in Definition 6, for any polynomial $p \in R$, one has $p \in \sqrt{[\Sigma]} \iff \mathsf{NF}(p, [C_1, \ldots, C_f]) = [0, \ldots, 0]$. Moreover, the normal form modulo a decomposition is $\mathbb{K}$-linear.*

4.4 First Integrals in Differential Algebra

Definition 7 (First integral modulo an ideal). *Let p be a differential polynomial and $\mathfrak{A}$ be an ideal. One says p is a first integral modulo $\mathfrak{A}$ if $\delta p \in \mathfrak{A}$.*

For any ideal $\mathfrak{A}$, the set of the first integrals modulo $\mathfrak{A}$ contains the ideal $\mathfrak{A}$. If $\mathfrak{A}$ is a proper ideal, the inclusion is strict since any element of $\mathbb{K}$ is a first integral. In practice, the first integrals taken in $\mathbb{K}$ are obviously useless.

Example 7 (Pendulum). Take $\mathbb{K} = \mathbb{Q}(m, l, g)$. Consider the ranking $\cdots > \dddot{l} > \dddot{x} > \dddot{y} > \ddot{l} > \dot{x} > \dot{y} > l > x > y$. Recall the pendulum equations Σ in Equations (1). Algorithm RosenfeldGroebner [4] shows that $\sqrt{[\Sigma]} = [C_1] : H_{C_1}^\infty \cap [C2] : H_{C2}^\infty$ where C_1 and C_2 are given by:

- $C_1 = [\dot{\lambda} = -3\frac{\dot{y}gm}{l^2}, \ddot{y}^2 = -\frac{-\lambda y^2 l^2 + \lambda l^4 - y^3 gm + ygml^2}{ml^2}, x^2 = -y^2 + l^2]$;
- $C_2 = [\lambda = -\frac{ygm}{l^2}, x = 0, y^2 = l^2]$.

Remark that the differential regular chain C_2 corresponds to a degenerate case, where the pendulum is vertical since $x = 0$. Further computations show that $\mathsf{NF}(\delta(\frac{m}{2}(\dot{x}^2 + \dot{y}^2) - mg\,y), [C_1, C_2]) = [0, 0])$, proving that $p = \frac{m}{2}(\dot{x}^2 + \dot{y}^2) - mg\,y$ is a first integral modulo $\sqrt{[\Sigma]}$. Remark that $x^2 + y^2$ is also a first integral. This is immediate since $\delta(x^2 + y^2) = \delta(x^2 + y^2 - 1) \in \sqrt{[\Sigma]}$.

Definition 7 is new to our knowledge. It is expressed in a differential algebra context. The end of Section 4.4 makes the link with the definition of first integral in a analysis context, through analytic solutions using results from [5].

Definition 8 (Formal power solution of an ideal). *Consider a n-uple $\bar{u} = (\bar{u}_1(t), \ldots, \bar{u}_n(t))$ of formal power series in t over $\mathbb{K}$. For any differential polynomial, one defines $p(\bar{u})$ as the formal power series in t obtained by replacing each u_i by $\bar{u}_i$ and interpreting the derivation δ as the usual derivation on formal power series. The n-uple $\bar{u} = (\bar{u}_1, \ldots, \bar{u}_n)$ is called a solution of an ideal $\mathfrak{A}$ if $p(\bar{u}) = 0$ for all $p \in \mathfrak{A}$.*

Lemma 4. *Take a differential polynomial p and n-uple $\bar{u} = (\bar{u}_1(t), \ldots, \bar{u}_n(t))$ of formal power series. Then $(\delta p)(\bar{u}) = \frac{\mathrm{d}p(\bar{u})}{\mathrm{d}t}$. If p is a first integral modulo an ideal $\mathfrak{A}$ and $\bar{u}$ is a solution of $\mathfrak{A}$, then the formal power series $p(\bar{u})$ satisfies $\frac{\mathrm{d}p(\bar{u})}{\mathrm{d}t} = 0$.*

Proof. Since δ is a derivation, $(\delta p)(\bar{u}) = \frac{\mathrm{d}p(\bar{u})}{\mathrm{d}t}$ is proved if one proves it when p equals any u_i. Assume that $p = u_i$. Then $(\delta p)(\bar{u}_i) = (\delta u_i)(\bar{u}_i) = \frac{\mathrm{d}p(\bar{u})}{\mathrm{d}t}$. Assume p is a first integral modulo $\mathfrak{A}$ and $\bar{u}$ is a solution of $\mathfrak{A}$. Then $\delta p \in \mathfrak{A}$ and $(\delta p)(\bar{u}) = 0$. Using $(\delta p)(\bar{u}) = \frac{\mathrm{d}p(\bar{u})}{\mathrm{d}t}$, one has $\frac{\mathrm{d}p(\bar{u})}{\mathrm{d}t} = 0$ □

Take a system of differential polynomials Σ. By [5, Theorem and definition 3], a differential polynomial p in R vanishes on all analytic solutions (over some open set with coordinates in the algebraic closure of the base field) of Σ if and only if $p \in \sqrt{[\Sigma]}$. Applying this statement to $p = \delta q$ for some first integral q w.r.t. $\sqrt{[\Sigma]}$, then δq vanishes on all analytic solutions of $\sqrt{[\Sigma]}$ so p is a first integral in the context of analysis, if one excludes the non analytic solutions.

4.5   Algorithm findAllFirstIntegrals

In this section, one considers a proper ideal $\mathfrak{A}$ given as $\mathfrak{A} = [C_1] : H_{C_1}^\infty \cap \cdots \cap [C_f] :$ $H_{C_f}^\infty$ where the C_i are differential regular chains. Denote $M = [C_1, \ldots, C_f]$. Take

---

Input: A list of differential regular chains $M = [C_1, \ldots, C_f]$
Input: A finite iterator J which outputs differential monomials
Output: A $\mathbb{K}$-basis of the linear combinations of momomials of J which are first integrals w.r.t. $[C_1] : H_{C_1}^\infty \cap \cdots \cap [C_f] : H_{C_f}^\infty$

```
1  begin
2  |   result ← the empty list [] ;   D ← the empty list [] ;
3  |   for i = 1 to f do
4  |   |   I ← the inverses of the initials and separants of Cᵢ modulo Cᵢ ;
5  |   |   append to D the denominators of the elements of I ;
6  |   Y ← the list of derivatives occurring in D ;
7  |   X ← the list of dummy variables [d₁, …, d_f] ;
8  |   M ← the 0 × 0 matrix ;   Q ← the empty list [] ;
9  |   E ← the empty list [] ;   F ← the empty list [];
10 |   while the iterator J is not empty do
11 |   |   μ ← getNewDerivativeMonomial(J);
12 |   |   q ← ∑ᶠᵢ₌₁ dᵢNF(δμ, Cᵢ);
13 |   |   append to X the variables of the numerator of q,
   |   |            which are neither in X nor Y ;
14 |   |   r ← incrementalFindDependence(Y, X, D, Q, E, M, q);
15 |   |   if r is a linear dependence then
16 |   |   |   // A new first integral has been found
17 |   |   |   (α₁, …, α_e, 1) ← r ;
18 |   |   |   append α₁μ₁ + ⋯ + α_eμ_e + w to result, where F = (μ₁, …, μ_e);
19 |   |   else
20 |   |   |   append μ to the end of F ;   append q to the end of Q ;
21 |   |   |   M′, (s, w) ← r ;   M ← M′ ;   append (s, w) to the end of E;
22 |   return result ;
```

Algorithm 5. findAllFirstIntegrals

a first integral modulo $\mathfrak{A}$ of the form $p = \sum \alpha_i \mu_i$, where the α_i's are in $\mathbb{K}$ and the μ_i's are monomials in the derivatives. Computing on lists componentwise, we have $0 = \mathsf{NF}(\delta p, M) = \sum \alpha_i \mathsf{NF}(\delta \mu_i, M)$. Consequently, a candidate $\sum \alpha_i \mu_i$ is a first integral modulo $\mathfrak{A}$ if and only if the $\mathsf{NF}(\delta \mu_i, M)$ are linearly dependent over $\mathbb{K}$.

Since the μ_i have no denominators, every irreducible factor of the denominator of any $\mathsf{NF}(w_i, C_j)$ necessarily divides (by Proposition 5) the denominator of the inverse of a separant or initial of C_j. As a consequence, the algorithms presented in the previous sections can be applied since we can precompute factors which should not be cancelled.

Algorithm findAllFirstIntegrals is very close to Algorithm findFirstDependence and only requires a few remarks. Instead of stopping at the first found dependence, it continues until the iterator has been emptied and stores all first dependences encountered. It starts by precomputing a safe set D for avoiding cancelling the denominators of any $\mathsf{NF}(\delta w_i, C_j)$. The algorithm introduces some dummy variables $d_1, \ldots, d_f$ for storing the normal form $\mathsf{NF}(\delta \mu, M)$, which is by definition a list, as the polynomial $d_1 \mathsf{NF}(\delta \mu, C_1) + \cdots + d_f \mathsf{NF}(\delta \mu, C_f)$. This alternative storage allows us to directly reuse Algorithm incrementalFindDependence which expects a polynomial.

Example 8 (Pendulum). Take the same notations as in Example 7. Take an iterator J enumerating the monomials 1, y, x, $\dot{y}$, $\dot{x}$, y^2, xy, y^2, $\dot{y}y$, $\dot{y}x$, $\dot{y}^2$, $\dot{x}y$, $\dot{x}x$, $\dot{x}\dot{y}$ and $\dot{x}^2$. Then Algorithm findAllFirstIntegrals returns the list $[1, x^2 + y^2, \dot{y}y + \dot{x}x, -2gy + \dot{x}^2 + \dot{y}^2]$. Note the presence of $\dot{y}y + \dot{x}x$ which is in the ideal since it is the derivative of $(x^2 + y^2 - 1)/2$.

The intermediate computations are too big be displayed here. As an illustration, one gives the normal forms of δy, $\delta(\dot{y}^2)$, $\delta(\dot{x}\dot{y})$ and $\delta(\dot{x}^2)$ modulo $[C_1, C_2]$ which are respectively

$$[\dot{y}, 0], \quad \left[\frac{2(\lambda y \dot{y} + mg\dot{y})}{m}, 0 \right], \quad \left[\frac{x\dot{y}(\lambda(2y^2 - l^2) + mgy)}{m(y^2 - l^2)}, 0 \right] \text{ and } \left[-\frac{2\lambda y \dot{y}}{m}, 0 \right].$$

When increasing the degree bound, one finds more and more spurious first integrals like $\dot{y}y^2 + \dot{x}xy$ (which is in the ideal) or some powers of the first integral $-2gy + \dot{x}^2 + \dot{y}^2$.

Example 9 (Lotka-Volterra equations).

$$C \begin{cases} \dot{x}(t) &= a\,x(t) - b\,x(t)\,y(t) \\ \dot{y}(t) &= -c\,y(t) + d\,x(t)\,y(t) \end{cases} \qquad \begin{matrix} x(t)\,\dot{u}(t) = \dot{x}(t) \\ y(t)\,\dot{v}(t) = \dot{y}(t) \end{matrix} \qquad (2)$$

Take $\mathbb{K} = \mathbb{Q}(a, b, c, d)$ and the ranking $\cdots > \dot{u} > \dot{v} > \dot{x} > \dot{y} > u > v > x > y$. One can show that C is a differential regular chain for the chosen ranking. The two leftmost equations of C corresponds to the classical Lotka-Volterra equations, and the two rightmost ones encode the logarithms of $x(t)$ and $y(t)$ in a polynomial way. A call to findAllFirstIntegrals with the monomials of degree at most 1 built over x, y, u, v yields $[1, -\frac{av}{d} - \frac{cu}{d} + \frac{by}{d} + x]$ which corresponds to the usual first integral $-a\ln(y(t)) - c\ln(x(t) + by(t) + dx(t)$.

Remark 2. The choice of the degree bounds and the candidate monomials in the first integrals is left to the user, through the use of the iterator J. This makes our algorithm very flexible especially if the user has some extra knowledge on the first integrals or is looking for specific ones. Finally, this avoids the difficult problem of estimating degree bounds, which can be quite high. Indeed, the simple system $\dot{x} = x, \dot{y} = -ny$, where n is any positive integer, admits $x^n y$ as a first integral, which is minimal in terms of degrees.

4.6 Complexity

The complexity for the linear algebra part of Algorithm findAllFirstIntegrals is the same as for Algorithm findFirstDependence: it is $O(e^3)$, where e is the cardinal of the iterator J, if one uses the LU-decomposition variant. However, e can be quite large in practice. For example, if one considers the monomials of total degree at most d involving s derivatives, then e is equal to $\binom{s+d}{s}$.

References

1. Boulier, F.: The BLAD libraries (2004).
   http://cristal.univ-lille.fr/~boulier/BLAD
2. Boulier, F., Lemaire, F.: A normal form algorithm for regular differential chains. Mathematics in Computer Science **4**(2–3), 185–201 (2010)
3. Boulier, F.: Efficient computation of regular differential systems by change of rankings using Kähler differentials. Technical report, Université Lille I, 59655, Villeneuve d'Ascq, France (November 1999) Ref. LIFL 1999–14, presented at the MEGA 2000 conference. http://hal.archives-ouvertes.fr/hal-00139738
4. Boulier, F., Lazard, D., Ollivier, F., Petitot, M.: Computing representations for radicals of finitely generated differential ideals. Applicable Algebra in Engineering, Communication and Computing 20(1), 73–121 (2009); (1997 Techrep. IT306 of the LIFL)
5. Boulier, F., Lemaire, F.: A computer scientist point of view on Hilbert's differential theorem of zeros. Submitted to Applicable Algebra in Engineering, Communication and Computing (2007)
6. Boulier, F., Lemaire, F., Sedoglavic, A.: On the Regularity Property of Differential Polynomials Modulo Regular Differential Chains. In: Gerdt, V.P., Koepf, W., Mayr, E.W., Vorozhtsov, E.V. (eds.) CASC 2011. LNCS, vol. 6885, pp. 61–72. Springer, Heidelberg (2011)
7. Bunch, J.R., Hopcroft, J.E.: Triangular factorization and inversion by fast matrix multiplication. Mathematics of Computation **28**(125), 231–236 (1974)
8. Faugère, J.C., Gianni, P., Lazard, D., Mora, T.: Efficient computation of Gröbner bases by change of orderings. Journal of Symbolic Computation **16**, 329–344 (1993)
9. Gathen, J.V.Z., Gerhard, J.: Modern Computer Algebra, 3rd edn. Cambridge University Press, New York (2013)
10. Kolchin, E.R.: Differential Algebra and Algebraic Groups. Academic Press, New York (1973)
11. Ritt, J.F.: Differential Algebra. Dover Publications Inc., New York (1950)
12. Stoutemyer, D.R.: Multivariate partial fraction expansion. ACM Commun. Comput. Algebra **42**(4), 206–210 (2009)

Simplification of Cylindrical Algebraic Formulas

Changbo Chen[1] and Marc Moreno Maza[1,2]

[1] Chongqing Key Laboratory of Automated Reasoning and Cognition, Chongqing Institute of Green and Intelligent Technology, Chinese Academy of Sciences, Chongqing, China
[2] ORCCA, University of Western Ontario, London, Canada
changbo.chen@hotmail.com, moreno@csd.uwo.ca

Abstract. For a set S of cells in a cylindrical algebraic decomposition of $\mathbb{R}^n$, we introduce the notion of generalized cylindrical algebraic formula (GCAF) associated with S. We propose a multi-level heuristic algorithm for simplifying the cylindrical algebraic formula associated with S into a GCAF. The heuristic strategies are motivated by solving examples coming from the application of automatic loop transformation. While the algorithm works well on these examples, its effectiveness is also illustrated by examples from other application domains.

1 Introduction

Cylindrical algebraic decomposition (CAD), introduced by G.E. Collins [8], is a fundamental tool in real algebraic geometry. One of its main applications, also the initial motivation, is to solve quantifier elimination problems in the first order theory of real closed fields. Since its introduction, CAD has been improved by many authors and applied to numerous applications. The implementation of CAD is now available in different software, such as QEPCAD, Mathematica, Redlog, SyNRAC, `RegularChains`, and many others.

A CAD of $\mathbb{R}^n$ decomposes $\mathbb{R}^n$ into disjoint connected semi-algebraic sets, called cells, such that any two cells are cylindrically arranged, which implies that the projections of any two cells onto any $\mathbb{R}^k$, $1 \leq k < n$, are either identical or disjoint. For a given semi-algebraic set S, one can compute a CAD $\mathcal{C}$ such that S can be written as a union of cells in $\mathcal{C}$. Each cell is represented by a cylindrical algebraic formula (CAF) [14], whose zero set is the cell.

The CAF $\phi_c(x_1, \ldots, x_n)$ representing a CAD cell c of $\mathbb{R}^n$ is a conjunction of finitely many atomic formulas of the form $x_i \ \sigma \ \mathrm{Root}_{x_i,k}(p)$, where $p \in \mathbb{R}[x_1, \ldots, x_i]$ and $\mathrm{Root}_{x_i,k}(p)$ denotes the k-th real root (counting multiplicities) of p treated as a univariate polynomial in x_i. The precise definition of CAF is given in Section 2. The CAF $\phi_c(x_1, \ldots, x_n)$ has a very nice property, namely the projection of c onto $\mathbb{R}^j$, $1 \leq j < n$, is exactly the zero set of the sub-formula of ϕ_c, which is obtained by taking the conjunction of all atomic formulas in ϕ_c involving only the variables $x_1, \ldots, x_j$. Let S be a set of cells $c_1, \ldots, c_t$ from a CAD $\mathcal{C}$. Denote by ϕ_S the zero set of S, thus we have $\phi_S := \vee_{i=1}^t \phi_{c_i}$. The formula ϕ_S is also called a *cylindrical algebraic formula*.

© Springer International Publishing Switzerland 2015
V.P. Gerdt et al. (Eds.): CASC 2015, LNCS 9301, pp. 119–134, 2015.
DOI: 10.1007/978-3-319-24021-3_9

A CAF is a special extended Tarski formula, see [2]. While a Tarski formula is often the default output of quantifier elimination procedures, a CAF is also important for several reasons. Firstly, computing CAFs can be done by means of a CAD procedure without introducing additional augmented projection factors, which can bring substantial savings in terms of computation resources. Secondly, CAFs have a nice structure: the projection of a CAF onto any lower-dimensional space can be easily read off from the CAF itself, as mentioned before. Moreover, since a CAF is used to describe CAD cells, it naturally exhibits a polychotomous structure. This property is usually not true for Tarski formula output. Thirdly, each atomic formula of a CAF has the convenient format $x \, \sigma \, E$, where E is an indexed root expression. This explicit expression is particularly useful in applications which care about the specific value of each coordinate, like in loop transformations of computer program [12,11]. Last but not least, performing set-theoretical operations on CAFs can be done efficiently, without explicit conversion to Tarski formulas [14]. This latter property supports an incremental algorithm for computing CADs [15].

While CAFs have many advantages, they have also their own drawbacks. Firstly, indexed root expressions are usually less handy to manipulate than polynomial expressions. This is because a polynomial function is defined everywhere while an indexed root expression is usually defined on a particular set. Secondly, due to numerous CAD cells being generated, a CAD-based QE solver usually outputs very lengthy CAFs, which could make the output formula not easy to use. For the particular application of loop transformation of computer programs, too many case distinctions might substantially increase the arithmetic cost of evaluating the transformed program as well as the number of misses in accessing cache memories. Therefore, simplification of CAFs is clearly needed. However, we have not seen much literature devoted to this topic except Chapter 8 of Brown's PhD thesis [2], The differences between Brown's approach and the one proposed is the present paper are discussed in Section 6. We remark that the `Reduce` command of `Mathematica` does perform some simplification before outputting CAFs. See Section 5 for an experimental comparison with `Mathematica`.

Since CAFs are generated from CAD cells, it is a natural idea to make use of the CAD data structure to simplify CAFs. In this paper, we produce a multi-level merging procedure for simplifying CAFs by exploiting structural properties of the CAD from which those CAFs are being generated. Although this procedure aims at improving the output of CAD solvers based on regular chains, it is also applicable to other CAD solvers. The merging procedure, presented formally in Section 4, consists of several reasonable and workable heuristics, most of which are motivated by solving examples taken from [12,7]. See Section 3 for details. The simplification procedure has four levels, where an upper level never produces more conjunctive clauses than the lower levels. The first two levels merge adjacent CAD cells, whereas the last two levels attempt to simplify a CAF into a single conjunctive clause, which is usually expected in the application of loop transformation. Thus the first two levels are expected to be effective for general QE problems whereas the last two are expected to be effective for QE problems arising from loop transformation. This expectation is justified also by the experimentation in Section 5.

The method has been implemented and new options are added to both the CylindricalAlgebraicDecompose and QuantifierElimination commands of the Regular Chains library. The effectiveness of this algorithm is illustrated by examples in Sections 3 and 5. The experimentation shows that our heuristics work well. The running time overhead of simplification compared to the running time of the quantifier elimination procedure itself is negligible in the first two levels and acceptable in the advanced levels of the proposed heuristics.

There have already been a few works on the simplification of Tarski formulas, see for example [10,4,3,13]. Our work is concerned with the simplification of extended Tarski formulas, which allow indexed root expressions besides polynomial constraints. Moreover, the simplification goal here is to reduce as much as possible the number of conjunctive CAF clauses while still maintaining the feature of case distinctions. We emphasize that the motivation and the main targeting application of this work is to unify the CAFs generated in the application of loop transformation. In such applications, explicit bounds of loop indices are needed and the number of case distinctions is expected to be as small as possible in order to reduce the code size.

2   Preliminary

In this section, we first review the notion of cylindrical algebraic decomposition and cylindrical algebraic formula (CAF). Then we define the notion of generalized CAF in order to represent the combination of CAFs.

Real Algebraic Function. Let $S \subset \mathbb{R}^n$. Let $f(x_1, \ldots, x_n, y) \in \mathbb{R}[x_1, \ldots, x_n, y]$. Let k be a positive integer. Assume that for every point α of S, the univariate polynomial $f(\alpha, y)$ has at least k real roots ordered by increasing value, counting multiplicities. Let $\mathrm{Root}_{y,k}(f)$ be a function which maps every point α of S to the k-th real root of $f(\alpha, y)$. The function $\mathrm{Root}_{y,k}(f)$ is called a real algebraic function defined on S.

Stack over a Semia-algebraic Set. Let S be a connected semi-algebraic subset of $\mathbb{R}^{n-1}$. The *cylinder* over S in $\mathbb{R}^n$ is defined as $Z_{\mathbb{R}}(S) := S \times \mathbb{R}$. Let $\theta_1 < \cdots < \theta_s$ be continuous real algebraic functions defined on S. Denote $\theta_0 = -\infty$ and $\theta_{s+1} := \infty$. The intersection of the graph of θ_i with $Z_{\mathbb{R}}(S)$ is called the θ_i-*section* of $Z_{\mathbb{R}}(S)$. The set of points between θ_i-section and θ_{i+1}-section, $0 \leq i \leq s$, of $Z_{\mathbb{R}}(S)$ is a connected semi-algebraic subset of $\mathbb{R}^n$, called a (θ_i, θ_{i+1})-*sector* of $Z_{\mathbb{R}}(S)$. The sequence (θ_0, θ_1)-sector, θ_1-section, (θ_1, θ_2)-sector, $\ldots$, θ_s-section, (θ_s, θ_{s+1})-sector form a disjoint decomposition of $Z_{\mathbb{R}}(S)$, called a *stack* over S, which is uniquely defined for given functions $\theta_1 < \cdots < \theta_s$.

Cylindrical Algebraic Decomposition. Let π_{n-1} be the standard projection from $\mathbb{R}^n$ to $\mathbb{R}^{n-1}$ mapping $(x_1, \ldots, x_{n-1}, x_n)$ onto $(x_1, \ldots, x_{n-1})$. A finite partition $\mathcal{D}$ of $\mathbb{R}^n$ is called a *cylindrical algebraic decomposition* (CAD) of $\mathbb{R}^n$ if one of the following properties holds.

- either $n = 1$ and $\mathcal{D}$ is a stack over $\mathbb{R}^0$,
- or the set of $\{\pi_{n-1}(D) \mid D \in \mathcal{D}\}$ is a CAD of $\mathbb{R}^{n-1}$ and each $D \in \mathcal{D}$ is a section or sector of the stack over $\pi_{n-1}(D)$.

When this holds, the elements of $\mathcal{D}$ are called *cells*. The set $\{\pi_{n-1}(D) \mid D \in \mathcal{D}\}$ is called the *induced* CAD of $\mathcal{D}$. A CAD $\mathcal{D}$ of $\mathbb{R}^n$ can be encoded by a tree, called a CAD tree (denoted by T), as below. The root node, denoted by r, is $\mathbb{R}^0$. The children nodes of r are exactly the elements of the stack over $\mathbb{R}^0$. Let T_{n-1} be a CAD tree of the induced CAD of $\mathcal{D}$ in $\mathbb{R}^{n-1}$. For any leaf node C of T_{n-1}, its children nodes are exactly the elements of the stack over C.

Cylindrical Algebraic Formula. Let c be a cell in a CAD of $\mathbb{R}^n$. A cylindrical algebraic formula associated with c, denoted by ϕ_c, is defined recursively.
(i) The case for $n = 1$. If $c = \mathbb{R}$, then $\phi_c := true$. If c is a point α, then define $\phi_c := x_1 = \alpha$. If c is an open interval $(\alpha, \beta) \neq \mathbb{R}$, then $\phi_c := c > \alpha \wedge c < \beta$. For the special case that $\alpha = -\infty$, then ϕ_c is simply written as $c < \beta$. Similarly if $\beta = +\infty$, ϕ_c is simply written as $c > \alpha$.
(ii) The case for $n > 1$. Let c_{n-1} be the projection of c onto $\mathbb{R}^{n-1}$. If $c = c_{n-1} \times \mathbb{R}$, then define $\phi_c := \phi_{c_{n-1}}$. If c is an θ_i-section, then $\phi_c := \phi_{c_{n-1}} \wedge x_n = \theta_i$. If c is an (θ_i, θ_{i+1})-sector, then $\phi_c := \phi_{c_{n-1}} \wedge x_n > \theta_i \wedge x_n < \theta_{i+1}$. If $\theta_i = -\infty$, then ϕ_c is simply written as $\phi_{c_{n-1}} \wedge x_n < \theta_{i+1}$. If $\theta_{i+1} = +\infty$, then ϕ_c is simply written as $\phi_{c_{n-1}} \wedge x_n > \theta_i$.
If ϕ_c is the CAF associated with c, its zero set is defined as $Z_{\mathbb{R}}(\phi_c) := c$. Let S be a set of disjoint cells in a CAD. If $S = \emptyset$, $\phi_S := false$. Otherwise, a CAF associated with S is defined as $\phi_S := \vee_{c \in S} \phi_c$. Its zero set is $Z_{\mathbb{R}}(\phi_S) := \cup_{c \in S} c$.

Example 1. *Consider the closed unit disk S defined by $x^2 + y^2 \leq 1$. Then a CAF associated with S is as below.*

$$
\begin{aligned}
(x = -1 \wedge y = 0) &\vee (-1 < x \wedge x < 1 \wedge y = -\sqrt{1 - x^2}) \\
&\vee (-1 < x \wedge x < 1 \wedge -\sqrt{1 - x^2} < y \wedge y < \sqrt{1 - x^2}) \\
&\vee (-1 < x \wedge x < 1 \wedge y = \sqrt{1 - x^2}) \\
&\vee (x = 1 \wedge y = 0)
\end{aligned}
$$

Extended Tarski formula [2]. A (restricted) *extended Tarski formula* (ETF) is a Tarski formula, with possibly the addition of atomic formulas of the form $x_i \, \sigma \, \text{Root}_{x_i,k}(f)$, where $\text{Root}_{x_i,k}(f)$, $1 \leq i \leq n$, is a real algebraic function (defined on some set), and $\sigma \in \{=, \neq, >, <, \geq, \leq\}$. Given an ETF Φ, we can always write it in a disjunctive normal form $\Phi = \vee_{i=1}^{s} \wedge_{j=1}^{s_i} \phi_{i,j}$. Let $\Phi_i := \wedge_{j=1}^{s_i} \phi_{i,j}$. Assume that the variables $x_1, \ldots, x_n$ are ordered as $x_1 < \cdots < x_n$. Let $v(\phi_{i,j})$ be the biggest variable appearing in $\phi_{i,j}$. Then we can always arrange the order of atomic formulas appearing in each Φ_i such that for any ϕ_{i,j_1} and ϕ_{i,j_2}, where $j_1 < j_2$, we have $v(\phi_{i,j_1}) \leq v(\phi_{i,j_2})$. Let $w \in \{x_1, \ldots, x_n\}$. Denote by $\Phi_i^{<w} := \wedge_{v(\phi_{i,j}) < w} \phi_{i,j}$. We say Φ_i is *proper* if for any $j = 1, \ldots, s_i$, if $\phi_{i,j} = v \, \sigma \, \text{Root}_{v,k}(f)$, then $\text{Root}_{v,k}(f(\alpha))$ is defined for all α satisfying $\Phi_i^{<v}$. We say Φ is proper if every Φ_i is proper. It is clear that a CAF is a proper restricted ETF.

Generalized Cylindrical Algebraic Formula (GCAF). Let S be a set of disjoint cells in a CAD of $\mathbb{R}^n$. A GCAF associated with S, denoted by Φ, is a proper restricted ETF $\Phi = \vee_{i=1}^{s} \wedge_{j=1}^{s_i} \phi_{i,j}$ such that

- the zero set of Φ is exactly $\cup_{c \in S} c$,
- the zero set of $\Phi_i := \wedge_{j=1}^{s_i} \phi_{i,j}$ is a union of some cells in S,
- the zero sets of Φ_i and Φ_j are disjoint for $1 \leq i < j \leq s$,
- each $\phi_{i,j}$ is of the form $v = \text{Root}_{v,k}(f)$, where $v \in \{x_1, \ldots, x_n\}$,
- for every $w \in \{x_1, \ldots, x_n\}$, we have $\pi_{<w}(\Phi_i^{\leq w}) = (\Phi_i^{<w})$, where $\Phi_i^{\leq w} := \wedge_{v(\phi_{i,j}) \leq w} \phi_{i,j}$ and $\Phi_i^{<w} := \wedge_{v(\phi_{i,j}) < w} \phi_{i,j}$.

A GCAF clearly has a cylindrical structure justifying its name. Note that both a CAF and a GCAF can be naturally encoded in a tree data structure, with each node representing an atomic formula $\phi_{i,j}$. This tree together with the CAD cells information is used in algorithms for simplifying CAFs in Section 4.

Example 2. *Both* $(-1 \leq x \wedge x \leq 1 \wedge -\sqrt{1-x^2} \leq y \wedge y \leq \sqrt{1-x^2})$ *and*

$$(x = -1 \wedge y = 0) \vee (-1 < x \wedge x < 1 \wedge -\sqrt{1-x^2} \leq y \wedge y \leq \sqrt{1-x^2})$$
$$\vee (x = 1 \wedge y = 0)$$

are GCAFs equivalent to the CAF in Example 1.

The simplification procedure presented in this paper turns a CAF ϕ (in disjunctive normal form) into an equivalent GCAF Φ (in disjunctive normal form). We say Φ is simpler than ϕ if the number of conjunctive clauses in Φ is strictly less than that of ϕ.

3 Motivating Examples

In this section, we go through several examples, coming from the application of automatic loop transformation, so as to introduce the heuristic strategies for combing CAFs. Those strategies are formally presented in Section 4. We refer the reader to [12,7] for the application background of these examples.

Example 3. We consider polynomial multiplication with synchronous scheduling. It is formulated as the following quantifier elimination problem. If no simplification is applied, the result consists of 10 conjunctive clauses, as shown below.

```
ff := &E([i,j]), (0 <= i) &and (i <= n) &and (0 <= j) &and
                 (j <= n) &and (t = n - j) &and  (p = i + j);
R := PolynomialRing([i,j,p,t,n]);
sols := QuantifierElimination(ff,R,output=rootof,simplification=false);
'&or'('&and'(n = 0,t = n,p = 0),'&and'(0 < n,t = 0,p = n),
'&and'(0 < n,t = 0,n < p,p < 2*n),'&and'(0 < n,t = 0,p = 2*n),
'&and'(0 < n,0 < t,t < n,p = -t+n),
'&and'(0 < n,0 < t,t < n,-t+n < p,p < 2*n-t),
'&and'(0 < n,0 < t,t < n,p = 2*n-t),'&and'(0 < n,t = n,p = 0),
'&and'(0 < n,t = n,0 < p,p < n),'&and'(0 < n,t = n,p = n))
```

We observe that some conjunctive clauses can be merged into one. For instance, consider the subformula

$$(0 < n \wedge t = 0 \wedge p = n) \vee (0 < n \wedge t = 0 \wedge n < p \wedge p < 2n) \\ \vee (0 < n \wedge t = 0 \wedge p = 2n)$$ (1)

Note that $(0 < n \wedge t = 0)$ is common to all three conjunctive clauses. Applying the distributivity law, the above formula is equivalent to

$$(0 < n \wedge t = 0) \wedge ((p = n) \vee (n < p \wedge p < 2n) \vee (p = 2n)).$$

Observe that $(p = n) \vee (n < p \wedge p < 2n) \vee (p = 2n)$ can be combined into one conjunctive clause, namely $n \leq p \wedge p \leq 2n$. Thus, the above sub-formula of Equation (1) is equivalent to

$$0 < n \wedge t = 0 \wedge n \leq p \wedge p \leq 2n.$$

The above transformation can be explained in the language of CAD. Here $(0 < n \wedge t = 0)$ represents a CAD cell of $\mathbb{R}^2$ while $p = n$, $n < p \wedge p < 2n$, and $p = 2n$ represent respectively three adjacent children of it, which makes the combination straightforward. This observation forms our first idea. Applying this strategy to the entire expression in Equation (1) yields the following simplified formula:

```
'&or'('&and'(n = 0,t = n,p = 0),
      '&and'(0 < n,t = 0,n <= p,p <= 2*n),
      '&and'(0 < n,0 < t,t < n,-t+n <= p,p <= 2*n-t),
      '&and'(0 < n,t = n,0 <= p,p <= n)
```

Let us look at the last three conjunctive clauses. At first glance, it seems that they cannot be combined into one. A key observation is that if we specialize $-t + n \leq p \wedge p \leq 2n - t$ at $t = 0$ and $t = n$, we obtain $n \leq p \wedge p \leq 2n$ and $0 \leq p \wedge p \leq n$. Thus the last three conjunctive clauses can be combined into one:

```
'&and'(0 < n,0 <= t,t <= n,-t+n <= p,p <= 2*n-t).
```

Applying this *specialization technique* again, we obtain the final output:

```
'&and'(0 <= n,0 <= t,t <= n,-t+n <= p,p <= 2*n-t).
```

Combining together the two simplification techniques introduced in Example 3 forms the basis for the first level of our simplification procedure. Here is the result produced by the QuantifierElimination command of the RegularChains library at level L1:

```
QuantifierElimination(ff, R, output=rootof, simplification='L1');
((((0 <= n) &and (0 <= t)) &and (t <= n)) &and
(-t + n <= p)) &and (p <= 2 n - t)
```

Example 4. In this example, we consider polynomial multiplication with asynchronous scheduling. The following shows the output without using simplification. The output has 12 conjunctive clauses.

```
ff := &E([i,j]), (0 <= i) &and (i <= n) &and (0 <= j) &and (j <= n) &and
              (t = n - j) &and  (p = i + j);
R :=  PolynomialRing([i,j,t,p,n]);
sols := QuantifierElimination(ff,R,output=rootof,simplification=false);
'&or'('&and'(n = 0,p = 0,t = n), '&and'(0 < n,p = 0,t = n),
      '&and'(0 < n,0 < p,p < n,t = -p+n),
      '&and'(0 < n,0 < p,p < n,-p+n < t,t < n),
      '&and'(0 < n,0 < p,p < n,t = n),
      '&and'(0 < n,p = n,t = 0),'&and'(0 < n,p = n,0 < t,t < n),
      '&and'(0 < n,p = n,t = n),'&and'(0 < n,n < p,p < 2*n,t = 0),
      '&and'(0 < n,n < p,p < 2*n,0 < t,t < -p+2*n),
      '&and'(0 < n,n < p,p < 2*n,t = -p+2*n),'&and'(0 < n,p = 2*n,t = 0))
```

Here is the result produced by the `QuantifierElimination` command of the
`RegularChains` library with the simplification level set to **L1**:

```
((((n = 0) &and (p = 0)) &and (t = 0)) &or (((((0 < n) &and (0 <= p))
&and (p <= n)) &and (-p + n <= t)) &and (t <= n))) &or (((((0 < n)
&and (n < p)) &and (p <= 2 n)) &and (0 <= t)) &and (t <= -p + 2 n))
```

The situation here is a bit interesting. It seems that it is impossible to combine
the three conjunctive clauses into one. This is because to make $-p + n \leq t \wedge t \leq
n \iff 0 \leq t \wedge t \leq -p + 2n$ hold, we must have $-p + n = 0$ and $-p + 2n = n$,
that is $n = p$ must hold. This obviously does not always hold under either the
condition $0 < n \wedge 0 \leq p \wedge p \leq n$ or the condition $0 < n \wedge n < p \wedge p \leq 2n$.

However, if we look more closely at the example, we find that the formulas

$$\forall n, p, t \quad 0 < n \wedge 0 \leq p \wedge p \leq n \wedge -p + n \leq t \wedge t \leq n \implies 0 \leq t \wedge t \leq -p + 2n.$$

and

$$\forall n, p, t \quad 0 < n \wedge n < p \wedge p \leq 2n \wedge 0 \leq t \wedge t \leq -p + 2n \implies -p + n \leq t \wedge t \leq n.$$

are always true. Thus the last two can be combined into one

$$0 < n \wedge 0 \leq p \wedge p \leq 2n \wedge -p + n \leq t \wedge t \leq n \wedge 0 \leq t \wedge t \leq -p + 2n.$$

This third simplification technique forms the foundation of option 3 of the algorithm presented in the next section. Now, the simplified output consists of the
following single conjunction: The following is the simplified output.

```
((((((0 <= n) &and (0 <= p)) &and (p <= 2 n)) &and (n - p <= t))
&and (t <= n)) &and (0 <= t)) &and (t <= 2 n - p)
```

Example 5. Consider a more advanced example from [7]. Without simplification, the output cannot be displayed completely. Below, we only display the first
2 and the last 2 conjunctive clauses. There are 223 conjunctive clauses, in total.

```
R := PolynomialRing([i, j, t, p, u, b, B, n]);
ff := &E([i,j]), (0 < n) &and (0 <= i) &and (i <= n) &and (0 <= j) &and
                 (j <= n) &and  (t = n - j) &and  (p = i + j) &and
                 (b>=0) &and (0<=u) &and (u<B) &and (p=b*B+u);
QuantifierElimination(ff,R,partial=true,output=rootof,simplification
=false);
'&or'('&and'(0 < n,0 < B,B < n,b = 0,u = 0,p = b*B+u,t = n),
'&and'(0 < n,0 < B,B < n,b = 0,0 < u,u < B,p = b*B+u,t = -u+n),
...

'&and'(0 < n,2*n < B,n/B < b,b < 2*n/B,u = -B*b+2*n,p = b*B+u,t = 0),
'&and'(0 < n,2*n < B,b = 2*n/B,u = 0,p = b*B+u,t = 0))
```

If the first two simplification techniques (options 1 or 2 in Section 4) are used, there are 29 conjunctive clauses. We only display the first 2 and the last 2 ones.

```
'&or'(
'&and'(0 < n, 0 < B, B < n, 0 <= b, b <= -(B-n)/B, 0 <= u, u < B,
       p = b*B+u, -B*b+n-u <= t, t <= n),
'&and'(0 < n, 0 < B, B < n, -(B-n)/B < b, b < n/B, 0 <= u, u <= -B*b+n,
       p = b*B+u, -B*b+n-u <= t, t <= n),
...

'&and'(0 < n, 2*n < B, 0 < b, b < n/B, -B*b+n < u, u <= -B*b+2*n,
       p = b*B+u, 0 <= t, t <= -B*b+2*n-u),
'&and'(0 < n, 2*n < B, n/B <= b, b <= 2*n/B, 0 <= u, u <= -B*b+2*n,
       p = b*B+u, 0 <= t, t <= -B*b+2*n-u))
```

If we use Option 3, the output consist only 5 conjunctive clauses:

```
'&or'(
'&and'(0 < n,0 < B,B < 2*n,0 <= b,b <= 2*n/B,0 <= u,u < B,u <= -B*b+2*n,
       p = B*b+u, -B*b+n-u <= t, t <= n, 0 <= t, t <= -B*b+2*n-u),
'&and'(0 < n, B = 2*n, b = 0, 0 <= u, u < 2*n,
       p = u, n-u <= t, t <= n, 0 <= t, t <= 2*n-u),
'&and'(0 < n, B = 2*n, 0 < b, b < 1/2, 0 <= u, u <= -2*b*n+2*n,
       p = 2*b*n+u, -2*b*n+n-u <= t, t <= n, 0 <= t, t <= -2*b*n+2*n-u),
'&and'(0 < n, B = 2*n, 1/2 <= b, b <= 1, 0 <= u, u <= -2*b*n+2*n,
       p = 2*b*n+u, 0 <= t, t <= -2*b*n+2*n-u),
'&and'(0 < n, 2*n < B, 0 <= b, b <= 2*n/B, 0 <= u, u <= -B*b+2*n,
       p = B*b+u, -B*b+n-u <= t, t <= n, 0 <= t, t <= -B*b+2*n-u))
```

The situation here is more subtle. It can be shown that the third one and the fourth one can be combined together using the technique shown in option 3. But it can not be combined with the second one by specialization. This is because in the second one we have $0 \leq u \wedge u < 2n$ while in the third one we have $0 \leq u \wedge u \leq -2bn + 2n$. When $b = 0$, the latter is equivalent to $0 \leq u \wedge u \leq 2n$. So a more general "pivot formula" is needed, which can be found in the first and the fifth conjunctive clauses. The technical details are covered in Section 4. This idea form the basis of option 4, which now brings a single conjunctive clause.

```
'&and'(0 < n,0 < B,0 <= b,b <= 2*n/B,0 <= u,u < B,u <= -B*b+2*n,
       p = B*b+u, -B*b+n-u <= t, t <= n, 0 <= t, t <= -B*b+2*n-u)
```

4   Algorithm

In this section, we present a heuristic algorithm Merge for combining cylindrical algebraic formulas, motivated by the examples shown in last section. In our algorithm, we have several levels of simplification. The most advanced level requires to decide the truth value of a quantifier free formula, which can be accomplished by a special QE procedure such as the one in [5] for computing triangular decomposition of semi-algebraic systems. In the following, we provide the pseudo code for the algorithm and explain the subroutines in detail. The explanation supplies a loose proof of the correctness of the algorithms.

The function Merge takes a CAF tree T as input and returns an equivalent GCAF tree T. In below, we say a child node is *even* if it represents a section cell and *odd* if it represents a sector cell. Merge is called recursively to merge its odd children subtree first. The reason for doing this is that the representation of the odd nodes might be used as a reference to merge the even nodes. At last, it calls BasicMerge to merge the children of r, each of which is a rooted GCAF tree.

In BasicMerge, one type of subtree is treated in a special manner. It is a subtree rooted at a node c in T consisting of a single path. Let Γ be such a subtree of height h. It can be represented by a list of nodes $\{c_0, c_1, \ldots, c_h\}$, where $c_0 := c$ and c_i is the child of c_{i-1}, $i = 1, \ldots, h$. Each c_i is uniquely identified by an atomic formula $\psi(c_i)$ in a CAD tree data structure. In the following subroutines, we use $c.nodeRep$ to denote $\psi(c_0)$, $c.rep$ to denote $\wedge_{i=0}^h \psi(c_i)$, $c.desRep$ to denote $\wedge_{i=1}^h \psi(c_i)$, and $c.desDesRep$ to denote $\wedge_{i=2}^h \psi(c_i)$. Denote by $c.ancRep$ the conjunction of c's ascendants' representing atomic formulas.

In the context of the subroutines, these objects are all well defined. The top level function Merge always compresses such a tree Γ into a node. The procedure starts by dividing children of r into blocks s.t. children in different blocks are not adjacent. If opt is 1, the child node which has children is not processed. If opt is 3 or 4, the procedure tries to combine all the children nodes in a block into one. If it fails, opt 2 is adapted to combine as much adjacent children nodes in a block as possible. Note that we can also merge the nodes of the type $x_i < Root_{x_i,k}f$ and $x_i > Root_{x_i,k}f$ into one $x_i \neq Root_{x_i,k}f$ even the two nodes are not adjacent. This justifies the call of IneqMerge.

The procedure BlockMergable is used to test if the children of r in a block can be merged into one. Note that this function only handle the case that none of the children nodes in the block have children. To merge the children nodes which themselves have children, extra cost might be paid, since each subtree rooted at a child node is for sure not a tree consisting of a single path. The conjunction of the odd children is selected first as a pivot. If it fails and opt is 4, the conjunction of odd siblings of r is selected as a pivot.

When the procedure BlockMergable fails, the function NextMergable takes over to merge adjacent nodes in a block incrementally, which calls SameDesRep to test if the representation of the descendants of two adjacent nodes are equivalent. The equivalence is conducted by some simple test. In the case that the adjacent nodes have children, one checks if the subtrees rooted at them have physically

---

Algorithm 1. Merge(r, T, opt)

Input: A CAF tree T rooted at r, an option 'opt' on simplification level.
Output: An equivalent GCAF tree T.

1 **begin**
2   **if** r *has no children* **then** return
      `// The recursive call is first made for the odd children`
3   **for** *each odd child c of r* **do** Merge(c, T, opt)
4   **for** *each even child c of r* **do** Merge(c, T, opt)
5   BasicMerge(r, T, opt);

---

the same representation by calling ExactSameDesRep. Otherwise, some simple specialization is conducted to check the equivalence.

5   Experimentation

In this section, we apply the algorithms of Section 4 to more examples. A comparison with the Reduce command of Mathematica is also provided.

Example 6. [11] This example is about scanning index sets generated by applying a non-linear schedule. Without simplification, the output has 58 clauses. With option 1, the output has 10 conjunctive clauses. With option 3, the output has exactly one conjunctive clause, although it takes about 200 more seconds.

```
ff := (n>=7) &and (2<=x) &and (x<=n) &and (4<=y) &and (y<=n)
            &and (n-x<=y) &and (t=(n-3)*x+y);
R := PolynomialRing([y, x, t, n]);
QuantifierElimination(ff,R,output=rootof,simplification='L3');
'&and'(7 <= n,3*n-8 <= t,t <= n^2-2*n,2 <= x,x <= -(n-t)/(n-4),
     -(n-t)/(n-3) <= x,x <= (t-4)/(n-3),x <= n,y = -n*x+t+3*x)
```

Example 7. [11] This example is about scanning index sets generated by normalizing loop strides. Without simplification, the output has 8 conjunctive clauses. The option 1 is sufficient to simplify it into one piece.

```
ff := (i>=2) &and (i^2<=n) &and (k>=i) &and (k*i<=n) &and (j=k*i);
R := PolynomialRing([k, i, j, n]);
QuantifierElimination(ff,R,output=rootof,simplification='L1');
'&and'(4 <= n,4 <= j,j <= n,2 <= i,i <= j^(1/2),k = j/i)
```

Example 8. [2] An example illustrating simple CAD. Without simplification, there are 51 clauses. After simplification with option 1, we have

```
R := PolynomialRing([z, y, x]);
qff := &E([z]), (19*z - 10*x + 10*y < 0) &and ((x^2+y^2+(z-3)^2<9)
                &or (2*x+19*z+10*y-11>=0));
QuantifierElimination(qff, R, output=rootof, simplification='L1');
```

Algorithm 2. BasicMerge(r, T, opt)

Input: A GCAF tree T rooted at r, an option 'opt' on simplification level.
Output: A simplified GCAF tree T.

```
 1  begin
 2  |   if r has no children then
 3  |   |   return;
 4  |   else if r has only one child then
 5  |   |   let c be the only child of r;
 6  |   |   if c has children then return r.rep := r.nodeRep ∧ c.rep;
    |   |   r.children := ∅, return;
 7  |   put the adjacent children of r into the same block;
 8  |   if opt = 1 then put each child having children in a separate block
    |       NCL := ∅; // NCL means new children list
 9  |   for each block B do
10  |   |   if opt = 3 or opt = 4 then
11  |   |   |   bool, pivotRep := BlockMergable(B, opt);
12  |   |   |   if bool then
13  |   |   |   |   key := BlockMerge(B, pivotRep);
14  |   |   |   |   NCL.append(key); next;
15  |   |   MGL := {B[1]};// MGL means a merge list
16  |   |   pivot := −1; m := |B|;
17  |   |   for i from 1 to m do
18  |   |   |   if NextMergable(B, i, m, pivot, opt) then
19  |   |   |   |   MGL := MGL.append(B[i + 1]);
20  |   |   |   else
21  |   |   |   |   if |MGL| = 1 then
    |   |   |   |   |   // no new children created
22  |   |   |   |   |   NCL.append(B[i]);
23  |   |   |   |   else
24  |   |   |   |   |   key := BlockMerge(MGL, B[pivot].rep);
25  |   |   |   |   |   NCL.append(key);
26  |   |   |   |   if i < m then  MGL := {B[i + 1]}
27  |   if opt ≠ 1 then
28  |   |   if |NCL| = 2 or |NCL| = 3 then
29  |   |   |   NCL := IneqMerge(NCL);
30  |   if |NCL| = 1 then
31  |   |   let c be the only element of NCL;
32  |   |   if c has children then
33  |   |   |   r.children := NCL;
34  |   |   else
35  |   |   |   r.rep := r.nodeRep ∧ c.rep; r.children := ∅;
36  |   else
37  |   |   r.children := NCL;
```

Algorithm 3. BlockMergable(B, T, opt)

Input: A block of adjacent children in one stack.
Output: If the children can be combined into one by the heuristic strategy
            below, return true and the combined representation; otherwise return
            false and empty.

begin
    if *at least one node in B has children* **then** return false, $\emptyset$ let L be the odd
    nodes of B; let $pivotRep := \wedge_{b \in L} b.desRep$; $res := true$;
    for $b \in B$ **do**
        set res to the truth value of $\forall \mathbf{x}, (b.ancRep \wedge b.nodeRep \wedge b.desRep \Leftrightarrow$
        $b.ancRep \wedge b.nodeRep \wedge pivotRep)$;
        if *not res* **then** break
    if *res* **then** return res, pivotRep **if** $opt = 4$ **then**
        let p be the parent of nodes in B;
        if *p is not root and p is even node* **then**
            let sL be the odd siblings of p;
            if $|sL| \neq 0$ *and none of sL has children* **then**
                $pivotRep := \wedge_{s \in sL} s.desDesRep$; $res := true$;
                for $b \in B$ **do**
                    $res := \forall \mathbf{x}, (b.ancRep \wedge b.nodeRep \wedge b.desRep \Leftrightarrow$
                    $b.ancRep \wedge b.nodeRep \wedge pivotRep)$;
                  if *not res* **then** break
                if *res* **then** return res, pivotRep
    return false, $\emptyset$

Algorithm 4. BlockMerge(B, pivotRep)

Input: A block B of adjacent nodes in a stack. A pivot representation pivotRep.
Output: A conjunctive ETF clause equivalent to $(\vee_{c \in B} c) \wedge pivotRep$.
begin
    create a new node c;
    let a and b be respectively the first and the last element of B;
    // Next we abuse the notation of intervals to represent a node
       for simplicity.
    if $a = (a_1, a_2)$ **then**
        if $b = (b_1, b_2)$ **then** $r := (a_1, b_2)$ **else if** $b = [b_1, b_1]$ **then** $r := (a_1, b_1]$
    else if $a = [a_1, a_1]$ **then**
        if $b = (b_1, b_2)$ **then** $r := [a_1, b_2)$ **else if** $b = [b_1, b_1]$ **then** $r := [a_1, b_1]$
    let $c := r \wedge pivotRep$;
    return c

Algorithm 5. NextMergable(A, i, m, pivot, opt)

Input: A block of adjacent nodes A, current node $A[i]$, final node $A[m]$ and pivot node *pivot*.
Output: Determine if $A[i]$ can be combined with its right adjacent sibling.
begin

   if $i = m$ **then**
      return false;

   if $pivot = -1$ **then**
      if $A[i]$ *is even node* **then**
         $pivot := i + 1$;
         if $SameDesRep(A[pivot], A[i], opt)$ **then** return true **else** return false;
      else
         $pivot = i$;
         if $SameDesRep(A[pivot], A[i + 1]), opt$ **then** return true **else** $pivot := -1$; return false;

   else
      if $A[i]$ *is even node* **then**
         if $SameDesRep(A[pivot], A[i + 1], opt)$ **then** return true **else** $pivot := i + 1$; return false;
      else
         if $SameDesRep(A[pivot], A[i + 1], opt)$ **then** return true **else** $pivot := -1$; return false;

Algorithm 6. SameDesRep(pivot, i, opt)

Input: A pivot node *pivot* and a node i in the same block.
Output: Test if the representation of their descendants are the same by some simple heuristics.
begin

   if *opt is* $2, 3, 4$ **then**
      if *both pivot and i have children* **then**
         return ExactSameDesRep(pivot, i);

   if i *is odd* **then** return the truth value of $i.desRep = pivot.desRep$ **else** return the truth value of $i.desRep = subs(i.nodeRep, pivot.desRep)$

Algorithm 7. IneqMerge(L, opt)

Input: A list L of nodes in a stack.
Output: Return a list NL of merged nodes.
begin

   if $|L| = 2$ **then** $a := L[1]$; $b := L[2]$ **else if** $|L| = 3$ **then** $a := L[1]$; $b := L[3]$ **else** return L **if** $a.nodeRep = (-\infty, \alpha)$ and $b.nodeRep = (\alpha, +\infty)$ and $SameDesRep(a, b, opt)$ **then**
      create a new node c; $c.rep := x \neq \alpha \wedge a.desRep$;
      if $|L| = 2$ **then** return $[c]$ **else if** $|L| = 3$ **then** return $[c, L[2]]$

```
'&or'('&and'(190/187-10/187*922^(1/2) < x, x < 11/12,
         100/461*x-570/461-19/461*(-561*x^2+1140*x+900)^(1/2) < y,
         y < 100/461*x-570/461+19/461*(-561*x^2+1140*x+900)^(1/2)),
    '&and'(x = 11/12,-1435/1383-19/5532*212199^(1/2) < y,
         y < -1435/1383+19/5532*212199^(1/2)), 11/12 < x)
```

Example 9. An example from program termination analysis. Without simplification, the output has 246 conjunctive clauses. After simplification with option 2, the output has 14 conjunctive clauses.

```
R := PolynomialRing([v1,v2,v3,labda,a11,a21,a22,a33,b12,b22,b23]);
f:= &E([v1,v2,v3,labda]), (labda>0) &and (a11*v1=labda*v1) &and
                   (a21*v1+a22*v2=labda*v2) &and (a33*v3=labda*v3) &and
                   (b12*v2>0) &and (b22*v2+b23*v3>0);
QuantifierElimination(f,R,output=rootof,partial=true,simplification
='L2');
'&or'('&and'(b23 <> 0, b22 < 0, b12 < 0, a22 <= 0, a21 <> 0, 0 < a11),
'&and'(b23 <> 0, b22 < 0, b12 < 0, 0 < a22),
'&and'(b23 <> 0, b22 < 0, 0 < b12, 0 < a33, a22 <> a33, a21 <> 0,
 a11 = a33), '&and'(b23 <> 0, b22 < 0, 0 < b12, 0 < a33, a22 = a33),
'&and'(b23 <> 0, b22 = 0, b12 <> 0, 0 < a33, a22 <> a33, a21 <> 0,
 a11 = a33), '&and'(b23 <> 0, b22 = 0, b12 <> 0, 0 < a33, a22 = a33),
'&and'(b23 <> 0, 0 < b22, b12 < 0, 0 < a33, a22 <> a33, a21 <> 0,
a11 = a33), '&and'(b23 <> 0, 0 < b22, b12 < 0, 0 < a33, a22 = a33),
'&and'(b23 <> 0, 0 < b22, 0 < b12, a22 <= 0, a21 <> 0, 0 < a11),
'&and'(b23 <> 0, 0 < b22, 0 < b12, 0 < a22),
'&and'(b23 = 0, b22 < 0, b12 < 0, a22 <= 0, a21 <> 0, 0 < a11),
'&and'(b23 = 0, b22 < 0, b12 < 0, 0 < a22),
'&and'(b23 = 0, 0 < b22, 0 < b12, a22 <= 0, a21 <> 0, 0 < a11),
'&and'(b23 = 0, 0 < b22, 0 < b12, 0 < a22))
```

Example 10. Consider computing a CAF of $p_1 \leq 0 \wedge p_2 \leq 0$, where p_1, p_2 are random polynomials of given degree d and with n variables. We have tested the case for $n = 2, d = 2, 3, 4$ and $n = 3, d = 2, 3$ and $n = 4, d = 2$. The experimentation shows that the number of conjunctive clauses after simplification using option 2 is about 3 to 5 times less than without applying simplification.

We have also selected some examples from [16], generated automatically from QEPCADexamplebank_v4.txt, to test the simplification procedure. The experimental results are summarized in Table 1. The examples in previous sections are included. The QuantifierElimination command is called with options partial='true', output='rootof' and different simplification flavors. In the table, L0 corresponds to the option simplification='false'. Two columns are dedicated to each simplification option, as well as to the Reduce command of Mathematica: the left column contains the time spent on calling the QE procedure while the right column gives the number of conjunctive GCAF clauses in the output.

From the table, we derive the following observations. For all the examples, there is almost no simplification overhead for the levels L1 and L2. This is because almost no algebraic computations are needed for those two levels. There

Table 1. Comparison between different simplification options and **Reduce**.

Example	L0		L1		L2		L3		L4		Reduce	
ArcSin-B	0.899	10	1.192	8	0.986	6	1.004	6	0.983	6	0.096	12
A-Real-Implicitization	0.977	6	1.021	5	1.013	2	1.264	2	1.309	2	0.069	5
Collision-A-from-B-H	2.340	13	2.344	1	2.273	1	2.328	1	2.374	1	0.180	1
Cyclic-4	0.851	2	0.868	2	0.880	1	0.874	1	0.861	1	0.055	2
Edges-Square-Product	57.378	249	57.507	5	58.158	5	58.188	5	58.668	5	4.753	7
Ellipse-A-from-B-H	2.305	12	2.352	2	2.345	2	2.439	2	2.376	2	0.400	2
Intersection-dagger-B	1.797	113	1.837	84	1.810	78	2.160	78	2.145	78	0.289	104
Kahan-A	1.034	17	1.414	10	1.107	10	1.129	10	1.113	10	0.116	10
Kahan-B	1.056	13	1.135	9	1.152	9	1.133	9	1.136	9	0.145	9
Loop-Norm	1.418	8	1.399	1	1.414	1	1.402	1	1.441	1	0.059	2
Loop-Termination	44.383	246	43.795	186	44.128	14	44.215	14	44.547	14	0.161	58
NL-Schedule	2.915	58	3.039	10	3.018	10	222.654	1	236.145	1	0.073	12
Off-Center-Ellipse	1.847	4	1.917	2	1.893	2	1.923	2	1.914	2	0.129	2
Parametric-Parabola	0.839	7	0.871	5	0.874	4	1.057	4	1.110	4	0.075	5
Poly-Mul-ASyn	1.191	12	1.232	3	1.257	3	1.672	1	1.692	1	0.054	6
Poly-Mul-Syn	0.959	10	1.028	1	1.015	1	1.379	1	1.360	1	0.052	2
Poly-Mul-Tile	12.001	223	12.522	29	12.333	29	152.901	5	207.635	1	0.123	50
Positivity-of-Quartic	1.480	10	1.497	4	1.575	3	1.473	3	1.468	3	0.094	4
Putnum-Example	2.872	76	2.897	8	3.016	8	2.934	8	2.941	8	0.430	6
Random-dagger-A	2.106	214	2.150	160	2.162	148	2.209	148	2.146	148	0.155	189
Range-of-Lower-Bounds	1.057	2	1.121	1	1.119	1	1.091	1	1.157	1	0.075	1
Sphere-Half-Space	2.297	9	2.319	3	2.298	3	2.302	3	2.423	3	0.076	2
Whitney-Umbrella	0.908	8	0.964	7	1.003	3	1.257	3	1.455	3	0.066	7
X-axis-Ellipse-Problem	12.099	104	12.322	24	13.070	24	12.552	24	12.398	24	0.466	16
YangXia	2.225	8	2.245	2	2.257	2	2.897	2	2.850	2	0.101	2

are only two examples, namely NL-Schedule and Poly-Mul-Tile, for which the simplification levels L3 and L4 incur significant overhead but also reduce significantly the amount of conjunctive clauses. For examples coming from the application of loop transformation, the levels L3 and L4 are effective. But for examples from other application domains, the level L2 is the best to choose considering the computation overhead, which is the default option of our QE procedure. W.r.t. the **Reduce** command, our QE procedure can generate less number of conjunctive clauses for 16 out of the 25 examples. For the three examples for which our QE procedure generates more conjunctive clauses, we have checked that the reason is that our routine ExactSameDesRep only checks whether two subtrees are physically the same while the **Reduce** command is more aggressive. Finding cost-effective heuristic strategies to handle this problem is left for future work.

6   Conclusions

We have presented a multi-level heuristic algorithm for simplifying cylindrical algebraic formulas, motivated by applications like automatic loop transformation in computer programs. The experimentation shows that the method can reduce significantly the number of conjunctive clauses of a CAF. Nevertheless, more work is required to obtain more compact output in less computing time. In particular, the following related work needs to be investigated. In Chapter 8 of his thesis [2], Brown presented algorithms for constructing cylindrical formulas, which are proper restricted extended Tarski formulas having cylindrical properties, but not necessarily GCAFs. An essential idea in his approach is the concept of "polynomially compatible". But it is not always easy to determine if two sections in different stacks are polynomially compatible, especially for cylindrical algebraic decomposition computed by the partial CAD approach [9] or the regular chain approach [6]. Moreover, the simplification techniques of [2] cannot handle the case taken care of by options 3 and 4 of our method. Nevertheless, the ideas of "polynomially compatible" and "truth-boundary cells" deserve to be investigated further and related to our approach. The adjacency algorithm [1] might also help determining whether two intro-stack sections can be combined.

Acknowledgments. Supported by the NSFC (11301524,11471307,61202131).

References

1. Arnon, D.S., Collins, G.E., McCallum, S.: Cylindrical algebraic decomposition II: an adjacency algorithm for the plane. SIAM J. Comput. **13**(4), 878–889 (1984)
2. Brown, C.W.: Solution Formula Construction for Truth Invariant CAD's. PhD thesis, University of Delaware (1999)
3. Brown, C.W.: Fast simplifications for tarski formulas based on monomial inequalities. Journal of Symbolic Computation **47**(7), 859–882 (2012)
4. Brown, C.W., Strzeboński, A.: Black-box/white-box simplification and applications to quantifier elimination. In: Proc. of ISSAC 2010, pp. 69–76 (2010)
5. Chen, C., Davenport, J.H., May, J., Moreno Maza, M., Xia, B., Xiao, R.: Triangular decomposition of semi-algebraic systems. In: Watt, S.M. (ed.) Proceedings ISSAC 2010, pp. 187–194 (2010)
6. Chen, C., Moreno Maza, M.: An incremental algorithm for computing cylindrical algebraic decompositions. In: Computer Mathematics: Proc. of ASCM 2012, pp. 199–222 (2014)
7. Chen, C., Moreno Maza, M.: Quantifier elimination by cylindrical algebraic decomposition based on regular chains. In: Proc. of ISSAC 2014, pp. 91–98 (2014)
8. Collins, G.E.: Quantifier elimination for real closed fields by cylindrical algebraic decomposition. Springer Lecture Notes in Computer Science **33**, 515–532 (1975)
9. Collins, G.E., Hong, H.: Partial cylindrical algebraic decomposition. Journal of Symbolic Computation **12**(3), 299–328 (1991)
10. Dolzmann, A., Sturm, T.: Simplification of quantifier-free formulas over ordered fields. Journal of Symbolic Computation **24**, 209–231 (1995)
11. Größlinger, A.: Scanning index sets with polynomial bounds using cylindrical algebraic decomposition. Number MIP-0803 (2008)
12. Größlinger, A., Griebl, M., Lengauer, C.: Quantifier elimination in automatic loop parallelization. J. Symb. Comput. **41**(11), 1206–1221 (2006)
13. Iwane, H., Higuchi, H., Anai, H.: An effective implementation of a special quantifier elimination for a sign definite condition by logical formula simplification. In: Gerdt, V.P., Koepf, W., Mayr, E.W., Vorozhtsov, E.V. (eds.) CASC 2013. LNCS, vol. 8136, pp. 194–208. Springer, Heidelberg (2013)
14. Strzeboński, A.: Computation with semialgebraic sets represented by cylindrical algebraic formulas. In: Proc. of ISSAC 2010, pp. 61–68. ACM (2010)
15. Strzeboński, A.: Solving polynomial systems over semialgebraic sets represented by cylindrical algebraic formulas. In: Proc. of ISSAC 2012, pp. 335–342. ACM (2012)
16. Wilson, D.J.: Real geometry and connectedness via triangular description: Cad example bank (2013)

Quasi-Steady State – Intuition, Perturbation Theory and Algorithmic Algebra

Alexandra Goeke[1], Sebastian Walcher[2], and Eva Zerz[3]

[1] Lehrstuhl A für Mathematik, RWTH Aachen, 52056 Aachen, Germany
{alexandra.goeke,walcher}@matha.rwth-aachen.de
[2] Lehrstuhl D für Mathematik, RWTH Aachen, 52056 Aachen, Germany
eva.zerz@math.rwth-aachen.de

Abstract. This survey of mathematical approaches to quasi-steady state (QSS) phenomena provides an analytical foundation for an algorithmic-algebraic treatment of the associated (parameter-dependent) ordinary differential systems, in particular for reaction networks. Topics include an ad hoc reduction procedure, singular perturbations, and methods to identify suitable parameter regions.

MSC (2010): 92C45, 34E15, 80A30, 13P10.

1 Introduction

The notion quasi-steady state (QSS) characterizes the behavior of certain reaction networks with slow and fast species, or slow and fast reactions. It was introduced in the early 20th century, with arguments based on scientific intuition. Mathematically, an intuitive approach prevailed for several decades as well, one reason being that the appropriate mathematical tools did not yet exist. With the emergence of singular perturbation theory, a possible appropriate translation from (bio-) chemical phenomenon to mathematical terms became available, and much of the subsequent work referred to singular perturbations. However, alternative interpretations and approaches do exist.

In the present paper we will survey two possible interpretations of QSS, and highlight the (perhaps surprising) fact that their implementation naturally leads to algorithmic algebra. "Naturally" in this context means that the familiar modelling of reaction networks by mass-action kinetics, and the subsequent reductions based on mathematical arguments, yield differential equations with polynomial or rational right-hand sides that are defined on algebraic varieties. (It is not necessary to invoke any additional assumptions.) From this perspective, QSS represents a very fitting example for "algebraic biology".

We will survey some recent work (mostly by the authors and co-workers) discussing the passage from (biological to) analytical to algebraic concepts. The analytical results lead to problems which, at least initially, are amenable to standard methods of algorithmic algebra.

© Springer International Publishing Switzerland 2015
V.P. Gerdt et al. (Eds.): CASC 2015, LNCS 9301, pp. 135–151, 2015.
DOI: 10.1007/978-3-319-24021-3_10

To illustrate the notions and arguments, we choose the Michaelis-Menten re-action system. (We do so with some reluctance, but this system is very relevant and very well-suited for brief illustrations.) In the Michaelis-Menten network, substrate S and enzyme E reversibly combine to a complex C, which in turn degrades – reversibly or irreversibly – to E and product P; thus one has the reaction scheme

$$E + S \underset{k_{-1}}{\overset{k_1}{\rightleftharpoons}} C \underset{k_{-2}}{\overset{k_2}{\rightleftharpoons}} E + P.$$

Mass action kinetics and conservation laws yield the differential system

$$
\begin{aligned}
\dot{s} &= -\, k_1 e_0 s + (k_1 s + k_{-1})c, \\
\dot{c} &= k_1 e_0 s - (k_1 s + k_{-1} + k_2)c + k_{-2}(e_0 - c)(s_0 - s - c),
\end{aligned}
\tag{1}
$$

for the concentrations, usually with initial values $s(0) = s_0 > 0$ and $c(0) = 0$. Here $k_{-2} = 0$ defines the irreversible scenario, while a network with $k_{-2} > 0$ is called reversible; all other parameters are > 0.

We instantly employ this system to give readers with little or no background in biochemistry an impression of the arguments and problems arising in the practice of quasi-steady state.

As for a first example, consider the irreversible Michaelis-Menten system

$$
\begin{aligned}
\dot{s} &= -\, k_1 e_0 s + (k_1 s + k_{-1})c, \\
\dot{c} &= k_1 e_0 s - (k_1 s + k_{-1} + k_2)c
\end{aligned}
\tag{2}
$$

and assume (based on intuition or experiments) that the complex concentration c has a negligible rate of change, thus $0 \approx \dot{c} = k_1 e_0 s - (k_1 s + k_{-1} + k_2)c$, for an extended duration of time. This gives rise to a heuristic approach: Set $k_1 e_0 s - (k_1 s + k_{-1} + k_2)c = 0$ (i.e., make the stronger assumption that $\dot{c} = 0$ holds exactly), solve this relation for c and substitute into the first equation of (2). Using some high school algebra one obtains the so-called Michaelis-Menten equation

$$\dot{s} = -\frac{k_1 k_2 e_0 s}{k_1 s + k_{-1} + k_2}.$$

This heuristics clearly needs a justification.

As for a second example, assume (based, again, on intuition or experimental data) that the rate constants k_2 and k_{-2} in the reversible system (1) are very small, thus the "right half" of the reaction scheme proceeds at a much slower pace than the "left half". Here it is natural to consider the limit $k_2 \to 0$ and $k_{-2} \to 0$. Thus one obtains the system

$$
\begin{aligned}
\dot{s} &= -\, k_1 e_0 s + (k_1 s + k_{-1})c, \\
\dot{c} &= k_1 e_0 s - (k_1 s + k_{-1})c.
\end{aligned}
$$

This system admits non-isolated stationary points, viz. a curve Z of equilibria corresponding to the fast reactions. In this approach the (presumed) interesting dynamics near Z is not being accounted for, hence more delicate limiting processes are necessary.

It goes without saying that a solid mathematical underpinning is necesssary in both examples. One purpose of the present contribution is to sketch the underlying theory, and to set the heuristics on firm ground.

2   Transferring Scientific to Mathematical Notions

The following short historical sketch of relevant contributions is necessarily incomplete, but it should serve to give an impression to non-expert readers.

Quasi-steady state assumptions for the irreversible reaction system (2) go back (among others) to Henri [19] in 1903, and Michaelis and Menten [25] in 1913. In the early stages, two incarnations of QSS materialized: One may say that Henri and Michaelis/Menten discussed QSS assuming slow and fast reactions (also known as partial equilibrium assumption, briefly PEA), while Briggs and Haldane [6] in 1925 considered QSS with a slow variable, and gave a heuristic derivation of the familiar Michaelis-Menten equation starting from the irreversible system (2). There was much work in the following decades, with arguments generally based on (bio-) chemical considerations; one representative is Laidler [22]. On the mathematical side, Tikhonov's work [36] on singular perturbations provided a solid mathematical foundation for slow-fast phenomena, as well as solid results. Heineken, Tsuchiya and Aris [18] were among the first to consider QSS (for irreversible Michaelis-Menten) from the perspective of singular perturbation theory; Schauer and Heinrich [30] gave a general discussion of slow-fast reactions from this point of view. One of Fenichel's [13] fundamental contributions to singular perturbation theory was a characterization with no reference to special coordinates. This was employed (and partly derived in an alternative way) by Stiefenhofer [35], as well as several others. The authors' work in [15,16,17] is also based on Fenichel's results. As for recent papers on slow/fast reactions one could mention Lee and Othmer [24] who included a discussion of the initial phase. A (quite efficient) numerical approach related to singular perturbation results, called computational singular perturbation (CSP), was introduced by Lam and Goussis [23] in the 1990s.

In addition to reducing a system with given small parameters, QSS also involves finding parameter regions where such phenomena occur (briefly, "finding small parameters"). Such lines of reasoning also go back to Henri, Michaelis/ Menten, and Briggs/Haldane. Segel and Slemrod [32] in 1989 introduced an approach to determine appropriate small parameters via time scale heuristics, and their work triggered a large number of follow-up publications, such as Borghans et al. [5]. A different approach, to be discussed below, was recently introduced in [15,17].

Generally, algebraic-algorithmic techniques for biological (and chemical) systems have been in the focus of attention since about 2000. An overview of early developments and an impression of the range of applications is provided in the conference proceedings [1,20]. We mention only a few publications that are of relevance for reaction equations. Gatermann and Huber [14], as well as Shiu and Sturmfels [34] make use of the special structure of mass action systems to discuss the variety of stationary points, resp. siphons. Conditions for Hopf bifurcations were considered by Niu and Wang [26], and by Errami et al. [9], among others. Parameter reduction from an algebraic perspective was discussed by Sedoglavic [31], and by Hubert and Labahn [21]. It seems that there exists relatively little work on algorithmic algebra aspects of quasi-steady state phenomena and reduction; one should, however, mention Boulier et al. [2,3,4].

The interpretation of QSS as a singular perturbation phenomenon is widely accepted but this does not seem to be a foregone conclusion. While a singular perturbation approach is natural for slow and fast chemical reactions, with the "small parameter" appearing in rate constants of certain reactions, it is less straightforward for QSS involving chemical species, and actually there exist alternative mathematical interpretations. Thus, Heinrich and Schauer [29] emphasized the approximate invariance of the set defined by $\dot{c} = 0$ in equation (1). This approximate invariance is implicitly assumed by practitioners who use the ad hoc reduction method as in Briggs/Haldane [6]. We will discuss both interpretations.

3   Preliminaries and Notation

In this section we fix notation and recall a few notions and results. Throughout the paper we will consider a parameter-dependent ordinary differential equation

$$\dot{x} = h(x, \pi), \quad x \in U \subseteq \mathbb{R}^n, \quad \pi \in \Pi \subseteq \mathbb{R}^m \tag{3}$$

with U open and the right-hand side h smooth in the variable (x, π). (The case $m = 0$ describes parameter-independent systems.) Our principal interest lies in polynomial or rational systems; this is a natural assumption in the setting of chemical reaction equations. When appropriate, we will pass to the complexification.

3.1   Lie Derivatives and Invariance Criteria

Given a smooth function $\psi : U \times \Pi \to \mathbb{R}$, we call $L_h(\psi)$, with

$$L_h(\psi)(x, \pi) = D_1 \psi(x, \pi) h(x, \pi)$$

(note that only the partial derivative with respect to x is involved) the *Lie derivative* of ψ with respect to h. The Lie derivative describes the rate of change for ψ along solutions of (3); it is useful for invariance criteria such as the following.

Lemma 1. *(a) Let $\psi_1, \ldots, \psi_s$ be smooth on $U \times \Pi$, and assume that there are smooth functions μ_{jk} such that*

$$L_h(\psi_j) = \sum_{k=1}^{s} \mu_{jk} \psi_k, \quad 1 \leq j \leq s. \tag{4}$$

Then the common zero set Y of the ψ_j is an invariant set of (3); i.e., whenever $y \in Y$ then the solution trajectory through y is contained in Y.

(b) *For complex polynomial functions and vector fields, the following converse holds: If the ψ_j generate a radical ideal, then invariance of the set Y will imply a relation (4), with polynomials μ_{jk}.*

3.2 Singular Perturbations

Singular perturbation theory for ODEs starts with Tikhonov's theorem (for details see the monograph by Verhulst [37]; Theorem 8.1). We specialize it to smooth autonomous equations. Consider a system in *Tikhonov standard form* in "slow time"

$$y_1' = f(y_1, y_2) + \varepsilon \dots, \quad y_1 \in D,$$
$$\varepsilon y_2' = g(y_1, y_2) + \varepsilon \dots, \quad y_2 \in G, \tag{5}$$

with small parameter $\varepsilon \geq 0$, defined on an open set $D \times G \subset \mathbb{R}^r \times \mathbb{R}^{n-r}$. Under some technical assumptions, the theorem guarantees that solutions of this system converge to solutions of a reduced system on the r-dimensional asymptotic slow manifold

$$\widetilde{Z} := \left\{ (y_1, y_2)^{\mathrm{tr}} \in D \times G; \, g(y_1, y_2) = 0 \right\};$$

the reduced system being given by

$$y_1' = f(y_1, y_2), \quad g(y_1, y_2) = 0, \tag{6}$$

on a suitable time interval as $\varepsilon \to 0$. A crucial technical assumption is satisfied whenever a uniform linear stability condition holds for the eigenvalues of the Jacobian $D_2 g(y_1, y_2)$ with respect to y_2. Generalizations for systems that are not in standard form, as well as less restrictive eigenvalue conditions (normal hyperbolicity), are due to Fenichel [13].

4 The Ad Hoc Approach

The following classical reduction heuristic (which we call the *ad hoc reduction*) is directly related to an intuitive quasi-steady state assumption for chemical species, such as by Briggs and Haldane [6]: In the differential equation, set the negligible rates of change equal to zero, and use the subsequent algebraic relations to obtain a reduced system. The procedure may be formalized by introducing the notion of *enforced invariant sets*.

Definition 1. *Given system (3), let $\psi_1, \dots, \psi_{n-r}$ be smooth in a neighborhood of some $(x_0, \pi_0) \in U \times \Pi$, and assume that the rank of the Jacobian of $\Psi := (\psi_1, \dots, \psi_{n-r})^{\mathrm{tr}}$ equals $n - r$ on this neighborhood. Let Y be (locally) the set of common zeros of these functions. Assume furthermore (w.l.o.g. upon relabelling) that the rank of $(x_1, \dots, x_r, \psi_1, \dots, \psi_{n-r})$ equals n on this neighborhood. Partition $x = (x^{[1]}, x^{[2]})$, with $x^{[1]} = (x_1, \dots, x_r)^{\mathrm{tr}}$ and $x^{[2]} = (x_{r+1}, \dots, x_n)^{\mathrm{tr}}$. Then we will call any system*

$$\dot{x}^{[1]} = h^{[1]}(x, \pi) + \sum_j m_j^{[1]} \psi_j$$
$$\dot{x}^{[2]} = -D_2 \Psi(x, \pi)^{-1} D_1 \Psi(x, \pi) h^{[1]}(x, \pi) + \sum_j m_j^{[2]} \psi_j \tag{7}$$

with arbitrary smooth $m_j^{[1]}$ and $m_j^{[2]}$, a system associated to (3) with enforced invariant set Y.

For the standard QSS setting, assuming "slow" variables $x_{r+1}, \ldots, x_n$ and $\psi_j = h_{r+j} = L_h(x_{r+j})$, this definition is applicable for ad hoc reduction. The common strategy is to consider only the first r equations, and to replace $x_{r+1}, \ldots, x_n$ in $h^{[1]}$ via $\Psi = 0$. System (7) on the invariant manifold Y provides an alternative to this (generally non-constructive) approach. In the polynomial or rational setting there remains to discuss a polynomial or rational system on an invariant algebraic variety; this seems more amenable to algebraic techniques. Boulier et al. [3,4] present an algorithmic approach to ad hoc reduction using elimination; in [4] the authors specifically introduce a variant to describe the dynamics on invariant manifolds arising from slow and fast reactions.

One must note that such procedures are consistent only if Y is actually an invariant set. We record a few facts; the proof is straightforward with Lemma 1.

Lemma 2. *(a) The set Y is invariant for system (7).*
(b) If Y is an invariant set for system (3) then its solutions on Y coincide with those of (7).

Example 1. Ad hoc approach for the irreversible Michaelis-Menten system, assuming QSS for the variable s. A system with enforced invariant set Y defined by $\psi = L_h(s) = -k_1 e_0 s + (k_1 s + k_{-1})c = 0$ has the form

$$\dot{s} = \frac{k_1 s + k_{-1}}{k_1(e_0 - c)}\left(k_1 e_0 s + (k_1 s + k_{-1} + k_2)c\right),$$
$$\dot{c} = k_1 e_0 s - (k_1 s + k_{-1} + k_2)c; \tag{8}$$

here – in the notation of (7) – we chose $m^{[1]} = m^{[2]} = 0$. The choice $m^{[2]} = 1$ provides a more convenient version of the second equation on Y, viz. $\dot{c} = -k_2 c$.

We will return later to the crucial condition of invariance (or "approximate invariance") of the QSS set Y, which is fundamental for the ad hoc reduction.

5   Reduction in the SPT Setting

In this section we discuss the (standard) singular perturbation approach to reduction. Thus we specialize system (3) to a smooth system with one small parameter, viz.

$$\dot{x} = h(x, \varepsilon) = h^{(0)}(x) + \varepsilon h^{(1)}(x) + \cdots \tag{9}$$

with ε in some neighborhood of 0. (One may think of other parameters as being "frozen".)

5.1   Conditions

Assume that system (9) is – in principle, after a coordinate change – amenable to reduction by Tikhonov's theorem. An obvious problem is to cast equations (9) into standard form (5). Fenichel [13] discussed this, and the following local characterization of systems which admit a coordinate transformation to Tikhonov standard form is a consequence of his results. (An elementary proof is given in [28].)

Proposition 1. *Let system* (9) *be given, and denote by* $Z = \mathcal{V}(h^{(0)})$ *the zero set of* $h^{(0)}$. *Let* $a \in Z$ *and assume that there exists a neighborhood* U *of* a *in* $\mathbb{R}^n$ *such that* $Z \cap U$ *is an* r-*dimensional submanifold. Then there exists an invertible local coordinate transformation to standard form* (5), *satisfying a linear stability condition in some neighborhood of* a, *if and only if the following hold.*

(i) The rank of $Dh^{(0)}(a)$ *is equal to* $n - r$, *and there exists a direct sum decomposition*

$$\mathbb{R}^n = \operatorname{Ker} Dh^{(0)}(a) \oplus \operatorname{Im} Dh^{(0)}(a).$$

(ii) The nonzero eigenvalues of $Dh^{(0)}(a)$ *have real part* < 0.

Adopting the nomenclature from subsection 3.2, we will refer to Z as the (asymptotic) slow manifold of (9).

As pointed out in [28], Proposition 1 guarantees the existence of a transformation to Tikhonov standard form, but generally one cannot determine such a transformation explicitly, even for polynomial systems. The main obstacle lies in the explicit determination of first integrals of system (9) when $\varepsilon = 0$. (There is an important exception to this rule for slow and fast reactions, if and when stoichiometry provides sufficiently many linear first integrals; see Schauer and Heinrich [30], Lee and Othmer [24] and the algorithmic implementation by Boulier et al. [2].) But even if such a transformation cannot be determined, for rational sytems one can compute a reduced system on Z explicitly.

5.2   Reduction of Rational Systems

The main result of this section (taken from [15,16]) provides an algorithmic approach to the computation of reduced equations for general systems (9) with rational right-hand side, in particular for reaction equations with mass action kinetics. We will freely use some notions and results from commutative algebra and algebraic geometry. Underlying the reduction theorem is a fact from classical algebraic geometry; viz., a variety of dimension r can be represented as the zero set of $n - r$ regular functions in a Zariski-open neighborhood of a simple point (see e.g. Shafarevich [33], Ch.II, §3).

Theorem 1. (See [16]) *Consider system* (9) *with rational right-hand side* h, *and let* $a \in \mathbb{R}^n$ *be a simple point of* $\mathcal{V}(h^{(0)})$, *with* $n - r = \operatorname{rank} Dh^{(0)}(a)$. *(Thus locally the dimension of* $\mathcal{V}(h^{(0)})$ *equals* r.) *Assume moreover a direct sum decomposition*

$$\mathbb{R}^n = \operatorname{Ker} Dh^{(0)}(a) \oplus \operatorname{Im} Dh^{(0)}(a).$$

Then the following hold.

(a) *There exist a Zariski-open neighborhood U_a of a in $\mathbb{R}^n$ and a product decomposition with matrices $\mu(x) \in \mathbb{R}(x)^{(n-r)\times 1}$, $P(x) \in \mathbb{R}(x)^{n\times(n-r)}$, such that*

$$h^{(0)}(x) = P(x)\mu(x), \quad x \in U_a \tag{10}$$

with $\operatorname{rank} P(a) = n - r$, $\operatorname{rank} D\mu(a) = n - r$, *and*

$$\mathcal{V}(h^{(0)}) \cap U_a = \mathcal{V}(\mu) \cap U_a$$

is an r–dimensional submanifold.

(b) *The following system is defined on a Zariski-open neighborhood of a in $\mathbb{R}^n$, and admits a relatively Zariski-open neighborhood $\mathcal{U}_a \subset \mathcal{V}(h^0)$ as an invariant set:*

$$x' = \left[I_n - P(x)A(x)^{-1}D\mu(x)\right] h^{(1)}(x), \tag{11}$$

with

$$A(x) := D\mu(x)P(x) \in \mathbb{R}(x)^{(n-r)\times(n-r)}.$$

(c) *If all the nonzero eigenvalues of $Dh^{(0)}(a)$ have negative real part then system (11), restricted to $\mathcal{U}_a$, corresponds to the reduced system (6) in Tikhonov's theorem.*

If the condition in part (c) is not satisfied then formula (11) still defines a system with invariant set $\mathcal{U}_a$; we will call this a *formal reduction* of (9). (Fenichel's theory [13] implies convergence under mild hyperbolicity conditions.)

Example 2. In the Michaelis-Menten system, let $k_{-1} = \varepsilon\kappa_{-1}$ and $k_2 = \varepsilon\kappa_2$. Then

$$h^{(0)} = P \cdot \mu \text{ with } P = \begin{pmatrix} -k_1 s \\ k_1 s + k_{-2}(s_0 - s - c) \end{pmatrix} \text{ and } \mu = e_0 - c,$$

and $h^{(1)} = (\kappa_{-1}c, -(\kappa_{-1} + \kappa_2)c)^{\mathrm{tr}}$; thus the slow manifold is given by $c = e_0$. With the notation from Theorem 1 one has, for instance,

$$I_2 - P \cdot A^{-1} \cdot D\mu = \begin{pmatrix} 1 & -k_1 s/(k_1 s + k_{-2}(s_0 - s - c)) \\ 0 & 0 \end{pmatrix},$$

and the reduced equation on the slow manifold reads

$$\dot{s} = k_{-1}e_0 + k_1 s(k_{-1} + k_2)e_0/(k_1 s + k_{-2}(s_0 - e_0 - s))$$

together with $\dot{c} = 0$. This setting corresponds to slow degradation of complex; for biologically relevant parameter values the solution will eventually escape from the slow manifold.

The reduction of the irreversible Michaelis-Menten system with small k_2 is discussed by Schauer and Heinrich [30], who (essentially) employ a linear transformation to Tikhonov standard form, and by Boulier et al. [4] who take an approach via enforced invariant sets.

5.3   Algorithmic Aspects

The decomposition in Theorem 1 (a) is found by inspection in many applications, but we point out that it can be determined constructively. First, according to Shafarevich [33], one may choose (any) $n - r$ functionally independent entries of $h^{(0)}$ for μ; thus a generator system of the vanishing ideal in the local ring of a is known. An algorithm to determine P (in effect, to express the remaining entries of $h^{(0)}$ by μ) uses standard bases and Mora's algorithm (see Decker and Lossen [11], Lecture 9). There is an implementation of Mora's algorithm in the computer algebra system SINGULAR [10]. Thus, to a great extent the reduction of rational systems is manageable by customary algorithms. Some restrictions apply, however, since inequations (to ensure rank conditions) and inequalities (to ensure negative real parts for eigenvalues) also play a role in the discussion. Moreover, one usually deals with semi-algebraic sets in phase and parameter space, due to nonnegativity conditions.

6   Identifying "Small Parameters"

In the previous section we assumed that a "small parameter" was known a priori for system (9). In some applications (such as networks with slow and fast reactions) this is the case, but in others it is not. In particular, for a quasi-steady state assumption for chemical species the designation of small parameters is not straightforward. In their influential paper on the irreversible Michaelis-Menten system, Segel and Slemrod [32] introduced a heuristics to identify and compare time scales, from which they derived appropriate "small parameters"; this laid the foundation for many publications on quasi-steady state phenomena. As it seems, all approaches in the literature to identify "small parameters" require some intuition (or initial assumption) about the reaction network. In recent work [15,17] the authors proceeded differently, just assuming the existence of a singular perturbation scenario. A mathematical motivation for this approach was obtained from numerically oriented work, such as Lam and Goussis [23], Duchêne and Rouchon [12].

6.1   Definition and Basic Properties

Looking at a system in Tikhonov standard form (5) shows that "small parameters" ε are in fact distinguished by properties of the system at $\varepsilon = 0$, such as the existence of non-isolated stationary points.

Definition 2. *A* $\pi^* \in \Pi$ *will be called a* Tikhonov-Fenichel parameter value (TFPV) *for dimension* r *(* $1 \leq r \leq n - 1$*) of system (3) whenever the following hold:*

(i) *The zero set* $\mathcal{V}(h(\cdot, \pi^*))$ *of* $x \mapsto h(x, \pi^*)$ *contains a local submanifold* Z *of dimension* r.

(ii) There is a point $x_0 \in Z$ such that rank $D_1h(x_0, \pi^*) = n - r$ *and*

$$\mathbb{R}^n = \text{Ker } D_1h(x, \pi^*) \oplus \text{Im } D_1h(x, \pi^*), \quad \text{all } x \in Z \text{ near } x_0.$$

(iii) The nonzero eigenvalues of $D_1h(x_0, \pi^*)$ have real part < 0.

If only conditions (i) and (ii) hold then we will call π^* a weak Tikhonov-Fenichel parameter value for dimension r.

A straightforward application of Theorem 1 shows:

Remark 1. Let $\pi^* \in \Pi$ be a Tikhonov-Fenichel parameter value for dimension r of system (3), and let $x_0 \in V(h(\cdot, \pi^*))$ be such that the conditions in Definition 2 are satisfied. Then for any smooth curve $\gamma \colon \mathbb{R} \to \Pi$, $\delta \mapsto \gamma(\delta)$ in parameter space with $\gamma(0) = \pi^*$, the system

$$\dot{x} = h(x, \gamma(\delta)) = h(x, \pi^*) + \delta \cdot D_2h(x, \pi^*)\gamma'(0) + O(\delta^2) \tag{12}$$

admits a Tikhonov-Fenichel reduction for $\delta \to 0$.

6.2   Structure of the TFPV Set

We turn to the computation of TFPV's for polynomial (or rational) vector fields; at the same time we will clarify the structure of the Tikhonov-Fenichel parameter value set. Thus we consider system (3) with a polynomial (or rational) right-hand side, and we will also assume that the domain of interest is a Zariski-open subset Δ of a semi-algebraic set in $\mathbb{R}^{n+m}$. (All proofs, as well as further details, may be found in [17].)

Given system (3) and $(x, \pi)^{\text{tr}} \in U \times \Pi$, we denote by

$$\chi(\tau) = \chi_{x,\pi}(\tau) := \tau^n + \sigma_{n-1}(x, \pi)\tau^{n-1} + \cdots + \sigma_1(x, \pi)\tau + \sigma_0(x, \pi) \tag{13}$$

the characteristic polynomial of the Jacobian $D_1h(x, \pi)$, in the indeterminate τ. The coefficients σ_i are polynomial (resp. rational) functions in x and π. We first list a few technical facts.

Lemma 3. *Let $(x_0, \pi^*) \in \Delta$ be such that $h(x_0, \pi^*) = 0$, and let the characteristic polynomial of $D_1h(x_0, \pi^*)$ be given by (13). Then $\pi^* \in \Pi$ is a Tikhonov-Fenichel parameter value for dimension r, and x_0 lies in the local slow manifold of π^*, only if the following hold.*

(i) One has $\sigma_0(x_0, \pi^*) = \cdots = \sigma_{r-1}(x_0, \pi^*) = 0$.
(ii) The polynomial

$$\widetilde{\chi}(\tau) = \tau^{n-r} + \sigma_{n-1}(x_0, \pi^*)\tau^{n-r-1} + \cdots + \sigma_r(x_0, \pi^*)$$

has only zeros with negative real part.

Note that the subset of Δ satisfying (i) and (ii) is defined by polynomial equations and inequalities (Routh-Hurwitz conditions for $\widetilde{\chi}$). The next auxiliary result is proven via Poincaré-Dulac normal form theory.

Lemma 4. *Let $(x_0, \pi^*) \in \Delta$ satisfy the conditions in Lemma 3. Then π^* is a TFPV for dimension r, and x_0 lies in the local slow manifold of π^*, if and only if the system $\dot{x} = h(x, \pi^*)$ admits r functionally independent analytic (equivalently, formal) first integrals in a neighborhood of x_0. The lowest-degree terms of these first integrals may be chosen as (linearly independent) linear forms in $x - x_0$.*

There remains the task to restate these conditions by polynomial equalities or inequalities. Assuming $h(x_0, \pi^*) = 0$, consider the Taylor expansion with respect to $y := x - x_0$:

$$h(x, \pi^*) = \sum_{k \geq 1} h_k(x_0, \pi^*, y), \tag{14}$$

with h_k homogeneous of degree k in y, and in particular $h_1(x_0, \pi^*, y) = Dh(x_0, \pi^*)y$. Every h_k is rational in (x_0, π^*), since h is rational. A formal power series

$$\psi(y) = \sum_{j \geq 0} \psi_j(y), \quad \psi_j \text{ homogeneous of degree } j$$

is a first integral of h near $y = 0$ if and only if

$$L_h(\psi)(y) := D\psi(y) \, h(x_0, \pi^*, y) = 0,$$

equivalently (comparing homogeneous parts with respect to y) if

$$\sum_{j=1}^{k} L_{h_j}(\psi_{k-j}) = 0 \text{ for all } k.$$

Now denote by S_k the space of homogeneous polynomials of degree k in y. For any $d \geq 1$ define the linear map

$$L_h^{(d)} : S_1 + \cdots + S_d \to S_1 + \cdots + S_d$$

by sending $\psi = \psi_1 + \cdots + \psi_d$ to the truncation of $L_h(\psi)$ at degree d.

Proposition 2. *Let $(x_0, \pi^*) \in \Delta$ satisfy the conditions in Lemma 3. There exist r independent first integrals for $\dot{x} = h(x, \pi^*)$ near x_0 if and only if*

$$\dim \operatorname{Ker} L_h^{(d)} = \vartheta_{r,d} := \sum_{j=1}^{d} \binom{r+j-1}{j} \quad \text{for all } d \geq 1.$$

One knows that the kernel of $L_h^{(d)}$ has dimension at most $\vartheta_{r,d}$, hence the conditions may be restated, in a manner similar to Lemma 3(i), via coefficients of the characteristic polynomial of $L_h^{(d)}$. (The condition corresponding to (ii) is automatically satisfied.) Due to Hilbert's *Basissatz*, it is sufficient to check the dimension condition for some suitable $d = d^*$, which, however, is not known a priori. Hence we have the foundation for a pseudo-algorithm, rather than an algorithm. The above observations, in conjunction with Tarski-Seidenberg, also lead to a proof of the following structure theorem.

Theorem 2. *The Tikhonov-Fenichel parameter values of (3) for dimension r form a semialgebraic subset of $\mathbb{R}^m$.*

6.3 Algorithmic Aspects

In this subsection we further restrict attention to systems (3) with polynomial right-hand side; for rational systems matters are similar, but the presentation is more cumbersome. Moreover we will only consider the setting of Lemma 3 (which also corresponds to the $d = 1$ case of Proposition 2), and we will only discuss the equations necessary for a TPFV (disregarding inequalities). This represents the first step in the general procedure.

Remark 2. If $\pi^* \in \Pi$ is a Tikhonov-Fenichel parameter value for dimension r (with $1 \leq r \leq n-1$) for system (3) then there exists $x_0 \in \mathbb{R}^n$ with the following properties.

(i) $h(x_0, \pi^*) = 0$;
(ii) the Jacobian $D_1 h(x_0, \pi^*)$ has rank $\leq n-r$, thus for any $k > n-r$, all $k \times k$ minors vanish;
(iii) $\sigma_r(x_0, \pi^*) \neq 0$.

There is some redundancy in the statement of part (ii) (clearly $k = n - r + 1$ suffices), which is harmless and sometimes even welcome for computations in small systems. To obtain TFPV's, it is natural to employ elimination ideals (see e.g. Cox, Little, O'Shea [8], pp. 24–26). Once more standard methods suffice for the initial computations.

Proposition 3. *Let $\pi^* \in \Pi$ be a TFPV of the polynomial system (3) for dimension r, $1 \leq r \leq n - 1$. Let $\gamma_1, \ldots, \gamma_{\ell_r} \in \mathbb{R}[x, \pi]$ denote all the $k \times k$ minors of $D_1 h(x, \pi)$, $n \geq k > n - r$, and let*

$$\mathcal{I} = \langle h_1, \ldots, h_n, \gamma_1, \ldots, \gamma_{\ell_r} \rangle \subseteq \mathbb{R}[x, \pi].$$

Then π^* is a zero of the elimination ideal $\mathcal{I}_\pi = \mathcal{I} \cap \mathbb{R}[\pi]$.

Corollary 1. *For the polynomial system (3), denote by*

$$\mathcal{J} = \langle h_1, \ldots, h_n, \det D_1 h(x, \pi) \rangle$$

the ideal generated by the entries of h and its Jacobian determinant. Then for every $r \geq 1$, a TFPV π^* of system (3) for dimension r is a zero of $\mathcal{J}_\pi := \mathcal{J} \cap \mathbb{R}[\pi]$.

Example 3. Consider the irreversible Michaelis-Menten equation (2). The components h_1 and h_2 and the Jacobian determinant d of h generate the ideal

$$\mathcal{I} = \langle h_1, h_2, d \rangle,$$

in $\mathbb{R}[x, \pi]$. With respect to lexicographic order, SINGULAR finds the reduced Groebner basis

$$g_1(x, \pi) = e_0 k_1 k_2,$$
$$g_2(x, \pi) = k_2 c,$$
$$g_3(x, \pi) = -k_1 s e_0 + (k_1 s + k_{-1}) c,$$

with the elimination ideal $\mathcal{I}_\pi = \mathcal{I} \cap \mathbb{R}[\pi]$ generated by g_1. The condition $g_1 = 0$ shows that the only possible "small parameters" in this system are e_0, k_1 and k_2. (One verifies that these actually yield Tikhonov-Fenichel reductions; furthermore g_2 and g_3 provide conditions on the slow manifold.)

7   The Ad Hoc Approach Revisited

We return to ad hoc QSS assumptions for chemical species, see Section 4. One motivation for this (as noted) lies in a possible different mathematical interpretation of QSS. But the strategy is also useful for finding certain TFPV's (keeping in mind that knowing partial solutions may be helpful when solving large polynomial systems). As noted above, constructing a differential equation with an enforced invariant set from a given one will only be relevant for the original when the set in question is "approximately invariant". Essentially this observation goes back to Schauer and Heinrich [29], who employed it to determine "small parameters" for the irreversible Michaelis-Menten system. In [27], the notion was investigated further, and an infinitesimal criterion was obtained. Recently in [7], the focus was on parameter values which force invariance (whence small perturbations force "approximate invariance"). These parameter values are algorithmically accessible for polynomial or rational vector fields.

7.1   Basics and Approximation Properties

Given the parameter-dependent system (3), we search for criteria to identify parameter regions where certain sets (defined by pre-imposed QSS conditions on certain species) are "approximately invariant" in the specific sense that a suitable nearby parameter will assure invariance. The following basic observation is a variant of [7], Prop. 4.1, with some details omitted.

Proposition 4. *Let $\pi^* \in \mathbb{R}^m$ such that the equations*

$$\psi_1(x, \pi^*) = \cdots = \psi_{n-r}(x, \pi^*) = 0$$

(with smooth functions and full rank) define a local r-dimensional submanifold Y_{π^*} of $\mathbb{R}^n$ which is invariant for the system (3) at $\pi = \pi^*$; moreover let $y^* \in Y_{\pi^*}$. Then there exist a compact neighborhood K of y^* and a neighborhood V of π^* such that for every $\pi \in V$ the set defined by the equations

$$\psi_1(x, \pi) = \cdots = \psi_{n-r}(x, \pi) = 0$$

contains an r-dimensional local submanifold Y_π which has nonempty compact intersection with K. As $\pi \to \pi^*$, solutions of (3) with initial value in $Y_\pi \cap K$ converge to solutions of any associated system (7) with enforced invariant set Y_{π^*}.

We will call π^* a *critical parameter value* with respect to $\psi_1, \ldots, \psi_{n-r}$ if the system of functions has full rank and defines an invariant r-dimensional local submanifold near some point (y^*, π^*).

7.2   Polynomial Systems and Algorithmic Aspects

The algorithmic-algebraic implementation of critical parameter values has not been previously discussed. (The approach by Boulier et al. [4] to enforce certain invariant manifolds, and to investigate the dynamics on these manifolds, has a different starting point, and uses different arguments.) The outline presented here (although based on arguments similar to section 6) may thus be considered new. Specializing to systems with polynomial right hand side, and keeping the maximal rank condition from Proposition 4, it is again natural to work in the local setting, as in subsection 5.2, and once more one obtains algebraic conditions on the parameters.

Proposition 5. *Given system* (3) *with polynomial right-hand side, and polynomials* $\psi_1, \ldots, \psi_{n-r}$ *as in Proposition 4, a critical parameter value lies in the ideal generated by all* $(n - r + 1) \times (n - r + 1)$ *minors of the Jacobians of*

$$\begin{pmatrix} \psi_1 \\ \vdots \\ \psi_{n-r} \\ L_h(\psi_j) \end{pmatrix}, \quad 1 \le j \le n - r.$$

Proof. We pass to the complexification, and choose $y \in Y_{\pi^*}$. By the rank condition, $\psi_1, \ldots, \psi_{n-r}$ locally define a variety of dimension r, and invariance, together with the local version of Hilbert's *Nullstellensatz* (see Shafarevich [33], Ch. III, §3) implies the existence of rational μ_{jk}, regular at y, such that

$$L_h(\psi_j) = \sum_{k=1}^{n-r} \mu_{jk} \psi_k, \quad 1 \le j \le n - r.$$

Differentiation of these identities with respect to x and using $\psi_j(z) = 0$ for $z \in Y_{\pi^*}$ shows that $D_1(L_h(\psi_j))(z, \pi^*)$ is a linear combination of the $D_1(\psi_k)(z, \pi^*)$ for all $z \in Y_{\pi^*}$. Therefore the Jacobian has rank $\le n - r$.

Together with the $n - r + 1$ equations $\psi_1 = \cdots = \psi_{n-r} = L_h(\psi_j) = 0$, these determinant conditions (of which there are at least r, so one has more than n equations in total) again allow to employ elimination ideals, with elimination of the variables. We discuss a small example to illustrate the procedure.

Example 4. We continue the example from section 4, searching for critical parameter values with respect to $\psi = -k_1 e_0 s + (k_1 s + k_{-1})c$, but we will consider the reversible case here. Elimination of s and c from the ideal generated by ψ, $L_h(\psi)$ and their Jacobian determinant yields an elimination ideal with one generator

$$g := e_0 k_{-1} \cdot p,$$

with

$$p = (k_1 k_2 + k_{-1} k_{-2})^2 + 2k_1 k_{-2}((e_0 + s_0)k_1 k_2 + k_{-1} k_{-2}) + k_1^2 k_{-2}^2 (e_0 - s_0)^2.$$

In view of nonnegativity conditions, this provides the following cases (and only these).

- $e_0 = 0$ (see Briggs-Haldane [6]);
- $k_{-1} = 0$; the invariant set is given here by $s = 0$;
- $k_1 = k_{-2} = 0$ (slow formation of complex);
- $k_2 = k_{-2} = 0$ (slow formation and degradation of product).

A comparison with [17], subsection 5.1 shows that (only) the second parameter value is not a TFPV. Thus, generally there exist critical parameter values which are not TFPV's.

A straightforward algorithmic search for critical parameter values via Proposition 5 seems to be more involved than a search for TFPV's. But a more detailed analysis will yield further (more convenient) conditions. This is the subject of ongoing work.

8 Conclusion

Readers with expertise in algorithmic algebra will have noticed that we did not discuss algorithmic-algebraic aspects in any depth or detail. Indeed, for a complete analysis of the relatively low-dimensional standard systems (involving relatively few parameters) from biochemistry that were discussed in [15,16,17], a combination of standard algorithms and case-by-case inspection of intermediate results turned out to be sufficient.

The principal purpose of the present note was to present and describe the basic analytical results which provide a natural foundation for algorithmic-algebraic work in the case of reaction equations. Indeed for this class all analytical conditions can be transferred to (semi-) algebraic ones, and the pertinent results of [15,16,17] were outlined here.

The (semi-) algebraic conditions we gave are quite likely not stated in an optimal manner, and it may be advisable for further analysis to use more special properties of reaction systems in a general discussion. (We essentially only required polynomiality of the systems and certain positivity conditions.) Certainly a general approach will take a substantially bigger effort to become feasible, and much work remains to be done. The authors hope that the present paper will provide a stimulus for such work.

References

1. Anai, H., Horimoto, K., Kutsia, T.: AB 2007. LNCS, vol. 4545. Springer, Heidlberg (2007)
2. Boulier, F., Lemaire, F., Sedoglavic, A., Ürgüplü, A.: Towards an Automated Reduction Method for Polynomial ODE Models of Biochemical Reaction Systems. Mathematics in Computer Science **2**, 443–464 (2009)
3. Boulier, F., Lefranc, M., Lemaire, F., Morant, P.E.: Model Reduction of Chemical Reaction Systems using Elimination. Mathematics in Computer Science **5**, 289–301 (2011)

4. Boulier, F., Lemaire, F., Petitot, M., Sedoglavic, A.: Chemical reaction systems, computer algebra and systems biology. In: Gerdt, V.P., Koepf, W., Mayr, E.W., Vorozhtsov, E.V. (eds.) CASC 2011. LNCS, vol. 6885, pp. 73–87. Springer, Heidelberg (2011)
5. Borghans, J.A.M., de Boer, R.J., Segel, L.A.: Extending the quasi-steady state approximation by changing variables. Bull. Math. Biol. **58**, 43–63 (1996)
6. Briggs, G.E., Haldane, J.B.S.: A note on the kinetics of enzyme action. Biochem. J. **19**, 338–339 (1925)
7. Cicogna, G., Gaeta, G., Walcher, S.: Side conditions for ordinary differential equations. J. Lie Theory **25**, 125–146 (2015)
8. Cox, D.A., Little, J., O'Shea, D.: Using algebraic geometry. Graduate Texts in Mathematics, vol. 185, 2nd edn. Springer, New York (2005)
9. Errami, H., Eiswirth, M., Grigoriev, D., Seiler, W.M., Sturm, T., Weber, A.: Efficient methods to compute Hopf bifurcations in chemical reaction networks using reaction coordinates. In: Gerdt, V.P., Koepf, W., Mayr, E.W., Vorozhtsov, E.V. (eds.) CASC 2013. LNCS, vol. 8136, pp. 88–99. Springer, Heidelberg (2013)
10. Decker, W., Greuel, G.-M., Pfister, G., Schönemann, H.: Singular 3-1-3 – A computer algebra system for polynomial computations (2011).
    http://www.singular.uni-kl.de
11. Decker, W., Lossen, Ch.: Computing in algebraic geometry. Algorithms and computation in mathematics, vol. 16. Springer, Berlin (2006)
12. Duchêne, P., Rouchon, P.: Kinetic scheme reduction via geometric singular perturbation techniques. Chem. Eng. Sci. **12**, 4661–4672 (1996)
13. Fenichel, N.: Geometric singular perturbation theory for ordinary differential equations. J. Differential Equations **31**(1), 53–98 (1979)
14. Gatermann, K., Huber, B.: A family of sparse polynomial systems arising in chemical reaction systems. J. Symbolic Comput. **33**, 275–305 (2002)
15. Goeke, A.: Reduktion und asymptotische Reduktion von Reaktionsgleichungen. Doctoral dissertation, RWTH Aachen (2013)
16. Goeke, A., Walcher, S.: A constructive approach to quasi-steady state reduction. J. Math. Chem. **52**, 2596–2626 (2014)
17. Goeke, A., Walcher, S., Zerz, E.: Determining "small parameters" for quasi-steady state. J. Diff. Equations **259**, 1149–1180 (2015)
18. Heineken, F.G., Tsuchiya, H.M., Aris, R.: On the mathematical status of the pseudo-steady state hypothesis of biochemical kinetics. Math. Biosci. **1**, 95–113 (1967)
19. Henri, V.: Lois générales de l'action des diastases. Hermann, Paris (1903)
20. Horimoto, K., Regensburger, G., Rosenkranz, M., Yoshida, H.: AB 2008. LNCS, vol. 5147. Springer, Heidelberg (2008)
21. Hubert, E., Labahn, G.: Scaling Invariants and Symmetry Reduction of Dynamical Systems. Found. Comput. Math. **13**, 479–516 (2013)
22. Laidler, K.J.: Theory of the transient phase in kinetics, with special reference to enzyme systems. Can. J. Chem. **33**, 1614–1624 (1955)
23. Lam, S.H., Goussis, D.A.: The CSP method for simplifying kinetics. Int. J. Chemical Kinetics **26**, 461–486 (1994)
24. Lee, C.H., Othmer, H.G.: A multi-time-scale analysis of chemical reaction networks: I Deterministic systems. J. Math. Biol. **60**, 387–450 (2009)
25. Michaelis, L., Menten, M.L.: Die Kinetik der Invertinwirkung. Biochem. Z **49**, 333–369 (1913)
26. Niu, W., Wang, D.: Algebraic analysis of bifurcations and limit cycles for biological systems. In: [20], pp. 156–171

27. Noethen, L., Walcher, S.: Quasi-steady state and nearly invariant sets. SIAM J. Appl. Math. **70**(4), 1341–1363 (2009)
28. Noethen, L., Walcher, S.: Tikhonov's theorem and quasi-steady state. Discrete Contin. Dyn. Syst. Ser. B **16**(3), 945–961 (2011)
29. Schauer, M., Heinrich, R.: Analysis of the quasi-steady-state approximation for an enzymatic one-substrate reaction. J. Theoret. Biol. **79**, 425–442 (1979)
30. Schauer, M., Heinrich, R.: Quasi-steady-state approximation in the mathematical modeling of biochemical networks. Math. Biosci. **65**, 155–170 (1983)
31. Sedoglavic, A.: Reduction of algebraic parametric systems by rectification of their affine expanded Lie symmetries. In: [1], pp. 277–291
32. Segel, L.A., Slemrod, M.: The quasi-steady-state assumption: A case study in perturbation. SIAM Review **31**, 446–477 (1989)
33. Shafarevich, I.R.: Basic algebraic geometry. Springer, New York (1977)
34. Shiu, A., Sturmfels, B.: Siphons in chemical reaction networks. Bull. Math. Biol. **72**, 1448–1463 (2010)
35. Stiefenhofer, M.: Quasi-steady-state approximation for chemical reaction networks. J. Math. Biol. **36**, 593–609 (1998)
36. Tikhonov, A.N.: Systems of differential equations containing a small parameter multiplying the derivative (in Russian). Math. Sb. **31**, 575–586 (1952)
37. Verhulst, F.: Methods and Applications of Singular Perturbations. Boundary Layers and Multiple Timescale Dynamics. Springer, New York (2005)

Polynomial Complexity Recognizing a Tropical Linear Variety

Dima Grigoriev

CNRS, Mathématiques, Université de Lille, Villeneuve d'Ascq, 59655, France,
Dmitry.Grigoryev@math.univ-lille1.fr
http://en.wikipedia.org/wiki/Dima_Grigoriev

Abstract. A polynomial complexity algorithm is designed which tests whether a point belongs to a given tropical linear variety.

Keywords: polynomial complexity, recognizing tropical linear variety.

Introduction

Consider a linear system

$$A \cdot X = b \tag{1}$$

with the $m \times n$ matrix $A = (a_{i,j})$ and the vector $b = (b_i)$ defined over the field $K = \mathbb{C}((t^{1/\infty})) = \{c = c_0 t^{i_0/q} + c_1 t^{(i_0+1)/q} + \cdots\}$ of Puiseux series where $i_0 \in \mathbb{Z}$, $1 \le q \in \mathbb{Z}$. Consider the map $Trop(c) = i_0/q \in \mathbb{Q}$ and $Trop(0) = \infty$. Denote by $P \subset K^n$ the linear plane determined by the system (1). The closure in the Euclidean topology $\overline{Trop(P)} \subset \mathbb{R}^n$ is called a *tropical linear variety* [14] (for the basic concepts of the tropical geometry see [12], [13]).

More generally, the *tropical variety* attached to an ideal $I \subset K[X_1, \ldots, X_n]$ is defined as $\overline{Trop(U)} \subset \mathbb{R}^n$ where $U \subset K^n$ is the variety determined by I. A *tropical basis* of I is a finite set $f_1, \ldots, f_k \in I$ such that $\overline{Trop(U)} = \overline{Trop(V(f_1))} \cap \cdots \cap \overline{Trop(V(f_k))}$ where $V(f_1) \subset K^n$ denotes the variety of all the zeroes of f_1. In [3] an algorithm is devised which produces a tropical basis of an ideal. Having a tropical basis, one can easily test, whether a point $v \in \mathbb{R}^n$ belongs to the tropical variety $\overline{Trop(U)}$ since $\overline{Trop(V(f))} = V(Trop(f))$ (due to [7]) where the tropicalization $Trop(f)$ of a polynomial $f \in K[X_1, \ldots, X_n]$ is defined coefficientwise, and $V(Trop(f))$ is the tropical hypersurface of all the zeroes of the tropical polynomial $Trop(f)$. But on the other hand, in [3] an example is exhibited of a tropical linear variety with any its tropical basis having at least an exponential number of elements, while the algorithm recognizing a tropical linear variety designed in the present paper has the polynomial complexity.

We study the problem of the complexity of recognizing $\overline{Trop(P)}$. In other words, we design a polynomial complexity algorithm which for a given vector $v = (v_1, \ldots, v_n) \in (\mathbb{R} \cap \overline{\mathbb{Q}})^n$ with real algebraic coordinates tests, whether a system (1) has a solution $x = (x_1, \ldots, x_n) \in K^n$ with $Trop(x) = v$. Obviously,

© Springer International Publishing Switzerland 2015
V.P. Gerdt et al. (Eds.): CASC 2015, LNCS 9301, pp. 152–157, 2015.
DOI: 10.1007/978-3-319-24021-3_11

this captures also the case $v \in ((\mathbb{R} \cap \overline{\mathbb{Q}}) \cup \{\infty\})^n$, so we can w.l.o.g. assume below that $v \in (\mathbb{R} \cap \overline{\mathbb{Q}})^n$.

Observe that the problem of recognizing, whether just the zero vector belongs to a given tropical (non-linear) variety, is NP-hard (see [8]) since the solvability of a system of polynomial equations from $\overline{\mathbb{Q}}[X_1, \ldots, X_n]$ is equivalent to that the zero vector belongs to the tropical variety determined by this system.

We mention also that testing emptiness of a tropical *non-linear* prevariety (i. e. an intersection of a few tropical hypersurfaces) is NP-complete [15], while testing the emptiness of a tropical *linear* prevariety belongs to $NP \cap coNP$ ([1], [4], [9]).

Algorithm Lifting a Point to a Tropical Linear Variety

We assume that the entries $a_{i,j}, b_i \in K$, $1 \leq i \leq m$, $1 \leq j \leq n$ are provided in the following way (cf. [5], [10], [2]). A *primitive element* $z \in K$ is given as a root of a polynomial equation $h(t, z) = 0$ where $h \in \mathbb{Z}[t, Z]$, and by means of a further specifying a beginning of the expansion of z as a Puiseux series over the field $\overline{\mathbb{Q}}$ of algebraic numbers (to make a root of h to be unique with this beginning of the expansion). Also we are supplied with rational functions $h_{i,j}, h_i \in \mathbb{Q}(t)[Z]$ such that $a_{i,j} = h_{i,j}(z)$, $b_i = h_i(z)$, $1 \leq i \leq m$, $1 \leq j \leq n$. We suppose that $\deg(h), \deg(h_{i,j}), \deg(h_i) \leq d$. In addition, we assume that each rational coefficient of the polynomials $h, h_{i,j}, h_i$ is given as a quotient of a pair of integers with absolute values less than 2^M. The latter means that the bit-size of this rational number is bounded by $2M$.

Here and below to develop a Puiseux series with coefficients from $\overline{\mathbb{Q}}$ within the polynomial (in the bit-size of the input and in the number of terms of the expansion) complexity, we exploit the algorithm from [6]. The algorithm makes use of a presentation of the field of the coefficients as a finite extension of $\mathbb{Q}$ via its primitive element (see e. g. [5], [10], [2]) similar to the presentation of $a_{i,j}, b_i$ above (we don't dwell here on the details since it does not influence the main body of the algorithm, for the sake of simplifying the exposition a reader can suppose that the coefficients of the Puiseux series of z are rational).

First, the algorithm cleans the denominator in the exponents of the Puiseux series of z replacing $t^{1/q}$ by t for a suitable $q \leq d$ to make z to be a Laurent series with integer exponents (and keeping the same notation for z, h, $h_{i,j}$, h_i). The coordinates of the vector v we also multiply by s and keep the same notation for v. We say that two coordinates v_{j_1}, v_{j_2} of v are *congruent* if $v_{j_1} - v_{j_2} \in \mathbb{Z}$. Consider any solution x of (1). For each congruence class $\alpha \in \overline{\mathbb{Q}} \cap \mathbb{R}$ of v select from x all the monomials with the exponents which belong to α, denote by $x^{(\alpha)} := (x_1^{(\alpha)}, \ldots, x_n^{(\alpha)})$ the resulting vector consisting of these selected subsums of $x_1, \ldots, x_n$. Then $x^{(0)}$ which corresponds to the congruence class of the integers satisfies (1), and any other $x^{(\alpha)}$ with $\alpha \notin \mathbb{Z}$ satisfies the homogeneous linear system $A \cdot x^{(\alpha)} = 0$, hence $A \cdot (t^{-\alpha} \cdot x^{(\alpha)}) = 0$ and $t^{-\alpha} \cdot x^{(\alpha)}$ is a Laurent series. Thus, we get the following

Lemma 1. *Vector* $v \in \overline{Trop(P)}$ *iff the conjunction of the following statements for all the congruence classes* $\alpha \in \overline{\mathbb{Q}} \cap \mathbb{R}$ *holds. System* (1) *when* $\alpha \in \mathbb{Z}$ *(or respectively, the homogeneous system* $A \cdot X = 0$ *when* $\alpha \notin \mathbb{Z}$*) has a solution* $x = (x_1, \dots, x_n)$ *in Laurent series* $x_1, \dots, x_n$ *satisfying the conditions either* $Trop(x_j) + \alpha = v_j$ *when* v_j *belongs to the congruence class of* α, *or* $Trop(x_j) + \alpha > v_j$ *otherwise,* $1 \leq j \leq n$.

We assume that the vector v is provided in the following way (cf. [11], [2] and also above). A primitive real algebraic element $u \in \overline{\mathbb{Q}} \cap \mathbb{R}$ is given as a root of a polynomial $g \in \mathbb{Z}[Y]$ together with specifying a rational interval $[e_1, e_2]$ which contains the unique root u of g. In addition, certain polynomials $g_j \in \mathbb{Q}[Y]$, $1 \leq j \leq n$ are given such that $v_j = g_j(u)$. We suppose that $\deg(g), \deg(g_j) \leq d$ and that the absolute values of the numerators and denominators of the (rational) coefficients of g, g_j and of e_1, e_2 do not exceed 2^M.

To detect whether for a pair of the coordinates the congruence $v_{j_1} - v_{j_2} \in \mathbb{Z}$ holds, the algorithm computes an integer approximation $e \in \mathbb{Z}$ of $|v_{j_1} - v_{j_2} - e| < 1/2$ (provided that it does exist) with the help of e. g. the algorithm from [2] and then verifies whether $v_{j_1} - v_{j_2} = e$ exploiting [5], [10] or [2]. This supplies us with the partition of the coordinates $v_1, \dots, v_n$ into the classes of congruence. Thus, for the time being we fix a congruence class α. The algorithm searches for vectors $x = (x_1, \dots, x_n)$ satisfying the conditions in Lemma 1. Denote by $J \subset \{1, \dots, n\}$ the set of j such that v_j belongs to the fixed congruence class. For every $j \in J$ we replace $a_{i,j}$ by $t^{v_j - \alpha} \cdot a_{i,j}$, $1 \leq i \leq m$. For every $j \notin J$ let $\alpha + s - 1 < v_j < \alpha + s$ for a suitable (unique) integer s, then we replace $a_{i,j}$ by $t^s \cdot a_{i,j}$, $1 \leq i \leq m$. After this replacement the algorithm searches for vectors $x = (x_1, \dots, x_n)$ satisfying the properties $Trop(x_j) = 0$, $j \in J$ and $Trop(x_j) \geq 0$, $j \notin J$.

Then by elementary transformations with the rows of matrix A over the quotient-ring $\mathbb{Q}(t)[Z]/h$ (making use of the basic algorithms e. g. from [2]) and an appropriate permutation of columns, the algorithm brings A to the form $a_{i,i} = 1$, $a_{i,j} = 0$, $1 \leq i \neq j \leq m$ (one can assume w.l.o.g. that $rk(A) = m$).

For $m < j \leq n$ denote $r_j := -\min_{1 \leq i \leq m}\{Trop(a_{i,j})\}$. If $r_j < 0$ we put the coordinate $x_j = 1$. Else if $r_j \geq 0$ we put $x_j = y_{j,0} + y_{j,1} \cdot t + \cdots + y_{j,r_j} \cdot t^{r_j}$ with the indeterminates $y_{j,0}, \dots, y_{j,r_j}$ over $\overline{\mathbb{Q}}$.

Below w.l.o.g. we carry out the calculations for the case of the congruence class of integers $\alpha \in \mathbb{Z}$. When $\alpha \notin \mathbb{Z}$ one should put below $b_i = 0$, $1 \leq i \leq m$ (cf. Lemma 1).

For $1 \leq i \leq m$ denote $s_i = \min_{m < j \leq n}\{Trop(a_{i,j}), Trop(b_i)\}$. The i-th equation of (1) one can rewrite as

$$x_i + \sum_{m < j \leq n} a_{i,j} \cdot x_j = b_i. \tag{2}$$

For every $s_i \leq k \leq 0$ one can express the coefficient of $\sum_{m < j \leq n} a_{i,j} \cdot x_j - b_i$ at the power t^k as a linear function $L_{i,k}$ over $\overline{\mathbb{Q}}$ in the indeterminates $Y := \{y_{j,l}, m < j \leq n, 0 \leq l \leq r_j\}$.

Consider the linear system

$$L_{i,k} = 0, \ 1 \leq i \leq m, \ s_i \leq k < 0 \tag{3}$$

in the indeterminates Y. The algorithm solves (3) and tests whether each of n linear functions from the family

$$L := \{L_{i,0}, \ i \in J, 1 \leq i \leq m; \ y_{j,0}, \ j \in J, m < j \leq n\}$$

does not vanish identically on the space of solutions of (3). If all of them do not vanish identically then take any values of Y which fulfil (3) with non-zero values of all the linear functions from the family L. Then the equation (2) determines uniquely x_i with $Trop(x_i) = 0$ when $i \in J$ and $Trop(x_i) \geq 0$ when $i \notin J$. This provides a solution x of the system (1) satisfying $Trop(x_j) = 0$ when $j \in J$ and $Trop(x_j) \geq 0$ when $j \notin J$ (cf. Lemma 1). Otherwise, if some of the linear functions from the family L vanishes identically on the space of solutions of (3) then the system (1) has no solutions satisfying the conditions of Lemma 1.

To test the above requirement of identically non-vanishing of the linear functions from the family L the algorithm finds a basis $w_1, \ldots, w_r \in (\overline{\mathbb{Q}})^N$ and a vector $w \in (\overline{\mathbb{Q}})^N$ where $N = |Y|$ such that the r-dimensional space of solutions of (3) is the linear hull of the vectors $w_1, \ldots, w_r$ shifted by the vector w. If each linear function from the family L does not vanish identically on this space then all of them do not vanish on at least one of the vectors from the family

$$F := \{w + \sum_{1 \leq l \leq r} p^l \cdot w_l, \ 1 \leq p \leq |J|r + 1 \leq nr + 1\}$$

because any linear function can vanish on at most of r vectors from F due to the non-singularity of the Vandermonde matrices. So, the algorithm substitutes each of the vectors of F into the linear functions from L and either finds a required one Y or discovers that (1) has no solution satisfying the conditions of Lemma 1.

To estimate the complexity of the designed algorithm observe that it solves the linear system (3) of the size bounded by a polynomial in n, d with the coefficients from a finite extension of $\mathbb{Q}$ having the bit-size less than linear in M and polynomial in n, d (again for the sake of simplifying the exposition a reader can think just of the rational coefficients, cf. the remark above at the beginning of the present Section). Thus, the algorithm solves this system within the complexity polynomial in M, n, d (see e. g. [2]), by a similar magnitude one can bound the complexity of the executed substitutions, and finally we can summarize the obtained results in the following theorem.

Theorem 1. *There is an algorithm which for a tropical linear variety $V := \overline{Trop(P)}$ defined by a linear system (1) over the field K of Puiseux series, recognizes whether a given real algebraic vector $v \in ((\mathbb{R} \cap \overline{\mathbb{Q}}) \cup \{\infty\})^n$ belongs to V. If yes then the algorithm yields a solution $x \in K^n$ of (1) with $Trop(x) = v$. The complexity of the algorithm is polynomial in the bit-sizes of the system (1) and of the vector v.*

A tropical linear variety $V := \overline{Trop(P)}$ lies in the real space $\mathbb{R}^n$. For a given real vector $v = (v_1, \ldots, v_n)$ one can test whether it belongs to V following the described above algorithm, provided that one is able to test whether $v_i - v_j$ is an integer and find it in this case.

Further Research

Let a point $v \in \mathbb{R}^n$ don't lie in a tropical linear variety $V = \overline{Trop(P)}$ (cf. the Introduction). Theorem 1 implies that one can verify the latter within the polynomial complexity. Owing to [14], [3] there exists a finite tropical basis $f_1, \ldots, f_k \in I$ of the ideal $I \subset K[X_1, \ldots, X_n]$ of P. This allows one to find $f \in \{f_1, \ldots, f_k\} \subset I$ such that v is not a tropical zero of the tropical polynomial $Trop(f)$. But the number k can be exponential ([3]). Is it possible to construct $f_0 \in I$ for which v is not a tropical zero of $Trop(f_0)$, within the polynomial complexity?

What is the complexity to detect whether a given tropical linear prevariety $V(R_1, \ldots, R_p)$, with $R_1, \ldots, R_p$ being tropical linear polynomials, is a tropical variety?

Acknowledgements. The author is grateful to the Max-Planck Institut für Mathematik, Bonn for its hospitality during writing this paper and to Labex CEMPI (ANR-11-LABX-0007-01).

References

1. Akian, M., Gaubert, S., Guterman, A.: Tropical polyhedra are equivalent to correspondence mean payoff games. Int. J. Algebra Comput. **22**(1), 1250001, 43 p. (2012)
2. Basu, S., Pollack, R., Roy, M.-F.: Algorithms in Real Algebraic Geometry. Springer, Berlin (2006)
3. Bogart, T., Jensen, A.N., Speyer, D., Sturmfels, B., Thomas, R.R.: Computing tropical varieties. J. Symb. Comput. **42**, 54–73 (2007)
4. Butkovic, P.: Min-plus Systems: Theory and Algorithms. Springer, London (2010)
5. Chistov, A.: An algorithm of polynomial complexity for factoring polynomials, and determination of the components of a variety in a subexponential time. J. Soviet Math. **34**, 1838–1882 (1986)
6. Chistov, A.L.: Polynomial complexity of Newton-Puiseux algorithm. In: Gruska, J., Rovan, B., Wiedermann, J. (eds.) Mathematical Foundations of Computer Science 1986. LNCS, vol. 233, pp. 247–255. Springer, Heidelberg (1986)
7. Einsiedler, M., Kapranov, M., Lind, D.: Non-archemidean amoebas and tropical varieties. J. Reine Angew. Math. **601**, 139–157 (2007)
8. Garey, M., Johnson, D.: Computers and Intractability: a Guide to the Theory of NP-Completeness. W.H. Freeman and Company, San Francisco (1979)
9. Grigoriev, D.: Complexity of solving tropical linear systems. Computational Complexity **22**, 71–88 (2013)
10. Grigoriev, D.: Polynomial factoring over a finite field and solving systems of algebraic equations. J. Soviet Math. **34**, 1762–1803 (1986)

11. Grigoriev, D., Vorobjov, N.: Solving systems of polynomial inequalities in subexponential time. J. Symp. Comput. **5**, 37–64 (1988)
12. Itenberg, I., Mikhalkin, G., Shustin, E.: Tropical algebraic geometry. Oberwolfach Seminars, vol. 35. Birkhauser, Basel (2009)
13. Maclagan, D., Sturmfels, B.: Introduction to Tropical Geometry. Graduate Studies in Math., vol. 161. AMS, Providence (2015)
14. Speyer, D., Sturmfels, B.: The tropical Grassmanian. Adv. Geom. **4**, 389–411 (2004)
15. Theobald, T.: On the frontiers of polynomial computations in tropical geometry. J. Symbolic Comput. **41**, 1360–1375 (2006)

Computing Highest-Order Divisors for a Class of Quasi-Linear Partial Differential Equations

Dima Grigoriev[1] and Fritz Schwarz[2]

[1] CNRS, Mathématiques, Université de Lille, Villeneuve d'Ascq, 59655, France
Dmitry.Grigoryev@math.univ-lille1.fr
http://en.wikipedia.org/wiki/Dima_Grigoriev
[2] Fraunhofer Gesellschaft, Institut SCAI 53754 Sankt Augustin, Germany
fritz.schwarz@scai-extern.fraunhofer.de
http://www.scai.fraunhofer.de/schwarz.html

Abstract. A differential polynomial G is called a divisor of a differential polynomial F if any solution of the differential equation $G = 0$ is a solution of the equation $F = 0$. We design an algorithm which for a class of quasi-linear partial differential polynomials of order $k + 1$ finds its quasi-linear divisors of order k.

Keywords: quasi-linear differential polynomial, divisor, algorithm.

Introduction

The problem of factoring linear ordinary differential operators $L = T \circ Q$ was studied in [15]. Algorithms for this problem were designed in [8], [16] (in [8] a complexity bound better than for the algorithm from [15] was established). An algorithm is exhibited in [10] for factoring a partial linear differential operator in two variables with a separable symbol. In [9], an algorithm is constructed for finding all first-order factors of a partial linear differential operator in two variables. A generalization of factoring for D-modules (in other words, for systems of linear partial differential operators) was considered in [11,17]. A particular case of factoring for D-modules is the Laplace problem [6,19] (one can find a short exposition of the Laplace problem in [12]).

The meaning of factoring for search of solutions is that any solution of operator Q is a solution of operator L, thus, factoring allows one to diminish the order of operators.

Much less is known for factoring non-linear (even ordinary) differential equations.

We note that our definition of divisors is in the frame of differential ideals [14], rather than the definition of factorization from [18,4] being in terms of a composition of nonlinear ordinary differential polynomials. In [4], a decomposition algorithm is designed.

We consider partial differential polynomials viewing them as polynomials in independent variables $x_1, \ldots, x_n$ and in derivatives

© Springer International Publishing Switzerland 2015
V.P. Gerdt et al. (Eds.): CASC 2015, LNCS 9301, pp. 158–165, 2015.
DOI: 10.1007/978-3-319-24021-3_12

$$\frac{d^{i_1+\cdots+i_n} u}{dx_1^{i_1} \cdots dx_n^{i_n}}$$

[14]. We study a class of *quasi-linear* differential polynomials in which the coefficients at all its highest derivatives, i. e., with the biggest value of the order $i_1 + \cdots + i_n$, are constants.

We design an algorithm which for a given quasi-linear differential polynomial F of order $k + 1$ finds the algebraic variety of all its quasi-linear divisors G of order k. Moreover, we show that in this case, $\deg G \leq \deg F$ (treating F and G as algebraic polynomials). This result generalizes [13] where an algorithm was designed for finding quasi-linear divisors for quasi-linear *ordinary* differential polynomials F of order $k = 2$.

In Section 1, we bound the degree of a divisor, and in Section 2, we describe the algorithm.

It would be interesting to find divisors of F of arbitrary orders (rather than just of k) even in the case of ordinary differential equations. Also an extension to arbitrary differential polynomials (rather than quasi-linear) looks as a challenge.

Another issue to be studied is constructing a common multiple of a pair of partial differential polynomials, i. e., a differential polynomial whose solutions contain the solutions of both differential polynomials; for the case of quasi-linear ordinary differential polynomials, an algorithm was designed in [13].

1   Bound on a Degree of a Divisor

We study partial differential polynomials, i. e., polynomials of the form

$$F(\ldots, \frac{d^{i_1+\cdots+i_n} u}{dx_1^{i_1} \cdots dx_n^{i_n}}, \ldots, x_1, \ldots, x_n)$$

with coefficients over $\overline{\mathbb{Q}}$ where the maximal value of $i_1 + \cdots + i_n$ is denoted by $\operatorname{ord} F$ (the order of F) [14]. We denote the differential ring of all partial differential polynomials by D.

Definition 1. *A differential polynomial G is a divisor of F if any solution u from the universal extension (see, e. g., p. 133 [14]) of the field of rational functions $\overline{\mathbb{Q}}(x_1, \ldots, x_n)$ of equation $G = 0$ is a solution of $F = 0$ as well.*

Due to the differential Nullstellensatz (see, e. g., Corollary 1 p. 148 [14]) a differential polynomial G is a divisor of F iff F belongs to the radical differential ideal generated by G. We mention that a bound being in general not primitive-recursive, for the differential Nullstellensatz was established in [5].

We say that F of order $k + 1$ is *quasi-linear* if

$$F = \sum_{i_1+\cdots+i_n=k+1} a_{i_1,\ldots,i_n} \cdot \frac{d^{k+1} u}{dx_1^{i_1} \cdots dx_n^{i_n}} + f$$

where coefficients $a_{i_1,\ldots,i_n} \in \overline{\mathbb{Q}}$ and $\operatorname{ord} f \leq k$.

In the present section we provide an algebraic criterion for a quasi-linear G of order k to be a divisor of F and bound the degree of G.

For the sake of simplifying notations we will assume that there are just two independent variables x, y, i. e., $n = 2$. Denote a quasi-linear differential polynomial

$$F = \sum_{0 \le i \le k+1} a_i \cdot \frac{d^{k+1}u}{dx^i dy^{k+1-i}} + f. \tag{1}$$

Let a quasi-linear differential polynomial

$$G = \sum_{0 \le i \le k} b_i \cdot \frac{d^k u}{dx^i dy^{k-i}} + g \tag{2}$$

be a divisor of F where $\operatorname{ord} g \le k - 1$ and $b_0, \ldots, b_k \in \overline{\mathbb{Q}}$. Making a $\overline{\mathbb{Q}}$-linear transformation of the independent variables x, y one can assume w.l.o.g. that $b_0 = 1$.

Theorem 1. *i) A quasi-linear differential polynomial G of order k is a divisor of a quasi-linear differential polynomial F of order $k + 1$ (with $\deg F = d$) iff G divides (as polynomials)*

$$\left(F - c_1 \cdot \frac{dG}{dx} - c_2 \cdot \frac{dG}{dy} \right)^d$$

where

$$\operatorname{ord}\left(F - c_1 \cdot \frac{dG}{dx} - c_2 \cdot \frac{dG}{dy} \right) \le k$$

for suitable (unique) $c_1, c_2 \in \overline{\mathbb{Q}}$.
ii) In this case, $\deg G \le \deg F$.

Introduce the *highest order derivatives forms* being homogeneous polynomials

$$A := \sum_{0 \le i \le k+1} a_i \cdot v^i \cdot w^{k+1-i}, \ B := \sum_{0 \le i \le k} b_i \cdot v^i \cdot w^{k-i} \in \overline{\mathbb{Q}}[v, w]$$

of the differential polynomials F and G, respectively.

Lemma 1. *If a quasi-linear differential polynomial G with $\operatorname{ord} G = k$ is a divisor of a quasi-linear differential polynomial F with $\operatorname{ord} F = k + 1$ then there exist unique $c_1, c_2 \in \overline{\mathbb{Q}}$ such that $(c_1 \cdot v + c_2 \cdot w) \cdot B = A$, in other words $B | A$. Moreover, in this case $\operatorname{ord}(F - c_1 \cdot \frac{dG}{dx} - c_2 \cdot \frac{dG}{dy}) \le k$.*

Proof of Lemma. Due to the differential Nullstellensatz we have for suitable integer m

$$F^m = \sum_q H_q \cdot G_q \tag{3}$$

where G_q are certain partial derivatives of G and $H_q \in D$. Introduce variables $u_{i,j}$ for $\frac{d^{i+j}u}{dx^i dy^j}$ and making use repeatedly of relations $\frac{du_{i,j}}{dx} = u_{i+1,j}$, $\frac{du_{i,j}}{dy} = u_{i,j+1}$ we can consider (3) as an equality of polynomials in the variables $\{u_{i,j}\}_{i,j}$, x, y. Let a derivative of G of an order higher than 1 occur in (3) and denote by $s \geq 2$ the highest order of derivatives of G occurring in (3).

Taking appropriate $\overline{\mathbb{Q}}$-linear combinations of the equations

$$\frac{d^s G}{dx^i dy^{s-i}} = 0, \, 0 \leq i \leq s,$$

and considering their highest order derivatives one can express the variables

$$u_{j,s+k-j} = \sum_{s < l \leq s+k} c_l \cdot u_{l,s+k-l} + g_j, \, 0 \leq j \leq s \qquad (4)$$

for suitable coefficients $c_l \in \overline{\mathbb{Q}}$ and differential polynomials g_j with ord $g_j < s+k$. Substituting expressions (4) into (3) we get rid of all the derivatives G_q of G of order s. Observe that these substitutions do not change the left-hand side of (3). After that substitute 0 in all H_q for variables $u_{l,s+k-l}$, $s < l \leq s+k$ and for all variables $u_{i,j}$ with $i+j > s+k$, we obtain a formula similar to (3) with orders of derivatives G_q of G less than s and with variables $u_{i,j}$ occurring in G_q and H_q satisfying $i+j < s+k$.

Continuing in this way, we get rid of all the variables $u_{i,j}$ in the right-hand side of (3) with $i+j > k+1$.

After that we employ formulae (4) with $s = 1$ to achieve that the differential polynomial $F_0 := F - c_1 \cdot \frac{dG}{dx} - c_2 \cdot \frac{dG}{dy}$ does not contain derivatives $u_{0,k+1}$, $u_{1,k}$ for suitable $c_1, c_2 \in \overline{\mathbb{Q}}$. Then (3) implies that

$$F_0^m = H^{(1)} \cdot \frac{dG}{dx} + H^{(2)} \cdot \frac{dG}{dy} + H^{(0)} \cdot G \qquad (5)$$

for some differential polynomials $H^{(1)}$, $H^{(2)}$, $H^{(0)}$ of orders at most $k+1$. Now substitute formulae (4) with $s = 1$ in formula (5), this results in

$$F_0^m = H \cdot G \qquad (6)$$

for appropriate $H \in D$. Therefore, since F_0 contains derivatives of order $k+1$ with constant coefficients, all these coefficients vanish, thus, ord $F_0 \leq k$, hence ord $H \leq k$. Consequently,

$$F_0 = f - c_1 \cdot \frac{dg}{dx} - c_2 \cdot \frac{dg}{dy} \qquad (7)$$

(see (1), (2)) and $(c_1 \cdot v + c_2 \cdot w) \cdot B = A$. The Lemma is proved. $\qquad \square$

Proof of Theorem. Substitute formulae

$$\frac{dg}{dx} = \sum_{i+j \leq k-1} \frac{\partial g}{\partial u_{i,j}} \cdot u_{i+1,j} + \frac{\partial g}{\partial x}; \quad \frac{dg}{dy} = \sum_{i+j \leq k-1} \frac{\partial g}{\partial u_{i,j}} \cdot u_{i,j+1} + \frac{\partial g}{\partial y} \qquad (8)$$

in (7), and we substitute the obtained expression for F_0 in the left-hand side of (6), then we substitute in the resulting formula the expression for $u_{0,k} = -\sum_{1\le i<k} b_i \cdot u_{i,k-i} - g$ from (2). After the latter substitution, the right-hand side of (6) vanishes, and we deduce (taking into account (2)) the equality

$$0 = f|_{(u_{0,k}=-\sum_{1\le i<k} b_i \cdot u_{i,k-i}-g)} - c_1 \cdot \left(\sum_{i+j\le k-1} \frac{\partial g}{\partial u_{i,j}} \cdot u_{i+1,j} + \frac{\partial g}{\partial x} \right) + \qquad (9)$$

$$c_2 \cdot \left(\frac{\partial g}{\partial u_{0,k-1}} \left(\sum_{1\le i\le k} b_i u_{i,k-i} + g \right) - \sum_{i+j\le k-1,\,(i,j)\ne(0,k-1)} \frac{\partial g}{\partial u_{i,j}} u_{i,j+1} - \frac{\partial g}{\partial y} \right) \quad (10)$$

One can rewrite

$$f|_{(u_{0,k}=-\sum_{1\le i<k} b_i \cdot u_{i,k-i}-g)} = f|_{(u_{0,k}=-\sum_{1\le i<k} b_i \cdot u_{i,k-i})} + h \cdot g$$

for suitable $h \in D$. Therefore, (9) and (10) imply the following divisibility relation

$$g | (f|_{(u_{0,k}=-\sum_{1\le i<k} b_i \cdot u_{i,k-i})} - c_1 \cdot \left(\sum_{i+j\le k-1} \frac{\partial g}{\partial u_{i,j}} \cdot u_{i+1,j} + \frac{\partial g}{\partial x} \right) + \qquad (11)$$

$$c_2 \cdot \left(\frac{\partial g}{\partial u_{0,k-1}} \cdot \sum_{1\le i\le k} b_i \cdot u_{i,k-i} - \sum_{i+j\le k-1,\,(i,j)\ne(0,k-1)} \frac{\partial g}{\partial u_{i,j}} \cdot u_{i,j+1} - \frac{\partial g}{\partial y})) \quad (12)$$

Denote the polynomial in the variables $\{u_{i,j}\}_{i,j}$, x, y in the right-hand side of (11), (12) by P.

Our goal is to prove that $\deg g \le \deg f$. Suppose the contrary. Then (11), (12) entail that $\deg P \le \deg g$ (taking into account that $\deg f \le \deg g$ by the supposition) and whence $P = c \cdot g$ for appropriate $c \in \overline{\mathbb{Q}}$. Consider a linear deglex ordering $\prec$ of monomials in $\{u_{i,j}\}_{i+j\le k-1}$, x, y in which $u_{i,j} \prec u_{l,s}$ when $i + j > l + s$ (the remaining requirements on the ordering do not matter). We observe that the highest (w.r.t. $\prec$) monomial in g cannot occur in P since $\deg f < \deg g$ by the supposition. This leads to a contradiction with the equality $P = c \cdot g$ which proves inequality $\deg g \le \deg f$. Summarizing, we conclude Theorem 1 ii).

To prove Theorem 1 i) in the direction when G is a divisor of F we apply Lemma 1 and note that one can take $m = d$ in (6) owing to Theorem 1 ii) because if $G|F_0^m$ for some m then $G|F_0^{\deg G}$. To prove the converse we observe that $G|(F_0 - c_1 \cdot \frac{dg}{dx} - c_2 \cdot \frac{dg}{dy})^d$ implies (3) (with $m = d$). $\qquad \square$

We present the following simple example just to illustrate the notations.

Example 1. Here we use the notations $u_x = \frac{\partial u}{\partial x}$ and so on.

$$G = u_x + u_y + g(x, y);$$

$$F = u_{xx} + 5u_{xy} + 6u_{yy} + \frac{\partial g}{\partial x} + 3\frac{\partial g}{\partial y} + H(x, y, u, u_x, u_y) \cdot (u_x + 2u_y + g);$$

$$c_1 = 1, \ c_2 = 3, \ A = v^2 + 5vw + 6w^2, \ B = v + 2w;$$

$$F_0 = F - u_{xx} - 2u_{xy} - \frac{\partial g}{\partial x} - 3(u_{xy} + 2u_{yy} + \frac{\partial g}{\partial y}) = H \cdot G.$$

2   Algorithm to Find the Algebraic Variety of All the Divisors

Now we proceed to an algorithm which for a quasi-linear $F \in D$ with $\operatorname{ord} F = k + 1$, $\deg F = d$ yields the algebraic variety of all its divisors of order k (let $k \geq 1$). Making a $\overline{\mathbb{Q}}$-linear transformation of independent variables x, y one can assume w.l.o.g. that the coefficient $a_0 = 1$ (see (1)), this is compatible with the assumption $b_0 = 1$ due to Lemma 1.

First the algorithm factorizes the highest order derivatives form $A = \sum_{0 \leq i \leq k+1} a_i \cdot v^i \cdot w^{k+1-i} \in \overline{\mathbb{Q}}[v, w]$ (see Lemma 1), say with the help of [2], [7]. Pick one of its at most of $k+1$ factors with degree k as a candidate for the highest order derivatives form $B = \sum_{0 \leq i \leq k} b_i \cdot v^i \cdot w^{k-i} \in \overline{\mathbb{Q}}[v, w]$ of a divisor G of F. One can assume w.l.o.g. that $b_0 = 1$ (if $b_0 = 0$ we discard this candidate). Hence $(c_1 \cdot v + c_2 \cdot w) \cdot B = A$ for some $c_1, c_2 \in \overline{\mathbb{Q}}$ (actually, $c_2 = 1$ since $a_0 = b_0 = 1$).

Due to Theorem 1 ii) $\deg G \leq \deg F$, and we write a candidate for G as a polynomial with indeterminate coefficients over $\overline{\mathbb{Q}}$. In view of Theorem 1 i) one has to verify whether G divides $(F - c_1 \cdot \frac{dG}{dx} - c_2 \cdot \frac{dG}{dy})^d$ employing (8) (cf. also (9), (10), (11), and (12)). For this goal we introduce H (see (6)) with indeterminate coefficients over $\overline{\mathbb{Q}}$ and verify the condition

$$\left(F - c_1 \cdot \frac{dG}{dx} - c_2 \cdot \frac{dG}{dy} \right)^d = H \cdot G \qquad (13)$$

as a system of polynomial equations invoking the quantifier elimination algorithm from [3] (eliminating the indeterminate coefficients of H). The latter algorithm finds the irreducible components of the algebraic variety of all divisors G.

To estimate the complexity of the designed algorithm one has to specify how does the algorithm represent the coefficients of F from $\overline{\mathbb{Q}}$. A customary way to this end is to represent them as elements from an appropriate finite extension of $\mathbb{Q}$ (see e. g. [1,2,7,3]). Denote by L a bound on the bit-size of such a representation (say, in a particular case of rational numbers p/q its bit-size is defined as $\lceil \log_2(p + 1)(q + 1) \rceil$).

Denote

$$N_0 := \binom{k+n}{n} + n, \ N := \binom{N_0 + d^2}{d^2}.$$

The complexity of the designed algorithm is majorated by the complexity of solving (13) which leads to the quantifier elimination for a system of polynomials in at most of N indeterminates being the coefficients at the monomials of degrees d for polynomial G and of degrees $d^2 - d$ for polynomial H in N_0 variables $\{u_{i_1,\ldots,i_n} : i_1 + \cdots + i_n \le k\} \cup \{x_1, \ldots, x_n\}$. The degrees of these polynomials do not exceed d, and their number is bounded by N. The bit-sizes of the coefficients of these polynomials are less than $L + O(\log N)$. The complexity of the quantifier elimination algorithm [3] applied to this system does not exceed a polynomial in L, d^{N^2}. Summarizing and utilizing the notations introduced above, we conclude with

Theorem 2. *There is an algorithm which for a given quasi-linear differential polynomial of an order $k + 1$ produces the irreducible components of the algebraic variety of all its quasi-linear divisors of order k. The complexity of the algorithm can be bounded by a polynomial in L, d^{N^2}.*

Acknowledgements. The first author is grateful to the Max-Planck Institut für Mathematik, Bonn for its hospitality during writing this paper and to Labex CEMPI (ANR-11-LABX-0007-01). The authors appreciate the valuable remarks of the anonymous referees which encouraged an improvement of the exposition.

References

1. Basu, S., Pollack, R., Roy, M.-F.: Algorithms in real algebraic geometry. Springer, Berlin (2006)
2. Chistov, A.: An algorithm of polynomial complexity for factoring polynomials, and determination of the components of a variety in a subexponential time. J. Soviet Math. **34**, 1838–1882 (1986)
3. Chistov, A., Grigoriev, D.: Complexity of quantifier elimination in the theory of algebraically closed fields. In: Chytil, M.P., Koubek, V. (eds.) Mathematical Foundations of Computer Science 1984. LNCS, vol. 176, pp. 17–31. Springer, Heidelberg (1984)
4. Gao, X.S., Zhang, M.: Decomposition of ordinary differential polynomials. Appl. Alg. Eng. Commun. Comput. **19**, 1–25 (2008)
5. Golubitsky, O., Kondratieva, M., Ovchinnikov, A., Szanto, A.: A bound for orders in differential Nullstellensatz. J. Algebra **322**, 3852–3877 (2009)
6. Goursat, E.: Leçons sur L'intégration des Équations aux Dérivées Partielles, vol. II. A. Hermann, Paris (1898)
7. Grigoriev, D.: Polynomial factoring over a finite field and solving systems of algebraic equations. J. Soviet Math. **34**, 1762–1803 (1986)
8. Grigoriev, D.: Complexity of factoring and GCD calculating of ordinary linear differential operators. J. Symp. Comput. **10**, 7–37 (1990)
9. Grigoriev, D.: Analogue of Newton-Puiseux series for non-holonomic D-modules and factoring. Moscow Math. J. **9**, 775–800 (2009)
10. Grigoriev, D., Schwarz, F.: Factoring and solving linear partial differential equations. Computing **73**, 179–197 (2004)
11. Grigoriev, D., Schwarz, F.: Loewy and primary decompositions of D-modules. Adv. Appl. Math. **38**, 526–541 (2007)

12. Grigoriev, D., Schwarz, F.: Non-holonomic ideal in the plane and absolute factoring. In: Proc. Intern. Symp. Symbol. Algebr. Comput., pp. 93–97. ACM, Munich (2010)
13. Grigoriev, D., Schwarz, F.: Computing divisors and common multiples of quasi-linear ordinary differential equations. In: Gerdt, V.P., Koepf, W., Mayr, E.W., Vorozhtsov, E.V. (eds.) CASC 2013. LNCS, vol. 8136, pp. 140–147. Springer, Heidelberg (2013)
14. Kolchin, E.: Differential Algebra and Algebraic Groups. Academic Press, New York (1973)
15. Schlesinger, L.: Handbuch der Theorie der linearen Differentialgleichungen II. Teubner, Leipzig (1897)
16. Schwarz, F.: A factorization algorithm for linear ordinary differential equations. In: Proc. Intern. Symp. Symbol. Algebr. Comput., Portland, pp. 17–25. ACM (1989)
17. Schwarz, F.: Loewy Decomposition of Linear Differential Equations. Springer, Vienna (2012)
18. Tsarev, S.: On factorization of nonlinear ordinary differential equations. In: Proc. Symp. Symbol. Algebr. Comput., Vancouver, pp. 159–164. ACM (1999)
19. Tsarev, S.: Generalized laplace transformations and integration of hyperbolic systems of linear partial differential equations. In: Proc. Intern. Symp. Symbol. Algebr. Comput., pp. 325–331. ACM, Peking (2005)

Symbolic Algorithm for Generating Irreducible Bases of Point Groups in the Space of SO(3) Group

A.A. Gusev[1], V.P. Gerdt[1], S.I. Vinitsky[1],
V.L. Derbov[2], A. Góźdź[3], and A. Pędrak[3]

[1] Joint Institute for Nuclear Research, Dubna, Russia
[2] Saratov State University, Saratov, Russia
[3] Institute of Physics, Maria Curie-Skłodowska University, Lublin, Poland
gooseff@jinr.ru, vinitsky@theor.jinr.ru, andrzej.gozdz@umcs.lublin.pl

Abstract. A symbolic algorithm which can be implemented in computer algebra systems for generating bases for irreducible representations of the laboratory and intrinsic point symmetry groups acting in the rotor space is presented. The method of generalized projection operators is used. First the generalized projection operators for the intrinsic group acting in the space $L^2(SO(3))$ are constructed. The efficiency of the algorithm is investigated by calculating the bases for both laboratory and intrinsic octahedral groups irreducible representations for the set of angular momenta up to $J = 10$.

Keywords: intrinsic point symmetry groups, method of generalized projection operators, octahedral group, irreducible representations in the space $L^2(SO(3))$.

1 Introduction

The rotational motion is one of the most important motions in Nature. To treat the rotational degrees of freedom in quantum objects like molecules and nuclei an intrinsic frame is commonly introduced [7,8,10,11].

In this case, the rotational motion is described on the manifold of the rotation group SO(3) which is usually parameterized by Euler angles. Molecules and nuclei can possess point symmetries [3,6]. This fact requires the construction of irreducible representations and their bases in terms of Wigner functions for arbitrary angular momentum. Particular examples of such states were already constructed "by hand" for study of the nuclear collective models [9].

However, a universal exact method is required to allow the analysis of rotational states in arbitrary molecules and nuclei [4,5]. The method of generalized projection operators (GPOs), proposed here for solving this problem, is, in principle, well known [1]. The novel idea is to use it for constructing the bases of irreducible representations of the *intrinsic* point groups. It means that we construct these bases in the intrinsic frame instead of the laboratory one, which is,

© Springer International Publishing Switzerland 2015
V.P. Gerdt et al. (Eds.): CASC 2015, LNCS 9301, pp. 166–181, 2015.
DOI: 10.1007/978-3-319-24021-3_13

in fact, required for the description of these objects. For this purpose we use the notion of the so-called intrinsic group [2].

In this paper, the symbolic algorithm implemented in computer algebra systems (CAS) for generating irreducible bases of intrinsic point groups in the group space of the SO(3) group is formulated. To perform noncommutative operator multiplication the appropriate tools of the Reduce are applied, though the Maple or Mathematica can be used as well. The efficiency of the algorithm is investigated by calculating the bases of the irreducible representations of the intrinsic octahedral group $\overline{O}$ for the values of angular momentum up to $J = 10$.

The paper is organized as follows. In Section 2, the action of the rotation group in the geometrical space and the functional spaces is defined. In Section 3, the definition of intrinsic groups is given. In Section 4, the algorithm for calculating the GPOs and the bases of irreducible representations of the intrinsic finite groups are presented. In Sections 5 and 6, the algorithm is applied to the octahedral group. In the Conclusion, the results and perspectives of applications of this algorithm are summarized.

2   Rotations

We start from the precise definition of the rotation group action in both the geometrical space and the functional spaces spanned by the quantum states of a certain quantum system. Because of some discrepancies in the literature related to the definition of rotation in 3D space, we first specify the precise description of rotational angles, rotational matrix elements, and the action of rotation operators in the space $L^2(SO(3))$ used in the present paper.

2.1   Geometric Rotation

According to Refs. [8,10,11], the rotation of the right-handed Cartesian frame $S(x, y, z)$ in three-dimensional geometric space R^3 through the Euler angles (α, β, γ) can be defined as follows:

1° the initial frame $S(x, y, z)$ is rotated about the axis Oz through the angle α; we get the new frame $S(x', y', z' = z)$;
2° the next step is the rotation of $S(x', y', z' = z)$ through the angle β about the axis Oy'; we get the new frame $S(x'', y'' = y', z'')$;
3° finally we perform the last rotation of $S(x'', y'', z'')$ through the angle γ about the axis Oz'' and get the frame $S(x''', y''', z''' = z'')$.

The appropriate rotation matrix has the form (see [10,11]):

$$R = \begin{bmatrix} \cos\alpha\,\cos\beta\,\cos\gamma - \sin\alpha\,\sin\gamma & -\cos\alpha\,\cos\beta\,\sin\gamma - \sin\alpha\,\cos\gamma & \cos\alpha\,\sin\beta \\ \sin\alpha\,\cos\beta\,\cos\gamma + \cos\alpha\,\sin\gamma & -\sin\alpha\,\cos\beta\,\sin\gamma + \cos\alpha\,\cos\gamma & \sin\alpha\,\sin\beta \\ -\sin\beta\,\cos\gamma & \sin\beta\,\sin\gamma & \cos\beta \end{bmatrix}. \quad (1)$$

The basic vectors of the rotated frame e'_i are related to those of the initial frame e_i as

$$e'_i = \sum_k R_{ki}(\alpha, \beta, \gamma) e_k, \quad i, k = x, y, z \tag{2}$$

and the resulting transformations of the vector coordinates that follow from Eqs. (1), (2) can be written as:

$$A'_i = \sum_k R_{ki}(\alpha, \beta, \gamma) A_k, \tag{3}$$

where A'_i are the components of a vector in the rotated frame, and A_i are the components of the same vector in the laboratory frame. Note that there is an inconsistency between the English edition [11] of the book by Varshalovich et al and the Russian edition [10]. As the reference book we use the English edition.

2.2　Rotation in Functional Spaces

Another problem is to define the action of the rotation group on the Cartesian basis to be consistent with the corresponding action on the space of functions depending on Euler angles, $L^2(SO(3))$. The Cartesian coordinates in the three-dimensional Euclidean space are known to be related to the Wigner functions of the rotation group $SO(3)$. First, the Cartesian components (A_x, A_y, A_z) of the vector $\boldsymbol{A}$ are expressed in terms of the spherical functions (see [8,11]):

$$A_x = \sqrt{2\pi/3}\,|\boldsymbol{A}|\{Y_{1-1}(\theta, \varphi) - Y_{11}(\theta, \varphi)\},$$
$$A_y = i\sqrt{2\pi/3}\,|\boldsymbol{A}|\{Y_{1-1}(\theta, \varphi) + Y_{11}(\theta, \varphi)\}, \quad A_z = \sqrt{4\pi/3}\,|\boldsymbol{A}|Y_{10}(\theta, \varphi). \tag{4}$$

In turn, the spherical functions can be expressed in terms of Wigner functions

$$Y_{JM}(\theta, \varphi) = \sqrt{(2J+1)/(4\pi)}\,D^{J*}_{M0}(\varphi, \theta, \chi). \tag{5}$$

Let us denote by $r^J_{MK}(\varphi, \theta, \chi)$ the eigenvectors of the angular momentum operators in the space $L^2(SO(3))$ with orthonormalization conditions:

$$r^J_{MK}(\varphi, \theta, \chi) = \sqrt{2J+1}\,D^{J*}_{MK}(\varphi, \theta, \chi), \quad \langle r^J_{MK} | r^J_{M'K'} \rangle = \delta_{MM'}\delta_{KK'}. \tag{6}$$

Then

$$Y_{JM}(\theta, \varphi) = \sqrt{1/4\pi}\,r^J_{M0}(\varphi, \theta, \chi). \tag{7}$$

To simplify the notation, let us denote the set of Euler angles by a single letter $\Omega = (\varphi, \theta, \chi)$. The action of the laboratory group $SO(3)$ in the space $L^2(SO(3))$ is defined by the left shift operation:

$$\hat{g}\psi(\Omega) = \psi(g^{-1}\Omega). \tag{8}$$

Applying this operation to the components (4) treated as functions of Euler angles, we get (3), i.e., the left shift operation yields the same rotation of the vector as produced by the geometric rotation of Cartesian coordinates. Thus, acting on functions of Euler angles we arrive at the transformation of coordinates.

The same is valid for an arbitrary spherical tensor, it transforms as coordinates rather than as a function, where the standard action on a function of coordinates is assumed to be:

$$\hat{g}f(\xi) = f(g^{-1}\xi). \tag{9}$$

Note that the operation (8) applied to the functions (4) of Euler angles is consistent with the definition of spherical tensors, but not with the operation (9) applied to (4) as functions of Cartesian coordinates.

The SO(3) action (9) on a function of Cartesian coordinates yields ($x_1 = x, x_2 = y, x_3 = z$):

$$\hat{g}f(x,y,z) = f\left(\hat{g}^{-1}x,\, \hat{g}^{-1}y,\, \hat{g}^{-1}z\right)$$
$$= f\left(\sum_k R_{k1}(g^{-1})x_k,\, \sum_k R_{k2}(g^{-1})x_k,\, \sum_k R_{k3}(g^{-1})x_k\right)$$
$$= f\left(\sum_k R_{1k}(g)x_k,\, \sum_k R_{2k}(g)x_k,\, \sum_k R_{3k}(g)x_k\right). \tag{10}$$

Obviously, the matrices of the rotation group representations calculated in the Cartesian basis, consisting of some functions $f_k(x,y,z)$, are different from the matrices calculated in the space $L^2(SO(3))$. It can be shown that these matrices are transposed with respect to each other. For example, in the special case of 3D representations with the basis functions $f_k(x_1, x_2, x_3) = x_k$, where $k = 1, 2, 3$, Eq.(10) yields

$$\hat{g}f_k(x_1, x_2, x_3) = \sum_{k'} R^T_{k'k}(g)f_{k'}(x_1, x_2, x_3), \tag{11}$$

where $R^T(g)$ is the transposed three-dimensional rotation matrix (1) and $g = g(\alpha, \beta, \gamma)$, while the geometric rotation of coordinates is

$$x'_k = \hat{g}x_k = \sum_{k'} R_{k'k}(g)x_{k'}. \tag{12}$$

3   Intrinsic Group

The definition of intrinsic group $\overline{G}$ is related to the so called right shift operation acting on the group manifold of a given group G. The right shift operator is chosen in the form that allows one to obtain, for convenience, the intrinsic group anti-isomorphic to the group G. This property of intrinsic groups is good for physical applications because it allows one to use most of the standard results obtained for standard groups. The idea of intrinsic groups is described in detail in Ref. [2]. Some applications of these groups to the description of quantum systems in the intrinsic frame are presented in Ref. [4].

For each element g of the group G, one can define the corresponding operator $\overline{g}$ acting in the group linear space $\mathcal{L}_G$ as:

$$\overline{g}S = Sg, \quad \text{for all } S \in \mathcal{L}_G. \tag{13}$$

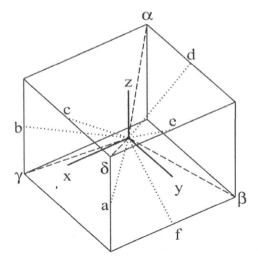

	x	y	z	$\sin\theta$	$\cos\theta$	φ
C_x	1	0	0	1	0	0
C_y	0	1	0	1	0	$\pi/2$
C_z	0	0	1	0	1	0
C_α	-1	-1	1	$\frac{\sqrt{2}}{\sqrt{3}}$	$\frac{1}{\sqrt{3}}$	$(5\pi)/4$
C_β	-1	1	-1	$\frac{\sqrt{2}}{\sqrt{3}}$	$-\frac{1}{\sqrt{3}}$	$(3\pi)/4$
C_γ	1	-1	-1	$\frac{\sqrt{2}}{\sqrt{3}}$	$-\frac{1}{\sqrt{3}}$	$(7\pi)/4$
C_δ	1	1	1	$\frac{\sqrt{2}}{\sqrt{3}}$	$\frac{1}{\sqrt{3}}$	$\pi/4$
C_a	1	1	0	1	0	$\pi/4$
C_b	1	-1	0	1	0	$(7\pi)/4$
C_c	1	0	1	$\frac{1}{\sqrt{2}}$	$\frac{1}{\sqrt{2}}$	0
C_d	-1	0	1	$\frac{1}{\sqrt{2}}$	$\frac{1}{\sqrt{2}}$	π
C_e	0	1	1	$\frac{1}{\sqrt{2}}$	$\frac{1}{\sqrt{2}}$	$\pi/2$
C_f	0	1	-1	$\frac{1}{\sqrt{2}}$	$\frac{-1}{\sqrt{2}}$	$\pi/2$

Fig. 1. The rotational axes of the octahedral group

The group $\overline{G}$ consisting of the operators $\bar{g}$ is called the intrinsic group related to the group G. It is easy to see some implications of the definition (13), for example,

$$\bar{g}h = h\bar{g} \tag{14}$$

for any $\bar{g} \in \overline{G}$ and $h \in G$. This brings us to the important property

$$[G, \overline{G}] = 0, \tag{15}$$

i.e., the actions of the laboratory and intrinsic group commute. This property fulfils the physical requirement of independence between the laboratory and intrinsic frames.

Another noteworthy fact is the anti-isomorphism between the groups G and $\overline{G}$, i.e.,

$$\psi_G : \overline{G} \to G, \text{ where } \psi_G(\bar{g}) = g, \quad \psi_G(\bar{g}_1 \bar{g}_2) = g_2 g_1. \tag{16}$$

This property makes it possible to apply all already known properties of the laboratory group, such as the representations, Clebsch–Gordan coupling coefficients, etc., to the intrinsic groups.

The action of the intrinsic group $\overline{G}$ on complex functions $\Psi : H \to \mathcal{C}$, where G is a subgroup of the group H, e.g., H = SO(3), is defined as [2]:

$$\hat{\bar{g}}\psi(h) = \psi(hg^{-1}). \tag{17}$$

According to [2], the representation $\bar{\Delta}^\Gamma$ of the intrinsic group $\overline{G}$ is equal to the transposed representation Δ^Γ of the laboratory group G:

$$\bar{\Delta}^\Gamma = (\Delta^\Gamma)^T. \tag{18}$$

Table 1. Values of n_1, n_2 and n_3 of the rotational angles of $R(C_*)$ from Eq. (29)

N	C_*	n_1 n_2 n_3	N	C_*	n_1 n_2 n_3	N	C_*	n_1 n_2 n_3	N	C_*	n_1 n_2 n_3
1	E	0 0 0	2	C_{2x}	0 2 2	3	C_{2y}	0 2 0	4	C_{2z}	0 0 2
5	$C_{3\alpha}$	2 1 3	6	$C_{3\beta}$	0 1 3	7	$C_{3\gamma}$	2 1 1	8	$C_{3\delta}$	0 1 1
9	$C_{3\alpha}^{-1}$	3 1 0	10	$C_{3\beta}^{-1}$	3 1 2	11	$C_{3\gamma}^{-1}$	1 1 0	12	$C_{3\delta}^{-1}$	1 1 2
13	C_{4x}	3 1 1	14	C_{4y}	0 1 0	15	C_{4z}	0 0 1	16	C_{4x}^{-1}	1 1 3
17	C_{4y}^{-1}	2 1 2	18	C_{4z}^{-1}	0 0 3	19	C_{2a}	0 2 1	20	C_{2b}	0 2 3
21	C_{2c}	0 1 2	22	C_{2d}	2 1 0	23	C_{2e}	1 1 1	24	C_{2f}	3 1 3

It means that to construct representations of intrinsic groups in the space $L^2(H)$ one can use the representation matrices Δ^Γ obtained for the group $G \subset H$, but in the other space. For example, for $H = SO(3)$, one can use the appropriate Cartesian basis $f_k(x, y, z)$, where $k = 1, 2, \ldots, \dim(\Gamma)$.

4   Generalized Projection Operators

The technique of projecting on the spaces of irreducible representation of a given group G is described, e.g., in [1]. The idea is based on using the generalized projection operators (GPOs) for the irreducible representations of locally compact groups. In this paper, we restrict ourselves to finite groups. The projection method is very useful for obtaining orthogonal bases of irreducible representations in the required spaces of functions. It can be applied to both laboratory groups and intrinsic groups.

The main idea is to decompose the Hilbert space $\mathcal{K} = L^2(G)$ of square integrable functions, defined in the group manifold of the group G, into a set of orthogonal irreducible subspaces that are invariant with respect to the action of the group G, irreducible subspaces $\mathcal{K}_\kappa$:

$$\mathcal{K} = \bigoplus_\kappa \mathcal{K}_\kappa. \qquad (19)$$

This technique requires using the GPOs defined as:

$$\hat{P}_{ab}^\Gamma = \frac{\dim(\Gamma)}{\mathrm{card}(G)} \sum_{g \in G} \left(\Delta_{ab}^\Gamma(g) \right)^* \hat{g}, \qquad (20)$$

where $\dim(\Gamma)$ denotes the dimension of the irreducible representation Γ of the group G, $\mathrm{card}(G)$ is the order of the group G (i.e., the number of its elements), $\Delta_{ab}^\Gamma(g)$ denotes the matrix elements of the irreducible representation Γ for the element g. The symbol $\hat{g}$ denotes a unitary operator of the group G, acting in the space $\mathcal{K}$.

The operators $\hat{P}_{ab}^\Gamma$ have the following properties (see [1], Chap. 7, §3):

$$\left(\hat{P}_{ab}^\Gamma \right)^\dagger = \hat{P}_{ba}^\Gamma, \qquad \hat{P}_{ab}^\Gamma \hat{P}_{a'b'}^{\Gamma'} = \delta_{\Gamma\Gamma'} \delta_{ba'} \hat{P}_{ab'}^{\Gamma'}. \qquad (21)$$

Table 2. The multiplication table of operators $C_1,...,C_{24}$ numbered in Table 1

1	2	3	4	5	6	7	8	9	10	11	12	13	14	15	16	17	18	19	20	21	22	23	24
1	2	3	4	5	6	7	8	9	10	11	12	13	14	15	16	17	18	19	20	21	22	23	24
2	1	4	3	7	8	5	6	12	11	10	9	16	21	20	13	22	19	18	15	14	17	24	23
3	4	1	2	8	7	6	5	10	9	12	11	24	17	19	23	14	20	15	18	22	21	16	13
4	3	2	1	6	5	8	7	11	12	9	10	23	22	18	24	21	15	20	19	17	14	13	16
5	8	6	7	9	12	10	11	1	4	2	3	15	16	22	19	24	17	14	21	23	13	20	18
6	7	5	8	11	10	12	9	4	1	3	2	18	24	14	20	16	21	22	17	13	23	19	15
7	6	8	5	12	9	11	10	2	3	1	4	20	13	17	18	23	22	21	14	24	16	15	19
8	5	7	6	10	11	9	12	3	2	4	1	19	23	21	15	13	14	17	22	16	24	18	20
9	11	12	10	1	3	4	2	5	7	8	6	22	19	13	14	18	24	16	23	20	15	21	17
10	12	11	9	3	1	2	4	8	6	5	7	21	15	24	17	20	13	23	16	18	19	22	14
11	9	10	12	4	2	1	3	6	8	7	5	14	20	23	22	15	16	24	13	19	18	17	21
12	10	9	11	2	4	3	1	7	5	6	8	17	18	16	21	19	23	13	24	15	20	14	22
13	16	23	24	17	14	22	21	19	20	15	18	2	8	10	1	7	9	12	11	6	5	4	3
14	22	17	21	15	20	18	19	24	13	23	16	9	3	8	11	1	6	5	7	2	4	12	10
15	19	20	18	24	16	13	23	14	21	22	17	8	11	4	5	10	1	3	2	12	9	7	6
16	13	24	23	22	21	17	14	18	15	20	19	1	6	11	2	5	12	9	10	8	7	3	4
17	21	14	22	19	18	20	15	13	24	16	23	10	1	5	12	3	7	8	6	4	2	11	9
18	20	19	15	16	24	23	13	22	17	14	21	7	9	1	6	12	4	2	3	10	11	8	5
19	15	18	20	13	23	24	16	17	22	21	14	5	12	2	8	9	3	1	4	11	10	6	7
20	18	15	19	23	13	16	24	21	14	17	22	6	10	3	7	11	2	4	1	9	12	5	8
21	17	22	14	20	15	19	18	23	16	24	13	12	4	6	10	2	8	7	5	1	3	9	11
22	14	21	17	18	19	15	20	16	23	13	24	11	2	7	9	4	5	6	8	3	1	10	12
23	24	13	16	21	22	14	17	20	19	18	15	3	7	12	4	8	11	10	9	5	6	1	2
24	23	16	13	14	17	21	22	15	18	19	20	4	5	9	3	6	10	11	12	7	8	2	1

These properties imply that the operator $\hat{P}_{bb}^{\Gamma}$ is a true projection operator:
$(\hat{P}_{bb}^{\Gamma})^{\dagger} = \hat{P}_{bb}^{\Gamma}, \quad \hat{P}_{bb}^{\Gamma}\hat{P}_{bb}^{\Gamma} = \hat{P}_{bb}^{\Gamma}.$

Let us describe step by step the procedure of constructing the orthogonal basis of irreducible representation using the concept of GPOs (see [1], Chap. 7, §4A). This approach is rather general, but for the purpose of this paper we restrict our consideration to the subgroup $\overline{G}$ of the intrinsic rotation group $\overline{G} \subset \overline{SO}(3)$.

Let us introduce the orthogonal basis in the space $\mathcal{K} = L^2(SO(3))$ as the set of Wigner functions (6). For fixed J and M, the vectors $|r_{MK}^J\rangle$, where $K = -J, -J+1, \ldots, J$, span the orthonormal basis in the subspace $\mathcal{K}_J$.

Note that for any subgroup of $\overline{SO}(3)$, the subspaces related to different angular momentum quantum numbers J and J' are orthogonal reducible subspaces. For this reason, our procedure can be applied to each subspace $\mathcal{K}_J$ separately.

The GPOs in this example are constructed in accordance with Eq. (20). The algorithm [1] consists of the following steps:

Step 1. *Choose one of the subspaces $\mathcal{K}_b^{\Gamma} = \hat{P}_{bb}^{\Gamma}\mathcal{K}$, where $\hat{P}_{bb}^{\Gamma}$ is the projection operator. The label Γ denotes the required irreducible representation, b is an arbitrary but fixed index.*

Table 3. Elements of octahedral group

C_*	$R(C_*)$	C_*	$R(C_*)$	C_*	$R(C_*)$	C_*	$R(C_*)$
E	$\begin{pmatrix}1&0&0\\0&1&0\\0&0&1\end{pmatrix}$	C_{2x}	$\begin{pmatrix}1&0&0\\0&-1&0\\0&0&-1\end{pmatrix}$	C_{2y}	$\begin{pmatrix}-1&0&0\\0&1&0\\0&0&-1\end{pmatrix}$	C_{2z}	$\begin{pmatrix}-1&0&0\\0&-1&0\\0&0&1\end{pmatrix}$
$C_{3\alpha}$	$\begin{pmatrix}0&0&-1\\1&0&0\\0&-1&0\end{pmatrix}$	$C_{3\beta}$	$\begin{pmatrix}0&0&1\\-1&0&0\\0&-1&0\end{pmatrix}$	$C_{3\gamma}$	$\begin{pmatrix}0&0&-1\\-1&0&0\\0&1&0\end{pmatrix}$	$C_{3\delta}$	$\begin{pmatrix}0&0&1\\1&0&0\\0&1&0\end{pmatrix}$
$C_{3\alpha}^{-1}$	$\begin{pmatrix}0&1&0\\0&0&-1\\-1&0&0\end{pmatrix}$	$C_{3\beta}^{-1}$	$\begin{pmatrix}0&-1&0\\0&0&-1\\1&0&0\end{pmatrix}$	$C_{3\gamma}^{-1}$	$\begin{pmatrix}0&-1&0\\0&0&1\\-1&0&0\end{pmatrix}$	$C_{3\delta}^{-1}$	$\begin{pmatrix}0&1&0\\0&0&1\\1&0&0\end{pmatrix}$
C_{4x}	$\begin{pmatrix}1&0&0\\0&0&-1\\0&1&0\end{pmatrix}$	C_{4z}	$\begin{pmatrix}0&-1&0\\1&0&0\\0&0&1\end{pmatrix}$	C_{4y}	$\begin{pmatrix}0&0&1\\0&1&0\\-1&0&0\end{pmatrix}$	C_{4x}^{-1}	$\begin{pmatrix}1&0&0\\0&0&1\\0&-1&0\end{pmatrix}$
C_{4y}^{-1}	$\begin{pmatrix}0&0&-1\\0&1&0\\1&0&0\end{pmatrix}$	C_{4z}^{-1}	$\begin{pmatrix}0&1&0\\-1&0&0\\0&0&1\end{pmatrix}$	C_{2a}	$\begin{pmatrix}0&1&0\\1&0&0\\0&0&-1\end{pmatrix}$	C_{2b}	$\begin{pmatrix}0&-1&0\\-1&0&0\\0&0&-1\end{pmatrix}$
C_{2c}	$\begin{pmatrix}0&0&1\\0&-1&0\\1&0&0\end{pmatrix}$	C_{2d}	$\begin{pmatrix}0&0&-1\\0&-1&0\\-1&0&0\end{pmatrix}$	C_{2e}	$\begin{pmatrix}-1&0&0\\0&0&1\\0&1&0\end{pmatrix}$	C_{2f}	$\begin{pmatrix}-1&0&0\\0&0&-1\\0&-1&0\end{pmatrix}$

One can directly check the orthogonality of the spaces $\mathcal{K}_b^\Gamma$ for different b.
Let $|\hat{P}_{bb}^\Gamma u\rangle \in \mathcal{K}_b^\Gamma$ and $|\hat{P}_{cc}^\Gamma u\rangle \in \mathcal{K}_c^\Gamma$, $\langle\hat{P}_{bb}^\Gamma u|\hat{P}_{cc}^\Gamma u\rangle = \langle u|\hat{P}_{bb}^\Gamma\hat{P}_{cc}^\Gamma u\rangle = \delta_{bc}\langle u|\hat{P}_{bc}^\Gamma u\rangle.$

Step 2. *Construct in an arbitrary manner the basis in the subspace $\mathcal{K}_b^\Gamma$, we denote it by $u_1, u_2, \ldots, u_{\dim(\mathcal{K}_b^\Gamma)}$.*

Since the vectors $|r_{MK}^J\rangle$ are basic vectors in $\mathcal{K}_J$, the projected vectors $|K\rangle = \hat{P}_{bb}^\Gamma|r_{MK}^J\rangle$ for $K = -J, -J+1, \ldots, J$ span the the projected subspace $\mathcal{K}_{Jb}^\Gamma$. The vectors $|K\rangle$ do not form an orthonormal basis in the space $\mathcal{K}_{Jb}^\Gamma$ because they can be neither orthogonal, nor even linearly independent.

To solve the problem we use the symmetric orthonormalization procedure. The overlaps

$$\langle K|K'\rangle = \langle r_{MK}^J|\hat{P}_{bb}^\Gamma|r_{MK'}^J\rangle \tag{22}$$

form a finite $(2J+1) \times (2J+1)$ dimensional Hermitian matrix referred to as Gramm matrix, which we denote here by $\mathcal{N}$. Solving the eigenvalue problem for the matrix $\mathcal{N}$ allows one to find the orthonormal basis, in which this matrix is diagonal:

$$\mathcal{N}w_t(K) = \lambda_t w_t(K). \tag{23}$$

The coefficients $w_t(K)$ allow for the construction of the states, which in the Generator Coordinate Method [7] are named the "natural states", and which furnish the required basis in our space. For $\lambda_t \neq 0$ one gets the following basis of orthonormal states:

$$|u_t\rangle = A_t \sum_K w_t(K)\hat{P}_{bb}^\Gamma|r_{MK}^J\rangle, \tag{24}$$

where J, M, Γ, and b are fixed, and A_t is the normalization coefficient.

Table 4. Classes, Cartesian bases, and character table for octahedral group O

Irr\Classes	$\mathcal{C}_1$	$\mathcal{C}_2$	$\mathcal{C}_3$	$\mathcal{C}_4$	$\mathcal{C}_5$	Cartesian bases
$\Gamma_1 = A_1$	1	1	1	1	1	$v_1 = R^2 = x^2 + y^2 + z^2$
$\Gamma_2 = A_2$	1	1	1	-1	-1	$v_1 = xyz$
$\Gamma_3 = E$	2	-1	2	0	0	$v_1 = \sqrt{3}(x^2 - y^2),\quad v_2 = (2z^2 - x^2 - y^2)$
$\Gamma_4 = T_2$	3	0	-1	-1	1	$v_1 = xy,\quad v_2 = yz,\quad v_3 = xz$
$\Gamma_5 = T_1$	3	0	-1	1	-1	$v_1 = x,\quad v_2 = y,\quad v_3 = z$

$\mathcal{C}_1 = \{E\}$; $\mathcal{C}_2 = \{C_{3\alpha}, C_{3\beta}, C_{3\gamma}, C_{3\delta}, C_{3\alpha}^{-1}, C_{3\beta}^{-1}, C_{3\gamma}^{-1}, C_{3\delta}^{-1}\}$; $\mathcal{C}_3 = \{C_{2x}, C_{2y}, C_{2z}\}$;
$\mathcal{C}_4 = \{C_{4x}, C_{4y}, C_{4z}, C_{4x}^{-1}, C_{4y}^{-1}, C_{4z}^{-1}\}$; $\mathcal{C}_5 = \{C_{2a}, C_{2b}, C_{2c}, C_{2d}, C_{2e}, C_{2f}\}$

Let us calculate the scalar products:

$$\langle u_{t'} | u_t \rangle = A_{t'} A_t \sum_{K',K} w_{t'}(K')^\star w_t(K) \langle r_{MK'}^J | \hat{\bar{P}}_{bb}^\Gamma | r_{MK}^J \rangle =$$

$$= A_{t'} A_t \sum_K w_{t'}(K')^\star \lambda_t w_t(K) = \delta_{t't} A_{t'} A_t \lambda_t \sum_K |w_t(K)|^2. \qquad (25)$$

This result means that the states $|u_t\rangle$, for which the eigenvalue $\lambda_t = 0$ are the zero vectors, and for the cases when $\lambda_t \neq 0$ the normalization coefficient is equal to $A_t = (\lambda_t \sum_K |w_t(K)|^2)^{-1/2}$. Note that since the eigenequation (23) is, in fact, the eigenequation for the matrix representation of the projection operator $\hat{\bar{P}}_{bb}^\Gamma$, the eigenvalues are equal to $\lambda_t = 0, 1$. Thus, we obtain the required basis in the projected subspace $\mathcal{K}_{Jb}^\Gamma$.

Step 3. *Apply successively the operators $\hat{\bar{P}}_{ab}^\Gamma$ to each of the vectors u_t, i.e., $\hat{\bar{P}}_{ab}^\Gamma u_t$, where Γ and b are fixed and $a = 1, 2, \ldots, \dim(\Gamma)$.*

As a result, we get the required basis states of the irreducible representation Γ of the group $\overline{G}$ in the rotor space:

$$|\Gamma a; tJ\rangle = A_t \sum_K w_t(K) \hat{\bar{P}}_{ab}^\Gamma | r_{MK}^J \rangle. \qquad (26)$$

The projection of the angular momentum M and the quantum number b can be chosen in an arbitrary manner. These states satisfy the usual conditions

$$\langle \Gamma' a'; t'J' | \Gamma a; tJ \rangle = \delta_{\Gamma'\Gamma} \delta_{a'a} \delta_{t't} \delta_{J'J}. \qquad (27)$$

The last but not the least issue is how to construct the representations Δ^Γ to be used in the GPOs. These representations have to be consistent with the action of group operators, the elements of GPOs. This problem was solved in Ref. [9]. In table (28), the relations between the action of the rotation group and the appropriate matrix representations are shown. Δ^Γ denotes the irreducible representation of the laboratory group, obtained using the Cartesian basis $f_k(x, y, z)$, where $k = 1, 2, \ldots, \dim(\Gamma)$. The right column of table (28) shows which representation should be used in the GPOs while acting in the space spanned by the basis shown in the left column [9]:

Table 5. The operators in the octahedral group representations Γ_1, Γ_2 and Γ_3

C_*	$\Gamma_1(C_*)$	$\Gamma_2(C_*)$	$\Gamma_3(C_*)$
E , C_{2x}, C_{2y}, C_{2z}	$(\ 1\ )$	$(\ 1\ )$	$\begin{pmatrix} 1 & 0 \\ 0 & 1 \end{pmatrix}$
$C_{3\alpha}, C_{3\beta}, C_{3\gamma}, C_{3\delta}$	$(\ 1\ )$	$(\ 1\ )$	$\begin{pmatrix} -1/2 & \sqrt{3}/2 \\ -\sqrt{3}/2 & -1/2 \end{pmatrix}$
$C_{3\alpha}^{-1}, C_{3\beta}^{-1}, C_{3\gamma}^{-1}, C_{3\delta}^{-1}$	$(\ 1\ )$	$(\ 1\ )$	$\begin{pmatrix} -1/2 & -\sqrt{3}/2 \\ \sqrt{3}/2 & -1/2 \end{pmatrix}$
$C_{4x}, C_{4x}^{-1}, C_{2e}, C_{2f}$	$(\ 1\ )$	(-1)	$\begin{pmatrix} 1/2 & -\sqrt{3}/2 \\ -\sqrt{3}/2 & -1/2 \end{pmatrix}$
$C_{4y}, C_{4y}^{-1}, C_{2c}, C_{2d}$	$(\ 1\ )$	(-1)	$\begin{pmatrix} 1/2 & \sqrt{3}/2 \\ \sqrt{3}/2 & -1/2 \end{pmatrix}$
$C_{4z}, C_{4z}^{-1}, C_{2a}, C_{2b}$	$(\ 1\ )$	(-1)	$\begin{pmatrix} -1 & 0 \\ 0 & 1 \end{pmatrix}$

rotation group action	matrices required in projection operators
$\hat{g} f_k(x,y,z) = f_k(g^{-1}(x,y,z))$	$\Delta^{\Gamma}(g)$
$\hat{g} r_{MK}^J(\Omega) = r_{MK}^J(g^{-1}\Omega)$	$(\Delta^{\Gamma})^T(g)$
$\hat{g} r_{MK}^J(\Omega) = r_{MK}^J(\Omega g^{-1})$	$\Delta^{\Gamma}(g)$

$$\tag{28}$$

Here $\bar{g}$ denotes an element of the intrinsic group $\overline{SO}(3)$.

5   Example of Using the Algorithm for the Octahedral Group

Below we consider only one example, the octahedral point group. However, the formalism and the algorithm is quite general and can be applied to any point group. In principle, all considerations of the previous sections can be applied to a wider class of groups, namely, to any finite group.

5.1   Construction of Elements of the Octahedral Group from Its Generators

The octahedral group O consists of 24 rotations grouped into five classes C_n. The element $C_{n\zeta}$ denotes the right-handed rotation through the angle $2\pi/n$ around the axis ζ. The direction of the required rotation axes and the directional angles of their unit vectors are shown in Fig. 1. The notation corresponds to Appendix D in [3]. **Step O1.** All three-dimensional matrices representing the elements $R(C_*)$ of the group O were calculated using the matrix $R(\alpha, \beta, \gamma)$, Eq. (1)

$$R(C_*) = R(\alpha = \pi n_1/2, \beta = \pi n_2/2, \gamma = \pi n_3/2), \tag{29}$$

with the values n_1, n_2 and n_3 specified in Table 1. The angles $\alpha = \pi n_1/2$, $\beta = \pi n_2/2$ and $\gamma = \pi n_3/2$ were calculated using the formulas [11]

$$\sin(\beta/2) = \sin\theta\sin(\omega/2), \quad \tan((\alpha+\gamma)/2) = \cos\theta\tan(\omega/2), \quad (\alpha-\gamma)/2 = \varphi - \pi/2,$$

Table 6. Operators of the octahedral group representation Γ_4

C_*	$\Gamma_4(C_*)$	C_*	$\Gamma_4(C_*)$	C_*	$\Gamma_4(C_*)$	C_*	$\Gamma_4(C_*)$
E	$\begin{pmatrix} 1 & 0 & 0 \\ 0 & 1 & 0 \\ 0 & 0 & 1 \end{pmatrix}$	C_{2x}	$\begin{pmatrix} -1 & 0 & 0 \\ 0 & 1 & 0 \\ 0 & 0 & -1 \end{pmatrix}$	C_{2y}	$\begin{pmatrix} -1 & 0 & 0 \\ 0 & -1 & 0 \\ 0 & 0 & 1 \end{pmatrix}$	C_{2z}	$\begin{pmatrix} 1 & 0 & 0 \\ 0 & -1 & 0 \\ 0 & 0 & -1 \end{pmatrix}$
$C_{3\alpha}$	$\begin{pmatrix} 0 & 0 & -1 \\ -1 & 0 & 0 \\ 0 & 1 & 0 \end{pmatrix}$	$C_{3\beta}$	$\begin{pmatrix} 0 & 0 & -1 \\ 1 & 0 & 0 \\ 0 & -1 & 0 \end{pmatrix}$	$C_{3\gamma}$	$\begin{pmatrix} 0 & 0 & 1 \\ -1 & 0 & 0 \\ 0 & -1 & 0 \end{pmatrix}$	$C_{3\delta}$	$\begin{pmatrix} 0 & 0 & 1 \\ 1 & 0 & 0 \\ 0 & 1 & 0 \end{pmatrix}$
$C_{3\alpha}^{-1}$	$\begin{pmatrix} 0 & -1 & 0 \\ 0 & 0 & 1 \\ -1 & 0 & 0 \end{pmatrix}$	$C_{3\beta}^{-1}$	$\begin{pmatrix} 0 & 1 & 0 \\ 0 & 0 & -1 \\ -1 & 0 & 0 \end{pmatrix}$	$C_{3\gamma}^{-1}$	$\begin{pmatrix} 0 & -1 & 0 \\ 0 & 0 & -1 \\ 1 & 0 & 0 \end{pmatrix}$	$C_{3\delta}^{-1}$	$\begin{pmatrix} 0 & 1 & 0 \\ 0 & 0 & 1 \\ 1 & 0 & 0 \end{pmatrix}$
C_{4x}	$\begin{pmatrix} 0 & 0 & -1 \\ 0 & -1 & 0 \\ 1 & 0 & 0 \end{pmatrix}$	C_{4y}	$\begin{pmatrix} 0 & 1 & 0 \\ -1 & 0 & 0 \\ 0 & 0 & -1 \end{pmatrix}$	C_{4z}	$\begin{pmatrix} -1 & 0 & 0 \\ 0 & 0 & 1 \\ 0 & -1 & 0 \end{pmatrix}$	C_{4x}^{-1}	$\begin{pmatrix} 0 & 0 & 1 \\ 0 & -1 & 0 \\ -1 & 0 & 0 \end{pmatrix}$
C_{4y}^{-1}	$\begin{pmatrix} 0 & -1 & 0 \\ 1 & 0 & 0 \\ 0 & 0 & -1 \end{pmatrix}$	C_{4z}^{-1}	$\begin{pmatrix} -1 & 0 & 0 \\ 0 & 0 & -1 \\ 0 & 1 & 0 \end{pmatrix}$	C_{2a}	$\begin{pmatrix} 1 & 0 & 0 \\ 0 & 0 & -1 \\ 0 & -1 & 0 \end{pmatrix}$	C_{2b}	$\begin{pmatrix} 1 & 0 & 0 \\ 0 & 0 & 1 \\ 0 & 1 & 0 \end{pmatrix}$
C_{2c}	$\begin{pmatrix} 0 & -1 & 0 \\ -1 & 0 & 0 \\ 0 & 0 & 1 \end{pmatrix}$	C_{2d}	$\begin{pmatrix} 0 & 1 & 0 \\ 1 & 0 & 0 \\ 0 & 0 & 1 \end{pmatrix}$	C_{2e}	$\begin{pmatrix} 0 & 0 & -1 \\ 0 & 1 & 0 \\ -1 & 0 & 0 \end{pmatrix}$	C_{2f}	$\begin{pmatrix} 0 & 0 & 1 \\ 0 & 1 & 0 \\ 1 & 0 & 0 \end{pmatrix}$

with θ and φ shown in Fig. 1 for the rotation C_{n*} through the angle $\omega = 2\pi/n$. As a result, we obtain the required $R(C_{n*})$ presented in Table 3. The multiplication table of the elements $C_1,...,C_{24}$ of the octahedral group (Table 2) uses the direct numeration/coding $1, 2, .., 24$ of Table 1.

5.2   Construction of Irreducible Representations of the Octahedral Group O in the Cartesian Bases

Step O2. Let us describe the construction of irreducible representations in the Cartesian bases. There are five classes and five irreducible representations of the octahedral group O, the notations of which, similar to those of Ref. [3], are presented in Table 4 together with the Cartesian basis vectors taken from Ref. [6], where the appropriate action is applied according to the first row of Eq. (28). The matrices $\Gamma_*(C_*)$ of the operators C_* describing the action of the representation Γ_* on the Cartesian basis vectors v_i of Table 4, are sought in the form:

$$C_* v_j(x,y,z) = \sum_i (\Gamma_*(C_*))_{ij} v_i(x,y,z). \qquad (30)$$

The unknown coefficients $(\Gamma_*(C_*))_{ij}$ in Eq. (30) are calculated by direct substitution of this relation

Table 7. Operators of the octahedral group representation Γ_5

C_*	$\Gamma_5(C_*)$	C_*	$\Gamma_5(C_*)$	C_*	$\Gamma_5(C_*)$	C_*	$\Gamma_5(C_*)$
E	$\begin{pmatrix} 1 & 0 & 0 \\ 0 & 1 & 0 \\ 0 & 0 & 1 \end{pmatrix}$	C_{2x}	$\begin{pmatrix} 1 & 0 & 0 \\ 0 & -1 & 0 \\ 0 & 0 & -1 \end{pmatrix}$	C_{2y}	$\begin{pmatrix} -1 & 0 & 0 \\ 0 & 1 & 0 \\ 0 & 0 & -1 \end{pmatrix}$	C_{2z}	$\begin{pmatrix} -1 & 0 & 0 \\ 0 & -1 & 0 \\ 0 & 0 & 1 \end{pmatrix}$
$C_{3\alpha}$	$\begin{pmatrix} 0 & 0 & -1 \\ 1 & 0 & 0 \\ 0 & -1 & 0 \end{pmatrix}$	$C_{3\beta}$	$\begin{pmatrix} 0 & 0 & 1 \\ -1 & 0 & 0 \\ 0 & -1 & 0 \end{pmatrix}$	$C_{3\gamma}$	$\begin{pmatrix} 0 & 0 & -1 \\ -1 & 0 & 0 \\ 0 & 1 & 0 \end{pmatrix}$	$C_{3\delta}$	$\begin{pmatrix} 0 & 0 & 1 \\ 1 & 0 & 0 \\ 0 & 1 & 0 \end{pmatrix}$
$C_{3\alpha}^{-1}$	$\begin{pmatrix} 0 & 1 & 0 \\ 0 & 0 & -1 \\ -1 & 0 & 0 \end{pmatrix}$	$C_{3\beta}^{-1}$	$\begin{pmatrix} 0 & -1 & 0 \\ 0 & 0 & -1 \\ 1 & 0 & 0 \end{pmatrix}$	$C_{3\gamma}^{-1}$	$\begin{pmatrix} 0 & -1 & 0 \\ 0 & 0 & 1 \\ -1 & 0 & 0 \end{pmatrix}$	$C_{3\delta}^{-1}$	$\begin{pmatrix} 0 & 1 & 0 \\ 0 & 0 & 1 \\ 1 & 0 & 0 \end{pmatrix}$
C_{4x}	$\begin{pmatrix} 1 & 0 & 0 \\ 0 & 0 & -1 \\ 0 & 1 & 0 \end{pmatrix}$	C_{4y}	$\begin{pmatrix} 0 & 0 & 1 \\ 0 & 1 & 0 \\ -1 & 0 & 0 \end{pmatrix}$	C_{4z}	$\begin{pmatrix} 0 & -1 & 0 \\ 1 & 0 & 0 \\ 0 & 0 & 1 \end{pmatrix}$	C_{4x}^{-1}	$\begin{pmatrix} 1 & 0 & 0 \\ 0 & 0 & 1 \\ 0 & -1 & 0 \end{pmatrix}$
C_{4y}^{-1}	$\begin{pmatrix} 0 & 0 & -1 \\ 0 & 1 & 0 \\ 1 & 0 & 0 \end{pmatrix}$	C_{4z}^{-1}	$\begin{pmatrix} 0 & 1 & 0 \\ -1 & 0 & 0 \\ 0 & 0 & 1 \end{pmatrix}$	C_{2a}	$\begin{pmatrix} 0 & 1 & 0 \\ 1 & 0 & 0 \\ 0 & 0 & -1 \end{pmatrix}$	C_{2b}	$\begin{pmatrix} 0 & -1 & 0 \\ -1 & 0 & 0 \\ 0 & 0 & -1 \end{pmatrix}$
C_{2c}	$\begin{pmatrix} 0 & 0 & 1 \\ 0 & -1 & 0 \\ 1 & 0 & 0 \end{pmatrix}$	C_{2d}	$\begin{pmatrix} 0 & 0 & -1 \\ 0 & -1 & 0 \\ -1 & 0 & 0 \end{pmatrix}$	C_{2e}	$\begin{pmatrix} -1 & 0 & 0 \\ 0 & 0 & 1 \\ 0 & 1 & 0 \end{pmatrix}$	C_{2f}	$\begin{pmatrix} -1 & 0 & 0 \\ 0 & 0 & -1 \\ 0 & -1 & 0 \end{pmatrix}$

$$C_* v_j(x, y, z) = v_j \Big(R(C_*)_{11}x + R(C_*)_{12}y + R(C_*)_{13}z, \tag{31}$$

$$R(C_*)_{21}x + R(C_*)_{22}y + R(C_*)_{23}z, \; R(C_*)_{31}x + R(C_*)_{32}y + R(C_*)_{33}z \Big),$$

with the appropriate permutations of the indexes i and j:

$$\Gamma_{ij}(C_*) = R_{ji}(C_*) = R_{ij}(C_*)^T.$$

From Eqs. (30) and (31) we get the following expression:

$$v_j \Big(R(C_*)_{11}x + R(C_*)_{12}y + R(C_*)_{13}z, \; R(C_*)_{21}x + R(C_*)_{22}y + R(C_*)_{23}z,$$

$$R(C_*)_{31}x + R(C_*)_{32}y + R(C_*)_{33}z \Big) - \sum_i (\Gamma_*(C_*))_{ij} v_i(x, y, z) = 0. \tag{32}$$

After equating in Eq. (32) the coefficients of monomials $x^{i_1} y^{i_2} z^{i_3}$ to zero we arrive at the set of equations with respect to the required coefficients $(\Gamma_*(C_*))_{ij}$. The irreducible representations $\Gamma_1, ..., \Gamma_5$ of the octahedral group found with this algorithm are presented in Tables 5–7.

Step O3. The characters $\chi^\Gamma(C_*)$ of the representation $\Gamma_*(C_*)$ are calculated as

$$\chi^\Gamma(C_*) = \mathrm{Tr}(\Gamma_*(C_*)) = \sum_{k=1}^{\dim(\Gamma_*)} (\Gamma_*(C_*))_{kk} \tag{33}$$

and are presented in Table 4.

Remark 1. The obtained octahedral group rotation matrices $R(C_*)$ and representations $\Gamma_1,...,\Gamma_5$ differ from the analogous results by Cornwell [3]. They differ by transposition because our rotation matrix $R(\alpha,\beta,\gamma)$ from Eq. (1) and those of Cornwell's paper differ by transposition, too.

5.3   GPOs Implementation for Standard and Intrinsic Point Groups

Now we proceed to the explicit form of the GPOs required for point groups. Though in the rest of this paper we consider only intrinsic groups, we start from the standard (laboratory) groups for comparison.

Laboratory Groups. To construct the bases in the space of rotational functions for the laboratory point groups we start from calculating the action of the operators (20) belonging to the *standard point group*, using the appropriate representation from (28), on the functions (6):

$$
R^{JK}_{npqM}(\Omega) = \frac{\dim(\Gamma_n)}{\mathrm{card}(G)} \sum_{g\in G} (\Gamma_n(g)^T)^\star_{pq}\, \hat{g}\, r^J_{MK}(\Omega) = \sum_{M'=-J}^{J} {}^{\mathrm{lab}}A^{JK}_{npqM'M}\, r^J_{M'K}(\Omega),
$$

$$
{}^{\mathrm{lab}}A^{JK}_{npqM'M} = \frac{\dim(\Gamma_n)}{\mathrm{card}(G)} \sum_{g\in G} (\Gamma_n(g)^T)^\star_{pq}\, D^J_{M'M}(g),
\tag{34}
$$

where the appropriate matrices of the representations Γ_n, obtained from the action in the space of functions of the Cartesian coordinates, are substituted instead of the general symbols Δ^Γ. Here $M = -J, -J+1, \ldots, J$ that provides full decomposition of the rotor space for a given J.

Intrinsic Groups. To construct the appropriate bases in the space of rotational functions for intrinsic point groups we have to use the action of the operators (20) belonging to the *intrinsic* group, using the appropriate representation from (28), on the functions (6):

$$
R^{JM}_{npqK}(\Omega) = \frac{\dim(\Gamma_n)}{\mathrm{card}(G)} \sum_{g\in G} \Gamma_n(g)^\star_{pq}\, \hat{\hat{g}}\, r^J_{MK}(\Omega) = \sum_{K'=-J}^{J} {}^{\mathrm{int}}A^{JM}_{npqKK'}\, r^J_{MK'}(\Omega),
$$

$$
{}^{\mathrm{int}}A^{JM}_{npqKK'} = \frac{\dim(\Gamma_n)}{\mathrm{card}(G)} \sum_{g\in G} \Gamma_n(g)^\star_{pq} D^J_{KK'}(g),
\tag{35}
$$

where the representation matrices Γ_n, obtained from the action in the space of functions of the Cartesian coordinates, are used to replace the general symbols Δ^Γ. Here K runs over the full range $K = -J, -J+1, \ldots, J$ that provides decomposition of the rotor space for a given J into irreducible representations.

Remark 2. From the relations $\Gamma_n = \Gamma_n^\star$, $(\Gamma_n(g))^T = \Gamma_n(g^{-1})$ and $D^J_{KK'}(g) = (D^J_{K'K}(g))^\star$ (see (36)) for the octahedral group it follows that the constituent matrices (34) and (35) of the rotational functions for the laboratory and intrinsic groups are hermitian conjugate with respect to each other ${}^{\mathrm{int}}A^{JM}_{npq} = \left({}^{\mathrm{lab}}A^{JM}_{npq}\right)^{\star T}$.

Consider step by step the algorithm from Section 4 for explicit construction of irreducible bases for the intrinsic octahedral group.

Step 1. First, one needs to organize the loop, running over all irreducible representation $\Gamma_n = \Gamma_1, \Gamma_2, ..., \Gamma_{n_{\max}}$ of the group G. In Eq. (35) let us choose a single (though we are doing the loop over all q_0) fixed $q = q_0$, $q_0 = 1, ..., \dim(\Gamma_n)$ such that there exists K, for which $R^{JM}_{nq_0q_0K}(\Omega)$ is not identically equal to zero, if possible. It can happen so that either for given q_0, or for given representation Γ_n all $R^{JM}_{nq_0q_0K}(\Omega) = 0$.

Step 2. Given the set of vectors $\tilde{u}_{nK}(\Omega) = R^{JM}_{nq_0q_0K}(\Omega)$, where $K = -J, -J + 1, ..., J$ one needs to choose among them a set of linearly independent vectors. Let us assume the quantum number $K = K_1, K_2, ..., K_s$, where $s \leq 1, 2, ..., 2J + 1$, to number the linearly independent vectors $\tilde{u}_{nq_0K}(\Omega)$

$$\tilde{u}_{nq_0K}(\Omega) = R^{JM}_{nq_0q_0K}(\Omega) = \sum_{K'=-J}^{J} A^{JM}_{nq_0KK'}r^J_{MK'}(\Omega), \quad K = K_1, ..., K_s(q_0),$$

$$A^{JM}_{nq_0KK'} = \frac{\dim(\Gamma_n)}{\text{card}(G)} \sum_{g \in G} \Gamma_n(g)^*_{q_0q_0} D^J_{KK'}(g) \text{ at } q = q_0 = p.$$

Here the Wigner functions $D^J_{mm'}(g) = D^J_{mm'}(\alpha = \pi n_1/2, \beta = \pi n_2/2, \gamma = \pi n_3/2)$ for the rotational operators $R(C_*)$ of the octahedral group O are calculated using the values n_1, n_2, n_3 from Table 1 as

$$D^J_{mm'}(g) = \begin{cases} (-1)^{-m(n_1+n_3)/2}\delta_{mm'}, & n_2 = 0, \\ (-1)^J(-1)^{-m(n_1-n_3+2)/2}\delta_{m,-m'}, & n_2 = 2, \\ (-1)^{-mn_1/2}d^J_{mm'}(\pi/2)(-1)^{-m'n_3/2}, & n_2 = 1, \\ (-1)^{-mn_1/2}(-1)^{m-m'}d^J_{mm'}(\pi/2)(-1)^{-m'n_3/2}, & n_2 = 3, \end{cases} \quad (36)$$

where for calculation of $d^J_{mm'}(\pi/2) = D^J_{mm'}(\alpha = 0, \beta = \pi/2, \gamma = 0)$ we use [10,11]

$$d^J_{mm'}(\pi/2) = \frac{(-1)^{m-m'}}{2^J}\sqrt{\frac{(J+m)! \ (J-m)!}{(J+m')! \ (J-m')!}}$$

$$\times \sum_{k=\max(0,m'-m)}^{\min(J+m',J-m)} \frac{(-1)^k(J+m')!(J-m')!}{k!(k+m-m')!(J+m'-k)!(J-m-k)!}.$$

Following section 4, one needs to orthogonalize this set of vectors. The result of orthogonalization using the Gram–Schmidt procedure can be written as:

$$u_{nq_0t}(\Omega) = \sum_{K=-J}^{J} \bar{B}^{JM}_{nq_0tK}\tilde{u}_{nq_0K}(\Omega).$$

After the orthogonalization for every $n = 1, ..., n_{\max}$, where $n_{\max}$ is the number of representations of the point group, we obtain the set of s orthonormalized vectors which we denote by $u_{nq_01}(\Omega), u_{nq_02}(\Omega), \ldots, u_{nq_0t}(\Omega), \ldots, u_{nq_0s}(\Omega), t = 1, \ldots, s(q_0)$. These vectors $u_{nq_0t}(\Omega)$ are decomposed using the basis $r_{MK}^{J}(\Omega)$

$$u_{nq_0t}^{JM}(\Omega) = \sum_{K'=-J}^{J} B_{nq_0tK'}^{JM} r_{MK'}^{J}(\Omega), \quad B_{nq_0tK'}^{JM} = \sum_{K=-J}^{J} \bar{B}_{nq_0tK}^{JM} A_{nq_0KK'}^{JM}, \quad (37)$$

where $t = 1, \ldots, s(q_0)$, $q_0 = 1, \ldots, \dim(\Gamma_n)$.

Step 3. Now for the irreducible representation Γ_n of the group G and for every vector $u_{nq_0t}^{JM}(\Omega), t = 1, \ldots, s(q_0)$ at fixed $q_0 = 1, \ldots, \dim(\Gamma_n)$ we apply the projection operator (35) for every $p = 1, 2, \ldots, \dim(\Gamma_n)$:

$$\bar{v}_{ntpq_0}^{JM}(\Omega) = \bar{P}_{pq_0}^{\Gamma_n} u_{nq_0t}^{JM}(\Omega) = \frac{\dim(\Gamma_n)}{\mathrm{card}(G)} \sum_{g \in G} \Gamma_n(g)_{pq_0}^{\star} \hat{g}\, u_{nq_0t}^{JM}(\Omega)$$

$$\bar{v}_{ntpq_0}^{JM}(\Omega) = \sum_{K'=-J}^{J} \check{A}_{ntpq_0 K'}^{JM} r_{MK'}^{J}(\Omega), \quad t = 1, \ldots, s(q_0), \quad (38)$$

$$\check{A}_{ntpq_0 K'}^{JM} = \sum_{K=-J}^{J} B_{nq_0tK}^{JM} \frac{\dim(\Gamma_n)}{\mathrm{card}(G)} \sum_{g \in G} \Gamma_n(g)_{pq_0}^{\star} D_{KK'}^{J}(g),$$

where $K = K_1, K_2, \ldots, K_s(q_0)$, $D_{KK'}^{J}(g)$ and $B_{nq_0tK}^{JM}$ are calculated by formulas (37). Using the orthonormalization conditions (6) for the basis functions $r_{MK}^{J}(\Omega)$, we calculate the normalization factor $N_{ntpq_0}^2$ and the required set of orthonormalized vectors $\bar{v}_{ntpq_0}(\Omega)$

$$v_{ntpq_0}^{JM}(\Omega) = \bar{v}_{ntpq_0}^{JM}(\Omega)/N_{ntpq_0}, \quad N_{ntpq_0}^2 = \langle \bar{v}_{ntpq_0}^{JM} | \bar{v}_{ntpq_0}^{JM} \rangle = \sum_{K=-J}^{J} (\check{A}_{ntpq_0 K}^{JM})^2. \quad (39)$$

The above algorithm was realized in the form of the program implemented in the computer algebra system Reduce. The typical running time of calculating the irreducible representations $\Gamma_1, \ldots, \Gamma_5$ for the octahedral group for $J \leq 10$ is 980 seconds using the PC Inter Pentium CPU 1.50 GHz 4 GB 64 bit Windows 8.

Results of Step 3. In the following, the output of the program for the irreducible representation Γ_5 is presented. The first set of vectors (40) corresponds to the basis furnishing representations of the laboratory group the second one (41) of the intrinsic group. The index t distinguishes among equivalent representations.

For example, the orthonormalized vectors $v_{ntpq_0}^{JM}(\Omega)$ given by Eqs. (38), (39) after the projection (37) at $q_0 = 3$ read as:

$$v_{5113}^{5K} = -(\sqrt{42} r_{1K}^{-5}(\Omega) + 9 r_{3K}^{-5}(\Omega) + \sqrt{5} r_{5K}^{-5}(\Omega))/\sqrt{128},$$

$$v_{5213}^{5K} = -(\sqrt{30} r_{1K}^{-5}(\Omega) - \sqrt{35} r_{3K}^{-5}(\Omega) + \sqrt{63} r_{5K}^{-5}(\Omega))/\sqrt{128},$$

$$v_{5123}^{5K} = i(\sqrt{42} r_{1K}^{+5}(\Omega) - 9 r_{3K}^{+5}(\Omega) + \sqrt{5} r_{5K}^{+5}(\Omega))/\sqrt{128}, \quad (40)$$

$$v_{5223}^{5K} = i(\sqrt{30} r_{1K}^{+5}(\Omega) + \sqrt{35} r_{3K}^{+5}(\Omega) + \sqrt{63} r_{5K}^{+5}(\Omega))/\sqrt{128},$$

$$v_{5133}^{5K} = (r_{4K}^{+5}(\Omega)), \quad v_{5233}^{5K} = (r_{0K}^{5}(\Omega));$$

$$v_{5113}^{5M}=-(\sqrt{42}r_{M1}^{5-}(\Omega)+9r_{M3}^{5-}(\Omega)+\sqrt{5}r_{M5}^{5-}(\Omega))/\sqrt{128},$$
$$v_{5213}^{5M}=-(\sqrt{30}r_{M1}^{5-}(\Omega)-\sqrt{35}r_{M3}^{5-}(\Omega)+\sqrt{63}r_{M5}^{5-}(\Omega))/\sqrt{128},$$
$$v_{5123}^{5M}=-i(\sqrt{42}r_{M1}^{5+}(\Omega)-9r_{M3}^{5+}(\Omega)+\sqrt{5}r_{M5}^{5+}(\Omega))/\sqrt{128}, \qquad (41)$$
$$v_{5223}^{5M}=-i(\sqrt{30}r_{M1}^{5+}(\Omega)+\sqrt{35}r_{M3}^{5+}(\Omega)+\sqrt{63}r_{M5}^{5+}(\Omega))/\sqrt{128},$$
$$v_{5133}^{5M}=(r_{M4}^{5+}(\Omega)), \qquad v_{5233}^{5M}=(r_{M0}^{5}(\Omega)),$$

where

$$r_{MK}^{J\pm}(\Omega)=(r_{MK}^{J}(\Omega)+r_{M-K}^{J}(\Omega))/\sqrt{2}, \; r_{MK}^{\pm J}(\Omega)=(r_{MK}^{J}(\Omega)+r_{-MK}^{J}(\Omega))/\sqrt{2}.$$

6   Conclusion

We present the symbolic algorithm for calculating irreducible representations of intrinsic as well as for laboratory point groups in the rotor space $L^2(SO(3))$, which can be implemented in computer algebra systems. The bases of these representations are required for calculating spectra and electromagnetic transitions in molecular and nuclear physics. The program is now prepared for calculating both the laboratory and the intrinsic octahedral groups, which is typically considered as the highest rotation point group (without the space inversion). It can be adopted for any subgroup of the octahedral group.

The work was partially supported by the Russian Foundation for Basic Research (RFBR) (grants Nos. 14-01-00420 and 13-01-00668), the Bogoliubov-Infeld and the Hulubei-Meshcheryakov JINR-Romania programs.

References

1. Barut, A., Rączka, R.: Theory of Group Representations and Applications. PWN, Warszawa (1977)
2. Chen, J.Q., Ping, J., Wang, F.: Group Representation Theory for Physicists. World Sci., Singapore (2002)
3. Cornwell, J.F.: Group Theory in Physics. Academic Press, London (1984)
4. Góźdź, A., Pędrak, A.: Hidden symmetries in the intrinsic frame. Phys. Scr. **154**, 014025 (2013)
5. Góźdź, A., Szulerecka, A., Dobrowolski, A., Dudek, J.: Symmetries in the intrinsic nuclear frame. Int. J. Mod. Phys. E **20**, 199–206 (2011)
6. Koster, G.F., Dimmock, J.O., Wheeler, R.G., Statz, H.: Properties of the thirty-two point groups. The M.I.T. Press, Cambridge (1936)
7. Ring, P., Schuck, P.: The Nuclear Many-Body Problem. Springer, N.Y. (1980)
8. Rose, M.E.: Elementary Theory of Angular Momentum. Wiley, New York (1957)
9. Szulerecka, A., Dobrowolski, A., Góźdź, A.: Generalized projection operators for intrinsic rotation group and nuclear collective model. Phys. Scr. **89**, 054033 (2014)
10. Varshalovich, D.A., Moskalev, A.N., Khersonskii, V.K.: Kvantovaja teorija uglovogo momenta. Nauka, Moscow (1975). (in Russian)
11. Varshalovich, D.A., Moskalev, A.N., Khersonskii, V.K.: Quantum Theory of Angular Momentum. World Sci., Singapore (1989)

Symbolic-Numeric Solution of Boundary-Value Problems for the Schrödinger Equation Using the Finite Element Method: Scattering Problem and Resonance States

A.A. Gusev[1], L. Le Hai[1,2], O. Chuluunbaatar[1,3], V. Ulziibayar[4],
S.I. Vinitsky[1], V.L. Derbov[5], A. Góźdź[6], and V.A. Rostovtsev[1]

[1] Joint Institute for Nuclear Research, Dubna, Russia
gooseff@jinr.ru
[2] Belgorod State University, Belgorod, Russia
[3] National University of Mongolia, UlaanBaatar, Mongolia
[4] Mongolian University of Science and Technology, UlaanBaatar, Mongolia
[5] Saratov State University, Saratov, Russia
[6] Institute of Physics, Maria Curie-Skłodowska University, Lublin, Poland

Abstract. We present new symbolic-numeric algorithms for solving the Schrödinger equation describing the scattering problem and resonance states. The boundary-value problems are formulated and discretized using the finite element method with interpolating Hermite polynomials, which provide the required continuity of the derivatives of the approximated solutions. The efficiency of the algorithms and programs implemented in the Maple computer algebra system is demonstrated by analysing the scattering problems and resonance states for the Schrödinger equation with continuous (piecewise continuous) real (complex) potentials like single (double) barrier (well).

1 Introduction

High-accuracy efficient algorithms and programs for solving boundary-value problems are presently indispensable for studying important mathematical models, describing wave propagation in smoothly irregular waveguides, tunnelling and channelling of compound quantum systems through multidimensional potential barriers, photoionization, photoabsorption, and transport in atomic, molecular, and quantum-dimensional semiconductor systems [1–15].

For this class of problems not only the solution itself, but also its first derivative must be continuous, which is of particular importance in the case of quantum-dimensional semiconductor systems and smoothly irregular waveguides, described by partial differential equations with piecewise-continuous coefficient functions [2, 16–18]. As shown by the example of solving an eigenvalue problem for the Schrödinger equation [19], the required continuity of the derivatives can be efficiently implemented in the approximating numerical solution on a finite-element grid using the Hermite interpolating elements [17, 20]. The reduction of the initial

© Springer International Publishing Switzerland 2015
V.P. Gerdt et al. (Eds.): CASC 2015, LNCS 9301, pp. 182–197, 2015.
DOI: 10.1007/978-3-319-24021-3_14

boundary-value problems to the corresponding algebraic problems is a cumbersome problem of the Finite Element Method using high-order approximation. The generation of the local functions using the high-order Hermite interpolation polynomials and the elements of mass and stiffness matrices is performed in the analytic form using the algorithm elaborated by the authors and implemented in CAS Maple. Using CAS Maple is a key point of the approach. Now it is possible to work with multiprocessor computers that implement parallel computations of algebraic problem with high-dimension matrices using the LinearAlgebra package of CAS Maple. Moreover, in our previous paper [19] we also used the symbolic algorithm to generate Fortran routines that allow the solution of the generalized algebraic eigenvalue problem with high-dimension matrices for real-valued potentials. Further development of this approach for solving the scattering problem and calculating the resonance metastable states for real-valued and complex potentials is an important problem that constitutes the goal of the present paper.

In this paper we present a new approach to the study of the resonance scattering problem and the metastable states for both continuous and piecewise continuous real-valued and complex potentials. The discretization of the corresponding boundary-value problem reformulated in terms of symmetric quadratic functionals is implemented using the Hermite interpolation polynomials which provide the required continuity of the derivatives of the approximated solutions. The continuity of the approximate solutions derivatives is the key point in the problems of quantum mechanics, waveguide theory, etc. For the scattering problem with the fixed real energy value $E=\Re E$, $\Re E>0$ we formulate the boundary-value problem for the Schrödinger equation in the finite interval $|z|\leq|z^{\max}|$ with the conditions of the third kind at the boundary points of the interval and construct the appropriate variational functional. The asymptotic solutions of the scattering problem at $|z^{\max}| \leq |z| < \infty$ comprise the incident wave and the unknown amplitudes of transmitted $T(E)$ and reflected $R(E)$ waves, which are calculated together with the desired numerical solution in the finite interval and its logarithmic derivatives at the boundary points of the interval. To calculate the resonance state with the unknown complex eigenvalue of energy $E_r=\Re E_r+i\Im E_r$, $\Re E_r>0$, $\Im E<0$ we formulate the boundary-value problem for the Schrödinger equation in the finite interval with the conditions of the third kind at the boundary points of the interval and construct the appropriate variational functional. In contrast to the scattering problem, in the asymptotic solutions of this problem the amplitude of the incident waves is zero, i.e., only the outgoing waves are present, $\exp(i\sqrt{E_r}|z|)$, that meet the radiation condition [23] and are considered within the sufficiently large but finite interval $|z|\leq|z^{\max}|$.

The constructed stiffness and mass matrices for the variational functionals, comprising the boundary conditions of the first, second, or third kind, are used to formulate the generalized algebraic eigenvalue problem. To calculate the resonance states with unknown complex energy eigenvalues E_r we use the Newton iteration scheme, in which the initial approximation is chosen as the solution of the scattering problem with the boundary conditions of the third kind and the real values of energy $E=\Re E>0$, close to the resonance ones, $E_r=\Re E_r+i\Im E_r$, and corresponding to the maximal value of the transmission coefficient $|T(E)|^2$.

We also used the appropriate solutions of the eigenvalue problem with the boundary conditions of the first or the second kind.

The efficiency (the order of approximation with respect to the finite element grid step) and the capability of time saving (the execution time for the Maple algorithms for banded matrices with the dimension up to 300) is demonstrated by the test calculations of scattering and resonance states for the Schrödinger equation with continuous (piecewise continuous) real (complex) barrier (double barrier) or well (double well) potential functions.

The paper is organized as follows. In Section 2 the formulation of the boundary-value problems with the boundary conditions of the first, second, and third kind is presented, as well as the appropriate variational functionals. Sections 3 presents the finite-element scheme with the Hermite interpolating polynomials and describes the algorithm of reducing the boundary-value problems to the algebraic ones. Section 4 is devoted to test calculations that demonstrate the efficiency and time-saving capability of the proposed computational schemes, implemented as a Maple program. In the Conclusion we discuss the results and the possible applications of the proposed computational schemes and computer programs.

2   Formulation of Boundary-Value Problems

Consider the second-order differential equation with respect to the unknown function $\Phi(z)$ in the interval $z \in \Omega_z = (z^{\min}, z^{\max})$ [19]

$$(D-2E)\,\Phi(z) = 0, \quad D = -\frac{1}{f_1(z)}\frac{\partial}{\partial z}f_2(z)\frac{\partial}{\partial z} + V(z). \tag{1}$$

The coefficient functions $f_1(z) > 0$, $f_2(z) > 0$ and the real or complex potential function $V(z)$ are assumed to be continuous and to possess derivatives up to the order $\kappa^{\max}-1 \geq 1$ in the domain $z \in \bar{\Omega}_z = [z^{\min}, z^{\max}]$. Alternative assumptions for piecewise continuous functions will be also considered below.

Depending on the physical problem, the desired solution is to obey the appropriate boundary conditions at the end points $z^{\min}$ and $z^{\max}$ of the interval $\bar{\Omega}_z$:

$$\text{(I)}: \quad \Phi_m(z^t) = 0, \quad t = \min \text{ and/or max}, \tag{2}$$

$$\text{(II)}: \quad d\Phi_m(z)/dz\big|_{z=z^t} = 0, \quad t = \min \text{ and/or max}, \tag{3}$$

$$\text{(III)}: \quad d\Phi_m(z)/dz\big|_{z=z^t} = \mathcal{R}(z^t)\Phi_m(z^t), \quad t = \min \text{ and/or max}. \tag{4}$$

The solution of the boundary-value problem can be reduced to the determination of the stationary point (or minimal value) of the variational functional [21]

$$\Xi(\Phi, E, z^{\min}, z^{\max}) \equiv \int_{z^{\min}}^{z^{\max}} \Phi(z)\,(D-2E)\,\Phi(z)f_1(z)dz = \Pi(\Phi, E, z^{\min}, z^{\max}) + C,$$

$$C = -f_2(z^{\max})\Phi(z^{\max})\mathcal{R}(z^{\max})\Phi(z^{\max}) + f_2(z^{\min})\Phi(z^{\min})\mathcal{R}(z^{\min})\Phi(z^{\min}),$$

$$\tag{5}$$

where $\Pi(\Phi, E, z^{\min}, z^{\max})$ is the symmetric functional

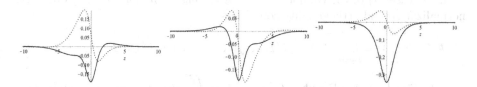

Fig. 1. Real (solid line) and imaginary (dotted line) parts of the eigenfunctions $\Phi_1^+(z)$, $\Phi_1^-(z)$ and $\Phi_1(z)$ with eigenvalues $E_1^\pm$ and E_1, respectively, given in Table 1.

$$\Pi(\Phi, E, z^{\min}, z^{\max}) = \int_{z^{\min}}^{z^{\max}} \left[f_2(z) d\Phi(z)/dz d\Phi(z)/dz \right. \tag{6}$$
$$\left. + f_1(z)\Phi(z)V(z)\Phi(z) - f_1(z)2E\Phi(z)\Phi(z) \right] dz.$$

Problem 1. For bound states the eigenfunctions are considered that obey the boundary conditions of the second kind (3) or the first kind (2) for $\mathcal{R}(z) = 0$ or $\mathcal{R}(z) \to \infty$ in the functional (5), (6), respectively.

In the case (a) of the complex potential and complex eigenvalues $E_m = \Re E_m + \imath \Im E_m$ the eigenfunctions $\Phi_m(z)$ obey the normalization and orthogonality conditions

$$\langle \Phi_m | \Phi_{m'} \rangle = \int_{z^{\min}}^{z^{\max}} \Phi_m(z)\Phi_{m'}(z)f_1(z)dz = \delta_{mm'}. \tag{7}$$

In the case (b) of the real eigenvalues E_m, i.e., $E_m = E_m^*$, $\Im E = 0$, the left-hand function $\Phi_m(z)$ in the scalar product (7) and the functional (5), (6) is replaced with the complex conjugate function $\Phi_m^*(z)$, corresponding to the same eigenvalue $E_m^* = E_m$.

Problem 2. For solving the scattering problem with fixed real eigenvalues E the eigenfunctions $\Phi(E, z)$ are to satisfy the boundary conditions of the third kind (4). The asymptotic solutions of the scattering problem at $|z^{\max}| \le |z| < \infty$ comprise the incident wave and the unknown amplitudes of transmitted $T(E)$ and reflected $R(E)$ waves, which are calculated together with the desired numerical solution in the finite interval and its logarithmic derivatives $\mathcal{R}(z^t)$ at the boundary points of the interval. The unknown eigenvalues $\mathcal{R}(z^{\min})$ (or $\mathcal{R}(z^{\max})$) are determined by solving the problem (19) with the boundary conditions (4) taken into account in a way similar to [9]. The parameter $\mathcal{R}(z^t)$, $t = \max$ (or $t = \min$), in the functional (5), (6) is determined from the asymptotic boundary conditions, $\mathcal{R}(z^t) = \left. \dfrac{d\Phi_{as}(E, z)}{dz} \right|_{z=z^t} \dfrac{1}{\Phi_{as}(E, z^t)}$, where the asymptotic solutions $\Phi_{as}(E, z)$ are δ-function normalized.

In the case (a) of the complex potential the eigenfunctions $\Phi(E, z)$ obey the normalization and orthogonality conditions

$$\langle \Phi(E) | \Phi(E') \rangle = \int_{z^{\min}}^{z^{\max}} \Phi(E, z) \Phi(E', z) f_1(z) dz + C(E, E') = 2\pi \delta(E - E'), \quad (8)$$

$$C(E, E') = \int_{-\infty}^{z^{\min}} \Phi_{as}(E, z) \Phi_{as}(E', z) f_1(z) dz + \int_{z^{\max}}^{+\infty} \Phi_{as}(E, z) \Phi_{as}(E', z) f_1(z) dz.$$

In the case (b) of the real potential the left-hand function $\Phi(E, z)$ in the scalar product (8) and in the functional (5), (6) is replaced with the complex conjugate eigenfunction $\Phi^*(E, z)$. The detailed consideration of the asymptotic functions $\Phi_{as}(E, z)$ will be presented below.

Problem 3. For metastable states the solution satisfies the boundary conditions of the third kind (4), where the parameter $\mathcal{R}(z^t)$ depends upon the complex energy value $E = \Re E + \imath \Im E$ in the lower semiplane: $\mathcal{R}(z^{\min}) = -\sqrt{-2E}$, $\mathcal{R}(z^{\max}) = \sqrt{-2E}$, with $\Re E > 0$ and $\Im E < 0$. In this case for the real (b) and complex (a) potentials (provided that the real and imaginary parts of the latter are specifically chosen, see [13]) the solution satisfies the normalization condition

$$(\Phi_m | \Phi_m) = 2\sqrt{-2E_m} \left(\int_{z^{\min}}^{z^{\max}} \Phi_m(z) \Phi_m(z) f_1(z) dz - 1 \right) + C_{mm} = 0, \quad (9)$$

$$C_{mm} = -f_2(z^{\max}) \Phi_m(z^{\max}) \Phi_m(z^{\max}) + f_2(z^{\min}) \Phi_m(z^{\min}) \Phi_m(z^{\min}),$$

and the orthogonality condition

$$(\Phi_m | \Phi_{m'}) = (\sqrt{-2E_m} + \sqrt{-2E_{m'}}) \int_{z^{\min}}^{z^{\max}} \Phi_m(z) \Phi_{m'}(z) f_1(z) dz + C_{mm'} = 0, \quad (10)$$

$$C_{mm'} = -f_2(z^{\max}) \Phi_m(z^{\max}) \Phi_{m'}(z^{\max}) + f_2(z^{\min}) \Phi_m(z^{\min}) \Phi_{m'}(z^{\min}),$$

that follows from calculating the difference of the functionals (5) with the eigenvalues E_m, $E_{m'}$ and the corresponding eigenfunctions $\Phi_m(z)$, $\Phi_{m'}(z)$ substituted into them and with $\imath p_m = \imath \sqrt{2E_m}$ and $\imath p_{m'} = \imath \sqrt{2E_{m'}}$ substituted into the parameter $\mathcal{R}(z^{\max})$ and with inverse sign into the parameter $\mathcal{R}(z^{\min})$, respectively. Similar orthogonality condition for real potentials was derived earlier using the Green function of the semiaxis [22].

2.1   Scattering Problem: The Physical Asymptotic Solutions in Longitudinal Coordinates and the Scattering Matrix

The solutions of the scattering problem with the fixed energy value $E > 0$ normalized by the condition (8) on the axis $z \in (-\infty, +\infty)$ possess the "incident wave + outgoing waves" asymptotic form

$$\Phi_v(z \to \pm\infty) = \begin{cases} \begin{cases} X^{(+)}(z) T_v, & z > 0, \\ X^{(+)}(z) + X^{(-)}(z) R_v, & z < 0, \end{cases} & v = \to, \\ \begin{cases} X^{(-)}(z) + X^{(+)}(z) R_v, & z > 0, \\ X^{(-)}(z) T_v, & z < 0, \end{cases} & v = \leftarrow, \end{cases} \quad (11)$$

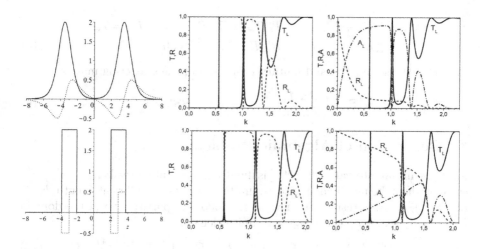

Fig. 2. The system of two complex Scarf potentials with $V_1 = 2$, $V_2 = 1$ separated by the distance $d = 7/2$, and the system of two complex rectangular potential barriers. The solid line shows the real part and the dotted line shows the imaginary part (left-hand panel). The coefficients of transmission $T_L = |T_\rightarrow|^2$ (solid line), reflection $R_L = |R_\rightarrow|^2$ (dotted line), and absorption A_L (dash-dotted line) versus the wave number $k = \sqrt{2E}$ for the systems of two purely real potentials (center panel) and complex potentials (right-hand panel).

where T_v and R_v are the transmission and reflection amplitudes, v is the initial direction of the particle motion along the z axis. For example, for $f_1(z) = f_2(z) = 1$ and rapidly decreasing $V(z \rightarrow \pm\infty) \rightarrow 0$ the asymptotic solutions $X^{(\pm)}(z) \equiv X^{(\pm)}(z, E)$ have the form

$$X^{(\pm)}(z) \rightarrow (p)^{-1/2} \exp\left(\pm\imath pz\right), p = \sqrt{2E} \qquad (12)$$

with the normalization condition

$$\int_{-\infty}^{\infty} (X^{(\pm)}(z, E'))^* X^{(\pm)}(z, E) dz = 2\pi\delta(E' - E). \qquad (13)$$

Generally, the functions $X^{(\pm)}(z)$ satisfy the conditions with the Wronskian

$$\mathrm{Wr}(X^{(\mp)}(z), X^{(\pm)}(z)) = \pm 2\imath, \quad \mathrm{Wr}(X^{(\pm)}(z), X^{(\pm)}(z)) = 0, \qquad (14)$$
$$\mathrm{Wr}(a(z), b(z)) = f_2(z)\left(a(z)db(z)/dz - da(z)/dzb(z)\right).$$

For real-valued potentials the Wronskian is constant, which yields the following properties of the reflection and transmission amplitudes

$$T_{\to}^* T_{\to} + R_{\to}^* R_{\to} = T_{\leftarrow}^* T_{\leftarrow} + R_{\leftarrow}^* R_{\leftarrow} = 1, \quad T_{\to} = T_{\leftarrow}$$
$$T_{\to}^* R_{\leftarrow} + R_{\to}^* T_{\leftarrow} = R_{\leftarrow}^* T_{\to} + T_{\leftarrow}^* R_{\to} = 0, \tag{15}$$

as well as the symmetric and unitary properties of the scattering matrix

$$\mathbf{S} = \begin{pmatrix} R_{\to} & T_{\leftarrow} \\ T_{\to} & R_{\leftarrow} \end{pmatrix}, \qquad \mathbf{S}^\dagger \mathbf{S} = \mathbf{S}\mathbf{S}^\dagger = 1. \tag{16}$$

3   Generation of Algebraic Problems

First, the initial interval $[z^{\min}, z^{\max}]$ is divided into n' subintervals $\tilde{\Omega}_i = [z'_{i-1}, z'_i]$, each of them being divided into n_i finite elements of different length $h_i = (z'_i - z'_{i-1})/n_i$. As a result we arrive at the following partitioning of the domain into $n = n_1 + \ldots + n_i + \ldots + n_{n'} \geq n'$ finite elements

$$\Omega_{h_j(z)}^p[z^{\min}, z^{\max}] = \cup_{j=1}^n \Omega_j = \cup_{i=1}^{n'} \tilde{\Omega}_i, \quad \tilde{\Omega}_i = \cup_{j=n_1+\ldots+n_{i-1}+1}^{n_1+\ldots+n_{i-1}+n_i} \Omega_j, \tag{17}$$
$$\Omega_j = [z_j^{\min}, z_j^{\max} \equiv z_{j+1}^{\min}], \quad j = 0, \ldots, n,$$
$$z_{j=i'+n_1+\ldots+n_{i-1}}^{\max} = (z'_{i-1}(n_i-i')+z'_i i')/n_i, \quad i' = 0, \ldots, n_i, \quad i = 1, \ldots, n'.$$

Each of the finite elements is then divided into p similar intervals, thus forming the finite-element grid $\Omega_{h_j(z)}^p[z^{\min}, z^{\max}] = \{z_0, z_1, \ldots, z_{np}\}$, where $z_{p(j-1)+r} = (z_j^{\min}(p-r) + z_j^{\max}r)/p$, $r = 0, \ldots, p$.

The solutions $\hat{\Phi}(z)$ are sought for in the form of a finite sum over the basis of local functions $N_\mu^g(z)$ at each nodal point $z = z_k$ of the grid $\Omega_{h_j(z)}^p[z^{\min}, z^{\max}]$:

$$\hat{\Phi}(z) = \sum_{\mu=0}^{L-1} \Phi_\mu^h N_\mu^g(z), \quad \hat{\Phi}(z_l) = \Phi_{l\kappa^{\max}}^h, \quad d^\kappa \hat{\Phi}(z)/dz^\kappa \big|_{z=z_l} = \Phi_{l\kappa^{\max}+\kappa}^h \tag{18}$$

where $L = (pn+1)\kappa^{\max}$ is the number of local functions and Φ_μ^h at $\mu = l\kappa^{\max}+\kappa$ are the nodal values of the κ-th derivatives of the function $\hat{\Phi}(z)$ (including the function $\hat{\Phi}(z)$ itself for $\kappa = 0$) at the points z_l.

The local functions $N_\mu^g(z) \equiv N_{l\kappa^{\max}+\kappa}^g(z)$ are piecewise polynomials of the given order $p' = \kappa^{\max}(p+1)-1$, constructed in our previous paper [19]. Their derivative of the order κ at the node z_l equals one, and the derivative of the order $\kappa' \neq \kappa$ at this node equals zero, while the values of the function $N_\mu^g(z)$ with all its derivatives up to the order $(\kappa^{\max}-1)$ equal zero at all other nodes $z_{l'} \neq z_l$ of the grid $\Omega_{h_j(z)}^p[z^{\min}, z^{\max}]$, i.e., $d^\kappa N_{l'\kappa^{\max}+\kappa'}^g/dz^\kappa \big|_{z=z_l} = \delta_{ll'}\delta_{\kappa\kappa'}$, $l = 0, \ldots, np$, $\kappa = 0, \ldots, \kappa^{\max}-1$.

The substitution of the expansion (18) into the variational functional (5), (6) reduces the solution of the eigenvalue *problem 1 or 3* (1)–(4) with the normalization condition (7)) or (9), or the scattering *problem 2* (1)–(4) with the fixed energy E to the solution of the algebraic problem with respect to the desired set $\boldsymbol{\Phi}^h = \{\Phi_\mu^h\}_{\mu=0}^{L-1}$:

$$(\mathbf{A} - \mathbf{M}_{\max} + \mathbf{M}_{\min} - 2E\,\mathbf{B})\boldsymbol{\Phi}^h = 0. \tag{19}$$

Here $\mathbf{A}$ and $\mathbf{B}$ are the symmetric $L \times L$ stiffness and mass matrices, $L = \kappa^{\max}(np+1)$,

$$A_{\mu_1;\mu_2} = \int_{z^{\min}}^{z^{\max}} f_2(z) \frac{dN_{\mu_1}^g(z)}{dz} \frac{dN_{\mu_2}^g(z)}{dz} dz + \int_{z^{\min}}^{z^{\max}} f_1(z) dz N_{\mu_1}^g(z) V(z) N_{\mu_2}^g(z),$$

$$B_{l_1;l_2} = \int_{z^{\min}}^{z^{\max}} f_1(z) N_{\mu_1}^g(z) N_{\mu_2}^g(z) dz,$$

$\mathbf{M}_{\max}$ and $\mathbf{M}_{\min}$ are $L \times L$ the matrices with zero elements except $M_{11} = f_2(z^{\min}) \mathcal{R}(z^{\min})$ and $M_{L+1-\kappa^{\max}, L+1-\kappa^{\max}} = f_2(z^{\max}) \mathcal{R}(z^{\max})$, respectively. The unknown eigenvalues $\mathcal{R}(z^{\min})$ or $\mathcal{R}(z^{\max})$ are determined by solving the problem (19) with the boundary conditions (4) taken into account in a way similar to that of Ref. [9].

The theoretical estimate for the $\mathbf{H}^0$ norm of the difference between the exact solution $\Phi_m(z) \in \mathcal{H}_2^2$ and the numerical one $\Phi_m^h(z) \in \mathbf{H}^{\kappa^{\max}}$ has the order of

$$|E_m^h - E_m| \le c_1 h^{2p'}, \quad \left\| \Phi_m^h(z) - \Phi_m(z) \right\|_0 \le c_2 h^{p'+1}, \tag{20}$$

where $h = \max_{1<j<n} h_j$ is the maximal step of the grid [21].

Remark. To obtain the eigenvalue estimate of the order $2p'$ the integrals are to be calculated with the same order of accuracy $2p'$. If the integrals are calculated with the accuracy $p'+1$, then we get the estimate of the same order $p'+1$ both for eigenvalues and for eigenfunctions.

3.1   The Calculation Scheme for the Solution Matrix $\Phi^h = \underset{\leftarrow}{\Phi^h}$

In this case Eq. (19) can be written in the following form

$$(\mathbf{G} + \mathbf{M}_{\min}) \begin{pmatrix} \underset{\leftarrow}{\Phi^a} \\ \underset{\leftarrow}{\Phi^b} \end{pmatrix} \equiv \begin{pmatrix} \underset{\leftarrow}{\mathbf{G}^{aa}} & \underset{\leftarrow}{\mathbf{G}^{ab}} \\ \underset{\leftarrow}{\mathbf{G}^{ba}} & \underset{\leftarrow}{\mathbf{G}^{bb}} \end{pmatrix} \begin{pmatrix} \underset{\leftarrow}{\Phi^a} \\ \underset{\leftarrow}{\Phi^b} \end{pmatrix} = \begin{pmatrix} 0 & 0 \\ 0 & \mathbf{G}(z^{\max}) \end{pmatrix} \begin{pmatrix} \underset{\leftarrow}{\Phi^a} \\ \underset{\leftarrow}{\Phi^b} \end{pmatrix}, \tag{21}$$

where $(\mathbf{M}_{\min})_{11} = M_{11} = f_2(z^{\min}) \mathcal{R}(z^{\min})$, $\mathcal{R}(z^{\min}) = \imath \sqrt{2E}$, the solutions $\underset{\leftarrow}{\Phi^a}$ and $\underset{\leftarrow}{\Phi^b} \equiv \Phi_{\leftarrow}(z^{\max})$ are vectors with the dimension $(L-1)$ and 1, respectively.

Hence the explicit expressions follow

$$\underset{\leftarrow}{\Phi^a} = -(\underset{\leftarrow}{\mathbf{G}^{aa}})^{-1} \underset{\leftarrow}{\mathbf{G}^{ab}} \underset{\leftarrow}{\Phi^b}, \quad \mathbf{G}(z^{\max}) = \underset{\leftarrow}{\mathbf{G}^{bb}} - \underset{\leftarrow}{\mathbf{G}^{ba}} (\underset{\leftarrow}{\mathbf{G}^{aa}})^{-1} \underset{\leftarrow}{\mathbf{G}^{ab}}. \tag{22}$$

From Eqs. (21) and (22) the relation between $\underset{\leftarrow}{\Phi^b}$ and its derivative follows

$$d\underset{\leftarrow}{\Phi^b}/dz = \mathcal{R}(z^{\max}) \underset{\leftarrow}{\Phi^b}, \quad \mathcal{R}(z^{\max}) = \mathbf{G}(z^{\max}). \tag{23}$$

Note, that the matrix $\mathbf{G}(z^{\max})$ is defined as the inverse of the submatrix $\underset{\leftarrow}{\mathbf{G}^{aa}}$, the calculation of which requires significant computer resources. To solve Eq. (23) without inverting $\underset{\leftarrow}{\mathbf{G}^{aa}}$, let us consider the set of algebraic equations with respect to the vectors $\underset{\leftarrow}{\mathbf{F}^a}$ and $\underset{\leftarrow}{\mathbf{F}^b}$

$$\begin{pmatrix} \underset{\leftarrow}{\mathbf{G}^{aa}} & \underset{\leftarrow}{\mathbf{G}^{ab}} \\ \underset{\leftarrow}{\mathbf{G}^{ba}} & \underset{\leftarrow}{\mathbf{G}^{bb}} \end{pmatrix} \begin{pmatrix} \underset{\leftarrow}{\mathbf{F}^a} \\ \underset{\leftarrow}{\mathbf{F}^b} \end{pmatrix} = f_2(z^{\max}) \begin{pmatrix} 0 \\ \mathbf{I} \end{pmatrix}. \tag{24}$$

Since the determinant of the matrix $\mathbf{G} + \mathbf{M}_{\min}$ is nonzero, the set of equations has the unique solution

$$\mathbf{F}^a_{\leftarrow} = -(\mathbf{G}^{aa}_{\leftarrow})^{-1}\mathbf{G}^{ab}_{\leftarrow}\mathbf{F}^b_{\leftarrow}, \quad \mathbf{F}^b_{\leftarrow} = f_2(z^{\max})\left(\mathbf{G}^{bb}_{\leftarrow} - \mathbf{G}^{ba}_{\leftarrow}(\mathbf{G}^{aa}_{\leftarrow})^{-1}\mathbf{G}^{ab}_{\leftarrow}\right)^{-1}. \quad (25)$$

Then the expression for $\mathcal{R}(z^{\max})$ follows

$$\mathcal{R}(z^{\max}) = \left(\mathbf{F}^b_{\leftarrow}\right)^{-1}. \quad (26)$$

From Eqs. (23) and (11) we get the equation for the reflection amplitude $R_{\leftarrow}$:

$$Y^{(+)}_{\leftarrow}(z^{\max})R_{\leftarrow} = -Y^{(-)}_{\leftarrow}(z^{\max}), \quad Y^{(\pm)}_{\leftarrow}(z) = dX^{(\pm)}(z)/dz - \mathcal{R}(z)X^{(\pm)}(z). \quad (27)$$

Having solved this equation, we find the reflection amplitude $R_{\leftarrow}$

$$R_{\leftarrow} = -(Y^{(+)}_{\leftarrow}(z^{\max}))^{-1}Y^{(-)}_{\leftarrow}(z^{\max}). \quad (28)$$

Then the desired solution $\boldsymbol{\Phi}^h_{\leftarrow}$ is calculated from Eqs. (11), (22), and (25)

$$\boldsymbol{\Phi}^b_{\leftarrow} = X^{(-)}(z^{\max}) + X^{(+)}(z^{\max})R_{\leftarrow}, \quad \boldsymbol{\Phi}^a_{\leftarrow} = F^a_{\leftarrow}\left(F^b_{\leftarrow}\right)^{-1}\boldsymbol{\Phi}^b_{\leftarrow}. \quad (29)$$

The transmission amplitude $T_{\leftarrow}$ is determined by solving the equation

$$X^{(-)}(z^{\min})T_{\leftarrow} = \boldsymbol{\Phi}^h_{\leftarrow}(z^{\min}), \quad T_{\leftarrow} = \left(X^{(-)}(z^{\min})\right)^{-1}\boldsymbol{\Phi}^h_{\leftarrow}(z^{\min}).$$

3.2   The Calculation Scheme for the Solution Matrix $\boldsymbol{\Phi}^h = \boldsymbol{\Phi}^h_{\rightarrow}$

In this case Eq. (19) can be written as follows:

$$(\mathbf{G} - \mathbf{M}_{\max})\begin{pmatrix}\boldsymbol{\Phi}^a_{\rightarrow} \\ \boldsymbol{\Phi}^b_{\rightarrow}\end{pmatrix} \equiv \begin{pmatrix}\mathbf{G}^{aa}_{\rightarrow} & \mathbf{G}^{ab}_{\rightarrow} \\ \mathbf{G}^{ba}_{\rightarrow} & \mathbf{G}^{bb}_{\rightarrow}\end{pmatrix}\begin{pmatrix}\boldsymbol{\Phi}^a_{\rightarrow} \\ \boldsymbol{\Phi}^b_{\rightarrow}\end{pmatrix} = \begin{pmatrix}-\mathbf{G}(z^{\min}) & 0 \\ 0 & 0\end{pmatrix}\begin{pmatrix}\boldsymbol{\Phi}^a_{\rightarrow} \\ \boldsymbol{\Phi}^b_{\rightarrow}\end{pmatrix}, \quad (30)$$

where $(\mathbf{M}^p_{\max})_{LL} = M_{L+1-\kappa^{\max},L+1-\kappa^{\max}} = f_2(z^{\max})\mathcal{R}(z^{\max})$, $\mathcal{R}(z^{\max}) = -i\sqrt{2E}$, the solutions $\boldsymbol{\Phi}^a_{\rightarrow}$ and $\boldsymbol{\Phi}^b_{\rightarrow} \equiv \boldsymbol{\Phi}_{\rightarrow}(z_{\min})$ are vectors with the dimension 1 and $(L-1)$, respectively.

The desired matrix $\mathbf{G}(z^{\min}) = \mathcal{R}(z^{\min})$ is expressed as

$$\mathcal{R}(z^{\min}) = (\mathbf{F}^a_{\rightarrow})^{-1}, \quad (31)$$

and the desired solution $\boldsymbol{\Phi}^h_{\rightarrow}$ is calculated as

$$\boldsymbol{\Phi}^b_{\rightarrow} = \mathbf{F}^b_{\rightarrow}\left(\mathbf{F}^a_{\rightarrow}\right)^{-1}\boldsymbol{\Phi}^a_{\rightarrow}, \quad \boldsymbol{\Phi}^a_{\rightarrow} = X^{(+)}(z^{\min}) + X^{(-)}(z^{\min})R_{\rightarrow}. \quad (32)$$

Here $\boldsymbol{\Phi}^a_{\rightarrow} \equiv \boldsymbol{\Phi}_{\rightarrow}(z^{\min})$ and $\boldsymbol{\Phi}^b_{\rightarrow}$ are vectors with the dimension 1 and $(L-1)$. The column vectors $\mathbf{F}^a_{\rightarrow}$ and $\mathbf{F}^b_{\rightarrow}$ with the dimension 1 and $(L-1)$ are solutions of the sets of algebraic equations

$$(\mathbf{G} - \mathbf{M}_{\max})\begin{pmatrix}\mathbf{F}^a_{\rightarrow} \\ \mathbf{F}^b_{\rightarrow}\end{pmatrix} \equiv \begin{pmatrix}\mathbf{G}^{aa}_{\rightarrow} & \mathbf{G}^{ab}_{\rightarrow} \\ \mathbf{G}^{ba}_{\rightarrow} & \mathbf{G}^{bb}_{\rightarrow}\end{pmatrix}\begin{pmatrix}\mathbf{F}^a_{\rightarrow} \\ \mathbf{F}^b_{\rightarrow}\end{pmatrix} = -f_2(z^{\min})\begin{pmatrix}\mathbf{I} \\ 0\end{pmatrix}. \quad (33)$$

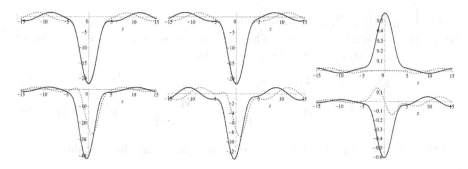

Fig. 3. Wave functions of the scattering problem for the first resonance value of energy $2E_1^{\max T}$, corresponding to the full transparency, i.e., the maximal transmission coefficient, for $\Phi_\rightarrow$ (left-hand panels) and $\Phi_\leftarrow$ (central panels); the functions of resonance metastable states with the energies $2E_1^T$ (right-hand panels), respectively, given in Table 3. The upper panels refer to the system of two real Scarf potentials with $V_1 = 2$, $V_2 = 0$, the lower panels refer to the system of two complex Scarf potentials with $V_1 = 2$, $V_2 = 1$. Solid and dotted lines show the real and imaginary parts of the wave functions, respectively.

Finally, we arrive at the following equations for the amplitudes of reflection $R_\rightarrow$ and transmission $T_\rightarrow$:

$$Y_\rightarrow^{(-)}(z^{\min})R_\rightarrow = -Y_\rightarrow^{(+)}(z^{\min}), \quad X^{(+)}(z_{\max})T_\rightarrow = \Phi_\rightarrow^h(z_{\max}), \qquad (34)$$

$$Y_\rightarrow^{(\pm)}(z) = \frac{dX^{(\pm)}(z)}{dz} - \mathcal{R}(z)X^{(\pm)}(z). \qquad (35)$$

The amplitudes of reflection $R_\rightarrow$ and transmission $T_\rightarrow$ take the form

$$R_\rightarrow = -\left(Y_\rightarrow^{(-)}(z^{\min})\right)^{-1} Y_\rightarrow^{(+)}(z^{\min}), \quad T_\rightarrow = \left(X^{(+)}(z^{\max})\right)^{-1} \Phi_\rightarrow^h(z^{\max}).$$

3.3 Algorithm for Calculating the Complex Eigenvalues and Eigenfunctions of Metastable States

To calculate a complex eigenvalue and the corresponding eigenfunction a discrete problem is solved for the equation

$$\mathcal{F}(u) = 0, \quad \Leftrightarrow \quad \{\mathcal{F}_1(u) = 0, \ \mathcal{F}_2(u) = 0\} \qquad (36)$$

with respect to the pair of unknowns $u = \{E^h, \Phi^h\}$, where $\mathcal{F}_1(u)$ and $\mathcal{F}_2(u)$ are given by the expressions

$$\mathcal{F}_1(u) = [\mathbf{A} - 2E^h\mathbf{B} + \mathbf{M}_{\min}(E^h) - \mathbf{M}_{\max}(E^h)] \Phi^h, \quad \mathcal{F}_2(u) = (\Phi^h)^T\mathcal{F}_1(u).$$

The transition from the approximate solution u_k to the approximate solution u_{k+1} is given by the formulas

$$2E_{k+1}^h = 2E_k^h + \mu_k\tau_k, \quad \Phi_{k+1}^h = \Phi_k^h + \mathbf{v}_k\tau_k, \qquad (37)$$

$$\mathbf{v}_k = \mathbf{v}_k^{(1)} + \mathbf{v}_k^{(2)}\mu_k, \quad \Phi_{k+1}^h = \Phi_{k+1}^h((\Phi_{k+1}^h)^T\mathbf{B}\Phi_{k+1}^h)^{-1/2},$$

where $2E^h_{k=0} = 2E_0$, $\mathbf{\Phi}^h_{k=0} = \mathbf{\Phi}_0$ is the initial approximation from the vicinity of the solution $2E = 2E_*$, $\mathbf{\Phi}^h = \mathbf{\Phi}_*$. The iteration corrections $\mathbf{v}^{(1)}_k$, $\mathbf{v}^{(2)}_k$ are found by solving the inhomogeneous algebraic problems

$$\mathcal{F}_1(E^h_k, \mathbf{v}^{(1)}_k) = -\mathcal{F}_1(E^h_k, \mathbf{\Phi}^h_k) = -\mathcal{F}_1(u_k), \quad \Rightarrow \quad \mathbf{v}^{(1)}_k = -\mathbf{\Phi}^h_k, \qquad (38)$$

$$\mathcal{F}_1(E^h_k, \mathbf{v}^{(2)}_k) = \left(\mathbf{B} - \frac{d\mathbf{M}_{\min}(E^h_k)}{2dE^h_k} + \frac{d\mathbf{M}_{\max}(E^h_k)}{2dE^h_k}\right)\mathbf{\Phi}^h_k, \qquad (39)$$

and the correction μ_k to the eigenvalue E^h_k is found using the formula

$$\mu_k = \frac{\mathcal{F}_2(E^h_k, \mathbf{\Phi}^h_k)}{(\mathbf{\Phi}^h_k)^T \mathbf{B} \mathbf{\Phi}^h_k} \equiv \frac{(\mathbf{\Phi}^h_k)^T \mathcal{F}_1(E^h_k, \mathbf{\Phi}^h_k)}{(\mathbf{\Phi}^h_k)^T \mathbf{B} \mathbf{\Phi}^h_k},$$

that follows from Eq. (36). The expressions for nonzero elements of $\mathbf{M}_{\min}(E^h_k)$, $\mathbf{M}_{\max}(E^h_k)$, and their derivatives by $2E^h_k$ have the form $(L' = L + 1 - \kappa^{\max})$

$$(\mathbf{M}_{\min}(E^h_k))_{11} = -f_2(z^{\min})\sqrt{-2E^h_k}, \quad (\mathbf{M}_{\max}(E^h_k))_{L',L'} = f_2(z^{\max})\sqrt{-2E^h_k},$$

$$\frac{d(\mathbf{M}_{\min}(E^h_k))_{11}}{d(2E^h_k)} = \frac{f_2(z^{\min})}{2\sqrt{-2E^h_k}}, \quad \frac{d(\mathbf{M}_{\max}(E^h_k))_{L',L'}}{d(2E^h_k)} = -\frac{f_2(z^{\max})}{2\sqrt{-2E^h_k}}.$$

The iteration step τ_k in the vicinity of the solution is equal to one, and the optimal step τ_k is calculated using the formula [24]

$$\tau_k = \max\left(\theta, \delta_k(0)/(\delta_k(0) + \delta_k(1))\right), \quad \theta = 0.1.$$

Here $\delta_k(0) = |\mathcal{F}_1(E^h_k, \mathbf{\Phi}^h_k)|^2$ and $\delta_k(1) = |\mathcal{F}_1(E^h_{k+1}, \mathbf{\Phi}^h_{k+1})|^2$ are the residuals and E_{k+1} and $\mathbf{\Phi}^h_{k+1}$ are calculated using Eqs. (37) at $\tau_k = 1$. In all cases $\theta < \tau_k < 1$. The iteration process (37) is terminated, when the condition $|\mathcal{F}_2(E^h_k, \mathbf{\Phi}^h_k)|^2 < \varepsilon$ becomes valid, where $\varepsilon > 0$ is the predetermined accuracy of the approximate solution calculation.

4   Benchmark Calculations

As an example, let us consider the Schrödinger equation (1) at $f_1(z) = f_2(z) = 1$ with the complex Scarf potential on the axis $z \in (-\infty, +\infty)$:

$$V_{Scarf}(z) = V_1 \cosh^{-2} z + \imath V_2 \sinh z \cosh^{-2} z. \qquad (40)$$

Problem 1. For $V_1 < 0$ and $V_2^2 \in \mathcal{R}$ the bound state problem has a finite set of known analytic solutions [12]. At $|V_2| < 1/4 - V_1$ the eigenvalues are essentially complex conjugate pairs:

$$E^\pm_n = -\left(n - (g^*_+ \pm \imath g^*_- - 1)/2\right)^2, \; g^*_\pm = \sqrt{1/4 - V_1 \mp V_2}, \; n = 0, 1, \ldots < (g^*_+ - 1)/2. \,(41)$$

At $|V_2| > 1/4 - V_1$ (or when V_2 is imaginary) the eigenvalues are real:

$$E_n = -\left(n - (g^*_+ + g^*_- - 1)/2\right)^2, \quad n = 0, 1, \ldots < (g^*_+ + g^*_- - 1)/2. \qquad (42)$$

Table 1. Eigenvalues $E_1^{\pm}$, E_1 and their differences from the corresponding analytic values calculated using the grid $(-20(N_1) - 4(N_2)4(N_3)20)$ with the number $N_1=N_2=N_3$ of the eighth-order finite elements ($\kappa_{max}=3$, $p=2$) in each of the subintervals, depending on N_1. The last row presents the analytic values and the Runge coefficient (43).

N_1	$V_1 = -2,\ V_2 = -3$		$V_1 = -2,\ V_2 = -1$	
4	$-0.229080666\pm0.559461207{*}I$	2.8E− 4∓3.1E−4*I	-0.921836165	5.4E−4+6E−14*I
8	$-0.229357025\pm0.559142713{*}I$	−9.5E− 7±1.3E−6*I	-0.922378370	−9.6E−7−6E−13*I
16	$-0.229356076\pm0.559144037{*}I$	3E−10±2.5E−9*I	-0.922377406	−6 E−10−3E−11*I
ext	$-0.229356076\pm0.559144040{*}I$	$Ru = 8.005$	-0.922377405	$Ru = 9.130$

Table 2. Dependence of the coefficients of transmission T, reflection R, and absorption A, calculated using the grid $(-20(N_1)20)$ upon the number N_1 of the eighth-order finite elements ($\kappa_{max} = 3$, $p = 2$) for $V_1 = 2$, $V_2 = 2$, $k = 2E = 1$. The last two rows present the analytical solution and the Runge coefficient (43).

N_1	Digits	$T_\rightarrow$	$R_\rightarrow$	$A_\rightarrow$
20	16	0.6005954018870188	0.0007394643169153872	0.3986651337960658
40	16	0.5984475588608321	0.0007498888028424546	0.4008025523363254
80	16	0.5984514912751766	0.0007498689244704378	0.4007986398003530
80	8	0.59845983	0.00074979961	0.4007903704
ext		0.5984515130037975	0.0007498688034693990	0.4007986181927332
Ru	16	9.088	9.029	9.088

The numerical experiments using the finite-element grid $\Omega_{h_j(z)}^p[z^{min}, z^{max}]$ demonstrated strict correspondence to the theoretical estimations (20) for both eigenvalues and eigenfunctions. In particular, we calculated the Runge coefficients

$$\beta_l = \log_2 \left| (\sigma_l^h - \sigma_l^{h/2})/(\sigma_l^{h/2} - \sigma_l^{h/4}) \right|, \quad l = 1, 2, \tag{43}$$

on three twice condensed grids with the absolute errors

$$\sigma_1^h = |F(E_m^{exact}) - F(E_m^h)|, \quad \sigma_2^h = \max_{z \in \Omega^h(z)} |\Phi_m^{exact}(z) - \Phi_m^h(z)| \tag{44}$$

for the eigenvalues and eigenfunctions, respectively. From Eq. (44) we obtained the numerical assessment of the convergence order $Ru \sim 8 \div 9$ of the proposed numerical schemes (shown for $F(E) = E$ in Table 1 and for $F(E) = T_\rightarrow, R_\rightarrow, A_\rightarrow$ in Table 2), the theoretical estimates being $\beta_1 = p' + 1$ and $\beta_2 = p' + 1$, in accordance with the *Remark* following Eq. (20).

Problem 2. For the scattering problem with fixed real-valued energy $2E = k^2 > 0$ and the complex Scarf potential (40) the coefficients of transmission $|T|^2$ and reflection $|R|^2$ are expressed as

$$|R_\rightarrow|^2 = D_\rightarrow/D, \quad |R_\leftarrow|^2 = D_\leftarrow/D, \quad |T_\rightarrow|^2 = |T_\leftarrow|^2 = \sinh^2(2\pi k)/D, \tag{45}$$
$$D_\rightarrow = (2\cosh(\pi g_+)\cosh(\pi g_-) + \cosh^2(\pi g_+)e^{-2\pi k} + \cosh^2(\pi g_-)e^{2\pi k}),$$
$$D_\leftarrow = (2\cosh(\pi g_+)\cosh(\pi g_-) + \cosh^2(\pi g_+)e^{2\pi k} + \cosh^2(\pi g_-)e^{-2\pi k}),$$
$$D = \sinh^2(2\pi k) + 2\cosh(2\pi k)\cosh(\pi g_+)\cosh(\pi g_-) + \cosh^2(\pi g_+) + \cosh^2(\pi g_-).$$

Here the notation $g_\pm = \sqrt{V_1 \pm V_2 - 1/4}$ is used. It has been proved [11] that when the potential is complex and spatially non-symmetric, the reflectivity depends on whether the particle is incident from the left or the right side. For the complex potential scattering with the fixed real $E > 0$ the conditions (15) are modified as follows:

$$|R_\rightarrow|^2 + |T_\rightarrow|^2 = 1 - A_\rightarrow, \quad |R_\leftarrow|^2 + |T_\leftarrow|^2 = 1 - A_\leftarrow, \quad T_\rightarrow = T_\leftarrow \equiv T.$$

For the complex Scarf potential $A_\rightarrow$ and $A_\leftarrow$ are expressed as

$$A_\rightarrow = \frac{s_+ s_- - s_-^2}{1 + s_+ s_-}, \quad A_\leftarrow = \frac{s_+ s_- - s_+^2}{1 + s_+ s_-}, \quad s_\pm = \frac{\cosh(\pi g_+)e^{\pm\pi k} + \cosh(\pi g_-)e^{\mp\pi k}}{\sinh(2\pi k)}. \tag{46}$$

Here we consider only positive values $A_\rightarrow > 0$ (or $A_\leftarrow > 0$), commonly interpreted as the probability of absorption [11, 13].

The *Problem 1* of determining the eigenvalues E_m^h and the corresponding eigenfunctions $\Phi_m^h(z)$ for Eq. (19) was solved using the built-in package LinearAlgebra of the Maple system. Table 1 presents the dependence of the eigenvalues calculated using the grid $(-20(N_1) - 4(N_2)4(N_3)20)$ with the number $N_1 = N_2 = N_3$ of the eighth-order finite elements ($\kappa_{max} = 3$, $p = 2$) in each of the subintervals upon N_1. One can see that these sequences converge to the analytical results (41) and (42). The behaviour of the eigenfunctions $\Phi_m^h(z)$ is illustrated by Fig. 1. The time of computing the auxiliary integrals is nearly 42 seconds, the time of constructing the matrices and solving the algebraic eigenvalue problem at $N = 16$ amounts to 4.5 seconds. Table 2 illustrates the dependence of the coefficients of transmission T, reflection R, and absorption A calculated using the grid $(-20(N_1)20)$ upon the number N_1 of the eighth-order finite elements ($\kappa_{max} = 3$, $p = 2$). One can see that these sequences converge to the analytical results (45) and (46) . The time of constructing the matrices and solving the algebraic problem for $N_1 = 20$ and $N_1 = 80$ (Digits:=16) amounts to 5 and 22 seconds, respectively, and for $N_1 = 80$ (Digits:=8) this time is 5 seconds.

For a system of multiple-barrier Scarf potentials separated from each other the approximate analytic expressions for the coefficients of transmission, reflection, and absorption are also available. In particular, for a system of two Scarf potential the analytic expressions are presented in Ref. [13]).

The scattering *Problem 2* with Eqs. (21) and (30) was solved following the algorithm of Sections 3.1 and 3.2 and using the built-in package LinearAlgebra of the Maple system using the finite-element grid $(-8(N_1 = 40)8)$ c N_1 with Hermite eighth-order finite elements ($\kappa_{max} = 3$, $p = 2$). The dependence upon k for the coefficients of transmission, reflection, and absorption, calculated with

Table 3. The first resonance energy values $2E_i^{\max T}$ for the maximal transmission coefficient (full transparency) and the eigenvalues $2E_i^r$ of resonance metastable states.

Scarf	$V_1 = 2$	$2E_1^{\max T} = 0.310918$	$2E_1^r = 0.31093782 - \imath 0.00069129$
	$V_2 = 0$	$2E_2^{\max T} = 1.025359$	$2E_2^r = 1.02413913 - \imath 0.01733149$
	$V_1 = 2$	$2E_1^{\max T} = 0.360240$	$2E_1^r = 0.36025570 - \imath 0.00103794$
	$V_2 = 1$	$2E_2^{\max T} = 1.036324$	$2E_2^r = 1.03383748 - \imath 0.02383030$
Steps/	$V_1 = 2$	$2E_1^{\max T} = 0.329476$	$2E_1^r = 0.32921557 - \imath 0.00247662$
wells	$V_2 = 0$	$2E_2^{\max T} = 1.254400$	$2E_2^r = 1.25175270 - \imath 0.03351010$
	$V_1 = 2$	$2E_1^{\max T} = 0.331776$	$2E_1^r = 0.33292316 - \imath 0.00247662$
	$V_2 = 1/2$	$2E_2^{\max T} = 1.263376$	$2E_2^r = 1.26054650 - \imath 0.03359483$

the absolute accuracy 0.001, in the system of two Scarf potentials with $V_1 = 2$, $V_2 = 1$ separated by the interval $d = 7/2$, is presented in the upper panel of Fig. 2. The resonance structure of the transmission coefficient is due to the presence of metastable states submerged in the continuous spectrum.

Problem 3. The complex eigenvalues and the corresponding eigenfunctions of the metastable states are calculated by means of the Newton iteration algorithm of Section 3.3 using the built-in package LinearAlgebra of the Maple system. For the initial approximation we used both the solutions of the bound-state *Problem 1* and the solutions of the scattering *Problem 2* with the resonance values of energy $E = E^r$, corresponding to the peaks of the transmission coefficient.

For the system of two real- and complex-valued Scarf potentials Fig. 3 presents the wave functions of the scattering problem for the first resonance state, corresponding to the maximal transmission coefficient (full transparency), and the functions of a resonance metastable state. The first resonance energy values $2E_i^{\max T}$ corresponding to the maximal transmission coefficient (full transparency) and the eigenvalues $2E_i^r$ of the resonance metastable states are shown in Table 3. The calculations were performed using the grid $(-8(N_1 = 40)8)$ with N_1 Hermite eighth-order finite elements ($\kappa_{\max} = 3$, $p = 2$).

In a similar way the piecewise continuous potentials are considered, in particular, the systems of potential steps/wells with rectangular-shaped walls. The latter problem can be solved analytically. The lower panel of Fig. 2 presents the approximation of the system of two Scarf potentials with a system of potential steps/wells. As seen from Fig. 2, with the increase of the wave number k the transmission, reflection, and absorption coefficients differ stronger. The calculations were performed using the grid $(-7(N_1{=}10){-}2{-}7/4(N_2{=}3){-}2{-}7/8(N_3{=}3)$ $-2(N_4{=}20)2(N_5{=}3)2{+}7/8(N_6{=}3)2{+}7/4(N_7{=}10)7)$ with N_i ($i = 1, ..., 7$) Hermite seventh-order finite elements ($\kappa_{\max}{=}2$, $p = 3$). The eigenvalues for the system of real and complex potential steps/wells (see Table 3) qualitatively agree with the results presented in the above paragraph. The scattering wave functions for the first two resonance energy values and for the resonance metastable states behave qualitatively similar to those of the system of Scarf potentials, and for this reason are not presented here.

5   Conclusion

The presented analysis of solving the eigenvalue problem, the scattering problem, and the calculation of resonance metastable states for the Schrödinger equation with continuous and piecewise continuous real-valued and complex potentials demonstrated the efficiency of the developed algorithms and programs, implemented in the Maple computer algebra system. The algorithm conserves the derivative continuity property, inherent in the desired solution, in the approximating numeric solution, defined on the finite-element grid using the Hermite interpolating elements.

Further development of the proposed algorithms and programs is targeted at the solution of the problems that describe the scattering processes in the quantum-dimensional semiconductor systems and smoothly irregular waveguides with piecewise continuous real-valued and complex coefficient functions in the partial differential equations, which require the continuity of not only the solution itself, but also of its first derivative.

The authors thank Prof. V.P. Gerdt for collaboration and support of this work. The work was partially supported by the Russian Foundation for Basic Research (RFBR) (grants No. 14-01-00420 and 13-01-00668), the Bogoliubov-Infeld and the Hulubei-Meshcheryakov JINR-Romania programs.

References

1. Kotlyar, V.V., Kovalev, A.A., Nalimov, A.G.: Gradient microoptical elements for achieving superresolution. Kompyuternaya optika **33**, 369–378 (2009). (in Russian)
2. Rezanur Rakhman, K.M., Sevastyanov, L.A.: One-dimensional scattering problem at stepwise potential with non-coincident asymptotic forms. Vestnik RUDN, ser. Fizika No. 5 (1), 35–38 (1997) (in Russian)
3. Sevastyanov, L.A., Sevastyanov, A.L., Tyutyunnik, A.A.: Analytical calculations in maple to implement the method of adiabatic modes for modelling smoothly irregular integrated optical waveguide structures. In: Gerdt, V.P., Koepf, W., Seiler, W.M., Vorozhtsov, E.V. (eds.) CASC 2014. LNCS, vol. 8660, pp. 419–431. Springer, Heidelberg (2014)
4. Chuluunbaatar, O., Gusev, A.A., Gerdt, V.P., Kaschiev, M.S., Rostovtsev, V.A., Samoylov, V., Tupikova, T., Vinitsky, S.I.: A symbolic-numerical algorithm for solving the eigenvalue problem for a hydrogen atom in the magnetic field: cylindrical coordinates. In: Ganzha, V.G., Mayr, E.W., Vorozhtsov, E.V. (eds.) CASC 2007. LNCS, vol. 4770, pp. 118–133. Springer, Heidelberg (2007)
5. Gusev, A.A., Chuluunbaatar, O., Gerdt, V.P., Rostovtsev, V.A., Vinitsky, S.I., Derbov, V.L., Serov, V.V.: Symbolic-numeric algorithms for computer analysis of spheroidal quantum dot models. In: Gerdt, V.P., Koepf, W., Mayr, E.W., Vorozhtsov, E.V. (eds.) CASC 2010. LNCS, vol. 6244, pp. 106–122. Springer, Heidelberg (2010)
6. Gusev, A.A., Vinitsky, S.I., Chuluunbaatar, O., Gerdt, V.P., Rostovtsev, V.A.: Symbolic-numerical algorithms to solve the quantum tunneling problem for a coupled pair of ions. In: Gerdt, V.P., Koepf, W., Mayr, E.W., Vorozhtsov, E.V. (eds.) CASC 2011. LNCS, vol. 6885, pp. 175–191. Springer, Heidelberg (2011)

7. Vinitsky, S., Gusev, A., Chuluunbaatar, O., Rostovtsev, V., Le Hai, L., Derbov, V., Krassovitskiy, P.: Symbolic-numerical algorithm for generating cluster eigenfunctions: tunneling of clusters through repulsive barriers. In: Gerdt, V.P., Koepf, W., Mayr, E.W., Vorozhtsov, E.V. (eds.) CASC 2013. LNCS, vol. 8136, pp. 427–442. Springer, Heidelberg (2013)

8. Vinitsky, S., Gusev, A., Chuluunbaatar, O., Le Hai, L., Góźdź, A., Derbov, V., Krassovitskiy, P.: Symbolic-numeric algorithm for solving the problem of quantum tunneling of a diatomic molecule through repulsive barriers. In: Gerdt, V.P., Koepf, W., Seiler, W.M., Vorozhtsov, E.V. (eds.) CASC 2014. LNCS, vol. 8660, pp. 472–490. Springer, Heidelberg (2014)

9. Gusev, A.A., Chuluunbaatar, O., Vinitsky, S.I., Abrashkevich, A.G.: KANTBP 3.0: New version of a program for computing energy levels, reflection and transmission matrices, and corresponding wave functions in the coupled-channel adiabatic approach. Comput. Phys. Commun. **185**, 3341–3343 (2014)

10. Molinàs-Mata, P., Molinàs-Mata, P.: Electron absorption by complex potentials: One-dimensional case. One-dimensional case. Phys. Rev. A **54**, 2060–2065 (1996)

11. Ahmed, Z.: Schrödinger transmission through one-dimensional complex potentials. Phys. Rev. A **64**, 042716 (2001)

12. Ahmed, Z.: Real and complex discrete eigenvalues in an exactly solvable one-dimensional complex PT -invariant potential. Phys. Lett. A **282**, 343–348 (2001)

13. Cerveró, J.M., Rodríguez, A.: Absorption in atomic wires. Phys. Rev. A **70**, 052705 (2004)

14. Muga, J.G., Palao, J.P., Navarro, B., Egusquiza, I.L.: Complex absorbing potentials. Phys. Reports **395**, 357–426 (2004)

15. Cannata, F., Dedonder, J.-P., Ventura, A.: Scattering in PT-symmetric quantum mechanics. Annals of Physics **322**, 397–433 (2007)

16. Becker, E.B., Carey, G.F., Oden, T.J.: Finite elements. An introduction, vol. I. Prentice-Hall Inc., Englewood Cliffs (1981)

17. Ram-Mohan, R.L.: Finite Element and Boundary Element Aplications in Quantum Mechanics. Oxford University Press, New York (2002)

18. Amodio, P., Blinkov, Y., Gerdt, V., La Scala, R.: On consistency of finite difference approximations to the navier-stokes equations. In: Gerdt, V.P., Koepf, W., Mayr, E.W., Vorozhtsov, E.V. (eds.) CASC 2013. LNCS, vol. 8136, pp. 46–60. Springer, Heidelberg (2013)

19. Gusev, A.A., Chuluunbaatar, O., Vinitsky, S.I., Derbov, V.L., Góźdź, A., Le Hai, L., Rostovtsev, V.A.: Symbolic-numerical solution of boundary-value problems with self-adjoint second-order differential equation using the finite element method with interpolation hermite polynomials. In: Gerdt, V.P., Koepf, W., Seiler, W.M., Vorozhtsov, E.V. (eds.) CASC 2014. LNCS, vol. 8660, pp. 138–154. Springer, Heidelberg (2014)

20. Berezin, I.S., Zhidkov, N.P.: Computing Methods, vol. I. Pergamon Press, Oxford (1965)

21. Strang, G., Fix, G.J.: An Analysis of the Finite Element Method. Prentice-Hall, Englewood Cliffs (1973)

22. Kukulin, V.I., Krasnopol'sky, V.M., Horáček, J.: Theory of Resonances, pp. 107–112. Academia, Praha (1989)

23. Siegert, A.J.F.: On the derivation of the dispersion formula for nuclear reactions. Phys. Rev. **56**, 750–752 (1939)

24. Ermakov, V.V., Kalitkin, N.N.: The optimal step and regularization for Newton's method. USSR Computational Mathematics and Mathematical Physics **21**, 235–242 (1981)

Application of Computer Algebra Methods to Investigation of Influence of Constant Torque on Stationary Motions of Satellite

Sergey A. Gutnik[1], Anna Guerman[2], and Vasily A. Sarychev[3]

[1] Moscow State Institute of International Relations (University) 76, Prospekt Vernadskogo, Moscow, 119454, Russia
s.gutnik@inno.mgimo.ru
[2] University of Beira Interior, 6200-001 Covilha, Portugal
anna@ubi.pt
[3] Keldysh Institute of Applied Mathematics (Russian Academy of Sciences) 4, Miusskaya Square, Moscow, 125047, Russia
vas31@rambler.ru

Abstract. Methods of computer algebra are used to study the properties of a nonlinear algebraic system that determines equilibrium orientations of a satellite moving along a circular orbit under the action of gravitational and constant torques. An algorithm for the construction of a Groebner basis is proposed for determining the equilibrium orientations of a satellite with a given constant torque and given principal central moments of inertia. The number of equilibria depending on the parameters of the problem is found by the analysis of real roots of algebraic equation of degree 6 from constructed Groebner basis. The domains with different numbers of equilibria are specified by the discriminant hyper surface given by discriminant of 6 degree polynomial, which was computed symbolically. The equations of boundary curves of two-dimensional section of the discriminant hypersurface are determined in function of values of the components of constant torque. Classification of domains with different number of equilibria from 24 to 0 is carried out for arbitrary values of the parameters.

1 Introduction

Celestial mechanics and astrodynamics are popular domains of application of symbolic computation methods. The important aspect of the development of astrodynamics and space engineering is the design of systems of orientation of the satellites. Among the various types of attitude control systems of orientation, the most widespread are the gravity orientation systems of the satellite. These systems are based on the fact that a satellite with different moments of inertia in the central Newtonian force field in the circular orbit has 24 equilibrium orientations, and four of them are stable [1], [2] and [3]. An important property of gravity orientation systems is that these systems can operate for a long time without spending energy. The problem to be analyzed in the present

© Springer International Publishing Switzerland 2015
V.P. Gerdt et al. (Eds.): CASC 2015, LNCS 9301, pp. 198–209, 2015.
DOI: 10.1007/978-3-319-24021-3_15

work is related to the behavior of the satellite acted upon by the gravity gradient and constant torques. The constant torque may be produced actively or caused, for example, by gas or fuel escape from the satellite. The action of some constant torque changes the orientations of the satellite and can destroy some or even all equilibria. Therefore, it is necessary to study the joint action of gravitational and constant torques and, in particular, to analyze all possible satellite's equilibria in a circular orbit. Such solutions can be used in practical space technology in the design of control systems of the satellites. In [4], the existence of equilibria for the satellite under the action of the gravity gradient and constant disturbing torques in some particular cases was indicated. For general values of constant torque, this problem was studied in [5], using aircraft angles approach for determining the satellite equilibrium orientations. It was shown that for small constant torque there exist 24 equilibria, and the number of equilibria decreases with the increase of constant torque. Using the above approach, the classification of different distributions of the number of equilibria as the function of parameters of the problem, namely, the components of the constant torque, the inertial parameters of the satellite and the angular velocity of its orbital motion was done in [6].

In the present work, the problem of determination of the classes of equilibrium orientations for the general values of constant torque is considered. The equilibrium orientations are determined by real roots of the system of algebraic equations. The investigation of equilibria was possible due to application of Computer Algebra Groebner basis and resultant methods. Evolution of domains with a fixed number of equilibria is investigated by the analysis of the singular points of the discriminant hypersurface in dependence of three dimensionless system parameters. Bifurcation values of the system parameters corresponding to the qualitative change of these domains were determined.

2   Equations of Motion

The motion of the satellite subjected to gravitational and constant torques in a circular orbit is considered. We assume that 1) the gravity field of the Earth is central and Newtonian, 2) the satellite is a triaxial rigid body, 3) the satellite is subjected to the gravity gradient torque and the torque that is fixed with respect to the body of satellite, so the components of this torque in the body fixed frame are constant. To write the equations of motion we introduce two right Cartesian coordinate systems with origin in the satellite's center of mass O. $OXYZ$ is the orbital coordinate system whose OZ axis is directed along the radius vector connecting the centers of mass of the Earth and of the satellite; the OX axis is directed along the vector of linear velocity of the center of mass O. $Oxyz$ is the satellite body coordinate system; Ox, Oy, and Oz are the principal central axes of inertia of the satellite. The orientation of the satellite body coordinate system $Oxyz$ with respect to the orbital coordinate system is determined by means of the aircraft angles of pitch (α), yaw (β) and roll (γ), and the direction cosines in transformation matrix between the orbital coordinate system $OXYZ$ and $Oxyz$ are represented by the following expressions [1]:

$$
\begin{aligned}
a_{11} &= \cos(x, X) = \cos\alpha\cos\beta, \\
a_{12} &= \cos(y, X) = \sin\alpha\sin\gamma - \cos\alpha\sin\beta\cos\gamma, \\
a_{13} &= \cos(z, X) = \sin\alpha\cos\gamma + \cos\alpha\sin\beta\sin\gamma, \\
a_{21} &= \cos(x, Y) = \sin\beta, \\
a_{22} &= \cos(y, Y) = \cos\beta\cos\gamma, \\
a_{23} &= \cos(z, Y) = -\cos\beta\sin\gamma, \\
a_{31} &= \cos(x, Z) = -\sin\alpha\cos\beta, \\
a_{32} &= \cos(y, Z) = \cos\alpha\sin\gamma + \sin\alpha\sin\beta\cos\gamma, \\
a_{33} &= \cos(z, Z) = \cos\alpha\cos\beta - \sin\alpha\sin\beta\sin\gamma.
\end{aligned}
\tag{1}
$$

Then equations of the satellite's attitude motion can be written in the Euler form [5]:

$$
\begin{aligned}
A\dot{p} + (C - B)qr - 3\omega_0^2(C - B)a_{32}a_{33} - \tilde{a} &= 0, \\
B\dot{q} + (A - C)rp - 3\omega_0^2(A - C)a_{31}a_{33} - \tilde{b} &= 0, \\
C\dot{r} + (B - A)pq - 3\omega_0^2(B - A)a_{31}a_{32} - \tilde{c} &= 0,
\end{aligned}
\tag{2}
$$

$$
\begin{aligned}
p &= (\dot{\alpha} + \omega_0)a_{21} + \dot{\gamma}, \\
q &= (\dot{\alpha} + \omega_0)a_{22} + \dot{\beta}\sin\gamma, \\
r &= (\dot{\alpha} + \omega_0)a_{23} + \dot{\beta}\cos\gamma.
\end{aligned}
\tag{3}
$$

In equations (2) and (3), A, B, and C are the principal central moments of inertia of the satellite; p, q, and r are the projections of the angular velocity of the satellite onto the Ox, Oy, and Oz axes; ω_0 is the angular velocity of the orbital motion of the satellite center of mass, while $\tilde{a}$, $\tilde{b}$, $\tilde{c}$ are the components of the constant torque in the satellite body coordinate system. The dot designates differentiation with respect to time t.

3   Equilibrium Orientations

Setting in (2) and (3) $\alpha = \alpha_0 = $ const, $\beta = \beta_0 = $ const, $\gamma = \gamma_0 = $ const, we obtain at $A \neq B \neq C$ the equations

$$
\begin{aligned}
a_{22}a_{23} - 3a_{32}a_{33} &= a, \\
a_{21}a_{23} - 3a_{31}a_{33} &= b, \\
a_{21}a_{22} - 3a_{31}a_{32} &= c,
\end{aligned}
\tag{4}
$$

allowing us to determine the satellite equilibrium orientations in the orbital coordinate system. Here

$$
a = \frac{\tilde{a}}{\omega_0^2(C - B)}, \quad b = \frac{\tilde{b}}{\omega_0^2(A - C)}, \quad c = \frac{\tilde{c}}{\omega_0^2(B - A)}
$$

are the constants that characterize the dimensionless components of the constant torque.

Substituting the expressions for the direction cosines from (1) into Eqs. (4), we arrive at the system of three equations in the three unknowns α, β, and γ. Another, more convenient, way to close Eqs. (4) is to add the following three conditions of orthogonality of the direction cosines

$$a_{21}^2 + a_{22}^2 + a_{23}^2 = 1,$$
$$a_{31}^2 + a_{32}^2 + a_{33}^2 = 1, \tag{5}$$
$$a_{21}a_{31} + a_{22}a_{32} + a_{23}a_{33} = 0.$$

System (4), (5) includes six algebraic equations for six unknown direction cosines, which allow us to determine the satellite equilibria in the orbital reference frame. After a_{21}, a_{22}, a_{23}, a_{31}, a_{32}, and a_{33} are found, the direction cosines a_{11}, a_{12} and a_{13} can be determined from the conditions of orthogonality. For the system of equations (4) and (5), we state the following problem: for given a, b, and c find all nine direction cosines, i.e., to find all the equilibrium orientations of the satellite. It is possible to see at once that for any given attitude of the satellite there will always exist a, b, and c such that this attitude is an equilibrium orientation.

In [5], using the direction cosines in terms of orientation angles (1) and concept of resultant, is shown that the system of equations (4), (5) can be reduced to a single algebraic equation of sixth degree with real coefficients, which represent polynomials depending on three dimensionless parameters of the system. Four equilibrium orientations of a satellite correspond to every real root of this algebraic equation. Since the number of real roots of the algebraic equation of degree 6 does not exceed 6, the satellite subjected to gravitational and constant torques can have no more than 24 equilibrium orientations in a circular orbit. In the case $a = b = c = 0$, it has been proved that the system (4), (5) has 24 solutions describing the equilibrium orientations of the satellite-rigid body [1], [2] and [3].

To solve algebraic system (4), (5) we applied the algorithm of constructing the Groebner bases [7]. The method of constructing the Groebner bases is an algorithmic procedure for complete reduction of the problem in the case of the system of polynomials in many variables to the polynomial of one variable. Using the Groebner[gbasis] Maple package [8] for constructing Groebner bases with linear ordering with respect to *tdeg* powers, we constructed the Groebner basis for system of six polynomials (4), (5) with six variables a_{ij} $(i = 2, 3; j = 1, 2, 3)$ under the ordering on the total power of the variables.

In the list of variables in the Maple Groebner package, we use six direction cosines, and we include in the list of polynomials the polynomials from the left-hand sides f_i $(i = 1, 2, ...6)$ of the algebraic equations (4), (5):

G:=map(factor, Groebner[gbasis]([f1,f2,...,f6], tdeg(a21,a22,a23, a31,a32,a33))).

Below are the polynomials from the constructed Groebner basis that depend only on five variables a_{22}, a_{23}, a_{31}, a_{32} and a_{33}:

$$ba_{22}^2 - 3ba_{32}^2 + 12a_{31}a_{33} - ac + 3b = 0,$$
$$144a_{32}^2a_{33}^2 + (84a - 24bc)a_{32}a_{33} + 12a^2(a_{32}^2 + a_{33}^2)$$
$$+a^2c^2 + a^2b^2 + b^2c^2 - 7abc = 0,$$
$$12ca_{33}^3 + 12ba_{32}a_{33}^2 - b(c^2 + 12)a_{32} - c(b^2 + 12)a_{33} \qquad (6)$$
$$+12ca_{33}a_{32}^2 + 12ba_{32}^3 + (ac^2 + ab^2 - 7bc)a_{31} = 0,$$
$$aba_{23}^2 - 3aba_{33}^2 - 12aa_{31}a_{33} - 12ba_{32}a_{33} + ca^2 + cb^2 - 4ab = 0,$$
$$12aa_{31}a_{33}^2 + 12ba_{32}a_{33}^2 + a^2ba_{32} + ab^2a_{31} - (cb^2 + ca^2 - 7ab)a_{33} = 0.$$

To calculate the sixth-order polynomial from the constructed Groebner basis that depends only on one variable a_{33}^2 the lexicographic monomial order was chosen: map(factor,Groebner[Basis](G,plex(a31, a32, a33, a21, a22, a23))). Another way to have such a polynomial is possible by calculating the resultant from the second and third polynomials of Groebner basis in (6), which involve only two unknowns a_{32} and a_{33}. Such resultant also yields the same sixth-degree polynomial in a_{33}^2. A simpler sixth-order algebraic equation from the constructed Groebner basis (6) that depends only on one variable $x = a_{23}^2$ is possible to construct by changing the order in the lexicographic list –
    plex(a21, a22, a23, a31, a32, a33). This polynomial has the form

$$P(x) = p_0x^6 + p_1x^5 + p_2x^4 + p_3x^3 + p_4x^2 + p_5x + p_6 = 0, \qquad (7)$$

where

$$p_0 = 4096,$$
$$p_1 = -8192,$$
$$p_2 = 256(16 - 6u^2 + 20uc + 17v),$$
$$p_3 = -128(u^2v + 8c^2u^2 - 8u^2 + 40cu + 5cuv + 34v - c^2v^2),$$
$$p_4 = 16(9u^4 + 257u^2 - 43u^2v + 16v^2 + 210cuv - 20cuv^2 + 20cu^3$$
$$\qquad + 17v^2c^2 + 64c^2u^2),$$
$$p_5 = 8(4u^2v^2 - 34u^2v - 38u^4 + 3u^4v - 130cu^3 + 5cu^3v - 20cuv^2 - 4c^2v^3$$
$$\qquad - 68c^2u^2v + 3c^2u^2v^2 - 10c^3uv^2),$$
$$p_6 = (4u^2 + u^2v + c^2v^2 + 5cuv)^2, \quad v = a^2 + b^2, \quad u = ab.$$

Equation (7) together with (6) and (4) can be used to determine all the equilibrium orientations of the satellite under the influence of gravitational and constant torques. The number of real roots of the algebraic equation (7) is even and not greater than 6. For each solution one can find two values of a_{23} and, then, their respective values a_{31}, a_{32} and a_{33} from the equations (6). For each set of values a_{31}, a_{32}, and a_{33}, one can define from the first equation (6) two values of a_{22} and then from original system (4) the respective value a_{21}. After a_{21}, a_{22}, a_{23}, a_{31}, a_{32}, and a_{33} are found, the direction cosines a_{11}, a_{12}, and a_{13} can be unambiguously determined from the conditions of orthogonality. Thus, each real root of the algebraic equation (7) is matched with four sets of values a_{ij} (four equilibrium orientations). Since the number of real roots of equation (7) does not exceed 6, the satellite at the circular orbit can have no more than 24 equilibrium orientations. Analysis of special cases when only one parameter has a nonzero value (the constant torque vector coincides with the principal axes of inertia of the satellite), say $a \neq 0$, $b = c = 0$, was studied in [5], another special cases, one of them, for example, $b = c \neq 0$ presented in [6].

Equation (7) determines the hypersurface in (a, b, c)−space that satisfies the following system of equations:

$$P(x) = 0, \quad P'(x) = 0. \tag{8}$$

The variable x can be eliminated from (8) using the method of resultant. Expanding the determinant of resultant matrix of equations (8) with the help of Maple symbolic matrix function, we obtain the algebraic equation of the hypersurface in the form

$$(a^2 - b^2)^6 P_1^2(a, b, c) P_2(a, b, c) = 0. \tag{9}$$

Here $P_1(a, b, c)$ and $P_2(a, b, c)$ are the polynomials of the 15th degree and 18th degree, respectively.

To select regions in the parameter space with the same number of real roots of equation (7) Meiman's theorem was used [9]. It follows from the Meiman's theorem that partition of the space of parameters on the domains with the same number of real roots is given by discriminant hypersurface. This hypersurface contains a component of codimension 1, which is the boundary of these domains. The discriminant of polynomial $P(x)$ is given by the determinant of resultant of polynomials (8). Now we have to verify whether the number of equilibria really changes when we cross one of the surfaces (9). This can be done directly, finding the number of equilibria numerically at a single point on each one of the domains in (a, b, c) – space limited by these surfaces. This analysis showed that only the surface

$$P_2(a, b, c) = 0. \tag{10}$$

separates domains with different number of equilibria.

The polynomial $P_2(a, b, c)$ has rather cumbersome form:

$$
\begin{aligned}
P_2(a, b, c) = {} & 729a^{10}(b^2 + c^2)^4 + 15120a^9 bc(b^2 + c^2)^2 \\
& + a^8(729b^{10} + 243b^8(24c^2 - 13) + 27b^6(513c^4 - 1638c^2 - 676) \\
& + b^4(27376 + 91260c^2 - 82134c^4 + 13851c^6) \\
& + b^2(5832c^8 - 44226c^6 + 91260c^4 - 7456c^2) \\
& + c^4(27376 - 18252c^2 - 3159c^4 + 729c^6)) \\
& + 70a^7 bc(405b^6 + 9b^4(75c^2 - 338) + b^2(675c^4 - 3276c^2 + 2704) \\
& + 405c^6 - 3042c^4 + 2704c^2 - 1152) \\
& + a^6(2916b^{10}c^2 + 27b^8(513c^4 - 1638c^2 - 676) + b^6(21870c^6 \\
& - 167427c^4 + 314847c^2 + 132619) + b^4(13851c^8 - 167427c^6
\end{aligned}
$$

$$+ 771147c^4 - 903619c^2 - 118976) + b^2(2916c^{10} - 44226c^8$$
$$+ 314847c^6 - 903619c^4 + 522288c^2 - 97344)$$
$$- 18252c^8 + 132619c^6 - 118976c^4 - 97344c^2 + 20736)$$
$$+ 70a^5bc(216b^8 + 9b^6(75c^2 - 338) + b^4(891c^4 - 3276c^2 + 15041)$$
$$+ b^2(675c^6 - 3276c^4 + 11492c^2 - 18440)$$
$$+ 7488 - 18440c^2 + 15041c^4 - 3042c^6 + 216c^8)$$
$$+ a^4(4374b^{10}c^4 + b^8(27376 + 91260c^2 - 82134c^4 + 13851c^6)$$
$$+ b^6(-118976 - 903619c^2 + 771147c^4 - 167427c^6 + 13851c^8)$$
$$+ b^4(-170183 + 3452995c^2 - 2647224c^4 + 771147c^6 - 82134c^8$$
$$+ 4374c^{10})$$
$$+ b^2(648288 - 3298542c^2 + 3452995c^4 - 903619c^6 + 91260c^8)$$
$$- 134784 + 648288c^2 - 170183c^4 - 118976c^6 + 27376c^8)$$
$$+ 70a^3bc(432b^8c^2 + b^6(2704 - 3276c^2 + 675c^4)$$
$$+ b^4(-18440 + 11492c^2 - 3276c^4 + 675c^6)$$
$$+ b^2(29497 - 11306c^2 + 11492c^4 - 3276c^6 + 432c^8)$$
$$- 12168 + 29497c^2 - 18440c^4 + 2704c^6)$$
$$+ a^2(2916b^{10}c^6 + b^8(-7456c^2 + 91260c^4 - 44226c^6 + 5832c^8)$$
$$+ b^6(-97344 + 522288c^2 - 903619c^4 + 314847c^6 - 44226c^8$$
$$+ 2916c^{10})$$
$$+ b^4(648288 - 3298542c^2 + 3452995c^4 - 903619c^6 + 91260c^8)$$
$$+ b^2(-1045044 + 4826593c^2 - 3298542c^4 + 522288c^6 - 7456c^8)$$
$$- 36(-6084 + 29029c^2 - 18008c^4 + 2704c^6))$$
$$+ 70abc(216b^8c^4 + b^6(-1152 + 2704c^2 - 3042c^4 + 405c^6)$$
$$+ b^4(7488 - 18440c^2 + 15041c^4 - 3042c^6 + 216c^8)$$
$$+ b^2(-12168 + 29497c^2 - 18440c^4 + 2704c^6)$$
$$- 72(c^2(4c^2 - 13)^2 - 54))$$
$$+ b^8c^4(27376 - 18252c^2 - 3159c^4 + 729c^6)$$
$$+ b^6(20736 - 97344c^2 - 118976c^4 + 132619c^6 - 18252c^8)$$
$$+ b^4(-134784 + 648288c^2 - 170183c^4 - 118976c^6 + 27376c^8)$$
$$- 36b^2(-6084 + 29029c^2 - 18008c^4 + 2704c^6)$$
$$+ 1296(c^2(4c^2 - 13)^2 - 36).$$

Evolution of domains with the fixed number of equilibria in dependence of three system parameters a, b, c is possible to investigate by the analysis of hypersurface (10). The regions of the (a, b, c)–space, where equilibria exist, are limited by the inequalities [5]

$$a^2 + b^2 + c^2 \leq 5, \quad a^2 + b^2 \leq 4, \quad c^2 + b^2 \leq 4, \quad a^2 + c^2 \leq 4. \tag{11}$$

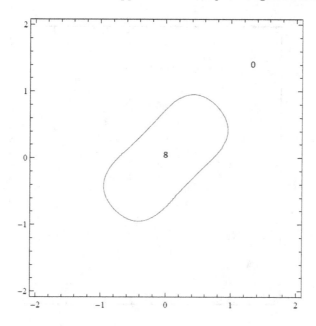

Fig. 1. The regions with the fixed number of equilibria for $a = (-35 + \sqrt{193})/12$

¿From the form of coefficients of equation (7), it follows that the parameters a and b occur symmetrically in it. In the terms of the coefficients p_i, we can separate out the factor abc so that $a, b,$ and c occur only in even powers, so the transformation $a \to -a$ leads to a distribution of number of equilibria symmetric with respect to either b- or c-axis. For the numerical investigation of equation (7), it is sufficient to consider the domain of the parameters delimited by inequalities (11) and the inequalities

$$-2 \le a \le 2, \quad -2 \le b \le 2, \quad -2 \le c \le 2; \tag{12}$$

in the plane (b, c), for the fixed a the parameters can be confined to the sector between the lines $c = b$ and $c = 0$ by virtue of symmetry. On the diagonal of the square $|b| \le 2$, $|c| \le 2$, when $|b| = |c|$, there is additional restriction $|b| \le \sqrt{2}$. The dependence of the number of real roots of equation (7) have been analyzed in space of parameters delimited by inequalities (11), (12). Some classification results [6] of different distributions of number of real roots for values of parameters in the domain (11), (12) were used.

The numerical analysis of the boundary curve (10) as well as the direct calculations of number of equilibria within the cube (12) resulted in the set of pictures of the distribution of the number of equilibria in the plane (b, c) for $a = \text{const}$ (Figs. 1–5). In fact, we calculated the two-dimensional section of the discriminant hypersurface that is given by the algebraic equation of two parameters b and c. It was found that most of the regions with specific values of number of equilibria arise or vanish on the diagonal of the square $|b| \le 2$, $|c| \le 2$, that is,

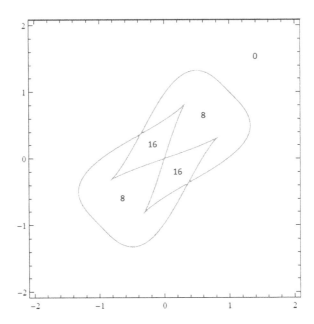

Fig. 2. The regions with the fixed number of equilibria for $a = -3/2$

when $|b| = |c|$. In this condition, we use the special cases results of [6] to calculate analytically the critical values of a, which correspond to qualitative changes of the distribution of number of equilibria as well as the coordinates (b, c) of points giving rise to new region (or ones where regions disappear). In some cases, the critical values of a were found numerically.

Figures 1–5 present some essentially different distributions of the number of equilibrium orientations in the planes $a = $ const for negative a. The classification was made for $-2 \leq a \leq 0$, because the transformation $a \to -a$ leads to a distribution of number of equilibria symmetric with respect to either b– or c– axis. Each figure corresponds to a specific value of a and shows the regions with different number of equilibria in (b, c)–plane (b–axis is horizontal, and c–axis is vertical) and boundary curves obtained as a cross section of surface (10) by the plane $a = $ const. The value of a in the space of parameters, where the picture changes qualitatively, changing the number of regions with specific values of the number of equilibria as well as their mutual disposition was defined as bifurcational. Figures 1–5 present the pictures where the principal changes appear. Here are indicated regions with 24, 16, 8, and 0 number of equilibria.

The first critical value is $a = -2$. The parameter values $b = c = 0$ are the only ones which correspond to equilibria (there are four of them). In the interval $-2 < a < (-35 + \sqrt{193})/12$, there is one region with the number of equilibria equal to 8. The second critical value is $(-35 + \sqrt{193})/12$ (Fig. 1). There is only one region that corresponds to 8 equilibrium orientations. For the next interval $(-35 + \sqrt{193})/12 < a < -3/2$, there exist 8 and 16 equilibria. For $a = -3/2$

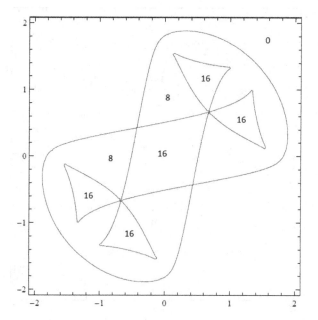

Fig. 3. The regions with the fixed number of equilibria for $a = -2/3$

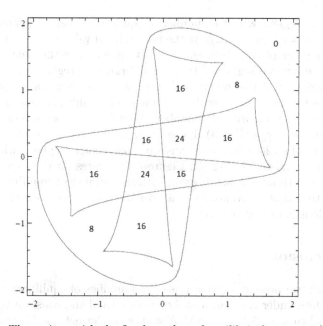

Fig. 4. The regions with the fixed number of equilibria for $a = -1/2$

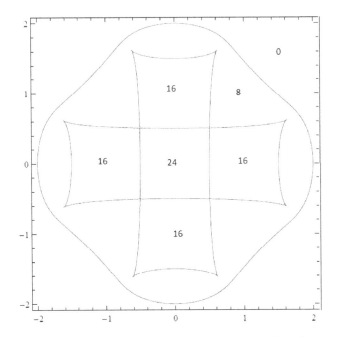

Fig. 5. The regions with the fixed number of equilibria for $a = 0$

there are two regions with the number of equilibria equal to 16 and they merge at the point $b = c = 0$ (Fig. 2). At the next critical value $a = -7/6$, four regions with the number of equilibria equal to 24 appear. For the next critical value $a = -1$ there are two regions with 24 equilibria, nine regions with 16 equilibria and two regions and two points with the number of equilibria equal to 8. For $-1 < a < -2/3$, two regions with the number of equilibria equal to 0 appear. At $a = -2/3$, the regions with the number of equilibria equal to 0 degenerate into the point $b = c = 2/3$ (Fig. 3). For $a = -1/2$, two regions with the number of equilibria equal to 24 merge and constitute one central region with the number of equilibria equal to 24 (Fig. 4). Figure 5 illustrates for $a = 0$ a cross-shaped cross section with large central region with the number of equilibria equal to 24.

The picture that corresponds to a positive value of a is obtained by one of the reflections: b to $-b$, or c to $-c$.

4   Conclusion

In this work, we present the analysis of the number of equilibrium orientations of the satellite under the action of the gravitational and constant torque in general case when $a \neq 0, b \neq 0$, and $c \neq 0$ with the help of Computer Algebra methods. The Computer Algebra system Maple is applied to reduce the satellite stationary motion system of six algebraic equations with six variables to a single algebraic equation of sixth degree in one variable, using the Groebner package for the construction of the Groebner basis. We have obtained the following new results. For general values of the components of the constant torque, we have indicated the analytic equation of the discriminant hypersurfaces that limit regions with different number of equilibrium orientations. The hypersurface equation was computed symbolically using the resultant approach. In this case, both the classification of different distributions of the number of equilibria and the coordinates of the bifurcation points were obtained combining analytical and numerical analysis.

These results permit us to describe the change of the number of equilibrium orientations of the satellite as a function of the parameters a, b, and c. When the constant torque is small enough, there exist 24 equilibria; when it is large enough, there are none. The evolution of domains with the fixed number of equilibrium orientations was investigated both analytically and numerically in the plane of two parameters b and c for different values of parameter a. The bifurcation values of a corresponding to the qualitative change of domains with the fixed number of equilibria were determined. From these results, we conclude that the satellite can have no more than 24 equilibrium orientations in a circular orbit.

The results of the study can be used at the stage of preliminary design of the satellite with gravitational control system.

Acknowledgements. The authors thank Doctor of Science V. Varin for his help in the Groebner basis calculation and Professor V. Gerdt and Professor Y. Blinkov for advising on the effectiveness of methods and algorithms of Groebner basis construction.

References

1. Sarychev, V.A.: Problems of Orientation of Satellites, Itogi Nauki i Tekhniki. Ser. Space Research, vol. 11. VINITI, Moscow (1978)
2. Beletsky, V.V.: Attitude Motion of Satellite in Gravitational Field. MGU Press, Moscow (1975)
3. Likins, P.W., Roberson, R.E.: Uniqueness of equilibrium attitudes for earth-pointing satellites. J. Astronaut. Sci. **13**, 87–88 (1966)
4. Garber, T.B.: Influence of constant disturbing torques on the motion of gravity-gradient stabilized satellites. AIAA J. **1**, 968–969 (1963)
5. Sarychev, V.A., Gutnik, S.A.: Equilibria of a satellite subjected to gravitational and constant torques. Cosmic Research **32**, 43–50 (1994)
6. Sarychev, V.A., Paglione, P., Guerman, A.: Influence of constant torque on equilibria of a satellite in a circular orbit. Celestial Mechanics and Dynamical Astronomy **87**, 219–239 (2003)
7. Buchberger, B.: Theoretical basis for the reduction of polynomials to canonical forms. SIGSAM Bulletin, 19–29 (1976)
8. Char, B.W., Geddes, K.O., Gonnet, G.H., Monagan, M.B., Watt, S.M.: Maple Reference Manual. Watcom Publications Limited, Waterloo (1992)
9. Meiman, N.N.: Some problems on the distribution of the zeros of polynomials. Uspekhi Mat. Nauk **34**, 154–188 (1949)

Bounds for the Condition Number
of Polynomials Systems with Integer Coefficients
(Invited Talk)

Aaron Herman[1] and Elias Tsigaridas[2,3]

[1] Department of Mathematics, North Carolina State University
aherman@ncsu.edu
[2] INRIA, Paris-Rocquencourt Center, PolSys Project
[3] UPMC, Univ Paris 06, LIP6, CNRS, UMR 7606, LIP6, Paris, France
elias.tsigaridas@inria.fr

Abstract. Polynomial systems of equations are a central object of study in computer algebra. Among the many existing algorithms for solving polynomial systems, perhaps the most successful numerical ones are the homotopy algorithms. The number of operations that these algorithms perform depends on the condition number of the roots of the polynomial system. Roughly speaking the condition number expresses the sensitivity of the roots with respect to small perturbation of the input coefficients.

A natural question to ask is how can we bound, in the worst case, the condition number when the input polynomials have integer coefficients? We address this problem and we provide effective bounds that depend on the number of variables, the degree and the maximum coefficient bitsize of the input polynomials. Such bounds allows to estimate the bit complexity of the algorithms that depend on the separation bound, like the homotopy algorithms, for solving polynomial systems.

1 Introduction

The study of algorithms for solving polynomial systems are in the center of study of computational algebra and computational mathematics in general. In this context, it is of great importance to define measures of hardness to express the difficulty to compute the solutions of a polynomial system. By "compute" we mean to approximate, up to any desired precision, one or all the roots of a polynomial system.

The condition number of the roots of a polynomial system could be considered, among other things, as such a measure of hardness. It expresses the sensitivity of the roots of a polynomial system, when we allow perturbations in the coefficients of the input polynomials. We refer to the fundamental work of Shub and Smale [11], see also [4], or to the recent book of Bürgisser and Cucker [5] for a detailed exposition. The complexity of numerical algorithms for polynomial system solving depends on the condition number. Among these numerical algorithms, the most successful, in theory and in practice, are the homotopy methods, e.g. [5,2].

© Springer International Publishing Switzerland 2015
V.P. Gerdt et al. (Eds.): CASC 2015, LNCS 9301, pp. 210–219, 2015.
DOI: 10.1007/978-3-319-24021-3_16

When the coefficients of the input polynomials are rational numbers, then besides the number of variables and degree of the polynomials, we have one more input parameter; the bitsize of the coefficients. We consider the maximum bitsize of the coefficients of a polynomial as the bitsize of the polynomial. Then, we should be able to express or to bound the measure of hardness for solving a polynomial system, or especially the condition number of the roots, with respect to these three parameters. That is to provide effective bounds as a function of the number of variables, the degree, and the bitsize of the input polynomials [9].

From a, first glance completely, different point of view, when the input coefficients are rational numbers, then a fundamental question of great importance is the following: What is the number of bits up to which we need to approximate the roots of a polynomial system, to distinguish them from each other? Can we provide such a bound as a function in the number of variables, the degree, and bitsize of the polynomials? Separation bounds, that is lower bounds on the minimum distance between the isolated roots of a polynomial system provide an answer to this question, in the worst case. We refer the reader to the DMM bound [6], that is the best known such bound. It is a natural question to ask if the bounds on the condition number and the separation bounds are connected. We will provide a positive answer to this question.

1.1   Our Results

We consider the problem of bounding the condition number of the roots of square-free univariate polynomials and 0-dimensional polynomial systems with a smooth zero set, when the input polynomials have integer coefficients. We also introduce an aggregate version of the condition number and we prove bounds of the same order of magnitude as in the case of the condition number of a single root.

In the univariate case we improve the currently known bounds [9, Theorem 2.4] by a factor of d (Proposition 1), where d is the degree of the polynomial. For the multivariate case the previous bounds [9, Theorem 2.5], which like ours are single exponential with respect to the number of variables, do not specify the constant in the exponent. We provide precise bounds (Theorem 1) and our approach leads to better bounds than the ones we can obtain by performing the calculations using the previously known approach [9]. The exact constants in the exponents can be useful in many applications e.g. [1,7,8]. Such bounds are also needed to establish a connection between Turing machines and the Blum-Cucker-Shub-Smale model and to certify and analyze the Boolean complexity of algorithms based on homotopy techniques [3].

The aggregate versions of the condition numbers we introduce (Proposition 2 and Theorem 2) encapsulate the condition number of all the roots. Contrary to what is expected as a bound in this case, that is the number of roots times the worst case bound for the condition number at a single root, our aggregate version saves a factor equal to the number of roots. As a consequence, in the multivariate case, we gain a factor of d^n, where d is the degree of the polynomials and n the number of variables.

1.2   Notation

In what follows $\mathcal{O}_B$, resp. $\mathcal{O}$, means bit, resp. arithmetic, complexity and $\widetilde{\mathcal{O}}_B$, resp. $\widetilde{\mathcal{O}}$, means that we are ignoring logarithmic factors. For a polynomial $A = \sum_{i=0}^{d} a_i\, x^i \in \mathbb{Z}[x]$, $\deg(A) = d$ denotes its degree. We consider the height function $\mathsf{H}(\cdot)$ which is defined as follows. If $a \in \mathbb{Z}$ then $\mathsf{H}(a) = |a|$. For $a, b \in \mathbb{Z}$, $\mathsf{H}(\frac{a}{b}) = \max\{\mathsf{H}(a), \mathsf{H}(b)\}$. For a polynomial A, we have $\mathsf{H}(A) = \max_k |a_k|$. Finally, for a matrix $M \in \mathbb{Z}^{n \times n}$, $\mathsf{H}(M) = \max_{i,j} |M_{i,j}|$. The logarithmic height is defined as $\mathsf{h}(\cdot) = \lg \mathsf{H}(\cdot)$, where $\lg(\cdot)$ is the logarithm of base 2. The Mahler bound (or measure) of A is $\mathcal{M}(A) = a_d \prod_{|\alpha| \geq 1} |\alpha|$, where α runs through the complex roots of A, e.g. [10,12]. If $A \in \mathbb{Z}[x]$ and $\mathsf{H}(A) = 2^\tau$, then $\mathcal{M}(A) \leq \|A\|_2 \leq \sqrt{d+1}\mathsf{H}(A) = 2^\tau \sqrt{d+1}$.

2   Condition Number for Univariate Polynomials

Let $A = \sum_{k=0}^{d} a_k X^k \in \mathbb{C}[X]$ and α be one of its roots. The condition number of A at α is defined as follows

$$\mu(A, \alpha) = \frac{\left(\sum_{i=0}^{d} |\alpha|^{2i}\right)^{\frac{1}{2}}}{|A'(\alpha)|} \tag{1}$$

where A' is the derivative of A. We define the condition number of A as

$$\mu(A) = \max_{\substack{\alpha \in \mathbb{C} \\ A(\alpha)=0}} \mu(A, \alpha) \tag{2}$$

If A is a square-free integer polynomial such that $\mathsf{H}(A) = 2^\tau$, Malajovich [9] provided the following bounds for the condition number at a root α,

$$\mu(A, \alpha) \leq 2^{2d^2-2} d^{2d} 2^{2\tau d^2}$$

which in turn leads to the following estimation for the condition number of A

$$\mu(A) = 2^{\mathcal{O}(\tau d^2)},$$

or

$$\lg(\mu(A)) \in \mathcal{O}(\tau d^2).$$

The following proposition improves this bound by a factor of d.

Proposition 1. *Consider the square-free polynomial* $A = \sum_{i=0}^{d} a_i X^i \in \mathbb{Z}[X]$ *with* $\mathsf{H}(A) \leq 2^\tau$ *and* $a_0 \neq 0$. *Let* $\alpha \neq 0$ *be a root of* A. *Then*

$$\mu(A, \alpha) \leq \sqrt{d+1}^{15d+1} 2^{15d\tau + \tau + 18d \lg(d)}.$$

Hence $\lg(\mu(A)) \in \mathcal{O}(d\tau + d \lg(d))$.

Proof: First we bound the numerator of formula (1) as follows

$$(\sum_{i=0}^{d} |\alpha|^{2i})^{\frac{1}{2}} = \|(1, \alpha, \dots, \alpha^d)\|_2 = \sqrt{d+1}\, \|(1, \alpha, \dots, \alpha^d)\|_\infty \leq \sqrt{d+1}\, 2^d\, \mathsf{H}(A)^d$$

$$(3)$$

To bound the denominator we need the following result, e.g. [6]. For $A \in \mathbb{Z}[X]$, let Ω be a set of k pairs of indices of non-zero roots of A. Then

$$\prod_{(i,j)\in\Omega} |\alpha_i - \alpha_j| \geq d^{-18d}(d+1)^{-15d/2}\mathsf{H}(A)^{-15d} \geq 2^{-30d\lg d}\, \mathsf{H}(A)^{-15d}.$$

Using the previous bound we can bound $A'(\alpha)$. We notice that

$$|A'(\alpha)| = |a_d \prod_{\substack{\gamma\neq\alpha \\ f(\gamma)=0}} (\alpha - \gamma)| \geq \prod_{\substack{\gamma\neq\alpha \\ f(\gamma)=0}} |\alpha - \gamma| \geq 2^{-30d\lg d}\, \mathsf{H}(A)^{-15d}$$

as $|a_d| \geq 1$. Finally, by combining the equations

$$\mu(A,\alpha) = \frac{\left(\sum_{i=0}^{d} |\alpha|^{2i}\right)^{\frac{1}{2}}}{|A'(\alpha)|} \leq 2^{2d}\, \mathsf{H}(A)^d\, 2^{30d\lg d}\, \mathsf{H}(A)^{15d} \leq 2^{32d\lg d}\, \mathsf{H}(A)^{16d}$$

$$\leq 2^{\mathcal{O}(d\tau + d\lg d)} = 2^{\tilde{\mathcal{O}}(d\tau)}.$$

$$\square$$

The condition number of A, Eq. (2), expresses the maximum condition of all the roots. Hence, one might suggest that if we are interested in a notion of the condition number that accounts for all the roots, then we have to multiply the worst case bound by their number; in our case d. However, it turns out that this is not the case. We consider the following aggregate version of the condition number

$$\tilde{\mu}(A) = \prod_{i=1}^{d} \mu(A, \alpha_i)$$

where $\{\alpha_i\}_{1\leq i\leq d}$ is the set of roots of f. We prove that a bound similar to the one of Proposition 1 holds for $\tilde{\mu}$.

Proposition 2. *Consider the square-free polynomial $A = \sum_{k=0}^{d} a_k X^k \in \mathbb{Z}[X]$ with $\mathsf{H}(A) = 2^\tau$. Then*

$$\tilde{\mu}(A) \leq \sqrt{d+1}^{2d} 2^{\tau d}.$$

Hence $\lg(\tilde{\mu}(A)) \in \mathcal{O}(d\tau + d\lg(d))$.

Proof: We obtain the bound using the properties of the Mahler measure and the discriminant, $\mathsf{disc}(A)$.

$$
\tilde{\mu}(A) = \prod_{i=1}^{d} \mu(A, \alpha_i) = \prod_{i=1}^{d} \frac{||(1, \alpha_i, \dots, \alpha_i^d)||_2}{|A'(\alpha_i)|} = \prod_{i=1}^{d} \frac{||(1, \alpha_i, \dots, \alpha_i^d)||_2}{a_d \prod_{j \neq i} |\alpha_i - \alpha_j|}
$$

$$
= \frac{\prod_{i=1}^{d} ||(1, \alpha_i, \dots, \alpha_i^d)||_2}{a_d^d \prod_{i \neq j} |\alpha_i - \alpha_j|} = \frac{\prod_{i=1}^{d} ||(1, \alpha_i, \dots, \alpha_i^d)||_2}{a_d^d \left(\frac{|\mathsf{disc}(A)|}{a_d^{2d-2}} \right)}
$$

$$
= a_d^{d-2} \frac{\prod_{i=1}^{d} ||(1, \alpha_i, \dots, \alpha_i^d)||_2}{|\mathsf{disc}(A)|} \leq a_d^{d-2} \frac{\prod_{i=1}^{d} \sqrt{d+1} ||(1, \alpha_i, \dots, \alpha_i^d)||_\infty}{|\mathsf{disc}(A)|}
$$

$$
= a_d^{d-2} \sqrt{d+1}^d \frac{\prod_{i=1}^{d} ||(1, \alpha_i, \dots, \alpha_i^d)||_\infty}{|\mathsf{disc}(A)|}
$$

$$
= a_d^{d-2} \sqrt{d+1}^d \frac{\prod_{i=1}^{d} \max\{1, |\alpha_i|^d\}}{|\mathsf{disc}(A)|} = a_d^{d-2} \sqrt{d+1}^d \frac{\left(\prod_{i=1}^{d} \max\{1, |\alpha_i|\} \right)^d}{|\mathsf{disc}(A)|}
$$

$$
= \frac{\sqrt{d+1}^d}{a_d^2} \frac{\left(a_d \prod_{i=1}^{d} \max\{1, |\alpha_i|\} \right)^d}{|\mathsf{disc}(A)|} = \frac{\sqrt{d+1}^d}{a_d^2} \frac{(\mathcal{M}(A))^d}{|\mathsf{disc}(A)|}
$$

$$
\leq \frac{\sqrt{d+1}^d}{a_d^2} \frac{(||A||_2)^d}{|\mathsf{disc}(A)|} \leq \frac{\sqrt{d+1}^d}{a_d^2} \frac{(\sqrt{d+1}\, H(A)))^d}{|\mathsf{disc}(A)|}
$$

$$
= \frac{\sqrt{d+1}^{2d}}{a_d^2} \frac{H(A)^d}{|\mathsf{disc}(A)|} \leq \sqrt{d+1}^{2d} H(A)^d.
$$

$\square$

3   Condition Number for Polynomial Systems

In this section we generalize the bounds of Propositions 1 and 2 to the case of polynomial systems. The definition of the condition number of a root of a polynomial system is given in equation (4). We assume that the polynomial systems are 0-dimensional and their zero set is smooth, that is the Jacobian of the system is invertible.

First we need to introduce additional notation, which follows closely [3]. Let $\mathcal{H}_d^n$ be the vector space of homogeneous polynomials in $n + 1$ variables, $X_0, X_1, \dots, X_n$, of degree d. If $f \in \mathcal{H}_d^n$ then

$$
f = \sum_{|\boldsymbol{\alpha}|=d} f_{\boldsymbol{\alpha}} \boldsymbol{X}^{\boldsymbol{\alpha}} = \sum_{|\boldsymbol{\alpha}|=d} f_{\boldsymbol{\alpha}} X_0^{\alpha_0} X_1^{\alpha_1} \cdots X_n^{\alpha_n}.
$$

For $f, g \in \mathcal{H}_d^n$ we consider the following inner product

$$
\langle f, g \rangle = \sum_{|\boldsymbol{\alpha}|=d} f_{\boldsymbol{\alpha}} g_{\boldsymbol{\alpha}} \binom{d}{\boldsymbol{\alpha}}^{-1} = \sum_{|\boldsymbol{\alpha}|=d} f_{\boldsymbol{\alpha}} g_{\boldsymbol{\alpha}} \binom{d}{\alpha_0, \alpha_1, \dots, \alpha_n}^{-1}
$$

and the corresponding norm

$$\|f\|_b^2 = \langle f, f \rangle = \sum_{|\alpha|=d} |f_\alpha|^2 \binom{d}{\alpha}^{-1}.$$

We consider $f = (f_1, \ldots, f_n) \in \mathcal{H}_{d_1}^n \times \cdots \times \mathcal{H}_{d_n}^n = \mathcal{H}$ to be a 0-dimensional polynomial system of n homogeneous equations in $n+1$ variables, with a smooth zero set. For a system of equations, f, we have the following definition of the norm

$$\|f\|^2 = \sum_{i=1}^{n} \|f_i\|_b^2.$$

The condition number of a polynomial system f at a number $z \in \mathbb{C}^n$ is defined in [4] as

$$\mu(f, z) = \|f\| \, \|(Df(z)|_{z^\perp})^{-1} \, \mathrm{Diag}(\|z\|^{d_i-1} d_i^{1/2})\|. \tag{4}$$

However, to bound the various quantities that appear we use an equivalent definition, Eq. (5), from Malajovich [9]. Moreover, we follow the notation from [3] to bound condition number of a polynomial system of polynomials having integer coefficients. In this case we assume that $\mathsf{H}(f_i) \leq 2^\tau$ for all i.

Let $f \in \mathcal{H}$ be a polynomial system and let $z \in \mathbb{P}(\mathbb{C}^{n+1})$. Let $\chi_1 = \chi_1(f, z)$ defined by

$$\chi_1 = \left\| \begin{pmatrix} Df(z) \\ z^* \end{pmatrix}^{-1} \begin{pmatrix} \sqrt{d_1}\,\|f\|\,\|z\|^{d_1-1} & & \\ & \ddots & \\ & & \sqrt{d_n}\,\|f\|\,\|z\|^{d_n-1} \\ & & & \|z\| \end{pmatrix} \right\|$$

$$= \|M_1^{-1} \cdot M_2\| \tag{5}$$

Note that these formulas do not depend on the representative of z and thus are well defined. Their value is also invariant under multiplication of f by a non–zero complex number $\lambda \in \mathbb{C}$. Our goal is to estimate a bound for $\chi_1(f, \zeta)$, where ζ is a root of f.

Recall that for any matrix M it holds $\|M\| \leq \|M\|_F$. The second norm is the Frobenius norm, that is $\|M\|_F = \sqrt{\sum_{i,j} |M_{i,j}^2|}$.

First we consider bounds for the norm of M_2. To bound $\|f\|$, assuming $\mathsf{H}(f_i) \leq 2^\tau$, we proceed as follows:

$$\|f\| = \sqrt{\sum_{i=1}^{n} \|f_i\|_b^2} \leq \sqrt{\sum_{k=1}^{n} 2^{2\tau + d_k \lg(nd_k)}} \leq 2^{\tau + d \lg(nd)}. \tag{6}$$

To bound $\|\zeta\|$ we use the DMM bounds [6]. The DMM is defined for sparse systems but we can also use it for the homogeneous case. To see this notice that we consider all the possible dehomogenizations of the system and we apply to each of them DMM. Then we take the worst bound.

For any root $\boldsymbol{\zeta} = (\zeta_0, \zeta_1, \ldots, \zeta_n)$ of the system it holds [6, Cor. 4]

$$\lg(\max_{0 \leq k \leq n} |\zeta_k|) \leq 1 + \prod_{i=1}^{n} d_i + \sum_{i=1}^{n} \prod_{j \neq i} d_i \left(\tau + \lg(2\, d_i^n)\right)$$

$$= \eta_1 = \mathcal{O}(d^n + n d^{n-1}\tau + n^2 d^{n-1} \lg d). \tag{7}$$

Now we are ready to bound $\|M_2\|$ by combining equations (6) and (7). The bound is as follows:

$$\|M_2\|_F^2 \leq \sum_{i=1}^{n} \left(\sqrt{d_i} \, \|\boldsymbol{f}\| \, \|\boldsymbol{\zeta}\|^{d_i-1}\right)^2 + \|\boldsymbol{\zeta}\|^2 \leq 2^{2\tau + 3d \lg(nd) + d\,\eta_1},$$

which simplifies to

$$\lg\|M_2\|_F \leq \mathcal{O}(d^{n+1} + n d^n \tau + n^2 d^n \lg d) = \widetilde{\mathcal{O}}(d^{n+1} + d^n \tau). \tag{8}$$

To bound M_1^{-1} it suffices to bound $\|M_1\|$. It holds $\|M_1^{-1}\| \leq n^n\, \mathsf{H}(M_1)$, e.g. [9, Lemma 4.5]. To obtain a bound for $\mathsf{H}(M_1)$, first we need an estimation on the evaluation of the derivatives $G_{i,j}(\boldsymbol{X}) = \frac{\partial}{\partial X_j} f_i(\boldsymbol{X})$ at the roots of the system, $\boldsymbol{\zeta}$.

Let $f_{n+1}^{(i,j)}(\boldsymbol{X}, Y) = Y - G_{i,j}(\boldsymbol{X}))$ and consider the polynomial system

$$(\Sigma_{i,j}) \quad \{f_1(\boldsymbol{X}) = \cdots f_n(\boldsymbol{X}) = f_{n+1}^{(i,j)}(\boldsymbol{X}, Y) = 0\}. \tag{9}$$

This is a system in $n + 1$ equations in $n + 1$ variables. It holds $\deg(f_{n+1}^{(i,j)}) = \deg(G_{i,j}) \leq d_i - 1$ and $\mathsf{H}(f_{n+1}^{(i,j)}) = \mathsf{H}(G_{i,j}) \leq d\, \mathsf{H}(f_i) \leq \tau + \lg d_i$. The resultant of $(\Sigma_{i,j})$ that eliminates the variables $X_1, \ldots, X_n$, is

$$R_{i,j} = \mathsf{Res}_{d_1,\ldots,d_n}(f_1(\boldsymbol{X}), \ldots, f_n(\boldsymbol{X}), y - G_{i,j}(\boldsymbol{X})) \in \mathbb{Z}[y]$$

where $R_{i,j} \in \mathbb{Z}[Y]$. The roots of $R_{i,j}$ correspond to the evaluations of $G_{i,j}$ at the roots of the system $\boldsymbol{f} = 0$. Therefore, an upper bound on the roots of $R_{i,j}$ provides an upper bound on the evaluation. We should notice that $R_{i,j}$ is not identically zero.

Hence, to obtain the required bounds we can consider the system $(\Sigma_{i,j})$. From this point of view we need to provide lower bounds on the coordinates of solutions of the system. For this we use DMM [6, Thm. 3 and Cor. 4] directly.

First, we need to define (bound) various quantities, see [6, Eq. (3)]. The mixed volume(s) $\mathsf{M}_0 = d_1 \cdots d_n (d_i - 1) \leq d^n (d - 1) \leq d^{n+1}$, $\mathsf{M}_k = d_1 \ldots d_{k-1} d_{k+1} \cdots d_n (d_i - 1) \leq d^{n-1}(d - 1) \leq d^n$ for $1 \leq k \leq n$, and $\mathsf{M}_{n+1} = d_1 \cdots d_n \leq d^n$; and the integer coefficients that appear in the resultant polynomial

$$\varrho = \prod_{k=1}^{n+1} (\#Q_k)^{\mathsf{M}_k} \leq 2^{\sum_{k=1}^{n+1} \mathsf{M}_i} \prod_{i=1}^{n} d_k^{n \mathsf{M}_k} (d_i - 1)^{n \mathsf{M}_{n+1}} \leq 2^{2nd^n} d^{2n^2 d^n}.$$

Finally, we bound the weighted heights of the input polynomials $C = \prod_{k=1}^{n+1} \mathsf{H}(f_k)^{M_k} \leq 2^{(n+\lg d)\tau d^n}$. An isolated root of the system with Y coordinate equal to y follows the bound $|y| \leq 2^{M_0} \varrho C$. Thus

$$|G_{i,j}(\varsigma)| \leq 2^{M_0} \varrho C \leq 2^{d^{n+1}+8n^2 d^n \lg d + (n+\lg d)\tau d^n}$$

for any i, j and for any root ς of the system. For ς^* it holds that $\mathsf{H}(\varsigma^*) \leq \mathsf{H}(\varsigma)$ and so we can use the bound from (7). Putting all these together we have the bound

$$\mathsf{H}(M_1) \leq 2^{\eta_2}$$

where

$$\eta_2 = \mathcal{O}(d^{n+1} + n^2 d^n \lg d + (n+\lg d)\tau d^n) = \tilde{\mathcal{O}}(d^{n+1} + n^2 d^n + n\tau d^n)$$

and so $\|M_1^{-1}\| \leq n^n \mathsf{H}(M_1) \leq 2^{\eta_2} \leq 2^{\tilde{\mathcal{O}}(d^{n+1}+n^2 d^n+n\tau d^n)}$.

Combining the bounds for $\|M_1^{-1}\|$ and $\|M_2\|$ we obtain the following bound for χ_1 which also a bound for the condition number of a complex root of the system.

$$\chi_1 \leq 2^{\eta_2} \leq 2^{\mathcal{O}(d^{n+1}+n^2 d^n \lg d + (n+\lg d)\tau d^n)}. \tag{10}$$

The previous discussion leads to the following theorem

Theorem 1. *Let $f = (f_1, \ldots, f_n) \in \mathcal{H}$ be a 0-dimensional polynomial system such that its zero set consists of smooth points. Assume $f_i \in \mathbb{Z}[X_0, X_1, \ldots, X_n]$ such that they have degrees bounded by d and $\mathsf{H}(f_i) \leq 2^\tau$. Then, we have the following bound for the condition number of any root ς of the system*

$$\mu(f, \varsigma) \leq 2^{\mathcal{O}(d^{n+1}+n^2 d^n \lg d + (n+\lg d)\tau d^n)}.$$

3.1   Multivariate Aggregate Condition Number

In this section we sketch the proof of an aggregate version of Theorem 1. It provides bounds similar to the ones of Proposition 2 and to the aggregate nature of the DMM bounds [6, Theorem 3].

In the view of Theorem 1 if we wanted to consider a bound on the condition number for all the roots of the system, then we have to multiply $\mu(f, \varsigma)$ by their number. There are d^n roots in the worst case, by the Bézout bound. This leads to a bound of $\tilde{\mathcal{O}}_B(d^{2n+1} + d^{2n}\tau)$.

In the sequel we will improve this bound to $\tilde{\mathcal{O}}_B(d^{n+1}+d^n\tau)$ using aggregation. Some elementary properties are in place.

$$\|M\| \leq \|M\|_F \leq \sqrt{n^2 \mathsf{H}(M)^2} \leq n \mathsf{H}(M).$$

If the entries of the matrix M depend on a root ς then we write $M(\varsigma)$ to emphasize this. In this context it holds

$$\chi_1(\varsigma) \leq \|M_1^{-1}(\varsigma) M_2(\varsigma)\| \leq (n+1)^2 \mathsf{H}(M_1(\varsigma))^{n+1} \mathsf{H}(M_2(\varsigma))$$

and

$$\tilde{\chi}_1(\zeta) = \prod_\zeta \chi_1(\zeta) \le (n+1)^{2d^n} \prod_\zeta \mathsf{H}(M_1(\zeta))^{n+1} \prod_\zeta \mathsf{H}(M_2(\zeta)).$$

We have to bound each factor independently. We sketch the approach for the second one. For the first factor we work similarly.

To bound $\prod_\zeta \mathsf{H}(M_2(\zeta))$ we can apply directly Eq. (5) or (8). However, this approach gives an exponent of d^{2n+1}, which is a big overestimation; by a factor of d^n.

We rely on aggregation bounds of polynomial system, provided by the DMM bounds [6]. Consider the polynomial $f_{n+1}(\boldsymbol{X}, Y) = Y - X_1^2 - \cdots - X_n^2$ and the polynomial system

$$(\Sigma_{i,j}) \quad \{f_1(\boldsymbol{X}) = \cdots f_n(\boldsymbol{X}) = f_{n+1}(\boldsymbol{X}, Y) = 0\}. \tag{11}$$

The resultant of the system encapsulates (all) the evaluations of f_{n+1} over the roots of $\boldsymbol{f}$. Therefore, it suffices to bound the height of the resultant. The bounds that we get are similar to the ones of the previous section. The calculations lead to the following theorem

Theorem 2. *Let $\boldsymbol{f} = (f_1, \ldots, f_n) \in \mathcal{H}$ be a 0-dimensional polynomial system. Assume $f_i \in \mathbb{Z}[X_0, X_1, \ldots, X_n]$ such that they have degrees bounded by d and $\mathsf{H}(f_i) \le 2^\tau$. Then, if ζ runs over all the solutions of the system, it holds*

$$\tilde{\chi}_1(\boldsymbol{f}) = \prod_\zeta \chi_1(\zeta) \le 2^{\tilde{\mathcal{O}}(d^{n+1} + d^n \tau)}.$$

Acknowledgments. ET is partially supported by GeoLMI (ANR 2011 BS03 011 06), HPAC (ANR ANR-11-BS02-013), and an FP7 Marie Curie Career Integration Grant.

References

1. Basu, S., Roy, M.: Bounding the radii of balls meeting every connected component of semi-algebraic sets. J. Symb. Comp. **45**, 1270–1279 (2010)
2. Bates, D.J., Hauenstein, J.D., Sommese, A.J., Wampler, C.W.: Numerically solving polynomial systems with Bertini, vol. 25. SIAM (2013)
3. Beltrán, C., Leykin, A.: Robust certified numerical homotopy tracking. Foundations of Computational Mathematics **13**(2), 253–295 (2013)
4. Blum, L., Cucker, F., Shub, M., Smale, S.: Complexity and Real Computation. Springer-Verlag (1998)
5. Bürgisser, P., Cucker, F.: Condition: The geometry of numerical algorithms, vol. 349. Springer Science & Business Media (2013)
6. Emiris, I.Z., Mourrain, B., Tsigaridas, E.P.: The DMM bound: multivariate (aggregate) separation bounds. In: Watt, S. (ed.) Proc. 35th ACM Int'l Symp. on Symbolic & Algebraic Comp. (ISSAC), pp. 243–250. ACM, Munich, July 2010
7. Hansen, K.A., Koucky, M., Lauritzen, N., Miltersen, P.B., Tsigaridas, E.P.: Exact algorithms for solving stochastic games. In: Proc. 43rd Annual ACM Symp. Theory of Computing (STOC) (2011)

8. Hansen, K.A., Koucky, M., Miltersen, P.B.: Winning concurrent reachability games requires doubly-exponential patience. In: Proc. 24th Annual IEEE Symposium on Logic in Computer Science (LICS), pp. 332–341. IEEE Computer Society, Washington, DC (2009)
9. Malajovich, G.: Condition number bounds for problems with integer coefficients. Journal of Complexity 16(3), 529–551 (2000)
10. Mignotte, M.: Mathematics for Computer Algebra. Springer-Verlag, New York (1991)
11. Shub, M., Smale, S.: Complexity of bezout's theorem: Iii. condition number and packing. J. Complexity 9(1), 4–14 (1993)
12. Yap, C.: Fundamental Problems of Algorithmic Algebra. Oxford University Press, New York (2000)

On Invariant Manifolds and Their Stability in the Problem of Motion of a Rigid Body under the Influence of Two Force Fields

Valentin Irtegov and Tatiana Titorenko

Institute for System Dynamics and Control Theory SB RAS,
134, Lermontov str., Irkutsk, 664033, Russia
irteg@icc.ru

Abstract. On the basis of the Routh–Lyapunov method and its generalizations, we study the structure of the phase space of the conservative system which describes the motion of a rigid body in gravitational and magnetic fields. Within the framework of this study, the invariant manifolds of various dimension have been found, their simplest classification has been performed, and sufficient conditions of stability have been obtained for the stationary invariant manifolds.

1 Introduction

In [1], the non-integrable, in general case, conservative system which describes the rotation of a rigid body around a fixed point in two uniform potential force fields – gravitational and magnetic – is considered. Such problems arise in many applications, e.g., space dynamics [2]. In this work, we study the particular integrable case of the given system which was found by Bogoyavlenskii (see [1]). When the equations of motion of the body are written on one of their manifolds, then they possess an additional first integral. Thus, the system under consideration becomes completely Liouville integrable on this manifold. Its qualitative analysis is the topic of the present paper.

We study the special solutions of the motion equations of the body which can be obtained from stationary conditions for the elements of algebra of problem's first integrals. Special attention is given to the families of invariant manifolds (IMs). Two procedures for their finding are proposed. Specificities which arise in investigation of stability of IMs families by the 2nd Lyapunov method are also discussed.

As research techniques we apply the Routh–Lyapunov method [3] and its generalizations (see [4]). This method allows one not only to find the desired solutions, but also to investigate their stability. We use the methods of computer algebra system "Mathematica", in particular, Gröbner bases method, as computing ones.

Most of the results presented in this paper for the differential equations written on the manifold can be extended to the initial equations of motion (in the original phase space). It concerns the families of IMs found and their classification.

© Springer International Publishing Switzerland 2015
V.P. Gerdt et al. (Eds.): CASC 2015, LNCS 9301, pp. 220–232, 2015.
DOI: 10.1007/978-3-319-24021-3_17

2   Formulation of the Problem

The rotation of a rigid body around a fixed point in two uniform potential force fields – gravitational and magnetic – is considered. The distribution of mass in the body corresponds to the Kowalewski integrable case [5].

The equations of motion of the body in a coordinate system which is rigidly attached to the body and its center is at the fixed point can be written as:

$$
\begin{aligned}
2\dot{p} = b\delta_3 + q\,r, & \quad \dot{\gamma}_1 = \gamma_2 r - \gamma_3 q, & \quad \dot{\delta}_1 = \delta_2 r - \delta_3 q, \\
2\dot{q} = x_0\gamma_3 - p\,r, & \quad \dot{\gamma}_2 = \gamma_3 p - \gamma_1 r, & \quad \dot{\delta}_2 = \delta_3 p - \delta_1 r, \\
\dot{r} = -b\delta_1 - x_0\gamma_2, & \quad \dot{\gamma}_3 = \gamma_1 q - \gamma_2 p, & \quad \dot{\delta}_3 = \delta_1 q - \delta_2 p.
\end{aligned}
\tag{1}
$$

Here p, q, r are the projections of the angular velocity onto the axes related to the body, $\gamma_1, \gamma_2, \gamma_3$ are the direction cosines of the upward vertical, $\delta_1, \delta_2, \delta_3$ are the direction cosines of the vector of constant magnetic moment, parameters x_0, b are proportional to the coordinate of the mass center of the body and the coordinate of the vector of constant magnetic moment, respectively.

Equations (1) admit the following first integrals:

$$
\begin{aligned}
2H &= 2(p^2 + q^2) + r^2 + 2(x_0\gamma_1 - b\,\delta_2) = 2h, \\
V_1 &= (p^2 - q^2 - x_0\gamma_1 - b\,\delta_2)^2 + (2p\,q - x_0\gamma_2 + b\,\delta_1)^2 = c_1, \\
V_2 &= \gamma_1^2 + \gamma_2^2 + \gamma_3^2 = 1, \; V_3 = \delta_1^2 + \delta_2^2 + \delta_3^2 = 1, \\
V_4 &= \gamma_1\delta_1 + \gamma_2\delta_2 + \gamma_3\delta_3 = c_2.
\end{aligned}
\tag{2}
$$

In general case, system (1) is non-integrable. In [1], several particular integrable cases of this system are given. When $b = 0$, it corresponds to the Kowalewski integrable case. When system (1) is written on the invariant manifold of codimension 2

$$
p^2 - q^2 - x_0\gamma_1 - b\,\delta_2 = 0, \; 2p\,q - x_0\gamma_2 + b\,\delta_1 = 0,
\tag{3}
$$

it has an additional cubic integral and is completely Liouville integrable. In the present work, we study the latter integrable case.

On IM (3), equations (1) can be written as

$$
\begin{aligned}
2\dot{p} = q\,r + b\,\delta_3, & \quad x_0\dot{\gamma}_3 = -[(p^2 + q^2)\,q + b\,(p\,\delta_1 + q\,\delta_2)], \\
2\dot{q} = x_0\gamma_3 - p\,r, & \quad \dot{\delta}_1 = r\,\delta_2 - q\,\delta_3, \\
\dot{r} = -2(p\,q + b\,\delta_1), & \quad \dot{\delta}_2 = \delta_3 p - \delta_1 r, \\
& \quad \dot{\delta}_3 = \delta_1 q - \delta_2 p.
\end{aligned}
\tag{4}
$$

They are derived from equations (1) by elimination of the variables γ_1, γ_2 from them with the aid of (3).

The first integrals of equations (4) are given by

$$
2\tilde{H} = 4p^2 + r^2 - 4b\,\delta_2 = 2\tilde{h},
$$

$$\tilde{V}_2 = \gamma_3^2 + \frac{(2p\,q + b\,\delta_1)^2}{x_0^2} + \frac{(q^2 - p^2 + b\delta_2)^2}{x_0^2} = 1,$$

$$V_3 = \delta_1^2 + \delta_2^2 + \delta_3^2 = 1, \tag{5}$$

$$\tilde{V}_4 = \frac{2p\,q\,\delta_2 + (p^2 - q^2)\,\delta_1}{x_0} + \gamma_3\,\delta_3 = \tilde{c}_2,$$

$$2V_5 = (p^2 + q^2)\,r - 2x_0 p\,\gamma_3 + 2bq\,\delta_3 = m.$$

Here the top four integrals are obtained from integrals (2) by elimination of the variables γ_1, γ_2 from them with the aid of (3), V_5 is the integral found by Bogoyavlenskii.

Within the framework of the study of the phase space structure of system (4), we state the problem of finding IMs of this system for their simplest classification and investigation of stability.

3   Finding Invariant Manifolds

According to the Routh–Lyapunov method, the IMs of equations under study can be found by solving the extremum problem for the elements of algebra of the first integrals of these equations. For this purpose, we take, e.g., the complete linear combination of first integrals (5)

$$K = \lambda_0 \tilde{H} - \frac{1}{2}\lambda_1 \tilde{V}_2 - \frac{1}{2}\lambda_2 V_3 - \lambda_3 \tilde{V}_4 - \lambda_4 V_5 \tag{6}$$

and write down the necessary conditions for the integral K to have an extremum with respect to the phase variables $p, q, r, \gamma_3, \delta_1, \delta_2, \delta_3$:

$$\partial K/\partial p = 4\lambda_0 p - \frac{2\lambda_1[(p^2 + q^2)\,p + b(q\,\delta_1 - p\,\delta_2)]}{x_0^2} - \frac{2\lambda_3(p\,\delta_1 + q\,\delta_2)}{x_0}$$
$$+\lambda_4\,(x_0\gamma_3 - p\,r) = 0,$$

$$\partial K/\partial q = -\frac{2\lambda_1\,[(p^2 + q^2)\,q + b\,(p\,\delta_1 + q\,\delta_2)]}{x_0^2} + \frac{2\lambda_3\,(q\,\delta_1 - p\,\delta_2)}{x_0}$$
$$-\lambda_4\,(q\,r + b\,\delta_3) = 0,$$

$$\partial K/\partial r = \lambda_0\,r - \frac{1}{2}\lambda_4\,(p^2 + q^2) = 0, \tag{7}$$

$$\partial K/\partial \gamma_3 = -\lambda_1\gamma_3 - \lambda_3\,\delta_3 + \lambda_4 x_0 p = 0,$$

$$\partial K/\partial \delta_1 = -\frac{\lambda_1 b\,(2p\,q + b\,\delta_1)}{x_0^2} - \lambda_2\,\delta_1 - \frac{\lambda_3\,(p^2 - q^2)}{x_0} = 0,$$

$$\partial K/\partial \delta_2 = -2b\,\lambda_0 - \lambda_2\,\delta_2 - \frac{\lambda_1 b\,(q^2 - p^2 + b\,\delta_2)}{x_0^2} - \frac{2\lambda_3 p\,q}{x_0} = 0,$$

$$\partial K/\partial \delta_3 = -\lambda_2\,\delta_3 - \lambda_3\,\gamma_3 - \lambda_4\,bq = 0.$$

Here λ_i are the parameters of the family of the integrals K.

The solutions of system (7) in the case when its equations are dependent allow one to define the IMs and the IMs families for differential equations (4) which correspond to the family of the first integrals K. Below, we seek such solutions with two procedures.

The first procedure is based on solving the stationary equations for a family of first integrals with respect to some part of phase variables and parameters of the family. It gives a possibility to obtain both the conditions of dependence for the equations and the desired solutions themselves. This technique was already applied by authors earlier (see, e.g., [6]). In this way, one can find the IMs embedded in one another. The second procedure enables us to obtain new IMs by eliminating the family parameters from the IMs families found earlier. In this way, one can find enveloping IMs (which contain the initial ones).

3.1   Finding the Invariant Manifolds Embedded in One Another

We shall seek the IMs of various dimension for equations (4). Since first integrals are IMs of codimension 1, let us begin with IMs of codimension 2.

Take as unknowns, e.g., $\delta_1, \delta_2, \lambda_1, \lambda_2, \lambda_0, \lambda_4$ and construct a Gröbner basis with respect to the lexicographic ordering $\delta_1 > \delta_2 > \lambda_1 > \lambda_2 > \lambda_0 > \lambda_4$ for the polynomials of system (7). As a result, we have a system of equations which is equivalent to the initial one:

$$\lambda_4 \, g_1(p, q, r, \gamma_3, \lambda_3, \lambda_4) = 0, \qquad g_2(p, q, r, \lambda_0, \lambda_4) = 0,$$
$$g_3(q, \gamma_3, \delta_3, \lambda_2, \lambda_3, \lambda_4) = 0, \qquad g_4(p, \gamma_3, \delta_3, \lambda_1, \lambda_3, \lambda_4) = 0,$$
$$g_5(p, q, r, \gamma_3, \delta_2, \delta_3, \lambda_3, \lambda_4) = 0, \quad g_6(p, q, r, \gamma_3, \delta_1, \delta_3, \lambda_3, \lambda_4) = 0, \qquad (8)$$

where $g_j (j = 1, \ldots, 6)$ are the polynomials of the basis. Because the resulting system is bulky, it is not represented explicitly here.

The first equation of system (8) is factorized. Hence, we can decompose the given system into two subsystems. Below, they are represented in an explicit form.

Subsystem 1:

$$\lambda_4 \, b \, x_0 [\varrho - 2(p^2 + q^2) p \, q] - \lambda_3 \, [x_0 \, \gamma_3 \, (2p \, (p^2 + q^2) + x_0 \gamma_3 \, r)$$
$$+ b \, (b \, \delta_3 r - 2q \, (p^2 + q^2)) \, \delta_3] = 0,$$
$$2\lambda_0 \, b \, x_0 \, [\varrho - 2(p^2 + q^2) \, p \, q] \, r - \lambda_3 \, (p^2 + q^2) \, [x_0 \, \gamma_3 \, (2p \, (p^2 + q^2)$$
$$+ x_0 \gamma_3 \, r) + b \, (b \, \delta_3 \, r - 2q \, (p^2 + q^2)) \, \delta_3] = 0, \qquad (9)$$
$$\lambda_2 \, x_0 \, [2(p^2 + q^2) \, p \, q - \varrho] + \lambda_3 \, b \, [2(p^2 + q^2) q^2 - \varrho_2] = 0,$$
$$\lambda_1 \, b \, [2(p^2 + q^2) \, p \, q - \varrho] + \lambda_3 \, x_0 \, [2(p^2 + q^2) p^2 + \varrho_2] = 0,$$

$$2b \, (p^2 + q^2) \, r \, \delta_2 + b \, [b \, r \, \delta_3 r + q(r^2 - 2(p^2 + q^2))] \, \delta_3 - (p \, r - x_0 \, \gamma_3)$$
$$\times [2p \, (p^2 + q^2) + x_0 \, \gamma_3 \, r] = 0,$$
$$-2b \, (p^2 + q^2) \, \delta_1 - p \, [2q \, (p^2 + q^2) + b \, \delta_3 \, r] - x_0 \gamma_3 \, q \, r = 0. \qquad (10)$$

Subsystem 2:

$$\lambda_4 = 0, \quad \lambda_0 = 0, \quad -(\lambda_2 \delta_3 + \lambda_3 \gamma_3) = 0, \quad -(\lambda_1 \gamma_3 + \lambda_3 \delta_3) = 0, \qquad (11)$$
$$(x_0^2 \gamma_3^2 + b^2 \, \delta_3^2) \, \delta_2 - [2 x_0 \gamma_3 p \, q + b \, (p^2 - q^2) \, \delta_3] \, \delta_3 = 0,$$
$$-(x_0^2 \gamma_3^2 + b^2 \, \delta_3^2) \, \delta_1 - [2b \, \delta_3 p \, q - x_0 \, \gamma_3 (p^2 - q^2)] \, \delta_3 = 0. \qquad (12)$$

Here $\varrho = (b \, \delta_3 p - x_0 \gamma_3 \, q) \, r$, $\varrho_2 = (b \, \delta_3 q + x_0 \gamma_3 p) \, r$.

Let us analyse the equations of subsystem 1.

It can easily be verified by IM definition that equations (10) define the IM of codimension 2 of differential equations (4): the derivative of expressions (10) calculated by virtue of equations (4) must vanish on these expressions.

Indeed, rewrite expressions (10) in the form

$$\delta_1 = -\frac{2(p^2 + q^2)\,p\,q + \varrho_2}{2b\,(p^2 + q^2)},$$

$$\delta_2 = \frac{b\,[(2(p^2 + q^2) - r^2)\,q - b\,r\,\delta_3]\,\delta_3 + (p\,r - x_0\gamma_3)(2p\,(p^2 + q^2) + x_0\gamma_3\,r)}{2b\,(p^2 + q^2)\,r} \quad (13)$$

and calculate the derivative of them by virtue of equations (4):

$$-\frac{(b\delta_3 p + x_0\gamma_3 q)\,y_1}{p^2 + q^2} + \frac{r y_2}{2} = 0,$$

$$-\frac{[(2(p^2 + q^2) + r^2)r + 4(b\delta_3 p - x_0\gamma_3 q)]\,y_1}{2r^2} - \frac{(b\delta_3 p + x_0\gamma_3 q)\,y_2}{p^2 + q^2} = 0. \quad (14)$$

Here $y_1 = \delta_1 + \dfrac{2(p^2 + q^2)\,p\,q + \varrho_2}{2b\,(p^2 + q^2)},$

$$y_2 = \delta_2 - \frac{b[(2(p^2 + q^2) - r^2)q - b\,r\,\delta_3]\delta_3 + (p\,r - x_0\gamma_3)(2p\,(p^2 + q^2) + x_0\gamma_3\,r)}{2b\,(p^2 + q^2)\,r}.$$

Because equalities (14) become identities when $y_1 = y_2 = 0$, hence the latter proves the desired property of invariance of solution (10).

The equations of vector field on IM (10) are given by

$$2\dot{p} = q\,r + b\,\delta_3, \quad 2\dot{q} = x_0\gamma_3 - p\,r, \quad \dot{r} = \frac{\varrho_2}{p^2 + q^2},$$

$$\dot{\gamma}_3 = \frac{b\,[b q\,r\,\delta_3 - (p^2 + q^2)(2q^2 - r^2)]\,\delta_3}{2x_0\,(p^2 + q^2)\,r} + \frac{p\,\gamma_3\,q}{r} + \frac{(x_0^2\gamma_3^2 - 2(p^2 + q^2)^2)\,q}{2x_0\,(p^2 + q^2)},$$

$$\dot{\delta}_3 = \frac{[b\,r\,\delta_3 - 2(p^2 + q^2)\,q]\,p\,\delta_3}{2\,(p^2 + q^2)\,r} - \frac{1}{2b} + \frac{x_0\gamma_3\,p\,(2p\,(p^2 + q^2) + x_0\gamma_3\,r)}{2b\,(p^2 + q^2)\,r}. \quad (15)$$

They are derived from equations (4) by elimination of the variables δ_1, δ_2 from them with the aid of (13).

From (9) we find the values $\lambda_0, \lambda_1, \lambda_2, \lambda_4$:

$$\lambda_1 = -\frac{\lambda_3 x_0(2(p^2 + q^2)\,p^2 + \varrho_2)}{b(2(p^2 + q^2)\,p\,q - \varrho)}, \quad \lambda_2 = -\frac{b\lambda_3(2(p^2 + q^2)q^2 - \varrho_2)}{x_0(2(p^2 + q^2)pq - \varrho)},$$

$$\lambda_0 = \frac{\lambda_3(p^2 + q^2)[b\,(b\,\delta_3 r - 2q\,(p^2 + q^2))\,\delta_3 + x_0\,(2p\,(p^2 + q^2) + x_0 r\,\gamma_3)\,\gamma_3]}{2b\,r x_0\,[\varrho - 2(p^2 + q^2)\,p\,q]},$$

$$\lambda_4 = \lambda_3\left(\left(\frac{1}{x_0 p} - \frac{(p^2 + q^2)\,r\,\gamma_3}{(2(p^2 + q^2)\,p\,q - \varrho)\,p\,q}\right)\delta_3 - \frac{\gamma_3}{b\,q}\right). \quad (16)$$

The derivatives of right-hand sides of (16) which are computed by virtue of equations (15) are equal to zero. The latter means that relations (16) are the first integrals of equations (15).

In a similar manner, we have established that equations (12) define the IM of codimension 2 for differential equations (4), and the values of λ_1, λ_2 found from the two latter expressions of (11) are the first integrals for the equations of vector field on this IM. Obviously, these integrals will be dependent.

Applying the described technique, it is not difficult to obtain the IMs of dimension less than the above. Let us seek the IMs of codimension 3. For this purpose, we shall increase the number of the phase variables and decrease the number of the parameters in the above list of the unknowns.

Take as unknowns $\delta_1, \delta_2, \delta_3, \lambda_2, \lambda_0, \lambda_4$ and construct a Gröbner basis with respect to the lexicographic ordering $\delta_1 > \delta_2 > \delta_3 > \lambda_2 > \lambda_0 > \lambda_4$ for the polynomials of system (7). As a result, we have a system of equations which is decomposed into two subsystems:

Subsystem 1:

$$\lambda_4 = 0, \ \lambda_0 = 0, \ \lambda_3^2 - \lambda_1\lambda_2 = 0, \tag{17}$$

$$
\begin{aligned}
&-\lambda_3\,\delta_3 - \lambda_1\gamma_3 = 0, \\
&-(\lambda_1^2 b^2 + \lambda_3^2 x_0^2)\,\delta_2 + \lambda_1(\lambda_1 b\,(p^2 - q^2) - 2\lambda_3 x_0 p\,q) = 0, \\
&-(\lambda_1^2 b^2 + \lambda_3^2 x_0^2)\,\delta_1 - \lambda_1(\lambda_3 x_0(p^2 - q^2) - 2\lambda_1 bp\,q) = 0.
\end{aligned}
\tag{18}
$$

Subsystem 2:

$$
\begin{aligned}
&\lambda_4 b\, x_0\,\sigma_1 - \lambda_1 b\,(2\lambda_3(p^2 + q^2)\,q + \lambda_1 b\gamma_3 r) - \lambda_3^2 x_0\,(2p\,(p^2 + q^2) + x_0\gamma_3 r) = 0, \\
&2\lambda_0\,bx_0\,r\,\sigma_1 - (p^2 + q^2)[\,2\lambda_1\lambda_3 b\,(p^2 + q^2)\,q + (\lambda_1^2 b^2 + \lambda_3^2 x_0^2)\,\gamma_3 r \\
&\quad + 2\lambda_3^2 x_0\,(p^2 + q^2)\,p\,] = 0, \\
&-\lambda_2 x_0\,[2p\,(p^2 + q^2) + x_0\gamma_3 r] - b\,[2\lambda_3(p^2 + q^2)\,q + \lambda_1 b\,\gamma_3 r] = 0,
\end{aligned}
\tag{19}
$$

$$
\begin{aligned}
&-b\,\sigma_1\,\delta_3 + (2p\,(p^2 + q^2) + x_0\,\gamma_3 r)(b\lambda_1 q + x_0\lambda_3 p) = 0, \\
&-2b\,(p^2 + q^2)\,\sigma_1^2\,\delta_2 - \lambda_1\lambda_3 b\,x_0\,[2(p^2 + q^2)^2\,q + (3p^2 - q^2)r^2]\,\sigma_2 \\
&\quad + \lambda_1^2 b^2\,[(q^2 - p^2)\,p\,r + x_0\,(p^2 + q^2)\,\gamma_3 + \lambda_3^2 x_0^2\,[2p\,((p^2 + q^2)^2 - q^2 r^2) \\
&\quad + x_0(p^2 + q^2)\,r]\,]\,\gamma_3\,r = 0, \\
&-2b\,(p^2 + q^2)\,\delta_1\,\sigma_1 - (2\lambda_1 b\,p\,q + \lambda_3 x_0\,(p^2 - q^2))\,\sigma_2 = 0.
\end{aligned}
\tag{20}
$$

Here $\sigma_1 = (b\lambda_1 p - \lambda_3 x_0 q)\,r$, $\sigma_2 = 2p\,(p^2 + q^2) + x_0\,\gamma_3 r$.

¿From the analysis of the subsystems, one can conclude that equations (18) define the family of IMs of codimension 3 for differential equations (4). The family parameters are λ_1, λ_3. Unlike the above case, the values of λ_i found here do not contain the phase variables. The integral $2\bar{K} = -\lambda_1\tilde{V}_2 - \lambda_3^2/\lambda_1\,V_3 - 2\lambda_3\tilde{V}_4$ takes a stationary value on the elements of the IMs family. We call such IMs the stationary ones.

Equations (20) also define the family of IMs of codimension 3 for equations (4) which is parameterized by λ_1, λ_3. The values for $\lambda_0, \lambda_2, \lambda_4$ obtained from relations (19) are the first integrals for the equations of vector field on the elements of this family.

We have also found the families of IMs of codimension 4 and 5. Let us represent the equations of the latter.

Take as unknowns $\delta_1, \delta_2, \delta_3, \gamma_3, r, \lambda_0$ and construct a Gröbner basis with respect to the lexicographic ordering $\delta_1 > \delta_2 > \delta_3 > \gamma_3 > r > \lambda_0$ for the polynomials of system (7). The system of equations which is equivalent to the initial one writes:

$$-4\lambda_0\alpha_2 + \lambda_4^2\alpha_1 = 0, \qquad (21)$$

$$\lambda_4\,\alpha_1\,r - 2\alpha_2\,(p^2 + q^2) = 0, \quad -\alpha_2\,\gamma_3 - \lambda_4\,(\lambda_2 x_0 p + \lambda_3\,b\,q) = 0,$$

$$\alpha_2\,\delta_3 - \lambda_4\,(\lambda_3 x_0\,p + \lambda_1\,b\,q) = 0,$$

$$2\alpha_1\,\alpha_2\,\delta_2 - 2\lambda_1\,\alpha_2\,b\,(p^2 - q^2) + x_0(4\lambda_3\,\alpha_2\,p\,q + \lambda_4^2\,\alpha_1\,b\,x_0) = 0, \qquad (22)$$

$$-\alpha_1\,\delta_1 - 2\lambda_1 b\,p\,q - \lambda_3 x_0\,(p^2 - q^2) = 0,$$

where $\alpha_1 = \lambda_1 b^2 + \lambda_2 x_0^2$, $\alpha_2 = \lambda_3^2 - \lambda_1\lambda_2$.

Equations (22) define the family of IMs of codimension 5 for differential equations (4) which is parameterized by $\lambda_1, \lambda_2, \lambda_3, \lambda_4$. It is the family of stationary IMs: the integral K takes a stationary value on the elements of the family when $\lambda_0 = \lambda_4^2\alpha_1/(4\alpha_2)$ (this value is found from equation (21)).

Choosing as the unknowns another combinations of the phase variables and the parameters λ_i, one can obtain the other IMs families of the same dimension. Having taken $\delta_1, \delta_2, \delta_3, \gamma_3, r, \lambda_4$ as the unknowns, we have found two IMs families of codimension 5 which differ from IMs family (22). Their equations are given by

$$2\lambda_0 r - \lambda_4\,(p^2 + q^2) = 0, \quad -\alpha_2\gamma_3 - \lambda_4(\lambda_2 x_0 p + b\lambda_3 q) = 0,$$

$$\alpha_2\delta_3 - \lambda_4\,(\lambda_3 x_0 p + b\lambda_1 q) = 0,$$

$$-\alpha_1\delta_2 + b\lambda_1\,(p^2 - q^2) - 2x_0(\lambda_3 p q + b\lambda_0 x_0) = 0, \qquad (23)$$

$$-\alpha_1\delta_1 - 2b\lambda_1 p q - \lambda_3 x_0(p^2 - q^2) = 0,$$

where $\lambda_4 = \pm\sqrt{\lambda_0\,\alpha_2/\alpha_1}$. The families are parameterized by $\lambda_0, \lambda_1, \lambda_2, \lambda_3$. The elements of the families provide a stationary value for the integral K under the above values λ_4.

Obviously, the solutions found by the described technique will be related. Indeed, on substituting expressions (22) (resolved with respect to $\delta_1, \delta_2, \delta_3, \gamma_3, r$) into equations (20), the latter become identities. Next, on substituting expressions (20) (resolved with respect to $\delta_1, \delta_2, \delta_3$) into equations (10), they become identities. Hence, one can conclude that the elements of IMs family (22) are submanifolds of IMs family (20), and the elements of IMs family (20) are submanifolds of IM (10).

Analogously, we have established that the elements of IMs families (23) are submanifolds of IMs family (20), and the elements of IMs family (18) are submanifolds of IM (12).

Thus, applying the procedure presented above, we obtain the IMs families embedded in one another. The latter enables us to classify IMs on the basis of their embedding and degree of their degeneration.

The IMs families found for differential equations written on IM (3) can be "lifted up" as invariant into the phase space of system (1). To this end, it is sufficient to add equations (3) to the equations of these IMs families. For example, the equations of IMs family (22) "lifted up" into the original phase space write:

$$p^2 - q^2 - x_0\gamma_1 - b\,\delta_2 = 0, \quad 2pq - x_0\gamma_2 + b\,\delta_1 = 0,$$
$$\lambda_4\,\alpha_1\,r - 2\alpha_2\,(p^2 + q^2) = 0, \quad -\alpha_2\,\gamma_3 - \lambda_4\,(\lambda_2 x_0 p + \lambda_3\,b\,q) = 0,$$
$$\alpha_2\,\delta_3 - \lambda_4\,(\lambda_3 x_0\,p + \lambda_1\,b\,q) = 0,$$
$$2\alpha_1\,\alpha_2\,\delta_2 - 2\lambda_1\,\alpha_2\,b\,(p^2 - q^2) + x_0(4\lambda_3\,\alpha_2\,pq + \lambda_4^2\,\alpha_1\,b\,x_0) = 0,$$
$$-\alpha_1\,\delta_1 - 2\lambda_1 b\,p\,q - \lambda_3 x_0\,(p^2 - q^2) = 0.$$

One can verify by IM definition that the latter equations define the IMs family of codimension 7 of system (1).

3.2   Finding the Enveloping Invariant Manifolds

Let us consider another technique for finding IMs. We shall seek an IMs family which includes the family of IMs (22).

Eliminate parameter λ_4 from equations (22) with the aid of one of the equations, e.g., the first. The value of λ_4 found from this equation is

$$\lambda_4 = \frac{2\alpha_2\,(p^2 + q^2)}{\alpha_1\,r}. \tag{24}$$

Next, construct a lexicographic Gröbner basis with respect to the lexicographic ordering $\delta_1 > \delta_2 > \delta_3 > \gamma_3 > r$ for the polynomials of a resulting system (after eliminating λ_4 from equations (22)). The system obtained

$$\alpha_1\,\gamma_3\,r + 2(p^2 + q^2)(\lambda_3 bq + \lambda_2 x_0 p) = 0,$$
$$\alpha_1\,r\,\delta_3 - 2(p^2 + q^2)(\lambda_1 bq + \lambda_3 x_0 p) = 0,$$
$$\alpha_1^2 r^2\,\delta_2 + \alpha_1\,[\lambda_1 b\,(q^2 - p^2) + 2\lambda_3\,x_0\,p\,q]\,r^2 - 2\alpha_2(p^2 + q^2)^2 = 0,$$
$$-\alpha_1\,\delta_1 - 2b\lambda_1 p q + \lambda_3 x_0(q^2 - p^2) = 0 \tag{25}$$

defines the IMs family of codimension 4 for differential equations (4) which is parameterized by $\lambda_1, \lambda_2, \lambda_3$. The latter is easily verified by IM definition. Expression (24) is the first integral for the equations of vector field on the elements of IMs family (25).

The elements of IMs family (22) are submanifolds of the IMs family found: on substituting expressions (22) (resolved with respect to $\delta_1, \delta_2, \delta_3, \gamma_3, r$) into equations (25), the latter become identities.

After eliminating, e.g., the parameter λ_2 from system (25) with the aid of the first equation of this system we have the equations of IMs family (20), and a value found for λ_2 is the first integral for the equations of vector field on the elements of this family.

Finally, after eliminating the parameter λ_1 from equations (20) we have the equations of IM (10).

Thus, the above examples show that the presented procedure provides a possibility to find enveloping IMs families by eliminating the family parameters from the equations of IMs families obtained earlier.

3.3   On Invariant Manifolds under Restrictions on the Constants of the Problem Integrals

From geometrical considerations, there exist some restrictions on the constants of first integrals $\tilde{V}_2, V_3$ (5) of the problem under study. To take into account these restrictions, it is sufficient to add expressions $\tilde{V}_2 = 1$, $V_3 = 1$ to the equations of the IM and IMs families found above. So, their codimension will be, generally speaking, by two greater than the indicated one.

Adding the integrals $\tilde{V}_2, V_3$ to the equations of IMs family, we separate in such a way some IMs subfamily from this family. Below, the given problem is considered for IMs family (22).

The equations of vector field on the elements of IMs family (22) are given by

$$2\dot{p} = \frac{2\alpha_2(p^2 + q^2)\,q}{\alpha_1\lambda_4} + \frac{b\lambda_4(b\lambda_1 q + \lambda_3 x_0 p)}{\alpha_2},$$

$$2\dot{q} = -\frac{2\alpha_2(p^2 + q^2)\,p}{\alpha_1\lambda_4} - \frac{\lambda_4 x_0(b\lambda_3 q + \lambda_2 x_0 p)}{\alpha_2}.$$

These equations admit two first integrals with the fixed constants:

$$\bar{V}_2 = \frac{(2\lambda_2 x_0 p\,q - b\lambda_3(p^2 - q^2))^2}{\alpha_1^2} + \frac{\lambda_4^2(b\lambda_3 q + \lambda_2 x_0 p)^2}{\alpha_2^2}$$

$$+ \left(\frac{2b\lambda_3 p\,q + \lambda_2 x_0(p^2 - q^2)}{\alpha_1} + \frac{b^2\lambda_4^2 x_0}{2\alpha_2}\right)^2 = 1,$$

$$\bar{V}_3 = \frac{2b\lambda_1 p\,q + \lambda_3 x_0(p^2 - q^2))^2}{\alpha_1^2} + \frac{\lambda_4^2(b\lambda_1 q + \lambda_3 p x_0)^2}{\alpha_2^2}$$

$$+ \left(\frac{b\lambda_1(p^2 - q^2) - 2\lambda_3 x_0 p\,q}{\alpha_1} - \frac{b\lambda_4^2 x_0^2}{2\alpha_2}\right)^2 = 1.$$

When $b = x_0, \lambda_1 = \lambda_2$, these integrals coincide and become:

$$\tilde{V} = \frac{\sigma_2(p^2 + q^2)^2}{4\lambda_2^2 x_0^2} + \frac{\lambda_4^2\sigma_2 x_0^2\,(2\lambda_3 p\,q + \lambda_2(p^2 + q^2))}{2\lambda_2\sigma_1^2} + \frac{\lambda_4^4 x_0^6}{4\sigma_1^2} = 1. \quad (26)$$

Here $\sigma_1 = \lambda_2^2 - \lambda_3^2, \sigma_2 = \lambda_2^2 + \lambda_3^2$.

After substitution of the above values for b, λ_1 into equations (22), a resulting system together with integral (26) defines the family of one-dimensional IMs for differential equations (4). It will be a subfamily of IMs family (22). Its equations can be written as:

$$\sigma_1^2\,[4\lambda_2^2 x_0^2 - (p^2 + q^2)^2\sigma_2] - x_0^4\lambda_2\lambda_4^2\sigma_2\,[4\lambda_3 p\,q + \lambda_2(2p^2 + q^2)] - x_0^8\lambda_2^2\lambda_4^4 = 0,$$
$$\lambda_2\lambda_4 x_0^2 r + \sigma_1(p^2 + q^2) = 0, \quad \sigma_1\gamma_3 - \lambda_4 x_0(\lambda_2 p + \lambda_3 q) = 0,$$
$$-\sigma_1\delta_3 - \lambda_4 x_0(\lambda_3 p + \lambda_2 q) = 0, \quad (27)$$
$$-2\delta_2\lambda_2\sigma_1 x_0 + \lambda_2\sigma_1(p^2 - q^2) - 2\lambda_3\sigma_1 p\,q + \lambda_2\lambda_4^2 x_0^4 = 0,$$
$$-2\delta_1\lambda_2 x_0 - \lambda_3(p^2 - q^2) - 2\lambda_2 p\,q = 0,$$

The found IMs family is stationary: the integral

$$4\hat{K} = -\frac{\lambda_2\lambda_4^2 x_0^2}{2\sigma_1}\,\tilde{H} - 2\lambda_2\,(\tilde{V}_2 + V_3) - 2\lambda_3\tilde{V}_4 - 2\lambda_4 V_5 \quad (28)$$

takes a stationary value on the elements of this family. Geometrically, the elements of IMs family (27) correspond to curves in R^7.

Let us find an enveloping IMs family for family (27). To this end, we eliminate the parameter λ_4 from equations (27) with one of its equations, e.g., $-\sigma_1\delta_3 - \lambda_4 x_0(\lambda_3 p + \lambda_2 q) = 0$, and construct a lexicographic Gröbner basis with respect to the variables $\delta_1, \delta_2, \delta_3, \gamma_3, r$ for the polynomials of a resulting (after eliminating λ_4) system. The system obtained

$$
\begin{aligned}
&(\lambda_2^2(p^2 + q^2)^2\sigma_2 - 4\lambda_2^4 x_0^2)r^4 + 2\lambda_2(p^2 + q^2)^2(2\lambda_3 pq + \lambda_2(p^2 + q^2))\sigma_2\, r^2 \\
&\quad + (p^2 + q^2)^4\sigma_1^2 = 0, \\
&\sigma_1^2 x_0\,(p^2 + q^2)^3\gamma_3 - 2(\lambda_2 p + \lambda_3 q)(p^2 + q^2)^2(2\lambda_3 pq + \lambda_2(p^2 + q^2))\sigma_2\, r \\
&\quad -\lambda_2(\lambda_2 p + \lambda_3 q)((p^2 + q^2)^2\sigma_2\, r^3 - 4\lambda_2^2 x_0^2) = 0, \\
&\sigma_1^2 x_0\,(p^2 + q^2)^3\, \delta_3 + 2(\lambda_3 p + \lambda_2 q)(p^2 + q^2)^2(2\lambda_3 pq + \lambda_2(p^2 + q^2))\,\sigma_2 r \quad (29) \\
&\quad +\lambda_2(\lambda_3 p + \lambda_2 q)\sigma_2((p^2 + q^2)^2 r^3 - 4\lambda_2^2 x_0^2) = 0, \\
&2\lambda_2\sigma_1 x_0(p^2 + q^2)^2\, \delta_2 + (p^2 + q^2)^2(6\lambda_2^2\lambda_3 pq + 2\lambda_3^3 pq + \lambda_2\lambda_3^2(3p^2 + q^2) \\
&\quad +\lambda_2^3(p^2 + 3q^2)) + (\lambda_2(p^2 + q^2)^2\sigma_2 - 4\lambda_2^3 x_0^2)r^2 = 0, \\
&-2\lambda_2 x_0\delta_1 - 2\lambda_2 pq - \lambda_3(p^2 - q^2) = 0
\end{aligned}
$$

defines the IMs family of codimension 5 for differential equations (4).

The elements of family (27) are submanifolds of IMs family (29): on substituting expressions (27) (resolved with respect to $\delta_1, \delta_2, \delta_3, \gamma_3, r, q$) into equations (29), the latter become identities. Geometrically, the elements of IMs family (29) correspond to surfaces of codimension 5 in R^7.

4   On the Stability of Stationary Invariant Manifolds

In this section, we investigate the stability of the stationary IMs families found above. For this purpose, the Routh–Lyapunov method is applied.

First, we analyse IM (3). It is not difficult to show that integral V_1 (2) takes a stationary value on the given IM. Therefore, in the case under study, the investigation of stability is reduced to verifying the sign-definiteness conditions for the variation of the integral V_1 obtained in the neighbourhood of IM (3). Introducing the deviations

$$
y_1 = p^2 - q^2 - x_0\gamma_1 - b\,\delta_2, \quad y_2 = 2pq - x_0\gamma_2 + b\,\delta_1
$$

of the IM, we have the sign-definite variation of the integral $\Delta V_1 = y_1^2 + y_2^2$. Whence, one can conclude that IM (3) is stable in the phase space of system (1).

Below, we investigate the stability of the elements of stationary IMs families (18) and (22) in the phase space of system (4).

The integral $2\bar{K} = -\lambda_1\tilde{V}_2 - \lambda_3^2/\lambda_1\,V_3 - 2\lambda_3\tilde{V}_4$ takes a stationary value on the elements of IMs family (18). The variation of the integral $\bar{K}$ in the deviations

$$
y_1 = \delta_1 + \frac{\lambda_1(\lambda_3 x_0\,(p^2 - q^2) + 2b\lambda_1 p\,q)}{\lambda_1^2\,b^2 + \lambda_3^2 x_0^2},
$$

$$
y_2 = \delta_2 - \frac{\lambda_1(\lambda_1\,b\,(p^2 - q^2) - 2\lambda_3 x_0 p\,q)}{\lambda_1^2\,b^2 + \lambda_3^2 x_0^2}, \quad y_3 = \delta_3 + \frac{\gamma_3\lambda_1}{\lambda_3}
$$

of the elements of the given IMs family can be written as:

$$\Delta \bar{K} = -\left(\frac{\lambda_3^2}{2\lambda_1} + \frac{b^2\lambda_1}{2x_0^2}\right)(y_1^2 + y_2^2) - \frac{\lambda_3^2 y_3^2}{2\lambda_1}. \tag{30}$$

The conditions for quadratic form (30) to be positive definite are given by $b \neq 0$, $x_0 \neq 0$, $\lambda_1 < 0$, $\lambda_3 \neq 0$. Hence, the elements of IMs family (18) are stable under the above restrictions on the parameters λ_1, λ_3. Note that the obtained sufficient conditions of stability impose the restrictions on the family parameters λ_1, λ_3, instead of the problem parameters. Thus, a subfamily (the elements of which are stable) was separated from the IMs family studied.

Let us investigate IMs family (22). Integral K (6) takes a stationary value on the elements of this family when $\lambda_0 = -\lambda_4^2(\lambda_1 b^2 + \lambda_2 x_0^2)/(4(\lambda_1\lambda_2 - \lambda_3^2))$. The variation of the integral in the deviations

$$y_1 = \delta_1 + \frac{2\lambda_1 bp\,q + \lambda_3 x_0(p^2 - q^2)}{\lambda_1 b^2 + \lambda_2 x_0^2},$$

$$y_2 = \delta_2 + \frac{b\lambda_4^2 x_0^2}{2(\lambda_3^2 - \lambda_1\lambda_2)} - \frac{b\lambda_1(p^2 - q^2) - 2\lambda_3 x_0 p\,q}{\lambda_1 b^2 + \lambda_2 x_0^2}, \quad y_3 = \delta_3 - \frac{\lambda_4(b\lambda_1 q + \lambda_3 x_0 p)}{\lambda_3^2 - \lambda_1\lambda_2},$$

$$y_4 = \gamma_3 + \frac{\lambda_4(b\lambda_3 q + \lambda_2 x_0 p)}{\lambda_3^2 - \lambda_1\lambda_2}, \quad y_5 = r + \frac{2(\lambda_1\lambda_2 - \lambda_3^2)(p^2 + q^2)}{\lambda_4(\lambda_1 b^2 + \lambda_2 x_0^2)}$$

of the elements of the IMs family can be written as

$$\Delta K = -\frac{\lambda_1 b^2 + \lambda_2 x_0^2}{2x_0^2}(y_1^2 + y_2^2) - \frac{\lambda_2}{2}y_3^2 - \lambda_3 y_3 y_4 - \frac{\lambda_1}{2}y_4^2 - \frac{\lambda_4^2(\lambda_1 b^2 + \lambda_2 x_0^2)y_5^2}{8(\lambda_1\lambda_2 - \lambda_3^2)}. \tag{31}$$

The conditions for quadratic form (31) to be positive definite are given by $b \neq 0$, $x_0 \neq 0$, $\lambda_2 < 0$, $\lambda_1\lambda_2 < \lambda_3^2$. Hence, the elements of IMs family (22) are stable under the above restrictions on the parameters $\lambda_1, \lambda_2, \lambda_3$. Likewise as above, the sufficient conditions of stability separate a subfamily (the elements of which are stable) from IMs family (22).

Finally, we investigate the stability of the IMs families, which correspond to IMs families (18) and (22) in the phase space of system (1).

A family of stationary IMs corresponds to IMs family (18) in the phase space of system (1). Its equations write:

$$p^2 - q^2 - x_0\gamma_1 - b\,\delta_2 = 0, \quad 2p\,q - x_0\gamma_2 + b\,\delta_1 = 0,$$
$$-\lambda_3\,\delta_3 - \lambda_1\gamma_3 = 0,$$
$$-(\lambda_1^2 b^2 + \lambda_3^2 x_0^2)\,\delta_2 + \lambda_1(\lambda_1 b\,(p^2 - q^2) - 2\lambda_3 x_0 p\,q) = 0,$$
$$-(\lambda_1^2 b^2 + \lambda_3^2 x_0^2)\,\delta_1 - \lambda_1(\lambda_3 x_0(p^2 - q^2) - 2\lambda_1 bp\,q) = 0.$$

They are derived by addition of equations (3) to (18). The elements of this family provide a stationary value for the following family of integrals (2):

$$2K_1 = -\left[\mu_1 V_1 + \mu_4\left(\frac{\lambda_1}{\lambda_3}V_2 + \frac{\lambda_3}{\lambda_1}V_3 + 2V_4\right)\right].$$

Here μ_1, μ_4 are the family parameters.

Introducing the deviations

$$y_1 = \delta_1 + \frac{\lambda_1(\lambda_3 x_0\,(p^2 - q^2) + 2b\lambda_1 p\,q)}{\lambda_1^2\,b^2 + \lambda_3^2 x_0^2},$$

$$y_2 = \delta_2 - \frac{\lambda_1(\lambda_1\,b\,(p^2 - q^2) - 2\lambda_3 x_0 p\,q)}{\lambda_1^2\,b^2 + \lambda_3^2 x_0^2},$$

$$y_3 = \delta_3 + \frac{\gamma_3\lambda_1}{\lambda_3},\; y_4 = \gamma_1 - \frac{\lambda_1(\lambda_3 x_0\,(p^2 - q^2) + 2b\lambda_1 p\,q)}{\lambda_1^2\,b^2 + \lambda_3^2 x_0^2},$$

$$y_5 = \gamma_2 + \frac{\lambda_1(\lambda_1\,b\,(p^2 - q^2) - 2\lambda_3 x_0 p\,q)}{\lambda_1^2\,b^2 + \lambda_3^2 x_0^2}$$

of the elements of the IMs family under consideration, we have the sign-definite variation of the integral

$$\Delta K_1 = -\frac{\lambda_1\mu_1 b^2 + \lambda_3\mu_4}{2\lambda_1}\,(y_1^2 + y_2^2) - \mu_4(y_1 y_4 + y_2 y_5) + \mu_1 b x_0(y_1 y_5 - y_2 y_4)$$

$$-\frac{\lambda_1\mu_4 + \lambda_3\mu_1 x_0^2}{2\lambda_3}\,(y_4^2 + y_5^2)$$

under the following conditions imposed on the parameters b, x_0, λ_1, λ_3, μ_1, μ_4:

$$b \neq 0, x_0 \neq 0, \mu_1 = \lambda_1, \mu_4 = \lambda_3, \lambda_3 \neq 0, \lambda_1 < 0.$$

The elements of IMs family under study are stable under the above restrictions on the parameters. The interpretation of the sufficient conditions obtained is similar to the previous cases considered. We can also see that the sufficient conditions of stability for IMs family (18) and the solution which corresponds to it in the original phase space coincide.

A stationary IMs family also corresponds to IMs family (22) in the phase space of system (1). Its equations are derived by addition of equations (3) to (22). The elements of this family provide a stationary value for integral V_1 (2).

Introducing the deviations

$$y_1 = \delta_1 + \frac{2\lambda_1 b p\,q + \lambda_3 x_0(p^2 - q^2)}{\lambda_1 b^2 + \lambda_2 x_0^2},$$

$$y_2 = \delta_2 + \frac{b\lambda_4^2 x_0^2}{2(\lambda_3^2 - \lambda_1\lambda_2)} - \frac{b\lambda_1(p^2 - q^2) - 2\lambda_3 x_0 p\,q}{\lambda_1 b^2 + \lambda_2 x_0^2},\; y_3 = \delta_3 - \frac{\lambda_4(b\lambda_1 q + \lambda_3 x_0 p)}{\lambda_3^2 - \lambda_1\lambda_2},$$

$$y_4 = \gamma_3 + \frac{\lambda_4(b\lambda_3 q + \lambda_2 x_0 p)}{\lambda_3^2 - \lambda_1\lambda_2},\; y_5 = r + \frac{2(\lambda_1\lambda_2 - \lambda_3^2)(p^2 + q^2)}{\lambda_4(\lambda_1 b^2 + \lambda_2 x_0^2)},$$

$$y_6 = \gamma_1 + \frac{b^2 x_0\lambda_4^2}{2\lambda_1\lambda_2 - 2\lambda_3^2} - \frac{2b\lambda_3 p\,q + x_0\lambda_2(p^2 - q^2)}{b^2\lambda_1 + x_0^2\lambda_2},$$

$$y_7 = \gamma_2 + \frac{b\lambda_3(p^2 - q^2) - 2x_0\lambda_2 p\,q}{b^2\lambda_1 + x_0^2\lambda_2}$$

of the elements of the family, we have the sign-definite variation of the integral $\Delta V_1 = \zeta^2 + \xi_2^2$. Here $\zeta_1 = by_2 + x_0 y_4$, $\xi = by_1 - x_0 y_5$. So, the elements of the IMs family are stable with respect to the variables ζ, ξ.

5   Conclusion

The conservative system which describes the rotation of a rigid body around a fixed point in gravitational and magnetic force fields was considered. The structure of the phase space of this system was studied by solving the extremum problem for the system's first integrals. Special attention was given to the IMs families which can be obtained from stationary conditions for the integrals. Two procedures to seek such solutions have been proposed. They allow one to find IMs families embedded in one another and enveloping ones. A series of the IMs families of various dimension have been found, and their simplest classification has been performed.

Some parametric analysis and the investigation of stability for the stationary IMs families have been carried out. Some specificities of properties of these families have been revealed. In particular, we have established that the sufficient conditions of stability for the IMs families impose the restrictions on the parameters of these families, instead of the problem parameters.

The IMs families found for the integrable system on the manifold can be "lifted up" as invariant into the phase space of the original system. In some cases, they keep the property to be stationary IMs families.

We have found and analyzed some part of possible IMs families of the problem only. For the exhaustive analysis of the problem on the base of the presented approach, it is necessary to investigate in detail the algebra of the problem first integrals. In the given work, we restricted our consideration to the linear combination of the integrals.

References

1. Bogoyavlenskii, O.I.: Two integrable cases of a rigid body dynamics in a force field. USSR Acad. Sci. Doklady **275**(6), 1359–1363 (1984)
2. Sarychev, V.A., Mirer, S.A., Degtyarev, A.A., Duarte, E.K.: Investigation of equilibria of a satellite subjected to gravitational and aerodynamic torques. Celestial Mechanics and Dynamical Astronomy **97**(4), 267–287 (2007)
3. Lyapunov, A.M.: On Permanent Helical Motions of a Rigid Body in Fluid. Collected Works, vol. 1. USSR Acad. Sci., Moscow-Leningrad (1954)
4. Irtegov, V.D., Titorenko, T.N.: The invariant manifolds of systems with first integrals. J. Appl. Math. Mech. **73**(4), 379–384 (2009)
5. Kowalewski, C.V.: Scientific Works. USSR Acad. Sci., Moscow (1948)
6. Irtegov, V., Titorenko, T.: Invariant manifolds in the classic and generalized Goryachev–Chaplygin problem. In: Gerdt, V.P., Koepf, W., Seiler, W.M., Vorozhtsov, E.V. (eds.) CASC 2014. LNCS, vol. 8660, pp. 218–229. Springer, Heidelberg (2014)

Homotopy Analysis Method for Stochastic Differential Equations with Maxima

Maciej Janowicz[1,2], Joanna Kaleta[1], Filip Krzyżewski[2], Marian Rusek[1], and Arkadiusz Orłowski[1]

[1] Faculty of Applied Informatics and Mathematics (WZIM),
Warsaw University of Life Sciences (SGGW),
ul. Nowoursynowska 159, 02-775 Warsaw, Poland
[2] Institute of Physics of the Polish Academy of Sciences,
Aleja Lotników 32/46, 02-668 Warsaw, Poland

Abstract. The homotopy analysis method known from its successful applications to obtain quasi-analytical approximations of solutions of ordinary and partial differential equations is applied to stochastic differential equations with Gaussian stochastic forces. It has been found that the homotopy analysis method yields excellent agreement with exact results (when the latter are available) and appears to be a very promising approach in the calculations related to quantum field theory and quantum statistical mechanics. Using a computer algebra system Maxima has considerably influenced and simplified the calculations and results.

Keywords: homotopy analysis method, stochastic differential equations, quantum scalar field, computer algebra systems, Maxima.

1 Introduction

The unprecedented recent growth of computing power of modern machines as well as the development of the software has made it possible to substantially enhance the quantitative predictive capabilities of physical research. Nevertheless, it appears that the role played by analytical techniques to solve mathematical problems of science and engineering has by no means diminished. And the development of symbolic algebra systems like Mathematica, Maple or Maxima has enabled the analytical methods to become more reliable and efficient. Among those analytical methods one can mention a particularly efficient approach recently developed by Liao [14,9,11,13,10,12] called by that author the "homotopy analysis method". One of the most significant features of the homotopy analysis method is its apparent independence of the presence of any small parameter. The approximate solution has the form of a series expansion. Usually, it is required that the series is truncated at a rather high order. However, skillful choice — e.g., on using variational tools — of some additional control parameters (the presence of which the method admits) allows one to obtain meaningful results even in low orders, as reported by Marinca and coworkers [17,15,16].

© Springer International Publishing Switzerland 2015
V.P. Gerdt et al. (Eds.): CASC 2015, LNCS 9301, pp. 233–244, 2015.
DOI: 10.1007/978-3-319-24021-3_18

The purpose of this paper is to demonstrate usefulness of the homotopy analysis method in the case of stochastic differential equations. To illustrate the quality of approximation provided by Liao's method, we use a single and very simple stochastic differential equation, all statistical properties of its solution being easily obtained from the corresponding Fokker–Planck equation. Our choice of working example has been, however, also motivated by the far-reaching analogy with the finite-temperature quantum field theory in the stochastic quantization framework.

The main body of this work is organized as follows. In Section 2, we describe the model which is to be solved using the homotopy analysis method and provide main motivation by showing connections of that simple model with the quantum field theory. Section 3 contains a short description of the Liao as well as Marinca approaches to obtain series solutions of the differential equations. In Section 4, such a series is obtained for our model, i.e., a non-linear stochastic differential equation. Section 5 contains several final remarks.

2   The Model

We consider the following non-linear stochastic ordinary differential equation:

$$\frac{d\phi(s)}{ds} = -\alpha\phi(s) - \frac{1}{6}\lambda\phi(s)^3 + f(s), \tag{1}$$

where the stochastic "force" $f(s)$ is a Gaussian and Markovian, and its statistical properties are given by:

$$\langle f(s)\rangle = 0,$$

$$\langle f(s)f(s')\rangle = 2v\delta(s - s'),$$

where v is a real parameter larger than zero, and the sharp brackets denote expectation values.

Physically, the above equation describes heavily overdamped anharmonic oscillations of a Brownian particle. It has been used by Bender and coworkers to introduce strong-coupling expansion in classical statistical mechanics [1]. However, we would like to emphasize here another motivation: Eq.(1) can be obtained as:

$$\frac{d\phi(s)}{ds} = -\frac{dS_E}{d\phi} + f(s), \tag{2}$$

where

$$S_E = \frac{1}{2}\phi a\phi + \frac{1}{4!}\phi^4 \tag{3}$$

is sometimes called the action of a "zero-dimensional" scalar meson field theory with quartic self-interaction. This is because the full four-dimensional action (in imaginary or Euclidean time $\tau = -it$) has the form:

$$S_E = \int d\tau d^3 r \left(\frac{1}{2} \Phi(\mathbf{r}, \tau) L_{KG} \Phi(\mathbf{r}, \tau) + \frac{1}{4!} \Phi(\mathbf{r}, t)^4 \right), \qquad (4)$$

where L_{KG} is the linear Klein–Gordon operator written in the imaginary time.

The fact that all statistical properties of the quantized meson field thery can be obtained from stationary $(s \to \infty)$ solutions of the system:

$$\frac{d\Phi(s)}{ds} = -\frac{\delta S_E}{\delta \Phi} + f(\mathbf{r}, \tau, s), \qquad (5)$$

where $\delta/\delta\Phi$ is the variational derivative, has been established in [19,3,18,6,5]. The transition from (4) to (5) is usually called "stochastic quantization".

By stationary solution of (1) or (5) we mean the solution obtained for large s so that all the initial correlations have died out, and it is sufficient to take into account only the special solution of the inhomogeneous equation.

The fact that (3) is the zero-dimensional caricature of (4) (S becomes merely a simple function instead of being a functional of the field Φ) is evident, and the same is true about (1) and (5). The way any special method is applied to solve (5) is a direct generalization of its application to (1).

Let us now observe that Eq. (1) admits an exact solution in the sense that all moments of ϕ can be easily obtained. Indeed, the corresponding Fokker-Planck equation for the distribution function $\rho(\phi, s)$ takes the form:

$$\frac{\partial \rho(\phi, s)}{\partial s} = v \frac{\partial^2 \rho(\phi, s)}{\partial \phi^2} + \frac{\partial}{\partial \phi} \left[\left(a\phi + \frac{1}{6} \lambda \phi^3 \right) \rho(\phi, s) \right], \qquad (6)$$

which admits the stationary (s-independent) solution:

$$\rho = \rho(\phi) = N \exp \left[-\frac{1}{v} \left(\frac{1}{2} a\phi^2 + \frac{1}{4!} \lambda \phi^4 \right) \right], \qquad (7)$$

where N is a normalization constant obtained from the normalization condition:

$$\int_{-\infty}^{\infty} \rho(\phi) d\phi = 1,$$

so that

$$N = \left[\sqrt{3} \exp \left(\frac{3\alpha^2}{4\lambda v} \right) \sqrt{\frac{\alpha}{\lambda}} K_{1/4} \left(\frac{3\alpha^2}{4\lambda v} \right) \right]^{-1}.$$

This way we obtain, in particular:

$$\langle \phi^2 \rangle_{st} = \frac{e^{-\frac{3\alpha^2}{4\lambda v}} \left(16\sqrt{3}\sqrt{\lambda v} \Gamma \left(\frac{7}{4} \right) \,_1F_1 \left(\frac{3}{4}; \frac{1}{2}; \frac{3\alpha^2}{2\lambda v} \right) - 9\sqrt{2}\alpha \Gamma \left(\frac{1}{4} \right) \,_1F_1 \left(\frac{5}{4}; \frac{3}{2}; \frac{3\alpha^2}{2\lambda v} \right) \right)}{32^{3/4} \sqrt[4]{3} \sqrt{\alpha} \lambda^{3/4} K_{\frac{1}{4}} \left(\frac{3\alpha^2}{4\lambda v} \right)},$$

where the braces $\langle ... \rangle st$ denote stationary expecation value, $\,_1F_1$ is the hypergeometric function, Γ is the Euler Gamma function, and $K_{1/4}$ is the modified Bessel function of the second kind.

Below, we pretend that we do not know the above exact expressions. In order to test whether the homotopy analysis method is suitable for the stochastic differential equations, we shall attempt to solve (1) approximately.

3   Homotopy Analysis Method

Let us consider an arbitrary system of differential equations (linear or non-linear, ordinary or partial, homogeneous or not) of the form

$$\mathcal{N}_a(u, t) = 0, \tag{8}$$

where the index a enumerates the equations of the system, u is the vector of dependent variables, and t represents the set of independent variables. The family of solutions will of course not change if we multiply the left-hand side by a parameter q, $q \neq 0$. Alongside (8) let us also consider a one-parameter family of systems of the form:

$$[(1 - q)\mathcal{L}(u(t) - u^{(0)}(t))]_a + q\mathcal{N}_a(u, t) = 0, \tag{9}$$

where $\mathcal{L}$ is a linear operator such that $\mathcal{L}u = 0$ can easily be solved while $u^{(0)}$ is an initial guess of the solution. Very often $u^{(0)}$ is chosen either to satisfy $\mathcal{L}u^{(0)} = 0$ or simply $u^{(0)} = 0$. Thus, for $q = 0$ we have to do with an easy to solve linear system while for $q = 1$ we obtain the initial system. The idea of Liao has been to solve system (9) by expanding u into a power series with respect to q

$$u = u^{(0)} + qu^{(1)} + q^2 u^{(2)} + \dots$$

and set $q = 1$ at the end of calculations. Obviously, if that were all, we would merely obtain a variant of perturbation expansion with artificial "small" parameter introduced by hand, even though $\mathcal{L}$ need not be contained in $\mathcal{N}$ and $u^{(0)}$ need not be a solution to $\mathcal{L}u = 0$. However, Liao observed that $q\mathcal{N}_a(u, t)$ can still be multiplied by a parameter ξ and function $h(q)$, the latter being analytical in q without changing the solution of (9) for $q = 1$. This gives the system:

$$[(1 - q)\mathcal{L}(u(t) - u^{(0)}(t))]_a + \xi q h(q)\mathcal{N}_a(u, t) = 0. \tag{10}$$

The function $h(q)$ can be chosen in such a way as to, e.g., redefine a "real" expansion parameter (which is not explicitly visible in (8) to make it smaller, while ξ can be chosen at the end of calculations to improve the overall convergence. In fact, the proper choice of the parameter ξ can and usually does improve the quality of approximate solution quite dramatically, and its apparently insignificant presence facilitated the spectacular success of the method. Further considerable progress has been achieved by Marinca and Herisanu who proposed to optimize the choice of the coefficients in the expansion

$$h(q) = h_0 + h_1 q + h_2 q^2 + \dots$$

by minimization of residual error. Even with those amendments, the method, in generic case, suffers from the presence of secular terms if applied to resonant systems so that the expansion is not uniformly valid. However, any known mechanism to eliminate the secular terms can be employed to augment the power of Liao's approach, e.g., the Lindstedt–Poincaré method. Here, we will use a variant of the Lindstedt–Poincaré method by observing that the operator $\mathcal{L}$ can itself be an analytical function of q. This will allow us to eliminate (at least in the sense of expectation values, please see below) the terms which would correspond to the secular terms in ordinary differential equations.

4 Homotopy Analysis Method and Stochastic Differential Equation with Third-Order Nonlinearity

Following Liao, alongside (1) we consider the equation:

$$(1 - q)(\frac{d}{ds} + \nu(q))\phi(q, s) + qc(q)\left[\frac{d\phi(q, s)}{ds} + \alpha\phi(q, s) + \frac{1}{6}\lambda\phi(q, s)^3\right] = 0, \quad (11)$$

in which we expand ϕ, ν, c as:

$$\phi(q, s) = \sum_{n \geq 0} \phi_n q^n,$$

$$\nu(q) = \sum_{n \geq 0} \nu_n q^n,$$

$$c(q) = \sum_{n \geq 0} c_n q^n.$$

However, in order not to overburden the expansions, we will equate all c_n with non-zero n to zero. In the end of calculation, we shall set $q = 1$ and

$$\sum_{n \geq 0} \nu_n q^n|_{q=1} = \alpha. \quad (12)$$

The last condition means that for $\lambda = 0$, $q = 0$, we obtain the "unperturbed", linear problem

$$\frac{d\phi}{ds} = -\alpha\phi + f.$$

Let us notice here that the choice of the constraints in Eq. (12) is only the simplest of many possibilities. In specific cases, other choices may be preferable. As mentioned before, there is no obvious way to get ν_1, ν_2, etc. In the case of deterministic resonance problems, we would choose ν_n, $n \geq 1$ in such a way as to avoid the secular terms. Here, no secular terms appear. Below we propose a scheme to choose ν_n which seems to us simple and intuitive. Another choice is discussed in Section 4.

In the zeroth-order we obtain, of course:

$$\left(\frac{d}{ds} + \nu_0\right)\phi_0(s) = f(s), \tag{13}$$

so that

$$\phi_0(s) = \phi_0(0)e^{-\nu_0 s} + \int_0^s e^{-\nu_0(s-s')}f(s')ds'. \tag{14}$$

For large s we have

$$\phi_0(s) = \int_0^s e^{-\nu_0(s-s')}f(s')ds' \tag{15}$$

so that, for large s, s' the correlation function $\langle\phi_0(s)\phi_0(s')\rangle$ is given by:

$$\langle\phi_0(s)\phi_0(s')\rangle = \frac{v}{\nu_0}\left(\exp(-\nu_0|s-s'|) - \exp(-\nu_0(s+s'))\right), \tag{16}$$

which for $s = s'$ becomes:

$$\langle\phi_0(s)^2\rangle = \frac{v}{\nu_0}\left(1 - \exp(-2\nu_0 s)\right), \tag{17}$$

and, naturally, $\langle\phi^2\rangle_{st} = v/\nu_0$. In the first order we need to solve:

$$\left(\frac{d}{ds} + \nu_0\right)\phi_1(s) + (\nu_1 + c_0(a - \nu_0))\phi_0 + \frac{1}{6}c_0\lambda\phi_0(s)^3 = 0. \tag{18}$$

The solution for large s is given by:

$$\phi_1(s) = -\int_0^s e^{-\nu_0(s-s')}\left[(\nu_1 + c_0(\alpha - \nu_0))\phi_0(s') + \frac{1}{6}c_0\lambda\phi_0(s')^3\right]ds'. \tag{19}$$

In order to obtain ν_1, we require that:

$$\langle\phi_0\phi_1\rangle_{st} = 0.$$

That is, up to the first order in q, the second moment of ϕ is equal to second order of ϕ_0,

$$\langle\phi^2\rangle_{st} = \langle\phi_0^2\rangle_{st}.$$

We have chosen that way to establish ν_1 because it resembles somewhat the method of consecutive elimination of perturbing terms in the near-identity transformation methods in perturbation theory, please see Chapter 5 of [8]. We find without difficulties that

$$\langle\phi_0\phi_1\rangle_{st} = \frac{v}{2\nu_0^2}\left[\nu_1 + c_0(\alpha - \nu_0) + \frac{1}{2}c_0\lambda\frac{v}{\nu_0}\right],$$

and, therefore,

$$\nu_1 = c_0 \left(\nu_0 - \alpha - \frac{1}{2}c_0 \lambda \frac{v}{\nu_0} \right). \tag{20}$$

If we want to risk a seemingly crude approximation and stop the expansion already after the first order, we need to set

$$\nu_0 + \nu_1 = \alpha,$$

which implies the following self-consistent equation for ν_0:

$$\nu_0 = \alpha + \frac{1}{2} \frac{c_0}{c_0 + 1} \lambda \frac{v}{\nu_0}. \tag{21}$$

Its solution gives:

$$\nu_0 = \frac{1}{2} \left(\alpha \pm \sqrt{\alpha^2 + 2 \frac{c_0}{c_0 + 1} \lambda v} \right). \tag{22}$$

We need to take the "+" before the square root; otherwise ν_0 would be smaller than zero.

The question now appears how we can choose c_0. We could, in principle, follow the route known from the linear delta expansion method [4]. This method is inherently variational in its nature, containing usually an artificial parameter, say, μ, which, roughly speaking, plays the role analogous to c_0. One imposes there the reasonable condition that one of the physical quantities of interest (let us call it F) should exhibit "minimal sensitivity" for the change of μ. That is, one should have $\partial F / \partial \mu = 0$. If we were to follow that route, our quantity of interest, namely $\langle \phi^2 \rangle$ would have to have a vanishing derivative with respect to c_0. However, this cannot be the case, as the above derivative never vanishes. We could either simply proceed to the second order and then check the derivatives, or try to establish another way to find c_0. A very attractive technique to establish c_0 is that proposed by Marinca and coworkers; it consists in minimizing the "residual error" in ϕ_1. Let us observe here, however, that equally well one can impose conditions based on other error-related criteria, similarily as in the statistical linearization [7,2,20,21]. While we appreciate valuable insights of Marinca and coworkers, in this work we shall use the known fact that in the quantum field theory and statistical physics, the so-called strong coupling limit is very often available in addition to the weak-coupling one. In our language, this means that both limits $\lambda / \alpha \to 0$ and $\alpha / \lambda \to 0$ are available. In our simple system, this is of course the case, and we have in the limit $\alpha / \lambda \to 0$:

$$\langle \phi^2 \rangle_{st} = \frac{2\sqrt{6}\Gamma(\frac{3}{4})}{\Gamma(\frac{1}{4})} \sqrt{\frac{v}{\lambda}}, \tag{23}$$

where $\Gamma(x)$ is the Euler Gamma function. On the other hand, our approximate $\langle \phi^2 \rangle$ gives, for $\alpha = 0$, $\sqrt{2(c_0 + 1)} / \sqrt{c_0} \sqrt{v/l}$. Hence

$$c_0 = \frac{\Gamma(\frac{1}{4})^2}{12\Gamma(\frac{3}{4})^2 - \Gamma(\frac{1}{4})^2} \approx 2.6965826\ldots$$

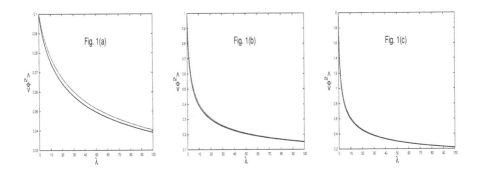

Fig. 1. Comparison of exact and first-order approximate results for the dependence of the second moment of the variable ϕ on λ for three different values of v; (a) $v = 0.1$; (b) $v = 1.0$; (c) $v = 2.0$. The parameter α is set to 1. The solid line represents exact solution, and the dotted line - the approximate one.

A comparison of exact and approximate first-order results with the above c_0 is contained in Fig. 1. The agreement seems quite spectacular if one takes into account the very low order of approximation.

Let us observe that for v of the order of 1 or larger, the exact and first-order approximate solutions are almost indistinguishable. Predictably, the results for the fourth moments are less impressive, but still remarkable, as shown in Fig. 2.

The results for the fourth moments vary from only qualitatively correct for small values of v to good agreement with exact values for $v > \alpha$. Let us proceed to the second order. Now, we need to solve:

$$\left(\frac{d}{ds} + \nu_0\right)\phi_2(s) = (\nu_1 + c_0(\alpha - \nu_0))\phi_1(s) + \tag{24}$$

$$+ (\nu_2 - c_0\nu_1 + c_0(1 - c_0)(\alpha - \nu_0))\phi_0(s) +$$

$$+ \frac{1}{6}c_0(1 - c_0)\lambda\phi_0(s)^3 + \frac{1}{2}c_0\lambda\phi_0(s)^2\phi_1(s).$$

The solution is again quite simple:

$$\phi_2(s) = -\int_0^s e^{-\nu_0(s-s')}\left[(\nu_1 + c_0(\alpha - \nu_0))\phi_1(s') + \tag{25}\right.$$

$$+ (\nu_2 + c_0(\alpha - \nu_0) - c_0(\nu_1 + c_0(\alpha - \nu_0)))\phi_0(s') +$$

$$\left. + \frac{1}{6}c_0(1 - c_0)\lambda\phi_0(s')^3 + \frac{1}{2}c_0\lambda\phi_0(s')^2\phi_1(s')\right]ds'.$$

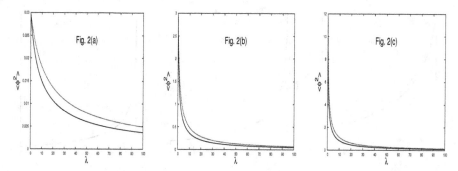

Fig. 2. Comparison of exact and first-order approximate results for the dependence of the fourth moment of the variable ϕ on λ for three different values of v; (a) $v = 0.1$; (b) $v = 1.0$; (c) $v = 2.0$. The parameter α is set to 1. The solid line represents exact solution, and the dotted line - the approximate ones

We require that:

$$\langle \phi^2 \rangle_{st} \approx \langle \phi_0^2 \rangle_{st}, \tag{26}$$

which means that

$$2\langle \phi_0 \phi_2 \rangle_{st} + \langle \phi_1^2 \rangle_{st} = 0.$$

The following algebra is simple but somewhat boring because ϕ_1 and ϕ_2 are no longer Gaussian random variables even though ϕ_0 is. It leads to the following self-consistent equations for ν_0:

$$\nu_0 = \alpha + \frac{c_0}{2c_0 + 1} \frac{v\lambda}{\nu_0} - \frac{1}{6} c_0^2 \lambda^2 \frac{v^2}{\nu_0^3}. \tag{27}$$

All higher-order moment can, of course, be expressed with the help of v and ν_0 because ϕ_0 is Gaussian. Now, the following difficulty appears. If we try to impose the condition: $\nu_0 + \nu_1 + \nu_2 = \alpha$, which is suitable in the second order, we are confronted with the algebraic equation of the fourth degree. While we of course can solve it, it is by no means clear whether the solutions contain a subset of real ones for sufficiently broad range of the parameter c_0. It is actually *not* the case for Eq. (27). In addition, it is not possible to reach the limit $\lambda/\alpha \to \infty$ in the sense that there are no real solutions for c_0. Therefore, we have adopted the following approach. The function of ν_0 which stands on the right-hand side of (27) can be understood as consisting just first few terms of a series in powers of ν_0^{-1}. We attempted to sum the series using the Pade $[1/2]$ approximant. The result is

$$\nu_0 = \frac{3 c_0{}^2 l \nu_0 v + \left(6 \alpha c_0{}^2 + 3 \alpha c_0\right) \nu_0{}^2 + \left(-4 \alpha^2 c_0{}^2 - 4 \alpha^2 c_0 - \alpha^2\right) \nu_0}{\left(2 c_0{}^2 + c_0\right) l v + \left(6 c_0{}^2 + 3 c_0\right) \nu_0{}^2 + \left(-4 \alpha c_0{}^2 - 4 \alpha c_0 - \alpha\right) \nu_0}. \tag{28}$$

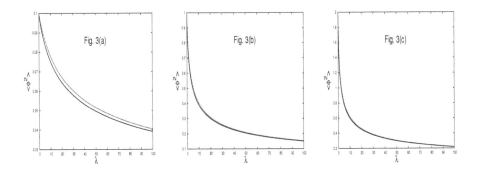

Fig. 3. Comparison of exact and second-order approximate results for the dependence of the second moment of the variable ϕ on λ for three different values of v; (a) $v = 0.1$; (b) $v = 1.0$; (c) $v = 2.0$. The parameter α is set to 1. The solid line represents exact solution, and the dotted line - the approximate one

This way we obtain for ν_0 an algebraic equation of the *third* degree. What is more, one solution is trivial, and it is fairly easy to pick up the proper one from the remaining two (of course, it is to be positive). Then the strong-coupling limit is also achievable. Interestingly, we have found a *negative* value of c_0, namely, -1.762159205046485, as that for which the limit of large λ for the second moment is obtained. The resulting dependence of the second moments on λ is shown in Fig. 3.

5   Concluding Remarks

In this paper, we have applied the homotopy analysis method to a nonlinear stochastic differential equation with Gaussian–Markovian stochastic.

The value of an artificial parameter the presence of which is characteristic for the method has been fixed by taking the limit of strong nonlinearity. Approximate solutions have been obtained and compared with the exact ones. It has been demonstrated that the second moments of the dependent stochastic variables as given by the homotopy analysis method agree very remarkably with exact moments. Broad perspectives of applications for the method seem to be open; for, despite the simplicity of the considered model, it possesses some characteristics of the stochastic differential equations of statistical mechanics and quantum field theory in the so-called stochastic quantization. As far as we know, there exist no other attempts to solve non-linear stochastic differential equations with the help of homotopy analysis method.

We would like to stress that the computer algebra systems appear to be ideally suited to enhance capabilities of both standard perturbation expansion and homotopy-related methods. In fact, the calculations by Liao mentioned above heavily rely on his use of Mathematica notebooks. In this paper, we have used

Mathematica, too, to obtain the explicit expressions which follow Eq. 10 as well as those in Eq. (23). However, it is the Maxima package which has been been most important for us. Indeed, writing the expansion in terms of q, gathering the terms, calculation of ν_0 in the first and second orders, writing down the Pade approximation and calculation of c_0 have all been done within Maxima. Last but not least, the gnuplot-made figures have also been obtained from within the Maxima environment.

Work is in progress on application of both those techniques in the physics of cold atomic gases and quantum optics.

References

1. Bender, C., Cooper, F., Guralnik, G., Rose, H.A., Sharp, D.H.: Strong coupling expansion for classical statistical dynamics. J. Stat. Phys. **22**, 647–660 (1980)
2. Caughey, T.K.: Equivalent linearization techniques. J. Acoust. Soc. Am. **35**, 1706–1711 (1963)
3. Damgaard, P.H., Hüffel, H.: Stochastic quantization. Phys. Rep. **152**, 227–398 (1987)
4. Farias, R.L.S., Krein, G., Ramos, R.O.: Applicability of the linear expansion for the field theory at finite temperature in the symmetric and broken phases. Phys. Rev. D **78**, 065046 (2008)
5. Gozzi, E.: Langevin simulation in Minkowski space. Phys. Lett. B **150**, 119–124 (1985)
6. Hüffel, H., Rumpf, H.: Stochastic quantization in Minkowski space. Phys. Lett. B **148**, 104–110 (1984)
7. Kazakov, I.E.: Approximate Methods for the Statistical Analysis of Nonlinear Systems. Trudy VVIA 394 (1954)
8. Kevorkian, J., Cole, J.D.: Multiple Scale and Singular Perturbation Method. Springer, Berlin (1996)
9. Liao, S.J.: An explicit, totally analytic approximate solution for Blasius viscous flow problems. Int. J. Non-Linear Mech. **34**, 759–778 (1999)
10. Liao, S.J.: An optimal homotopy-analysis approach for strongly nonlinear differential equations. Commun. Nonlinear Sci. Numer. Simul. **14**, 983–997 (2009)
11. Liao, S.J.: Beyond perturbation: introduction to the homotopy analysis method. Chapman and Hall/CRC Press, Boca Raton (2003)
12. Liao, S.J.: On the homotopy analysis method for nonlinear problems. Appl. Math. Comput. **147**, 499–513 (2004)
13. Liao, S.J., Tan, Y.: A general approach to obtain series solutions of nonlinear differential equations. Stud. Appl. Math. **119**, 297–355 (2007)
14. Liao, S.J.: The Proposed Homotopy Analysis Technique for the Solution of Nonlinear Problems. Thesis, Shanghai Jiao Tong University, Shanghai (1992)
15. Marinca, V., Herisanu, N.: Application of optimal homotopy asymptotic method for solving nonlinear equations arising in heat transfer. Int. Commun. Heat Mass Transfer **35**, 710–715 (2008)
16. Marinca, V., Herisanu, N., Bota, C., Marinca, B.: An optimal homotopy asymptotic method applied to the steady flow of a fourth-grade fluid past a porous plate. Appl. Math. Lett. **22**, 245–251 (2009)
17. Marinca, V., Herisanu, N., Nemes, I.: Optimal homotopy asymptotic method with application to thin film flow. Central Eur. J. Phys. **6**, 648–653 (2008)

18. Namiki, M.: Stochastic Quantization. Lect. Notes Phys., vol. m 9, X, 217 pages. Springer, Berlin (1992)
19. Parisi, G., Wu, Y.-S.: Perturbation theory without gauge fixing. Sci. Sinica **24**, 483–496 (1981)
20. Roberts, J.B., Spanos, P.D.: Random Vibrations and Statistical Linearization. Dover, Mineola (2003)
21. Socha, L.: Linearization Methods for Stochastic Dynamic Systems. Springer, Berlin (2007)

On the Topology and Visualization of Plane Algebraic Curves

Kai Jin[1], Jin-San Cheng[2], and Xiao-Shan Gao[2]

[1] School of Mathematics and Statistics, Central China Normal University,
Wuhan, China
[2] Key Lab of Mathematics Mechanization, AMSS, CAS, Beijing, China.
{jinkai,jcheng}@amss.ac.cn, xgao@mmrc.iss.ac.cn

Abstract. In this paper, we present a symbolic algorithm to compute the topology of a plane curve. The algorithm mainly involves resultant computations and real root isolation for univariate polynomials. The novelty of this paper is that we use a technique of interval polynomials to solve the system $\{f(\alpha, y) = \frac{\partial f}{\partial y}(\alpha, y) = 0\}$ and at the same time, get the simple roots of $f(\alpha, y) = 0$ on the α fiber. It greatly improves the efficiency of the lifting step since we need not compute the simple roots of $f(\alpha, y) = 0$ any more. After the topology is computed, we use a revised Newton's method to compute the visualization of the plane algebraic curve. We ensure that the meshing is topologically correct. Many nontrivial examples show our implementation works well.

Keywords: Plane curve, topology, interval polynomial, visualization, root candidate.

1 Introduction

The determination of the topology of a given algebraic plane curve and the computation of line segments which approximate the graph of the curve are basic operations in computer graphics and geometric modeling. A visualization of a plane curve could be used to display the curve correctly, and is the foundation for further displaying space curves and surfaces. We consider an implicit plane curve defined by $f(x, y) = 0$, denoted by $C_f = \{(x, y) \in \mathbb{R}^2 | f(x, y) = 0\}$, where $f(x, y)$ is a square-free polynomial with rational coefficients. Typically, the topology of C_f is given in terms of a plane graph G embedded in $\mathbb{R}^2$ that is ambient isotopic to C_f. For a geometric-topological analysis, we further require the vertices of G to be located on C_f (ignoring the representation error). There exists a large amount of work on computing the topology of plane algebraic curves (see [2, 10, 15, 17, 18, 21, 27, 30, 36, 40] and references therein). As far as we know, there are mainly two types of algorithms among these references.

The first type of algorithm is based on symbolic computation. Most symbolic methods proceed in three steps: (a) Projection step: Compute $R(x)$, the resultant of f and f_y with respect to y, and determine its real roots α_i, $i = 1, \ldots, l_0$. (b) Lifting step: For every α_i, compute the real roots of $f(\alpha_i, y) = 0$, i.e., the set

© Springer International Publishing Switzerland 2015
V.P. Gerdt et al. (Eds.): CASC 2015, LNCS 9301, pp. 245–259, 2015.
DOI: 10.1007/978-3-319-24021-3_19

$P_i = \{p_{i,1}, \ldots, p_{i,l_i}\}$. (c) Connection step: For each i and each $p \in P_i$, determine the left branch number and right branch number of p and connect the points related to P_i. Among the algorithms based on symbolic methods, there are two different classes.

The first class is based on solving a triangular system $\{R(x) = f(x, y) = 0\}$ (see [5–7, 17, 18, 21] and references therein). The classical example of this algorithm is the cylindrical algebraic decomposition (CAD) approach (see [5–7, 12, 17, 18, 20, 21, 27] and references therein). It divides the plane into cylindrical cells such that the given polynomial $f(x, y)$ has the same sign on each cell. Then, to determine the topology of the curve, one merely needs to give the adjacency information between adjacent cells [4, 6, 7]. The CAD-based methods need to solve a system of $O(n^3)$ roots, where n is the degree of $f(x, y)$. The second class is based on solving the system $\{f(x, y) = \frac{\partial f}{\partial y}(x, y) = 0\}$ which has $O(n^2)$ roots (see [10, 15, 30, 34, 40] and references therein).

In [30], González-Vega and Necula use Sturm-Habicht sequences to express the y-coordinate of each x-critical point (see the definition of x-critical points in Section 2.1) as a rational function of its x-coordinate. When the curve is in generic position, it is easy to determine the connection relationships without computing the numbers of left and right branches since there is only one x-critical point on each fiber. Additionally, there exist algorithms that handle curves that are not in generic position [34]. This approach, however, requires substantial additional time-consuming symbolic computations such as computing Sturm sequences.

In [40], Seidel and Wolpert present an alternate approach to compute the critical points which avoids the most costly algebraic computation, but it computes several projections of the critical points. First, the authors project the x-critical points onto the x and y-axes, respectively, which leads to a set of boxes containing the solutions of the system $\{f(x, y) = \frac{\partial f}{\partial y}(x, y) = 0\}$. Then, the authors project the x-critical points on a third random axis to discard the redundant boxes. In their algorithm, all computations involve only rational numbers, but the authors still need to compute Sturm-Habicht sequences for refining the boxes containing the x-critical points until each box intersects only with the branches originating from the x-critical point. Their approach assumes that the curve is in generic position by a pre-processing phase in which the curve is sheared, if needed.

More recently, Berberich *et al.* [10] present a certified algorithm to compute the topology of plane curves. They project the x-critical points on both x and y-axes to get a set of boxes. Instead of taking a third projection as in Seidel 2005 [40], they compute a bound on the boundary of each box. Based on this bound, they give a criterion to test whether a box certainly contains a real solution of the system $\{f(x, y) = \frac{\partial f}{\partial y}(x, y) = 0\}$. Furthermore, they also present an efficient, numerical method to solve the system $\{f(x, y) = \frac{\partial f}{\partial y}(x, y) = 0\}$. If the numerical method fails in some special geometric situation, they reuse the first algorithm to keep the certification of the algorithm.

The other type of algorithms are numerical algorithms which rely on subdivision of the original domain. The famous example is the family of marching cube algorithms (Lorensen and Cline [29]). This approach avoids the expensive algebraic computations of computing the x-critical points. The method does not give any guarantee on the topological correctness of its output since the method fails if singular points exist in the domain. But it has inspired many algorithms [2, 11, 13, 22, 28, 31, 36, 37]. In [36], Plantinga and Vegter give an efficient numerical subdivision algorithm that is certified when the curve C_f is bounded and has no singular points. They use a computational model that relies only on function evaluation and interval arithmetic. An excellent survey of numerical methods based on interval arithmetic was given in [31]. In order to get the correct output, the authors in [11] and [2] use evaluation bounds and the topological degree, respectively, to guarantee the correct topology near singular points. These methods are very fast in the regular domain, however, they become very slow near singular points of the curve, if certification is required.

In our paper, a CAD based method is presented which is a combination of the above two symbolic algorithms. Explicitly, we first use the efficient algorithm in [14] with a little modification to solve the system $\{f(x, y) = \frac{\partial f(x,y)}{\partial y} = 0\}$, but, meanwhile, we get many simple roots of $f(\alpha, y) = 0$ in the sense that the simple roots of $f(\alpha, y) = 0$ are either isolated or contained in the isolating intervals of the multiple roots of $f(\alpha, y) = 0$. The method of our algorithm is very efficient since it involves only resultant computations and real root isolation of univariate polynomials. We compute the number of branches of each point on the α fiber by the properties of multiplicity on the point. The idea is not new (e.g. see [10, 15, 40]).

After the topology is obtained, we also compute a meshing of the curve. We separate the curve into regular curve segments then trace the curve segments by a revised Newton's method which guarantees the efficiency and correctness. Different from our pure numerical method, the authors of the papers [20, 25, 35] approximate regular curve segments with splines.

We implement the algorithm in Maple and compare our method with some efficient methods. The visualization of plane curves with our software show a good result.

The rest of paper is organized as follows: In Section 2, necessary preparations and notations are given. In Section 3, we give an algorithm to compute the topology of an algebraic plane curve. An isotopic meshing for the plane curve is computed in Section 4 using an improved Newton's method. Many experiments are presented in Section 5, and finally, we conclude the paper in Section 6.

2 Topology Computation for Plane Algebraic Curves

Let $\mathbb{Q}, \mathbb{R}, \mathbb{C}$ be the fields of rational, real and complex numbers, respectively. For $\forall f \in \mathbb{Q}[x,y]$, let C_f denote the plane real algebraic curve defined by the polynomial $f(x,y) \in \mathbb{Q}[x,y]$.

Given a square-free polynomial $f(x, y) \in \mathbb{Q}[x, y]$, we assume that C_f has no vertical lines. If $f = f^*(x, y)t(x)$, we first remove $t(x)$ and analyse the curve $C' = C_{f/t}$, then, we merge the vertical lines defined by $t = 0$ into the graph of C'.

Hence, in the reminder of the paper, we always assume that $\mathcal{C}_f$ has no vertical lines. We have the following main steps to compute the topology of $\mathcal{C}_f$. Because of the page limitation of the paper, we just give an outline of the steps.

The novelty of the algorithm presented below is we use the steps and properties of our techniques in solving bivariate systems to service for our topology computation algorithm. For example, we directly get simple zeros of $f(\alpha, y) = 0$ on the fiber $x = \alpha$ in Section 2.1. The multiplicities of the zeros of the given system can be use to separate the multiple zeros of $f(\alpha, y) = 0$ on the fiber $x = \alpha$ in Section 2.2.

2.1   Solving the System $\{f = f_y = 0\}$

Denote $h(x) = \text{Res}_y(f, f_y)$. We compute the real roots of the system with Algorithm 2 in [14]. Let $\alpha \in \mathbb{I}$ be a real root of $h(x) = 0$ and $\mathbb{I}$ its isolating interval. For the real roots of $f(\mathbb{I}, y) = 0$, some intervals, derived by Algorithm 1 in [14], we will not remove it even if they have no intersection with that of $f_y(\mathbb{I}, y) = 0$. We use the monotonous property (Lemma 4 in [14]) to certify some simple zeros of $f(\alpha, y) = 0$. Note that there may exist some simple zeros which is contained in the intersection of the real roots of $f(\mathbb{I}, y) = 0$ and $f_y(\mathbb{I}, y) = 0$. We will split them in the next step.

2.2   Computing the Fibers

The aim of this step is to separate the roots of $f(\alpha, y) = 0$. If we know the multiplicity of $f(\alpha, y) = 0$ at its root β, denoted as $\text{mult}(f(\alpha, y), \beta) = k$, we get $\frac{\partial^k f}{\partial y^k}(\alpha, \beta) \neq 0$ and $\frac{\partial^j f}{\partial y^j}(\alpha, \beta) = 0 (0 \leq j \leq k - 1)$. Then we can separate any multiple root β of $f(\alpha, y) = 0$ from its simple roots in an interval by refining the interval. Let $Int(p, q, z)$ denote the intersection multiplicities of two bivariate polynomials p, q at point z. From [19, 41], we have

$$\text{mult}(f(\alpha, y), \beta) = Int(f, f_y, p) - Int(f_x, f_y, p) + 1, p = (\alpha, \beta). \qquad (1)$$

Since the algorithm in [14] can get the intersection multiplicity of a root of a bivariate system, we can get $\text{mult}(f(\alpha, y), \beta)$ by solving the systems $\{f, f_y\}$ and $\{f_x, f_y\}$.

2.3   Branch Number Computation and Connection

In order to compute the topology of $\mathcal{C}_f$, we have to determine the branch number for each x-critical point computed before. First, we introduce the definition of the segregating box, which is similar as local box in [13] and some related references.

A box $\mathbf{B} = [a, b] \times [c, d]$ is called **segregating** w.r.t. $\mathcal{C}_f$ if

$$\mathcal{C}_f \cap [a, b] \times [c, c] = \mathcal{C}_f \cap [a, b] \times [d, d] = \emptyset.$$

If each point on the fiber α is segregating, we can count the number of real roots of $f(a, y) = 0$ $(f(b, y) = 0)$ in $[c, d]$. It is the number of left (right) branches of (α, β). Note that we can set the isolating interval for α of all the roots of $f(\alpha, y)$ as the same, thus we can count them once. If α is rational and $a = b$, we can set a small number $\epsilon > 0$ such that $[\alpha - \epsilon, \alpha - \epsilon]$ is an isolating interval for α. Connect all the points on the fibers, we can get the topological graph of a given plane curve.

3   Isotopic Meshing for a Plane Curve

In this section, we use a modified Newton's method to approximate the plane curve $\mathcal{C}_f$.

We say two graphs $\mathcal{G}_1$ and $\mathcal{G}_2$ are **isotopic** if there exists a continuous mapping $\gamma : \mathbb{R}^2 \times [0,1] \rightarrow \mathbb{R}^2$, such that for any fixed $t \in [0,1]$, $\gamma(\cdot, t)$ is a homeomorphism from $\mathbb{R}^2$ to itself, and γ continuously deforms $\mathcal{G}_1$ into $\mathcal{G}_2$: $\gamma(\mathcal{G}_1, 0) = \mathcal{G}_1$, $\gamma(\mathcal{G}_1, 1) = \mathcal{G}_2$.

For an arbitrary $\epsilon > 0$, an ϵ-**meshing**, for $\mathcal{C}$ is an isotopic meshing $\mathcal{G}$ for $\mathcal{C}$, which gives an ϵ-approximation for $\mathcal{C}$ in the sense $\| P - \gamma(P, 1) \| \leq \epsilon$ for all $P \in \mathcal{G}$.

A key difficulty for tracing a plane curve is to make the approximate curve isotopic to the curve, especially at the neighborhood of isolated singularities or complicated ones with high multiplicity of tangency. We first compute the topology of $\mathcal{C}_f$, then an efficient numerical method is used to approximate the regular monotonous curve segments. At first, we divide $\mathcal{C}_f$ into monotonous curve segments. Let

$$H(x) = \mathrm{sqrfree}(\mathrm{Res}_y(f, f_x)), \ h(x) = \frac{H(x)}{\gcd(H(x), r(x))},$$

where $r(x) = \mathrm{sqrfree}(\mathrm{Res}_y(f, f_y))$. Assume that $\Sigma = \{h(x), f(x, y)\}$, and Σ is a triangular system, but we do not need to use general algorithms to solve Σ because of the fact that each solution of Σ is a simple point of $\mathcal{C}_f$. Hence, to solve Σ, we use consecutive real roots isolation for univariate rational polynomials. Then, we insert the solutions of Σ into the topological graph of $\mathcal{C}_f$. Thus, the curve $\mathcal{C}_f$ is divided into monotonous curve segments and what we need to do is to visualize monotonous curve segments.

3.1   Tracing Regular Curve Segments

In this part, we show how to use a modified Newton's method to trace a regular monotonous curve segment. Newton's method is a classic and efficient method for solving nonlinear equations, but it is a local method. Thus, the choice of an initial value is crucial when using Newton's method. If the initial value is not proper, it may occur that the iteration either will not converge, or converge to another point, see Figure 3.

Based on the analysis above, The computation of reliable solutions is crucial when using Newton's method. A simple idea is to get an interval I in advance which contains the exact root x^*, if the final convergence value $\bar{x} \in I$, then we regard $\bar{x}$ as the exact root x^* (ignoring the floating point representation error); Otherwise, we abandon $\bar{x}$ and take another better initial value continuing to approximate the curve.

According to the preparation above, we can explain how to trace a curve segment. Consider a regular curve segment $C_1 = \widehat{AB}$ as in Figure 1 (similar to the projection in x direction in Figure 2 and other cases). Now that the coordinates of A are given, what we have to do is to sample a finite number of points on the curve segment, then connect each pair of adjacent points with line segments. We illustrate just one step to explain this process. That is, finding approximate coordinates for B. First, we step a fixed stepsize Δx along the **positive tangent**

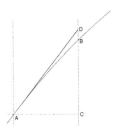
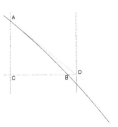

Fig. 1. Approximation in y-direction   **Fig. 2.** Approximation in x-direction

direction (that is, the first component of the tangent vector is positive) of $A = (x_A, y_A)$ to $D = (x_D, y_D)$, then we sketch a vertical line through D and a horizontal line through A, these two straight lines intersect at $C = (x_C, y_C)$. So we have the x-coordinate of B, leaving y_B undetermined, then we consider the solution of the univariate polynomial equation $g(y) = f(x_C, y) = 0$ in the interval $[y_C, y_D]$. We use Newton's method to compute y_B with the initial value $y_0 = y_D$. If the convergence value $\bar{y} \in [y_C, y_D]$, then we let $y_B = \bar{y}$; otherwise, it means that the initial value is not well enough, and we should choose a better initial value y_{CD}, it can be obtained from the following expression:

$$y_{CD} = y_C - \frac{g(y_C)(y_C - y_D)}{g(y_C) - g(y_D)} = \frac{g(y_D)y_C - g(y_C)y_D}{g(y_D) - g(y_C)} \qquad (2)$$

The geometric meaning of the expression above is illustrated in Figure 4. Where $y_C = Y_1$, $y_D = Y_2$, $y_{CD} = Y_{1,2}$. The new isolating interval for y_B is $[Y_1, Y_{1,2}]$, if $g(Y_1)g(Y_{1,2}) < 0$; and the interval is updated to $[Y_{1,2}, Y_2]$, if $g(Y_2)g(Y_{1,2}) < 0$. In the case of Figure 4, the new interval is $[Y_1, Y_{1,2}]$. Actually, we ignore the trivial case that the value of Formula (2) equals zero, because when this case occurs, we have already obtained the approximate value of y_B. Repeat above process until we get the coordinates of B.

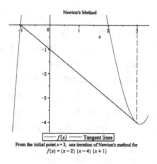

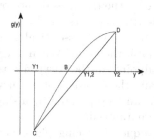

Fig. 3. The expected solution is $x = 2$, the process gets convergence after just one iteration, that is $x_1 = x_0 - \frac{f(x_0)}{f'(x_0)} = 3 - \frac{-4}{-1} = -1$, since $f(-1) = 0$. Hence, the method fails.

Fig. 4. Computing new initial point

In order to guarantee the meshing of the curve is isotopic, a key ingredient is that the isolating interval for undetermined coordinate contains exactly one real zero of the univariate polynomial. When the step size along the positive tangent direction is too big, it may happen that an interval contains two or more solutions of $g(y)$. See Figure 5, where the line segment B_2D_2 intersects to each point with curve segments C_2, C_3 of $\mathcal{C}$. So in this case, we have to decrease step size to make sure that triangles $\triangle A_iB_iD_i$, $i = 1, 2, 3$ don't meet (see Figure 6). Ensuring tracing the curve inside the triangles guarantees the topology of the curve for this case. This rule can avoid branch jumping for most of time.

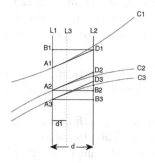

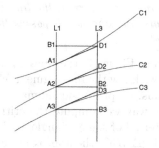

Fig. 5. One isolating interval contains several curve segments whose triangles intersect, in this case the Newton iteration may not be guaranteed.

Fig. 6. The related triangles of the curve segments are disjoint, the tracing is topology guaranteed.

In some case, it may happen that one triangle may contain two or more curve segments even if the triangles do not overlap. Since the curve C_f has no singular points or x-critical points in the domain between two adjacent fibers. Thus, when a singular point or x-critical point occurs, we know the number of branch originating form the point. When tracing the branches, we ensure that for each step, the endpoints of the Newton's iteration for each branch have the same x-coordinate and the approximation curve segments are disjoint in each step until each branch can be separated by a single triangle. Thus the meshing in the end is topology guaranteed.

The technique of tracing plane curve segments is much like the technique of homotopy continuation method (see [9], [42] and references therein). But our method focuses on the real components of plane curves. Furthermore, our method guarantees the topology of the curve and it is not difficult to implement.

3.2   Error Control for the Meshing of a Plane Curve

We show the error control for the plane approximate curve in this subsection. In geometry, the approximation error should be defined as the Hausdorff distance between the segment S and its approximation S_a,

$$e(S, S_a) = \mathrm{dis}(S, S_a) = \max_{P \in S} \min_{p' \in S_a} d(P, P').$$

However, such a distance is difficult to compute. In this paper, we compute an ϵ meshing for the plane curve segments, and, ϵ is an upper bound of the Hausdorff distance between the original curve and the approximate curve. First, we give the following result:

Proposition 1. *For any triangle ΔABC and a vector direction $\mathbf{n} = (n_1, n_2)$, consider all the lines which are parallel to $\mathbf{n}$ and have intersection points with ΔABC, each of those lines is cut into a line segment by the triangle ΔABC (we did not consider the degenerate case that the lines meet the triangle ΔABC at only one point, see Figure 7, the line L_6 is the degenerate case), among all the line segments, the one which passes through a vertex of ΔABC is the longest line segment.*

Proof: From Figure 7, the lines L_i, $i = 1, \ldots, 6$ are parallel to $\mathbf{n} = (n_1, n_2)$. Obviously, we know the line segment $\overline{M_3 N_3}$ is the longest one among $\overline{M_j N_j}$, $j = 1, \ldots, 5$. Thus, we prove the lemma.   □

From the way we compute the critical points as before, each curve segment is in the internal of some triangle as shown in Figure 8. The blue original curve segment is in ΔABC, the approximate curve segment is the line segment $\overline{ADC}$. According to Proposition 1, the length of $\overline{BD}$ is the longest line segments among all of them, which is absolutely an upper bound of the Hausdorff distance between the original and approximate curve segments. All the plane curve segments can be handled like this, and we use the largest error to denote the global error for the entire curve. Thus, we get the upper bound for the approximation curve although the bound can be greatly improved.

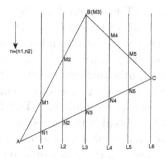

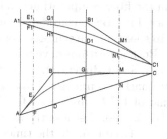

Fig. 7. The explanation of Lemma 1: BN_3 is the longest line segments under the direction n and intersect the triangle $\triangle ABC$

Fig. 8. The error control of the approximation: the error of the curve segment $AC(A_1C_1)$ is bounded by $|BD|$ (resp. $|B_1D_1|$)

4 Experiments

In this section, we list many nontrivial examples and the results show that our algorithm works well. We implemented our algorithm in Maple and tested the examples in Maple 15 on a PC with Intel(R)Core(TM)i3-2100 CPU @3.10GHz 3.10GHz, 2G memory and Windows 7 operating system.

Table 1. The computing time of random dense polynomials (in seconds). f^* is a dense polynomial f^* of degree 70 with coefficients ranging from -5 to 5. It has 2335 terms. This almost reaches the limit of the computational ability of our implementation.

degree	5	10	15	20	25	30	35	40	45	50	55	70(f^*)
T_1	0.0047	0.028	0.11	0.36	0.91	1.98	3.87	7.02	12.08	19.77	30.94	67.44
T_2	0.0031	0.0125	0.088	0.344	0.77	1.87	3.89	7.03	10.31	20.65	30.53	90.21
T	0.047	0.33	0.82	2.25	4.64	10.92	20.35	35.9	66.75	103.64	195.90	647.09

In Table 1, we consider random dense polynomials with coefficients ranging from -100 to 100. The first row is the degrees of polynomials. The second is the times in seconds of computing $R(x) = \text{Res}_y(f, f_y)$, denoted as T_1, the third is the times for isolating the real roots of $R(x)$, denoted as T_2. These two operations are the basic symbolic computation for many algorithms based on elimination technique. The last row is the total times (T) for computing the topology. For each column, the entries are averaged with 10 random polynomials. In Table 2 we also compute random polynomials in [10] and compare the times of our algorithm (LUR-top) with the algorithms GEOTOP-BS and GEOTOP in [10]. GEOTOP-BS exclusively uses LIFT-BS[1] for the fiber lifting. GEOTOP combines LIFT-NT and

---

[1] LIFT-BS and LIFT-NT are two methods in lifting process, the first is a symbolic and complete approach while the second a numeric but not certified one, for more details, please refer to [10].

LIFT-BS in the fiber computations and it uses LIFT-NT first, and if it fails for a certain fiber, LIFT-BS is considered for this fiber. Both of these two algorithms use GPU to speed up the symbolic computation, such as resultant and gcd computations. Both algorithms have to compute at least two resultants (projections in x-direction and y-direction), while LUR-top computes just one resultant for these benchmarks, in practice. From Table 2, we can see that most of the computing time of LUR-top is used for T_1 and T_2 operations while GEOTOP-BS and GEOTOP are not, because the authors use GPU to deal with these symbolic computations. Even though, the times of LUR-top in our machine are less than that of GEOTOP-BS in their machine when the bits equals 10. When bits equals to 512, LUR-top becomes slow due to the low efficiency of the symbolic computation in Maple. GEOTOP-BS and GEOTOP are two two algorithms in [10] for computing the topology of the curve, the corresponding columns are the times in their paper (their code is not available). Moreover they outsource many symbolic computations to the graphics hardware to reduce the computing time. For the GPU-part of the algorithm, they use the GeForce GTX580 Graphics card (Fermi Core). We implement our code in Maple and illustrate the times for the same benchmarks, where T_1 denotes the total time for computing $\mathrm{Res}_y(f, f_y)$, and T_2 denotes total time for isolating the real zeros of $\mathrm{Res}_y(f, f_y)$ which have the same meaning to Table 1. T denotes the total time for computing the topology of the curves.

Table 2. Timings for analysing the topology of five random curves from [10].

			LUR-top		
			T_1	T_2	T
degree, bits	GEOTOP-BS	GEOTOP			
Machine	Linux platform on a 2.8 GHz 8-Core Inter Xeon W3530 with 8MB of L2 cache		Win 7 on 3.1GHz dual Core i3-2100 CPU with 256KB of L2 cache		
Code language	C++		Maple language		
GPU speedup	YES		NO		

Sets of five random dense curves

degree, bits	GEOTOP-BS	GEOTOP	T_1	T_2	T
06, 10	0.71	0.14	0.046	0.016	0.655
06, 512	0.15	0.29	0.219	0.047	0.874
09, 10	1.50	0.23	0.124	0.016	1.155
09, 512	2.38	0.57	2.012	0.156	3.166
12, 10	4.54	0.65	0.281	0.187	2.854
12, 512	7.37	1.49	5.553	0.468	8.628
15, 10	5.81	0.92	0.686	0.312	3.931
15, 512	11.16	2.46	13.011	0.982	17.161

Sets of five random sparse curves

degree, bits	GEOTOP-BS	GEOTOP	T_1	T_2	T
06, 10	0.25	0.07	0.016	0	0.265
06, 512	0.42	0.13	0.032	0.031	0.235
09, 10	0.54	0.11	0.031	0.015	0.265
09, 512	0.78	0.20	0.499	0.157	0.899
12, 10	0.88	0.17	0.062	0.047	0.359
12, 512	1.73	0.42	1.311	0.888	2.543
15, 10	3.03	0.59	0.201	0.204	0.890
15, 512	5.88	1.22	5.054	3.401	9.126

Table 3. The degree and running time (in sec) for special curves with complicated singular points, see Figure 9,10,11,12. In these figures, we can see that the curve is divided into monotonous curve segments with different colors. We approximate its curve segments one by one and display all at last. As we have presented in Subsection 3.2, the error bound has a lot of redundancy which can be greatly improved. That is to say, the magnitude for veritable error of each curve should be much smaller than the values of last column in the Table above.

curve	degree	topology computation	times for visualization	total time	error
$FTT^2_{2,5}$	20	17.68	3.16	21.36	0.065
$FTT^2_{3,4}$	24	8.66	1.36	10.02	0.0015
$FTT^2_{3,5}$	30	35.68	9.14	44.82	0.00091
$FTT^2_{4,5}$	40	106.95	16.97	123.92	0.0041
$fl_{4,2}$	18	2.64	7.41	10.05	0.18
f_1	10	0.078	0.20	0.28	0.0029
f_2	8	0.078	0.19	0.27	0.0016
f_3	7	0.031	0.14	0.17	0.00098
f_4	9	0.032	0.11	0.14	0.0037
F^*	50	261.83	97.02	358.85	0.074

We also consider some special and difficult examples, see Figure 9 and Figure 10 which are introduced in [26]. As to their defining equations, we just give a simple description, for more details, please refer to Challenge 12 and Challenge 14 in [26]. In Figure 9, the two polynomials are $FTT^2_{3,5}$, $FTT^2_{4,5}$, and $FTT^2_{k,m}$ are defined by the following:

$$FTT^2_{k,m} = (T_m(y) - T_m(x)^k)^2 - T_m(y)^{2k}$$

for $k = 2, 3, \ldots$, $m = 2, 3, \ldots$, where $T_m(x)$ is the Tchebychev polynomial of the first kind with degree m. The small diamonds represent isolated singular points, and from this picture, we can see that the outputs approximate the curve with a small error. In Figure 10, the polynomial is $fl_{4,2}$ whose expression can be obtained below:

$$fl_{4,k} = ((y-x^k)^2 - y^{2k})((y-x^k)^2 - 2y^{2k})((x-y^k)^2 - x^{2k})((x-y^k)^2 - 2x^{2k}) - (xy)^{4k+1}, \ k \geq 2.$$

The curve $fl_{4,2}$ has exactly two tangent directions at singular point and 8 half branches for each of them, see in Figure 10. The curves f_i, $i = 1, \ldots, 4$ are also taken from [26], see Figure 11 for their visualization (we list only one of them for the sake of space), their expressions are given in Equation (3), for more information about these four curves, please refer Challenge 15 of [26].

$$f_1 = x^3 - x^2 y^2 + 5/7\, y^{10}\ , \quad f_2 = -x^4 + x^2 y^2 - 5/7\, y^8\ ,$$
$$f_3 = -x^2 y^2 + x^7 + 5/7\, y^5\ , \quad f_4 = x^3 - x^2 y^2 + 5/7\, y^9. \tag{3}$$

The curve $F^* = \mathrm{Res}_z(f, g)$ is the projection of a space curve defined by f and g, where $f = -y^3 - y^2 z + x^4 z + x^2 yz^2 - yz^5 - x^5 z^2 + x^4 yz^2 - x^3 y^3 z - x^7 y + x^7 z^2 + x^5 z^4 + x^3 y^6 + y^{10} + z^{10}$, $g = x - y - z + z^2 + x^2 y - x^2 z - xyz - x^4 + x^2 yz - xy^3 - yz^3 + y^5 + z^5$. The degree and computing time is presented in Table 3.

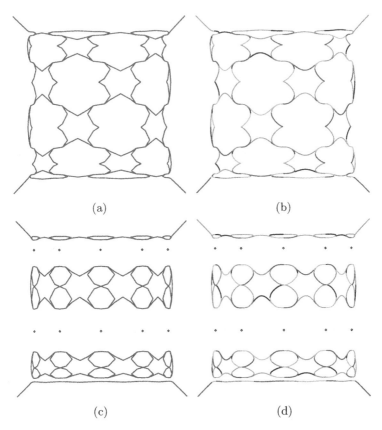

(a)   (b)

(c)   (d)

Fig. 9. The topology and visualization for 2 complicated curves, we list only two of them for the sake of space: (a) is the topology of $FTT^2_{3,5}$, (b) is the visualization of $FTT^2_{3,5}$, (c) is the topology of $FTT^2_{4,5}$, (d) is the visualization of $FTT^2_{4,5}$.

Fig. 10. The topology of $fl_{4,2}$ **Fig. 11.** The visualization of f_2

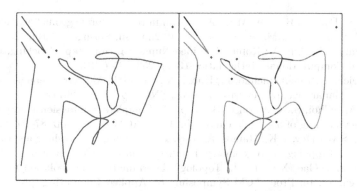

Fig. 12. The topology and visualization of the curve F^*, from the above picture, we can see, it has 11 singular points, of which 7 are isolated singular points.

5   Conclusions

Based on some existing ideas to compute the topology of plane algebraic curves, we use our bivariate system solving technique with revision, which is a symbolic method, to compute the topology of plane curves. Tracing the regular curve segments with a revised Newton method, we get a meshing of the curve. To get a better error control for a meshing of a curve is our future work.

Acknowledgement.   The   work   is   partially   supported   by   NKBRPC (2011CB302400), NSFC Grants (11471327, 11001258, 60821002), SRF for ROCS, SEM. The first author is also financially supported by self-determined research funds of CCNU15A05042 from the colleges' basic research and operation of MOE.

References

1. Akritas, A.G.: An implementation of Vincents Theorem. Numerische Mathematik **36**, 53–62 (1980)
2. Alberti, L., Mourrain, B., Wintz, J.: Topology and arrangement computation of semi-algebraic planar curves. Comp. Aid. Geom. Des. **25**(8), 631–651 (2008)
3. Alcázar, J.G., Schicho, J., Sendra, J.R.: A delineablity-based method for computing critical sets of algebraic surfaces. J. Symb. Comp. **42**, 678–691 (2007)
4. Arnon, D.S., McCallum, S.: A polynomial-time algorithm for the topological type of a real algebraic curve. J. Symb. Comp. **5**, 213–236 (1998)
5. Arnon, D.S., Collins, G., McCallum, S.: Cylindrical algebraic decomposition, I: the Basic Algorithm. SIAM Journal on Computing **13**(4), 865–877 (1984)
6. Arnon, D.S., Collins, G., McCallum, S.: Cylindrical algebraic decomposition, II: an adjacency algorithm for plane. SIAM Journal on Computing **13**(4), 878–889 (1984)
7. Arnon, D.S., Collins, G.: Cylindrical algebraic decomposition, III: an adjacency algorithm for 3D space. J. Symb. Comp. **5**(1,2), 163–187 (1988)

8. Basu, S., Pollack, R., Roy, M.F.: Algorithm in real algebraic geometry. Algorithms and Computation in Mathematics, vol. 10, 2nd edn. Springer, Berlin (2006)

9. Beltrán, C., Leykin, A.: Robust Certified Numerical Homotopy Tracking. Foundations of Computational Mathematics **13**(2), 253–295 (2013)

10. Berberich, E., Emeliyanenko, P., Kobel, A., Sagraloff, M.: Arrangement computation for planar algebraic curves. In: Proc. SNC, pp. 88–99 (2011)

11. Burr, M., Choi, S., Galehouse, B., Yap, C.: Complete subdivision algorithms, II: Isotopic meshing of algebraic curves. In: Proc. ACM ISSAC, pp. 87–94 (2008)

12. Cheng, S.W., Dey, T.K., Ramos, A., Ray, T.: Sampling and meshing a surface with guaranteed topology and geometry. In: Proc. Symp. on CG, pp. 280–289 (2004)

13. Cheng, J.S., Gao, X.S., Li, J.: Topology determination and isolation for implicit plane curves. In: Proc. ACM Symposium on Applied Computing, pp. 1140–1141 (2009)

14. Cheng, J.S., Jin, K.: A generic position based method for real root isolation of zero-dimensional polynomial systems. J. Symb. Comp. **68**, 204–224 (2015)

15. Cheng, J.S., Lazard, S., Peñaranda, L., Pouget, M., Rouillier, F., Tsigaridas, E.: On the topology of real algebraic plane curves. Mathematics in Computer Science **4**, 113–117 (2010)

16. Collins, G., Akritas, A.: Polynomial real roots isolation using Descartes' rule of signs. In: ISSAC, pp. 272–275 (1976)

17. Eigenwillig, A., Kerber, M.: Exact and efficient 2d-arrangements of arbitrary algebraic curves. In: Proc. 19th Annual ACM-SIAM Symposium on Discrete Algorithm (SODA 2008), San Francisco, USA, January 2008, pp. 122–131 (2008)

18. Eigenwillig, A., Kerber, M., Wolpert, N.: Fast and exact geometric analysis of real algebraic plane curves. In: Proc. ACM ISSAC 2007, pp. 151–158. ACM Press (2007)

19. Fulton, W.: Introduction to intersection theory in algebraic geometry. In: CBMS Regional Conference Series in Mathematics, vol. 54. Published for the Conference Board of the Mathematical Sciences, Washington, DC (1984)

20. Gao, X.S., Li, M.: Rational Quadratic Approximation to Real Algebraic Curves. Comp. Aid. Geom. Des. **21**, 805–828 (2004)

21. Hong, H.: An Efficient Method for Analyzing the Topology of Plane Real Algebraic Curves. Mathematics and Computers in Simulation **42**(4–6), 571–582 (1996)

22. Gomes, A.J.P., Morgado, J.F.M., Pereira, E.S.: A BSP-based algorithm for dimensionally nonhomogeneous planar implicit curves with topological guarantees. ACM Trans. Graph. **28**(2) (2009)

23. Kerber, M.: Geometric algorithms for algebraic curves and surfaces. Ph.D. Thesis, Universität des Saarlandes, Germany (2009)

24. Kerber, M., Sagraloff, M.: A worst-case bound for topology computation of algebraic curves. J. Symb. Comp. **47**, 239–258 (2012)

25. Kvasov, B.I.: Methods of Shape-Preserving Spline Approximation. World Scientific, Singapore (2000)

26. Labs, O.: A list of challenges for real algebraic plane curve visualization software In: Nolinear Computational Geometry. The IMA Volumes, vol. 151., pp. 137–164. Springer, New York (2010)

27. Lazard, D.: CAD and topology of semi-algebraic sets. Mathematics in Computer Science **4**(1), 93–112 (2010)

28. Liang, C., Mourrain, B., Pavone, J.P.: Subdivision methods for the topology of 2d and 3d implicit curves. In: Computational Methods for Algebraic Spline Surfaces. Springer-Verlag (2006)

29. Lorensen, W.E., Cline, H.E.: Marching cubes: a high resolution 3d surface construction algorithm. In: Proc. SIGGRAPH 1987. ACM Press (1987)
30. González-Vega, L., Necula, I.: Efficient topology determination of implicitly defined algebraic plane curves. Comp. Aid. Geom. Des. **19**, 719–743 (2002)
31. Martin, R., Shou, H., Voiculescu, I., Bowyer, A., Wang, G.: Comparison of interval methods for plotting algebraic curves. Comp. Aid. Geom. Des. **19**, 553–587 (2002)
32. Mccallum, S., Collins, G.E.: Local box adjacency a lgorithms for cylindrical algebraic decompositions. J. Symb. Comp. **33**, 321–342 (2002)
33. Moore, R.E., Kearfott, R.B., Cloud, M.J.: Introduction to Interval Analysis. Society for Industrial and Applied Mathematics, Philadelphia (2009)
34. Mourrain, B., Pion, S., Schmitt, S., Técourt, J.P., Tsigaridas, E.P., Wolpert, N.: Algebraic issues in computational geometry. In: Boissonnat, J.D., Teillaud, M. (eds.) Effective Computaional Geometry for Curves and Surfaces. Mathematics and Visualization, chapter 3. Springer, Berlin (2006)
35. Noakes, L., Kozera, R.: Cumulative chords, piecewise-quadratics and piecewise-cubics. In: Klette, R., et al. (eds.) Geometric Properties for Incomplete Data, pp. 59–75. Springer, Printed in the Netherlands (2006)
36. Plantinga, S., Vegter, G.: Isotopic meshing of implicit surfaces. Visual Computer **23**, 45–58 (2007)
37. Ratschek, H.: Scci-hybrid method for 2d curve tracing. International Journal of Image and Graphics World Scientific Publishing Company **5**(3), 447–479 (2005)
38. Rouillier, F., Zimmermann, P.: Efficient isolation of polynomial real roots. Journal of Computational and Applied Mathematics **162**(1), 33–50 (2003)
39. Sakkalis, T.: The topological configuration of a real algebraic curve. Bull. Aust. Math. Soc. **43**, 37–50 (1991)
40. Seidel, R., Wolpert, N., On the exact computation of the topology of real algebraic curves. In: Proceedings of Symposium on Computational Geometry, pp. 107–115 (2005)
41. Teissier, B.: Cycles évanescents, sections planes et conditions de Whitney. (french). Singularités à Cargèse. Astérisque 7 et 8, 285–362 (1973)
42. Li, T.Y.: Numerical solution of polynomial systems by homotopy continuation methods. Handbook of numerical analysis **11**, 209–30 (2003)

Piecewise-Quadratics and Reparameterizations for Interpolating Reduced Data

Ryszard Kozera[1] and Lyle Noakes[2]

[1] Warsaw University of Life Sciences - SGGW
Faculty of Applied Informatics and Mathematics
Nowoursynowska str. 159, 02-776 Warsaw, Poland
[2] Department of Mathematics and Statistics
The University of Western Australia
35 Stirling Highway, Crawley W.A. 6009, Perth, Australia
ryszard_kozera@sggw.pl, lyle.noakes@maths.uwa.edu.au

Abstract. This paper tackles the problem of interpolating reduced data $Q_m = \{q_i\}_{i=0}^m$ obtained by sampling an unknown curve γ in arbitrary euclidean space. The interpolation knots $\mathcal{T}_m = \{t_i\}_{i=0}^m$ satisfying $\gamma(t_i) = q_i$ are assumed to be unknown (*non-parametric interpolation*). Upon selecting a specific numerical scheme $\hat{\gamma}$ (here a *piecewise-quadratic* $\hat{\gamma} = \hat{\gamma}_2$), one needs to supplement Q_m with knots' estimates $\{\hat{t}_i\}_{i=0}^m \approx \{t_i\}_{i=0}^m$. A common choice of $\{\hat{t}_i^\lambda\}_{i=0}^m$ ($\lambda \in [0,1]$) frequently used in curve modeling and data fitting (e.g. in computer graphics and vision or in computer aided design) is called *exponential parameterizations* (see, e.g., [11] or [16]). Recent results in [8] and [14] show that $\hat{\gamma}_2$ combined with exponential parameterization yields (in trajectory estimation) either linear $\alpha(\lambda) = 1$ ($\lambda \in [0,1)$) or cubic $\alpha(1) = 3$ convergence orders, once Q_m gets progressively denser. The asymtotics proved in [8] relies on the extra assumptions requiring $\hat{\gamma}_2$ to be reparameterizable to the domain of γ. Indeed, as shown in [14], a natural candidate ψ for such a reparameterization meets this criterion only for $\lambda = 1$, whereas the latter (see [8]) may not hold for the remaining $\lambda \in [0,1)$ (which e.g. brings difficulty in length estimation of γ by using $\hat{\gamma}$). Our paper fills out this gap and establishes sufficient conditions imposed on $\mathcal{T}_m$ to render ψ a genuine reparameterization with $\lambda \in [0,1)$ (see Th. 4). The derivation of a such a condition involves theoretical analysis and symbolic computation, and this constitutes a novel contribution of the present work. The numerical tests verifying whether ψ indeed is a reparameterization (for $\lambda \in [0,1)$ and for more-or-less uniform samplings $\mathcal{T}_m$) are also performed. The sharpness of the asymptotics in question is additionally confirmed with the aid of numerical tests.

1 Introduction

Sampled data points $\gamma(t_i) = q_i$ with $(\{t_i\}_{i=0}^m, Q_m)$ form a pair of a so-called *non-reduced data*. It is required here that $t_i < t_{i+1}$ and $q_i \neq q_{i+1}$ hold. In addition, we assume that $\gamma : [0, T] \to E^n$ (with $0 < T < \infty$) is sufficiently smooth (specified

V.P. Gerdt et al. (Eds.): CASC 2015, LNCS 9301, pp. 260–274, 2015.
DOI: 10.1007/978-3-319-24021-3_20

later) and that it defines a regular curve (i.e. $\dot{\gamma}(t) \neq \mathbf{0}$). In order to approximate the curve γ with an arbitrary interpolant $\bar{\gamma}: [0, T] \to E^n$ it is necessary to assume that the knots fulfill the *admissibility condition* (denoted as $\{t_i\}_{i=0}^m \in V_G^m$):

$$\lim_{m \to \infty} \delta_m = 0, \quad \text{where} \quad \delta_m = \max_{0 \leq i \leq m-1} (t_{i+1} - t_i). \tag{1}$$

This paper discusses a special subfamily of admissible samplings $V_{mol}^m \subset V_G^m$ called *more-or-less uniform* samplings (see e.g. [7]) defined as:

$$\beta \delta_m \leq t_{i+1} - t_i \leq \delta_m, \tag{2}$$

for some $\beta \in (0, 1]$. Note that the left inequality in (2) excludes samplings with distance between consecutive knots smaller than $\beta \delta_m$. On the other hand the right inequality in (2) holds automatically due to (1).

For $\mathcal{T}_m$ unknown, a proper choice of guessed knots $\{\hat{t}_i\}_{i=0}^m \approx \{t_i\}_{i=0}^m$ is stipulated by enforcing the convergence of the selected interpolant $\hat{\gamma}$ (satisfying $\hat{\gamma}(\hat{t}_i) = q_i$) to the unknown curve γ with possibly fast orders. Recall that we choose here $\hat{\gamma} = \hat{\gamma}_2$ as a Lagrange piecewise-quadratic interpolant fitting consecutive triples of points from Q_m. At best the resulting asymptotics in trajectory estimation by any interpolant $\hat{\gamma}$ (and thus in particular by $\hat{\gamma}_2$) should match the asymptotics derived for the corresponding classical *parametric interpolant* $\bar{\gamma}$ (here a piecewise-quadratic $\bar{\gamma} = \tilde{\gamma}_2$) based on non-reduced data ($\{t_i\}_{i=0}^m, Q_m$), with $\mathcal{T}_m$ specified. The next section outlines the existing results on interpolating reduced data and formulates a research task for this paper (see Th. 4).

2   Problem Formulation and Motivation

Recall that the family $F_{\delta_m}: [0, T] \to E^n$ satisfies $F_{\delta_m} = O(\delta_m^\alpha)$ if $\|F_{\delta_m}\| = O(\delta_m^\alpha)$, where $\| \cdot \|$ denotes the euclidean norm and $\alpha \in \mathbb{R}$. By well-known convention, the latter guarantees the existence of constants $K > 0$ and $\bar{\delta} > 0$ (independent on m) such that $\|F_{\delta_m}\| \leq K \delta_m^\alpha$, for all $\delta_m \in (0, \bar{\delta})$.

A standard result for *non-reduced data* ($\{t_i\}_{i=0}^m, Q_m$) combined with piecewise r-degree polynomial $\bar{\gamma} = \tilde{\gamma}_r$ reads (see e.g. [7] or [1]):

Theorem 1. *Let $\gamma \in C^{r+1}$ be a regular curve $\gamma : [0, T] \to E^n$. Assume that the knot parameters $\{t_i\}_{i=0}^m \in V_G^m$ are given. Then a piecewise r-degree Lagrange polynomial $\tilde{\gamma}_r$ used with $\{t_i\}_{i=0}^m$ known, yields a sharp estimate for the trajectory estimation $\tilde{\gamma}_r = \gamma + O(\delta_m^{r+1})$.*

In particular, piecewise-quadratics $\tilde{\gamma}_2$ or piecewise-cubics $\tilde{\gamma}_3$ render *cubic* or *quartic* orders in Th. 1, respectively. In various applications in computer graphics and vision (e.g. image segmentation or curve modeling), engineering or physics (trajectory estimation) or medical image processing (e.g. in medical diagnosis) a common situation is to deal exclusively with the *reduced data* Q_m (see e.g. [11], [5], [16] or [3]). Any fitting scheme based on Q_m requires, in the first step to determine the respective substitutes $\{\hat{t}_i\}_{i=0}^m$ (i.e. external knots) approximating somehow the internal knots $\{t_i\}_{i=0}^m$. One particular family of $\{\hat{t}_i\}_{i=0}^m \approx \{t_i\}_{i=0}^m$ commonly used for curve modeling is the so-called *exponential parameterization* (see e.g. [11] or [12]):

$$\hat{t}_0 = 0, \qquad \hat{t}_{i+1} = \hat{t}_i + \|q_{i+1} - q_i\|^\lambda, \tag{3}$$

where $0 \leq \lambda \leq 1$ and $i = 0, 1, \ldots, m-1$. The special cases of $\lambda \in \{1, 0.5, 0\}$, yield the so-called *cumulative chords, centripetal, and blind uniform parameterizations* of external knots, respectively. We denote a piecewise degree-r polynomial based on (3) and Q_m as $\hat{\gamma} = \hat{\gamma}_r : [0, \hat{T}] \to E^n$, where $\hat{T} = \sum_{i=0}^{m-1} \|q_{i+1} - q_i\|^\lambda$. In order to establish the asymptotics in difference between the curve γ and any non-parametric interpolant $\hat{\gamma}$, a *reparameterization* $\psi : [0, T] \to [0, \hat{T}]$ synchronizing both domains of γ and $\hat{\gamma}_r$ is needed. As mentioned above, the case when $\lambda = 0$ transforms (3) into to blind *uniform* guesses (with no regard to the distribution of Q_m) with $\hat{t}_i = i$. For $r = 2$ and $\lambda = 0$ the linear convergence rate (i.e. $\alpha(0) = 1$) in trajectory estimation was originally proved in [15]. A faster convergence follows if $\lambda = 1$ is assumed to (3). This yields the so-called *cumulative chords* (see e.g. [11] or [16]) satisfying $\hat{t}_0 = 0$ and $\hat{t}_{i+1} = \hat{t}_i + \|q_{i+1} - q_i\|$ for $i = 0, 1, \ldots, m-1$. Such a choice of $\{\hat{t}_i\}_{i=0}^m$ incorporates the geometry of Q_m and consequently offers much better approximation orders $\alpha(1)$ (at least for $r = 2, 3$) as opposed to the case of $\lambda = 0$. Indeed the following holds (see [14]):

Theorem 2. *Suppose γ is a regular C^k curve in E^n, where $k \geqslant r + 1$ and $r = 2, 3$ sampled according to (1). Let $\hat{\gamma}_r : [0, \hat{T}] \to E^n$ be the cumulative chord piecewise degree-r interpolant defined by Q_m and (3) with $\lambda = 1$. Then there is a piecewise-C^r reparameterization $\psi : [0, T] \to [0, \hat{T}]$ with $\hat{\gamma}_r \circ \psi = \gamma + O(\delta_m^{r+1})$.*

Visibly, for $r = 2$ and $\lambda = 1$ the order $\alpha(1) = 3$ determined by Th. 2 improves by 2 the order $\alpha(0) = 1$. In addition, at least for $r = 2, 3$ both asymptotics established in Th. 2 (for $\lambda = 1$) and Th. 1 coincide. Recent result in [8] (for $r = 2$) extends the above two special cases of $\alpha(0) = 1$ and $\alpha(1) = 3$ to the entire family of exponential parameterizations (3), i.e. to all $\lambda \in [0, 1]$. As proved in [8], for arbitrary more-or-less uniform samplings (2) combined with (3) any choice of $\lambda \in [0, 1)$ does not improve the asymptotics in γ approximation by $\hat{\gamma}_2$. In fact a linear convergence order $\alpha(\lambda) = 1$ holds for all $\lambda \in [0, 1)$ and $r = 2$:

Theorem 3. *Suppose γ is a regular C^3 curve in E^n sampled more-or-less uniformly (2). Let $\hat{\gamma}_2 : [0, \hat{T}] \to E^n$ be the piecewise-quadratic interpolant defined by Q_m and (3) (with $\lambda \in [0, 1]$). Then for a special candidate of a piecewise-C^∞ reparameterization $\psi : [0, T] \to [0, \hat{T} = \sum_{i=0}^{m-1} \|q_{i+1} - q_i\|^\lambda]$ and $\lambda \in [0, 1)$ we have $\hat{\gamma}_2 \circ \psi = \gamma + O(\delta_m)$. In addition, for either $\{t_i\}_{i=0}^m$ uniform or $\lambda = 1$ and (1), ψ is a piecewise-C^∞ reparameterization for which we have $\hat{\gamma}_2 \circ \psi = \gamma + O(\delta_m^3)$.*

Th. 3 yields an unexpected left-hand side discontinuity in $\alpha(\lambda)$ at $\lambda = 1$. Indeed by Th. 3 the order $\alpha(\lambda) = 1$ (for $\lambda \in [0, 1)$) jumps abruptly to $\alpha(1) = 3$. As demonstrated in [8], opposite to the cumulative chords the natural candidate for a reparametrization, i.e a piecewise-quadratic $\psi = \{\psi_i\}_{i=0}^{m-2}$ with quadratic

$\psi_i : [t_i, t_{i+2}] \to [\hat{t}_i, \hat{t}_{i+2}]$ satisfying $\psi_i(t_{i+j}) = \hat{t}_{i+j}$ (for $j = 0, 1, 2$, see also (3)) may not render an injective function for the remaining $\lambda \in [0, 1)$. Note that $\psi : [0, T] \to [0, \hat{T}]$ invoked in Th. 3 is defined as a track-sum of ψ_i, where $i = 0, 2, 4, \ldots, m-2$. For various applications like e.g. length estimation of the 2D object in medical image processing or correct trajectory tracking it is vital that ψ is a reparameterization. Therefore in this paper, we formulate and substantiate sufficient conditions imposed on samplings $\{t_i\}_{i=0}^m$ to render ψ_i (and thus ψ) a genuine (in fact here a piecewise C^∞) reparameterization.

Note first that any admissible sampling (including a more-or-less uniform one) can be characterized by (for arbitrary m and $0 \le i \le m$):

$$t_{i+1} - t_i = M_{im}\delta_m \quad \text{and} \quad t_{i+2} - t_{i+1} = N_{im}\delta_m, \tag{4}$$

where $0 < M_{im}, N_{im} \le 1$. Our main result (complementing Th. 3) reads as follows:

Theorem 4. *Let the assumptions from Th. 3 hold. Suppose that sampling $\{t_i\}_{i=0}^m$ (see (4)) fulfills both inequalities determined by (12) and (13) for a fixed $\lambda_0 \in [0, 1)$. Then, each quadratic $\psi_i : [t_i, t_{i+2}] \to [\hat{t}_i, \hat{t}_{i+2}]$ is asymptotically (i.e. for sufficiently large m) a reparameterization, with $\{\hat{t}_i\}_{i=0}^m$ defined according to (3). In addition, the latter holds asymptotically for all exponential parameterizations (3) with $\lambda \in [\lambda_0, 1)$. Finally, if β determining a more-or-less uniformity (2) satisfies $\beta > \sqrt{2}-1$, then ψ is asymptotically a reparameterization for all $\lambda \in [0, 1]$.*

In this paper an analytical proof of Th. 4 is given. In addition, both (12) and (13) are interpreted with the aid of symbolic computation in *Mathematica* and geometrical argument. In practice the first condition (12) is not easy to check. Consequently, a stronger condition (16) is proposed, which can easily be checked by symbolic computation. The verification of (16) (as opposed to (12)) is simpler (at least for symbolic computation) and its satisfaction yields both (12) and (13). Again further geometrical insight is given. The entire procedure for determining whether a given ψ is a reparameterization is illustrated in the closing section of this paper. The experimental results (performed in *Mathematica*) of testing this procedure on various samplings and curves are also presented. Finally, the sharpness of the asymptotics derived in Th. 3 is numerically confirmed. Note also that sufficient conditions for ψ to be a reparameterization specified in Th. 4 complement Th. 3.

3   Main Result

Proof. (i) We prove now Th. 4 (by recalling first [8]). Let $\psi_i : [t_i, t_{i+2}] = I_i \to [\hat{t}_i, \hat{t}_{i+2}] = \hat{I}_i$, be the quadratic polynomial satisfying interpolation conditions $\psi_i(t_{i+j}) = \hat{t}_{i+j}$ (for $j = 0, 1, 2$) with $\hat{t}_i$ defined as in (3). The track-sum of $\{\psi_i\}_{i=0}^{m-2}$ (for $i = 0, 2, 4 \ldots m - 2$) defines a piecewise-C^∞ mapping $\psi : [0, T] \to [0, \hat{T}]$. The Newton Interpolation Formula (see [1]) over each I_i yields $\psi_i(t) = \psi_i[t_i] + \psi_i[t_i, t_{i+1}](t - t_i) + \psi_i[t_i, t_{i+1}, t_{i+2}](t - t_i)(t - t_{i+1})$, and therefore

$$\psi_i^{(1)}(t) = \psi_i[t_i, t_{i+1}] + (2t - t_{i+1} - t_i)\psi_i[t_i, t_{i+1}, t_{i+2}]. \tag{5}$$

By [8], for more-or-less uniform samplings $\{t_i\}_{i=0}^m \in V_{mol}^m$ we have (for $k = 0, 1$):

$$\psi_i[t_{i+k}, t_{i+k+1}] = (t_{i+k+1} - t_{i+k})^{-1+\lambda} + O((t_{i+k+1} - t_{i+k})^{1+\lambda}),$$

$$\psi_i[t_i, t_{i+1}, t_{i+2}] = \frac{(t_{i+2} - t_{i+1})^{-1+\lambda} - (t_{i+1} - t_i)^{-1+\lambda}}{t_{i+2} - t_i} + O(\delta_m^\lambda). \qquad (6)$$

(ii) We pass now to the *main contribution* of this paper (namely, a proof of Th. 4). For ψ_i to be a reparameterization we need $\psi_i^{(1)} > 0$ over I_i. Taking into account that $\psi_i^{(1)}(t)$ is affine, it is sufficient to show that both $\psi_i^{(1)}(t_i) > 0$ and $\psi_i^{(1)}(t_{i+2}) > 0$ hold, asymptotically. In doing so, by (5) we arrive at:

$$\psi_i^{(1)}(t_i) = \psi_i[t_i, t_{i+1}] + (t_i - t_{i+1})\psi_i[t_i, t_{i+1}, t_{i+2}],$$

$$\psi_i^{(1)}(t_{i+2}) = \psi_i[t_i, t_{i+2}] + [(t_{i+2} - t_{i+1}) + (t_{i+2} - t_i)]\psi_i[t_i, t_{i+1}, t_{i+2}]. \qquad (7)$$

We find now sufficient condition under which both inequalities $\psi_i^{(1)}(t_i) > 0$ and $\psi_i^{(1)}(t_{i+2}) > 0$ are satisfied asymptotically. By (1), the consecutive knots' differences determined in (4) yield $M_{im} = O(1)$ and $N_{im} = O(1)$ (in fact here $0 < M_{im} \leq 1$ and $0 < N_{im} \leq 1$). Coupling (4) with (6) gives

$$\psi_i[t_i, t_{i+1}] = M_{im}^{-1+\lambda}\delta_m^{-1+\lambda} + O(\delta_m^{1+\lambda}),$$

$$\psi_i[t_{i+1}, t_{i+2}] = N_{im}^{-1+\lambda}\delta_m^{-1+\lambda} + O(\delta_m^{1+\lambda}). \qquad (8)$$

Since $0 < M_{im}, N_{im} \leq 1$, we have $M_{im}^\theta = O(1)$ and $N_{im}^\theta = O(1)$ for each $\theta \geq 0$ (here $\theta = 1 + \lambda$). The latter (with $\theta = \lambda$) together with (4), (6), and $0 < (t_{i+j+1} - t_{i+j})(t_{i+2} - t_i) < 1$, for $j = 0, 1$ yield that $\psi_i[t_i, t_{i+1}, t_{i+2}]$

$$= \frac{N_{im}^{-1+\lambda}\delta_m^{-1+\lambda} - M_{im}^{-1+\lambda}\delta_m^{-1+\lambda}}{(M_{im} + N_{im})\delta_m} + \frac{O((t_{i+2} - t_{i+1})^{1+\lambda}) + O((t_{i+2} - t_{i+1})^{1+\lambda})}{t_{i+2} - t_i}$$

$$= \frac{N_{im}^{-1+\lambda}\delta_m^{-1+\lambda} - M_{im}^{-1+\lambda}\delta_m^{-1+\lambda}}{(M_{im} + N_{im})\delta_m} + O(\delta_m^\lambda). \qquad (9)$$

Combining (9) with (7), (8), $M_{im} = O(1)$ and $N_{im} = O(1)$ leads to

$$\psi_i^{(1)}(t_i) = \delta_m^{-1+\lambda}\left(M_{im}^{-1+\lambda} - \frac{M_{im}}{M_{im} + N_{im}}(N_{im}^{-1+\lambda} - M_{im}^{-1+\lambda})\right) + O(\delta_m^{1+\lambda}),$$

$$\psi_i^{(1)}(t_{i+2}) = \delta_m^{-1+\lambda}\left(M_{im}^{-1+\lambda} + \left(1 + \frac{N_{im}}{M_{im} + N_{im}}\right)(N_{im}^{-1+\lambda} - M_{im}^{-1+\lambda})\right)$$
$$+ O(\delta_m^{1+\lambda}). \qquad (10)$$

In order to ensure now that both $\psi_i^{(1)}(t_i) > 0$ and $\psi_i^{(1)}(t_{i+2}) > 0$ hold asymptotically, it is sufficient to assume *firstly* that:

$$\rho_1(M_{im}, N_{im}) = M_{im}^{-1+\lambda} - \frac{M_{im}}{M_{im} + N_{im}}(N_{im}^{-1+\lambda} - M_{im}^{-1+\lambda}),$$

$$\rho_2(M_{im}, N_{im}) = M_{im}^{-1+\lambda} + \left(1 + \frac{N_{im}}{M_{im} + N_{im}}\right)(N_{im}^{-1+\lambda} - M_{im}^{-1+\lambda}), \quad (11)$$

are positive and *secondly* that there exist constants $\alpha_1 < 2$, $\alpha_2 < 2$, $K_1 > 0$ and $K_2 > 0$ such that asymptotically we have:

$$\rho_1(M_{im}, N_{im}) \geq K_1\delta_m^{\alpha_1} \quad \text{and} \quad \rho_2(M_{im}, N_{im}) \geq K_2\delta_m^{\alpha_2}. \quad (12)$$

Condition (12) ensures that asymptotically the slowest components in (10) are of order less than $\lambda + 1$, which combined with the (11) results (for sufficiently large m) in positive derivatives of ψ_i at both ends of each I_i (and thus over entire I_i). As it turns out, the positivity of the expressions in (11) has a simple geometrical interpretation easily verifiable for any specific samplings $\{t_i\}_{i=0}^m$ and $\lambda \in [0, 1)$. Indeed, upon simple algebraic manipulations both inequalities from (11) are reducible to:

$$2 + \frac{N_{im}}{M_{im}} > \left(\frac{N_{im}}{M_{im}}\right)^{-1+\lambda} \quad \text{and} \quad \left(\frac{N_{im}}{M_{im}}\right)^{-1+\lambda} > 1 - \frac{1 + \frac{N_{im}}{M_{im}}}{2\frac{N_{im}}{M_{im}} + 1}. \quad (13)$$

The system of two non-linear inequalities (13) in two independent variables can be solved by *adopting one of two following geometrically driven approaches*:

a) The first method relies on *homogeneous substitution* $x = (N_{im}/M_{im}) > 0$ applied to both inequalities (13) which in turn leads into:

$$f(x) = 2 + x > g(x) = x^{-1+\lambda} \quad \text{and} \quad g(x) > h(x) = 1 - \frac{1 + x}{2x + 1}. \quad (14)$$

Plotting $f(x) > g(x)$ (e.g. in *Mathematica*) yields for each $\lambda \in [0, 1)$ an interval (a_λ, ∞), where $0 < a_\lambda < 1$ (here $f(a_\lambda) = g(a_\lambda)$). Similarly, upon plotting $g(x) > h(x)$ for each $\lambda \in [0, 1)$ we obtain an interval $(0, b_\lambda)$, where $b_\lambda > 1$ (here $g(b_\lambda) = h(b_\lambda)$). Thus a non-empty intersection of (a_λ, ∞) and $(0, b_\lambda)$ renders *an admissible interval* for samplings $\{t_i\}_{i=0}^m$ from (4) with $(N_{im}/M_{im}) \in (a_\lambda, b_\lambda)$ (see Figure 1, when $\lambda = 0.1$ is used). Table 1 (called here *a look-up table*) lists admissible intervals (a_λ, b_λ) (for $\lambda \in [0, 1)$) which are numerically computed by finding the sole roots of either $f(x) - g(x) = 0$ or $g(x) - h(x) = 0$, with the aid of *Mathematica's NSolve* function. Given $\lambda \in [0, 1)$, the existence of at least one pair of intersection points (a_λ, b_λ) results from the Darboux Theorem upon calculating the limits of f, g and h at the ends of the interval $(0, +\infty)$. On the other hand, the uniqueness of (a_λ, b_λ) follows from the strict monotonicity of f, g and h. Of course for $\lambda \in [0, 1)$ we have $a_\lambda < 1$, since $f(1) - g(1) = 2 > 0$ and $\lim_{x \to 0^+}(f(x) - g(x)) = -\infty$.

The next observation inferable from the graphs of f, g and h reads (for $\lambda \in [0, 1)$ as:

$$0 \leq \lambda_1 < \lambda_2 < 1 \quad \text{then} \quad 0 < a_{\lambda_2} < a_{\lambda_1} < 1 \quad \text{and} \quad 1 < b_{\lambda_1} < b_{\lambda_2}. \quad (15)$$

Note that it suffices to substantiate (15) only for a_λ as the other ones follow from the formula $a_\lambda = (1/b_\lambda)$ (hence also $a_\lambda < b_\lambda$) which is easily verifiable upon using

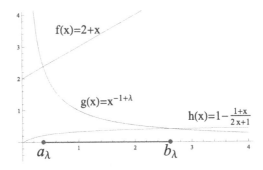

Fig. 1. An admissible interval for $(N_{im}/M_{im}) \in (a_\lambda, b_\lambda) = (0.381, 2.622)$ and $\lambda = 0.1$.

(14). The inequality $a_{\lambda_2} < a_{\lambda_1}$ follows either by an indirect argument or upon resorting to the implicit function theorem. Indeed for the function $i(\lambda, a) = 2 + a - a^{-1+\lambda}$ (with $(\lambda, a) \in [0, 1) \times (0, \infty)$), the equation $i(\lambda, a) = 0$ (since $(\partial i/\partial a)(\lambda, a) = 1 + (1-\lambda)a^{-2+\lambda} > 0$ does not vanish) determines $a_\lambda = a(\lambda)$ whose derivative satisfies $a'(\lambda) = (a^{-1+\lambda} \ln(a))/(1 + (1-\lambda)a^{-2+\lambda}) < 0$, as $0 < a < 1$. Thus $a_\lambda = a(\lambda)$ is strictly decreasing. Note also that if $\lambda_n = (1-c_n) \to 1^-$ (where $c_n \to 0^+$) then $a_{\lambda_n} \to 0^+$. Indeed by monotonicity of a_{λ_n} and $a_{\lambda_n} \geq 0$ the sequence a_{λ_n} is convergent with $\lim_{n\to\infty} a_{\lambda_n} = g \geq 0$. Taking into account (14) we have $2 + a_{\lambda_n} = e^{-c_n \ln(a_{\lambda_n})}$. The latter, if $g \neq 0$ yields $g = -1$, a contradiction. Thus since λ_n was chosen arbitrary, we arrived at (as $b_\lambda = 1/a_\lambda$)) $a_{\lambda=1^-} = \lim_{\lambda\to 1^-} a_\lambda = 0^+$ and $b_{\lambda=1^-} = \lim_{\lambda\to 1^-} b_\lambda = +\infty$. A simple inspection of (14) shows that $a_{\lambda=0} = \sqrt{2} - 1$ and $b_{\lambda=0} = (1/(\sqrt{2}-1)) = 1 + \sqrt{2}$. In addition as proved above $a_{\lambda=1^-} = 0^+$ and $b_{\lambda=1^-} = +\infty$.

Hence $x_0 \in (a_{\lambda_0}, b_{\lambda_0})$ ensures that $x_0 \in (a_\lambda, b_\lambda)$ for all $[\lambda_0, 1)$. The latter in fact follows for all $\lambda \in [\lambda_0, 1]$ as the case of $\lambda = 1$ always renders ψ a reparameterization (see [8] or [14]).

A particular candidate for λ_0 can be found e.g. upon inspecting *a look-up Table 1*. In order to accept vaster class of admissible exponential parameterizations yielding ψ as a reparameterization, the look-up table, should in practice contain the intervals (a_λ, b_λ) computed for a denser increment of $\lambda \in [0, 1)$, preferably equal to 0.01. This can be achieved by extending Table 1 to more entries upon again invoking *Mathematica* package.

Recall that more-or-less uniform samplings (2) satisfy $\beta \leq x = N_{im}/M_{im} \leq (1/\beta)$. Consequently, given the β is known a priori, if both inequalities $\beta > a_{\lambda_0}$ and $(1/\beta) < b_{\lambda_0}$ hold, then ψ is asymptotically a reparametrization for all $\lambda \in [\lambda_0, 1)$ (and also for $\lambda \in [\lambda_0, 1]$). In particular, if $\beta > a_{\lambda=0} = \sqrt{2} - 1$ and $(1/\beta) < b_{\lambda=0} = \sqrt{2} + 1$, which both hold for $\beta > \sqrt{2} - 1$, then ψ defines a genuine piecewise C^∞ reparameterization for each $\lambda \in [0, 1]$ defining (3). For such more-or-less uniform samplings no support of the look-up table is required.

b) Alternatively (to get more geometrical insight into (13)), for any fixed $\lambda \in [0, 1)$ one solves both inequalities from (11) over the domain $[0, 1] \times [0, 1]$ with the aid of symbolic computation (i.e. with the *Mathematica RegionPlot* function). The geometrical plots of *2D admissible zone* A_λ (see e.g. shaded areas

Table 1. Numerically computed admissible intervals (a_λ, b_λ) for various $\lambda \in [0,1)$

$\lambda = 0.0$	$\lambda = 0.1$	$\lambda = 0.2$	$\lambda = 0.3$	$\lambda = 0.4$
(0.414,2.414)	(0,381,2.622)	(0.345,2,901)	(0.304,3.294)	(0.257,3.885)
$\lambda = 0.5$	$\lambda = 0.6$	$\lambda = 0.7$	$\lambda = 0.8$	$\lambda = 0.9$
(0.216,4.865)	(0.148,6.761)	(0.086, 11.599)	(0.029,34.394)	(0.001,1028.990)

in Fig. 2) admits any sampling pairs $(x, y) = (M_{im}, N_{im}) \in A_\lambda$, for different $\lambda \in [0, 1]$ rendering ψ_i asymptotically a reparameterization. Note that for arbitrary more-or-less uniform samplings (and any fixed $\lambda \in [0, 1)$ in exponential parameterization (3)) an admissible zone for (M_{im}, N_{im}) is in fact a sub-square $[\beta, 1]^2 \subset [0, 1]^2$, where $0 < \beta \leq 1$ is defined as in (2). A straightforward verification shows that for $\lambda = 1$ both equations in (11) are satisfied and therefore $A_1 = [0, 1] \times [0, 1]$. The case when $\lambda = 0$ reduces (11) into $\frac{1}{x} - \frac{x}{x+y}\left(\frac{1}{y} - \frac{1}{x}\right) > 0$ and $\frac{1}{y} + \left(1 + \frac{x}{x+y}\right)\left(\frac{1}{y} - \frac{1}{x}\right) > 0$, which in turn become $y^2 - x^2 + 2xy > 0$ and $x^2 - y^2 + 2xy > 0$, thus yielding $A_0 = \{(x, y) \in [0, 1]^2 : y > (\sqrt{2} - 1)x, y < (1/(\sqrt{2} - 1))x = (\sqrt{2} + 1)x\}$. The latter corresponds to a_0 and b_0 computed above. The symmetry of A_λ for other $\lambda \in (0, 1)$ with respect to line $y = x$ (as expected due to $a_\lambda = (1/b_\lambda)$) is illustrated in Figure 2 for $\lambda \in \{0, 0.5, 0.7\}$. Noticeably symbolic computation by *RegionPlot* shows that $A_{\lambda_1} \subset A_{\lambda_2}$ for $\lambda_1 < \lambda_2$, where $\lambda_i \in [0, 1]$. The latter follows also independently from (15).

Since for samplings (2) we have $\beta \leq (x/y) \leq (1/\beta)$ and $\beta \leq (y/x) \leq (1/\beta)$, the pair (M_{im}, N_{im}) belongs to the set $B = \{(x, y) \in [0, 1]^2 : y \leq \beta^{-1}x, y \geq \beta x\}$. Hence if $B \subset A_{\lambda_0}$ then $B \subset A_\lambda$, for all $\lambda \in [\lambda_0, 1]$. This renders ψ asymptotically a piecewise C^∞ reparameteriztion for all $\lambda \in [\lambda_0, 1]$. In particular, the latter holds for all $\lambda \in [0, 1]$ given $B \subset A_0$ (since then $A_0 \subset A_\lambda$). $\qquad\square$

Unfortunately, the verification of condition (12) is harder and each time demands an extra analysis. To circumvent this difficulty, a stronger condition guaranteeing the fulfillment of both (11) and (12) is proposed below. More precisely, one assumes here the existence of two positive number (σ_1, σ_2) independent on

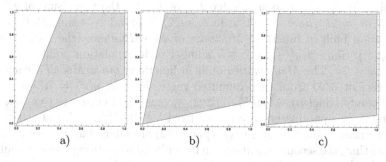

a)   b)   c)

Fig. 2. The plots of 2D admissible zones A_λ, for: a) $\lambda = 0$, b) $\lambda = 0.5$ and c) $\lambda = 0.7$

m, i and $\lambda \in [\lambda_0, 1]$ satisfying the following two inequalities (at least asymptotically):

$$\rho_1(M_{im}, N_{im}) \geq \sigma_1 > 0 \quad \text{and} \quad \rho_2(M_{im}, N_{im}) \geq \sigma_2 > 0. \tag{16}$$

Evidently, the satisfaction of (16) implies that (11) and (12) hold which in turn yields ψ_i as a reparameterization for $\lambda \in [\lambda_0, 1]$. Geometrically, any pair of $\sigma = (\sigma_1, \sigma_2)$ introduces an extra buffer zone in corrected 2D admissible regions A_λ^σ this time by making the left-hand sides in (11) stay away from zero, which is enforced by (16). The practical procedure to determine whether ψ_i indeed is a reparameterization relies on fixing $\sigma = (\sigma_1, \sigma_2)$ and performing *Mathematica* symbolic calculation with *RegionPlot* function to plot A_λ^σ for incremented λ. If there exists λ_0 such that a given family of more-or-less uniform samplings satisfies $(M_{im}, N_{im}) \in A_\lambda^\sigma$ for all $\lambda \in [\lambda_0, 1]$, then condition (16) follows. One can also vary σ and repeat the above procedure in a search for smaller value of λ_0.

The mathematical argument to solve (13) is extendable to (16) (with $\sigma = \sigma_1 = \sigma_2$) if extra assumptions on samplings $\mathcal{T}_m$ are made. This ultimately leads to the shifted inequalities $f(x) > g(x) + \sigma$ and $g(x) - \sigma > h(x)$ similar to those already discussed from (14). However, the latter exceeds the scope of this paper.

4   Experiments

The tests are conducted in *Mathematica 9.0* (see [17]) on a 2.4GHZ Intel Core 2 Duo computer with 8GB RAM. Note that since $T = \sum_{i=1}^{m}(t_{i+1} - t_i) \leq m\delta_m$ the following holds $m^{-\alpha} = O(\delta_m^\alpha)$ for $\alpha > 0$. Hence, the verification of the asymptotics expressed in terms of $O(\delta_m^\alpha)$ can be examined in terms of $O(1/m^\alpha)$ asymptotics. For a regular curve $\gamma : [0, T] \to E^n$, $\lambda \in [0, 1]$ and m varying between $m_{min} \leq m \leq m_{max}$, the error for γ estimation over $[t_i, t_{i+2}]$ reads

$$E_m^i = \sup_{t \in [t_i, t_{i+2}]} \|(\hat{\gamma}_{2,i} \circ \psi_i)(t) - \gamma(t)\| = max_{t \in [t_i, t_{i+2}]}\|(\hat{\gamma}_{2,i} \circ \psi_i)(t) - \gamma(t)\|, \tag{17}$$

as $\tilde{E}_m^i(t) = \|(\hat{\gamma}_{2,i} \circ \psi_i)(t) - \gamma(t)\| \geq 0$ is continuous over each sub-interval $[t_i, t_{i+2}] \subset [0, T]$. The maximal value E_m of $\tilde{E}_m(t)$ (the track-sum of $\tilde{E}_m^i(t)$), for each $m = 2k$ (here $k = 1, 2, 3, \ldots, m/2$) is found by using *Mathematica* optimization built-in functions *Maximize* or *FindMinimum* (the latter applied to $-\tilde{E}_m(t)$). Since $deg(\hat{\gamma}_2) = 2$, the number of interpolation points $\{q_i\}_{i=0}^{m}$ is odd i.e. $m = 2k$. The *Mathematica* built-in functions *LinearModelFit* calculates the coefficient $\bar{\alpha}(\lambda)$ from the computed regression line $y(x) = \bar{\alpha}(\lambda)x + b$ on pairs of points $\{(\log(m), -\log(E_m))\}_{m=m_{min}}^{m_{max}}$ (see [7]). In order to test that ψ_i is asymptotically a parameterization (for a fixed $\lambda \in [0, 1)$), it suffices to show that both constraints $(M_{im}/N_{im}) \in (a_\lambda, b_\lambda)$ and (12) are fulfilled for sufficiently large m. Recall that, the second condition can be replaced by a stronger one formulated in (16). In the event of difficulties in verification of the above two conditions, one may test in practice (for large $m = m_{max}$) both inequalities $\psi_i^{(1)}(t_i) > 0$

and $\psi_i^{(1)}(t_{i+2}) > 0$ which should hold over each sub-interval $[t_i, t_{i+2}]$. Evidently, such an approach only partially alleviates the problem, as it relies on the implicit assumption that both inequalities hold asymptotically i.e. for sufficiently large m. The next two examples confirm numerically the sharpness of Th. 3 and illustrate our procedure designed to determine when ψ is a piecewise C^∞ parameterization.

Example 1. Let a regular spatial curve in E^3 be *a quadratic elliptical helix*:

$$\gamma_h(t) = (2\cos(t), \sin(t), t^2), \tag{18}$$

with $t \in [0, 2\pi]$, be sampled more-or-less uniformly (2) (with $\beta = (1/3)$) as:

$$t_i = \begin{cases} \frac{2\pi i}{m} & \text{if } i \text{ even,} \\[2mm] \frac{2\pi i}{m} + \frac{\pi}{m} & \text{if } i = 4k+1, \\[2mm] \frac{2\pi i}{m} - \frac{\pi}{m} & \text{if } i = 4k+3. \end{cases} \tag{19}$$

Figure 3 shows a plot of γ_h sampled in accordance with (19) for $m = 22$. Note that here $\delta_m = (3\pi/m)$ and over each segment $[t_i, t_{i+2}]$ we either have $t_{i+1} - t_i = \delta_m$ and $t_{i+2} - t_{i+1} = (1/3)\delta_m$ or $t_{i+1} - t_i = (1/3)\delta_m$ and $t_{i+2} - t_{i+1} = \delta_m$. This results only in two pairs $(M_1, N_1) = (1, 1/3)$ and $(M_2, N_2) = (1/3, 1)$ which are independent of m. The inequality $\beta = (1/3) > \sqrt{2} - 1$ from Th. 4 does not hold and thus the look-up Table 1 is used which yields $(N_1/M_1) = (1/3) \in (a_{0.3}, b_{0.3}) = (0.304, 3.294)$ and $(N_2/M_2) = 3 \in (a_{0.3}, b_{0.3})$. Thus (11) holds for all $\lambda \in [0.3, 1)$. In fact there exists exactly one $\lambda_e \approx 0.3$ such that for all $\lambda \in [\lambda_e, 1)$ all intervals (a_λ, b_λ) are admissible. To test the first condition in (12) we solve (with respect to λ) two equations $\rho_1(M_1, N_1) = 0$ and $\rho_1(M_2, N_2) = 0$. Indeed, for $\rho_1(M_1, N_1) = 0$ we have

$$1^{\lambda-1} - \frac{1}{1 + \frac{1}{3}}\left(\left(\frac{1}{3}\right)^{\lambda-1} - 1^{\lambda-1}\right) = 0 \equiv \left(\frac{7}{9}\right) = \left(\frac{1}{3}\right)^\lambda.$$

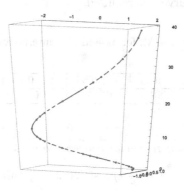

Fig. 3. The plot of the hellix γ_h from (18) sampled according to (19), for $m = 22$

The latter separates $\lambda_e = \frac{\ln(9)-\ln(7)}{\ln(3)} \approx 0.2287$ from the admissible interval $[0.3, 1)$ while $m \to \infty$. Similarly, for $\rho_1(M_2, N_2) = 0$ we obtain

$$\left(\frac{1}{3}\right)^{\lambda-1} - \frac{\frac{1}{3}}{\frac{1}{3}+1}\left(1^{\lambda-1} - \left(\frac{1}{3}\right)^{\lambda-1}\right) = 0 \equiv \left(\frac{1}{15}\right) = \left(\frac{1}{3}\right)^\lambda.$$

This gives another $\lambda_e = \frac{\ln(15)}{\ln(3)} \approx 2.465 \notin [0,1)$ separated from $[0.3, 1)$, while $m \to \infty$. Similarly, $\rho_2(M_1, N_1) = 0$ becomes

$$1^{\lambda-1} + \left(1 + \frac{\frac{1}{3}}{1+\frac{1}{3}}\right)\left(\left(\frac{1}{3}\right)^{\lambda-1} - 1^{\lambda-1}\right) = 0 \equiv \left(\frac{1}{15}\right) = \left(\frac{1}{3}\right)^\lambda.$$

As previously, $\lambda_e \approx 2.465 \notin [0,1)$ and also is separated from $[0.3, 1)$ for $m \to \infty$. Finally, $\rho_2(M_2, N_2) = 0$ transforms into:

$$\left(\frac{1}{3}\right)^{\lambda-1} + \left(1 + \frac{1}{1+\frac{1}{3}}\right)\left(1^{\lambda-1} - \left(\frac{1}{3}\right)^{\lambda-1}\right) = 0 \equiv \left(\frac{7}{9}\right) = \left(\frac{1}{3}\right)^\lambda.$$

Similarly, $\lambda_e \approx 0.2287 \in [0,1)$ is separated from $[0.3, 1)$, while $m \to \infty$. As both ρ_1 and ρ_2 are continuous over $(M_{im}, N_{im}) \in [\beta, 1] \times [\beta, 1]$, a small separation of (11) away from zero by introducing $\sigma = (\sigma_1, \sigma_2)$ (see (16)) should still keep the shifted λ_e^σ away from $[0.3, 1)$. This is illustrated in Fig. 4 with the aid of *Region-Plot* function visualizing 2D admissible zone $A_{0.3}$ against the buffer admissible zones A_λ^σ, with $\sigma_1 = \sigma_2 = 0.01$ and $\lambda \in \{0.3, 0.5, 0.7, 08\}$. Note that similarly to A_λ we also have $A_{\lambda_1}^\sigma \subset A_{\lambda_1}^\sigma$ for $\lambda_2 > \lambda_1$.

A linear regression to approximate $\alpha(\lambda)$ is used here with $m_{min} = 101 \leq m \leq m_{max} = 121$. The computed estimates $\bar{\alpha}(\lambda) \approx \alpha(\lambda)$ for various $\lambda \in [0,1]$ are shown in Table 2. Visibly the sharpness of Th. 3 is experimentally confirmed. Note that the asymptotics established in Th. 3 holds for any $\lambda \in [0,1]$ without actually assuming ψ_i to be a reparameterization. However, the latter is vital in approximating length of γ or if one-two-one local correspondence between interpolant $\hat{\gamma}_2$ and the curve γ is required.   □

Table 2. Estimated $\bar{\alpha}(\lambda) \approx \alpha(\lambda)$, for γ_h and (19) interpolated by $\hat{\gamma}_2$ with $\lambda \in [0, 1]$

$\lambda = 0.00$	$\lambda = 0.10$	$\lambda = 0.33$	$\lambda = 0.50$	$\lambda = 0.70$	$\lambda = 0.90$	$\lambda = 1.00$
1.001	1.000	1.001	0.999	1.001	1.058	2.931

Example 2. Consider *a planar regular convex spiral* $\gamma_{sp} : [0, 5\pi] \to E^2$

$$\gamma_{sp}(t) = ((6\pi - t)\cos(t), (6\pi - t)\sin(t)) \tag{20}$$

sampled first as in (19). Figure 5 a) illustrates the curve γ_{sp} sampled along (19) with Q_{22}. Evidently, the same conclusion as reached in the last example,

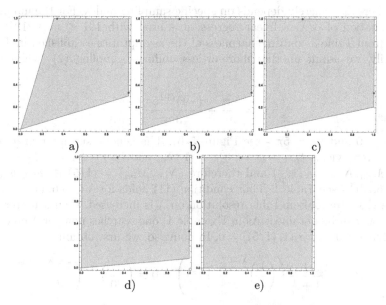

Fig. 4. The plots of admissible zones: a) $A_{0.3}$ b)-e) buffer $A_\lambda^{0.01}$ for $\lambda \in \{0.3, 0.5, 0.7, 0.9\}$ all containing samplings (19) (represented by two red dots)

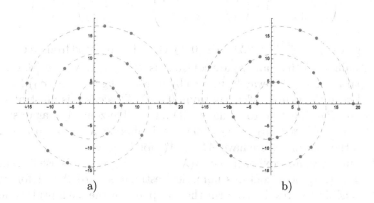

Fig. 5. The plot of γ_{sp} from (20) sampled according to: a) (19) and b) (21), for $m = 22$

Table 3. Estimated $\bar{\alpha}(\lambda) \approx \alpha(\lambda) = 1$ for γ_{sp} and (19) interpolated by $\hat{\gamma}_2$ with $\lambda \in [0, 1]$

$\lambda = 0.00$	$\lambda = 0.10$	$\lambda = 0.33$	$\lambda = 0.50$	$\lambda = 0.70$	$\lambda = 0.90$	$\lambda = 1.00$
0.980	0.982	1.012	1.021	1.009	1.619	2.997

applies here as it merely depends on specific sampling and λ. For the numerical approximation of $\alpha(\lambda)$ a linear regression is used with $101 \leq m \leq 121$. The results from Table 3 confirm sharpness of the asymptotics established in Th. 3.

Finally, we admit another more-or-less uniform sampling (2) (where $\beta = (1/5)$):

$$t_i = \frac{5\pi i}{m} + \frac{(-1)^{i+1}5\pi}{3m}, \tag{21}$$

with $t_0 = 0$ and $t_m = 5\pi$ - see Figure 5 b). The generic sub-interval $[t_i, t_{i+2}]$ (with i even) yields $t_{i+1} - t_i = \frac{5 \cdot 5\pi}{3m} = 1 \cdot \delta_m$ and $t_{i+2} - t_{i+1} = \frac{5\pi}{3m} = \frac{1}{5}\delta_m$. Thus $(M_{im}, N_{im}) = (1, \frac{1}{5})$ and therefore $(N_{im}/M_{im}) = (1/5) \in (a_{0.6}, b_{0.6}) = (0.148, 6.761)$ - see Table 1. Thus condition (11) holds for $\lambda \in [0.6, 1)$.

Note that if a look-up table resolution on λ is increased, then a better estimate of λ_0 can be obtained. As in Example 1, one searches now for λ enforcing $\rho_1(1, (1/5)) = 0$ and $\rho_2(1, (1/5)) = 0$. In doing so, we first obtain

$$1^{\lambda-1} - \frac{1}{1+\frac{1}{5}}\left(\left(\frac{1}{5}\right)^{\lambda-1} - 1^{\lambda-1}\right) = 0 \equiv \left(\frac{9}{20}\right) = \left(\frac{1}{5}\right)^{\lambda}.$$

Hence $\lambda_e = \frac{\ln(20)-\ln(9)}{\ln(5)} \approx 0.4961 \in [0, 1)$ is separated from $[0.6, 1)$ for $m \to \infty$. Similarly, $\rho_2(1, (1/5)) = 0$ amounts to

$$1^{\lambda-1} + \left(1 + \frac{\frac{1}{5}}{1+\frac{1}{5}}\right)\left(\left(\frac{1}{5}\right)^{\lambda-1} - 1^{\lambda-1}\right) = 0 \equiv \left(\frac{1}{35}\right) = \left(\frac{1}{5}\right)^{\lambda}.$$

The latter gives $\lambda_e = \frac{\ln(35)}{\ln(5)} \approx 2.2090 \notin [0, 1)$ which is separated from $[0.6, 1)$ when $m \to \infty$. Again, as both ρ_1 and ρ_2 are continuous over $(M_{im}, N_{im}) \in [\beta, 1] \times [\beta, 1]$ a small separation of (11) away from zero by introducing $\sigma = (\sigma_1, \sigma_2)$ (see (16)) should still keep the shifted λ_e^σ away from $[0.6, 1)$. This effect is illustrated with the aid *RegionPlot* function to visualize 2D admissible zone $A_{0.6}$ against buffer admissible zones A_λ^σ, with $\sigma_1 = \sigma_2 = 0.01$ and different $\lambda \in \{0.6, 0.7, 0.8, 0.9\}$. Again, as in the case of A_λ, we have $A_{\lambda_1}^\sigma \subset A_{\lambda_2}^\sigma$ for $\lambda_1 < \lambda_2$.

For the numerical approximation of $\alpha(\lambda)$ a linear regression is used again with $101 \leq m \leq 121$. Table 4 contains numerical estimates of $\alpha(\lambda) = 1$ for various $\lambda \in [0, 1)$. Visibly the results confirm the sharpness of the asymptotics derived in Th. 3.   □

Table 4. Estimated $\bar{\alpha}(\lambda, 0) \approx \alpha(\lambda, 0) = 1$ (with $\lambda \in [0, 1)$) and $\bar{\alpha}(1, 0) \approx \alpha(1, 0) = 3$ for γ_{sp} and sampling (21) interpolated by $\hat{\gamma}_2$ with $\lambda \in [0, 1]$

$\lambda = 0.00$	$\lambda = 0.10$	$\lambda = 0.33$	$\lambda = 0.50$	$\lambda = 0.70$	$\lambda = 0.90$	$\lambda = 1.00$
0.990	0.990	0.991	0.998	1.020	1.320	3.000

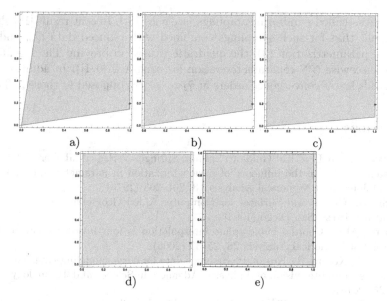

Fig. 6. The plots of admissible zones: a) $A_{0.6}$ b)-e) buffer $A_\lambda^{0.01}$ for $\lambda \in \{0.6, 0.7, 0.8, 0.9\}$ all containing samplings (21) (represented by single red dot)

5   Conclusions

This work supplements Th. 3 (see [8]) concerning the asymptotics in trajectory estimation via piecewise-quadratic interpolation $\hat{\gamma}_2 : [0, \hat{T}] \to E^n$ based on reduced data Q_m and exponential parameterization (3). The curve $\gamma : [0, T] \to E^n$ is assumed to be sampled more-or-less uniformly (2) for $\lambda \in [0, 1)$ and sampled along (1) for $\lambda = 1$. Th. 3 assumes the existence of a reparameterization $\psi : [0, T] \to [0, \hat{T}]$ between γ and $\hat{\gamma}_2$. This paper formulates sufficient conditions imposed on samplings (2) and $\lambda \in [0, 1)$ to guarantee that $\dot\psi > 0$ - see Th. 4 and (16). Recall that for $\lambda = 1$, a quadratic ψ defines a reparameterization (see [8]). The analysis performed here and the accompanying procedure is supported by the geometrical insight, symbolic computation and numerical verification of both Th. 3 and Th. 4. A possible extension of this work is to invoke smooth interpolation schemes (see [1]) combined with reduced data and exponential parameterization (see [11]). Certain clues may be found in [4], where complete C^2 splines [1] are dealt with $\lambda = 1$ to establish the fourth orders of convergence in trajectory and length estimation. In general for length estimation (as well as for the genuine trajectory tracking) the mapping ψ needs to be a reparameterization (see also [7]). The analysis of C^1 interpolation for reduced data with cumulative chords can be found in [7] or [9]. Some applications and theory on non-parametric interpolation can be found e.g., in [11], [7] or [5]. Related work on other knots' parameterizations are discussed in [11], [12], [6], [13] and [2]. Finally, we should point out that one can also consider the other subfamilies of admissible samplings (1) different than more-or-less uniform (2). One of them introduces so-called ε-uniform samplings (see e.g. [15]). Recent result [10] shows (among all) that for such samplings combined with reduced data Q_m and exponential parametrization (3), the quadratic ψ used in proving Th. 4 defines a genuine piecewise C^∞ reparameterization for each $\lambda \in [0,1]$). In addition, the latter yields faster convergence orders in $\hat{\gamma}_2 \circ \psi - \gamma$ as opposed to those in Th. 3.

References

1. de Boor, C.: A Practical Guide to Splines. Springer-Verlag, Heidelberg (2001)
2. Epstein, M.P.: On the influence of parameterization in parametric interpolation. SIAM Journal of Numerical Analysis **13**, 261–268 (1976)
3. Farin, G.: Curves and Surfaces for Computer Aided Geometric Design, 3rd edn. Academic Press, San Diego (1993)
4. Floater, M.S.: Chordal cubic spline interpolation is fourth order accurate. IMA Journal of Numerical Analysis **26**, 25–33 (2006)
5. Janik, M., Kozera, R., Kozioł, P.: Reduced data for curve modeling - applications in graphics, computer vision and physics. Advances in Science and Technology **7**(18), 28–35 (2013)
6. Kocić, L.M., Simoncelli, A.C., Della Vecchia, B.: Blending parameterization of polynomial and spline interpolants, Facta Universitatis (NIŠ). Series Mathematics and Informatics **5**, 95–107 (1990)
7. Kozera, R.: Curve modeling via interpolation based on multidimensional reduced data. Studia Informatica **25**(4B–61), 1–140 (2004)
8. Kozera, R., Noakes, L.: Piecewise-quadratics and exponential parametereziation for reduced data. Applied Mathematics and Computaion **221**, 1–19 (2013)
9. Kozera, R., Noakes, L.: C^1 interpolation with cumulative chord cubics. Fundamenta Informaticae **61**(3–4), 285–301 (2004)
10. Kozera, R., Noakes, L.: Exponential parameterization and ε-uniformly sampled reduced data. Applied Mathematics and Information Sciences. Accepted for publication
11. Kvasov, B.I.: Methods of Shape-Preserving Spline Approximation. World Scientific Publishing Company, Singapore (2000)
12. Lee, E.T.Y.: Choosing nodes in parametric curve interpolation. Computer-Aided Design **21**(6), 363–370 (1989)
13. Mørken, K., Scherer, K.: A general framework for high-accuracy parametric interpolation. Mathematics of Computaion **66**(217), 237–260 (1997)
14. Noakes, L., Kozera, R.: Cumulative chords piecewise-quadratics and piecewise-cubics. In: Klette, R., Kozera, R., Noakes, L., Weickert, J. (eds.) Geometric Properties of Incomplete Data, Computational Imaging and Vision, vol. 31, pp. 59–75. Kluver Academic Publishers, The Netherlands (2006)
15. Noakes, L., Kozera, R., Klette, R.: Length estimation for curves with different samplings. In: Bertrand, G., Imiya, A., Klette, R. (eds.) Digital and Image Geometry. LNCS, vol. 2243, pp. 339–351. Springer, Heidelberg (2002)
16. Piegl, L., Tiller, W.: The NURBS Book. Springer-Verlag, Heidelberg (1997)
17. Wolfram Mathematica 9, Documentation Center.
    http://reference.wolfram.com/mathematica/guide/Mathematica.html

Parametric Solvable Polynomial Rings and Applications

Heinz Kredel

IT-Center, University of Mannheim, Germany
kredel@rz.uni-mannheim.de

Abstract. We recall definitions and properties of parametric solvable polynomial rings and variants. For recursive solvable polynomial rings, i.e. solvable polynomial rings with coefficients from a solvable polynomial ring, also commutator relations between main and coefficient variables are introduced. From these rings solvable quotient and residue class rings can be constructed and used as coefficients of solvable polynomials. The resulting skew extension fields of the ground field can be applied for skew root finding or primary ideal decomposition. We present the design and implementation of these rings in the strongly typed, generic, object oriented computer algebra system JAS.

1 Introduction

Between commutative algebra and free non-commutative algebras lies an important area of solvable algebras, also called PBW algebras. Solvable algebras and polynomial rings still share many properties with the well studied polynomial rings, for example, being Noetherian rings or being tractable with computational means, like Gröbner bases, root finding, factorization or quantifier elimination. In contrast, free non-commutative polynomial rings are in general no more Noetherian and computational means suffer from infinite ideals with non-terminating Gröbner base computations and non-unique representations. Though, problems in commutative and solvable polynomial rings are not such easily computable as many algorithms have complexity bounds exponential in the number of variables.

In this paper we explore novel areas of solvable polynomial rings which have partly been addressed in theory, but lack implementations in computer algebra systems. Implementations of solvable polynomial ring algorithms have the advantage of being able to share the representation and many arithmetic algorithms with commutative polynomial rings but have to be efficient in the handling and application of commutator relations to carry out the non-commutative multiplications. The implementation of solvable polynomial rings should further be generic in the sense that various coefficient rings must be usable in a strongly type safe way with optimal compiling and performing code.

The new topic discussed in the article are parametric solvable polynomial rings which also define commutator relations between variables of the polynomial ring and the coefficient rings and new non-commutative coefficient rings composed

© Springer International Publishing Switzerland 2015
V.P. Gerdt et al. (Eds.): CASC 2015, LNCS 9301, pp. 275–291, 2015.
DOI: 10.1007/978-3-319-24021-3_21

of quotients of solvable polynomials and others. The new coefficient rings arise
in the investigation of primary ideal decomposition in solvable polynomial rings
as solvable extension fields of the base field.

The implementation is discussed in the framework of the Java Algebra System
(JAS), see [1, 2]. It facilitates the modeling of algebraic structures in a strongly
typed, generic, object oriented computer algebra software. JAS provides Java
libraries which provide several algorithm versions for greatest common divisor,
square-free decomposition, factorization and Gröbner bases computation in separate packages [3]. The libraries are enhanced for interactive usage with the help
of the Jython and JRuby scripting languages [4]. The implementations make use
of the multi-core CPUs, for example in greatest common divisor computations
[3] or parallel and distributed Gröbner base computations [5].

1.1   Related Work

The study of computer algebra aspects of enveloping fields of Lie algebras was
started by Apel and Lassner in [6, 7]. The solvable polynomial rings were first
studied by Kandri-Rodi and Weispfenning in [8]. Mora provided a more general
framework for non-commutative polynomial rings [9]. Comprehensive Gröbner
bases as a method to solve parametric (commutative) problems was introduced
by Weispfenning in [10]. Kredel and Weispfenning extended the method of Comprehensive Gröbner bases to parametric solvable polynomial rings [11, 12]. Ideal
theoretic properties, like primary ideal decomposition, of solvable and PBW algebras are studies by Gomez-Torrecillas [13]. An implementation of solvable and
PBW algebras in the Singular computer algebra system and further properties
of these algebras were studied by Levandovskyy [14, 15]. Today differential operator rings are implemented in Mathematica or Maple. Further related work is
mentioned in the paper as required.

1.2   Outline

The paper is organized as follows: Section 2 recalls some mathematical background on solvable polynomial rings. The design and implementation aspects of
solvable polynomial rings in JAS are presented in 3. Lastly, section 4 shows some
applications and examples.

2   Solvable Polynomial Rings

Recall from [12] the concept of solvable polynomial rings with commutator relations between variables and between variables and coefficients. Solvable polynomials rings S are associative rings $(S, 0, 1, +, -, *)$, with

$$S = \mathbf{K}\{X_1, \ldots, X_n; Q; Q'\} \tag{1}$$

and characterized as polynomial rings over skew fields $\mathbf{K}$ in variables $X_1, \ldots, X_n$,
$n \geq 0$, together with a new non-commutative product '$*$', defined by means of

commutator relations

$$Q = \{X_j * X_i = c_{ij} X_i X_j + p_{ij} : 0 \neq c_{ij} \in \mathbf{K}, X_i X_j > p_{ij} \in S, 1 \leq i < j \leq n\}$$

between the variables with respect to a $*$-compatible term order $>$ on $S \times S$ (extended from the order $>$ on the set of terms), and commutator relations

$$Q' = \{X_i * a = c_{ai} a X_i + p_{ai} : 0 \neq c_{ai} \in \mathbf{K}, p_{ai} \in \mathbf{K}, 1 \leq i \leq n, a \in \mathbf{K}\}$$

between the variables and the coefficients. In case, the commutator relations Q or Q' are empty, i.e. the respective relations are treated as commutative, and $\mathbf{K}$ is commutative, S is a commutative polynomial ring $S = \mathbf{K}[X_1, \ldots, X_n]$.

2.1   Parametric Solvable Polynomial Rings

In chapter 7, section 1, of [12], parametric coefficient rings were introduced to establish Comprehensive Gröbner bases [10] for solvable polynomial rings. In this case, the coefficient field $\mathbf{K}$ is replaced by a (commutative) ring R with commuting parameter variables $\{U_1, \ldots, U_m\}$ over a domain $\mathbf{R}$, $R = \mathbf{R}[U_1, \ldots, U_m]$. The parametric solvable polynomial ring $S = R\{X_1, \ldots, X_n; Q\}$, with empty Q', is denoted by

$$S = \mathbf{R}[U_1, \ldots, U_m]\{X_1, \ldots, X_n; Q\}. \tag{2}$$

With the help of the product lemma 1 the non-commutative multiplication of two solvable polynomials can be written as the commutative product of these polynomials times an element of R plus some rest, which is less than the product (in the given term order $<$).

Lemma 1 (7.1.2 in [12]). *Let $\mathbf{R}$ be a commutative Noetherian domain, $m \in \mathbb{N}$, $R = \mathbf{R}[U_1, \ldots, U_m]$. Let $S = R\{X_1, \ldots, X_n; Q\}$ be a parametric solvable polynomial ring as defined in Axioms 7.1.1 in [12] with respect to a $*$-compatible term order $<$. Let C be the multiplicative subset of R generated by the c_{ij} from the commutator relations Q. Then for $0 \neq f, g \in S$ one can compute $0 \neq c \in C$ and $p \in S$ with $p < f \cdot g$ such that*

$$f * g = c \cdot f \cdot g + p.$$

c and p are uniquely determined by these properties and the coefficients of p in R are polynomials in the c_{ij}, the coefficients of all p_{ij} from the commutator relations Q and of the coefficients of f, g. Furthermore these polynomials are formed uniformly, independently of the ring R.

2.2   Solvable Polynomial Coefficient Rings

In this section we define parametric solvable polynomial rings with coefficients from solvable polynomials and commutator relations between variables of the 'main' ring and the coefficient ring.

$$S = \mathbf{R}\{U_1, \ldots, U_m; Q_u\}\{X_1, \ldots, X_n; Q_x; Q'_{ux}\} \tag{3}$$

The coefficients are from a solvable polynomial ring $R = \mathbf{R}\{U_1, \ldots, U_m; Q_u\}$ in the variables $U_1, \ldots, U_m$, together with commutator relations Q_u between the U variables,

$$Q_u = \{ U_j * U_i = c_{uij} U_i U_j + p_{uij} :$$
$$0 \neq c_{uij} \in \mathbf{R}, U_i U_j > p_{uij} \in R, 1 \leq i < j \leq m \}.$$

Q'_u is assumed to be empty, i.e. the elements of $\mathbf{R}$ commute with the U variables.

For the main solvable polynomial ring $S = R\{X_1, \ldots, X_n; Q_x; Q'_{ux}\}$ in the variables $X_1, \ldots, X_n$ over R, there are commutator relations between the X variables Q_x and between the U and X variables Q'_{ux},

$$Q_x = \{ X_j * X_i = c_{xij} X_i X_j + p_{xij} :$$
$$0 \neq c_{xij} \in R, X_i X_j > p_{xij} \in S, 1 \leq i < j \leq n \}$$

and

$$Q'_{ux} = \{ X_j * U_i = c_{ij} U_i X_j + p_{ij} :$$
$$0 \neq c_{ij} \in \mathbf{R}, U_i X_j > p_{ij} \in S, 1 \leq i \leq m, 1 \leq j \leq n \}.$$

The p_{ij} are allowed to lie in S, and not only in R, provided that $p_{ij} < U_i X_j$. It is assumed that the elements of $\mathbf{R}$ commute with the U and X variables. The term orders $<$ are assumed to be $*$-compatible in the respective rings.

2.3   Recursive Solvable Polynomial Rings

The solvable polynomial ring from equation (3) does not completely match the situation of the ring from (1) as it still does not allow commutator relations between arbitrary base coefficients and main variables. However, it models *recursive* solvable polynomial rings, where main variables can be shifted to coefficient variables and vice versa as desired by an application.

$$S_k = \mathbf{R}\{X_1, \ldots, X_k; Q_k\}\{X_{k+1}, \ldots, X_n; Q_n; Q'_{kn}\}, \quad 0 \leq k \leq n \qquad (4)$$

The cases $k = 0$ and $k = n$ recover the usual non-parametric cases: $S_0 = S_n = \mathbf{R}\{X_1, \ldots, X_n; Q\}$.

2.4   Solvable Quotient and Residue Rings

With the help of Gröbner bases in solvable polynomial rings it is possible to define and construct various kinds of quotient rings or residue class rings.

Let $\mathcal{I} = \mathrm{ideal}(F)$ be a two-sided ideal for a (non-empty, finite) subset $F = \{f_1, \ldots, f_k\}, k \geq 1, F \subset R = \mathbf{R}\{U_1, \ldots, U_m; Q_u\}$. Then the following constructions are possible

1. solvable quotient ring, $\mathbf{R}(U_1, \ldots, U_m; Q_u)$, a skew field,
2. solvable residue class ring modulo $\mathcal{I}$, $\mathbf{R}\{U_1, \ldots, U_m; Q_u\}_{/\mathcal{I}}$,

3. solvable local ring, localized by ideal $\mathcal{I}$, $\mathbf{R}\{U_1, \ldots, U_m; Q_u\}_{\mathcal{I}}$,
4. solvable quotient and residue class ring modulo $\mathcal{I}$, $\mathbf{R}(U_1, \ldots, U_m; Q_u)_{/\mathcal{I}}$, it is a skew field if $\mathcal{I}$ is a completely prime ideal.

For the general theory of skew field constructions see [16]. The quotient ring from item 4) is the ring to be used in a prime or primary decomposition of ideals in solvable polynomial rings. This is different to commutative polynomial rings, where the (commutative) residue class ring from 2) can be used. As in this case it is possible to construct inverses modulo prime ideals, which are useless in the solvable case (or non-commutative case) [13].

The constructions are possible since the solvable polynomial rings R satisfy the left, and right Ore conditions, since they are Noetherian rings [12]. This means, that for any $a, b \in R$, there exist $c, d, c', d' \in R$ with

$$c * a = d * b \qquad \text{and} \qquad a * c' = b * d'$$

Such elements can be computed by left, respectively right syzygy computation in R, see section 3 and [6].

2.5 Solvable Quotient Rings as Coefficient Rings

We now consider solvable polynomial rings over the rings from the previous section. Interesting are the rings 1, 3, and 4, for example 1,

$$S = \mathbf{R}(U_1, \ldots, U_m; Q_u)\{X_1, \ldots, X_n; Q_x, Q'_{ux}\}. \tag{5}$$

The multiplication, in particular right multiplication between coefficients $\frac{n}{d} \in \mathbf{R}(U_1, \ldots, U_m; Q_u)$ and main variables $x \in \{X_1, \ldots, X_n\}$, has to be examined. For left fractions $d^{-1}*n$, let $x^e * d^{-1} * n = (x^e * \frac{1}{d}) * \frac{n}{1}$ and for right fractions $n * d^{-1}$, let $x^e * n * d^{-1} = (x^e * \frac{n}{1}) * \frac{1}{d}$. The multiplication by $\frac{n}{1}$ is just multiplication by n. The right multiplication by $\frac{1}{d}$ can be performed with the help of the following lemma. It shows moreover that the multiplication by $\frac{1}{d}$ is completely determined by the multiplication with d and the axioms of associative rings.

Lemma 2. *Assume $c_{ij} = 1$ in Q'_{ux}. Let $x^e * d = dx^e + p$, $dx^e > p \in S$, then*

$$x^e * \frac{1}{d} = \frac{1}{d}(x^e - (p * \frac{1}{d})). \tag{6}$$

Proof. The identity can be derived under the assumption that all $c_{ij} = 1$ in Q'_{ux}, as follows. If not all $c_{ij} = 1$ with some more care a corresponding factor c as product of c_{ij}'s can be established. Let $z = x^e$, then from $z * \frac{1}{d} * d = z$ and $p = z * d - dz$ together with the assumption $z * \frac{1}{d} = \frac{1}{d}z + q$ for some q, it follows $(\frac{1}{d}z + q) * d = z$ and $\frac{1}{d}(dz + p) + q * d = z$. Multiplying out, we get $\frac{1}{d}d * z + \frac{1}{d}p + q * d = z$ and so $\frac{1}{d}p + q * d = 0$ must hold. Multiplying with $\frac{1}{d}$ from right, we have $\frac{1}{d}p * \frac{1}{d} + q = 0$ and so $q = -\frac{1}{d}p * \frac{1}{d}$. With this, we get $z * \frac{1}{d} = \frac{1}{d}z - \frac{1}{d}p * \frac{1}{d} = \frac{1}{d}(z - p * \frac{1}{d})$. Since by assumption $p < z$ the claim follows by induction as finally $p \in R$ and the multiplication can be carried out in R. □

We are now prepared and turn to the design and implementation of parametric solvable polynomial rings from this section.

3   Implementation of Solvable Polynomial Rings

In this section we present the design and implementation of parametric solvable polynomial rings with solvable coefficient rings. We first summarize the existing classes and then turn to the design and implementation of the new variants.

Since its beginning, the JAS project [17], included polynomial rings and solvable polynomial rings. Polynomials are implemented using generic type parameters of Java 1.5 (or later). The implementation becomes type safe, interoperable and composable between various algebraic structures. We only discus relevant parts for this paper, for further details see [1, 17].

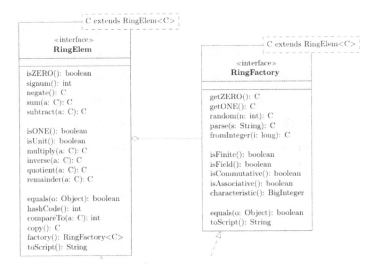

Fig. 1. Ring elements and ring factory interfaces.

3.1   Polynomial Rings

The coefficient rings must implement the `RingElem` interface, defining the main arithmetic operations, like + or ∗, and some utility methods, like `equals()`, mandated by the Java platform, see figure 1. Ring elements can be recursive by definition `C extends RingElem<C>` and know about the ring to which they belong by the `factory()` method. The rings, to which an ring element belongs, are implemented by a `RingFactory`. The ring factories handle most aspects of the creation of ring elements, like `getZERO()`, and methods to query mathematical properties of the ring, like `isFinite()`.

The class `GenPolynomial` implements `RingElem<GenPolynomial<C>>` and restricts the coefficients also to implementations of `RingElem`, see figure 2. Besides the methods mandated by the `RingElem` interface it implements methods peculiar to polynomials, like `leadingBaseCoefficient()` or `reductum()`. The polynomial ring factory `GenPolynomialRing` needs at least a factory for the coefficient ring and the number or names of the variables. Besides methods mandated

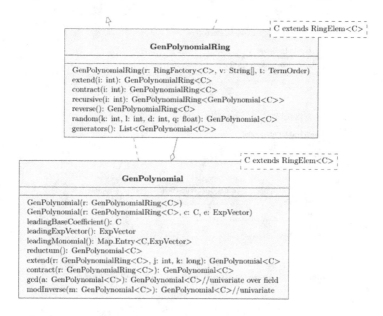

Fig. 2. Polynomial and polynomial ring.

by the `RingFactory` interface it implements methods for polynomial rings, like `extend()` or `contract()`, which add or remove variables.

3.2   Solvable Polynomial Rings

Solvable polynomials are implemented by classes `GenSolvablePolynomial` and `GenSolvablePolynomialRing`. Both extend the respective base class `GenPoly-nomial` and `GenPolynomialRing` to reuse common code, see figures 3 and 4. Solvable polynomials override the `multiply()` method with the new $*$ multipli-cation and rely on the implementation of other methods from `GenPolynomial`, like `sum()` or `leadingExpVector()`. The implementation ensures, that the re-sult of `sum()` correctly produces a solvable polynomial with the help of the ring factory for solvable polynomials available in the summand, respectively `this`.

The `GenSolvablePolynomialRing` maintains a table of commutator relations, class `RelationTable`, to implement the $*$-product. Missing relations between variables are treated as commutative relations. Relations between powers of variables are stored and used for faster computation of the product when they become available. The coefficients are assumed to commute with the polynomial variables but need not be from a commutative ring themselves.

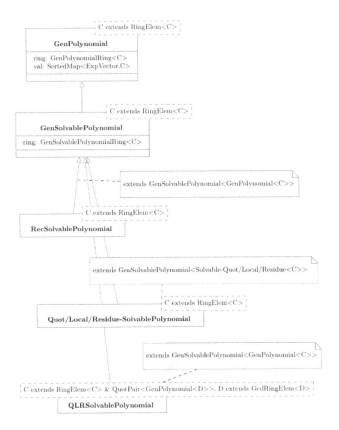

Fig. 3. Solvable polynomial classes overview.

Issues with Type Erasure. The implementation with Java type parameters has one caveat which has been accepted by the language design to be able to also use non typed library code together with typed code, namely "type erasure". This removes all nested type information and keeps only the outermost type. In our case we would like to let solvable polynomials implement the interface

```
RingElem<GenSolvablePolynomial<C>>
```

However, it can only implement `RingElem<GenPolynomial<C>>` because it extends `GenPolynomial`. Now type erasure reduces both more specific types of `RingElem<.>` just to `RingElem`, leading to a type clash. This has the consequence that we must use type casts sometimes. The run-time types of the polynomial objects are correctly constructed by the implementation. Type erasure makes the code less readable, but has otherwise no further undesirable issues [1].

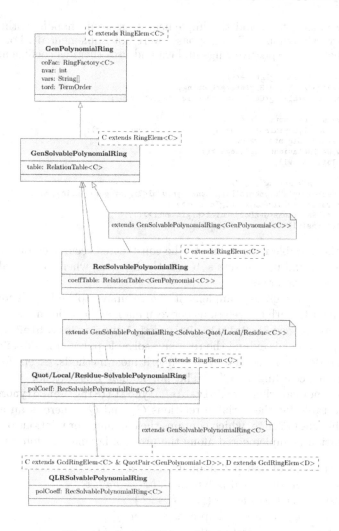

Fig. 4. Solvable Polynomial ring classes.

3.3 Recursive Solvable Polynomial Rings

By the design of the `RingElem` interface and the polynomial and solvable polynomial classes design it was possible to construct recursive polynomial rings from the beginning. For example, the ring from equation (2) has type

```
GenSolvablePolynomialRing<GenPolynomial<C>>
```

and the polynomial has type

```
GenSolvablePolynomial<GenPolynomial<C>>.
```

But also the recursive solvable polynomial ring from equation (3), only without the relations between main and coefficient variables Q'_{ux}, had been possible.

The type was as before, and the implementation would handle arithmetic operations correctly (though Gröbner bases were not implemented). For example, one constructs the respective rings in Java and uses them as coefficients.

```
BigInteger cfac = new BigInteger();
GenSolvablePolynomialRing<BigInteger> cring;
GenSolvablePolynomialRing<GenPolynomial<BigInteger>> ring;

String[] cvars = new String[] { "a", "b" };
cring = new GenPolynomialSolvableRing<BigInteger>(cfac, cvars);
RelationGenerator<BigInteger> cwl
      = new WeylRelations<BigInteger>();
cring.addRelations(cwl);

String[] vars = new String[] { "x", "y" };
ring = new GenSolvablePolynomialRing<GenPolynomial<BigInteger>>(cring, vars);
RelationGenerator<GenPolynomial<BigInteger>> wl
      = new WeylRelations<GenPolynomial<BigInteger>>();
ring.addRelations(wl);
```

Finally the variable `ring` points to the constructed recursive polynomial ring in variables `x,y` and parameters `a,b` over the integers, where the variables commute with the parameters.

It remains to design and implement the recursive solvable polynomial ring from equation (3), with the relations between main and coefficient variables Q'_{ux}. The classes are named `RecSolvablePolynomial` and `RecSolvablePolynomialRing`. The polynomials extend the recursive solvable polynomials `GenSolvablePolynomial<GenPolynomial<C>>` and inherit most methods, except for the new multiplication according to Q'_{ux}. The polynomial ring extends the recursive solvable polynomial ring `GenSolvablePolynomialRing<GenPolynomial<C>>`. Besides the table for the variable relations Q_x and Q_u, there is an additional relation table `coeffTable`, which contains the commutator relations of Q'_{ux}. The $*$-multiplication is implemented along the proof of lemma 1 as follows.

1. loop over terms of first polynomial: $ax^e = a'u^{e'}x^e$
2. loop over terms of second polynomial: $bx^f = b'u^{f'}x^f$
3. compute $(ax^e) * (bx^f)$ as $a * ((x^e * b) * x^f)$
   (a) $x^e * b = p_{eb}$, iterate lookup of $x_i * u_j$ in Q'_{ux}
   (b) $p_{eb} * x^f = p_{ebf}$, iterate lookup of $x_j * x_i$ in Q_x
   (c) $a * (p_{ebf}) = p_{aebf}$, in recursive coefficient ring lookup $u_j * u_i$ in Q_u
4. sum up the p_{aebf}

The implementation remembers computed powers of variables, for example $X_j^h * X_i^g = X_i^g X_j^h + p_{igjh}$, $i < j$. These relations are added to the respective relation table to avoid re-computation of the polynomials p_{igjh}. Arbitrary power products of variables are not stored but recomputed when needed. It could be a future optimization to track and store all products. The complexity of the multiplication seems to be very high. However, in cases when most variables commute, for example with Weyl differential polynomial rings, not too much expensive $*$-multiplications are executed. There is a new method `rightRecursivePolynomial` to represent the polynomial with coefficients on the right of the variables.

3.4   Solvable Quotient and Residue Rings

The solvable quotient and residue rings from section 2.4 are implemented by the following classes. Let the ideal $\mathcal{I}$ be defined as above, then

1. the solvable quotient ring, $\mathbf{R}(U_1, \ldots, U_m; Q_u)$, is implemented by classes `SolvableQuotient` and `SolvableQuotientRing`
2. the solvable residue class ring modulo $\mathcal{I}$, $\mathbf{R}\{U_1, \ldots, U_m; Q_u\}_{/\mathcal{I}}$, is implemented by classes `SolvableResidue` and `SolvableResidueRing`
3. the solvable local ring, localized by ideal $\mathcal{I}$, $\mathbf{R}\{U_1, \ldots, U_m; Q_u\}_{\mathcal{I}}$, is implemented by classes `SolvableLocal` and `SolvableLocalRing`
4. the solvable quotient and residue class ring modulo $\mathcal{I}$, $\mathbf{R}(U_1, \ldots, U_m; Q_u)_{/\mathcal{I}}$, is implemented by classes `SolvableLocalResidue` and `SolvableLocalResidueRing`

The left and right Ore conditions can be computed by left, respectively right syzygy computation in R, see e.g. [6]. Implementations can be found in classes which implement the `SolvableSyzygy` interface, namely `SolvableSyzygySeq`. The respective methods are called `leftOreCond()` and `rightOreCond()`.

The implementation of quotient construction suffers from the fact, that there is no suitable greatest common divisor implemented at the moment in solvable polynomial rings to be able to reduce the quotient to lower terms in contrast to commutative polynomial rings. There is a method `leftSimplifier()` after [7] to reduce the quotients to lower terms, using module Gröbner bases of syzygies of quotients. However, its complexity is very high and it can only be used for small degrees and small number of variables. A package for greatest common divisor computation in solvable polynomial rings (when being Ore domains) is under construction, but not yet in a useful state. It will be discussed elsewhere, see the source code in [2], package `edu.jas.fd` (factorization domain). However, greatest common divisors in solvable polynomial rings are not unique as in the commutative case.

3.5   Solvable Polynomial Rings with Non-commutative Coefficients

We can now discuss solvable polynomials with coefficients from the solvable quotient and residue rings in the previous section. The relation to solvable and recursive solvable polynomials and rings is shown in figures 3 and 4. We need two variants, one for residue class coefficients and one for quotient-like coefficients.[1] We try to use as much as possible the recursive solvable implementation.

The easy case is the residue class coefficient case. The implementation is in classes `ResidueSolvablePolynomial` and `ResidueSolvablePolynomialRing` which extend solvable polynomials and rings over `SolvableResidue<C>`. The ring contains a reference to a `RecSolvablePolynomialRing` named `polCoeff`. This ring has only solvable polynomials as coefficients. It is used to compute the product of polynomials with representatives of the multiplied residue classes.

---

[1] Actually there are five classes, three being obsolete by now and not discussed here.

From this, the product residue coefficients are constructed. So the polynomial multiplication implementation can be greatly simplified by delegation to the recursive solvable polynomial ring.

More difficult are the quotient-like coefficient cases. Instead of three separate implementations, we abstract the quotient-like appearance with an interface `QuotPair`. It defines methods to get the numerator and denominator of a quotient and to test if both are constants (as polynomials). The corresponding `QuotPairFactory` interface defines methods for creation of quotients `create(num)` and `create(num,den)` and the method `pairFactory()` to get a factory for the pair entries. The implementation is in classes `QLRSolvable-Polynomial` and `QLRSolvablePolynomialRing` which extend solvable polynomials and rings over `QuotPair<GenPolynomial<D>>`. The ring contains a reference to a `RecSolvablePolynomialRing` named `polCoeff` with only solvable polynomial coefficients. This restricts the commutator relations to polynomial relations, i.e. no denominators $\neq 1$ may appear. The ring has methods `fromPoly-Coefficients()` and `toPolyCoefficients()` to convert between polynomial coefficients (of the recursive solvable polynomials) and quotient-like coefficients of this class. In case the $p_{ij} \in R$ from the relations in Q'_{ux} it is sufficient to fill the `polCoeff.coeffTable` with relations. In case the $p_{ij} \in S$, we must fill also the `polCoeff.table` with relations.

We assume that all $c_{ij} = 1$ in Q'_{ux} for the rest of the section. The implementation of the multiplication computes the inner most product $(x^e * \frac{1}{d}) * \frac{n}{1}$ after lemma 2 using the following steps.

1. recursion base, denominator $= 1$: $x^e * \frac{n}{1}$. It computes $x^e * n$ from the recursive solvable polynomial ring, looking up $x^e * n$ in Q'_{ux}, and then converting the result to a polynomial with quotient coefficients.
2. recursion base, denominator $\neq 1$: $x^e * \frac{1}{d}$. Let p be computed by $x^e * d = dx^e + p$ then compute $x^e * \frac{1}{d}$ as $\frac{1}{d}(x^e - (p * \frac{1}{d}))$ by lemma 2. Since $p < x^e$, $p * \frac{1}{d}$ uses recursion on a polynomial with smaller head term, so the algorithm will terminate.
3. numerator $\neq 1$: let $p_{xed} = x^e * \frac{1}{d}$ and compute $p_{xed} * \frac{n}{1}$ by recursion.

We see that the polynomial multiplication is simplified by delegation to the recursive solvable polynomial ring.

4   Applications

We summarize some old and new applications of solvable polynomial rings and give an example for computation over solvable quotient coefficients. We conclude with a sketch of an extension to free non-commutative coefficients.

4.1   Comprehensive Gröbner Bases

Implementations of Comprehensive Gröbner bases in commutative parametric polynomial rings $S = \mathbf{R}[U_1, \dots, U_m][X_1, \dots, X_n]$ have been presented in [18–24].

The commutative version of [22, 24, 25] can be slightly modified to allow the computation of left comprehensive Gröbner bases in $\mathbf{R}[U_1, \ldots, U_m]\{X_1, \ldots, X_n, Q\}$ as presented and implemented in [26]. When the parameter ring is considered as transcendental field extension of $\mathbf{R}$, that is $S = \mathbf{R}(U_1, \ldots, U_m)\{X_1, \ldots, X_n, Q\}$, the well established theory and implementations of commutative or solvable polynomial rings over fields can be employed, see e.g. [1, 12, 27]. Also fraction free variants with coefficients from the ring $\mathbf{R}[U_1, \ldots, U_m]$ and taking primitive parts from the computed polynomials have been implemented.

4.2 Gröbner Bases and Applications

The solvable polynomials with residue or quotient-like coefficients can be used where normal solvable polynomials can be used. This includes left and right Gröbner bases. For two-sided Gröbner bases additionally right multiplications with coefficients have to be carried out, see Theorem 4.11.8 (corrected version) in [12]. In these cases the correct usage of left and right *-multiplication has been verified. Further left and right syzygies and left, right and two-sided module Gröbner bases are computable. For recursive solvable polynomial rings the Gröbner base algorithm is implemented using pseudo reduction (using left Ore conditions to adjust the multipliers).

4.3 Examples

We give some examples of computations in the ring

$$\mathbb{Q}(x, y, z, t; Q_x)/_{\mathcal{I}}\{r; Q_r\},$$

where $Q_x = \{z * y = yz + x, t * y = yt + y, t * z = zt - z\}$ and $\mathcal{I} = (t^2 + z^2 + y^2 + x^2 + 1)$. To avoid the verbose Java code we present the example using the JAS interface to jruby. Use `require "examples/jas";` to load the interface. The example will first use a solvable quotient with residue class in variables x,y,z,t and two-sided ideal [t**2 + z**2 + y**2 + x**2 + 1]. Then a solvable polynomial ring in variable r with coefficients from the just constructed coefficient ring is defined and then used to compute a few left common divisors.

```
pcz = PolyRing.new(QQ(),"x,y,z,t");
zrel = [z, y, ( y * z + x ), t, y, ( y * t + y ),
        t, z, ( z * t - z )];
pz = SolvPolyRing.new(QQ(), "x,y,z,t", PolyRing.lex, zrel);
ff = pz.ideal("", [t**2 + z**2 + y**2 + x**2 + 1]);
ff = ff.twosidedGB();
```

The output two-sided ideal $(x, y, z, t^2 + 1)$ in Ruby notation is

```
SolvIdeal.new(
  SolvPolyRing.new(QQ(),"x,y,z,t",PolyRing.lex,
```

```
    rel=[z, y,  ( y * z + x ), t, z,  ( z * t - z ),
         t, y,  ( y * t + y )]),
 "",[x, y, z, ( t**2 + 1 )])
```

The solvable quotient with residue class ring $\mathbb{Q}(x, y, z, t; Q_x)_{/\mathcal{I}}$ and elements in that ring can be set-up by the function SLR(ideal, numerator, denominator):

```
 f0 = SLR(ff, t + x + y + 1);
 f1 = SLR(ff, z**2+x+1  );
```

With this definitions we compute products and inverses in this ring.

```
 f2 = f1*f0: z**2 * t + x * t + t + y * z**2 + x * z**2 + z**2
           + 2 * x * z + x * y + y + x**2 + 2 * x + 1
 fi = 1/f1: 1 / ( z**2 + x + 1 )
 fi*f1 = f1*fi: 1
 f0*fi: ( x**2 * z * t**2 + ... ) / ( ... + 23 * x + 7 )
        ( 2 * t**2 + 7 ) / ( 2 * t + 7 )
```

One would like to have terms involving x, y and z to be simplified to 0, as shown in the last line, which is not possible in this ring. Next we construct an univariate solvable polynomial ring with this coefficient ring pzc = f0.ring.

```
 pt = SolvPolyRing.new(pzc, "r", PolyRing.lex);
```

We would like to define the same commutator relations for r as for t. However, in this ring we have $x = 0$, $y = 0$ and $z = 0$, so this is not possible and $Q_r = \emptyset$, i.e. r commutes with t. We look at the polynomial $r^2 + 1$ and check if it generates a two-sided Gröbner base, which turns out to be true.

```
 fr = r**2 + 1;
 iil = pt.ideal( "", [ fr ] );
 rgll = iil.twosidedGB();
 SolvIdeal.new(...,[( r**2 + 1 )])
```

Since $t^2 + 1 \in \mathcal{I}$ we see that t is a root of $r^2 + 1$.

```
 e = fr.evaluate( t );
 e: 0
```

Moreover $r - t$ and $r + t$ divide $r^2 + 1$;

```
 fp = (r-t);
 fr / fp: (r+t)
 fr % fp: 0
```

If we check the product $(r - t)(r + t)$, it is computed as $r^2 - t^2$ and not $r^2 + 1$ as expected. However, the two polynomials are equal modulo $\mathcal{I}$.

```
 frp = fp*(r+t);
 frp: ( r**2 - t**2 )
 frp-fr: 0
 frp == fr: true
```

Finally, we construct the extension field $\mathbb{Q}(x, y, z, t; Q_x)/\mathcal{I}(r; Q_r)/(r^2+1)$ with Q_x and Q_r as above, and check, that in this field $r^2 + 1 = 0$ (and also $t^2 + 1 = 0$).

```
rf = SLR(rgll, r);
rf**2 + 1: 0
ft = SLR(rgll, t);
ft**2 + 1: 0
(rf-ft)*(rf+ft): 0
```

In this field r^2+1 splits into linear factors as shown in the last line. This concludes the examples.

4.4   Extension to Free Non-commutative Coefficients

An extension to the rings presented are solvable polynomial rings with coefficients from a free non-commutative polynomial ring subject to some relations. Free non-commutative polynomial rings have been studied, for example in [9]. An implementation is contained in classes `GenWordPolynomial` and `GenWord-PolynomialRing`. Using such rings, it is possible to do computations in integro-differential algebras [28]. For example in the integro-differential Weyl algebra

$$\mathbf{K}\langle \ell, \partial \rangle_{/(\partial \ell = 1)}\{x; Q\}, \quad Q = \{x * \partial = \partial x - 1, \; x * \ell = \ell x + \ell^2\}.$$

The implementation of such rings is ongoing work and will be discussed elsewhere.

5   Conclusions

We presented parametric solvable polynomial rings, where the definition of commutator relations between polynomial variables and coefficient variables is possible. This enables the computation of recursive solvable polynomial rings and makes it possible to construct and compute in localizations with respect to two-sided ideals in such rings. Using these as coefficient rings of solvable polynomial rings makes computations of roots, common divisors and ideal constructions over skew fields feasible. The algorithms have been implemented in JAS in a type-safe, object oriented way with generic coefficients. Besides the high complexity of the solvable multiplication the lack of efficient simplifiers to reduce intermediate expression swell hinder practical computations.

Acknowledgments. We thank our colleagues Thomas Becker, Raphael Jolly, Wolfgang K. Seiler, Thomas Sturm, Axel Kramer, Victor Levandovskyy, Joachim Apel, Markus Aleksy and others for various discussions on the design and the requirements for JAS and its mathematical foundations. Thanks also to the referees for the insightful suggestions to improve the paper.

References

1. Kredel, H.: On a Java Computer Algebra System, its performance and applications. Science of Computer Programming **70**(2–3), 185–207 (2008)
2. Kredel, H.: The Java algebra system (JAS). Technical report (since 2000). http://krum.rz.uni-mannheim.de/jas/
3. Kredel, H.: Unique factorization domains in the Java computer algebra system. In: Sturm, T., Zengler, C. (eds.) ADG 2008. LNCS, vol. 6301, pp. 86–115. Springer, Heidelberg (2011)
4. Kredel, H., Jolly, R.: Algebraic structures as typed objects. In: Gerdt, V.P., Koepf, W., Mayr, E.W., Vorozhtsov, E.V. (eds.) CASC 2011. LNCS, vol. 6885, pp. 294–308. Springer, Heidelberg (2011)
5. Kredel, H.: Distributed Gröbner bases computation with MPJ. In: IEEE AINA Workshops, Barcelona, Spain, pp. 1429–1435 (2013)
6. Apel, J., Lassner, W.: An algorithm for calculations in eveloping fields of Lie algebras. In: Proc. Int. Conf. on Computer Algebra and its Applications in Theoretical Physics, JINR, D11-85-791, Dubna (1985)
7. Apel, J., Lassner, W.: Computation and simplification in Lie fields. In: EUROCAL 1987, pp. 468–478 (1987)
8. Kandri Rody, A., Weispfennning, V.: Non-commutative Gröbner bases in algebras of solvable type. J. Symb. Comp. **9**(1), 1–26 (1990)
9. Mora, T.: An introduction to commutative and noncommutative Gröbner bases. Theor. Comput. Sci. **134**(1), 131–173 (1994)
10. Weispfenning, V.: Comprehensive Gröbner bases. J. Symb. Comp. **14**(1), 1–29 (1992)
11. Kredel, H., Weispfenning, V.: Parametric Gröbner bases for non-commutative polynomials. In: Shirkov, D.V., Rostovtsev, V.A., Gerdt, V.P. (eds.) IV. Int. Conf. on Computer Algebra in Physical Research, Dubna, USSR, 1990, pp. 236–246. World Scientific Publishing (1991)
12. Kredel, H.: Solvable Polynomial Rings. Dissertation, Universität Passau, Passau (1992)
13. Bueso, J.L., Gómez-Torrecillas, J., Verschoren, A.: Algorithmic Methods in Non-Commutative Algebra: Applications to Quantum Groups. Kluwer Academic Publishers (2003)
14. Levandovskyy, V., Schönemann, H.: Plural: a computer algebra system for non-commutative polynomial algebras. In: Proc. Symbolic and Algebraic Computation, International Symposium ISSAC 2003, Philadelphia, USA, pp. 176–183 (2003)
15. Levandovskyy, V.: Non-commutative Computer Algebra for polynomial algebras: Gröbner bases, applications and implementation. Dissertation, University of Kaiserslautern, Kaiserslautern (2005)
16. Cohn, P.M.: Skew Fields, Theory of General Division Rings. Encyclopedia of Mathematics and its Applications, vol. 57. Cambridge University Press (1995)
17. Kredel, H.: On the design of a Java computer algebra system. In: Proc. PPPJ 2006, University of Mannheim, pp. 143–152 (2006)
18. Kredel, H., Pesch, M.: MAS: the modula-2 algebra system. In: Computer Algebra Handbook, pp. 421–428. Springer (2003)
19. Dolzmann, A., Sturm, T.: Redlog: Computer algebra meets computer logic. ACM SIGSAM Bull. **31**(2), 2–9 (1997)
20. Montes, A.: An new algorithm for discussing Gröbner basis with parameters. J. Symb. Comput. **33**(1–2), 183–208 (2002)

21. Suzuki, A., Sato, Y.: An alternative approach to comprehensive Gröbner bases. J. Symb. Comput. **36**(3–4), 649–667 (2003)
22. Nabeshima, K.: A speed-up of the algorithm for computing comprehensive Gröbner systems. In: Proc. ISSAC 2007, pp. 299–306 (2007)
23. Inoue, S., Sato, Y.: On the parallel computation of comprehensive Gröbner systems. In: Proc. PASCO 2007, pp. 99–101 (2007)
24. Kredel, H.: Comprehensive Gröbner bases in a Java computer algebra system. In: Proceedings ASCM 2009, Kyushu University, Fukuoka, Japan, pp. 77–90 (2009)
25. Kredel, H.: Comprehensive Gröbner bases in a Java computer algebra system. In: Feng, R., Lee, W.S., Sato, Y. (eds.) Computer Mathematics, pp. 93–108. Springer, Heidelberg (2014)
26. Nabeshima, K.: PGB: a package for computing parametric polynomial systems. In: Proceedings ASCM 2009, Kyushu University, Fukuoka, Japan, pp. 111–122 (2009)
27. Greuel, G., Pfister, G., Schönemann, H.: Singular - a computer algebra system for polynomial computations. In: Computer Algebra Handbook, pp. 445–450. Springer (2003)
28. Rosenkranz, M., Regensburger, G.: Solving and factoring boundary problems for linear ordinary differential equations in differential algebras. J. Symb. Comput. **43**(8), 515–544 (2008)

Triangular Decomposition of Matrices in a Domain

Gennadi Malaschonok and Anton Scherbinin

Tambov State University,
Internatsionalnaya 33, 392622 Tambov, Russia
malaschonok@gmail.com

Abstract. Deterministic recursive algorithms for the computation of matrix triangular decompositions with permutations like LU and Bruhat decomposition are presented for the case of commutative domains. This decomposition can be considered as a generalization of LU and Bruhat decompositions because they both may easily be obtained from this triangular decomposition. Algorithms have the same complexity as the algorithm of matrix multiplication.

1 Introduction

Traditionally, decomposition of matrices in a product of several triangular matrices and permutation matrices is investigated for matrices over fields.

Active use of functional and polynomial matrices in computer algebra systems has generated interest in matrix decomposition in commutative domains.

Matrix decomposition of the form $A = VwU$ is called the Bruhat decomposition, if V and U are nonsingular upper triangular matrices and w is a matrix of permutation. The generalized Bruhat decomposition was introduced by Dima Grigoriev [3,2].

Matrix decomposition of the form $A = LU$ or $A = PLUQ$ is called LU-decomposition of the matrix A, if L and U are lower and upper triangular matrices and P, Q are the matrices of permutations.

In general, for an arbitrary matrix, we cannot get the Bruhat decomposition based on the expansion LU. And vice versa.

In this paper, we suggest a new form of triangular decomposition. We call it LDU-decomposition. Here L and U are lower and upper triangular matrices, $D = PdQ$, where d is a diagonal matrix, P and Q are permutation matrices.

This decomposition can be considered as a generalization of LU and Bruhat decompositions. Since they both may easily be obtained from the LDU-decomposition.

We describe and prove the deterministic recursive algorithm for computation of such triangular decomposition. This algorithm has the same complexity as the algorithm of matrix multiplication (proof see in [8]). In [8], there were described particular cases of this algorithm when each of the main corner minor of a matrix A up to the rank is not equal to zero.

Now we present the complete algorithm for the general case and give its proof.

© Springer International Publishing Switzerland 2015
V.P. Gerdt et al. (Eds.): CASC 2015, LNCS 9301, pp. 292–306, 2015.
DOI: 10.1007/978-3-319-24021-3_22

2   Preliminary. Triangular Decomposition in Domain

Let R be a commutative domain, $A = (a_{i,j}) \in R^{n \times m}$ be a matrix of size $n \times m$, $\alpha_{i,j}^k$ be $k \times k$ minor of matrix A which disposed in rows $1, 2, \ldots, k - 1, i$ and columns $1, 2, \ldots, k - 1, j$. We denote $\alpha^0 = 1$ and $\alpha^k = \alpha_{k,k}^k$. And we use the notation δ_{ij} for Kronecker delta.

Let k and s be integers in the interval $0 \le k < s \le n$, $\mathcal{A}_{s,p}^k = (\alpha_{i,j}^{k+1})$ be the matrix of minors with size $(s - k) \times (p - k)$ which has elements $\alpha_{i,j}^{k+1}$, $i = k+1, \ldots, s - 1, s$, $j = k+1, \ldots, p - 1, p$ and $\mathcal{A}_n^0 = (\alpha_{i,j}^1) = A$. We denote $\mathcal{A}_{s,s}^k = \mathcal{A}_s^k$.

We shall use the following identity (see [1,9]):

Theorem 1 (Sylvester determinant identity)
Let $n = m$, k and s be the integers in the interval $0 \le k < s \le n$. Then it is true that

$$\det(\mathcal{A}_s^k) = \alpha^s(\alpha^k)^{s-k-1}. \tag{1}$$

Theorem 2 (LDU decomposition of the minors matrix)
Let $A = (a_{i,j}) \in R^{n \times m}$ be the matrix of rank r, $\alpha^i \ne 0$ for $i = k, k + 1, \ldots, r$, then the matrix of minors $\mathcal{A}_{s,p}^k$, $k < s \le n$, $k < p \le m$ is equal to the following product of three matrices:

$$\mathcal{A}_{s,p}^k = L_s^k D_{s,p}^k U_p^k = (a_{i,j}^j)(\delta_{ij}\alpha^k(\alpha^{i-1}\alpha^i)^{-1})(a_{i,j}^i). \tag{2}$$

The matrix $L_s^k = (a_{i,j}^j)$, $i = k+1 \ldots s$, $j = k+1 \ldots r$, is a lower triangular matrix of size $(s-k) \times (r-k)$, the matrix $U_p^k = (a_{i,j}^i)$, $i = k+1 \ldots r$, $j = k+1 \ldots p$, is an upper triangular matrix of size $(r - k) \times (p - k)$ and $D_{s,p}^k = (\delta_{ij}\alpha^k(\alpha^{i-1}\alpha^i)^{-1})$, $i = k + 1 \ldots r$, $j = k + 1 \ldots r$, is a diagonal matrix of size $(r - k) \times (r - k)$.

Proof. See proof in [8].

Remark 1. We can add the unit blocks to the matrices L_s^k and U_p^k, and we obtain nonsingular square matrices of size $(s - k)$ and $(p - k)$, respectively. We can add zero elements to the matrix $D_{s,p}^k$ and obtain the diagonal matrix of size $(s - k) \times (p - k)$. As a result, we obtain decomposition $\mathcal{A}_{s,p}^k = L_s^k D_{s,p}^k U_p^k$ with invertible square matrices L_s^k and U_p^k

Corollary 1 (LDU decomposition of matrix A in domain)
Let $A = (a_{i,j}) \in R^{n \times m}$ be the matrix of rank r, $\alpha^i \ne 0$ for $i = 1, 2, \ldots, r$, then matrix A is equal to the following product of three matrices:

$$A = L_n^0 D_{n,m}^0 U_m^0 = (a_{i,j}^j)(\delta_{ij}(\alpha^{i-1}\alpha^i)^{-1})(a_{i,j}^i). \tag{3}$$

Corollary 2 (Bruhat decomposition)
Let $A = LDU$ be the LDU-decomposition of a $n \times n$ matrix A with rank r, S be the "flipped" identity matrix, then $V = SLS$ and U are upper triangular matrices and

$$SA = V(SD)U \tag{4}$$

is the Bruhat decomposition of the matrix SA.

Corollary 3 (Cases of zero blocks). *Let* $\mathcal{A}^k_{s,p} = \begin{pmatrix} \mathbf{A} & \mathbf{B} \\ \mathbf{C} & \mathbf{D} \end{pmatrix}$, $S_2 = \begin{pmatrix} 0 & I \\ I & 0 \end{pmatrix}$,
$\mathbf{A} = \mathcal{A}^k_{r,r}$ *and* $\mathcal{A}^k_{s,p} = L^k_s D^k_{s,p} U^k_p$ *is a decomposition* (2) *for matrix* $\mathcal{A}^k_{s,p}$. *Then
this decomposition has the following properties:*

$$\text{if } \mathbf{C} = 0, \text{ then } L^k_s = \mathrm{diag}(L^k_r, L^r_s). \tag{5}$$

$$\text{if } \mathbf{B} = 0, \text{ then } U^k_p = \mathrm{diag}(U^k_r, U^r_p). \tag{6}$$

$$\text{if } \mathbf{A} = \mathbf{B} = \mathbf{C} = 0, \mathbf{D} = L^r_s D^r_{s,p} U^r_s \text{ then } \mathcal{A}^k_{s,p} = S_2 \begin{pmatrix} L^r_s D^r_{s,p} U^r_s & 0 \\ 0 & 0 \end{pmatrix} S_2. \tag{7}$$

Proof. Let $\mathbf{C} = (\alpha^k_{i,j})^{j=k+1,..r}_{i=r+1,..,s} = 0$. Then all the minors $(\alpha^j_{i,j})$ for $i = r+1, .., s$
and $j = k+1, ..r$ equal zero. It follows from the Sylvester Determinant Identity:

$$\alpha^j_{i,j} * (\alpha^k_{k,k})^{j-k-1} = \begin{vmatrix} \alpha^{k+1}_{k+1,k+1} & \cdots & \alpha^{k+1}_{k+1,j-1} & \alpha^{k+1}_{k+1,j} \\ \cdots & \cdots & \cdots & \cdots \\ \alpha^{k+1}_{j-1,k+1} & \cdots & \alpha^{k+1}_{j-1,j-1} & \alpha^{j-1}_{k+1,j} \\ \alpha^{k+1}_{i,k+1} & \cdots & \alpha^{k+1}_{i,j-1} & \alpha^{k+1}_{i,j} \end{vmatrix}, \quad k+1 < j \leq i.$$

The last row in this minor is a row in the block $\mathbf{C}$. Therefore, this row is zero
row. So all the elements in the lower left corner block of the matrix L^k_s equal
zero.

If $\mathbf{B} = 0$, then we can prove in the same manner that the upper right corner
block of the matrix U^k_p equals zero.

If $\mathbf{A} = \mathbf{B} = \mathbf{C} = 0$, then it is obvious that the last expression (7) for $\mathcal{A}^k_{s,p}$ is
true.

We use some block matrix notations: for any matrix A (or A^p_q), we denote by
$A^{i_1,i_2}_{j_1,j_2}$ (or $A^{p;i_1,i_2}_{q;j_1,j_2}$) the block which stands at the intersection of rows $i_1 + 1, \ldots, i_2$
and columns $j_1 + 1, \ldots, j_2$ of the matrix. We denote by $A^{i_1}_{i_2}$ the diagonal block
$A^{i_1,i_2}_{i_1,i_2}$.

2.1   LDU Algorithm for the Matrix with Nonzero Diagonal Minors Up to the Rank

Input: $(\mathcal{A}^k_n, \alpha^k)$, $0 \leq k < n$.
Output: $\{L^k_n, \{\alpha^{k+1}, \alpha^{k+2}, \ldots, \alpha^n\}, U^k_n, M^k_n, W^k_n\}$, where $M^k_n = \alpha^k (L^k_n D^k_n)^{-1}$,

$$D^k_n = \alpha^k \mathrm{diag}\{\alpha^k \alpha^{k+1}, \ldots, \alpha^{n-1} \alpha^n\}^{-1}, W^k_n = \alpha^k (D^k_n U^k_n)^{-1}.$$

1. If $k = n - 1$, $\mathcal{A}^{n-1}_n = (a^n)$ is a matrix of the first order, then we obtain

$$\{a^n, \{a^n\}, a^n, a^{n-1}, a^{n-1}\}, \quad D^{n-1}_n = (a^n)^{-1}.$$

2. If $k = n - 2$, $A_n^{n-2} = \begin{pmatrix} a^{n-1} & \beta \\ \gamma & \delta \end{pmatrix}$ is a matrix of second order, then we obtain

$$\left\{ \begin{pmatrix} a^{n-1} & 0 \\ \gamma & a^n \end{pmatrix}, \{a^{n-1}, a^n\}, \begin{pmatrix} a^{n-1} & \beta \\ 0 & a^n \end{pmatrix}, \begin{pmatrix} a^{n-2} & 0 \\ -\gamma & a^{n-1} \end{pmatrix}, \begin{pmatrix} a^{n-2} & -\beta \\ 0 & a^{n-1} \end{pmatrix} \right\}$$

where $a^n = (a^{n-2})^{-1} \begin{vmatrix} a^{n-1} & \beta \\ \gamma & \delta \end{vmatrix}$, $D_n^{n-2} = a^{n-2}\mathrm{diag}\{a^{n-2}a^{n-1}, a^{n-1}a^n\}^{-1}$.

3. If the order of the matrix A_n^k is more than two ($0 \le k < n - 2$), then we choose an integer s in the interval ($k < s < n$) and divide the matrix into blocks

$$A_n^k = \begin{pmatrix} A_s^k & \mathbf{B} \\ \mathbf{C} & \mathbf{D} \end{pmatrix}.$$

3.1. Recursive step: $\{L_s^k, \{a^{k+1}, a^{k+2}, \ldots, a^s\}, U_s^k, M_s^k, W_s^k\} = \mathbf{LDU}(A_s^k, a^k)$
3.2. We compute $\tilde{U} = (a^k)^{-1}M_s^k\mathbf{B}$, $\tilde{L} = (a^k)^{-1}\mathbf{C}W_s^k$,

$$A_n^s = (a^k)^{-1}a^s(\mathbf{D} - \tilde{L}D_s^k\tilde{U}).$$

3.3. Recursive step: $\{L_n^s, \{a^{s+1}, a^{s+2}, \ldots, a^n\}, U_n^s, M_n^s, W_n^s\} = \mathbf{LDU}(A_n^s, a^s)$
3.4 Result: $\{L_n^k, \{a^{k+1}, a^{k+2}, \ldots, a^n\}, U_n^k, M_n^k, W_n^k\}$, where

$$L_n^k = \begin{pmatrix} L_s^k & 0 \\ \tilde{L} & L_n^s \end{pmatrix}, \quad M_n^k = \begin{pmatrix} M_s^k & 0 \\ -M_n^s\tilde{L}D_s^kM_s^k/a^k & M_n^s \end{pmatrix},$$

$$U_n^k = \begin{pmatrix} U_s^k & \tilde{U} \\ 0 & U_n^s \end{pmatrix}, \quad W_n^k = \begin{pmatrix} W_s^k & -W_s^kD_s^k\tilde{U}W_n^s/a^k \\ 0 & W_n^s \end{pmatrix}.$$

Proof of the correctness of this algorithm you can fined in [8].

3   Triangular Matrix Decomposition

Now we are interested in the general case of initial matrix A, and we want to find decomposition in the form

$$A_{n,m}^k = P_n^k L_n^k D_{n,m}^k U_m^k Q_m^k, \tag{8}$$

with the additional

Property 1 (of triangular decomposition).
    If matrix $A_{n,m}^k$ has rank $\bar{r} - k$, then
(α) the lower and upper triangular matrices L_n^k and U_m^k are of the form

$$L_n^k = \begin{pmatrix} L_1 & 0 \\ L_2 & I_{n-\bar{r}} \end{pmatrix}, \quad U_m^k = \begin{pmatrix} U_1 & U_2 \\ 0 & I_{m-\bar{r}} \end{pmatrix}, \tag{9}$$

(β) the matrices $\mathcal{L} = P_n^k L_n^k (P_n^k)^T$ and $\mathcal{U} = (Q_m^k)^T U_m^k Q_m^k$ remain triangular after replacing in the L_n^k and U_m^k the unit block by arbitrary triangular block.

Example 1 (Non-trivial example for property β). We replace the right lower unit block I_2 by the block L_3: $\begin{pmatrix} 0 & I_1 \\ I_2 & 0 \end{pmatrix} \begin{pmatrix} L_1 & 0 \\ L_2 & \mathbf{L}_3 \end{pmatrix} \begin{pmatrix} 0 & I_2 \\ I_1 & 0 \end{pmatrix}$, with $L_2 = 0$.

Example 2 (Non-trivial example for property β).
$\begin{pmatrix} P & 0 \\ 0 & I \end{pmatrix} \begin{pmatrix} L_1 & 0 \\ L_2 & \mathbf{L}_3 \end{pmatrix} \begin{pmatrix} P^T & 0 \\ 0 & I \end{pmatrix}$, with PL_1P^T lower triangular matrix.

Definition 1 (of triangular decomposition of matrix in domain). *We call the matrix decomposition* (8) **triangular decomposition** *if it satisfies Property 1.*

In this case, the decomposition (8) takes the form $A_{n,m}^k = \mathcal{LDU}$ ($\mathcal{D} = P_n^k D_{n,m}^k Q_n^k$).

This decomposition gives the Bruhat decomposition for the matrix SA like it was shown in (4).

Let us note that property 1 is trivial if $k = n - 1$. Property (β) is trivial if $\bar{r} \geq n - 1$. Therefore, the recursive algorithm as set out below, we can prove by induction.

4   LDU Algorithm with Permutation Matrices

4.1   Left Upper Block A Is Not Zero Block

Input: $(\mathcal{A}_{n,m}^k, \alpha^k)$, $0 \leq k < n$, $0 \leq k < n$, $\mathcal{A}_{n,m}^k \neq 0$;
Output: $(P_n^k, L_n^k, \{\alpha^{k+1}, \alpha^{k+2}, \ldots, \alpha^{\bar{r}}\}, U_m^k, Q_m^k, M_{\bar{r}}^k, W_{\bar{r}}^k)$, where $\bar{r} - k$ is the rank of the matrix $\mathcal{A}_{n,m}^k$, $D_{\bar{r}}^k = \alpha^k \mathrm{diag}(\alpha^k \alpha^{k+1}, \ldots, \alpha^{\bar{r}-1} \alpha^{\bar{r}})^{-1}$, $D_{n,m}^k = \mathrm{diag}(D_{\bar{r}}^k, 0)$, $M_{\bar{r}}^k = \alpha^k (L_{\bar{r}}^k D_{\bar{r}}^k)^{-1}$, $W_{\bar{r}}^k = \alpha^k (D_{\bar{r}}^k U_{\bar{r}}^k)^{-1}$, $A_n^k = P_n^k L_n^k D_{n,m}^k U_m^k Q_m^k$.

1. If $k = \min(n, m) - 1$, $\mathcal{A}_{n,m}^k$ is a matrix of one row or one column, then

$$\{I_n, L_n^k, \{\alpha^{k+1}\}, U_m^k, I_m, (\alpha^{k+1}), (\alpha^{k+1})\} = \mathbf{LDU}(\mathcal{A}_{n,m}^k, \alpha^k).$$

If $\mathcal{A}_{n,m}^k = (a^{k+1}, .., a_{n,k+1}^{k+1})^T$ is a column-matrix then $m = k + 1$, $U_m^k = (a^{k+1})$ and $L_n^k = \begin{pmatrix} 0 \\ \mathcal{A}_{n,k+1}^k & I_{n-k} \end{pmatrix}$. If $\mathcal{A}_{n,m}^k = (a^{k+1}, .., a_{k+1,m}^{k+1})$ is a row-matrix then $n = k + 1$, $U_m^k = \begin{pmatrix} \mathcal{A}_{k+1,m}^k \\ 0 & I_{m-k} \end{pmatrix}$ and $L_n^k = (a^{k+1})$.

2. If the matrix $\mathcal{A}_{n,m}^k$ has more than one row and more than one column, then we choose an integer s in the interval $(k < s < n)$ and divide the matrix into blocks

$$\mathcal{A}_{n,m}^k = \begin{pmatrix} \mathcal{A}_s^k & \mathbf{B} \\ \mathbf{C} & \mathbf{D} \end{pmatrix}. \tag{10}$$

2.1. Recursive step

Let the block $\mathcal{A}_s^k$ has rank $r - k$, $r < s$, then we get

$$\{P_s^k, L_s^k, \{\alpha^{k+1}, \alpha^{k+2}, \ldots, \alpha^r\}, U_s^k, Q_s^k, M_r^k, W_r^k\} = \mathbf{LDU}(\mathcal{A}_s^k, \alpha^k)$$

$$L_s^k = \begin{pmatrix} L_r^k & 0 \\ L_1 & I_{s-r} \end{pmatrix}, \quad U_s^k = \begin{pmatrix} U_r^k & U_1 \\ 0 & I_{s-r} \end{pmatrix}, \quad D_s^k = \begin{pmatrix} D_r^k & 0 \\ 0 & 0 \end{pmatrix}, \quad M_r^k =$$
$$\alpha^k (L_r^k D_r^k)^{-1}, \; W_r^k = \alpha^k (D_r^k U_r^k)^{-1}, \; D_r^k = \alpha^k \mathrm{diag}\{\alpha^k \alpha^{k+1}, \ldots, \alpha^{r-1} \alpha^r\}^{-1}, \; A_s^k = P_s^k L_s^k D_s^k U_s^k Q_s^k.$$

Let us note that $\alpha_{i,j}^k$ and α^k are the minors of the matrix

$$(\mathrm{diag}(I_k, P_s^k)^T A (\mathrm{diag}(I_k, Q_s^k)^T).$$

2.2. We separate the matrix $\mathcal{A}_{n,m}^k$ into blocks another way

$$\mathrm{diag}^T(P_s^k, I_{n-s}) \mathcal{A}_{n,m}^k \mathrm{diag}^T(Q_s^k, I_{n-s}) = \begin{pmatrix} A_r^k & B \\ C & D \end{pmatrix}. \tag{11}$$

$$\tilde{U} = (\alpha^k)^{-1} M_r^k B, \tilde{L} = (\alpha^k)^{-1} C W_r^k, \tag{12}$$

$$\mathcal{A}_{n,m}^r = (\alpha^k)^{-1} \alpha^r (D - \tilde{L} D_r^k \tilde{U}). \tag{13}$$

2.3. Recursive step.

If $A_{n,m}^r \neq 0$, then

$$\{P_n^r, L_n^r, \{\alpha^{r+1}, \alpha^{r+2}, \ldots, \alpha^{\bar{r}}\}, U_m^r, Q_m^r, M_{\bar{r}}^r, W_{\bar{r}}^r\} = \mathbf{LDU}(\mathcal{A}_{n,m}^r, \alpha^r)$$

$$A_{n,m}^r = P_n^r L_n^r D_{n,m}^r U_m^r Q_m^r.$$

Otherwise($A_{n,m}^r = 0$), we do nothing.

2.4 Result:

$$\{P_n^k, L_n^k, \{\alpha^{k+1}, \alpha^{k+2}, \ldots, \alpha^{\bar{r}}\}, U_m^k, Q_m^k, M_{\bar{r}}^k, W_{\bar{r}}^k\},$$

If $A_{n,m}^r \neq 0$, then

$$P_n^k = \mathrm{diag}(P_s^k, I_{n-s})\mathrm{diag}(I_{r-k}, P_n^r), Q_m^k = \mathrm{diag}(I_{r-k}, Q_m^r)\mathrm{diag}(Q_s^k, I_{n-s})$$

$$L_n^k = \begin{pmatrix} L_r^k & 0 \\ P_n^{rT} \tilde{L} & L_n^r \end{pmatrix}, \quad U_m^k = \begin{pmatrix} U_r^k & \tilde{U} Q_n^{rT} \\ 0 & U_n^r \end{pmatrix}, \tag{14}$$

$$M_{\bar{r}}^k = \begin{pmatrix} M_r^k & 0 \\ -M_{\bar{r}}^r P_n^{rT} \tilde{L} D_r^k M_r^k / \alpha^k & M_{\bar{r}}^r \end{pmatrix}, \tag{15}$$

$$W_{\bar{r}}^k = \begin{pmatrix} W_r^k & -W_r^k D_r^k \tilde{U} Q_m^{rT} W_{\bar{r}}^r / \alpha^k \\ 0 & W_{\bar{r}}^r \end{pmatrix}. \tag{16}$$

Otherwise($A_{n,m}^r = 0$),

$$\bar{r} = r$$

$$P_n^k = \text{diag}(P_s^k, I_{n-s}), Q_m^k = \text{diag}(Q_s^k, I_{n-s}),$$

$$L_n^k = \begin{pmatrix} L_r^k & 0 \\ \tilde{L} & I_{n-r} \end{pmatrix}, \quad U_m^k = \begin{pmatrix} U_r^k & \tilde{U} \\ 0 & I_{m-r} \end{pmatrix}, \tag{17}$$

$$M_r^k = \alpha^k (L_r^k D_r^k)^{-1},$$

$$W_r^k = \alpha^k (D_r^k U_r^k)^{-1},$$

$$D_r^k = \alpha^k \text{diag}\{\alpha^k \alpha^{k+1}, \ldots, \alpha^{r-1} \alpha^r\}^{-1},$$

Theorem 3. *Algorithm* **4.1** *is correct.*

Proof. Proof of the correctness of this algorithm is like previous one. But now we have to prove that matrices $P_n^k, Q_m^k, L_n^k, U_m^k$ satisfy the Property of triangular decomposition.

Let matrices $P_s^k, Q_s^k, L_s^k, U_s^k$ and $P_n^r, Q_n^r, L_n^r, U_n^r$ satisfy the Property of triangular decomposition and $\mathcal{A}_{n,m}^r \neq 0$ at step 2.3.

Property β. We wish to prove the property (β) for $P_n^k L_n^k (P_n^k)^T$ and $(Q_m^k)^T U_m^k Q_m^k$.

$$P_n^k L_n^k (P_n^k)^T = \text{diag}(P_s^k, I_{n-s})\text{diag}(I_{r-k}, P_n^r)L_n^k \text{diag}^T(I_{r-k}, P_n^r)\text{diag}^T(P_s^k, I_{n-s})$$

$$= \text{diag}(P_s^k, I_{n-s}) \begin{pmatrix} L_r^k & 0 \\ \tilde{L} & P_n^r L_n^r P_n^{rT} \end{pmatrix} \text{diag}(P_s^k, I_{n-s})^T. \tag{18}$$

The blocks $P_n^r L_n^r P_n^{rT}$ and

$$P_s^k L_s^k (P_s^k)^T = P_s^k \begin{pmatrix} L_r^k & 0 \\ L_1 & I \end{pmatrix} (P_s^k)^T \tag{19}$$

are triangular by induction. The block

$$G = P_s^k \begin{pmatrix} L_r^k & 0 \\ \tilde{L}' & L'' \end{pmatrix} (P_s^k)^T \tag{20}$$

is located in the upper left corner of the matrix (18). The block $\tilde{L}'$ has size $(s-r) \times (r-k)$ and L'' has size $(s-r) \times (s-r)$. The block L'' is low triangular block and $\tilde{L}' = L_1$ by the construction. Due to the property (β) of triangular decomposition of A_s^k this corner block G is triangular too.

We can make a similar argument for the block $(Q_m^k)^T U_m^k Q_m^k$.

Property α. Due to the construction of the matrices L_n^k and U_m^k they can have the unit block in the lower right corner, but the size of this block is less than $n-r$. Therefore, this unit block is disposed in the blocks L_n^r and U_n^r. The property (α) of triangular decomposition of A_n^k is the consequence of such property of matrix A_n^r. The case when $\mathcal{A}_{n,m}^r = 0$ at step 2.3 is evident.

Corollary.

In section 6, we present a full example of triangular decomposition for a matrix of size 6×6. In this example, the (18)–(20) can be written as

$$(18): P_6^0 L_6^0 P_6^{0T} = \text{diag}(P_4^0, I_2)\text{diag}(I_2, P_6^2)L_6^0\text{diag}^T(I_2, P_6^2)\text{diag}^T(P_4^0, I_2) =$$

$$= \text{diag}(P_4^0, I_2)\begin{pmatrix} L_2^0 & 0 \\ \tilde{L}_2^0 & P_6^2 L_6^2 P_6^{2T} \end{pmatrix}\text{diag}(P_4^0, I_2)^T,$$

$$(19): P_4^0 L_4^0 (P_4^0)^T = P_4^0 \begin{pmatrix} L_2^0 & 0 \\ \tilde{L}_2^0 & I_2 \end{pmatrix}(P_4^0)^T,$$

$$(20): G = \begin{pmatrix} L_2^0 & 0 \\ \tilde{L}_2^0 & L_6^4 \end{pmatrix}.$$

4.2   Matrix A Has Full Rank, Matrices C and (or) B Are Zero Matrices

2.1. This step has no changes.

2.2. We compute $\tilde{U}$, $\tilde{L}$. One of these matrices or both equal zero. So for matrix $\mathcal{A}_n^s$, we obtain: $\mathcal{A}_n^s = (\alpha^k)^{-1}\alpha^s \mathbf{D}$.

2.3. Recursive step. We take matrix $\mathbf{D}$ instead of $\mathcal{A}_n^s$.

$$\{P_n^s, \bar{L}_n^s, \{\bar{\alpha}^{s+1}, \bar{\alpha}^{s+2}, \dots, \bar{\alpha}^n\}, \bar{U}_n^s, Q_n^s, \bar{M}_n^s, \bar{W}_n^s\} = \mathbf{LDU}(\mathbf{D}, \alpha^k)$$

Let us note that we can do these computations simultaneously, and the computations of the first step (3.1).

The value $\lambda = \alpha^s/\alpha^k$ is a coefficient in the equality $\mathcal{A}_n^s = (\alpha^k)^{-1}\alpha^s \mathbf{D}$. So it is easy to prove that $\lambda\bar{L}_n^s = L_n^s, \lambda\bar{\alpha}^{s+1} = \alpha^{s+1}, \lambda\bar{\alpha}^{s+2} = \alpha^{s+2}, \dots, \lambda\bar{\alpha}^n = \alpha^n,$ $\lambda^{-1}\bar{D}_n^s = D_n^s, \lambda\bar{U}_n^s = U_n^s, \lambda\bar{M}_n^s = M_n^s, \lambda\bar{W}_n^s = W_n^s$. This way we can find the factorization:

$$\{P_n^s, L_n^s, \{\alpha^{s+1}, \alpha^{s+2}, \dots, \alpha^n\}, U_n^s, Q_n^s, M_n^s, W_n^s\} = \mathbf{LDU}(\mathcal{A}_n^s, \alpha^s)$$

2.4 Result has no changes. But if $C = 0$ then $L_n^k = \text{diag}(L_s^k, L_n^s)$, if $B = 0$ then $U_m^k = \text{diag}(U_s^k, U_n^s)$.

4.3   Matrix A Has No Full Rank, Matrices C and (or) B Are Zero Matrices, $A \neq 0$

Let us assume that at some stage of algorithm 4.1, we obtain the matrix

$$\mathcal{A}_{n,m}^k = \begin{pmatrix} \mathbf{A} & \mathbf{B} \\ 0 & \mathbf{D} \end{pmatrix}.$$

Here $\mathbf{A} = \mathcal{A}_s^k$, and $\mathbf{D}$ is a block of size $(n-s) \times (m-s)$.

We can use the algorithm, which is set out in subsection 4.1. But in this case, the decomposition of block $\mathbf{A} = \mathcal{A}_s^k$ and $\mathbf{D}$ can be performed simultaneously and independently.

2.1. Recursive steps

$$\{P_s^k, L_s^k, \{\alpha^{k+1}, \alpha^{k+2}, \ldots, \alpha^r\}, U_s^k, Q_s^k M_r^k, \ W_r^k,\} = \mathbf{LDU}(\mathbf{A}, \alpha^k) \qquad (21)$$

$$\{P_n^s, \bar{L}_n^s, \{\bar{\alpha}^{s+1}, \bar{\alpha}^{s+2}, \ldots, \bar{\alpha}^{\bar{r}}\}, \bar{U}_m^s, Q_m^s, \bar{M}_{\bar{r}}^s, \ \bar{W}_{\bar{r}}^s\} = \mathbf{LDU}(\mathbf{D}, \alpha^k) \qquad (22)$$

$\mathbf{A} = P_s^k L_s^k D_s^k U_s^k Q_s^k, \ \mathbf{D} = P_m^s \bar{L}_n^s D_{n,m}^s \bar{U}_m^s Q_m^s.$

We suppose that matrix $\mathbf{A}$ has rank $r - k$, $k < r < s$, matrix $\mathbf{D}$ has rank $\bar{r} - s$, $s < \bar{r} \le \min(n, m)$. Then

$$L_s^k = \begin{pmatrix} L_0 & 0 \\ M_0 & I_{s-r} \end{pmatrix}, D_s^k = \begin{pmatrix} d_1 & 0 \\ 0 & 0 \end{pmatrix}, U_s^k = \begin{pmatrix} U_0 & V_0 \\ 0 & I_{s-r} \end{pmatrix},$$

2.2. We decompose the matrix $A_{n,m}^k$ into blocks another way

$$\mathrm{diag}^T(P_s^k, I_{n-s}) A_{n,m}^k \mathrm{diag}^T(Q_s^k, I_{n-s}) = \begin{pmatrix} A_r^k & B \\ C & D \end{pmatrix}.$$

$$\tilde{U} = (\alpha^k)^{-1} M_r^k B, \tilde{L} = (\alpha^k)^{-1} C W_r^k, \qquad (23)$$

$$A_{n,m}^r = (\alpha^k)^{-1} \alpha^r (D - \tilde{L} D_r^k \tilde{U}). \qquad (24)$$

Let us note that $\lambda = \alpha^r/\alpha^k$, $C = \begin{pmatrix} C_1 \\ 0 \end{pmatrix}$, $\tilde{L} = \begin{pmatrix} L_1 \\ 0 \end{pmatrix}$, $A_{n,m}^r = \lambda \begin{pmatrix} D_1 & D_2 \\ 0 & \mathbf{D} \end{pmatrix}$.

2.3. Recursive step.

$$\{P_n^r, L_n^r, \{\alpha^{r+1}, \alpha^{r+2}, \ldots, \alpha^{\bar{r}}\}, U_m^r, Q_m^r, M_{\bar{r}}^r, \ W_{\bar{r}}^r\} = \mathbf{LDU}(A_{n,m}^r, \alpha^r)$$

we can do with the help of block decomposition (22), and obtain

$A_{n,m}^r = P_n^r L_n^r D_{n,m}^r U_m^r Q_m^r.$

2.4 Result:

$$\{P_n^k, L_n^k, \{\alpha^{k+1}, \ldots, \alpha^r, \alpha^{r+1}, \ldots, \alpha^{\bar{r}}\}, U_m^k, Q_m^k, M_{\bar{r}}^k, W_{\bar{r}}^k\} = \mathbf{LDU}(A_{n,m}^k, \alpha^k),$$

is similar as in the section 4.1.

5   Matrix A is Zero Matrix

5.1   Matrix A Is Zero Matrix, C and (or) B Are Nonzero Matrices

Let us assume that at some stage of algorithm 4.1, we obtain the matrix

$$A_{n,m}^k = \begin{pmatrix} 0 & \mathbf{B} \\ \mathbf{C} & \mathbf{D} \end{pmatrix}, S = \begin{pmatrix} 0 & I \\ I & 0 \end{pmatrix}. \qquad (25)$$

Let $\mathbf{C} \ne 0$. Then for the matrix $SA_{n,m}^k$, we can make triangular decomposition as in section 4: $\begin{pmatrix} \mathbf{C} & \mathbf{D} \\ 0 & \mathbf{B} \end{pmatrix} = LDU$. Then $A_{n,m}^k = SLDU$. In this case, matrix L has the form $L = \mathrm{diag}(L_1, L_2)$ due to Corollary 3 of Theorem 1. Then we get $A_{n,m}^k = \mathrm{diag}(L_2, L_1)(SD)U$.

Let $\mathbf{B} \neq 0$. Then for the matrix $A_{n,m}^k S$, we can make triangular decomposition as in section 4: $\begin{pmatrix} \mathbf{B} & 0 \\ \mathbf{D} & \mathbf{C} \end{pmatrix} = LDU$. Then $A_{n,m}^k = LDUS$. In this case, matrix U has the form $U = \mathrm{diag}(U_1, U_2)$ due to Corollary 3 of Theorem 1. Then we get $A_{n,m}^k = L(DS)\mathrm{diag}(U_2, U_1)$.

Let $\mathbf{B} = \mathbf{C} = \mathbf{A} = 0$ and $\mathbf{D} \neq 0$, $\mathbf{D} = LDU$. Then for the matrix $A_{n,m}^k$ we can make triangular decomposition: $\begin{pmatrix} 0 & 0 \\ 0 & \mathbf{D} \end{pmatrix} = \begin{pmatrix} I & 0 \\ 0 & L \end{pmatrix}\begin{pmatrix} 0 & 0 \\ 0 & D \end{pmatrix}\begin{pmatrix} I & 0 \\ 0 & U \end{pmatrix}$.

5.2   All Cases When Half of the Matrix Is Equal to Zero

Theorem 4. *Let* $A_{n,m}^k = \begin{pmatrix} \mathbf{A} & \mathbf{B} \\ \mathbf{C} & \mathbf{D} \end{pmatrix}$.

If $\mathbf{C} = \mathbf{D} = 0$ *and* $(\mathbf{A}, \mathbf{B}) = PL(D, 0)UQ$ *is a triangular decomposition then* $\begin{pmatrix} P & 0 \\ 0 & I \end{pmatrix}\begin{pmatrix} L & 0 \\ 0 & I \end{pmatrix}\begin{pmatrix} D & 0 \\ 0 & 0 \end{pmatrix} UQ$ *is a triangular decomposition of* $A_{n,m}^k$.

If $\mathbf{A} = \mathbf{B} = 0$ *and* $(\mathbf{C}, \mathbf{D}) = PL(D, 0)UQ$ *is a triangular decomposition then* $\begin{pmatrix} 0 & I \\ P & 0 \end{pmatrix}\begin{pmatrix} L & 0 \\ 0 & I \end{pmatrix}\begin{pmatrix} D & 0 \\ 0 & 0 \end{pmatrix} UQ$ *is a triangular decomposition of* $A_{n,m}^k$.

If $\mathbf{B} = \mathbf{D} = 0$ *and* $(\mathbf{A}, \mathbf{C})^T = PL(D, 0)^T UQ$ *is a triangular decomposition then* $PL \begin{pmatrix} D & 0 \\ 0 & 0 \end{pmatrix}\begin{pmatrix} U & 0 \\ 0 & I \end{pmatrix}\begin{pmatrix} Q & 0 \\ 0 & I \end{pmatrix}$ *is a triangular decomposition of* $A_{n,m}^k$.

If $\mathbf{A} = \mathbf{C} = 0$ *and* $(\mathbf{B}, \mathbf{D})^T = PL(D, 0)^T UQ$ *is a triangular decomposition then* $PL \begin{pmatrix} D & 0 \\ 0 & 0 \end{pmatrix}\begin{pmatrix} U & 0 \\ 0 & I \end{pmatrix}\begin{pmatrix} 0 & I \\ Q & 0 \end{pmatrix}$ *is a triangular decomposition of* $A_{n,m}^k$.

Proof. To prove we have to check the Property of triangular decomposition for each case.

6   Example of Triangular Decomposition on $\mathbb{Z}$

Let a matrix A be given. We show how to obtain the decomposition $A = \mathbf{LDU}$:

$$\begin{bmatrix} 3 & 2 & 3 & 5 & 1 & 2 \\ 1 & 3 & 4 & 2 & 3 & 4 \\ 3 & 2 & 3 & 5 & 5 & 6 \\ 1 & 3 & 4 & 2 & 2 & 1 \\ 2 & 1 & 3 & 2 & 2 & 3 \\ 2 & 1 & 3 & 2 & 2 & 3 \end{bmatrix} = \begin{bmatrix} 3 & 0 & 0 & 0 & 0 & 0 \\ 1 & 7 & 0 & 0 & 0 & 0 \\ 3 & 0 & 40 & 0 & 0 & 0 \\ 1 & 7 & -10 & -80 & 0 & 0 \\ 2 & -1 & 0 & 0 & 10 & 0 \\ 2 & -1 & 0 & 0 & 10 & 1 \end{bmatrix}\begin{bmatrix} \frac{1}{3} & 0 & 0 & 0 & 0 & 0 \\ 0 & \frac{1}{21} & 0 & 0 & 0 & 0 \\ 0 & 0 & 0 & \frac{1}{400} & 0 & 0 \\ 0 & 0 & 0 & 0 & 0 & \frac{-1}{3200} \\ 0 & 0 & \frac{1}{70} & 0 & 0 & 0 \\ 0 & 0 & 0 & 0 & 0 & 0 \end{bmatrix}\begin{bmatrix} 3 & 2 & 3 & 5 & 1 & 2 \\ 0 & 7 & 9 & 1 & 8 & 10 \\ 0 & 0 & 10 & -9 & 12 & 15 \\ 0 & 0 & 0 & 1 & 0 & 0 \\ 0 & 0 & 0 & 0 & 40 & 40 \\ 0 & 0 & 0 & 0 & 0 & -80 \end{bmatrix}$$

Let the upper left block have the size of 4×4:

$$A_6^0 = \begin{pmatrix} A_4^0 & B_4^0 \\ C_4^0 & D_4^0 \end{pmatrix}, \quad A_4^0 = \begin{bmatrix} 3 & 2 & 3 & 5 \\ 1 & 3 & 4 & 2 \\ 3 & 2 & 3 & 5 \\ 1 & 3 & 4 & 2 \end{bmatrix}, \quad B_4^0 = \begin{pmatrix} 1 & 2 \\ 3 & 4 \end{pmatrix}, \quad C_4^0 = \begin{pmatrix} 2 & 1 & 3 & 2 \\ 2 & 1 & 3 & 2 \end{pmatrix}, \quad D_4^0 = \begin{pmatrix} 2 & 3 \\ 2 & 3 \end{pmatrix}.$$

6.1 Recursive Step of the First Level

Triangular Decomposition for $\mathcal{A}_4^0$. Let us split matrix $\mathcal{A}_4^0$ into blocks again.

$$\mathcal{A}_4^0 = \begin{pmatrix} \mathcal{A}_2^0 & \mathcal{B}_2^0 \\ \mathcal{C}_2^0 & \mathcal{D}_2^0 \end{pmatrix}, \text{ where } \mathcal{A}_2^0 = \begin{pmatrix} 3 & 2 \\ 1 & 3 \end{pmatrix}, \mathcal{B}_2^0 = \begin{pmatrix} 3 & 5 \\ 4 & 2 \end{pmatrix}, \mathcal{C}_2^0 = \begin{pmatrix} 3 & 2 \\ 1 & 3 \end{pmatrix}, \mathcal{D}_2^0 = \begin{pmatrix} 3 & 5 \\ 4 & 2 \end{pmatrix}.$$

6.1.1 Recursive Step

Triangular Decomposition for $\mathcal{A}_2^0$. We split matrix $\mathcal{A}_2^0$: $\mathcal{A}_2^0 = \begin{pmatrix} \mathcal{A}_1^0 & \mathcal{B}_1^0 \\ \mathcal{C}_1^0 & \mathcal{D}_1^0 \end{pmatrix}$,

where $\mathcal{A}_1^0 = (3)$, $\mathcal{B}_1^0 = (2)$, $\mathcal{C}_1^0 = (1)$, $\mathcal{D}_1^0 = (3)$.

6.1.1.1 Recursive Step

Triangular Decomposition for $\mathcal{A}_1^0$. Matrix $\mathcal{A}_1^0$ has size 1×1, then we can write result for this decomposition: $\mathcal{A}_1^0 = (3)$, $\alpha_0 = 1$,

$$\{P_1^0, L_1^0, \{\alpha_1\}, U_1^0, Q_1^0, M_1^0, W_1^0\} = \mathbf{LDU}(\mathcal{A}_1^0, \alpha_0),$$

$L_1^0 = U_1^0 = (3)$, $P_1^0 = Q_1^0 = M_1^0 = W_1^0 = (1)$, $\alpha_1 = 3$, $D_1^0 = (\frac{1}{3})$.

Calculations for Next Recursive Step

$$\overline{L_1^0} = (\alpha_0)^{-1} C_1^0 Q_1^0 W_1^0 = (1)^{-1}(1)(1)(1) = (1),$$
$$\overline{U_1^0} = (\alpha_0)^{-1} M_1^0 P_1^0 \mathcal{B}_1^0 = (1)^{-1}(1)(1)(2) = (2),$$
$$\mathcal{A}_2^1 = (\alpha_1/\alpha_0)(\mathcal{D}_1^0 - \overline{L_1^0} D_1^0 \overline{U_1^0}) = \frac{3}{1}((3) - (1)(1/3)(2)) = (7).$$

6.1.1.2 Recursive Step

Triangular Decomposition for $\mathcal{A}_2^1$. Matrix $\mathcal{A}_2^1$ has size 1×1: $\mathcal{A}_2^1 = (7)$, $\alpha_1 = 3$,

$$\{P_2^1, L_2^1, \{\alpha_2\}, U_2^1, Q_2^1, M_2^1, W_2^1\} = \mathbf{LDU}(\mathcal{A}_2^1, \alpha^1),$$

$L_2^1 = U_2^1 = (7)$, $P_2^1 = Q_2^1 = (1)$, $M_2^1 = W_2^1 = (3)$, $\alpha_2 = 7$, $D_2^1 = (1/7)$.

$$\overline{M_1^0} = -(\alpha_0)^{-1} M_2^1 \overline{L_1^0} D_1^0 M_1^0 = (1)^{-1}(3)(1)(1/3)(1) = (-1)$$
$$\overline{W_1^0} = -(\alpha_0)^{-1} W_1^0 D_1^0 \overline{U_1^0} W_2^1 = (1)^{-1}(1)(1/3)(2)(3) = (-2)$$

6.1.1.3 Result of Decomposition of Matrix $\mathcal{A}_2^0$

$$\{P_2^0, L_2^0, \{\alpha_1, \alpha_2\}, U_2^0, Q_2^0, M_2^0, W_2^0\} = \mathbf{LDU}(\mathcal{A}_2^0, \alpha_0),$$

$$P_2^0 = \begin{pmatrix} P_1^0 & 0 \\ 0 & P_2^1 \end{pmatrix} = Q_2^0 = \begin{pmatrix} Q_1^0 & 0 \\ 0 & Q_2^1 \end{pmatrix} = \begin{pmatrix} 1 & 0 \\ 0 & 1 \end{pmatrix}, L_2^0 = \begin{pmatrix} L_1^0 & 0 \\ \overline{L_1^0} & L_2^1 \end{pmatrix} = \begin{pmatrix} 3 & 0 \\ 1 & 7 \end{pmatrix},$$

$$U_2^0 = \begin{pmatrix} U_1^0 & \overline{U_1^0} \\ 0 & U_2^1 \end{pmatrix} = \begin{pmatrix} 3 & 2 \\ 0 & 7 \end{pmatrix}, M_2^0 = \begin{pmatrix} M_1^0 & 0 \\ \overline{M_1^0} & M_2^1 \end{pmatrix} = \begin{pmatrix} 1 & 0 \\ -1 & 3 \end{pmatrix}, \alpha_1 = 3, \alpha_2 = 7,$$

$$W_2^0 = \begin{pmatrix} W_1^0 & \overline{W_1^0} \\ 0 & W_2^1 \end{pmatrix} = \begin{pmatrix} 1 & -2 \\ 0 & 3 \end{pmatrix}, D_2^0 = \begin{pmatrix} 1/3 & 0 \\ 0 & 1/21 \end{pmatrix}.$$

Calculations for Next Recursive Step

$$\overline{L_2^0} = (\alpha_0)^{-1} \mathcal{C}_2^0 Q_2^0 W_2^0 = (1)^{-1} \begin{pmatrix} 3 & 2 \\ 1 & 3 \end{pmatrix} \begin{pmatrix} 1 & 0 \\ 0 & 1 \end{pmatrix} \begin{pmatrix} 1 & -2 \\ 0 & 3 \end{pmatrix} = \begin{pmatrix} 3 & 0 \\ 1 & 7 \end{pmatrix},$$

$$\overline{U_2^0} = (\alpha_0)^{-1} M_2^0 P_2^0 \mathcal{B}_2^0 = (1)^{-1} \begin{pmatrix} 1 & 0 \\ -1 & 3 \end{pmatrix} \begin{pmatrix} 1 & 0 \\ 0 & 1 \end{pmatrix} \begin{pmatrix} 3 & 5 \\ 4 & 2 \end{pmatrix} = \begin{pmatrix} 3 & 5 \\ 9 & 1 \end{pmatrix},$$

$$\mathcal{A}_4^2 = \frac{\alpha_2}{\alpha_0}(\mathcal{D}_2^0 - \overline{L_2^0} D_2^0 \overline{U_2^0}) = \frac{7}{1}\left(\begin{pmatrix} 3 & 5 \\ 4 & 2 \end{pmatrix} - \begin{pmatrix} 3 & 0 \\ 1 & 7 \end{pmatrix}\begin{pmatrix} \frac{1}{3} & 0 \\ 0 & \frac{1}{21} \end{pmatrix}\begin{pmatrix} 3 & 5 \\ 9 & 1 \end{pmatrix}\right) = \begin{pmatrix} 0 & 0 \\ 0 & 0 \end{pmatrix}.$$

6.1.2 Recursive Step

Triangular Decomposition for $\mathcal{A}_4^2$. Since the matrix $\mathcal{A}_4^2$ is zero, we can write the result of decomposition of matrix $\mathcal{A}_4^0$.

6.1.3 Result of Decomposition of Matrix $\mathcal{A}_4^0$

$$\{P_4^0, L_4^0, \{\alpha_1, \alpha_2\}, U_4^0, Q_4^0, M_2^0, W_2^0\} = \mathbf{LDU}(\mathcal{A}_4^0, \alpha^0),$$

$$P_4^0 = \begin{pmatrix} P_2^0 & 0 \\ 0 & I_2 \end{pmatrix}, \quad L_4^0 = \begin{pmatrix} L_2^0 & 0 \\ \overline{L_2^0} & I_2 \end{pmatrix}, \quad U_4^0 = \begin{pmatrix} U_2^0 & \overline{U_2^0} \\ 0 & I_2 \end{pmatrix},$$

$$Q_4^0 = \begin{pmatrix} Q_2^0 & 0 \\ 0 & I_2 \end{pmatrix}, \quad M_2^0 = \begin{pmatrix} 1 & 0 \\ -1 & 3 \end{pmatrix}, \quad W_2^0 = \begin{pmatrix} 1 & -2 \\ 0 & 3 \end{pmatrix},$$

$$D_4^0 = \begin{pmatrix} D_2^0 & 0 \\ 0 & 0 \end{pmatrix}, \quad \alpha_1 = 3, \quad \alpha_2 = 7, \quad (P_4^0, L_4^0, D_4^0, U_4^0, Q_4^0) =$$

$$\left(\begin{pmatrix} 1&0&0&0 \\ 0&1&0&0 \\ 0&0&1&0 \\ 0&0&0&1 \end{pmatrix}, \begin{pmatrix} 3&0&0&0 \\ 1&7&0&0 \\ 3&0&1&0 \\ 1&7&0&1 \end{pmatrix}, \begin{pmatrix} \frac{1}{3}&0&0&0 \\ 0&\frac{1}{21}&0&0 \\ 0&0&0&0 \\ 0&0&0&0 \end{pmatrix}, \begin{pmatrix} 3&2&3&5 \\ 0&7&9&1 \\ 0&0&1&0 \\ 0&0&0&1 \end{pmatrix}, \begin{pmatrix} 1&0&0&0 \\ 0&1&0&0 \\ 0&0&1&0 \\ 0&0&0&1 \end{pmatrix}\right).$$

6.2   Recursive Step of the First Level

Since $\mathcal{A}_4^0$ is not full rank, we should split matrix again: $P_4^0 \mathcal{A}_6^0 Q_4^0 = \begin{pmatrix} \mathcal{A}_2^0 & \mathcal{B}_2^0 \\ \mathcal{C}_2^0 & \mathcal{D}_2^0 \end{pmatrix}$,

where $\mathcal{A}_2^0 = \begin{pmatrix} 3 & 2 \\ 1 & 3 \end{pmatrix}$, $\mathcal{B}_2^0 = \begin{pmatrix} 3 & 5 & 1 & 2 \\ 4 & 2 & 3 & 4 \end{pmatrix}$, $\mathcal{C}_2^0 = \begin{pmatrix} 3 & 2 \\ 1 & 3 \\ 2 & 1 \\ 2 & 1 \end{pmatrix}$ $\mathcal{D}_2^0 = \begin{pmatrix} 3 & 5 & 5 & 6 \\ 4 & 2 & 2 & 1 \\ 3 & 2 & 2 & 3 \\ 3 & 2 & 2 & 3 \end{pmatrix}$.

$$\widetilde{L_2^0} = (\alpha_0)^{-1} \mathcal{C}_2^0 W_2^0 = (1)^{-1} \begin{pmatrix} 3 & 2 \\ 1 & 3 \\ 2 & 1 \\ 2 & 1 \end{pmatrix}\begin{pmatrix} 1 & -2 \\ 0 & 3 \end{pmatrix} = \begin{pmatrix} 3 & 0 \\ 1 & 7 \\ 2 & -1 \\ 2 & -1 \end{pmatrix}, \widetilde{U_2^0} = (\alpha_0)^{-1} M_2^0 \mathcal{B}_2^0 =$$

$$(1)^{-1} \begin{pmatrix} 1 & 0 \\ -1 & 3 \end{pmatrix}\begin{pmatrix} 3 & 5 & 1 & 2 \\ 4 & 2 & 3 & 4 \end{pmatrix} = \begin{pmatrix} 3 & 5 & 1 & 2 \\ 9 & 1 & 8 & 10 \end{pmatrix}, \mathcal{A}_6^2 = \frac{\alpha_2}{\alpha_0}(\mathcal{D}_2^0 - \widetilde{L_2^0} D_2^0 \widetilde{U_2^0}) =$$

$$\frac{7}{1}\left(\begin{pmatrix} 3&5&5&6 \\ 4&2&2&1 \\ 3&2&2&3 \\ 3&2&2&3 \end{pmatrix} - \begin{pmatrix} 3&0 \\ 1&7 \\ 2&-1 \\ 2&-1 \end{pmatrix}\begin{pmatrix} \frac{1}{3} & 0 \\ 0 & \frac{1}{21} \end{pmatrix}\begin{pmatrix} 3&5&1&2 \\ 9&1&8&10 \end{pmatrix}\right) = \begin{pmatrix} 0&0&28&28 \\ 0&0&-7&-21 \\ 10&-9&12&15 \\ 10&-9&12&15 \end{pmatrix}.$$

Split matrix $\mathcal{A}_6^2$ into blocks. $\mathcal{A}_6^2 = \begin{pmatrix} \mathcal{A}_4^2 & \mathcal{B}_4^2 \\ \mathcal{C}_4^2 & \mathcal{D}_4^2 \end{pmatrix}$, where $\mathcal{A}_4^2 = \begin{pmatrix} 0 & 0 \\ 0 & 0 \end{pmatrix}$, $\mathcal{B}_4^2 = \begin{pmatrix} 28 & 28 \\ -7 & -21 \end{pmatrix}$, $\mathcal{C}_4^2 = \begin{pmatrix} 10 & -9 \\ 10 & -9 \end{pmatrix}$, $\mathcal{D}_4^2 = \begin{pmatrix} 12 & 15 \\ 12 & 15 \end{pmatrix}$. Since matrix $\mathcal{A}_4^2$ is zero, we should use some permutation of blocks, and then we can find decomposition of $\mathcal{C}_4^2$ and $\mathcal{B}_4^2$ simultaneously.

$$\begin{pmatrix} 0 & I_2 \\ I_2 & 0 \end{pmatrix} \mathcal{A}_6^2 = \begin{pmatrix} 0 & I_2 \\ I_2 & 0 \end{pmatrix} \begin{pmatrix} 0 & \mathcal{B}_4^2 \\ \mathcal{C}_4^2 & \mathcal{D}_4^2 \end{pmatrix} = \begin{pmatrix} \mathcal{C}_4^2 & \mathcal{D}_4^2 \\ 0 & \mathcal{B}_4^2 \end{pmatrix}.$$

6.2.1 Recursive Step

Triangular Decomposition of Matrix

$$\mathcal{C}_4^2 = \begin{pmatrix} 10 & -9 \\ 10 & -9 \end{pmatrix}, \alpha_2 = 7$$

$$\{P_4^2, L_4^2, \{\alpha_3\}, U_4^2, Q_4^2, M_3^2, W_3^2\} = \mathbf{LDU}(\mathcal{C}_4^2, \alpha^2),$$

where $P_4^2 = \begin{pmatrix} 1 & 0 \\ 0 & 1 \end{pmatrix}$, $L_4^2 = \begin{pmatrix} L_0 & 0 \\ M_0 & I_1 \end{pmatrix} = \begin{pmatrix} 10 & 0 \\ 10 & 1 \end{pmatrix}$, $U_4^2 = \begin{pmatrix} U_0 & V_0 \\ 0 & I_1 \end{pmatrix} = \begin{pmatrix} 10 & -9 \\ 0 & 1 \end{pmatrix}$,

$Q_4^2 = \begin{pmatrix} 1 & 0 \\ 0 & 1 \end{pmatrix}$, $M_3^2 = (7)$, $W_3^2 = (7)$, $\alpha_3 = 10$.

6.2.2 Recursive Step

Triangular Decomposition of Matrix

$$\mathcal{B}_4^2 = \begin{pmatrix} 28 & 28 \\ -7 & -21 \end{pmatrix}, \alpha_2 = 7$$

$$\{P_6^4, \bar{L}_6^4, \{\bar{\alpha}_4, \bar{\alpha}_5\}, \bar{U}_6^4, Q_6^4, \bar{M}_6^4, \bar{W}_6^4\} = \mathbf{LDU}(\mathcal{B}_4^2, \alpha_2),$$

where $P_6^4 = \begin{pmatrix} 1 & 0 \\ 0 & 1 \end{pmatrix}$, $\bar{L}_6^4 = \begin{pmatrix} 28 & 0 \\ -7 & -56 \end{pmatrix} = \begin{pmatrix} 10 & 0 \\ 10 & 1 \end{pmatrix}$, $\bar{U}_6^4 = \begin{pmatrix} 28 & 28 \\ 0 & -56 \end{pmatrix}$, $Q_6^4 = \begin{pmatrix} 1 & 0 \\ 0 & 1 \end{pmatrix}$, $\bar{M}_6^4 = \begin{pmatrix} 7 & 0 \\ 7 & 28 \end{pmatrix}$, $\bar{W}_6^4 = \begin{pmatrix} 7 & -28 \\ 0 & 28 \end{pmatrix}$, $\bar{\alpha}_4 = 28, \bar{\alpha}_5 = -56.$

Calculations before Next Recursive Step

$$L_6^4 = \frac{\alpha_3}{\alpha_2} \bar{L}_6^4 = \frac{10}{7} \begin{pmatrix} 28 & 0 \\ -7 & -56 \end{pmatrix} = \begin{pmatrix} 40 & 0 \\ -10 & -80 \end{pmatrix},$$

$$U_6^4 = \frac{\alpha_3}{\alpha_2} \bar{U}_6^4 = \frac{10}{7} \begin{pmatrix} 28 & 28 \\ 0 & -56 \end{pmatrix} = \begin{pmatrix} 40 & 40 \\ 0 & -80 \end{pmatrix},$$

$$M_6^4 = \frac{\alpha_3}{\alpha_2} \bar{M}_6^4 = \frac{10}{7} \begin{pmatrix} 28 & 0 \\ -7 & -56 \end{pmatrix} = \begin{pmatrix} 10 & 0 \\ 10 & 40 \end{pmatrix},$$

$$W_6^4 = \frac{\alpha_3}{\alpha_2} \bar{W}_6^4 = \frac{10}{7} \begin{pmatrix} 7 & -28 \\ 0 & 28 \end{pmatrix} = \begin{pmatrix} 10 & -40 \\ 0 & 40 \end{pmatrix}, \tag{26}$$

$\alpha_4 = \frac{\alpha_3 \bar{\alpha}_4}{\alpha_2} = \frac{10 \times 28}{7} = 40$, $\alpha_5 = \frac{\alpha_3 \bar{\alpha}_5}{\alpha_2} = \frac{10 \times (-56)}{7} = -80$.

Since C_4^2 has no full rank, we should split into blocks matrix $\mathcal{D}_4^2$. $P_4^2 \mathcal{D}_4^2 Q_6^4 = \begin{pmatrix} D_1 \\ D_2 \end{pmatrix}$, where $D_1 = (12\ 15)$ and $D_2 = (12\ 15)$.

$$U^d = (\alpha_2)^{-1} M_3^2 D_1 = (7)^{-1}(7)(12\ 15) = (12\ 15),$$

$$L^d = (\alpha_2)^{-1}(D_2 - M_0(\frac{1}{\alpha_3})U^d)W_6^4 =$$

$$(7)^{-1}\left((12\ 15) - (10)(\frac{1}{10})(12\ 15)\right)\begin{pmatrix} 10 & -40 \\ 0 & 40 \end{pmatrix} = (0\ 0),$$

$$M^{bc} = -(\alpha_2)^{-1} M_6^4 0 (\frac{1}{\alpha_3}) M_3^2 = 0,$$

$$W^{bc} = -\frac{1}{\alpha_2} W_3^2 \frac{1}{\alpha_3} U^d W_6^4 = -\frac{7}{7}\frac{1}{10}(12\ 15)\begin{pmatrix} 10 & -40 \\ 0 & 40 \end{pmatrix} = (-12\ -12).$$

6.2.3. Result of Decomposition of Matrix $\mathcal{A}_6^2$

$$\{P_6^2, L_6^2, \{\alpha_3, \alpha_4, \alpha_5\}, U_6^2, Q_6^2, M_6^2, W_6^2\} = \mathbf{LDU}(\mathcal{A}_6^2, \alpha_2),$$

where $P_6^2 = \begin{pmatrix} 0 & I_2 \\ I_2 & 0 \end{pmatrix}\begin{pmatrix} P_4^2 & 0 \\ 0 & P_6^4 \end{pmatrix}\begin{pmatrix} I_1 & 0 & 0 \\ 0 & 0 & I_1 \\ 0 & I_2 & 0 \end{pmatrix}$, $Q_6^2 = \begin{pmatrix} I_1 & 0 & 0 \\ 0 & 0 & I_2 \\ 0 & I_1 & 0 \end{pmatrix}\begin{pmatrix} Q_4^2 & 0 \\ 0 & Q_6^4 \end{pmatrix}$,

$L_6^2 = \begin{pmatrix} L_0 & 0 & 0 \\ 0 & L_6^4 & 0 \\ M_0 & L_d & I_1 \end{pmatrix}$, $U_6^2 = \begin{pmatrix} U_0 & U^d & V_0 \\ 0 & U_6^4 & 0 \\ 0 & 0 & I_1 \end{pmatrix}$, $M_6^2 = \begin{pmatrix} M_3^2 & 0 \\ M^{bc} & M_6^4 \end{pmatrix} = \begin{pmatrix} 7 & 0 & 0 \\ 0 & 10 & 0 \\ 0 & 10 & 40 \end{pmatrix}$,

$W_6^2 = \begin{pmatrix} W_3^2 & W^{bc} \\ 0 & W_6^4 \end{pmatrix} = \begin{pmatrix} 7 & -12 & -12 \\ 0 & 10 & -40 \\ 0 & 0 & 40 \end{pmatrix}$, $\alpha_3 = 10, \alpha_4 = 40, \alpha_5 = -80$,

$D_6^2 = \mathrm{diag}(\frac{1}{\alpha_3}, \frac{\alpha_2}{\alpha_3\alpha_4}, \frac{\alpha_2}{\alpha_4\alpha_5}, 0)$, $(P_6^2, L_6^2, D_6^2, U_6^2, Q_6^2) =$

$$\left(\begin{bmatrix} 0 & 1 & 0 & 0 \\ 0 & 0 & 1 & 0 \\ 1 & 0 & 0 & 0 \\ 0 & 0 & 0 & 1 \end{bmatrix}, \begin{bmatrix} 10 & 0 & 0 & 0 \\ 0 & 40 & 0 & 0 \\ 0 & -10 & -80 & 0 \\ 10 & 0 & 0 & 1 \end{bmatrix}, \begin{bmatrix} \frac{1}{10} & 0 & 0 & 0 \\ 0 & \frac{7}{400} & 0 & 0 \\ 0 & 0 & \frac{-7}{3200} & 0 \\ 0 & 0 & 0 & 0 \end{bmatrix}, \begin{bmatrix} 10 & 12 & 15 & -9 \\ 0 & 40 & 40 & 0 \\ 0 & 0 & -80 & 0 \\ 0 & 0 & 0 & 1 \end{bmatrix}, \begin{bmatrix} 1 & 0 & 0 & 0 \\ 0 & 0 & 1 & 0 \\ 0 & 0 & 0 & 1 \\ 0 & 1 & 0 & 0 \end{bmatrix}\right)$$

6.3   Final Result

$$\{P_6^0, L_6^0, \{\alpha_1, \alpha_2\alpha_3, \alpha_4, \alpha_5\}, U_6^0, Q_6^0\} = \mathbf{LDU}(\mathcal{A}_6^0, \alpha_0),$$

$D_6^0 = \mathrm{diag}(\frac{\alpha_0}{\alpha_0\alpha_1}, \frac{\alpha_0}{\alpha_1\alpha_2}, \frac{\alpha_0}{\alpha_2\alpha_3}, \frac{\alpha_0}{\alpha_3\alpha_4}, \frac{\alpha_0}{\alpha_4\alpha_5}, 0) = \mathrm{diag}(\frac{1}{3}, \frac{1}{21}, \frac{1}{70}, \frac{1}{400}, \frac{-1}{3200}, 0)$

$P_6^0 = \begin{pmatrix} P_4^0 & 0 \\ 0 & I_2 \end{pmatrix}\begin{pmatrix} I_2 & 0 \\ 0 & P_6^2 \end{pmatrix}$, $L_6^0 = \begin{pmatrix} L_2^0 & 0 \\ P_6^{2T}\widetilde{L}_2^0 & L_6^2 \end{pmatrix}$, $U_6^0 = \begin{pmatrix} U_2^0 & \widetilde{U}_2^0 Q_6^{2T} \\ 0 & U_6^2 \end{pmatrix}$,

$$Q_6^0 = \begin{pmatrix} I_2 & 0 \\ 0 & Q_6^2 \end{pmatrix} \begin{pmatrix} Q_4^0 & 0 \\ 0 & I_2 \end{pmatrix}, (P_6^0, L_6^0, U_6^0, Q_6^0) =$$

$$\left(\begin{bmatrix} 1 & 0 & 0 & 0 & 0 & 0 \\ 0 & 1 & 0 & 0 & 0 & 0 \\ 0 & 0 & 0 & 1 & 0 & 0 \\ 0 & 0 & 0 & 0 & 1 & 0 \\ 0 & 0 & 1 & 0 & 0 & 0 \\ 0 & 0 & 0 & 0 & 0 & 1 \end{bmatrix}, \begin{bmatrix} 3 & 0 & 0 & 0 & 0 & 0 \\ 1 & 7 & 0 & 0 & 0 & 0 \\ 2 & -1 & 10 & 0 & 0 & 0 \\ 3 & 0 & 0 & 40 & 0 & 0 \\ 1 & 7 & 0 & -10 & -80 & 0 \\ 2 & -1 & 10 & 0 & 0 & 1 \end{bmatrix}, \begin{bmatrix} 3 & 2 & 3 & 1 & 2 & 5 \\ 0 & 7 & 9 & 8 & 10 & 1 \\ 0 & 0 & 10 & 12 & 15 & -9 \\ 0 & 0 & 0 & 40 & 40 & 0 \\ 0 & 0 & 0 & 0 & -80 & 0 \\ 0 & 0 & 0 & 0 & 0 & 1 \end{bmatrix}, \begin{bmatrix} 1 & 0 & 0 & 0 & 0 & 0 \\ 0 & 1 & 0 & 0 & 0 & 0 \\ 0 & 0 & 1 & 0 & 0 & 0 \\ 0 & 0 & 0 & 0 & 1 & 0 \\ 0 & 0 & 0 & 0 & 0 & 1 \\ 0 & 0 & 0 & 1 & 0 & 0 \end{bmatrix} \right)$$

$$\mathcal{L} = P_6^0 L_6^0 P_6^{0^T}, \ \mathcal{D} = P_6^0 D_6^0 Q_6^0, \ \mathcal{U} = Q_6^{0^T} U_6^0 Q_6^0.$$

7 Conclusion

We propose an algorithm for the matrix factorization, which has the complexity of matrix multiplication. It was implemented at http://mathpartner.com. An example can be obtained as follows: "A=[[1,2],[3,4]]; \LDU(A)".

References

1. Akritas, A.G., Akritas, E.K., Malaschonok, G.I.: Various proofs of Sylvester's (determinant) identity. In: Proc. Int. IMACS Symp. on Symbolic Computation, June 14–17, 1993, Lille, France, pp. 228–230 (1993)
2. Grigoriev, D.: Additive complexity in directed computations. Theoretical Computer Science **19**, 39–67 (1982)
3. Grigoriev, D.: Analogy of Bruhat decomposition for the closure of a cone of Chevalley group of a classical series. Soviet Math. Dokl. **23**, 393–397 (1981)
4. Malaschonok, G.I.: A fast algorithm for adjoint matrix computation. Tambov University Reports **5**(1), 142–146 (2000)
5. Malaschonok, G.I.: Effective matrix methods in commutative domains. In: Krob, D., Mikhalev, A.A., Mikhalev, A.V. (eds.) Formal Power Series and Algebraic Combinatorics, pp. 506–517. Springer, Berlin (2000)
6. Malaschonok, G.: Fast generalized bruhat decomposition. In: Gerdt, V.P., Koepf, W., Mayr, E.W., Vorozhtsov, E.V. (eds.) CASC 2010. LNCS, vol. 6244, pp. 194–202. Springer, Heidelberg (2010)
7. Malaschonok, G.I.: Fast matrix decomposition in parallel computer algebra. Tambov University Reports **15**(4), 1372–1385 (2010)
8. Malaschonok, G.: Generalized bruhat decomposition in commutative domains. In: Gerdt, V.P., Koepf, W., Mayr, E.W., Vorozhtsov, E.V. (eds.) CASC 2013. LNCS, vol. 8136, pp. 231–242. Springer, Heidelberg (2013)
9. Malaschonok, G.I.: Matrix Computational Methods in Commutative Rings. Tambov University Publishing House, Tambov (2002)
10. Malaschonok, G.I.: On the fast generalized Bruhat decomposition in domains. Tambov University Reports **17**(2), 544–550 (2012)

Automated Reasoning in Reduction Rings Using the *Theorema* System[*]

Alexander Maletzky

Doctoral College "Computational Mathematics" and RISC
Johannes Kepler University Linz, Austria
alexander.maletzky@dk-compmath.jku.at

Abstract. In this paper, we present the computer-supported theory exploration, including both formalization and verification, of a theory in commutative algebra, namely the theory of reduction rings. Reduction rings, introduced by Bruno Buchberger in 1984, are commutative rings with unit which extend classical Gröbner bases theory from polynomial rings over fields to a far more general setting.

We review some of the most important notions and concepts in the theory and motivate why reduction rings are a natural candidate for being explored with the assistance of a software system, which, in our case, is the *Theorema* system. We also sketch the special prover designed and implemented for the purpose of semi-automated, interactive verification of the theory, and outline the structure of the formalization.

Keywords: Gröbner bases, reduction rings, mathematical theory exploration, automated reasoning, formalized mathematics, Theorema.

1 Introduction

Automated reasoning, or, more precisely, computer-supported mathematical theory exploration, aims at the systematic creation of machine-checked, formal mathematics, either entirely by or at least with extensive support of software systems, where the meaning of "formal mathematics" may range from individual theorems over whole theories up to huge structured knowledge bases. In this paper, we demonstrate how a non-trivial theory in the realm of commutative algebra, namely the theory of reduction rings and Gröbner bases, can be formally developed in the *Theorema* software system, including semi-automated, interactive verification. Up to our knowledge, this is the first time this theory is the subject of computer-supported theory exploration in *any* software system, although it has to be mentioned that Gröbner bases in the *classical* setting (i.e., in polynomial rings over fields) have already undergone formal treatment in various flavors [15,11,5,6,7].

The theory of *reduction rings*, to be presented in more detail in Section 2, is a natural candidate for computer-supported exploration: a good deal of the notions

---

[*] This research was funded by the Austrian Science Fund (FWF): grant no. W1214-N15, project DK1.

© Springer International Publishing Switzerland 2015
V.P. Gerdt et al. (Eds.): CASC 2015, LNCS 9301, pp. 307–321, 2015.
DOI: 10.1007/978-3-319-24021-3_23

and concepts involved have rather lengthy and technical definitions, which quite often leads to comparatively technical proofs with many case distinctions, subgoals, etc. Although the correctness of all results has already been established more than 20 years ago using "pencil and paper", the theory has hardly been extended and generalized since then (e. g., to non-commutative rings; see also Section 6), which is clearly a non-trivial task, but perhaps also due the aforementioned manifold of complicated (from the formal point of view) definitions. Further developing the theory, finally, is the main motivation and the long term goal of the project.

Before this goal can be achieved, however, the existing part has to be formalized and formally verified first, which is what will be presented in the rest of the paper. Still, one must also mention that the exploration of reduction ring theory in *Theorema* was started only recently and, thus, many things have not been completed yet: although the formal representation of the theory by means of a structured *Theorema* document is mostly finished, only a few results have been mechanically proved so far. Nevertheless, the completed proofs (one of them being sketched in Section 5) give a good impression of how semi-automated theorem proving in *Theorema* usually proceeds.

The structure of this paper is as follows: Section 2 very briefly reviews reduction rings and their relation to Gröbner bases. Section 3 contains an overview of the *Theorema* system, and Section 4 presents the *special prover* designed and implemented for the computer-supported verification of the formalization of reduction ring theory in *Theorema*. Afterward, Section 5 describes the outline of this formalization and demonstrates how computer-supported theorem proving in *Theorema* usually proceeds, by means of a concrete example. Finally, Section 6 summarizes the contents of this paper and provides an outlook over possible extensions of the work presented here.

2   Reduction Rings

The notion of *reduction ring* was introduced by Bruno Buchberger [1] in 1984 and later generalized and extended by Sabine Stifter [12,13]. In short, a reduction ring R is a commutative ring with unit, where in addition a Noetherian order relation $\prec$ and, for each ring element c, a subset $M_c \subseteq R$ are defined. The M_c are the sets of *multipliers*, i. e., they consist of those elements c may be multiplied with when reducing an arbitrary ring element modulo c. If $\prec$ and M_c have certain properties, and, thus, R is a reduction ring, then Gröbner bases of finitely generated ideals in R can be computed algorithmically from any given finite ideal basis. Moreover, if R is a reduction ring, then also $R[X]$ (the polynomial ring in indeterminates $X = \{x_1, \ldots, x_n\}$ over R) and R^n (the n-fold direct product of R) can be made reduction rings by defining $\prec$ and M_c appropriately. This already indicates that reduction rings do not necessarily have any polynomial structure. Furthermore, reduction rings not even have to be free of zero divisors; for instance, $\mathbb{Z}_m$, the ring of residue classes modulo m, constitutes a reduction ring for *any* (not necessarily prime) m. Other examples of reduction rings are $\mathbb{Z}$, all fields, and, thus, also $\mathbb{Z}^n$, $K[X]$ (for fields K), etc.

It has to be remarked that many other generalizations of Gröbner bases theory, both in the commutative and non-commutative case, have been proposed by various authors (see, e. g., [1] for a review of commutative generalizations), and that the term "reduction ring" could, thus, also be understood in a broader sense, simply as a ring where a "reduction operation" is defined (in whatever way). However, the term "reduction ring" will always refer to reduction rings according to [13] in this paper.

2.1 Gröbner Bases in Reduction Rings

Many different characterizations of Gröbner bases exist in the literature, where most of them turn out to be equivalent if restricted to certain domains. For instance, in the classical setting of polynomials over a field, Gröbner bases can be characterized as sets G such that

(G1) every leading power-product of polynomials in $\langle G \rangle$ (i. e., the ideal generated by G) is a multiple of the leading power-product of at least one polynomial in G, or

(G2) the *reduction relation* modulo G is Church–Rosser, or

(G3) All S-polynomials of elements in G can be reduced to 0 modulo G.

It is clear that in reduction rings, there is no analogue of (G1): elements of reduction rings, in general, cannot be decomposed into a "leading part" and into a "rest". Hence, the only candidates remaining are (G2) and (G3), and actually (G2) is taken as the definition of Gröbner bases in reduction rings. In the sequel, R is always assumed to be a reduction ring.

Definition 1 (Gröbner basis). *Let $G \subseteq R$ be finite. G is a* Gröbner basis *of the ideal it generates iff the reduction relation modulo G, i. e., $\rightarrow_G$, is Church–Rosser.*

Definition 1 provides an algebraic characterization of Gröbner bases, but not an algorithmic one: given a finite set $B \subseteq R$, it is in general not possible to find out *algorithmically* whether B is a Gröbner basis or not, and furthermore, to compute a Gröbner basis G of the ideal generated by B (if it exists, which is not yet clear either). Hence, what is needed is an analogue of (G3) in reduction rings, and indeed there is one.

Theorem 1 (Main Theorem). *Let $G \subseteq R$ be finite. Then G is a Gröbner basis of the ideal it generates iff for all $g_1, g_2 \in G$ (not necessarily distinct), all $i, j \in I_{g_1}$, and all non-trivial common reducibles a of g_1 and g_2 w. r. t. (i, j) there exists a critical pair (b_1, b_2) for g_1 and g_2 w. r. t. a and (i, j), such that $b_1 \leftrightarrow_G^{\preceq a} b_2$.*

Unfortunately, due to the space limitations imposed on this paper, we cannot go further into the details of Theorem 1, nor provide the formal (and slightly technical) definitions of *non-trivial common reducibles*, *critical pair*, $\rightarrow_G$ and

$\leftrightarrow_G^{\prec a}$; the interested reader is referred to [1,12,13] instead. Constructing a completely formal *Theorema* proof of Theorem 1 is one of the goals of the formal treatment of reduction ring theory described in this paper.

As in the classical setting of polynomials over a field, Theorem 1 provides both an algorithmic criterion for deciding whether a given set G is a Gröbner basis, and also for computing a Gröbner basis for the ideal generated by G in case it is not. The algorithm is almost exactly the same critical-pair/completion algorithm as Buchberger's algorithm in the classical setting, with three minor modifications:

- Not only pairs of *distinct* elements g_1, g_2 of G have to be considered, but also pairs where both constituents are identical.
- For a given pair (g_1, g_2), *all* minimal non-trivial common reducibles a (for *all* indices $i \neq j \in I_{g_1}$ if $g_1 = g_2$) have to be considered, not just one. Still, there are only finitely many, and considering only *one* critical pair (b_1, b_2) for g_1 and g_2 w.r.t. a and (i, j) is sufficient.
- Instead of reducing the difference $b_1 - b_2$ of b_1 and b_2 to normal form, both b_1 and b_2 have to be reduced separately to normal form. If the two normal forms are not identical, their difference has to be added to the basis.

Of course, the algorithm is only an algorithm relative to the computability of the basic ring operations, as well as the computability of normal forms and minimal non-trivial common reducibles. This can either be required by the axioms of reduction rings, as it is done in [1,12,13], or left as an additional degree of freedom. In the latter case, which is how it is done in the *Theorema*-formalization underlying this paper, one then has to distinguish between *algorithmic* and *non-algorithmic* reduction rings. All examples of reduction rings mentioned at the beginning of this section, however, are algorithmic, and furthermore being an algorithmic reduction ring carries over from R to $R[X]$ and to R^n.

3   The *Theorema* System – A Short Overview

Theorema [4,19] is a system for computer-supported mathematical theory exploration, conceived in the mid-90s by Bruno Buchberger and now developed under his guidance at RISC. One of the major goals of the project has ever since been the seamless integration of *proving*, *computing* and *solving* within one single software system, and, thus, supporting the working mathematician in his everyday life.

Recently, a completely new version of *Theorema* was released, called *Theorema* 2.0 [18]. Although the main design principles have not changed, and it is still based on *Mathematica* [20], the software was entirely redesigned and reimplemented from scratch, with substantial improvements compared to the previous version at all levels of its architecture. Since the research presented in this paper was entirely carried out in *Theorema* 2.0, here and henceforth, *Theorema* will always refer to *Theorema* 2.0 unless explicitly stated otherwise.

Besides the formal treatment of the theory of reduction rings presented in this paper, there are also other theories in the area of computer algebra developed with *Theorema*, which are worth being mentioned here in order to illustrate the capabilities of the system. All of them were developed in *Theorema* 1, but can easily be updated to function under the new version as well: the symbolic treatment of linear boundary problems by means of Green's functions and Green's operators, for instance, was developed by Markus Rosenkranz [10] with (at least partial) support from *Theorema*, and moreover, the resulting algorithm was implemented in the system in a generic, highly-structured way (following the *Theorema*-functor approach), see [14]. A second example of a non-trivial problem solved with *Theorema* is the problem of automatically synthesizing Buchberger's algorithm for computing Gröbner bases only from its specification, following the principle of *Lazy Thinking* [3,5].

An important concept in *Theorema* is the concept of *domains – functors – categories* [17,2]. In order for no confusion to arise, it must be pointed out already here that the terms *functor* and *category* have a slightly different meaning than in classical category theory, to be explained below. Also, the term *domain* does not necessarily refer to a ring without zero divisors (as usually in algebra), but rather to a general algebraic structure formed by a carrier together with operations defined on it.

In *Theorema*, a functor is essentially a function mapping domains (and possibly other entities) to domains, where in turn a domain is characterized by a carrier and operations. Hence, a functor typically takes as input a domain $\mathcal{A}$ and constructs a new domain $\mathcal{B}$, by defining $\mathcal{B}$'s carrier and operations in terms of $\mathcal{A}$'s carrier and operations. Moreover, in order to make things more compact, the carrier of a domain is usually not represented as a set, but as a further operation, namely a unary decision predicate that decides for any given object x whether x belongs to the respective domain or not. One of the simplest examples of a functor is the functor $\mathcal{DP}$ which takes as input two domains $\mathcal{D}_1$ and $\mathcal{D}_2$ and constructs the direct product $\mathcal{P}$ of $\mathcal{D}_1$ and $\mathcal{D}_2$, where all operations of $\mathcal{P}$ are defined component-wise in terms of the operations of the $\mathcal{D}_i$. Categories, finally, are classes of domains sharing common properties, e. g., the category of all commutative rings with unit, the category of all fields, and so on.

The formal development of reduction ring theory presented in this paper also follows the functor approach. Natural candidates for functors are, of course, the $\mathcal{POLY}$ functor that constructs the ring of multivariate polynomials over a given coefficient domain and power-product domain. The most important category then, obviously, is the category of reduction rings as described in Section 2.

4   The REDUCTIONRINGPROVER **Special Prover**

As many other systems for mathematical theory exploration, *Theorema* is not specialized to work only in one single mathematical domain, e. g., in algebra or geometry, but its design allows for completely arbitrary mathematical content, formulated in the language of *higher-order predicate logic and set theory*, to be treated by the system. Still, this does not mean there is no specialization at all, but the specialization happens at a different level: an important aspect of the philosophy behind *Theorema*, and one that distinguishes it from other systems, is the idea that exploring a theory does not only happen at the object level, but also at the meta level. This means that when working in a certain theory T one should not have to fall back to the very elementary and general proving techniques of predicate logic all the time, but rather use more advanced techniques that eventually lead to more elegant and shorter proofs which ideally even resemble the way human mathematicians would proceed. These advanced techniques, however, might only be correct in T but not in general.

A concrete example for a theory T and special proving techniques is given by geometric theorem proving, i. e., where T can be thought of as the theory of real numbers with addition and multiplication. It is a well-known fact that Gröbner bases can be used for proving statements in T, simply by finding out whether a certain polynomial identity follows from a system of algebraic equations or not. Now, a mathematician working in this area most probably would like to automate this process, or more precisely, type in the statement he wants to prove into a computer system that internally automatically uses the method of Gröbner bases for proving or disproving it – and this is possible in *Theorema* by creating a *special prover* for geometric theorem proving. Such a prover will directly use Gröbner bases on the inference level and hence prove theorems in geometry automatically in a short and elegant way, precisely as desired by the human mathematician.

It has to be pointed out that a geometry-prover of the form sketched above does not yet exist in *Theorema* 2.0 (it was available in the old version of the system, though; see [8,9]). Instead, in the sequel, another *Theorema* special prover, created for the treatment of the theory of reduction rings, is presented.

In *Theorema*, a prover consists of two parts which are mainly independent of each other: a collection of *inference rules* and a proving *strategy*. The inference rules describe how a certain proof situation, characterized by a set of assumptions and a proof goal, can be transformed into one or more simpler proof situations. The proving strategy guides the application of the inference rules, i. e., it specifies in which order the rules are tried, what to do if more than one rules are applicable, etc.

The proving strategy used for verifying the theory of reduction rings is a *fully interactive* strategy. This means that in each step, the human operator has full control over the whole proof search: he decides which inference to perform (and in which way) and at which position in the proof to proceed. Nevertheless, if he feels that the current proof situation is simple enough for *Theorema* to automatically find a proof, it is still possible to trigger an automatic proof search as well.

Regarding inference rules, the REDUCTIONRINGPROVER actually does not consist of that many inference rules being special in the sense that they can only be used in the theory of reduction rings, but not elsewhere; they are presented in the next two subsections.

One remark is still in place: Most functions and relations, e. g., $\prec$, are only defined for arguments of a particular domain $\mathcal{D}$. Since *Theorema* is not typed, however, a proof situation may well contain functions/relations applied on arguments *not* in that particular domain, leading to undefined expressions. Therefore, in order to reason correctly, domain-membership of all terms involved in an inference step must always be checked explicitly. In *Theorema*, membership of x in $\mathcal{D}$ is usually denoted by $\underset{\mathcal{D}}{\in} [x]$, and $\underset{\mathcal{D}}{\in} [x_1, \ldots, x_n]$ abbreviates $\bigwedge_{i=1,\ldots,n} \underset{\mathcal{D}}{\in} [x_i]$.

4.1 Order Relations

One of the ubiquitous objects in the theory of reduction rings are order relations of all kinds: every reduction ring is ordered by an arbitrary *partial* order relation, power products are ordered by *admissible total* orderings, and the divisibility relation on power products is a *monotonic* (w. r. t. multiplication) partial ordering. Moreover, some of the orderings have been defined as *irreflexive* and *asymmetric*, others as *reflexive* and *antisymmetric*. Although both kinds of order relations are more or less equivalent to each other (one can always make a reflexive ordering out of an irreflexive one, and vice versa), from the point of view of theorem proving it is desirable to be able to handle both kinds directly, without the need for any conversion taking place beforehand.

Let in the sequel $\prec$ always be defined on the domain $\mathcal{D}$. The following are the three ordering inference rules:

OrderingGoal If the proof goal is of the form $x \prec y$ or $\neg x \prec y$, various attempts for simplifying the goal are made, depending on whether the ordering $\prec$ is partial or total, and whether it is reflexive or irreflexive. For instance, if $\prec$ is any order relation, we have

$$\frac{K \;\vdash\; \underset{\mathcal{D}}{\in} [x, y, z_1, \ldots, z_n]}{K, x \prec z_1, z_1 \prec z_2, \ldots, z_{n-1} \prec z_n, z_n \prec y \;\vdash\; x \prec y}$$

by transitivity. Similarly, if $\prec$ is asymmetric, we have

$$\frac{K \;\vdash\; \underset{\mathcal{D}}{\in} [x, y]}{K, y \prec x \;\vdash\; \neg x \prec y}$$

All inferences of that kind are incorporated in the *OrderingGoal* inference rule in the REDUCTIONRINGPROVER.

OrderingKB If the set of assumptions of the current proof situation contains a formula of the form $x \prec x$ and $\prec$ is an irreflexive ordering, then the assumptions are apparently contradictory, so the proof is finished. This gives rise to the following two inferences, both incorporated in the *OrderingKB* inference rule in the REDUCTIONRINGPROVER:

$$\frac{K \;\vdash\; \underset{\mathcal{D}}{\in} [x]}{K, x \prec x \;\vdash\; \Gamma}$$

if $\prec$ is irreflexive, and

$$\frac{K \;\vdash\; \underset{\mathcal{D}}{\in} [x]}{K, \neg x \prec x \;\vdash\; \Gamma}$$

if $\prec$ is reflexive.

OrderingEqualGoal If the proof goal is of the form $x = y$ and both x and y are elements of a domain that is totally ordered by $\prec$, then it suffices to prove both $\neg x \prec y$ and $\neg y \prec x$. Similarly, if $\prec$ is antisymmetric, it suffices to prove both of $x \prec y$ and $y \prec x$. This gives rise to the following two inferences, both incorporated in the *OrderingEqualGoal* inference rule in the REDUCTIONRINGPROVER:

$$\frac{\mathrm{K} \;\vdash\; \underset{\mathcal{D}}{\in} [x,y] \qquad \mathrm{K} \;\vdash\; \neg x \prec y \qquad \mathrm{K} \;\vdash\; \neg y \prec x}{\mathrm{K} \;\vdash\; x = y}$$

if $\prec$ is total, and

$$\frac{\mathrm{K} \;\vdash\; \underset{\mathcal{D}}{\in} [x,y] \qquad \mathrm{K} \;\vdash\; x \prec y \qquad \mathrm{K} \;\vdash\; y \prec x}{\mathrm{K} \;\vdash\; x = y}$$

if $\prec$ is antisymmetric.

4.2   Commutative Rings with Unit

Since every reduction ring by definition is a commutative ring with unit, it is desirable to have inference rules incorporating the logical axioms of the ring operations $+$ and $\cdot$ in a compact and easy-to-use way, such that it is not necessary to fall back to the very definitions of commutativity and distributivity to prove that, say, $x \cdot y + y \cdot z$ and $x \cdot (z + y)$ are equal. Rather, all this should happen fully automatically whenever the proof goal is an equality of two terms, where the outermost function symbol of at least one of the terms is $+$, $\cdot$, $-$ (the additive inverse) or $\sum$ (since also sums over several ring elements play an important role in reduction ring theory).

Fix now a commutative ring with unit $\mathcal{R}$ as the domain of discourse, i.e. $+$, $\cdot$, etc. are functions on $\mathcal{R}$. The following are the two special inference rules for commutative rings with unit (keep in mind that domain-membership in $\mathcal{R}$ of x and y in $x + y$ is not guaranteed and always has to be checked explicitly):

MembershipCommRings1 If one has to prove membership of a certain term in a commutative ring with unit, several properties of the functions $+$, $\cdot$, etc. are used in order to simplify the proof goal. Examples of inferences are

$$\frac{\mathrm{K} \;\vdash\; \underset{\mathcal{R}}{\in} [x] \qquad \mathrm{K} \;\vdash\; \underset{\mathcal{R}}{\in} [y]}{\mathrm{K} \;\vdash\; \underset{\mathcal{R}}{\in} [x+y]}$$

and

$$\frac{\mathrm{K} \;\vdash\; a \in \mathbb{Z} \qquad \mathrm{K} \;\vdash\; b \in \mathbb{Z} \qquad \mathrm{K} \;\vdash\; \underset{i=a,\dots,b}{\forall} \underset{\mathcal{R}}{\in} [f(i)]}{\mathrm{K} \;\vdash\; \underset{\mathcal{R}}{\in} [\;\sum_{i=a,\dots,b} f(i)\;]}$$

All inferences of that kind are incorporated in the *MembershipCommRings1* inference rule of the REDUCTIONRINGPROVER. In fact, since membership of certain terms in commutative rings with unit has to be proved very often, this is one of the most important special inference rules.

CommRing1Equal If the proof goal is of the form $a = b$, where the outermost function symbol of a or b is among $+, \cdot, -$ and $\sum$, both a and b are fully expanded using associativity and distributivity of the functions involved, and then the resulting terms are checked for being equal, further using commutativity of $+$ and $\cdot$. Of course, associativity/commutativity/distributivity of $+$ and $\cdot$ may only be exploited if the arguments of the respective functions belong to $\mathcal{R}$, which is checked analogously to the ordering-rules.

5 Formalized Reduction Rings

Theorema is a system for mathematical theorem exploration, opposed to isolated theorem proving. Thus, working in *Theorema* usually proceeds by developing a whole theory for what one wants to do, consisting of definitions, lemmas, theorems, computations, etc., all included in one or more *Theorema* notebooks and resembling the way how mathematical knowledge is presented in textbooks and articles. The theorems can be proved using definitions, lemmas, and special inference rules (see Section 4), and may then be used for carrying out sample computations or proving other theorems (although no restrictions are imposed on the order in which the theorems are proved: an unproved statement may well serve as knowledge for proving another statement).

5.1 Structure of the Formalization

We developed a formal *Theorema* theory for the theory of reduction rings. For this, we first introduced the category (in the sense of Section 3) **Reduction-Ring**. Having this category it is already possible to define the notion of Gröbner bases and state the Main Theorem of reduction ring theory, containing a finite criterion for checking whether a given set is a Gröbner basis or not.

We also introduced the categories **CommPPDomain** of (commutative) power-products and **ReductionPolynomialDomain** of polynomials over a co-efficient domain $\mathcal{R}$ and a power-product domain, where objects like the Noetherian ordering[1] $\prec$ are defined in terms of the respective objects in $\mathcal{R}$ in such a way that if $\mathcal{R}$ is a reduction ring, then so is the polynomial ring. Moreover, the theorem containing this claim and the various lemmas needed for its proof are already part of the formalization as well.

Complementing the purely theoretical concepts of categories and theorems, the formalization consists of computational parts, too. In particular, it contains several *Theorema* functors for constructing concrete reduction rings, e. g., a functor that turns an arbitrary field into a reduction ring by endowing it with a suitable Noetherian order relation $\prec$ and other objects needed in reduction rings. Another example of a functor is the functor that turns a reduction ring $\mathcal{R}$ and a power-product domain $\mathcal{T}$ into the polynomial ring over $\mathcal{R}$ and $\mathcal{T}$, again endowed with suitable reduction-ring-objects (note that the resulting domain

---

[1] Actually, the whole formalization is based upon the reverse relation $\succ$, but this should not lead to any confusion.

then belongs to category **ReductionPolynomialDomain**). Most importantly, however, there is a functor $\mathcal{GB}$ that takes as input an *algorithmic* reduction ring $\mathcal{R}$ and returns a new ring, where in addition also a function gb for computing Gröbner bases is defined. gb, of course, implements the critical-pair/completion algorithm sketched in Section 2.1 in terms of the operations of $\mathcal{R}$.

Regarding the formal verification of the formalization, we decided to begin with proving that if $\mathcal{P}$ is in category **ReductionPolynomialDomain**, the coefficient domain is in category **ReductionRing**, and the power-product domain is in category **CommPPDomain**, then also $\mathcal{P}$ is in category **ReductionRing**. Although this undertaking has been started only recently and, hence, it is not finished yet, first results, namely completely formal, machine-generated proofs of important lemmas, have already been achieved. One of them is presented in the following subsection.

Of course, the ultimate goal is the formal verification of *all* of reduction ring theory, not only of what is mentioned above. This is still future work.

5.2   Example: Noetherianity of $\prec$ in Polynomial Rings

As an example, we present here the formal statement of the theorem that $\prec$ in domains belonging to **ReductionPolynomialDomain** is Noetherian, and also sketch the main ideas behind its *Theorema*-generated proof.

Let in the sequel always $\mathcal{R}$ be a reduction ring, $\mathcal{T}$ a commutative power-product domain, and $\mathcal{P}$ the polynomial ring over $\mathcal{R}$ and $\mathcal{T}$. Since we have to refer to $\prec$ in $\mathcal{R}$, in $\mathcal{T}$ and in $\mathcal{P}$, we denote the relations by $\underset{\mathcal{R}}{\prec}$, $\underset{\mathcal{T}}{\prec}$ and $\underset{\mathcal{P}}{\prec}$, respectively, according to syntactic conventions regarding domain operations in *Theorema*. Before we can give the definition of $\underset{\mathcal{P}}{\prec}$, we need the auxiliary notions of H, lp and lc. The following is their definition in *Theorema* notation:

Definition 2 (H, lp, lc).

$$\underset{\underset{\mathcal{P}}{\in[p]}}{\forall}$$

$$\underset{\underset{\mathcal{T}}{\in[\tau,\sigma]}}{\forall} \ \mathrm{C}[\mathrm{H}[p,\tau],\sigma] \ = \ \begin{cases} \mathrm{C}[p,\sigma] \Leftarrow & \tau \underset{\mathcal{T}}{\prec} \sigma \\ 0 & \Leftarrow \text{otherwise} \end{cases} \tag{H}$$

$$p \neq 0 \Rightarrow$$

$$\mathrm{lp}[p] \ := \ \underset{\underset{\mathcal{T}}{\in[\tau]}}{\text{the}} \ (\mathrm{C}[p,\tau] \neq 0 \wedge \underset{\underset{\mathcal{T}}{\in[\sigma]}}{\forall} \ \tau \underset{\mathcal{T}}{\prec} \sigma \Rightarrow \mathrm{C}[p,\sigma] = 0) \tag{lp}$$

$$\mathrm{lc}[p] \ := \ \mathrm{C}[p,\mathrm{lp}[p]] \tag{lc}$$

Some remarks on Definition 2 are in place:

– $\mathrm{C}[p,\tau]$ denotes the coefficient of polynomial p at power-product τ. It cannot be defined in general for arbitrary polynomial domains, but rather has to be defined in each concrete domain.

- $H[p, \tau]$ denotes the *higher part* of p w. r. t. τ, i. e. the sum of those monomials of p whose power-product is strictly greater than τ. It is not defined explicitly, but only implicitly in terms of C.
- $lp[p]$ denotes the *leading power-product* of p, given that p is non-zero. It is defined using *Theorema*'s "the" quantifier.
- $lc[p]$, finally, denotes the *leading coefficient* of p.
- The first and the third line in the definition are so-called *global declarations*. They are simply prepended to all subsequent formulas, in order to ease reading and writing formulas in *Theorema*.
- Strictly speaking, not only the various versions of $\prec$ carry under-scripts in the formalization, but also all of C, H, etc. are under-scripted with $\mathcal{P}$, to explicitly state that they belong to domain $\mathcal{P}$. Here, these under-scripts are omitted for better readability.

Now we are able to define $\underset{\mathcal{P}}{\prec}$, again in *Theorema* notation:

Definition 3 ($\underset{\mathcal{P}}{\prec}$).

$$\underset{\underset{\mathcal{P}}{\in[p,q]}}{\forall}$$

$$p \underset{\mathcal{P}}{\prec} q \;:\Leftrightarrow\; \underset{\underset{\mathcal{T}}{\in[\tau]}}{\exists} \; (H[p, \tau] = H[q, \tau] \wedge C[p, \tau] \underset{\mathcal{R}}{\prec} C[q, \tau]) \qquad (\underset{\mathcal{P}}{\prec})$$

In the theorem claiming that $\underset{\mathcal{P}}{\prec}$ is Noetherian, one does not even need that $\mathcal{R}$ is a reduction ring, but only that $\underset{\mathcal{R}}{\prec}$ is a partial Noetherian ordering and that $\underset{\mathcal{T}}{\prec}$ is a total Noetherian ordering (which is implied by $\mathcal{R}$ being a reduction ring and $\mathcal{T}$ being a power-product domain, though).

Instead of presenting the theorem, which should be apparent, we present the key lemma for proving it, together with its proof:

Lemma 1 (Lemma for proving Noetherianity of $\underset{\mathcal{P}}{\prec}$).

$$\underset{\underset{\mathcal{T}}{\in[\tau]}}{\forall} \quad \underset{\underset{\text{DomainSets}[\mathcal{P}]}{\in}}{\forall}[A]$$

$$\left(A \neq \{\} \wedge \underset{p \in A}{\forall} \; p \neq 0 \Rightarrow lp[p] \underset{\mathcal{T}}{\prec} \tau\right) \Rightarrow \underset{p \in A}{\exists} \; isMin[p, A, \underset{\mathcal{P}}{\prec}] \qquad (1)$$

DomainSets is a general *Theorema* functor returning the domain of all sets of elements belonging to the input domain $\mathcal{D}$, without any operations other than $\underset{\text{DomainSets}[\mathcal{D}]}{\in}$. $isMin[p, A, \underset{\mathcal{P}}{\prec}]$ states that no element in A is strictly less than p w. r. t. $\underset{\mathcal{P}}{\prec}$. Since its formal definition should be obvious, it is spared in this paper.

The *Theorema*-generated proof of Lemma 1 essentially proceeds by Noetherian induction on the power-product τ (recall that $\underset{\mathcal{T}}{\prec}$ is Noetherian), to be explained now step-by-step:

1. Perform Noetherian induction on τ, i. e., choose $\overline{\tau}$ and $\overline{A}$ arbitrary but fixed, assume

$$\overline{A} \neq \{\} \qquad (A\#1)$$

$$\underset{p \in \overline{A}}{\forall} \; p \neq 0 \Rightarrow \mathrm{lp}[p] \underset{\mathcal{T}}{\prec} \overline{\tau} \tag{A\#2}$$

$$\underset{\underset{\mathcal{T}}{\in}[\tau]}{\forall} \; \tau \underset{\mathcal{T}}{\prec} \overline{\tau} \Rightarrow \underset{\underset{\mathrm{DomainSets}[\mathcal{P}]}{\in}}{\forall} [A] \qquad \cdots \tag{IH}$$

and prove

$$\underset{p \in A}{\exists} \; \mathrm{isMin}[\,p, A, \underset{\mathcal{P}}{\prec}] \tag{G\#1}$$

Formula (IH) of course is the induction hypothesis. Note that the principle of Noetherian induction does not have to be implemented as a special inference rule, but can be stated as a higher-order formula on the object level and then used like an inference rule on the meta level, thanks to the way how *Theorema* employs certain kinds of formulas (e. g. universally quantified implications) as rewrite-rules in proofs.

2. Distinguish two cases, based upon whether $0 \in \overline{A}$ or not. If $0 \in \overline{A}$ then 0 apparently witnesses the existential goal (G#1), so assume now $0 \notin \overline{A}$. From (A#2) we can readily infer

$$\underset{p \in \overline{A}}{\forall} \; p \neq 0 \wedge \mathrm{lp}[p] \underset{\mathcal{T}}{\prec} \overline{\tau} \tag{A\#3}$$

3. Consider the set $P := \{\mathrm{lp}[p] \;\; | \;\; \}$. P apparently is the non-empty (due to $p \in \overline{A}$ (A#2)) set consisting of all leading power-products of elements in $\overline{A}$, and, therefore, it contains a minimal element σ due to the Noetherianity of $\underset{\mathcal{T}}{\prec}$. We know

$$\mathrm{isMin}[\,\sigma, P, \underset{\mathcal{T}}{\prec}] \tag{A\#4}$$

as well as

$$\sigma \underset{\mathcal{T}}{\prec} \overline{\tau} \tag{A\#5}$$

from (A#3).

4. Consider the set $C := \{\mathrm{lc}[p] \;\; | \;\; \mathrm{lp}[p] = \sigma\}$. As can easily be seen, C is a $p \in \overline{A}$ non-empty set consisting of elements of the coefficient domain $\mathcal{R}$, and since $\underset{\mathcal{R}}{\prec}$ is Noetherian, it contains a minimal element c. Thus, we know

$$\mathrm{isMin}[\,c, C, \underset{\mathcal{R}}{\prec}] \tag{A\#6}$$

5. Consider the set $R := \{p - c \cdot \sigma \;\; | \;\; \mathrm{lp}[p] = \sigma \wedge \mathrm{lc}[p] = c\}$. R is again $p \in \overline{A}$ non-empty, consists of polynomials, and moreover satisfies

$$\underset{p \in R}{\forall} \; p \neq 0 \Rightarrow \mathrm{lp}[p] \underset{\mathcal{T}}{\prec} \sigma \tag{A\#7}$$

because all the leading monomials cancel by construction. This means that we can now use our induction hypothesis (IH), instantiated by $\tau \leftarrow \sigma$ and $A \leftarrow R$, and infer

$$\mathrm{isMin}[\overline{p}, R, \underset{\mathcal{P}}{\preceq}] \qquad (\mathrm{A}\#8)$$

from (A#5) and (A#7), for some $\overline{p} \in R$.

6. Instantiate the existentially quantified goal (G#1) by $p \leftarrow \overline{p} + c \cdot \sigma$, and prove

$$\mathrm{isMin}[\overline{p} + c \cdot \sigma, \overline{A}, \underset{\mathcal{P}}{\preceq}] \qquad (\mathrm{G}\#2)$$

which can be accomplished using Definition 3 together with (A#4), (A#6) and (A#8) (we spare the details).

The proof was generated interactively with the assistance of *Theorema*, following the six steps sketched above. Furthermore, ordering-rules, as described in Section 4.1, were used to abbreviate otherwise tedious inferences regarding, e. g., transitivity of $\underset{\mathcal{T}}{\preceq}$.

Please note that "interactive proving" in *Theorema* does not mean to write down the proof in sufficient detail and afterward let the system check all inferences and fill small gaps, but rather to initiate a proof attempt where the system simply asks the human user for support whenever it does not know how to proceed, in a dialog-oriented manner. Nevertheless, as soon as the proof is finished (either with success or with failure), a nicely-structured, human readable proof document describing each single step is generated fully automatically, where formal contents are interspersed with informal explanatory text in English (or any other language). In other words, the proof outlined above is rather the *output* than the *input* of interactive proving in *Theorema*.

6   Conclusion and Future Work

The preceding sections illustrated how the formal treatment of reduction ring theory in *Theorema* can be conducted. On the object level, this consists of providing suitable definitions for all notions and concepts of the theory. On the inference level, it is desirable to have specialized inference rules at one's disposal for efficiently proving statements about all kinds of order relations and equality of terms in commutative rings with unit, such that the resulting proofs become short and elegant.

One of the long-term goals of the project is to extend the theory of reduction rings, but also of Gröbner bases in general, in several ways, making use of the existing formalization:

- Introduce non-commutative reduction rings and prove an analogue of Theorem 1.
- Find further basic reduction rings and functors that conserve the property of being a reduction ring.

– Investigate the relation of Gröbner bases (in polynomial rings over fields) and generalized Sylvester matrices, such that Gröbner bases can effectively be computed by triangularizing coefficient-matrices [16].

These possible extensions are among the main motivations for a formal treatment of reduction ring theory. Proofs in this theory tend to be lengthy, tedious (with many case distinctions etc.) and very technical, but still comparatively straightforward. Therefore, having a software system supporting the human mathematician in all aspects of theory exploration, being in *automatically* proving simple lemmas, keeping track of cases still to be considered in long proofs, or even carrying out test-computations with notions just introduced, certainly is a great benefit.

Acknowledgments. I gratefully acknowledge the valuable discussions about formal mathematics and Gröbner bases with my PhD adviser Bruno Buchberger, and about *Theorema* with Wolfgang Windsteiger.

This research was funded by the Austrian Science Fund (FWF): grant no. W1214-N15, project DK1.

References

1. Buchberger, B.: A critical-pair/completion algorithm for finitely generated ideals in rings. In: Börger, E., Rödding, D., Hasenjaeger, G. (eds.) Rekursive Kombinatorik 1983. LNCS, vol. 171, pp. 137–161. Springer, Heidelberg (1984)
2. Buchberger, B.: Gröbner Rings in Theorema: A Case Study in Functors and Categories. Tech. Rep. 2003–49, RISC Report Series, Johannes Kepler University Linz, Austria (2003)
3. Buchberger, B.: Towards the automated synthesis of a Gröbner bases Algorithm. RACSAM (Revista de la Real Academia de Ciencias), Serie A: Mathematicas **98**(1), 65–75 (2004)
4. Buchberger, B., Craciun, A., Jebelean, T., Kovacs, L., Kutsia, T., Nakagawa, K., Piroi, F., Popov, N., Robu, J., Rosenkranz, M., Windsteiger, W.: Theorema: Towards computer-aided mathematical theory exploration. J. Applied Logic **4**(4), 470–504 (2006)
5. Craciun, A.: Lazy Thinking Algorithm Synthesis in Gröbner Bases Theory. PhD thesis, RISC, Johannes Kepler University Linz, Austria (2008)
6. Jorge, J.S., Guilas, V.M., Freire, J.L.: Certifying properties of an efficient functional program for computing Gröbner bases. J. Symbolic Computation **44**(5), 571–582 (2009)
7. Medina-Bulo, I., Palomo-Lozano, F., Ruiz-Reina, J.-L.: A verified Common Lisp implementation of Buchberger's algorithm in ACL2. J. Symbolic Computation **45**(1), 96–123 (2010)
8. Robu, J., Ida, T., Ţepeneu, D., Takahashi, H., Buchberger, B.: Computational origami construction of a regular heptagon with automated proof of its correctness. In: Hong, H., Wang, D. (eds.) ADG 2004. LNCS (LNAI), vol. 3763, pp. 19–33. Springer, Heidelberg (2006)

9. Robu, J.: Automated proof of geometry theorems involving order relation in the frame of the theorema project. In: Pop, H.F. (ed.) KEPT 2007 (Knowledge Engineering, Principles and Techniques). Special issue of Studia Universitatis "Babes-Bolyai", Series Informatica, pp. 307–315 (2007)

10. Rosenkranz, M.: A new symbolic method for solving linear two-point boundary value problems on the level of operators. J. Symbolic Computation **39**(2), 171–199 (2005)

11. Schwarzweller, C.: Gröbner bases – theory refinement in the mizar system. In: Kohlhase, M. (ed.) MKM 2005. LNCS (LNAI), vol. 3863, pp. 299–314. Springer, Heidelberg (2006)

12. Stifter, S.: Computation of Gröbner Bases over the Integers and in General Reduction Rings. Diploma thesis, Institut für Mathematik, Johannes Kepler University Linz, Austria (1985)

13. Stifter, S.: The reduction ring property is hereditary. J. Algebra **140**(89–18), 399–414 (1991)

14. Tec, L.: A Symbolic Framework for General Polynomial Domains in Theorema: Applications to Boundary Problems. PhD thesis, RISC, Johannes Kepler University Linz, Austria (2011)

15. Thery, L.: A machine-checked implementation of Buchberger's Aalgorithm. J. Automated Reasoning **26**, 107–137 (2001)

16. Wiesinger-Widi, M.: Gröbner Bases and Generalized Sylvester Matrices. PhD thesis, RISC, Johannes Kepler University Linz, Austria (2015 to appear)

17. Windsteiger, W.: Building up hierarchical mathematical domains using functors in theorema. In: Armando, A., Jebelean, T. (eds.) Calculemus 1999. ENTCS, vol. 23, pp. 401–419. Elsevier (1999)

18. Windsteiger, W.: Theorema 2.0: a system for mathematical theory exploration. In: Hong, H., Yap, C. (eds.) ICMS 2014. LNCS, vol. 8592, pp. 49–52. Springer, Heidelberg (2014)

19. The theorema System. http://www.risc.jku.at/research/theorema/software/

20. Wolfram Mathematica. http://www.wolfram.com/mathematica/

On the Partial Analytical Solution
of the Kirchhoff Equation

Dominik L. Michels[1], Dmitry A. Lyakhov[2], Vladimir P. Gerdt[3],
Gerrit A. Sobottka[4], and Andreas G. Weber[4]

[1] Computer Science Department, Stanford University, 353 Serra Mall, MC 9515,
Stanford, CA 94305, USA
michels@cs.stanford.edu
[2] Radiation Gaseous Dynamics Lab, A. V. Luikov Heat and Mass Transfer Institute
of the National Academy of Sciences of Belarus, P. Brovka St 15, 220072 Minsk,
Belarus
lyakhovda@hmti.ac.by
[3] Group of Algebraic and Quantum Computations, Joint Institute for Nuclear
Research, Joliot-Curie 6, 141980 Dubna, Moscow Region, Russia
gerdt@jinr.ru
[4] Multimedia, Simulation and Virtual Reality Group, Institute of Computer Science
II, University of Bonn, Friedrich-Ebert-Allee 144, 53113 Bonn, Germany
sobottka@cs.uni-bonn.de, weber@cs.uni-bonn.de

Abstract. We derive a combined analytical and numerical scheme to
solve the (1+1)-dimensional differential Kirchhoff system. Here the object
is to obtain an accurate as well as an efficient solution process. Purely
numerical algorithms typically have the disadvantage that the quality of
the solutions decreases enormously with increasing temporal step sizes,
which results from the numerical stiffness of the underlying partial differential equations. To prevent that, we apply a differential Thomas decomposition and a Lie symmetry analysis to derive explicit analytical
solutions to specific parts of the Kirchhoff system. These solutions are
general and depend on arbitrary functions, which we set up according to
the numerical solution of the remaining parts. In contrast to a purely numerical handling, this reduces the numerical solution space and prevents
the system from becoming unstable. The differential Kirchhoff equation
describes the dynamic equilibrium of one-dimensional continua, i.e. slender structures like fibers. We evaluate the advantage of our method by
simulating a cilia carpet.

Keywords: Differential Thomas Decomposition, Kirchhoff Rods, Lie
Symmetry Analysis, Partial Analytical Solutions, Partial Differential
Equations, Semi-analytical Integration.

1   Introduction

In general, to study algebraic properties of a polynomially nonlinear system of
partial differential equations (PDEs), the system has to be completed to involution [14]. In particular, by such a completion one can check the consistency of

© Springer International Publishing Switzerland 2015
V.P. Gerdt et al. (Eds.): CASC 2015, LNCS 9301, pp. 322–333, 2015.
DOI: 10.1007/978-3-319-24021-3_24

the system, detect arbitrariness in its analytical solution, eliminate a subset of dependent variables, and verify whether some another PDE is a differential consequence of the system, i.e. verify whether this PDE vanishes on all solutions of the system. However, algorithmically, a polynomially nonlinear PDE system may not admit completion. Instead, one can decompose such a system into finitely many involutive subsystems with disjoint sets of solutions. This is done by the Thomas decomposition [2].[1]

In the present paper, we apply first the differential Thomas decomposition to the equation system which comprises four linear PDEs in two independent and six dependent variables and also two quadratically nonlinear algebraic relations in the dependent variables. This partial differential-algebraic system is a parameter-free subsystem of the $(1+1)$-dimensional Kirchhoff equation. In doing so, we detect the arbitrariness in the general (analytical) solution to that subsystem. Then, by applying the classical theorem about conservative vector fields [5] and performing integration by parts we reduce the parameter-free subsystem to a system of three nonlinear PDEs in two independent variables that contains one arbitrary function.

As the next step, we perform the computer algebra based Lie symmetry analysis to the reduced system and discover that this system is equivalent (modulo Lie group transformations) to another system of three nonlinear PDEs in two independent variables (x, y) and two dependent variables f and g. The last system comprises two equations of zero Gaussian curvature for the surfaces $z = f(x, y)$ and $z = g(x, y)$ and one more equation that relates f and g. To construct the solution of the system we use the classical parametrization of a zero curvature surface [8]. As a result, we obtain a closed-form analytical solution to the parameter-free subsystem under consideration and prove that the solution is general. This opens up a wide range for applications of symbolic-numeric methods to the $(1+1)$-dimensional Kirchhoff equation. All symbolic computations related to the construction of the general analytical solution were done with Maple.

On that basis, we derive a combined analytical and numerical scheme to solve the $(1+1)$-dimensional Kirchhoff equation. This system of partial differential and algebraic equations describes the dynamic equilibrium of one-dimensional continua like slender structures, artificial fibers, human hair, cilia and flagella. Since their dynamical behavior act on different time-scales, numerical stiffness enters the problem and leads to huge instabilities in the case of basic purely numerical solution schemes. To prevent that, we exploit the explicit analytical expressions found for the solution to the parameter-free subsystem. In so doing, we set up the arbitrary functions in the exact solution according to the numerical solution of the Kirchhoff equation. In contrast to a purely numerical handling, this reduces the numerical solution space and prevents the Kirchhoff system from becoming unstable. The advantage is demonstrated by simulating a cilia carpet.

---

[1] For a more-general overview of the formal algorithmic elimination for PDEs, we additionally refer to [13].

2   Kirchhoff's Rod Theory

For a detailed derivation of the special Cosserat theory of rods, we refer to our last year's contribution [12]. According to the formulation presented therein, the rods in classical Cosserat theory can undergo shear and extension. We can avoid this by setting the linear strains to $\boldsymbol{\nu} := (\nu_1, \nu_2, \nu_3) = (0, 0, 1)$ in local coordinates. Geometrically this means, that the angle between the director $\boldsymbol{d}_3$ and the tangent to the centerline, $\partial_s \boldsymbol{r}$, always remains zero (no shear) and that the tangent to the centerline always has unit length (no elongation).[2] We further reduce the system to two dimensions by setting $\kappa_3 = \omega_3 = \nu_3 = 0$. Finally we separate the resulting equations into two PDE subsystems

$$\rho A \partial_t \boldsymbol{v} = \partial_s \boldsymbol{n} + \boldsymbol{f}, \tag{1}$$
$$\rho I \partial_t \boldsymbol{\omega} = \partial_s \boldsymbol{m} + \mathrm{adiag}(1, -1)\, \boldsymbol{n} + \boldsymbol{l} \tag{2}$$

and

$$\partial_t \boldsymbol{\kappa} = \partial_s \boldsymbol{\omega}, \tag{3}$$
$$\partial_s \boldsymbol{v} = \mathrm{adiag}(1, -1)\, \boldsymbol{\omega}, \tag{4}$$

and additionally a set of constraints

$$\boldsymbol{\omega}\, \mathrm{adiag}(1, -1)\, \boldsymbol{\kappa}^\mathsf{T} = 0, \tag{5}$$
$$\boldsymbol{v}\, \mathrm{adiag}(1, -1)\, \boldsymbol{\kappa}^\mathsf{T} = 0 \tag{6}$$

with an antidiagonal matrix

$$\mathrm{adiag}(1, -1) = \begin{pmatrix} 0 & 1 \\ -1 & 0 \end{pmatrix}.$$

Please note that equations (3)–(4) are parameter free, whereas equations (1)–(2) include parameters ρ, A, I.

3   Derivation of Explicit Analytical Solution

In this section, we derive an exact analytical solution to (3)–(4) under constraints (5)–(6).

3.1   Arbitrariness in General Solution

To determine the functional arbitrariness in the general analytical solution to equations (3)–(6) with non-vanishing values of the dependent variables, we apply

---

[2] For a more-general overview of the special Cosserat and Kirchhoff rod theory, we refer to [1].

to them the differential Thomas decomposition [2,13] by using its implementation in Maple[3] under the choice of a degree-reverse-lexicographical ranking with

$$s \succ t; \qquad \omega_2 \succ \omega_1 \succ \kappa_2 \succ \kappa_1 \succ v_2 \succ v_1.$$

Then, under the condition $\omega_1 \omega_2 \kappa_1 \kappa_2 v_1 v_2 \neq 0$, the decomposition algorithm outputs two involutive differential systems. One of them is given by the set of equations

$$v_1 \underline{\omega_2} - v_2 \omega_1 = 0,$$
$$\underline{\partial_s \omega_1} - \partial_t \kappa_1 = 0,$$
$$v_1 \underline{\kappa_2} - v_2 \kappa_1 = 0,$$
$$v_1^2 \omega_1^2 \underline{\partial_s \kappa_1} - \kappa_1^2 v_1 \omega_1 \partial_t v_1 - 2 \kappa_1 v_1^2 \omega_1 \partial_t \kappa_1 + \kappa_1 v_2 \omega_1^3 = 0,$$
$$\underline{\partial_s v_2} + \omega_1 = 0,$$
$$\kappa_1 v_1^2 \underline{\partial_t v_2} - \kappa_1 v_2 v_1 \partial_t v_1 + (v_2^2 + v_1^2) \omega_1^2 = 0,$$
$$v_1 \underline{\partial_s v_1} - v_2 \omega_1 = 0$$

and inequations

$$v_1 \neq 0, \ v_2 \neq 0, \ \kappa_1 \neq 0, \ \omega_2 \neq 0$$

with underlined *leaders*, i.e., the highest ranking derivatives occurring in the equations. The second involutive system in the output contains the equation $v_1^2 + v_2^2 = 0$ which has no real solution under our assumption of non-vanishing dependent variables.

The differential dimensional polynomial [11] computed for the left-hand sides of the above equation system

$$3 \binom{l+1}{l} + 1$$

shows that the general analytical solution depends on three arbitrary functions of one variable.

Remark 1. This functional arbitrariness can also be understood via well-posedness of the Cauchy (initial value) problem [7,14]. For the above involutive system, the Cauchy problem has a unique analytical solution in a vicinity of a given initial point (s_0, t_0) if $v_2(s_0, t_0)$ is an arbitrary constant and $\kappa_1(s_0, t), \omega_1(s_0, t), v_1(s_0, t)$ are arbitrary functions of t, analytical at the point $t = t_0$.

Remark 2. In the procedure described below, to construct a solution to (3)–(6) we have to solve a number of intermediate partial differential equations. In doing so, it suffices to take into account the functional arbitrariness in the solutions to those equations and to neglect the arbitrariness caused by arbitrary constants.

---

[3] The Maple package DIFFERENTIALTHOMAS is freely available and can be downloaded via http://wwwb.math.rwth-aachen.de/thomasdecomposition/.

3.2 Reduction to Two Dependent Variables

If D is an open simply connected subset of the real space $\mathbb{R}^2$ of the independent variables (s,t), then the partial differential equations (3) define conservative vector fields (κ_1, ω_1) and (κ_2, ω_2) on D.

By the classical theorem of calculus [5][4] there exist functions $p_1(s,t)$ and $p_2(s,t)$ such that

$$\kappa_1 = \partial_s p_1, \quad \omega_1 = \partial_t p_1, \quad \kappa_2 = \partial_s p_2, \quad \omega_2 = \partial_t p_2. \tag{7}$$

If one takes into account these relations, then the application of the same theorem to equation (4) transforms them into the equivalent form

$$v_1 = \partial_t f, \quad v_2 = \partial_t g, \quad p_1 = -\partial_s g, \quad p_2 = \partial_s f, \tag{8}$$

where $f = f(s,t)$ and $g = g(s,t)$ are some functions.

Therefore, the parameter-free system (3)–(6) is equivalent to (i.e., it has the same solution space as) the following system of two nonlinear PDEs:

$$f_{st} g_{ss} - g_{st} f_{ss} = 0, \tag{9}$$

$$g_{ss} g_t + f_{ss} f_t = 0. \tag{10}$$

Here and below we use the following notation: $f_s \equiv \partial_s f$, $f_t \equiv \partial_t f$, $f_{st} \equiv \partial_s \partial_t f$, et cetera. We look for a solution of (3)–(6) such that

$$\omega_1 \omega_2 \kappa_1 \kappa_2 v_1 v_2 \neq 0. \tag{11}$$

Equations (9)–(10) form a linear system in f_{ss} and g_{ss}. It has non-trivial solutions only if

$$g_{st} g_t + f_{st} f_t = 0 \tag{12}$$

what is equivalent to the equality $\omega_1 v_2 - \omega_2 v_1 = 0$ that follows from (5)–(6).

Obviously, equation (12) admits an integration by parts which yields

$$(g_t)^2 + (f_t)^2 - h(t) = 0,$$

where $h(t)$ is an arbitrary function. Therefore, system (3)–(6) is equivalent to the system

$$g_{ss} g_t + f_{ss} f_t = 0, \tag{13}$$

$$(g_t)^2 + (f_t)^2 - h = 0, \quad h_s = 0. \tag{14}$$

Now we apply the Thomas decomposition algorithm in its implementation in Maple [2] for the differential elimination of the function f from the system (13)–(14) and independently the differential elimination of the function g from the same system. In doing so, inequation (11) expressed in terms of f and g is taken into account. As a result of each these two eliminations, equation (13) is replaced with another equation whereas equations (14) are preserved by the elimination.

Consider now the PDE system which comprises two equations in (14) and two more equations[5] obtained by the differential elimination of the functions f and

---

[4] The proof is also given in `http://www.owlnet.rice.edu/~fjones/chap12.pdf`.

[5] Note that these equations are symmetric under the interchange of f with g.

g as explained above:

$$g_{ss}g_t h_t + 2\,h \cdot (g_{st})^2 - 2\,h\,g_{ss}g_{tt} = 0 \,, \tag{15}$$

$$f_{ss}f_t h_t + 2\,h \cdot (f_{st})^2 - 2\,h\,f_{ss}f_{tt} = 0 \,, \tag{16}$$

$$(g_t)^2 + (f_t)^2 - h = 0 \,, \quad h_s = 0 \,. \tag{17}$$

Remark 3. If one fixes the ranking and applies the differential Thomas decomposition to the equation systems (13)–(14) and (15)–(17) extended with inequation (11), then in both cases the same set of equations and inequations is output. This shows explicitly the equivalence of both PDE systems as having the same solution space.

3.3   Lie Symmetry Analysis

Since the PDE system (15)–(17) is symmetric with respect to f and g, it suffices to apply the methods of the Lie symmetry analysis [9] to (16)–(17). As in our paper [12], to perform related symbolic computations we use the Maple package DESOLVE [4] and its routine *gendef* to generate the determining linear PDE system for the infinitesimal point transformations. Then we apply the Maple package JANET[6] [3] to complete the determining system to involution[7].

As a result, the involutive determining system containing twenty-three one-term and two-term linear PDEs is easily solvable by the Maple routine *pdsolve*. This gives for the coefficients of the infinitesimal Lie symmetry generator

$$\mathcal{X} = \xi_1(s,t,f,h)\,\partial_s + \xi_2(s,t,f,h)\,\partial_t + \eta_1(s,t,f,h)\,\partial_f + \eta_2(s,t,f,h)\,\partial_h \tag{18}$$

the following expressions:

$$\eta_1 = c_1 s + c_2 f + c_3\,, \quad \eta_2 = -\frac{d\phi}{dt} + c_4\,, \quad \xi_1 = c_5 s + c_6 f + c_7\,, \quad \xi_2 = \phi\,,$$

where c_i $(1 \le i \le 7)$ are arbitrary constants, and $\phi = \phi(t)$ is an arbitrary function.

As we are looking for the functional arbitrariness only (see Remark 2), we put $c_i = 0$. The Lie symmetry group defined by the generator (18) is obtained by solving the Lie equations

$$\frac{ds}{d\alpha} = \xi_1\,, \quad \frac{dt}{d\alpha} = \xi_2\,, \quad \frac{df}{d\alpha} = \eta_1\,, \quad \frac{dh}{d\alpha} = \eta_2\,.$$

For the independent variables this gives the one-parameter transformation group

$$\tilde{s} = s\,, \quad \tilde{t} = \Phi(t,\alpha)\,.$$

Here α is a group parameter and Φ is an arbitrary function such that $\phi(t) = \partial_\alpha\Phi|_{\alpha=0}$. Now we can formulate and prove one of the main theoretical results of the present paper.

---

[6] The Maple package JANET is freely available and can be downloaded via http://wwwb.math.rwth-aachen.de/Janet/.

[7] The same output can also be generated using the package DIFFERENTIALTHOMAS.

Lemma 1. *Let $h(t) > 0$ be a given function. Then for any solution to equation (16) there exists an invertible transformation of variables $(s,t) \to (x,y)$ of the form*

$$x = s, \quad y = \Psi(t) \tag{19}$$

such that the solution is an image of a solution to the equation

$$f_{xx} f_{yy} - (f_{xy})^2 = 0. \tag{20}$$

Proof. First, we note that the transformation (19) is invertible if and only if $\Psi_t \neq 0$. Second, we apply this transformation to (20). This gives

$$\frac{f_{ss} f_{tt} - (f_{st})^2}{f_{ss} f_t} = \frac{\Psi_{tt}}{\Psi_t}. \tag{21}$$

On the other hand, equation (16) can be rewritten as

$$\frac{f_{ss} f_{tt} - (f_{st})^2}{f_{ss} f_t} = \frac{h_t}{2h}. \tag{22}$$

We see that at $(\Psi_t)^2 = h$ the right-hand sides of (21) and (22) coincide.   □

This lemma immediately implies the following proposition.

Proposition 1. *The PDE system (15)–(17) can be obtained from the system*

$$f_{xx} f_{yy} - (f_{xy})^2 = 0, \tag{23}$$
$$g_{xx} g_{yy} - (g_{xy})^2 = 0, \tag{24}$$
$$(f_y)^2 + (g_y)^2 - 1 = 0 \tag{25}$$

by the transformation (19) of the independent variables with $h = (\Psi_t)^2$.

Thus, we reduced the parameter-free subsystem (3)–(6) of the Kirchhoff equation to the equivalent system (23)–(25). In doing so, the independent variables of the latter system are related to the independent variables of the former one by the equalities (19). The advantages of the reduced system are:

1. there are only two dependent variables;
2. the system is symmetric with respect to the dependent variables;
3. they are separated in (23) and (24);
4. the differential equation of the form (23) is very well known in the theory of analytical surfaces and deeply studied in the literature [10]. It means that the surface $z = f(x, y)$ has zero Gaussian curvature. Such surfaces are locally isometric to a plane and admit a parametrization constructed in the classical paper [8]. In the following subsection, we use this parametrization to derive the general analytical solution to (3)–(6).

3.4 Parametrized Analytical Solution

In [8], a parametric form of a general analytical solution to an isolated zero curvature equation was found. For equation (23), this parametrization reads

$$f = a_1(u)v + b_1(u), \quad x = a_2(u)v + b_2(u), \quad y = a_3(u)v + b_3(u). \tag{26}$$

Here a_i and b_i are analytical functions of the parameter u.

Remark 4. In general, the function g may have the parameterizations of x and y such that $a_i(u), b_i(u)$ $(i = 2, 3)$ are different from that in (26). We assume, however, that both functions f and g have the same parametrizations of x and y so that $a_2 \neq 0$.

Our assumption allows to express the parameter v linearly in terms of x and leads to a rather compact parametrization of the functions f and g of the form

$$f = A_1(u)\,x + B_1(u), \quad y = A(u)\,x + B(u), \quad g = M(u)\,x + N(u)$$

with six functions. However, these functions are not arbitrary. By taking the first and second order derivatives of both sides in the equality

$$f\left(x, A(u)\,x + B(u)\right) = A_1(u)x + B_1(u)$$

with respect to x and u we obtain a linear equation system in $f_x, f_y, f_{xx}, f_{yy}, f_{xy}$. Its solution allows to express the partial derivatives f_{xx}, f_{yy}, f_{xy} in terms of the functions $A(u), A_1(u), B(u), B_1(u)$ and their first derivatives. Then, the substitution of these expressions into equation (23) shows that this equation is satisfied if

$$(A_1)_u = P(u)\,A_u, \quad (B_1)_u = P(u)\,B_u, \tag{27}$$

where $P(u)$ can be any function. Similarly, equation (24) is rewritten as two differential equalities

$$M_u = Q(u)\,A_u, \quad N_u = Q(u)\,B_u \tag{28}$$

with unspecified $Q(u)$. Next, equation (25) implies the algebraic equality $Q(u)^2 + P(u)^2 = 1$ or equivalently

$$Q(u) = \sin(C(u)), \quad P(u) = \cos(C(u)) \tag{29}$$

with arbitrary $C(u)$.

As the next step, we apply the following transformation of the independent variables, which are inverse to (19):

$$s = x, \quad t = F(y)$$

with $F = \Psi^{-1}$. For this purpose, we use the relations

$$f\left(s, F\left(A(u)\,s + B(u)\right)\right) = A_1(u)s + B_1(u),$$
$$g\left(s, F\left(A(u)\,s + B(u)\right)\right) = M(u)s + N(u)$$

together with (27)–(29) to compute all partial derivatives of the functions f and g with respect to s and t that occur in (7) and (8).

This gives the analytical solution

$$
\kappa = -\frac{A^2(u)C'(u)}{A'(u)s + B'(u)} \begin{pmatrix} \cos(C(u)) \\ \sin(C(u)) \end{pmatrix},
$$

$$
\omega = \frac{A(u)C'(u)}{F'(A(u)s + B(u))(A'(u)s + B'(u)))} \begin{pmatrix} \cos(C(u)) \\ \sin(C(u)) \end{pmatrix}, \tag{30}
$$

$$
v = \frac{1}{F'(A(u)s + B(u))} \begin{pmatrix} \cos(C(u)) \\ \sin(C(u)) \end{pmatrix}, \quad t = F\left(A(u)\,s + B(u)\right).
$$

Here $A(u), B(u), C(u)$ are arbitrary analytical functions of one variable, F is an arbitrary analytical function of the argument $A(u)\,s + B(u)$, and the prime mark denotes differentiation of a function with respect of its argument.

Substituting (30) in equations (3)–(6) shows that these equations are satisfied for any functions A, B, C, F. In Section 3.1, we saw that the general analytical solution depends on three arbitrary functions, each of one variable. The following theorem shows that the functional arbitrariness in (30) is superfluous.

Theorem 1. *Set $B(u) = u$ in (30). Then the analytical solution*

$$
\kappa = -\frac{A^2(u)C'(u)}{A'(u)s + 1} \begin{pmatrix} \cos(C(u)) \\ \sin(C(u)) \end{pmatrix},
$$

$$
\omega = \frac{A(u)C'(u)}{F'(A(u)s + u)(A'(u)s + 1)))} \begin{pmatrix} \cos(C(u)) \\ \sin(C(u)) \end{pmatrix}, \tag{31}
$$

$$
v = \frac{1}{F'(A(u)s + u)} \begin{pmatrix} \cos(C(u)) \\ \sin(C(u)) \end{pmatrix}, \quad t = F\left(A(u)\,s + u\right).
$$

to (3)–(6) is general.

Proof. Assume without loss of generality that the initial point is $(s_0, t_0) = (0, 0)$ and pose the Cauchy problem for (3)–(6) in a vicinity of this point. Then the well-posed Cauchy problem reads (see Remark 1):

$$
v_1(0, t) = f_1(t), \quad \omega_1(0, t) = f_2(t), \quad \kappa_1(0, t) = f_3(t), \quad v_2(0, 0) = C_1, \tag{32}
$$

where f_1, f_2, f_3 are analytic functions and C_1 is an arbitrary constant. Substitution of (31) into (32) implies the equations

$$
\frac{\cos(C(u))}{F'(u)} = f_1(F(u)),
$$

$$
\frac{C'(u)A(u)\cos(C(u))}{F'(u)} = f_2(F(u)),
$$

$$
-C'(u)A^2(u)\cos(C(u)) = f_3(F(u)), \tag{33}
$$

$$
\frac{\sin(C(0))}{F'(0)} = C_1,
$$

$$
F(0) = 0.
$$

Now we aim to show that for any predetermined functions satisfying

$$f_1(t) \neq 0, \quad f_2(t) \neq 0, \quad f_3(t) \neq 0, \quad v_2(0,0) = C_1 \neq 0$$

it is always possible to match up functions $C(u), A(u), F(u)$ such that in a vicinity of $(0,0)$, solution (31) satisfies the initial data (32). The squared second equation in (33) divided by the third equation gives the equality

$$-\frac{C'(u)\cos(C(u))}{(F'(u))^2} = \frac{f_2^2(F(u))}{f_3(F(u))}$$

which together with the first and third equations in (33) implies

$$
\begin{aligned}
C'(u) &= -\frac{f_2^2(F(u))\cos(C(u))}{f_3(F(u))f_1^2(F(u))}, \quad C(0) = \arctan\left(\frac{C_1}{f_1(0)}\right), \\
F'(u) &= \frac{\cos(C(u))}{f_1(F(u))}, \quad F(0) = 0, \\
A(u) &= -\frac{f_3(F(u))f_1(F(u))}{f_2(F(u))\cos(C(u))}.
\end{aligned}
\tag{34}
$$

If $f_1(0) \neq 0$, $f_2(0) \neq 0$, $f_3(0) \neq 0$, then in some vicinity of zero, problem (34) has the unique solution. Indeed, the first two equations in (34) define a Cauchy problem with regular right-hand sides, and the last equation sets up the function $A(u)$ explicitly. In doing so, $C(0) \in (\pi/2, -\pi/2)$ and, hence, $\cos(C(0)) \neq 0$. In a vicinity of $u = 0$ all factors occurring in $C'(u), F'(u), A(u)$ do not vanish, and the unique solution of this system satisfies (33) identically. The last statement is easily verified by substituting (34) in (33). □

4 Numerical Experiments

The resulting analytical solutions (31) for system (3)–(4) under constraints (5)–(6) contain four parameterization functions, which can be determined by the numerical integration of equations (1)–(2). The substitution of equations (31) into equations (1)–(2), the replacement of the spatial derivatives with central differences, and the replacement of the temporal derivatives according to the numerical scheme of a forward Euler integrator, leads to an explicit expression.[8] Iterating over this recurrence equation allows for the simulation of the dynamics of a Kirchhoff rod.

We present this in the context of the simulation of the beating pattern of a cilium. Such kind of scenarios are of interest in the context of simulations in biology and biophysics, since this kind of patterns occur widely in nature. For example, cilia carpets in the interior of the lung are responsible for the mucus transport, see [6]. The results are shown in Figure 1. Compared to a purely numerical handling using a discretization of equations (1)–(4) in a similar way,

---

[8] We do not explicitly write out the resulting equations here for brevity.

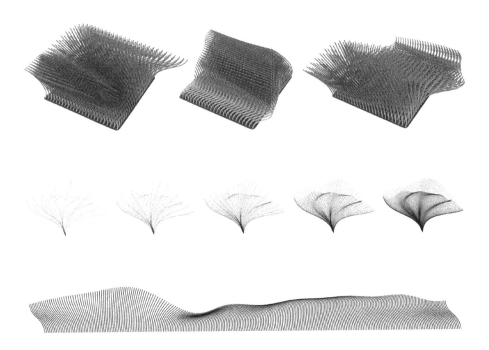

Fig. 1. Simulation of a cilia carpet (top) composed of multiple cilia beating in a metachronal rhythm (middle). This produces the appearance of a wave (below).

the step size can be increased by three orders of magnitude, which leads to an acceleration of two orders of magnitude. This is of special interest for complex systems like cilia carpets, in which multiple cilia are beating in parallel with phase differences in order to produce the appearance of a wave. This is illustrated in Figure 1.

5   Conclusion

In this contribution, we have studied the two-dimensional Kirchhoff system. To prevent from numerical problems which result from the numerical stiffness of the underlying differential equations, we developed a combined, partially symbolic and partially numeric, integration scheme. Within the symbolic part we constructed an explicit analytical solution to the parameter-free part of the Kirchhoff system and proved that this solution is general. The application of the analytical solution prevents from numerical instabilities and allows for highly accurate and efficient simulations. This was demonstrated for the scenario of a cilia carpet, which could be efficiently simulated with an acceleration of two orders of magnitude compared to a purely numerical handling.

Acknowledgements. This work was partially supported by the Max Planck Center for Visual Computing and Communication (D.L.M.) as well as by the grant No. 13-01-00668 from the Russian Foundation for Basic Research (V.P.G.). The authors thank Robert Bryant for useful comments on the solution of the PDE system (23)–(25) and the anonymous referees for their remarks and suggestions.

References

1. Antman, S.S.: Nonlinear Problems of Elasticity. Applied Mathematical Sciences, vol. 107. Springer-Verlag, Berlin (1995)
2. Bächler, T., Gerdt, V.P., Lange-Hegermann, M., Robertz, D.: Algorithmic Thomas decomposition of algebraic and differential systems. J. Symb. Comput. **47**, 1233–1266 (2012)
3. Blinkov, Y.A., Cid, C.F., Gerdt, V.P., Plesken, W., Robertz, D.: The maple package Janet: II. linear partial differential equations. In: Ganzha, V.G., Mayr, E.W., Vorozhtsov, E.V. (eds.) CASC 2003, pp. 41–54. Institut für Informatik, Technische Universität München, Garching (2003)
4. Carminati, J., Vu, K.: Symbolic computation and differential equations: Lie symmetries. J. Symb. Comput. **29**, 95–116 (2000)
5. Fikhtengol'ts, G.M.: A Course of Differential and Integral Calculus [in Russian], vol. 3. Nauka, Moscow (1966)
6. Fulford, G.R., Blake, J.R.: Muco-ciliary transport in the lung. J. Theor. Biol. **121**(4), 381–402 (1986)
7. Gerdt, V.P.: Algebraically simple involutive differential systems and the Cauchy problem. J. Math. Sci. **168**(3), 362–367 (2010)
8. Hartman, P., Nirenberg, L.: On spherical image maps whose Jacobians do not change sign. Am. J. Math. **81**(4), 901–920 (1959)
9. Ibragimov, N.H. (ed.): CRS Handbook of Lie Group Analysis of Differential Equations. New Trends in Theoretical Developments and Computational Methods, vol. 3. CRC Press, Boca Raton (1996)
10. Krivoshapko, S., Ivanov, V.N.: Encyclopedia of Analytical Surfaces. Springer, Cham (2015)
11. Lange-Hegermann, M.: The Differential Dimension Polynomial for Characterizable Differential Ideals. arXiv:1401.5959 (2014)
12. Michels, D.L., Lyakhov, D.A., Gerdt, V.P., Sobottka, G.A., Weber, A.G.: Lie symmetry analysis for cosserat rods. In: Gerdt, V.P., Koepf, W., Seiler, W.M., Vorozhtsov, E.V. (eds.) CASC 2014. LNCS, vol. 8660, pp. 324–334. Springer, Heidelberg (2014)
13. Robertz, D.: Formal Algorithmic Elimination for PDEs. Lecture Notes in Mathematics, vol. 2121. Springer, Cham (2014)
14. Seiler, W.M.: Involution - The Formal Theory of Differential Equations and its Applications in Computer Algebra. Algorithms and Computation in Mathematics, vol. 24. Springer, Heildelberg (2010)

Efficient Computation of Algebraic Local Cohomology Classes and Change of Ordering for Zero-Dimensional Standard Bases

Katsusuke Nabeshima[1] and Shinichi Tajima[2]

[1] Institute of Socio-Arts and Sciences, Tokushima University,
1-1 Minamijosanjima, Tokushima, JAPAN
`nabeshima@tokushima-u.ac.jp`
[2] Graduate School of Pure and Applied Sciences, University of Tsukuba,
1-1-1 Tennoudai, Tsukuba, JAPAN
`tajima@math.tsukuba.ac.jp`

Abstract. A new effective algorithm for computing a set of algebraic local cohomology classes is presented. The key ingredient of the proposed algorithm is to utilize a standard basis. As the application, an algorithm is given for the conversion of a standard basis of a zero-dimensional ideal with respect to any given local ordering into a standard basis with respect to any other local ordering, in the formal power series ring. The new algorithm always outputs a reduced standard basis.

Keywords: standard bases, algebraic local cohomology, singularities.

1 Introduction

Local cohomology is a key concept in algebraic geometry and commutative algebra, hence provides fundamental tools for applications in several fields both inside and outside mathematics [1, 6, 8, 12]. Thus, it is meaningful to provide an efficient algorithm for computing algebraic local cohomology classes.

In the previous work [18], computational methods for algebraic local cohomology classes and standard bases associated with Jacobi ideals of hypersurface isolated singularities were introduced. In this paper, new effective methods for computing algebraic local cohomology classes and standard bases associated with zero-dimensional ideals are considered.

In order to state precisely the problems and the results, let X be a neighborhood of the origin O of n-dimensional complex space $\mathbb{C}^n$ with coordinates $(x_1, \ldots, x_n)$, and assume that a set F of s polynomials $f_1, f_2, \ldots, f_s$ in $K[x_1, \ldots, x_n]$ satisfying $\{a \in X | f_1(a) = f_2(a) = \cdots = f_s(a) = 0\} = \{O\}$, is given where K is $\mathbb{Q}$ or $\mathbb{C}$. Let H_F denote a set of algebraic local cohomology classes supported at the origin that are annihilated by the ideal $\langle F \rangle$ generated, in the ring of formal power series, by F. The present paper contains two main results. One is an algorithm for computing a basis of the vector space H_F. The other one is a new algorithm for change of ordering for standard bases.

In the first part of this paper, we introduce a new algorithm for computing a basis of the vector space H_F. The idea of the new algorithm is an utilization

© Springer International Publishing Switzerland 2015
V.P. Gerdt et al. (Eds.): CASC 2015, LNCS 9301, pp. 334–348, 2015.
DOI: 10.1007/978-3-319-24021-3_25

of properties of standard basis of the ideal $\langle F \rangle$. We discuss how to compute a basis of the vector space H_F from a given standard basis of $\langle F \rangle$. It turns out, as is shown in section 3, that the resulting method is more effective than our previous algorithm [18].

In the second part of this paper, we introduce a new method for converting a standard basis of the ideal generated by F w.r.t. any given local ordering into a standard basis w.r.t. any other local ordering in a local ring.

The key ingredients of the algorithm, which is completely different from the famous FGLM algorithm [3], are algebraic local cohomology and the Grothendieck local duality theorem [5, 6].

There are two main advantages of the resulting algorithm of change of ordering. First, the algorithm always outputs the *reduced* standard bases. Other algorithm and implementation do not enjoy this property. Second, the resulting algorithm does not invoke Mora's reduction as in tangent cone algorithms [9, 10]. The computation consists of only linear algebra computation.

The organization of this paper is as follows. Related preparatory notation, definitions and former results used in this paper, are first reviewed in section 2. The main results are given in section 3.

2   Preliminaries

Throughout this paper, we use the notation x as the abbreviation of n variables $x_1, \ldots, x_n$. The set of natural numbers $\mathbb{N}$ includes zero. $\mathbb{C}$ and $\mathbb{Q}$ are the field of complex numbers and the field of rational numbers, respectively. K is $\mathbb{Q}$ or $\mathbb{C}$, and O is the origin of K^n.

In this section, we briefly review relations between algebraic local cohomology classes and standard bases. The details are in our previous works [15–18].

Let $H_{[O]}^n(K[x])$ denote the set of algebraic local cohomology classes supported at the origin O with coefficients in K, defined by

$$H_{[O]}^n(K[x]) := \lim_{k \to \infty} \mathrm{Ext}_{K[x]}^n(K[x]/\langle x_1, x_2, \ldots, x_n \rangle^k, K[x]),$$

where $\langle x_1, x_2, \ldots, x_n \rangle$ is the maximal ideal generated by $x_1, x_2, \ldots, x_n$.

Let X be a neighborhood of the origin O of K^n. Consider the pair $(X, X - O)$ and its relative Čech covering. Then, any element of $H_{[O]}^n(K[x])$ can be represented as an element of relative Čech cohomology. We represent an algebraic local cohomology class, given by finite sum of the form $\sum c_\lambda \left[\frac{1}{x^{\lambda+1}} \right]$, as a polynomial in n variables $\sum c_\lambda \xi^\lambda$ called "polynomial representation", where $c_\lambda \in K$, $\lambda \in \mathbb{N}^n$ and $\xi = (\xi_1, \ldots, \xi_n)$. The multiplication by x^α for polynomial representation is defined as

$$x^\alpha * \xi^\lambda := \begin{cases} \xi^{\lambda - \alpha}, & \lambda_i \geq \alpha_i, i = 1, \ldots, n, \\ 0, & \text{otherwise,} \end{cases}$$

where $\alpha = (\alpha_1, \ldots, \alpha_n) \in \mathbb{N}^n, \lambda = (\lambda_1, \ldots, \lambda_n) \in \mathbb{N}^n$, and $\lambda - \alpha = (\lambda_1 - \alpha_1, \ldots, \lambda_n - \alpha_n)$. We use " $*$ " as the multiplication.

Let fix a term ordering $\prec$ on $K[\xi]$. For a given algebraic local cohomology class of the form

$$\psi = c_\lambda \xi^\lambda + \sum_{\xi^{\lambda'} \prec \xi^\lambda} c_{\lambda'} \xi^{\lambda'}, \ c_\lambda \neq 0,$$

we call ξ^λ the **head term**, c_λ the **head coefficient**, $c_\lambda \xi^\lambda$ the **head monomial** and $\xi^{\lambda'}$ the **lower terms**. We denote the head term by $\mathrm{ht}(\psi)$, head coefficient by $\mathrm{hc}(\psi)$, head monomial by $\mathrm{hm}(\psi)$. Furthermore, we also denote the set of terms of ψ as $\mathrm{Term}(\psi) := \{\xi^\kappa | \psi = \sum_{\kappa \in \mathbb{N}^n} c_\kappa \xi^\kappa, c_\kappa \neq 0, c_\kappa \in K\}$ and the set of lower terms of ψ as $\mathrm{LL}(\psi) := \{\xi^\kappa \in \mathrm{Term}(\psi) | \xi^\kappa \neq \mathrm{ht}(\psi)\}$.

Let Ψ be a finite subset of $H_{[O]}^n(K[x])$. We denote the set of head terms of Ψ as $\mathrm{ht}(\Psi) := \{\mathrm{ht}(\psi) | \psi \in \Psi\}$, the set of terms of Ψ as $\mathrm{Term}(\Psi) := \bigcup_{\psi \in \Psi} \mathrm{Term}(\psi)$ and the set of lower terms of Ψ as $\mathrm{LL}(\Psi) := \bigcup_{\psi \in \Psi} \mathrm{LL}(\psi)$. Moreover, we denote the set of monomial elements of Ψ as $\mathrm{ML}(\Psi)$, and the set of linear combination elements of Ψ as $\mathrm{SL}(\Psi)$, i.e., $\Psi = \mathrm{ML}(\Psi) \cup \mathrm{SL}(\Psi)$.

Let ξ^λ be a term and let Φ be a set of terms in $K[\xi]$ where $\lambda = (\lambda_1, \ldots, \lambda_n) \in \mathbb{N}^n$. We call $\xi^\lambda \cdot \xi_i$ a **neighbor** of ξ^λ for each $i = 1, \ldots, n$. We define the neighbor of Φ as **Neighbor**(Φ), i.e., **Neighbor**$(\Phi) := \{\varphi \cdot \xi_i | \varphi \in \Phi, i = 1, \ldots, n\}$.

Note that, in this paper, for a polynomial and a set of polynomials in $K[x]$, we use the same notation as above, too.

Example 1. Let us consider algebraic local cohomology classes $\psi_1 = \xi_1^2 - \frac{3}{2}\xi_1\xi_2^4$, $\psi_2 = \xi_1^2\xi_1 - 3\xi_1\xi_2^3 + \xi_2$ and $\psi_3 = \xi_1^2\xi_2^3$, and $\Psi = \{\psi_1, \psi_2, \psi_3\}$ in $K[\xi_1, \xi_2]$. Let $\prec$ be the global lexicographical ordering s.t. $\xi_2 \prec \xi_1$. Then, $\mathrm{ht}(\psi_1) = \xi_1^2$, $\mathrm{Term}(\psi_1) = \{\xi_1^2, \xi_1\xi_2^4\}$ and $\mathrm{LL}(\psi_1) = \{\xi_1\xi_2^4\}$. For the set Ψ, $\mathrm{ML}(\Psi) = \{\psi_3\}$ and $\mathrm{SL}(\Psi) = \{\psi_1, \psi_2\}$. The set of neighbors of $\xi_1^2\xi_2^3$ is **Neighbor**$(\xi_1^2\xi_2^3) = \{\xi_1^3\xi_2^3, \xi_1^2\xi_2^4\}$.

We often use the following definitions.

Definition 1 (changing variables). *Let G be a set of polynomials in $K[x]$ and $g \in G$. For all $i \in \{1, \ldots, n\}$, a map $\mathcal{CV}$ is defined as changing variables x_i into ξ_i. The inverse map $\mathcal{CV}^{-1}$ is defined as changing variables ξ_i into x_i. That is, $\mathcal{CV}(g)$ is in $K[\xi]$. The set $\mathcal{CV}(G)$ is also defined as $\mathcal{CV}(G) = \{\mathcal{CV}(g) | g \in G\}$.*

For instance, $f = 3x_1^2 x_2 + \frac{1}{2}x_1 + 1 \in K[x_1, x_2]$ and $\psi = \frac{1}{2}\xi_1^3 - 3\xi_1\xi_2 + 2\xi_3 \in K[\xi_1, \xi_2, \xi_3]$. Then, $\mathcal{CV}(f) = 3\xi_1^2\xi_2 + \frac{1}{2}\xi_1 + 1$ and $\mathcal{CV}^{-1}(\psi) = \frac{1}{2}x_1^3 - 3x_1x_2 + 2x_3$.

Definition 2 (minimal bases). *A basis $\{x^{\gamma_1}, \ldots, x^{\gamma_l}\}$ for a monomial ideal I is said to be **minimal** if no x^{γ_i} in the basis divides another x^{γ_j} for $i \neq j$ where $\gamma_1, \ldots, \gamma_l \in \mathbb{N}^n$.*

Definition 3 (inverse orderings). *Let $\prec$ be a (local or global) term ordering. Then, the inverse ordering $\prec^{-1}$ of $\prec$ is defined by $x^\alpha \prec x^\beta \iff x^\beta \prec^{-1} x^\alpha$.*

If $\prec$ is a global term ordering (1 is the minimal term), then $\prec^{-1}$ is the local term ordering (1 is the maximal term). Conversely, if $\prec$ is a local term ordering, then $\prec^{-1}$ is the global term ordering.

Definition 4 (standard bases). *Let $\prec$ be a local term ordering and let R be a ring of formal power series. Let $I \subset R$ be an ideal. A standard basis of I w.r.t. $\prec$ is a set $\{g_1, \ldots, g_t\} \subset I$ such that $\mathrm{ht}(I) = \langle \mathrm{ht}(g_1), \ldots, \mathrm{ht}(g_t) \rangle$.*

If a basis $\{\mathrm{ht}(g_1), \ldots, \mathrm{ht}(g_t)\}$ is minimal for the monomial ideal $\mathrm{ht}(I)$, then $\{g_1, \ldots, g_t\}$ is said to be a minimal standard basis of I.

In the literature, the term "standard basis" is more common than "Gröbner basis" when working with local orderings and the local rings so we adopt that terminology in this paper.

Let F be a set of s polynomials $f_1, f_2, \ldots, f_s$ in $K[x]$ s.t. $\{a \in X | f_1(a) = f_2(a) = \cdots = f_s(a) = 0\} = \{O\}$. We define a set H_F to be the set of algebraic local cohomology classes in $H^n_{[O]}(K[x])$ that are annihilated by the ideal generated by F, where

$$H_F = \{\psi \in H^n_{[O]}(K[x]) | f_1 * \psi = f_2 * \psi = \cdots = f_s * \psi = 0\}.$$

In [11, 18], the set $H_f = \{\psi \in H^n_{[O]}(K[x]) | \frac{\partial f}{\partial x_1} * \psi = \frac{\partial f}{\partial x_2} * \psi = \cdots = \frac{\partial f}{\partial x_n} * \psi = 0\}$ is considered in the case where $f \in K[x]$ has an isolated singularity at O. As F satisfies the condition $\{a \in X | f_1(a) = f_2(a) = \cdots = f_s(a) = 0\} = \{O\}$, The discussion given in [11, 18] can be applied for computing H_F. In our works [16–18], algorithms for computing a basis of the vector space H_F, have been introduced and implemented in a computer algebra system Risa/Asir [14]. One can obtain a basis of the finite-dimensional vector space H_F by the algorithms and implementation.

Here, we recall a notation of transfer introduced in [11].

Definition 5 ([11]). *Let $\prec$ be a global ordering in $K[\xi]$ and let Ψ be a finite subset of $H^n_{[O]}(K[x])$. Let $c_{(\gamma, \kappa)} \in K$ denote the coefficient of the monomial ξ^κ in the polynomial representation*

$$\xi^\gamma + \sum_{\xi^\kappa \prec \xi^\gamma} c_{(\gamma, \kappa)} \xi^\kappa$$

of an element in $\mathrm{SL}(\Psi)$, where $\gamma \in \mathbb{N}^n$ is the exponent of its head term. Let Φ be a set of terms in $K[\xi]$, then, for $\xi^\lambda \in \Phi$, the transfer SB_Ψ is defined by the following:

$$\begin{cases} \mathrm{SB}_\Psi(\xi^\lambda) = x^\lambda - \displaystyle\sum_{\xi^\kappa \in \mathrm{ht}(\mathrm{SL}(\Psi))} c_{(\kappa, \lambda)} x^\kappa & \text{in } K[x], \text{ if } \xi^\lambda \in \mathrm{LL}(\Psi), \\ \mathrm{SB}_\Psi(\xi^\lambda) = x^\lambda & \text{in } K[x], \text{ if } \xi^\lambda \notin \mathrm{LL}(\Psi). \end{cases}$$

The set $\mathrm{SB}_\Psi(\Phi)$ is also defined by $\mathrm{SB}_\Psi(\Phi) = \{\mathrm{SB}_\Psi(\xi^\lambda) | \xi^\lambda \in \Phi\}$.

The next theorem shows the relation between a basis of the vector space H_F and a standard basis of $\langle F \rangle$.

Theorem 1 ([11, 18]). *Let $\prec$ be a global term ordering in $K[\xi]$ and Ψ be a basis of the vector space H_F such that, for all $\psi \in \Psi$, $\mathrm{hc}(\psi) = 1$ and $\mathrm{ht}(\psi) \notin \mathrm{ht}(\Psi \backslash \{\psi\})$. Let Φ be the minimal basis of $\langle \mathbf{Neighbor}(\mathrm{ht}(\Psi)) \backslash \mathrm{ht}(\Psi) \rangle$ in $K[\xi]$. Then, $\mathrm{SB}_\Psi(\Phi)$ is a minimal standard basis of $\langle F \rangle$ w.r.t. the local term ordering $\prec^{-1}$ in the ring of formal power series $K[[x]]$.*
Note that, $\mathcal{CV}(\Phi)$ is the set of head terms of a standard basis of $\langle F \rangle$ w.r.t. $\prec^{-1}$.

By this theorem, the minimal standard basis can be obtained from a basis of the vector space H_F in an efficient manner.

3   Main Results

In this section, we introduce two algorithms. One is an algorithm for computing a basis of the vector space H_F. The other one is an algorithm for change of ordering of standard bases. These two algorithms are the main results of this paper.

3.1   Computing a Basis of the Vector Space H_F from a Given Standard Basis of $\langle F \rangle$

Here, we illustrate how to compute a basis of the vector space H_F from a given standard basis of $\langle F \rangle$. Actually, we consider the opposite direction of Theorem 1.

Lemma 1. *Let T be the minimal basis of $\langle \mathrm{Term}(F) \rangle$ in $K[x]$ and M be the set of standard monomials of $\langle T \rangle$. Then, for all $\mathfrak{m} \in M$, $f_i * \mathcal{CV}(\mathfrak{m}) = 0$. Namely, the set $\mathcal{CV}(M)$ constitutes all monic monomial elements of a basis of the vector space H_F. We denote this set $\mathcal{CV}(M)$ as $\mathrm{MB}(H_F)$.*

Proof. Let $x_1^{\alpha_1} x_2^{\alpha_2} \cdots x_n^{\alpha_n} \in \langle T \rangle$ and $\xi_1^{\lambda_1} \xi_2^{\lambda_2} \cdots \xi_n^{\lambda_n} \in \mathcal{CV}(M)$ where $(\alpha_1, \alpha_2, \ldots, \alpha_n)$, $(\lambda_1, \lambda_2, \ldots, \lambda_n) \in \mathbb{N}^n$. As M is the set of standard monomials of $\langle T \rangle$, there always exists $j \in \{1, 2, \ldots, n\}$ such that $\alpha_j > \lambda_j$. Since $f_i \in \langle T \rangle$, hence, $f_i * (\xi_1^{\lambda_1} \xi_2^{\lambda_2} \cdots \xi_n^{\lambda_n}) = 0$ where $i \in \{1, \ldots, s\}$. $\square$

By computing a set of standard monomials of $\langle \mathrm{Term}(F) \rangle$, $\mathrm{MB}(H_F)$ can be immediately obtained.

Proposition 1. *Let G be a standard basis of $\langle F \rangle$ w.r.t. a local term ordering $\prec$ in $K[[x]]$ and let T be the set of standard monomials of $\langle \mathrm{ht}(G) \rangle$. Then, there exists a basis Ψ of the vector space H_F such that $\mathrm{ht}(\Psi) = \mathcal{CV}(T)$ w.r.t. $\prec^{-1}$.*

Proof. The minimal basis of the monomial ideal $\mathrm{ht}(\langle F \rangle)$ is unique and H_F is a dual space of $K[[x]]/\langle F \rangle$ [18]. Thus, by Theorem 1 (a relation of a standard basis), there exists a basis Ψ of the vector space H_F such that the minimal basis of the monomial ideal $\langle \mathbf{Neighbor}(\mathrm{ht}(\Psi)) \backslash \mathrm{ht}(\Psi) \rangle$ is equal to $\mathcal{CV}^{-1}(T)$. $\square$

The following theorem is a direct consequence of Theorem 1 and Proposition 1. This is the opposite direction of Theorem 1.

Theorem 2. *Let G be a minimal standard basis of $\langle F \rangle$ w.r.t. a local term ordering $\prec$ and let T be the set of standard monomials of $\langle \mathrm{ht}(G) \rangle$. Set $\{\xi^{\gamma_1}, \xi^{\gamma_2}, \ldots, \xi^{\gamma_l}\} = \mathcal{CV}(T) \backslash \mathrm{MB}(H_F)$ and let $c_{(\lambda, \kappa)} \in K$ denote the coefficient of the monomial x^κ in the polynomial representation*

$$x^\lambda + \sum_{x^\kappa \prec x^\lambda} c_{(\lambda, \kappa)} x^\kappa$$

of an element in G, where $\lambda \in \mathbb{N}^n$ is the exponent of its head term. Then, for each $i \in \{1, \ldots, l\}$, there exists $\varrho_i(\xi) \in K[\xi]$ such that

(i) $f_j * \psi_i = 0, j = 1, 2, \ldots, s$, where $\psi_i = \xi^{\gamma_i} - \displaystyle\sum_{x^{\gamma_i} \in \mathrm{Term}(G) \backslash \mathrm{ht}(G)} c_{(\lambda, \gamma_i)} \xi^\lambda + \varrho_i(\xi),$

(ii) $\mathrm{ht}(\varrho_i(\xi)) \prec^{-1} \xi^{\gamma_i}$.

Furthermore, $\{\psi_1, \ldots, \psi_l\} \cup \mathrm{MB}(H_F)$ is a basis of the vector space H_F. We call $\varrho_i(\xi)$ a sub-lower part of ψ_i w.r.t. $\prec^{-1}$.

We illustrate Theorem 2 with the following example.

Example 2. A polynomial $f = x_1^3 + x_1 x_2^5 + x_2^7 \in K[x_1, x_2]$ defines an isolated singularity at the origin O in K^2. Let us consider $F = \{\frac{\partial f}{\partial x_1}, \frac{\partial f}{\partial x_2}\} = \{3x_1^2 + x_2^5, 5x_1 x_2^4 + 7x_2^6\}$. Then, the computer algebra system SINGULAR [2] outputs the following as a standard basis of $\langle F \rangle$ w.r.t. the negative degree reverse lexicographical ordering[1] $\prec$ with the coordinate (x_1, x_2):

$$G = \{3x_1^2 + x_2^5, 5x_1 x_2^4 + 7x_2^6, x_2^8\}.$$

As the minimal basis of $\langle \mathrm{Term}(F) \rangle = \langle x_1^2, x_2^5, x_1 x_2^4, x_2^6 \rangle$ is $\{x_2^5, x_1 x_2^4, x_1^2\}$, a set of standard monomials of $\langle \mathrm{Term}(F) \rangle$ is

$$M = \{1, x_1, x_2, x_2^2, x_2^3, x_2^4, x_1 x_2, x_1 x_2^2, x_1 x_2^3\}.$$

Therefore, by Proposition 1, $\mathrm{MB}(H_F) = \{1, \xi_1, \xi_2, \xi_2^2, \xi_2^3, \xi_2^4, \xi_1 \xi_2, \xi_1 \xi_2^2, \xi_1 \xi_2^3\}$ where x_1 and x_2 are corresponding to ξ_1 and ξ_2, respectively. In Fig. 1 and Fig. 2, we represent an exponent of a term in $\mathrm{MB}(H_F)$ as •.

As $\mathrm{ht}(G) = \{x_2^8, x_1 x_2^4, x_1^2\}$, a set of standard monomials of $\langle \mathrm{ht}(G) \rangle$ is $T = M \cup \{x_2^5, x_2^6, x_2^7\}$. Hence, $\mathcal{CV}(T) \backslash \mathrm{MB}(H_F) = \{\xi_2^5, \xi_2^6, \xi_2^7\}$. In Fig. 2, we represent an exponent of a term in $\mathcal{CV}(T) \backslash \mathrm{MB}(H_F)$ as $\triangle$ w.r.t. $\prec^{-1}$.

By Theorem 2 and the standard basis G, the following set

$$\left\{ \xi_2^5 - \frac{1}{3}\xi_1^2 + \varrho_1(\xi), \xi_2^6 - \frac{7}{5}\xi_1 \xi_2^4 + \varrho_2(\xi), \xi_1^7 + \varrho_3(\xi) \right\}$$

is obtained where $\varrho_1(\xi), \varrho_2(\xi)$ and $\varrho_3(\xi)$ are sub-lower parts w.r.t. $\prec^{-1}$.□

---

[1] A definition of the negative degree reverse lexicographical ordering is in the appendix.

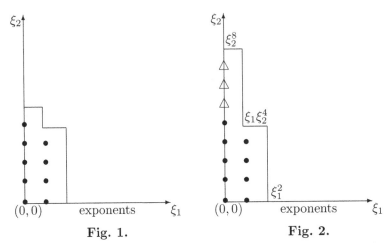

Fig. 1.   **Fig. 2.**

Next, we describe a method for computing sub-lower parts, namely a method to find candidates of unknown lower terms. The following corollary that immediately follows from Theorem 2, gives one of techniques for computing candidates of unknown lower terms.

Corollary 1. *Using the same notation as in Theorem 2, if G is the reduced standard basis, then no element in $CV(\mathrm{ht}(G))$ appears in the set* $\mathrm{Term}(\{\varrho_1(\xi), \ldots, \varrho_l(\xi)\})$.

The next lemma gives a property of lower terms.

Lemma 2 ([11, 18]). *Using the same notation as in Theorem 2, let $\varphi \in \Psi = \{\psi_1, \ldots, \psi_l\}$. Let $U = \{\psi' \in \Psi \mid \mathrm{ht}(\psi') \prec^{-1} \mathrm{ht}(\varphi)\}$. Then, if $\xi^\lambda \in \mathrm{LL}(\varphi)$ with $\lambda = (\lambda_1, \ldots, \lambda_n) \in \mathbb{N}^n$, then,*

(C): *for each $i = 1, 2, \ldots, n$, the term ξ^λ / ξ_i is in $\mathrm{Term}(U) \cup \mathrm{MB}(H_F)$, provided that $\lambda_i \geq 1$.*

The lemma implies that any element of

$$\mathbf{Neighbor}(\mathrm{Term}(U) \cup \mathrm{MB}(H_F)) \backslash (\mathrm{Term}(U) \cup \mathrm{MB}(H_F)),$$

that satisfies the condition **(C)** of Lemma 2 becomes a candidate of lower terms.

If a lower term of ψ_i is in $\mathrm{ht}(\{\psi_1, \ldots, \psi_l\}) \cup \mathrm{MB}(H_F)$ (the notation is from Lemma 2), then the lower term can be reduced by elements of $\{\psi_1, \ldots, \psi_l\} \cup \mathrm{MB}(H_F)$ in the vector space H_F. Namely, we do not need to consider any element of $\mathrm{ht}(\{\psi_1, \ldots, \psi_l\}) \cup \mathrm{MB}(H_F)$ as candidates of lower terms, for computing a basis of H_F. Furthermore, as a proper lower term ξ^β satisfies the condition **(C)** of Lemma 2, if a term ξ^α is a candidate of head terms such that ξ^α is greater than ξ^β w.r.t. a global term ordering, then the term ξ^β becomes a candidate of lower terms.

The following algorithm computes a basis of H_F from a (minimal) standard basis of $\langle F \rangle$ w.r.t. a local ordering $\prec$. After obtaining $\mathrm{MB}(H_F)$, the algorithm computes linear combination elements of a basis of H_F from bottom to up w.r.t. the inverse ordering $\prec^{-1}$.

Remark that if a given standard basis is minimal, then we can apply useful techniques devised from Theorem 2 and Corollary 1 for computing a basis of H_F. If a given standard basis is not minimal, we cannot apply these techniques. However, in such a case, a basis of H_F can be computed by applying Lemma 2 to decide candidates of lower terms.

In the following algorithm, the notation "LList" is the set of known lower terms that become candidate of lower terms.

Algorithm (LocalCohomology)

Specification: LocalCohomology$(F, \prec)$
Computing a basis of the vector space H_F.
Input: $F = \{f_1, \ldots, f_s\}$: a set of polynomials in $K[x]$ satisfying
$$\{a \in X \mid f_1(a) = f_2(a) = \cdots = f_s(a) = 0\} = \{O\}.$$
$\prec$: a local term ordering.
Output: $M \cup S$: a basis of H_F w.r.t. $\prec^{-1}$, where S is a set of linear combination elements and $M = \mathrm{MB}(H_F)$.
BEGIN

$G \leftarrow$ Compute a (minimal) standard basis of $\langle F \rangle$ w.r.t. $\prec$. Each element of G is represented in a form $x^\lambda + \sum_{x^\kappa \prec x^\lambda} c_{(\lambda,\kappa)} x^\kappa$, where $c_{(\lambda,\kappa)} \in K$.
$M \leftarrow$ Compute $\mathrm{MB}(H_F)$; $T \leftarrow$ Compute a standard monomial of $\langle \mathrm{ht}(G) \rangle$;
$S \leftarrow \emptyset$; $\Phi \leftarrow \mathcal{CV}(T) \backslash M$; $U \leftarrow$ **Neighbor**$(\mathcal{CV}(T)) \backslash \mathcal{CV}(T)$; LList $\leftarrow \emptyset$;

$$Q \leftarrow \begin{cases} \mathcal{CV}(\mathrm{ht}(G)) & \text{if } G \text{ is reduced,} \\ \emptyset & \text{if } G \text{ isn't reduced.} \end{cases} ;(*1) \qquad \text{/* Corollary 1*/}$$

$\mathrm{NC} \leftarrow \{\xi^\lambda \in U \mid \xi^\lambda \text{ satisfies } (\mathbf{C}) \text{ of Lemma 2}\} \backslash Q$;
while $\Phi \neq \emptyset$ **do**
$\xi^\gamma \leftarrow$ Select the smallest element from Φ w.r.t. $\prec^{-1}$; $\Phi \leftarrow \Phi \backslash \{\xi^\gamma\}$;

$$\psi \leftarrow \begin{cases} \xi^\gamma - \sum_{x^\gamma \in \mathrm{Term}(G) \backslash \mathrm{ht}(G)} c_{(\lambda,\gamma)} \xi^\lambda & \text{if } G \text{ is minimal,} \\ \xi^\gamma & \text{if } G \text{ isn't minimal.} \end{cases} ; (*2) \text{ /* Th.2*/}$$

$\mathrm{NC} \leftarrow \mathrm{NC} \backslash \mathrm{LL}(\psi)$;
$\mathrm{CL} \leftarrow \{\xi^\kappa \mid \xi^\kappa \prec^{-1} \xi^\gamma, \xi^\kappa \in \mathrm{NC}\} \cup \mathrm{LList}$; $(*3)$ /*candidates of lower terms*/
$\varphi \leftarrow \psi + \sum_{\xi^\lambda \in \mathrm{CL}} c_\lambda \xi^\lambda$, where c_λ's are indeterminates; $E \leftarrow \emptyset$;
for $i = 1$ **to** s **do**
$\eta \leftarrow f_i * \varphi$; /*check $f_i * \varphi = 0$. */
while $\eta \neq 0$ **do**
$E \leftarrow E \cup \{\mathrm{hc}(\eta) = 0\}$ w.r.t. $\prec^{-1}$; $\eta \leftarrow \eta - \mathrm{hm}(\eta)$;
end-while
end-for
$\varphi' \leftarrow$ Solve the system E of linear equations and substitute the solutions into c_λ of φ; $(*4)$
$\mathrm{LList} \leftarrow (\mathrm{LL}(\varphi') \backslash Q) \cup \mathrm{LList}$; $\mathrm{NC} \leftarrow \mathrm{NC} \backslash \mathrm{LL}(\varphi')$;

$\mathrm{NC} \leftarrow (\{\xi^\lambda \in \mathbf{Neighbor}(\mathrm{LL}(\varphi'))|\xi^\lambda \text{ satisfies } (\mathbf{C}) \text{ of Lemma 2}\}\backslash Q) \cup \mathrm{NC};$
$S \leftarrow S \cup \{\varphi'\};$
end-while
return $M \cup S$
END

Remark: If G is reduced, then the first one is selected at the box (∗1). If G is minimal, then the first one is selected at the box (∗2). Uniqueness of the solution c_λ at (∗4) was proved in our previous works [16, 18].

Theorem 3. *The algorithm* **LocalCohomology** *outputs a basis of the vector space* H_F *and terminates.*

Proof. Since H_F is a finite-dimensional vector space, the number of elements of T is finite. (Actually this number is the dimension of H_F.) Hence, as Φ is a finite set, the first **while-loop** terminates. Thus, this algorithm terminates.

By Lemma 1, the set M is a subset of a basis of H_F. Φ is the set of head terms of linear combination elements. By Theorem 2, Corollary 1 and Lemma 2, the candidates of lower terms are decided as CL at (∗3). At (∗4), each proper linear combination element φ of an algebraic local cohomology class, is determined as φ', and obviously $f_i * \varphi' = 0$ for each $i = 1, \dots, s$. Therefore, this algorithm outputs a basis of H_F. □

We illustrate the algorithm with the following example.

Example 3. Let us consider Example 2, again. $f = x_1^3 + x_1 x_2^5 + x_2^7$, $F = \{\frac{\partial f}{\partial x_1}, \frac{\partial f}{\partial x_2}\}$ and $\prec$ is the negative degree reverse lexicographical ordering s.t. $x_2 \prec x_1$. We set $\Phi = \mathcal{CV}(T)\backslash \mathrm{MB}(H_F)$ which is equal to $\{\xi_2^5, \xi_2^6, \xi_2^7\}$.

(1) Select the smallest element ξ_2^5 in Φ w.r.t. $\prec^{-1}$ and renew Φ as $\Phi\backslash\{\xi_2^5\}$. By Theorem 2, we get $\xi_2^5 - \frac{1}{3}\xi_1^2$. Next, we compute the set of candidates of unknown lower terms of ξ_2^5. Then, $\mathrm{NC} = \{\xi_1\xi_2, \xi_2^8\}$ and $\mathrm{CL} = \emptyset$. Thus, $\psi_1 = \xi_2^5 - \frac{1}{3}\xi_1^2$ is an element of a basis of H_F. Actually, $\frac{\partial f}{\partial x_1} * \psi_1 = 0$ and $\frac{\partial f}{\partial x_2} * \psi_1 = 0$. The set NC is renewed as $\{\xi_1^3, \xi_1^2\xi_2\}$ and LList $= \{\xi_1^2\}$.

(2) Select the smallest element ξ_2^6 in Φ w.r.t. $\prec^{-1}$ and renew Φ as $\Phi\backslash\{\xi_2^6\}$. By Theorem 2, we get $\xi_2^6 - \frac{7}{5}\xi_1\xi_2^4$. Next, we compute the set of candidates of unknown lower terms of ξ_2^6. Then, $\mathrm{CL} = \{\xi_1^3, \xi_1^2\xi_2, \xi_1^2\}$. Set $\psi_2 = \xi_2^6 - \frac{7}{5}\xi_1\xi_2^4 + c_{(3,0)}\xi_1^3 + c_{(2,1)}\xi_1^2\xi_2 + c_{(2,0)}\xi_1^2$ where $c_{(3,0)}$, $c_{(2,1)}$ and $c_{(2,0)}$ are indeterminates. Since $\frac{\partial f}{\partial x_1} * \psi_2 = 3c_{(2,1)}\xi_2 + 3c_{(3,0)}\xi_1 + \xi_2 + 3c_{(2,0)}$ and $\frac{\partial f}{\partial x_2} * \psi_2 = 0$, the system of equations is $E = [3c_{(2,1)} + 1 = 0, 3c_{(3,0)} = 0, 3c_{(2,0)} = 0]$. As the solution is $c_{(2,1)} = -\frac{1}{3}, c_{(3,0)} = c_{(2,0)} = 0$, $\psi_2 = \xi_2^6 - \frac{7}{5}\xi_1\xi_2^4 - \frac{1}{3}\xi_1^2\xi_2$ is an element of a basis of H_F.

The set NC is renewed as $\{\xi_1^3, \xi_1^2\xi_2^2, \xi_1\xi_2^5, \xi_2^8\}$ and the set LList is renewed as $\{\xi_1\xi_2^4, \xi_1^2\xi_2, \xi_1^2\}$.

(3) Select the smallest element ξ_2^7 in Φ w.r.t. $\prec^{-1}$ and renew Φ as $\Phi\backslash\{\xi_2^7\} = \emptyset$. In this case, since $x_2^7 \notin \mathrm{Term}(G)$, it is impossible to apply the technique of Theorem 2 for getting lower terms of ξ_2^7. We utilize only Lemma 2 to obtain candidates of lower terms of ξ_2^7. Then, $\mathrm{CL} = \{\xi_1\xi_2^5, \xi_1\xi_2^4, \xi_1^2\xi_2^2, \xi_1^3, \xi_1^2\xi_2, \xi_1^2\}$.

Set $\psi_3 = \xi_2^7 + c_{(1,5)}\xi_1\xi_2^5 + c_{(1,4)}\xi_1\xi_2^4 + c_{(2,2)}\xi_1^2\xi_2^2 + c_{(3,0)}\xi_1^3 + c_{(2,1)}\xi_1^2\xi_2 + c_{(2,0)}\xi_1^2$.
Check $\frac{\partial f}{\partial x_1} * \psi_3 = 0$, $\frac{\partial f}{\partial x_2} * \psi_3 = 0$ and solve the obtained system of linear equations. We get $c_{(1,5)} = -\frac{7}{5}, c_{(3,0)} = \frac{7}{15}, c_{(2,1)} = 0, \ c_{(2,2)} = -\frac{1}{3}, \ c_{(1,4)} = c_{(2,0)} = 0$. Therefore, $\psi_3 = \xi_2^7 - \frac{7}{5}\xi_1\xi_2^5 - \frac{1}{3}\xi_1^2\xi_2^2 + \frac{7}{15}\xi_1^3$ is an element of a basis of H_F.

Thus, $\{\psi_1, \psi_2, \psi_3\} \cup \mathrm{MB}(H_F)$ is a basis of the vector space H_F w.r.t. $\prec^{-1}$. $\square$

We have implemented the algorithm **LocalCohomology** in the computer algebra system Risa/Asir[14] and have executed some computation. Here, we give the results of benchmark tests. The following table shows a comparison of the implementation of **LocalCohomology** with our previous Risa/Asir-implementation [18]. As our previous implementation works for hypersurface isolated singularities, we consider a polynomial g_i that defines an isolated singularity at the origin, as a problem of the benchmark tests where $i = 1, \ldots, 8$. That is, $F = \{\frac{\partial g_i}{\partial x}, \frac{\partial g_i}{\partial y}\}$ or $F = \{\frac{\partial g_i}{\partial x}, \frac{\partial g_i}{\partial y}, \frac{\partial g_i}{\partial z}\}$ where x, y, z are variables. We used a PC [OS: Windows 7 (64bit), CPU: Intel(R) Core i-7-2600 CPU @ 3.40 GHz 3.40 GHz, RAM: 4 GB]. The time is given in seconds.

Problem (Milnor No.)	Algorithm [18]	LocalCohomology
$g_1 = (x^3 + yz^2)^2 + y^4 + z^4$ (45)	0.09	0.05
$g_2 = (x^3 + y^{10})^3 + xy^7$ (55)	0.02	0.02
$g_3 = (x^3 y + y^9 + 2xy^7)^2 + x^7$ (100)	1.29	0.76
$g_4 = (x^3 + yz^2)^2 + x^{10} + y^8 + z^4$ (105)	0.76	0.50
$g_5 = (x^4 + y^6 + x^2 y^3)^3 + x^8 y^6$ (187)	12.9	6.01
$g_6 = (y^4 + xz^3 + x^2)^2 + y^8 + z^9$ (168)	34.1	18.3
$g_7 = (x^3 y + y^7 + x^2 y^3)^4 + x^{14}$ (351)	198.4	84.3
$g_8 = (x^4 + y^9)^4 + 3x^{16}$ (525)	204.8	149.0

Since head terms of a basis of H_F are immediately decided by a standard basis of $\langle F \rangle$ by Proposition 1, the algorithm **LocalCohomology** is more efficient than the algorithm [18]. This is one of advantages.

3.2   Change of Ordering for Standard Bases and Reduced Standard Bases

Here, we introduce an algorithm for change of ordering for standard bases as an application of the algorithm **LocalCohomology**. The algorithm always outputs a reduced standard basis of $\langle F \rangle$ w.r.t. a given local term ordering. This is one of the advantages.

The next theorem, borrowed from [11, 18], describes how to compute a "reduced" standard basis from a basis of the vector space H_F. Required conditions of a basis of H_F are precisely stated in the theorem.

Theorem 4 ([11, 18]). *Let Ψ be a basis of the vector space H_F such that, for all $\psi \in \Psi$, $\mathrm{hc}(\psi) = 1$ and $\mathrm{ht}(\psi) \notin \mathrm{LL}(\Psi)$. Let $\prec$ be a global term ordering in*

$K[\xi]$ and Φ be the minimal basis of $\langle \mathbf{Neighbor}(\mathrm{ht}(\Psi)) \backslash \mathrm{ht}(\Psi) \rangle$ in $K[\xi]$. Then, $\mathrm{SB}_\Psi(\Phi)$ is a reduced standard basis of $\langle F \rangle$ w.r.t. the local term ordering $\prec^{-1}$ in the ring of formal power series $K[[x]]$.

Let Ψ' be an output of the **LocalCohomology**$(F, \prec)$ and G be a standard basis of $\langle F \rangle$ w.r.t. $\prec$. As Ψ' is a set of vectors, it is possible to compute a basis Ψ that satisfies the condition "for all $\psi \in \Psi$, $\mathrm{hc}(\psi) = 1$ and $\mathrm{ht}(\psi) \notin \mathrm{LL}(\Psi)$" w.r.t. a given ordering, from the basis Ψ'. The following algorithm outputs such Ψ and converts the standard basis G w.r.t. $\prec$ into the reduced standard basis w.r.t. another given local term ordering.

Algorithm (ChangeRSB)

Specification: ChangeRSB$(G, F, \prec_1, \prec_2)$
Converting a standard basis G of $\langle F \rangle$ w.r.t. $\prec_1$ into the reduced standard basis of $\langle F \rangle$ w.r.t. $\prec_2$.
Input: $F = \{f_1, \ldots, f_s\}$: a set of polynomials in $K[x]$ satisfying
$\quad \{a \in X | f_1(a) = f_2(a) = \cdots = f_s(a) = 0\} = \{O\}$.
$\quad \prec_1, \prec_2$: local term orderings.
$\quad G$: the standard basis of $\langle F \rangle$ w.r.t. $\prec_1$.

Output: S: a **reduced** standard basis of $\langle F \rangle$ w.r.t. $\prec_2$.
BEGIN
$\Psi' \leftarrow$ Compute a basis of H_F by **LocalCohomology** with G.
$v \leftarrow$ Sort and line up all elements of $\mathrm{Term}(\Psi') \backslash \mathrm{MB}(H_F)$ w.r.t. $\prec_2^{-1}$;
$A \leftarrow$ Make the coefficient matrix of $\Psi' \backslash \mathrm{MB}(H_F)$ w.r.t. v ;
$B \leftarrow$ Compute the row reduced echelon matrix of A;
${}^t(\psi_1, .., \psi_l) \leftarrow Bv$; /* ${}^t(\psi_1, .., \psi_l)$ is the transposed matrix of $(\psi_1, .., \psi_l)$*/
$\Psi \leftarrow \{\psi_1, \ldots, \psi_l\} \cup \mathrm{MB}(H_F)$;
$\Phi \leftarrow$ Compute the minimal basis of $\langle \mathbf{Neighbor}(\mathrm{ht}(\Psi)) \backslash \mathrm{ht}(\Psi) \rangle$ w.r.t. $\prec_2^{-1}$;
$S \leftarrow \mathrm{SB}_\Psi(\Phi)$;
return S
END

As the algorithm **LocalCohomology** terminates, the termination of this algorithm is obvious. The correctness of the algorithm is also obvious, since B is the row reduced echelon matrix.

Example 4. Let $f = x_1^3 + x_1 x_2^5 + x_2^7$ be a polynomial in $K[x_1, x_2]$ that defines an isolated singularity at the origin O in K^2. Set $F = \{\frac{\partial f}{\partial x_1}, \frac{\partial f}{\partial x_2}\}$. From Example 2, $G = \{3x_1^2 + x_2^5, 5x_1 x_2^4 + 7x_2^6, x_2^8\}$ is a standard basis of $\langle F \rangle$ w.r.t. the negative degree reverse lexicographical ordering $\prec_1$ with the coordinate (x_1, x_2).

Let us start computation of a standard basis of $\langle F \rangle$ w.r.t. the "**negative lexicographical**"[2] ordering $\prec_2$ with the coordinate (x_1, x_2).
The output of **LocalCohomology**$(F, \prec_1)$, already given in Example 3, is

$$\Psi' = \mathrm{MB}(H_F) \cup \left\{ \xi_2^5 - \frac{1}{3}\xi_1^2, \xi_2^6 - \frac{7}{5}\xi_1\xi_2^4 - \frac{1}{3}\xi_1^2\xi_2, \xi_2^7 - \frac{7}{5}\xi_1\xi_2^5 - \frac{1}{3}\xi_1^2\xi_2^2 + \frac{7}{15}\xi_1^3 \right\},$$

---

[2] A definition of the negative lexicographical ordering is in the appendix.

where $\mathrm{MB}(H_F) = \{1, \xi_1, \xi_2, \xi_1\xi_2, \xi_1^2, \xi_1^2\xi_2, \xi_2^3\}$.

Next, sort and line up all elements of $\mathrm{Term}(\Psi')\backslash \mathrm{MB}(H_F)$ w.r.t. $\prec_2^{-1}$ as

$$v = {}^t(\xi_1^3,\ \xi_1^2\xi_2^2,\ \xi_1^2\xi_2,\ \xi_1^2,\ \xi_1\xi_2^5,\ \xi_1\xi_2^4,\ \xi_2^7,\ \xi_2^6,\ \xi_2^5).$$

The coefficient matrix of $\Psi'\backslash \mathrm{MB}(H_F)$ w.r.t. v is

$$A = \begin{array}{c} \quad\xi_1^3\ \ \xi_1^2\xi_2^2\ \ \xi_1^2\xi_2\ \ \xi_1^2\ \ \xi_1\xi_2^5\ \ \xi_1\xi_2^4\ \ \xi_2^7\ \ \xi_2^6\ \ \xi_2^5 \\ \begin{pmatrix} 0 & 0 & 0 & -\frac{1}{3} & 0 & 0 & 0 & 0 & 1 \\ 0 & 0 & -\frac{1}{3} & 0 & 0 & -\frac{7}{5} & 0 & 1 & 0 \\ \frac{7}{15} & -\frac{1}{3} & 0 & 0 & -\frac{7}{5} & 0 & 1 & 0 & 0 \end{pmatrix} \end{array}.$$

The row reduced echelon matrix of A is

$$B = \begin{array}{c} \quad\xi_1^3\ \ \xi_1^2\xi_2^2\ \ \xi_1^2\xi_2\ \ \xi_1^2\ \ \xi_1\xi_2^5\ \ \xi_1\xi_2^4\ \ \xi_2^7\ \ \xi_2^6\ \ \xi_2^5 \\ \begin{pmatrix} 1 & -\frac{5}{7} & 0 & 0 & -3 & 0 & \frac{15}{7} & 0 & 0 \\ 0 & 0 & 1 & 0 & 0 & \frac{21}{5} & 0 & -3 & 0 \\ 0 & 0 & 0 & 1 & 0 & 0 & 0 & 0 & -3 \end{pmatrix} \end{array}.$$

Hence, a basis Ψ of the vector space H_F which satisfies $\mathrm{hc}(\psi) = 1$, $\mathrm{ht}(\psi) \notin \mathrm{ht}(\Psi\backslash\{\psi\})$ and $\mathrm{ht}(\psi) \notin \mathrm{LL}(\Psi)$ w.r.t. $\prec_2^{-1}$, for all $\psi \in \Psi$, is

$$\Psi = \mathrm{MB}(H_F) \cup \left\{\xi_1^3 - \frac{5}{7}\xi_1^2\xi_2^2 - 3\xi_1\xi_2^5 + \frac{15}{7}\xi_2^7, \xi_1^2\xi_2 + \frac{21}{5}\xi_1\xi_2^4 - 3\xi_2^6, \xi_1^2 - 3\xi_2^5\right\}.$$

The minimal basis of $\langle\mathbf{Neighbor}(\mathrm{ht}(\Psi))\backslash \mathrm{ht}(\Psi)\rangle$ w.r.t. $\prec_2^{-1}$ is

$$\Phi = \{\xi_1^4, \xi_1^3\xi_2, \xi_1^2\xi_2^2, \xi_1\xi_2^4, \xi_2^5\}.$$

In Fig. 3, an exponent of an element of $\mathrm{MB}(H_F)$ is represented as $\bullet$, and an exponent of an element of $\mathrm{ht}(\Psi)\backslash \mathrm{MB}(H_F)$ represented as $\triangle$ w.r.t. $\prec_2^{-1}$. As the set Φ plays a key role to construct a standard basis, we specially give the elements of Φ in the figure.

Then, a standard basis $\mathrm{SB}_\Psi(\Phi)$ of $\langle F\rangle$ w.r.t. $\prec_2$, is

$$\left\{x_1^4, x_1^3 x_2, x_1^2 x_2^2 + \frac{5}{7}x_1^3, x_1 x_2^4 - \frac{21}{5}x_1^2 x_2, x_2^5 + 3x_1^2\right\}. \quad \square$$

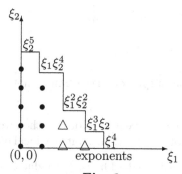

Fig. 3.

Example 5. Let $f = x^3y + xy^4 + x^2y^3$ be a polynomial in $K[x,y]$ that defines an isolated singularity at the origin O in K^2. Set $F = \{\frac{\partial f}{\partial x}, \frac{\partial f}{\partial y}\} = \{3x^2y + y^4 + 2xy^3, x^3 + 4xy^3 + 3x^2y^3\}$. Then, the computer algebra system SINGULAR outputs the following as a standard basis of $\langle F \rangle$ w.r.t. the negative degree reverse lexicographical ordering $\prec_1$ with the coordinate (x,y):

$$\{x^3 + 3x^2y^2 + 4xy^3, 3x^2y + 2xy^3 + y^4, 33xy^4 - 14xy^5 - 7y^6, y^7\}.^3$$

Then, the algorithm **LocalCohomology** outputs a basis of H_F as follows:

$$\left\{ 1, \xi_1, \xi_1^2, \xi_2, \xi_1\xi_2, \xi_2^2, \xi_1\xi_2^2, \xi_2^3, \xi_2^4 - \frac{1}{3}\xi_1^2\xi_2, \xi_1\xi_2^3 - 4\xi_1^3 - \frac{2}{3}\xi_1^2\xi_2, \right.$$
$$\left. \xi_2^5 - \frac{1}{3}\xi_1^2\xi_2^2 + \xi_1^3, \xi_2^6 - \frac{1}{3}\xi_1^2\xi_2^3 + \frac{7}{33}\xi_1\xi_2^4 + \frac{5}{33}\xi_1^3\xi_2 - \frac{14}{99}\xi_1^2\xi_2^2 + \frac{4}{3}\xi_1^4 + \frac{14}{33}\xi_1^3 \right\},$$

where x and y are corresponding to ξ_1 and ξ_2, respectively.

Let us consider a standard basis of $\langle F \rangle$ w.r.t. the negative lexicographical ordering $\prec_2$ with the coordinate (x,y). The algorithm **ChangeRSB** converts the basis above into a basis Ψ so that for all $\psi \in \Psi$, $\mathrm{hc}(\psi) = 1$, $\mathrm{ht}(\psi) \notin \mathrm{ht}(\Psi \backslash \{\psi\})$ and $\mathrm{ht}(\psi) \notin \mathrm{LL}(\Psi)$ w.r.t. $\prec_2^{-1}$ as follows:

$$\Psi = \left\{ 1, \xi_1, \xi_1^2, \xi_2, \xi_1\xi_2, \xi_2^2, \xi_1\xi_2^2, \xi_2^3, \xi_1^2\xi_2 - 3\xi_2, \xi_1^2\xi_2^2 - \frac{3}{4}\xi_1\xi_2^3 + \frac{3}{2}\xi_2^4 - 3\xi_2^5, \right.$$
$$\left. \xi_1^3 + \frac{1}{2}\xi_2^4 - \frac{1}{4}\xi_1\xi_2^3, \xi_1^4 + \frac{5}{44}\xi_1^3\xi_2 - \frac{1}{4}\xi_1^2\xi_2^3 + \frac{7}{44}\xi_1\xi_2^4 + \frac{3}{4}\xi_2^6 - \frac{7}{22}\xi_2^5 \right\}.$$

Therefore, by a basis Ψ, the reduced standard basis of $\langle F \rangle$ w.r.t. $\prec_2$ is

$$\left\{ x^5, x^3y - \frac{5}{44}x^4, xy^3 + \frac{3}{4}x^2y^2 + \frac{1}{4}x^3, y^4 + 3x^2y - \frac{3}{2}x^2y^2 - \frac{1}{2}x^3 \right\}.$$

The computer algebra system SINGULAR outputs the following as a standard basis of $\langle F \rangle$ w.r.t. $\prec_2$.

```
> ring A=0, (x,y),ls;
> poly f=x3y+xy4+x2y3;
> ideal I=diff(f,x),diff(f,y);
> std(I);
_[1]=y4+2xy3+3x2y
_[2]=4xy3+3x2y2+x3
_[3]=44x3y-15x3y2-5x4
_[4]=x5
```

It is easy to see that the output of SINGULAR is not the reduced standard basis of $\langle F \rangle$ w.r.t. $\prec_2$. □

Let us remark that if once we have a basis of H_F, then we are able to compute the reduced standard basis of $\langle F \rangle$ w.r.t. an arbitrary local ordering by computing

---

[3] This output is not the reduced standard basis. The reduced standard basis is: $\{x^3 + 4xy^3 - \frac{14}{33}y^6 - y^5, x^2y + \frac{2}{3}xy^3 + \frac{1}{3}y^4, xy^4 - \frac{7}{33}y^6, y^7\}$.

a row reduced echelon matrix of the coefficient matrix of the basis. That is, we need only the computation of the coefficient matrix to obtain a reduced standard basis of $\langle F \rangle$ w.r.t. an arbitrary local ordering. This computational step is not costly. Actually, for $g_1, \ldots, g_8$ presented in section 3.1., according to our implementation, the change of orderings from negative lexicographic ordering to negative degree lexicographical ordering, and the oposite direction, takes less than 0.5 seconds for getting a new standard basis from a basis of H_F.

The costliest part of the algorithm **LocalCohomology**, is to compute a basis of H_F. In our implementation, computing a basis of H_F is costlier than computing a standard basis of $\langle F \rangle$. In fact, for $g_1, \ldots, g_8$ presented in section 3.1., the command "std" of SINGULAR takes less than 0.5 seconds to output a standard basis w.r.t. negative lexicographical orderings and negative degree lexicographical orderings.

If one has a basis of H_F, then the algorithm **ChangeRSB** is a nice way to compute another standard basis of $\langle F \rangle$. The big advantage of the algorithm **ChangeRSB** is that the algorithm always outputs the *reduced* standard bases. Other algorithm and implementation do not enjoy this property. In the case that the *reduced* standard bases is not required, it is better to compute directly another standard basis by an algorithm for computing standard bases (Mora's tangent cone algorithms [9, 10], Lazard's homogenization technique [7] or "std" of SINGULAR) in speed and complexity.

We have introduced the new effective algorithm for computing a basis of H_F from a standard basis of $\langle F \rangle$. This algorithm is more efficient than the our previous algorithm [18]. We have also introduced the new algorithm for change of ordering of zero dimensional standard bases as the application of the algorithm **LocalCohomology**. The new algorithm does not invoke Mora's reduction as in tangent cone algorithms. Actually, in the algorithm **ChangeRSB**, we do not need to compute any reduction by a given standard basis. This is greatly different from FGLM algorithm.

Acknowledgment. This work has been partly supported by JSPS Grant-in-Aid for Young Scientists (B) (No.15K17513) and Grant-in-Aid for Scientific Research (C) (No.15K04891).

References

1. Brodmann, M.P., Sharp, R.Y.: Local Cohomology. Cambridge Univ. Press (1998)
2. Decker, W., Greuel, G.-M., Pfister, G., Schönemann, H.: SINGULAR 3-1-6 - A computer algebra system for polynomial computations (2012). http://www.singular.uni-kl.de
3. Faugère, J., Gianni, P., Lazard, D., Mora, T.: Efficient computation of zero-dimensional Gröbner bases by change of ordering. Journal of Symbolic Computation **16**, 329–344 (1993)
4. Greuel, G.-M., Pfister, G.: A Singular Introduction to Commutative Algebra, 2nd edn. Springer(2008)

5. Grothendieck, A.: Théorèmes de dualité pour les faisceaux algébriques cohérents. Séminaire Bourbaki **149** (1957)
6. Grothendieck, A.: Local Cohomology. Notes by Hartshorne, R. Lecture Notes in Math., vol. 41. Springer (1967)
7. Lazard, D.: Gröbner bases, Gaussian elimination, and resolution of systems of algebraic equations. In: van Hulzen, J.A. (ed.) ISSAC 1983 and EUROCAL 1983. LNCS, vol. 162, pp. 146–156. Springer, Heidelberg (1983)
8. Lyubeznik, G.: Local Cohomology and its Applications, Dekker (2002)
9. Mora, T.: An algorithm to compute the equations of tangent cones. In: Calmet, J. (ed.) EUROCAM 1982. LNCS, vol. 144, pp. 158–165. Springer, Heidelberg (1982)
10. Mora, T., Pfister, G., Traverso, C.: An introduction to the tangent cone algorithm. Adv. in Computing Research, Issued in Robotics and Nonlinear Geometry **6**, 199–270 (1992)
11. Nabeshima, K., Tajima, S.: On efficient algorithms for computing parametric local cohomology classes associated with semi-quasihomogeneous singularities and standard bases. In: Proc. ISSAC 2014, pp. 351–358. ACM-Press (2014)
12. Nabeshima, K., Tajima, S.: Logarithmic vector fields associated with parametric semi-quasihomogeneous hypersurface isolated singularities. In: Proc. ISSAC 2015. ACM-Press (2015) (to appear)
13. Nakamura, Y., Tajima, S.: On weighted-degrees for algebraic local cohomologies associated with semiquasihomogeneous singularities. Advanced Studies in Pure Mathematics **46**, 105–117 (2007)
14. Noro, M., Takeshima, T.: Risa/Asir - a computer algebra system. In: Proc. ISSAC 1992, pp. 387–396. ACM-Press (1992).
   http://www.math.kobe-u.ac.jp/Asir/asir.html
15. Tajima, S., Nakamura, Y.: Algebraic local cohomology class attached to quasi-homogeneous isolated hypersurface singularities. Publications of the Research Institute for Mathematical Sciences **41**, 1–10 (2005)
16. Tajima, S., Nakamura, Y.: Annihilating ideals for an algebraic local cohomology class. Journal of Symbolic Computation **44**, 435–448 (2009)
17. Tajima, S., Nakamura, Y.: Algebraic local cohomology classes attached to unimodal singularities. Publications of the Research Institute for Mathematical Sciences **48**, 21–43 (2012)
18. Tajima, S., Nakamura, Y., Nabeshima, K.: Standard bases and algebraic local cohomology for zero dimensional ideals. Advanced Studies in Pure Mathematics **56**, 341–361 (2009)

Appendix

In this paper, the following two concrete local orderings are utilized. The coordinate is $(x_1, x_2, \ldots, x_n)$ and $\alpha, \beta \in \mathbb{N}^n$.

(i) **Negative lexicographical ordering** $\succ$:
$x^\alpha \succ x^\beta : \iff \exists 1 \leq i \leq n, \alpha_1 = \beta_1, \ldots, \alpha_{i-1} = \beta_{i-1}, \alpha_i < \beta_i.$

(ii) **Negative degree lexicographical ordering** $\succ$:
$x^\alpha \succ x^\beta : \iff \deg(x^\alpha) < \deg(x^\beta),$ or $(\deg(x^\alpha) = \deg(x^\beta)$ and $\exists 1 \leq i \leq n : \alpha_1 = \beta_1, \ldots, \alpha_{i-1} = \beta_{i-1}, \alpha_i > \beta_i).$

These definitions are from [4].

Polynomial Real Root Isolation
by Means of Root Radii Approximation

Victor Y. Pan[1,2] and Liang Zhao[2]

[1] Departments of Mathematics and Computer Science,
Lehman College and the Graduate Center of the City University of New York,
Bronx, NY 10468, USA
[2] Ph.D. Programs in Mathematics and Computer Science,
The Graduate Center of the City University of New York,
New York, NY 10036, USA
victor.pan@lehman.cuny.edu,
lzhao1@gc.cuny.edu
http://comet.lehman.cuny.edu/vpan/

Abstract. Univariate polynomial root-finding is a classical subject, still important for modern computing. Frequently one seeks just the real roots of a real coefficient polynomial and is challenged to approximate them faster. The challenge is known for long time, and the subject has been intensively studied. The real roots can be approximated at a low computational cost if the polynomial has no non-real roots, but for high degree polynomials, non-real roots are typically much more numerous than the real ones. The bounds on the Boolean cost of the refinement of the simple and isolated real roots have been decreased to nearly optimal, but the success has been more limited at the stage of the isolation of real roots. By revisiting the algorithm of 1982 by Schönhage for the approximation of the root radii, that is, the distances between the roots and the origin, we obtain substantial progress: we isolate the simple and well conditioned real roots of a polynomial at the Boolean cost dominated by the nearly optimal bounds for the refinement of such roots. Our numerical tests with benchmark polynomials performed with the IEEE standard double precision show that the resulting nearly optimal real root-finder is practically promising. Our techniques are simple, and their power and application range may increase in combination with the known efficient methods. At the end we point out some promising directions to the isolation of complex roots and root clusters.

Keywords: Polynomials, real root-finding, root isolation, root radii.

1 Introduction

Assume a univariate polynomial of degree n with real coefficients,

$$p(x) = \sum_{i=0}^{n} p_i x^i = p_n \prod_{j=1}^{n} (x - x_j), \quad p_n \neq 0, \tag{1}$$

© Springer International Publishing Switzerland 2015
V.P. Gerdt et al. (Eds.): CASC 2015, LNCS 9301, pp. 349–360, 2015.
DOI: 10.1007/978-3-319-24021-3_26

which has r real roots $x_1, \ldots, x_r$ and $s = (n - r)/2$ pairs of non-real complex conjugate roots. In many applications, e.g., to algebraic and geometric optimization, one seeks only the real roots, which make up just a small fraction of all roots. This motivates a well studied subject of real root-finding (see, e.g., [28], [31], [37], and the extensive bibliography therein), but the most popular packages of subroutines for numerical root-finding such as MPSolve 2.0 [4], Eigensolve [7], and MPSolve 3.0 [5] still approximate the r real roots about as fast and as slow as all the n complex roots.

A typical fast real root-finder consists of two stages. At first one isolates all simple and well conditioned real roots (that is, the ones that admit isolation). Namely, one computes some complex discs, each covering a single real root and no other roots of the polynomial $p(x)$. Then the isolated roots are approximated fast by means of some specialized root refinement algorithms. The record and nearly optimal Boolean complexity at this stage has been obtained in the algorithm of [28], [31].

Presently we achieve progress, so far missing, at the former stage of the real root isolation. Our algorithm performs this stage and consequently approximates all the simple and well conditioned real roots within the same asymptotic Boolean complexity bounds of the papers [28], [31] (see our Theorem 7).

As in the papers [28], [31], we approximate only simple and well conditioned real roots, but not the multiple and ill conditioned roots. The latter roots little affect the complexity of our algorithm, e.g., our cost estimate do not depend on the minimal distance between the roots and do not include the terms like $\log(\mathrm{Discr}(p)^{-1})$.

Our overall cost bound also matches the nearly optimal one of the papers [20] and [23]. Their algorithm approximates all complex roots of a polynomial, but combines a number of advanced techniques. This makes it much harder to implement and even to comprehend than our algorithm, which is much less involved, more transparent, and more accessible for the implementation.

We have tested our algorithm applied with the IEEE standard double precision to some benchmark polynomials with small numbers of real roots. The test results are quite encouraging and are in good accordance with our formal study. The overall Boolean complexity of our isolation algorithm is dominated at the stage of performing the Dandelin's auxiliary root-squaring iterations, whose Boolean cost grows fast when their number increases, but in our tests, this number grew very slowly as we increased the degree of the input polynomials from 64 to 1024.

Technically, we achieve our progress by incorporating the old algorithm of [35], for the approximation of the root radii, that is, the distances of the roots to the origin (we refer the reader to [19], [8], [11], and [2] on the preceding works related to that algorithm). We feel that Schönhage in [35] used only a small part of the potential power of the algorithm. Namely he applied it to the rather modest task of the isolation of a single complex root, whereas we apply this simple but surprisingly efficient tool to the isolation of all simple and well conditioned real roots. We hope that this tool will be efficiently incorporated into other root-finders as well.

Our another basic sub-algorithm performs multi-point evaluation of the polynomial $p(x)$. The algorithm of [17], recently used in the root-finders of [28], [30], [31], and [37], solves this problem at a low Boolean cost, but is numerically unstable and only works with extended precision. Performed with double precision, it produces corrupted output already for polynomials of moderate degree (say, about 50 or so).

In some cases, we can avoid this drawback by applying the recent alternative numerically stable algorithms of [24] and [25], whose Boolean cost matches the one of [17] as long as the outputs are required within the relative approximation error bound $1/2^b$, for $b = O(\log(n))$. This is certainly the case at the initial stage of our algorithm, when we compute the sign of $p(x)$ at $2n$ points.

We organize our presentation as follows. In the next section, we cover some auxiliary results. In Section 3, we recall the Boolean complexity estimate for the approximation of the root radii. In Section 4, we describe our real root algorithm. In Section 5, we demonstrate it by working example. Section 6, the contribution of the second author, presents the results of our numerical tests. For some highly promising extension of our work, see Section 7 and [32].

2   Some Definitions and Auxiliary Results

Hereafter "flop" stands for "arithmetic operation". "lg" stands for "$\log_2$".

$O_B(\cdot)$ and $\tilde{O}_B(\cdot)$ denote the Boolean complexity up to some constant and poly-logarithmic factors, respectively.

τ is the overall bit-size of the coefficients of a polynomial $p(x)$ of (1).

Theorem 1. *Count the roots of a polynomial with their multiplicity. Then a polynomial $p(x)$ has an odd number of roots in the real line interval (α, β) if and only if $p(\alpha)p(\beta) < 0$.*

Definition 1. $D(z, \rho) = \{x : |x - z| \le \rho\}$ *denotes the closed disc with a complex center z and a radius ρ. Such a disc is γ-isolated, for $\gamma > 1$, if the disc $D(z, \gamma\rho)$ contains no other roots of the polynomial $p(x)$ of equation (1). A root x_j is γ-isolated if the disc $D(x_j, (\gamma - 1)|x_j|)$ contains no other roots of the polynomial $p(x)$ besides x_j.*

The following theorem states that Newton's iterations

$$y_0 = z, \quad y^{(h+1)} = y^{(h)} - p(y^{(h)})/p'(y^{(h)}), \quad h = 0, 1, \ldots \tag{2}$$

converge with quadratic rate globally, that is, right from the start, if they are initialized in a $3(n-1)$-isolated disc containing a single simple root of a polynomial $p(x)$ of (1).

Theorem 2. *Assume Newton's iteration (2) for a polynomial $p = p(x)$ of (1) and let $0 < 3(n-1)|y_0 - x_1| < |y_0 - x_j|$ for $j = 2, \ldots, n$. Then $|y_k - x_1| \le 2|y_0 - x_1|/2^{2^k}$ for $k = 0, 1, \ldots$.*

Proof. This is [38, Theorem 2.4], which strengthens [33, Corollary 4.5].

The following two theorems state some upper bounds on the Boolean cost of certain fundamental polynomial computations.

Theorem 3. Multi-point Polynomial Evaluation. *Assume a real $b \geq 1$, a polynomial $p(x)$ of (1), and k complex points $z_1, \ldots, z_k$ such that $k \geq n$. Write $l = \lg(\max_{j=1}^{n} |z_j|)$. Then, at the Boolean cost $\tilde{O}_B((b + \tau)k + lkn)$, one can compute the values v_j such that $|v_j - p(z_j)| \leq 1/2^b$ for $j = 1, \ldots, k$.*

This is [14, Theorem 3.9], also proved in [39], [15], and [29] (cf. [29, Lemma 21]. All proofs boil down to estimating the Boolean cost of the algorithm of [17], which relies on recursive polynomial division techniques of [6] and fast FFT-based polynomial division of [34].

Remark 1. Clearly, the same asymptotic bound applies to the Boolean complexity of a single Newton's iteration performed concurrently at k points.

Theorem 4. *Given a real $b \geq 1$, a polynomial $p(x)$ of (1), and the complex centers z_j and radii ρ_j of m discs $D_j = D(z_j, \rho_j)$, $j = 1, \ldots, m$, each covering a single simple root of the polynomial $p(x)$, write $L = \max_{j=1}^{m} \lg(|z_j| + \rho_j)$. Then, at the Boolean cost $\tilde{O}_B((b + L)n + \tau n^2)$, one can compute m new inclusion discs for the same roots, with the radii decreased by factors of at least 2^b provided that*
  (i) all the m roots and all the m centers $z_1, \ldots, z_m$ are real or
  (ii) all the m discs $D_1, \ldots, D_m$ are γ-isolated for a constant $\gamma > 1$.

Part (i) has been proved in [28] based on the algorithms of [26] and [27] (cf. also [31] and [37]). Part (ii) has been proved in [30] based on the approximation of the power sums of the roots of a polynomial.

3   Root Radii and Their Estimation

Definition 2. *List the absolute values of the roots of $p(x)$ in the non-increasing order, denote them $r_j = |x_j|$ for $j = 1, \ldots, n$, $r_1 \geq r_2 \geq \cdots \geq r_n$, and call them the root radii of the polynomial $p(x)$.*

The following result bounds the largest root radius r_1.

Theorem 5. *(See [40].) For a polynomial $p(x)$ of (1) and $r_1 = \max_{j=1}^{n} |x_j|$, it holds that*

$$0.5 r_1^+ / n \leq r_1 \leq r_1^+ \text{ for } r_1^+ = 2 \max_{i=1}^{n} |p_{n-i}/p_n|. \tag{3}$$

Remark 2. The theorem bounds all the root radii, but the paper [36] (extending the previous work in [16], [13], and [12]) presents stronger upper estimates for the positive roots of a polynomial $p(x)$. By applying these estimates to the reverse polynomial $p_{\text{rev}}(x) = x^n p(1/x)$ and to the polynomials $p(-x)$ and $p_{\text{rev}}(-x)$, one can bound the positive roots from below and the negative roots from both below and above.

Theorem 6. *Assume a polynomial $p = p(x)$ of (1) and a positive Δ. Then, within the Boolean cost bound $\tilde{O}_B(\tau n^2)$, one can compute approximations $\tilde{r}_j$ to all root radii r_j such that $1/\Delta \leq \tilde{r}_j/r_j \leq \Delta$ for $j = 1, \ldots, n$, provided that $\lg(\frac{1}{1-\Delta}) = O(\lg(n))$, that is, $|\tilde{r}_j/r_j - 1| \leq c/n^d$ for a fixed pair of constants $c > 0$ and d.*

This is [35, Corollary 14.3]. At first the root radii are approximated at a dominated cost in the case of $\Delta = 2n$. See some details of the algorithm in [21, Section 4]. [35, Section 14] and [21, Section 4] cite the related works [19], [8], [11], [1], and one can also compare the relevant techniques of the power geometry in [2] and [3], developed for the study of algebraic and differential equations.

In order to extend Theorem 6 to the case of $\Delta = (2n)^{1/2^k}$ for any positive integer k, at first apply k Dandelin's root-squaring iterations to the monic polynomial $q_0(x) = p(x)/p_n$ (cf. [10]), that is, compute recursively the polynomials

$$q_i(x) = (-1)^n q_{i-1}(\sqrt{x}\,)q_{i-1}(-\sqrt{x}\,) = \prod_{j=1}^{n}(x - x_j^{2^i}), \text{ for } i = 1, 2, \ldots \quad (4)$$

Then approximate the root radii $r_j^{(k)}$ of $q_k(x)$ by applying Theorem 6 for $\Delta = 2n$ and for $p(x)$ replaced by $q_k(x)$. Finally approximate the root radii r_j of $p(x)$ as $r_j = (r_j^{(k)})^{1/2^k}$. This enables us to decrease Δ from $2n$ to at most $1 + c/n^d = 1 + 2^{O(\lg(n))}$ for any fixed pair of constants $c > 0$ and d by using $k = O(\lg(n))$ Dandelin's iterations. The cost bound of Theorem 6 follows.

Remark 3. By using $k = \lceil \ln(s \lg(n)) \rceil$ Dandelin's root-squaring iterations, for $s > 1$, the same algorithm supporting Theorem 6 approximates the root radii within the relative error bound $\frac{1}{s \ln(2n)}$.

Corollary 1. *Assume a real $b \geq 1$, a polynomial $p(x)$ of (1), and a complex z. Then, within the Boolean cost bound $\tilde{O}_B((\tau + n(1 + \beta))n^2)$, for $\beta = \lg(2 + |z|)$, one can compute approximations $\tilde{r}_j \approx \bar{r}_j$ to the distances $\bar{r}_j = |z - x_j|$ from the point z to all roots x_j of the polynomial $p(x)$ such that $1/\Delta \leq \tilde{r}_j/\bar{r}_j \leq \Delta$, for $j = 1, \ldots, n$, provided that $\lg(\frac{1}{\Delta-1}) = O(\lg(n))$.*

Proof. Given a polynomial $p(x)$ of (1) and a complex scalar z, one can compute the coefficients of the polynomial $q(x) = p(x + z)$ by using $O(n \lg(n))$ flops (cf. [22]). The root radii of the polynomial $q(x)$ for a complex scalar z are equal to the distances $|x_j - z|$ from the point z to the roots x_j of $p(x)$.

Now let τ_q denote the bit-size of the coefficients of $q(x)$, and let $\bar{r}_j$ for $j = 1, \ldots, n$ denote its root radii, listed in the non-increasing order. Then, clearly, $\bar{r}_j \leq r_j + |z|$ for $j = 1, \ldots, n$. Furthermore represent the coefficients of the polynomial $q(x)$ with the bit-size $\tau + n$ by using representation of $p(x)$ with bit-size $\tilde{O}(\tau + n(1 + \beta))$ for $\beta = \lg(2 + |z|)$. By applying Theorem 6 to the polynomial $q(x)$, extend the cost bounds from the root radii to the distances.

4   Isolation and Approximation of Simple and Well-Conditioned Real Roots

Algorithm 1. Approximation of simple and well-conditioned real roots.

INPUT: *two positive integers n and r such that $0 < r < n$, a positive tolerance bound t, three real constants b, c and d such that $b \geq 1$ and $c > 0$, the coefficients of a polynomial $p(x)$ of equation (1), and an upper estimate r' for the unknown number of its real roots, $r' \leq n$.*

OUTPUT: *Approximations within the relative error bound $1/2^b$ to some or all real roots $x_1, \ldots, x_{r_+}$ of the polynomial $p(x)$, including all real roots that are both single and $(1 + c/n^d)$-isolated.*

COMPUTATIONS:

1. *Compute approximations $\tilde{r}_1, \ldots, \tilde{r}_n$ to the root radii of a polynomial $p(x)$ of (1) within the relative error bound c/n^d, $\tilde{r}_1 \leq \cdots \leq \tilde{r}_n$ (see Theorem 6). (Each approximation $\tilde{r}_i$ defines at most two real line intervals that can include real roots lying near $\pm\tilde{r}_i$. Overall we obtain $2n$ candidate inclusion intervals $\mathcal{I}_1, \ldots, \mathcal{I}_{2n}$, some of which can overlap or coincide with each other.)*
2. *Define $\mathbb{T} = \{t_1, \ldots, t_{n'}\}$, the set of the distinct endpoints of these intervals in non-decreasing order, for $n' \leq 4n$. Compute the sign of the polynomial $p(x)$ on this set. If the sign changes at two consecutive real points, $t' = t_j$ and $t'' = t_{j+1}$, select the line interval (t', t'').*
3. *Apply the algorithm of [28], [31], which supports part (i) of Theorem 4, concurrently to all selected line intervals. As soon as the algorithm decreases the length of a selected interval to or below $1/2^{b-1}$, output the midpoint as an approximation to a real root. If r' such approximations have been output or if the cost of performing the iterations of the algorithm of [28], [31] reaches the tolerance bound t, then stop the computations. (At stopping, there can remain some selected intervals of length exceeding $1/2^b$. We can output them with the label "intervals with real roots, apparently ill-conditioned".)*

Compare the error bound on the root radii at Stage 1 with the definition of a single $(1 + c/n^d)$-isolated real root of $p(x)$ in Definition 1 and conclude that one of the inclusion intervals $\mathcal{I}_j$, for $1 \leq j \leq 2n$, contains this root and no other roots of $p(x)$. Now the correctness of the algorithm follows from Theorem 1.

The Boolean cost at Stage 1 is $\tilde{O}_B((\tau + l)n^2)$, for $l = \max_{j=1}^{m} \lg(|x_j|)$ (by virtue of Theorem 6), at Stage 2 is $\tilde{O}_B(\tau n + ln^2)$ (by virtue of Theorem 3, for $b = 2$), and at Stage 3 is $\tilde{O}_B((\tau + l)n^2 + bn)$ (by virtue of part (i) of Theorem 4 for $L = O(l)$). In view of (3), we can write $l = O(\tau)$, and then the overall asymptotic cost bound, dominated at Stage 3, turns into $\tilde{O}_B(\tau n^2 + bn)$. This is the same cost bound as at Stage 1 if $b = O(\tau n)$ and is the same as Stage 2 if in addition $\tau = O(l)$. Summarizing we obtain the following estimate.

Theorem 7. *Suppose that we are given the coefficients of a polynomial $p(x)$ of equation (1) and three real constants b, c, and d such that $b \geq 1$ and $c > 0$.*

Then, at the cost $\tilde{O}_B(\tau n^2 + bn)$, we can approximate all the $(1 + c/n^d)$-isolated real roots of $p(x)$ within the relative error bound $1/2^b$.

Remark 4. Before we begin the computations of Stage 2 of Algorithm 1, we can narrow the ranges for positive and negative roots of the polynomial $p(x)$ by following the recipes of Remark 2. Then at Stage 2 we can avoid the evaluation of the polynomial at the points lying outside these ranges.

Remark 5. Before we started Stage 3, we can recompute the root radii for a smaller value c/n^d and then check if some intervals, selected under this value, lie strictly inside the old ones, sharing no endpoints, and hence are γ-isolated for some $\gamma > 1$. To such intervals we can apply a simplified version of the algorithm of [28], [31], which combines bisection and Newton's iteration and does not involve the more advanced algorithm of double exponential sieve (also called the bisection of the exponent).

Remark 6. For multi-point evaluation of the polynomial $p(x)$, we can apply the algorithm of [17] with extended precision, but if the output is only required with a low precision, then instead we can apply the algorithms of [24] and [25] with double precision at the same asymptotic Boolean cost. In particular, at Stage 2 of Algorithm 1 we only seek a single bit (of the sign of $p(x)$) per point of evaluation, and at Stage 3 at the initial Newton's iterations, crude values of $p(x)$ can be sufficient. At Stage 2 of contracting r suspect real intervals, we perform order of $O(r \lg(b))$ evaluations, and if r is small, which is quite typically the case in the applications to algebraic and geometric computations, then it can be non-costly even to apply the so called Horner's algorithm. It uses $2nr$ arithmetic operations, that is, less than the algorithms of [17], [24], and [25], if $r = o(\log^2(n))$.

Remark 7. If we only need to approximate the very well isolated real roots, we can apply Algorithm 1 at a lower cost for a larger isolation ratio. If we do not know how well the roots are isolated, we can at first apply Algorithm 1 assuming a larger ratio and, if the algorithm fails, re-apply it recursively, each time assuming stronger isolation.

5 Working Example

Consider the following polynomial

$$p(x) = 8x^7 + 16x^6 + 16x^5 + 16x^4 - 23x^3 - 30x^2 + 3x + 4.$$

This product of the 4th degree Chebyshev polynomial of the first kind and the cubic polynomial $x^3 + 2x^2 + 3x + 4$ has five real roots and two complex conjugate non-real roots. All seven roots are well-separated, namely,

$$0.3827 + 0.0000i, -0.3827 + 0.0000i, 0.9239 + 0.0000i, -0.9239 + 0.0000i,$$

$$-0.1747 + 1.5469i, -0.1747 - 1.5469i, -1.6506 + 0.0000i.$$

It took 14 root-squaring iterations to estimate the seven root-radii with a precision of at least 0.001:

$$[0.3827, 0.3827], [0.3826, 0.3828], [0.9237, 0.9240], [0.9238, 0.9241],$$

$$[1.5565, 1.5570], [1.5565, 1.5570], [1.6504, 1.6507].$$

The range $[1.5565, 1.5570]$ was for two roots. Taking the negations into account, we obtained 12 intervals containing all real roots, among which we counted the intervals $[-1.5565, -1.5570]$ and $[1.5565, 1.5570]$ twice:

$$[-1.6507, -1.6504], [-1.5570, -1.5565], [-0.9241, -0.9238],$$

$$[-0.9240, -0.9237], [-0.3828, -0.3826], [-0.3827, -0.3827], [0.3827, 0.3827],$$

$$[0.3826, 0.3828], [0.9237, 0.9240], [0.9238, 0.9241],$$

$$[1.5565, 1.5570], [1.6504, 1.6507].$$

Some of the ranges for the remaining root radii overlapped pairwise, but our algorithm evaluated the polynomial $p(x)$ at the endpoints of all the twelve intervals. Wherever there was the change of the sign, we concluded that the interval contains a root of the polynomial $p(x)$. We obtained nine (rather than five or even seven) intervals with the sign changes, namely:

$$[-1.6507, -1.6504][-0.9241, -0.9238][-0.9240, -0.9237][-0.3828, -0.3826]$$

$$[-0.3827, -0.3827][0.3827, 0.3827][0.3826, 0.3828][0.9237, 0.9240][0.9238, 0.9241]$$

Then our algorithm initiated Newton's iteration at the mid-point of each of these intervals. Whenever the Newton's map fell outside the interval, the algorithm switched to bisection method instead of Newton's. Very soon the algorithm converged to all real roots of $p(x)$.

The first Newton's iteration produced the following real root estimates:

$$-1.65062921, -0.92387954, -0.92387954, -0.38268343,$$

$$-0.38268343, 0.38268343, 0.38268343, 0.92387954, 0.92387954.$$

The second iteration produced the estimates

$$-1.65062919, -0.92387953, -0.92387953, -0.38268343, -0.38268343,$$

$$0.38268343, 0.38268343, 0.92387953, 0.92387953.$$

The maximum difference of the estimates output by the second and the third iterations was less than 10^{-8}. Thus the algorithm stopped and returned the values of five distinct real roots of the polynomial $p(x)$:

$$-1.65062919, -0.92387953, -0.38268343, 0.38268343, 0.92387953.$$

6   The Tests for Real Root-Finding with Algorithm 1

We tested a variant of our isolation Algorithm 1 described in Remarks 5–7. We applied it to polynomials of degrees $n = 64, 128, 256, 512, 1024$ having no clustered roots. In this variant, we simplified Stage 3: instead of the algorithm of [28], [31] we applied three Newton's iterations, initialized at the center of the isolation interval output by Algorithm 1. In the case of poor convergence, we performed a few bisections of the output interval of Algorithm 1 and then applied three Newton's iterations again. We performed all computations with double precision and estimated the output errors by comparing our results with the outputs of MATLAB function "roots()".

We generated our input polynomials as the products of the Chebyshev polynomials of the first kind (having degree r and, thus, having r simple real roots) with polynomials of degree $n - r$ of the following three families,

1. Polynomials with real standard Gaussian random coefficients.
2. Polynomials with complex standard Gaussian random coefficients.
3. Polynomials with integral consecutive coefficients starting at 1, i.e., $p(x) = 1 + 2x + \ldots + (n - 1)x^n$.

Table 1 displays the number of Dandelin's root-squaring iterations (in the column "Iter") and the maximum error (in the column "Error"). The tests showed very slow growth of the number of iterations, and consequently of the computational cost of Stage 1, as n increased from 64 to 1024.

Table 1. Real Root-finding with Using Root Radii Estimation

n	r	Type 1		Type 2		Type 3	
		Iter	Error	Iter	Error	Iter	Error
64	4	4	6.42E-11	4	6.42E-11	4	6.42E-11
64	8	6	3.16E-05	6	3.16E-05	6	3.16E-05
64	12	7	4.36E-03	7	4.36E-03	7	4.36E-03
128	4	4	6.42E-11	4	6.42E-11	4	6.42E-11
128	8	6	3.16E-05	6	3.16E-05	6	3.16E-05
128	12	7	4.36E-03	7	4.36E-03	7	4.36E-03
256	4	4	6.42E-11	4	6.42E-11	4	6.42E-11
256	8	6	3.16E-05	6	3.16E-05	6	3.16E-05
256	12	7	4.36E-03	7	4.36E-03	7	4.36E-03
512	4	4	6.42E-11	4	6.42E-11	4	6.42E-11
512	8	6	3.16E-05	6	3.16E-05	6	3.16E-05
512	12	7	4.36E-03	7	4.36E-03	7	4.36E-03
1024	4	4	6.42E-11	4	6.42E-11	4	6.42E-11
1024	8	6	3.16E-05	6	3.16E-05	6	3.16E-05
1024	12	7	4.36E-03	7	4.36E-03	7	4.36E-03

7   Conclusions

We plan to extend our techniques to the isolation of multiple roots and root clusters, by means of the elaboration upon our present constructions and of the incorporation of the techniques that extend part (ii) of Theorem 4 (cf. [32]). These inexpensive techniques would enable us to detect a multiple root or a root cluster isolated in a real interval and to refine its approximation. Having closely approximated such a cluster, we can split out the factor of $p(x)$ whose root set is precisely this cluster, and then we can work on root-finding for the remaining factor of smaller degree. Developing this approach, we are going to explore potential advantages of shifting the origin into the center of gravity of the roots, $-p_{n-1}/(np_n)$ (cf. [35]), or into the roots of higher order derivatives of $p(x)$ (cf. [18, Section 15]).

Another direction is the extension of our work to the isolation of the complex roots and root clusters (cf. [32]). Towards this goal, one can approximate closely enough, by means of computations with extended precision, the root radii of polynomials $q_0(x) = p(x - s_0)$, $q_1(x) = p(x - s_1)$ and $q_2(x) = p(x - s_2)$, for some appropriate shift values s_0, s_1, and s_2. This would define three families of narrow annuli (n annuli in each family, some of which can overlap or even coincide with other annuli), and then we would compute the intersections of the triples of the annuli, each triple coming from the three families. We could output the well isolated intersections as long as we determine that they contain roots or root clusters.

We plan to test our current, amended, and extended algorithms against the available real root-finders and to explore the chances for enhancing the efficiency of our algorithms by means of their combination with some of the other highly efficient techniques known for root-finding.

Acknowledgements.   This work has been supported by NSF Grant CCF 1116736 and PSC CUNY Award 67699-00 45. We are also grateful to the reviewers for helpful comments.

References

1. Alt, H.: Multiplication is the easiest non-trivial arithmetic function. Theoretical Computer Science **36**, 333–339 (1985)
2. Bruno, A.D.: The Local Method of non-linear Analysis of Differential Equations. Nauka, Moscow (1979) (in Russian). English translation: Soviet Mathematics, Springer, Berlin (1989)
3. Bruno, A.D.: Power Geometry in Algebraic and Differential Equations. Fizmatlit, Moscow (in Russian), English translation: North-Holland Mathematical Library, vol. 57. Elsevier, Amsterdam (2000), also reprinted in 2005, ISBN 0-444-50297
4. Bini, D.A., Fiorentino, G.: Design, analysis, and implementation of a multiprecision polynomial rootfinder. Numerical Algorithms **23**, 127–173 (2000)
5. Bini, D.A., Robol, L.: Solving secular and polynomial equations: a multiprecision algorithm. J. Computational and Applied Mathematics **272**, 276–292 (2014)

6. Fiduccia, C.M.: Polynomial evaluation via the division algorithm: the fast fourier transform revisited. In: Proc. 4th Annual ACM Symposium Theory of Computing, pp. 88–93 (1972)

7. Fortune, S.: An iterated eigenvalue algorithm for approximating roots of univariate polynomials. J. Symb. Comput. **33**(5), 627–646 (2002)

8. Graham, R.I.: An efficient algorithm for determining the convex hull of a finite planar set. Information Processing Letters **1**, 132–133 (1972)

9. Golub, G.H., Van Loan, C.F.: Matrix Computations, 4th edn. The Johns Hopkins University Press, Baltimore (2013)

10. Householder, A.S.: Dandelin, Lobachevskii, or Graeffe. Amer. Math. Monthly **66**, 464–466 (1959)

11. Henrici, P.: Applied and Computational Complex Analysis. Wiley, New York (1974)

12. Hong, H.: Bounds for absolute positiveness of multivariate polynomials. J. Symb. Comput. **25**, 571–585 (1998)

13. Kioustelidis, J.B.: Bounds for positive roots of polynomials. J. Comput. Appl. Math. **16**, 241–244 (1986)

14. Kirrinnis, P.: Partial fraction decomposition in $C(z)$ and simultaneous Newton iteration for factorization in $C[z]$. J. Complexity **14**, 378–444 (1998)

15. Kobel, A., Sagraloff, M.: On the complexity of computing with planar algebraic curves. J. Complexity, in press. Also arXiv:1304.8069v1 [cs.NA], April 30, 2013

16. Lagrange, J.L.: Traité de la résolution des équations numériques. Paris (1798). (Reprinted in Œuvres, t. VIII, Gauthier-Villars, Paris (1879))

17. Moenck, R., Borodin, A.: Fast modular transform via division. In: Proc. 13th Annual Symposium on Switching and Automata Theory, pp. 90–96, IEEE Comp. Society Press, Washington, DC (1972)

18. McNamee, J.M., Pan, V.Y.: Numerical Methods for Roots of Polynomials. Part 2 (XXII + 718 pages), Elsevier (2013)

19. Ostrowski, A.M.: Recherches sur la méthode de Graeffe et les zéros des polynomes et des series de Laurent. Acta Math. **72**, 99–257 (1940)

20. Pan, V.Y.: Optimal (up to polylog factors) sequential and parallel algorithms for approximating complex polynomial zeros. In: Proc. 27th Ann. ACM Symp. on Theory of Computing, pp. 741–750. ACM Press, New York (1995)

21. Pan, V.Y.: Approximating complex polynomial zeros: modified quadtree (Weyl's) construction and improved Newton's iteration. J. Complexity **16**(1), 213–264 (2000)

22. Pan, V.Y.: Structured Matrices and Polynomials: Unified Superfast Algorithms. Birkhäuser, Springer, Boston, New York (2001)

23. Pan, V.Y.: Univariate polynomials: nearly optimal algorithms for factorization and rootfinding. J. Symb. Computations **33**(5), 701–733 (2002). In: Proc. version in ISSAC'2001, pp. 253–267, ACM Press, New York (2001)

24. Pan, V.Y.: Transformations of matrix structures work again. Linear Algebra and Its Applications **465**, 1–32 (2015)

25. Pan, V.Y.: Fast Approximate Computations with Cauchy Matrices and Polynomials, arXiv:1506.02285 [math.NA], 32 p., 6 figures, 8 tables, June 7, 2015

26. Pan, V.Y., Linzer, E.: Bisection acceleration for the symmetric tridiagonal eigenvalue problem. Numerical Algorithms **22**(1), 13–39 (1999)

27. Pan, V.Y., Murphy, B., Rosholt, R.E., Qian, G., Tang, Y.: Real root-finding. In: Watt, S.M., Verschelde, J. (eds.) Proc. 2nd ACM International Workshop on Symbolic-Numeric Computation (SNC), pp. 161–169 (2007)

28. Pan, V.Y., Tsigaridas, E.P.: On the Boolean complexity of the real root refinement. In: M. Kauers (ed.) Proc. Intern. Symposium on Symbolic and Algebraic Computation (ISSAC 2013), pp. 299–306, Boston, MA, June 2013. ACM Press, New York (2013). Also arXiv 1404.4775 April 18, 2014

29. Pan, V.Y., Tsigaridas, E.P.: Nearly optimal computations with structured matrices. In: Proc. the Int. Conf. on Symbolic Numeric Computation (SNC 2014). ACM Press, New York (2014). Also April 18, 2014, arXiv:1404.4768 [math.NA]

30. Pan, V.Y., Tsigaridas, E.P.: Accelerated approximation of the complex roots of a univariate polynomial. In: the 2014 Proc. of the Int. Conf. on Symbolic Numeric Computation (SNC 2014), Shanghai, China, July 2014, pp. 132–134. ACM Press, New York (2014). Also April 18, 2014, arXiv : 1404.4775 [math.NA]

31. Pan, V.Y., Tsigaridas, E.: Nearly optimal refinement of real roots of a univariate polynomial. J. Symb. Comput. (in press)

32. Pan, V.Y., Zhao, L.: Polynomial root isolation by means of root radii approximation, arxiv, 1501.05386, June 15, 2015

33. Renegar, J.: On the worst-case arithmetic complexity of approximating zeros of polynomials. J. Complexity **3**(2), 90–113 (1987)

34. Sieveking, M.: An algorithm for division of power series. Computing **10**, 153–156 (1972)

35. Schönhage, A.: The fundamental theorem of algebra in terms of computational complexity. Math. Department, Univ. Tübingen, Germany (1982)

36. Stefanescu, D.: New bounds for positive roots of polynomials. Univ. J. Comput. Sci. **11**, 2125–2131 (2005)

37. Sagraloff, M., Mehlhorn, K.: Computing real roots of real polynomials - an efficient method based on Descartes' rule of signs and Newton iteration. J. Symb. Comput. (in press)

38. Tilli, P.: Convergence conditions of some methods for the simultaneous computations of polynomial zeros. Calcolo **35**, 3–15 (1998)

39. van der Hoeven, J.: Fast composition of numeric power series. Tech. Report 2008–09, Université Paris-Sud, Orsay, France (2008)

40. Van der Sluis, A.: Upper bounds on the roots of polynomials. Numerische Math. **15**, 250–262 (1970)

Randomized Circulant and Gaussian Pre-processing

Victor Y. Pan[1] and Liang Zhao[2]

[1] Departments of Mathematics and Computer Science,
Lehman College and the Graduate Center of the City University of New York,
Bronx, NY 10468 USA
victor.pan@lehman.cuny.edu
http://comet.lehman.cuny.edu/vpan/
[2] Ph.D. Programs in Mathematics and Computer Science,
The Graduate Center of the City University of New York,
New York, NY 10036 USA
lzhao1@gc.cuny.edu

Abstract. Circulant matrices have been extensively applied in Symbolic and Numerical Computations, but we study their new application, namely, to randomized pre-processing that supports Gaussian elimination with no pivoting, hereafter referred to as *GENP*. We prove that, with a probability close to 1, GENP proceeds with no divisions by 0 if the input matrix is pre-processed with a random circulant multiplier. This yields 4-fold acceleration (in the cases of both general and structured input matrices) versus pre-processing with the pair of random triangular Toeplitz multipliers, which has been the user's favorite since 1991. In that part of our paper, we assume computations with infinite precision, but in other parts with double precision, in the presence of rounding errors. In this case, GENP fails without pre-processing unless all square leading blocks of the input matrix are well-conditioned, but empirically GENP produces accurate output consistently if a well-conditioned input matrix is pre-processed with random circulant multipliers. We also support formally the latter empirical observation if we allow standard Gaussian random input and hence the average non-singular and well-conditioned input as well, but we prove that GENP fails numerically with a probability close to 1 in the case of some specific input matrix pre-processed with such multipliers. We also prove that even for the worst case well-conditioned input, GENP runs into numerical problems only with a probability close to 0, if a nonsingular and well-conditioned input matrix is multiplied by a standard Gaussian random matrix. All our results for GENP can be readily extended to the highly important block Gaussian elimination.

Keywords: Random circulant matrices, linear systems, pre-processing, pivoting, pre-conditioning, Gaussian random matrices.

© Springer International Publishing Switzerland 2015
V.P. Gerdt et al. (Eds.): CASC 2015, LNCS 9301, pp. 361–375, 2015.
DOI: 10.1007/978-3-319-24021-3_27

1   Introduction

Circulant matrices are among the most popular structured matrices. Such a matrix can be decomposed into the product of the matrix of the discrete Fourier transform (DFT), a diagonal matrix, and the matrix of the inverse DFT. This enables fast FFT-based multiplication and inversion of circulant matrices, prompting their numerous highly efficient applications to various Symbolic and Numerical Computations, but we study a new application, namely, to pre-processing Gaussian elimination with no pivoting (*GENP*). Unlike the celebrated circulant pre-conditioners that accelerate the convergence of iterative algorithms for structured linear systems of equations [7], [8], we use random circulant multipliers in order to avoid potential degeneracy and numerical instability of GENP applied to both general and structured linear systems.

At first we recall that the customary variant of Gaussian elimination with partial pivoting, hereafter referred to as *GEPP*, ensures that the computations are *safe*, that is, use no divisions by 0. This follows due to a proper policy of interchanging the rows of the input matrix, called *pivoting*. For an $n \times n$ input matrix, pivoting involves just the order of n^2 comparisons, but still takes quite a heavy toll. It interrupts the stream of arithmetic operations with foreign operations of comparison, involves book-keeping, compromises data locality, impedes parallelization of the computations, and increases communication overhead and data dependence.

GENP avoids these drawbacks of GEPP, but it is safe in the above sense if and only if the input matrix is *strongly non-singular*, that is, non-singular together with all its square leading block sub-matrices. Most of the matrices with entries from the infinite fields of complex, real, or rational numbers or the finite fields of large cardinality are strongly non-singular, but not so for the matrices in the fields of small cardinality, e.g., $GF(2)$. Over the fields of rational, real or complex numbers, the diagonally dominant, positive definite and totally positive matrices are strongly non-singular (cf. [15]), but for the other matrices, it is usually as hard to verify strong non-singularity as to perform GENP.

GENP preceded by randomized multiplications by a single structured matrix or by a pair of such matrices is an attractive alternative to GEPP. The frequently cited paper [17] proves that the $n \times n$ product UAL is strongly non-singular with probability close to 1 if A is a non-singular matrix and if U^T and L are random lower triangular Toeplitz matrices. One just needs to generate $2n - 1$ random parameters and to perform $O(n^2 \log(n))$ arithmetic operations in order to compute the product UAL.

We accelerate a little this pre-processing of [17], as well as of all its known variations (cf. [2, Section 2.13], [6] and [20, Sections 5.6 and 5.7]), by applying just a single circulant multiplier, but can we extend to this case the proof of [17] that GENP is safe? This is not straightforward at all, but we work out a proof in Section 3.

Namely we prove that, with a probability close to 1, any non-singular matrix becomes strongly non-singular after its multiplication by a random f-circulant multiplier, for any complex f. We recall the definition of f-circulant matrices in Section 2.3. This is a triangular Toeplitz matrix, for $f = 0$, and is called circulant, for $f = 1$, and skew circulant, for $f = -1$. If $f \neq 0$, we operate with an f-circulant matrix almost as fast as with a circulant one and twice as fast as with a triangular Toeplitz matrix.

All our results for GENP can be equally well applied to supporting block Gaussian elimination, whereas incorporation of pivoting into this highly important algorithm leads to additional difficulties and drawbacks. An important special case is the solution of a Toeplitz or Toeplitz-like linear system of equations, which again has widely known applications to Symbolic and Numerical Computations, e.g., to the computation of a polynomial GCD and an approximate GCD, Padé approximation, rational function reconstruction, and linear recurrence span.

The MBA superfast algorithm, by Morf [18], [19] and by Bitmead and Anderson [3], runs in nearly linear arithmetic time for both Toeplitz and Toeplitz-like inputs. This algorithm is precisely the recursive block Gaussian elimination, accelerated by means of exploiting the Toeplitz-like structure of the input matrix. Pivoting is not used for Toeplitz or Toeplitz-like inputs because it would immediately destroy their structure, but the structure is fully preserved by our randomized circulant pre-processing (cf. [20, Chapter 5]).

Extensive formal study of how randomized pre-processing counters singularity (the subject of our Section 3) goes back at least to the 1990s (e.g., see [20, Section 2.13], entitled "Regularization of a Matrix via Pre-conditioning with Randomization"), but in Section 4, we extend our formal study to the more recent research subject of randomized structured numerical pre-processing of GENP and block Gaussian elimination, in the presence of rounding errors, in which case both algorithms can readily fail [15], [1], and pre-processing is badly needed (see [23] on the history of this study).

Empirically, random circulant multipliers provide the desired remedy. In the extensive tests reported in [24] and [23], GENP with random circulant pre-processing has consistently produced accurate solutions for a great variety of inputs for which GENP failed without pre-processing.

These empirical observations may be sufficient for the user, but we also provide some formal support for them. Namely, we prove that for standard Gaussian random well-conditioned input and pre-processing with a non-singular and well-conditioned multiplier, GENP fails with a probability close to 0.

This motivates a research challenge of finding a small family of appropriate policies of structured pre-processing for which the intersection of such input sets is empty, but applying just a singe random circulant multiplier is not sufficient: we prove in Theorem 9 that in the presence of rounding errors GENP with such a multiplier fails with a probability near 1, for some specific input matrices of a large size pre-processed with such a multiplier.

Pre-processing with a standard Gaussian (unstructured) random multiplier is sufficient, however: we prove that even for the worst case input matrix (as long as it is well-conditioned and is pre-processed with such a multiplier) GENP and block Gaussian elimination are numerically stable with a probability close to 1.

The cost of the generation of the multiplier and the multiplication by it should be counted, but also discounted because pre-processing can be re-used and because modern technological trend motivates discounting the cost of arithmetic computations. Here is a relevant citation from [4]: "The traditional metric for the efficiency of a numerical algorithm has been the number of arithmetic operations it performs. Technological trends have long been reducing the time to perform an arithmetic operation, so it is no longer the bottleneck in many algorithms; rather, communication, or moving data, is the bottleneck".

We organize our paper as follows. We present some background material in the next section. In Section 3, we prove that GENP with a random f-circulant multiplier is expected to proceed with no divisions by 0. In Section 4, we study numerical GENP, performed with rounding errors.

To simplify our exposition, we present our results just for GENP, but for the sake of completeness, we link to them block Gaussian elimination in the Appendix.

We report our substantial further progress in [27].

2   Background

2.1   Safe GENP, Pre-processing and Strong Non-singularity

We call GENP *safe* whenever it proceeds to the end with no divisions by 0.
$A_{k,l}$ denotes the $k \times l$ leading (that is, northwestern) sub-matrix of a matrix A. A matrix is called *strongly non-singular* if all its square leading sub-matrices are non-singular.

Theorem 1. *(See [15], [20, Section 5.1], or Corollary 3 in our Appendix.) GENP is safe if and only if the input matrix is strongly non-singular.*

Suppose A is a non-singular, but not necessarily strongly non-singular $n \times n$ matrix, and seek a non-singular $n \times n$ matrix H such that the *pre-processed* matrix AH is strongly non-singular. If it is strongly non-singular, then we can safely apply to it GENP, output the inverse matrix $M = (AH)^{-1}$ or the solution $M\mathbf{f}$ to the linear system $AH\mathbf{y} = \mathbf{f}$, and readily recover the inverse $A^{-1} = HM$ or the solution $\mathbf{x} = H\mathbf{y}$ to the linear system $A\mathbf{x} = \mathbf{f}$. One can similarly use pre-processing $A \to FA$ or $A \to FAH$ where F is a non-singular $n \times n$ matrix.

2.2   Pre-processing by Means of Symmetrization
or Random Sampling

Hereafter A^H denotes the Hermitian transpose of a matrix A, which is just the usual transpose A^T of the matrix A if it is real. The matrices $A^H A$ and AA^H are Hermitian non-negative definite and, therefore, strongly non-singular if the matrix A is a complex, real or rational non-singular matrix (cf. [15]), but this does not work in finite fields; moreover, the maps $A \to A^H A$ and $A \to AA^H$ square the condition number, and this should be avoided in numerical computations with rounding errors.

We assume a non-singular matrix A and seek alternative pre-processing $A \rightarrow AH$ with a random matrix H such that GENP is expected to be safe for the matrix AH. Here we assume that the entries of the matrix H are linear combinations of finitely many independent identically distributed (hereafter referred to as *i.i.d.*) random variables, which are either standard Gaussian random variables (hereafter referred to just as *Gaussian*) or have their values uniformly sampled from a fixed finite set, that is, selected under the uniform probability distribution on that set. We call a matrix *Gaussian* if it is filled with independent Gaussian variables.

Theorem 2. *Assume a fixed $n \times n$ matrix A and an $n \times n$ matrix H whose entries are linear combinations of finitely many i.i.d. variables. Let $\det((AH)_{l,l})$ vanish identically in them for neither of the integers l, $l = 1, \ldots, n$.*

(i) If the values of the variables have been sampled uniformly at random from a set S of cardinality $|S|$, then the matrix $(AH)_{l,l}$ is singular with a probability at most $l/|S|$, for any l, and the matrix AH is strongly non-singular with a probability at least $1 - 0.5(n-1)n/|S|$.

(ii) If these i.i.d. variables are Gaussian, then the matrix AH is strongly non-singular with probability 1.

Proof. Part (i) of the theorem readily extends a celebrated lemma of [11], also known from [29] and [28]. The extension is specified, e.g., in [26]. Part (ii) follows because the equation $\det((AH)_{l,l})$ for any integer l in the range $[1, n]$ defines an algebraic variety of a lower dimension in the linear space of the input variables.

2.3 f-Circulant, Gaussian f-Circulant, and Toeplitz Matrices

In order to decrease the cost of sampling the matrix H and of multiplication by it, we choose it to be f-circulant and prove in the next sections that the assumptions of Theorem 2 hold for this matrix, provided that the matrix A is non-singular.

At first we recall some definitions. For a positive integer n and a complex scalar f, define the unit f-circulant matrix Z_f of size $n \times n$ as follows,

$$Z_f = \begin{pmatrix} 0 & \cdots & & \cdots & 0 & f \\ 1 & \ddots & & & 0 & 0 \\ \vdots & \ddots & \ddots & & \vdots & \vdots \\ \vdots & & \ddots & \ddots & 0 & 0 \\ 0 & \cdots & & \cdots & 1 & 0 \end{pmatrix}.$$

Note that $Z^n = fI$ for I denoting the identity matrix. An *f-circulant matrix*

$$Z_f(\mathbf{v}) = \sum_{i=0}^{n} v_i Z_f^i = \begin{pmatrix} v_0 & f v_{n-1} & \cdots & f v_1 \\ v_1 & v_0 & \ddots & \vdots \\ \vdots & \ddots & \ddots & f v_{n-1} \\ v_{n-1} & \cdots & v_1 & v_0 \end{pmatrix},$$

of size $n \times n$, is defined by its first column $\mathbf{v} = (v_i)_{i=0}^{n-1}$. $Z_f(\mathbf{v})$ is a lower triangular Toeplitz matrix for $f = 0$, *circulant* for $f = 1$, and skew-circulant for $f = -1$. We call an f-circulant matrix a *Gaussian f-circulant* (or just *Gaussian circulant* if $f = 1$) if its first column is filled with independent Gaussian variables. For every fixed f, the f-circulant matrices form an algebra in the linear space of $n \times n$ *Toeplitz matrices*

$$(t_{i-j})_{i,j=0}^{n-1} = \begin{pmatrix} t_0 & t_{-1} & \cdots & t_{1-n} \\ t_1 & t_0 & \ddots & \vdots \\ \vdots & \ddots & \ddots & t_{-1} \\ t_{n-1} & \cdots & t_1 & t_0 \end{pmatrix}. \tag{1}$$

Definition 1. *(i) Write $\omega = \exp(\frac{2\pi\sqrt{-1}}{n})$, $\Omega = (\omega^{ij})_{i,j=0}^{n-1}$, $\Omega^{-1} = \frac{1}{n}\Omega^H$, ω denotes a primitive n-th root of unity, Ω and Ω^{-1} denote the matrices of the discrete Fourier transform at n points and its inverse, respectively (hereafter referred to as DFT(n) and inverse DFT(n), respectively).*

(ii) Furthermore write $\mathbf{u} = (u_i)_{i,j=0}^{n-1}$, and $D(\mathbf{u}) = \mathrm{diag}(u_0, \ldots, u_{n-1})$, that is, $D(\mathbf{u})$ is the diagonal matrix with the diagonal entries $u_0, \ldots, u_{n-1}$

Remark 1. If $n = 2^k$ is a power of 2, we can apply the FFT algorithm and perform DFT(n) and inverse DFT(n) by using only $1.5n \log_2(n)$ and $1.5n \log_2(n) + n$ arithmetic operations, respectively, but if $2^{k-1} < n \le 2^k$ we can embed the matrix $\Omega = \Omega_n$ into the matrix Ω_{2^k} and obtain the vector $\Omega_n(v_i)_{i=0}^{n-1}$ readily from the vector $\Omega_{2^k}(v_i)_{i=0}^{2^k-1}$ where $v_i = 0$ for $i \ge n$, and similarly for Ω_n^{-1}.

Theorem 3. *(Cf. [9].) If $f \ne 0$, then f^n-circulant matrix $Z_{f^n}(\mathbf{v})$ of size $n \times n$ can be factored as follows,*

$$Z_{f^n}(\mathbf{v}) = U_f^{-1} D(U_f \mathbf{v}) U_f \text{ for } U_f = \Omega D(\mathbf{f}), \quad \mathbf{f} = (f^i)_{i=0}^{n-1}, \text{ and } f \ne 0.$$

In particular, for circulant matrices, $D(\mathbf{f}) = I$, $U_f = \Omega$, $Z_1(\mathbf{v}) = \Omega^{-1} D(\Omega \mathbf{v}) \Omega$.

The theorem implies that for $f \ne 0$, one can multiply an $n \times n$ f-circulant matrix by a vector by applying two DFT(n), an inverse DFT(n), and additionally n multiplications and $2\delta_f n$ multiplications and divisions where δ_f is 0 if $f = 1$ and is 1 otherwise. We cannot apply this theorem directly to 0-circulant, that is, a triangular Toeplitz matrix, but we can multiply this matrix by a vector at a double cost because we can represent such a matrix as the sum of a circulant matrix and a skew-circulant matrix.

Remark 2. Assume a Gaussian circulant matrix $Z_1(\mathbf{v})$. Its first column vector $\mathbf{v}$ is Gaussian, but then so is also the vector $\mathbf{u} = (u_i)_{i=1}^{n} = \frac{1}{\sqrt{n}}\Omega \mathbf{v}$, which defines the matrix $Z_1(\mathbf{v})$ by the diagonal matrix $D = \mathrm{diag}(u_i)_{i=1}^{n}$ of Theorem 3.

3   Pre-processing with an f-Circulant Multiplier

Theorem 4. *Suppose* $A = (a_{i,j})_{i,j=1}^n$ *is a non-singular matrix,* $T = (t_{i-j+1})_{i,j=1}^n$ *is a Gaussian f-circulant matrix,* $B = AT = (b_{i,j})_{i,j=1}^n$, f *is a fixed complex number,* $t_1, \ldots, t_n$ *are variables, and* $t_k = ft_{n+k}$ *for* $k = 0, -1, \ldots, 1-n$. *Let* $B_{l,l}$ *denote the l-th leading sub-matrices of the matrix B for* $l = 1, \ldots, n$, *and so* $\det(B_{l,l})$ *are polynomial in* $t_1, \ldots, t_n$, *for all* l, $l = 1, \ldots, n$. *Then neither of these polynomials vanishes identically in* $t_1, \ldots, t_n$.

Proof. Fix a positive integer $l \leq n$. With the convention $\alpha_{k \pm n} = f\alpha_k$ for $k = 1, \cdots, n$, we can write

$$B_{l,l} = \left(\sum_{k_1=1}^n \alpha_{k_1} t_{k_1}, \sum_{k_2=1}^n \alpha_{k_2+1} t_{k_2}, \ldots, \sum_{k_l=1}^n \alpha_{k_l+l-1} t_{k_l} \right), \qquad (2)$$

where α_j is the jth column of $A_{l,n}$. Let $a_{i,j+n} = fa_{i,j}$, for $k = 1, \cdots, n$, and readily verify that

$$b_{i,j} = \sum_{k=1}^n a_{i,j+k-1} t_k,$$

and so $\det(B_l)$ is a homogeneous polynomial in $t_1, \ldots, t_n$.

Now Theorem 4 is implied by the following lemma.

Lemma 1. *If* $\det(B_{l,l}) = 0$ *identically in all the variables* $t_1, \ldots, t_n$, *then*

$$\det(\alpha_{i_1}, \alpha_{i_2}, \ldots, \alpha_{i_l}) = 0 \qquad (3)$$

for all l-tuples of sub-scripts $(i_1, \ldots, i_l)$ *such that* $1 \leq i_1 < i_2 < \cdots < i_l \leq n$.

Indeed let $A_{l,n}$ denote the sub-matrix made up of the first l rows of A. Note that if (3) holds for all l-tuples of the sub-scripts $(i_1, \ldots, i_l)$ above, then the rows of the sub-matrix $A_{l,n}$ are linearly dependent, but they are the rows of the matrix A, and their linear dependence contradicts the assumption that the matrix A is non-singular.

In the rest of this section, we prove Lemma 1. At first we order the l-tuples $I = (i_1, \ldots, i_l)$, each made up of l positive integers written in non-decreasing order, and then we apply induction.

We order all l-tuples of integers by ordering at first their largest integers, in the case of ties by ordering their second largest integers, and so on.

We can define the classes of these l-tuples up to permutation of their integers and congruence modulo n, and then represent every class by the l-tuple of non-decreasing integers between 1 and n. Then our ordering of l-tuples of ordered integers takes the following form, $(i_1, \ldots, i_l) < (i_1', \ldots, i_l')$ if and only if there exist a sub-script j such that $i_j < i_j'$ and $i_k = i_k'$ for $k = j+1, \ldots, l$.

We begin our proof of Lemma 1 with the following basic result.

Lemma 2. *It holds that*

$$\det(B_{l,l}) = \sum_{1 \le i_1 \le i_2 \le \cdots \le i_l \le n} a_{\prod_{j=1}^l t_{i_j}} \prod_{j=1}^l t_{i_j}$$

where a tuple $(i_1, \ldots, i_l)$ may contain repeated elements,

$$a_{\prod_{j=1}^l t_{i_j}} = \sum_{(i_1', \ldots, i_l')} \det(\alpha_{i_1'}, \alpha_{i_2'+1}, \ldots, \alpha_{i_l'+l-1}), \tag{4}$$

and $(i_1', \ldots, i_l')$ ranges over all permutations of $(i_1, \ldots, i_l)$.

Proof. By using (2) we can expand $\det(B_{l,l})$ as follows,

$$
\begin{aligned}
\det(B_{l,l}) &= \det\left(\sum_{k_1=1}^n \alpha_{k_1} t_{k_1}, \sum_{k_2=1}^n \alpha_{k_2+1} t_{k_2}, \ldots, \sum_{k_l=1}^n \alpha_{k_l+l-1} t_{k_l} \right) \\
&= \sum_{i_1=1}^n t_{i_1} \det\left(\alpha_{i_1}, \sum_{k=1}^n \alpha_{k+1} t_k, \ldots, \sum_{k_l=1}^n \alpha_{k_l+l-1} t_{k_l} \right) \\
&= \sum_{i_1=1}^n t_{i_1} \sum_{i_2=1}^n t_{i_2} \det\left(\alpha_{i_1}, \alpha_{i_2+1}, \sum_{k_2=1}^n \alpha_{k_2+2} t_{k_2}, \ldots, \sum_{k_l=1}^n \alpha_{k_l+l-1} t_{k_l} \right) \\
&= \cdots \\
&= \sum_{i_1=1}^n t_{i_1} \sum_{i_2=1}^n t_{i_2} \cdots \sum_{i_l=1}^n t_{i_l} \det(\alpha_{i_1}, \alpha_{i_2+1}, \ldots, \alpha_{i_l+l-1}). \tag{5}
\end{aligned}
$$

Hence the coefficient $a_{\prod_{j=1}^l t_{i_j}}$ of any term $\prod_{j=1}^l t_{i_j}$ is the sum of all determinants $\det(\alpha_{i_1'}, \alpha_{i_2'+1}, \ldots, \alpha_{i_l'+l-1})$ where $(i_1', \ldots, i_l')$ ranges over all permutations of $(i_1, \ldots, i_l)$, and we arrive at (4).

In particular, the coefficient of the term t_1^l is $a_{t_1 \cdot t_1 \cdots t_1} = \det(\alpha_1, \alpha_2, \ldots, \alpha_l)$. This coefficient equals zero because $B_{l,l}$ is identically zero, by assumption of lemma 1, and we obtain

$$\det(\alpha_1, \alpha_2, \ldots, \alpha_l) = 0. \tag{6}$$

This is the basis of our inductive proof of Lemma 1. In order to complete the induction step, it remains to prove the following lemma.

Lemma 3. *Let $J = (i_1, \ldots, i_l)$ be a tuple such that $1 \le i_1 < i_2 < \cdots < i_l \le n$.*
Then J is a sub-script tuple of the coefficient of the term $\prod_{j=1}^l t_{i_j-j+1}$ in equation (4).
Moreover, J is the single largest tuple among all sub-script tuples.

Proof. Hereafter $\det(\alpha_{i'_1}, \alpha_{i'_2+1}, \ldots, \alpha_{i'_l+l-1})$ is said to be *the determinant associated with the permutation* $(i'_1, \ldots, i'_l)$ of $(i_1, \ldots, i_l)$ in (4). Observe that $\det(\alpha_{i_1}, \ldots, \alpha_{i_l})$ is the determinant associated with $\mathcal{I} = (i_1, i_2 - 1, \ldots, i_l - l + 1)$ in the coefficient $a_{\prod_{j=1}^{l} t_{i_j - j + 1}}$.

Let $\mathcal{I}'$ be a permutation of $\mathcal{I}$. Then $\mathcal{I}'$ can be written as $\mathcal{I}' = (i_{s_1} - s_1 + 1, i_{s_2} - s_2 + 1, \ldots, i_{s_l} - s_l + 1)$, where $(s_1, \ldots, s_l)$ is a permutation of $(1, \ldots, l)$. The determinant associated with $\mathcal{I}'$ has the sub-script tuple $\mathcal{J}' = (i_{s_1} - s_1 + 1, i_{s_2} - s_2 + 2, \ldots, i_{s_l} - s_l + l)$. j satisfies the inequality $j \leq i_j \leq n - l + j$ because by assumption $1 \leq i_1 < i_2 < \cdots < i_l \leq n$, for any $j = 1, 2, \ldots, l$. Thus, $i_{s_j} - s_j + j$ satisfies the inequality $j \leq i_{s_j} - s_j + j \leq n - l + j \leq n$, for any s_j. This fact implies that no sub-script of $\mathcal{I}'$ is negative or greater than n.

Let $\mathcal{J}'' = (i_{s_{r_1}} - s_{r_1} + r_1, i_{s_{r_2}} - s_{r_2} + r_2, \ldots, i_{s_{r_l}} - s_{r_l} + r_l)$ be a permutation of $\mathcal{J}$ such that its elements are arranged in the non-decreasing order. Now suppose $\mathcal{J}'' \geq \mathcal{J}$. Then we must have $i_{s_{r_l}} - s_{r_l} + r_l \geq i_l$. This implies that

$$i_l - i_{s_{r_l}} \leq r_l - s_{r_l}. \tag{7}$$

Observe that

$$l - s_{r_l} \leq i_l - i_{s_{r_l}} \tag{8}$$

because $i_1 < i_2 < \cdots < i_l$ by assumption. Combine bounds (7) and (8) and obtain that $l - s_{r_l} \leq i_l - i_{s_{r_l}} \leq r_l - s_{r_l}$ and hence $r_l = l$.

Apply this argument recursively for $l - 1, \ldots, 1$ and obtain that $r_j = j$ for any $j = 1, \ldots, l$. Therefore, $\mathcal{J} = \mathcal{J}'$ and $\mathcal{I}' = \mathcal{I}$. It follows that J is indeed the single largest sub-script tuple.

By combining lemmas 2 and 3, we support the induction step of the proof of lemma 1, which we summarize as follows:

Lemma 4. *Assume the class of l-tuples of l positive integers written in the increasing order in each l-tuple and write* $\det(I) = \det(\alpha_{i_1}, \alpha_{i_2}, \ldots, \alpha_{i_l})$ *if* $I = (\alpha_{i_1}, \alpha_{i_2}, \ldots, \alpha_{i_l})$.
Then $\det(I) = 0$ *provided that* $\det(J) = 0$ *for all* $J < I$.

Finally we readily deduce Lemma 2 by combining this result with equation (6). This completes the proof of Theorem 4.

Corollary 1. *Assume any non-singular* $n \times n$ *matrix* A *and a finite set* $\mathcal{S}$ *of cardinality* $|\mathcal{S}|$. *Sample the values of the n coordinates* $v_1, \ldots, v_n$ *of a vector* $\mathbf{v}$ *at random from this set. Fix a complex f and define the matrix* $H = Z_f(\mathbf{v})$ *of size* $n \times n$, *with the first column vector* $\mathbf{v} = (v_i)_{i=1}^{n}$. *Then GENP is safe for the matrix* AH
(i) with a probability of at least $1 - 0.5(n-1)n/|\mathcal{S}|$ *if the values of the n coordinates* $v_1, \ldots, v_n$ *of a vector* $\mathbf{v}$ *have been sampled uniformly at random from a finite set* $\mathcal{S}$ *of cardinality* $|\mathcal{S}|$ *or*
(ii) with probability 1 if these coordinates are i.i.d. Gaussian variables.
(iii) The same claims hold for the matrices FA *and* $F = H^T$.

Proof. Theorems 1, 2, and 4 together imply parts (i) and (ii) of the corollary. By applying transposition, extend them to part (iii).

4 Does Random Circulant Pre-processing Make GENP Numerically Stable?

Assume that GENP is performed numerically, with rounding to a fixed precision, e.g., IEEE standard double precision. Then we can naturally extend the concept of safe GENP to *numerically stable GENP* by requiring that the input matrix be strongly non-singular and *strongly well-conditioned*, that is, by requiring that the matrix itself and all its square leading sub-matrices be non-singular and well-conditioned.

Let us introduce some definitions.

$||M|| = ||M||_2$ denotes the spectral norm of a matrix M.

M^+ denotes its Moore–Penrose generalized inverse, $M^+ = M^{-1}$ for non-singular matrices M.

$\kappa(M) = ||M|| \, ||M^+||$ denotes its condition number.

The matrix M is called *ill-conditioned* if its condition number $\kappa(M)$ is large in context or equivalently if a matrix of smaller rank lies in an $\epsilon||M||$-neighborhood where ϵ is small in context. Otherwise the matrix is called *well-conditioned*.

Any inversion algorithm for a non-singular matrix is highly sensitive to both input and rounding errors if and only if the matrix is ill conditioned.

Likewise GENP is highly sensitive to the input and rounding errors if and only if some leading blocks are ill-conditioned (cf. [24, Theorem 5.1]).

Next we recall some estimates for the norms and condition numbers of Gaussian and Gaussian circulant matrices.

Theorem 5. *(Cf. [12, Theorem II.7].) Suppose G is a Gaussian $m \times n$ matrix, $h = \max\{m, n\}$, $t \geq 0$, and $z \geq 2\sqrt{h}$. Then*
Probability$\{||G|| > z\} \leq \exp(-(z - 2\sqrt{h})^2/2)$ *and*
Probability$\{||G|| > t + \sqrt{m} + \sqrt{n}\} \leq \exp(-t^2/2)$.

Theorem 6. *Suppose G is a Gaussian $m \times n$ matrix, $m \geq n$, and $x > 0$. Write $\Gamma(x) = \int_0^\infty \exp(-t)t^{x-1}dt$ and $\zeta(t) = 2\, t^{m-1}(\frac{m}{2})^{m/2}\exp(-\frac{m}{2}t^2)/\Gamma(\frac{m}{2})$. Then*
Probability $\{||G^+|| \geq m/x^2\} < \frac{x^{m-n+1}}{\Gamma(m-n+2)}$ *for $n \geq 2$*

and in particular

Probability $\{||G^+|| \geq n/x^2\} < x/2$ *for $m = n \geq 2$.*

Proof. See [5, Proof of Lemma 4.1]. ∎

The two theorems combined imply that an $m \times n$ Gaussian matrix is very well-conditioned if $m - n$ is large or even moderately large, but can be considered well-conditioned even if $m = n$. These properties are immediately extended to the sub-matrices because they are also Gaussian.

Theorem 7. *(Cf. [25].) Suppose $C = \Omega^H D\Omega$ is a non-singular circulant $n \times n$ matrix, where $D = \mathrm{diag}(g_i)_{i=1}^n$. Then $||C|| = \max_{i=1}^n |g_i|$, $||C^{-1}|| = \min_{j=1}^n |g_j|$, and $\kappa(C) = \max_{i,j=1}^n |g_i/g_j|$,*

In the case of a Gaussian circulant matrix C, the g_i are i.i.d. Gaussian variables, and $\kappa(C) = \max_{i,j=1}^n |g_i/g_j|$ is not large with a probability close to 1. This property is not extended to sub-matrices, unlike the case of a Gaussian matrix.

Now we come to the following question. Would we ensure, with a probability close to 1, numerically stable GENP for a non-singular and well-conditioned matrix A if we apply to it Gaussian multiplier G or if we apply Gaussian circulant multiplier?

The extensive experiments in [24] and [23] suggest the answers "yes" to both questions, but formal support for this heuristics has been missing so far. [23, Corollaries 5.2 and 5.3] imply the positive answer to the first question, but a flaw in the proof of [23, Corollary 4.1] invalidates the claimed estimates. Next we readily fix the flaw by proving our next Theorem 8. This validates the results of [23], except that one should substitute $\nu_{k,k}^+$ for $\nu_{n,k}^+$ in part (iii) of [23, Corollary 5.2]. (For simplicity we assume dealing with real matrices, but one can readily relax this assumption.)

Lemma 5. *Suppose that*
G is a Gaussian matrix, $G \in \mathbb{R}^{m \times n}$,
S and T are orthogonal matrices,
$S \in \mathbb{R}^{k \times m}$, and $T \in \mathbb{R}^{n \times k}$ for some k, m, and n.
Then SG and GT are Gaussian matrices.

Theorem 8. *(Cf. [23, Corollary 5.2].)*
Assume a non-singular and well-conditioned $n \times n$ matrix A and an $n \times n$ Gaussian matrix G. Then
(i) the matrix AG is strongly non-singular with a probability 1,
(ii) $\|(AG_{k,k})\| \leq \|A_{k,n}\| \, \|G_{n,k}\| \leq \|A\| \, \|G_{n,k}\|$, and
(iii) $\|((AG)_{k,k})^+\| \leq \|A^+\| \, \|H^+\|$ where H is a $k \times k$ Gaussian matrix.

Proof. We readily verify parts (i) and (ii).

Towards proving part (iii), note that $(AG)_{k,k} = A_{n,k}G_{n,k}$. Let $A = S\Sigma T^H$ be the SVD of A, where the matrices S and T are orthogonal and Σ is diagonal. Hence $A_{n,k} = S_{n,k}\Sigma T^H$ and $(AG)_{k,k} = BG'$ where $B = S_{n,k}\Sigma$, $G' = T^H G_{n,k}$. G' is a Gaussian matrix by virtue of Lemma 5.

Let $B = S_B \Sigma_B T_B^H$ be SVD, and so $(AG)_{k,k} = FH$, where $F = S_B \Sigma_B$, $H = T_B^H G'$, S_B, Σ_B, and H are $k \times k$ matrices, S_B and T_B are orthogonal, $\Sigma_B = \mathrm{diag}(\sigma_j)_{j=1}^k$, $\sigma_1 = \|B\| \leq \|\Sigma\| = \|A\|$, $\sigma_k = 1/\|B^+\| \geq 1/\|A^+\| = 1/\|F^+\|$ (implying that $\|F^+\| = \|A^+\|$).

Hence F is a non-singular matrix, H is a Gaussian matrix by virtue of Lemma 5, $(AG)_{k,k}$ and H are non-singular with probability 1. Assume that they are non-singular and then $\|((AG)_{k,k})^+\| = \|((AG)_{k,k})^{-1}\| \leq \|H^{-1}\| \, \|F^{-1}\| = \|H^{-1}\| \, \|A^+\|$. This completes the proof because H is a $k \times k$ Gaussian matrix.

Theorems 5, 6, and 8 together imply that the matrix AG is strongly non-singular with probability 1 and is strongly well-conditioned with a probability near 1, and we arrive at the following result.

Corollary 2. *GENP is numerically stable with a probability near 1 if it is applied to a non-singular and well-conditioned matrix pre-processed with a Gaussian multiplier.*

Seeking further insight into the subject, let us reverse Theorem 8 and the corollary by assuming that the input matrix A is Gaussian and a multiplier G is just non-singular and well-conditioned. The same results follow about the application of GENP to the matrix AG. We can view them as supporting GENP with any non-singular and well-conditioned pre-processing for an average matrix A, but one can argue that a Gaussian matrix poorly represents an average matrix in practice. Indeed for a fixed pre-processor GENP may fail numerically at a set of hard inputs, which can be most interesting inputs for the current application.

One is challenged to select a small family of pre-processors with no common hard inputs for GENP, but this may be a hard problem. We conclude with providing some insight: we exhibit specific inputs, which are hard for GENP with Gaussian circulant multiplier C and some of their natural variations.

At first we recall from [21] that, with a probability close to 1, GENP is numerically unstable for the $n \times n$ unitary matrix $A = \frac{1}{\sqrt{n}}\Omega$ and consequently for the inverse matrix $A^{-1} = \frac{1}{\sqrt{n}}\Omega^H$ as well, provided that n is a large integer. Scaling by $1/\sqrt{n}$ does not affect the condition numbers, but turns Ω and Ω^H into unitary matrices. Of course, one does not need to apply GENP to invert these matrices, but the result in [21] enables us to reveal some hard inputs for GENP pre-processed with a Gaussian circulant multiplier.

The proof in [21] can be readily extended to cover also the inputs ΩR and $R^H \Omega^H$ for a random permutation matrix R. Then, by extending these results, we prove that with a probability close to 1, GENP remains numerically unstable for the matrices ΩC, $\Omega C R$, $C\Omega^H$, and $R^H C \Omega^H$ where C is a Gaussian matrix and n is a large integer.

Theorem 9. *Assume a large integer n, the $n \times n$ DFT matrix Ω, a random $n \times n$ permutation matrix R, and the circulant $n \times n$ matrix $C = \Omega^{-1}D\Omega$ with Gaussian diagonal matrix $D = \mathrm{diag}(g_j)_{j=1}^n$ (having i.i.d. Gaussian diagonal entries $g_1, \ldots, g_n$). Then, with a probability close to 1, application of GENP to the matrices ΩC, $\Omega C R$, $C\Omega^H$, and $R^H C \Omega^H$ is numerically unstable.*

Proof. Theorem 3 implies that $\Omega C = D\Omega$. Moreover the Gaussian diagonal matrix D is expected to be well-conditioned, and GENP is expected to be numerically unstable for the matrix Ω. Conclude that the latter property applies to the matrix $\Omega C = D\Omega$ as well, and similarly to the matrices $\Omega C R = D\Omega R$, $C\Omega^H = \Omega^H D$, and $R^H C \Omega^H = R^H \Omega^H D$.

Finally note that up to scaling by a constant $D\,\Omega\,R$ is an $n \times n$ SRFT matrix from [16, Section 11.1] whose $n \times l$ and $l \times n$ sub-matrices are extensively used in various randomized matrix computations, for $l < n$ and usually for $l \ll n$. Theorem 9 shows that GENP with an $n \times n$ SRFT multiplier fails numerically even for the identity input matrix.

Acknowledgements. This work has been supported by NSF Grant CCF 1116736 and PSC CUNY Award 67699-00 45. We are also grateful to the reviewers for helpful comments.

References

1. Bunch, J.R.: Stability of methods for solving Toeplitz systems of equations. SIAM J. Sci. and Stat. Computing **6**(2), 349–364 (1985)
2. Bini, D., Pan, V.Y.: Polynomial and Matrix Computations, Fundamental Algorithms, vol. 1. Birkhäuser, Boston (1994)
3. Bitmead, R.R., Anderson, B.D.O.: Asymptotically fast solution of Toeplitz and related systems of linear equations. Linear Algebra and Its Applications **34**, 103–116 (1980)
4. Ballard, G., Carson, E., Demmel, J., Hoemmen, M., Knight, N., Schwartz, O.: Communication lower bounds and optimal algorithms for numerical linear algebra. Acta Numerica **23**, 1–155 (2014)
5. Chen, Z., Dongarra, J.J.: Condition numbers of Gaussian random matrices. SIAM. J. Matrix Analysis and Applications **27**, 603–620 (2005)
6. Chen, L., Eberly, W., Kaltofen, E., Saunders, B.D., Turner, W.J., Villard, G.: Efficient matrix preconditioners for black box linear algebra. Linear Algebra and Its Applications **343–344**, 119–146 (2002)
7. Chan, R.H., Ng, M.K.: Conjugate gradient methods for Toeplitz systems. SIAM Review **38**, 427–482 (1996)
8. Chan, R.H., Ng, M.K.: Iterative methods for linear systems with matrix structures. In: Kailath, T., Sayed, A.H. (eds.) Fast Reliable Algorithms for Matrices with Structure, pp. 117–152. SIAM, Philadelpha (1999)
9. Cline, R.E., Plemmons, R.J., Worm, G.: Generalized inverses of certain Toeplitz matrices. Linear Algebra and Its Applications **8**, 25–33 (1974)
10. Demmel, J.: The probability that a numerical analysis problem is difficult. Math. Comput. **50**, 449–480 (1988)
11. Demillo, R.A., Lipton, R.J.: A probabilistic remark on algebraic program testing. Information Processing Letters **7**(4), 193–195 (1978)
12. Davidson, K.R., Szarek, S.J.: Local operator theory, random matrices, and Banach spaces. In: Johnson, W.B., Lindenstrauss, J. (eds.) Handbook on the Geometry of Banach Spaces, pp. 317–368. North Holland, Amsterdam (2001)
13. Edelman, A.: Eigenvalues and condition numbers of random matrices. SIAM J. Matrix Analysis and Applications **9**(4), 543–560 (1988)
14. Edelman, A., Sutton, B.D.: Tails of condition number distributions. SIAM J. Matrix Anal. and Applications **27**(2), 547–560 (2005)
15. Golub, G.H., Van Loan, C.F.: Matrix Computations, 4th edn. Johns Hopkins University Press, Baltimore (2013)
16. Halko, N., Martinsson, P.G., Tropp, J.A.: Finding structure with randomness: probabilistic algorithms for constructing approximate matrix decompositions. SIAM Review **53**(2), 217–288 (2011)
17. Kaltofen, E., Saunders, B.D.: On Wiedemann's method for solving sparse linear systems. In: Mattson, H.F., Mora, T., Rao, T.R.N. (eds.) AAECC 1991. LNCS, vol. 539, pp. 29–38. Springer, Heidelberg (1991)
18. Morf, M.: Fast Algorithms for Multivariable Systems. Ph.D. Thesis, Department of Electrical Engineering, Stanford University, Stanford, CA (1974)

19. Morf, M.: Doubling algorithms for toeplitz and related equations. In: Proc. IEEE International Conference on ASSP, pp. 954–959. IEEE Press, Piscataway (1980)
20. Pan, V.Y.: Structured Matrices and Polynomials: Unified Superfast Algorithms. Birkhäuser/Springer, Boston/New York (2001)
21. Pan, V.Y.: How Bad Are Vandermonde Matrices? Available at arxiv: 1504.02118, April 8, 2015 (revised April 26, 2015 and June 2015)
22. Pan, V.Y.: Transformations of matrix structures work again. Linear Algebra and Its Applications **465**, 1–32 (2015)
23. Pan, V.Y., Qian, G., Yan, X.: Random multipliers numerically stabilize Gaussian and block Gaussian elimination: Proofs and an extension to low-rank approximation. Linear Algebra and Its Applications **481**, 202–234 (2015)
24. Pan, V.Y., Qian, G., Zheng, A.: Randomized preprocessing versus pivoting. Linear Algebra and Its Applications **438**(4), 1883–1899 (2013)
25. Pan, V.Y., Svadlenka, J., Zhao, L.: Estimating the norms of circulant and Toeplitz random matrices and their inverses. Linear Algebra and Its Applications **468**, 197–210 (2015)
26. Pan, V.Y., Wang, X.: Degeneration of integer matrices modulo an integer. Linear Algebra and Its Applications **429**, 2113–2130 (2008)
27. Pan, V.Y., Zhao, L.: Gaussian Elimination with Randomized Pre-processing, arxiv, 1501.05385, June 15, 2015
28. Schwartz, J.T.: Fast probabilistic algorithms for verification of polynomial identities. J. ACM **27**(4), 701–717 (1980)
29. Zippel, R.E.: Probabilistic algorithms for sparse polynomials. In: Ng, E.W. (ed.) EUROSAM 1979. LNCS, vol. 72, pp. 216–226. Springer, Berlin (1979)

Appendix

A   Block Gaussian Elimination and GENP

Hereafter I_k denotes the $k \times k$ identity matrix, $O_{k,l}$ denotes the $k \times l$ matrix filled with zeros.

For a non-singular 2×2 block matrix $A = \begin{pmatrix} B & C \\ D & E \end{pmatrix}$ of size $n \times n$ with non-singular $k \times k$ *pivot block* $B = A_{k,k}$, define $S = S(A_{k,k}, A) = E - DB^{-1}C$, the *Schur complement* of $A_{k,k}$ in A, and the block factorizations,

$$A = \begin{pmatrix} I_k & O_{k,r} \\ DB^{-1} & I_r \end{pmatrix} \begin{pmatrix} B & O_{k,r} \\ O_{r,k} & S \end{pmatrix} \begin{pmatrix} I_k & B^{-1}C \\ O_{k,r} & I_r \end{pmatrix} \tag{9}$$

and

$$A^{-1} = \begin{pmatrix} I_k & -B^{-1}C \\ O_{k,r} & I_r \end{pmatrix} \begin{pmatrix} B^{-1} & O_{k,r} \\ O_{r,k} & S^{-1} \end{pmatrix} \begin{pmatrix} I_k & O_{k,r} \\ -DB^{-1} & I_r \end{pmatrix}. \tag{10}$$

We verify readily that S^{-1} is the $(n-k) \times (n-k)$ trailing (that is, southeastern) block of the inverse matrix A^{-1}, and so the Schur complement S is non-singular since the matrix A is non-singular.

Factorization (10) reduces the inversion of the matrix A to the inversion of the leading block B and its Schur complement S, and we can recursively reduce the task to the case of the leading blocks and Schur complements of decreasing sizes as long as the leading blocks are non-singular. After sufficiently many recursive steps of this process of block Gaussian elimination, we only need to invert matrices of small sizes, and then we can stop the process and apply a selected black box inversion algorithm.

In $\lceil \log_2(n) \rceil$ recursive steps, all pivot blocks and all other matrices involved into the resulting factorization turn into scalars, all matrix multiplications and inversions turn into scalar multiplications and divisions, and we arrive at a *complete recursive factorization* of the matrix A. If $k = 1$ at all recursive steps, then the complete recursive factorization (10) defines GENP.

Actually, however, any complete recursive factorizations turn into GENP up to the order in which we consider its steps. This follows because at most $n - 1$ distinct Schur complements $S = S(A_{k,k}, A)$ for $k = 1, \ldots, n - 1$ are involved in all recursive block factorization processes for $n \times n$ matrices A, and so we arrive at the same Schur complement in a fixed position via GENP and via any other recursive block factorization (9). Hence we can interpret factorization step (9) as the block elimination of the first k columns of the matrix A, which produces the matrix $S = S(A_{k,k}, A)$. If the dimensions $d_1, \ldots, d_r$ and $\bar{d}_1, \ldots, \bar{d}_{\bar{r}}$ of the pivot blocks in two block elimination processes sum to the same integer k, that is, if $k = d_1 + \cdots + d_r = \bar{d}_1 + \cdots + \bar{d}_{\bar{r}}$, then both processes produce the same Schur complement $S = S(A_{k,k}, A)$. The following results extend this observation.

Theorem 10. *In the recursive block factorization process based on (9), every diagonal block of every block diagonal factor is either a leading block of the input matrix A or the Schur complement $S(A_{h,h}, A_{k,k})$ for some integers h and k such that $0 < h < k \leq n$ and $S(A_{h,h}, A_{k,k}) = (S(A_{h,h}, A))_{h,h}$.*

Corollary 3. *The recursive block factorization process based on equation (9) can be completed by involving no singular pivot blocks (and in particular no pivot elements vanish) if and only if the input matrix A is strongly non-singular.*

Proof. Combine Theorem 10 with the equation $\det A = (\det B) \det S$, implied by (9).

Symbolic Computation and Finite Element Methods
(Invited Talk)

Veronika Pillwein*

RISC, Johannes Kepler University, Linz, Austria
veronika.pillwein@risc.jku.at
https://www.risc.jku.at/home/vpillwei

Abstract. During the past few decades there have been many examples where computer algebra methods have been applied successfully in the analysis and construction of numerical schemes, including the computation of approximate solutions to partial differential equations. The methods range from Gröbner basis computations and Cylindrical Algebraic Decomposition to algorithms for symbolic summation and integration. The latter have been used to derive recurrence relations for efficient evaluation of high order finite element basis functions. In this paper we review some of these recent developments.

Keywords: computer algebra, symbolic summation, cylindrical algebraic decomposition, finite element methods.

1 Overview

Many problems in science and engineering are described by partial differential equations on non-trivial domains which – except in special cases – can not be solved analytically. Numerical methods such as finite difference methods (FDM) or finite element methods (FEM) are used to solve these equations. In the past decades, FEM have become the most popular tools for obtaining solutions of partial differential equations on complicated domains [9],[12,13]. The main advantage of finite element methods is their general applicability to a huge class of problems, including linear and nonlinear differential equations, coupled systems, varying material coefficients and boundary conditions.

Symbolic computation has in the last decades gained importance concerning ease of use and range of applicability in part due to the availability of bigger and faster computers and the development of efficient algorithms. There are hybrid symbolic-numeric algorithms for computing validated results, but there are also examples of collaborations where symbolic methods have been used to analyze numerical schemes, to develop new ones, and to simplify or speed up existing ones. We begin by giving a brief overview on some of these cooperations.

---

* Supported by the Austrian FWF grants F50-07 and W1214.

© Springer International Publishing Switzerland 2015
V.P. Gerdt et al. (Eds.): CASC 2015, LNCS 9301, pp. 376–390, 2015.
DOI: 10.1007/978-3-319-24021-3_28

In many numerical methods, large scale linear systems of equations need to be solved and one popular technique for speeding up the iterative solution process is using preconditioning. For this preconditioning matrices that are spectrally close to the given matrix need to be constructed. Langer, Schicho, and Reitzinger [24] carried this out by symbolically solving the optimization problem arising in this construction. For this type of problems Cylindrical Algebraic Decomposition (CAD) as well as Gröbner bases are applicable (in theory), but practically the issue of computational complexity has to be circumvented.

Levandovskyy and Martin [25] applied CAD to derive von Neumann stability of a given difference scheme. They also presented symbolic approaches to derive a finite difference scheme for a single PDE, where they focused on linear PDEs with constant coefficients as well as the computation of dispersion relations of a linear PDE in a symbolic, algorithmic manner. These tools were also implemented in the computer algebra system Singular.

CAD has been applied earlier by Hong, Liska, and Steinberg [18] in the analysis of (systems of) ordinary and partial differential-difference equations, where the necessary conditions for stability, asymptotic stability and well-posedness of the given systems were transformed into statements on polynomial inequalities using Fourier or Laplace transforms.

Very recently, Cluzeau, Dolean, Nataf and Quadrat [14,15] have used algebraic and symbolic techniques such as Smith normal forms and Gröbner basis techniques to develop new Schwarz-like algorithms and preconditioners for linear systems of partial differential equations.

High order FEM are often preferrable in the numerical solution of PDEs because of their fast convergence, but they require high computational effort. Hence, every simplification is welcome. In [2], two types of high order basis functions with good numerical properties were constructed. By means of symbolic summation algorithms recurrence relations for the fast evaluation of these basis functions were derived entirely automatically.

Koutschan, Lehrenfeld and Schöberl [23] were concerned with an efficient implementation of a numerical discretization of the time-domain Maxwell's equations. As part of this algorithm the tensor product structure of the basis functions and mixed difference-differential relations satisfied by the underlying orthogonal polynomials were exploited. Below we comment on this type of relations for the basis functions discussed next.

For the iterative solution of linear systems it is convenient, if the matrices are sparse. For different types of partial differential equations and the method of high order finite elements on simplices (i.e., triangles and tetrahedra), families of basis functions have been proposed [4] that yield sparse system matrices. The proof of sparsity was carried out using symbolic computation. We review some of these results and give some recurrences that can be used in fast computation of the matrix entries.

As another specific example we present symbolic local Fourier analysis (sLFA) below. Local Fourier analysis (LFA) has been introduced by Brandt [10] to analyze multigrid methods. Multigrid methods are common iterative solvers for the

large systems arising in FEM or FDM. LFA gives quantitative statements on
the methods under investigation, i.e., it leads to the determination of sharp convergence rates. Typically only convergence is proven for these methods and the
convergence rate is estimated using brute force numerical interpolation. In [27]
we introduced sLFA to derive a closed form upper bound for the convergence
rate of a model problem with a particular solver.

2   Brief Introduction to FEM

Before we give two specific examples of a symbolic-numeric cooperation we briefly
introduce the basic concept of FEM. Finite element methods are based on the
variational formulation of partial differential equations. Let us consider as an
example the following simple problem: given f, find u (both from appropriate
function spaces) s.t.

$$u(x) - \Delta u(x) = f(x),$$

in a given domain $\Omega \subseteq \mathbb{R}^d$, $d = 1, 2, 3$, where Δ denotes the standard Laplace
operator $\Delta u(x) = \sum_{j=1}^{d} \frac{\partial u}{\partial x_j}$. The PDE above is multiplied on both sides by a
smooth function v that vanishes at the boundary of Ω and then integrated over
the given domain. By partial integration (exploiting the compact support of v),
we obtain the variational formulation of the given PDE

$$\int_\Omega u(x)v(x)\,dx + \int_\Omega \nabla u(x) \cdot \nabla v(x)\,dx = \int_\Omega f(x)v(x), \qquad (2.1)$$

where ∇ denotes the gradient operator. In this variational problem the task is:
given f, find u (in a less restrictive function space) s.t. (2.1) is satisfied for all
test functions v, i.e., all smooth functions vanishing on the boundary of the given
domain.

For numerically determining an approximate solution to this problem, the
domain of interest is subdivided into simple geometrical objects such as triangles,
quadrilaterals, tetrahedra, or hexahedra. The approximate solution is expanded
in a (finite) basis of local functions $\{\phi_i\}_{i=1}^N$, each supported on a finite number
of elements in the subdivision. Continuing with our example (2.1), we replace

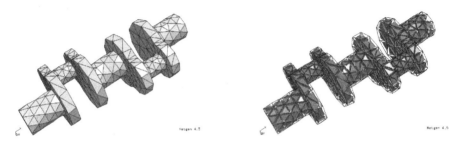

Fig. 1. FE mesh for a crankshaft, left: surface mesh; right: interior tetrahedral elements

the solution $u(x)$ by the approximate solution $u_h(x) = \sum_{i=1}^{N} u_i \phi_i(x)$ and use $\phi_j(x)$ as test functions. This yields the system of equations

$$\sum_{i=1}^{N} u_i \underbrace{\int_\Omega \left(\phi_i(x)\phi_j(x) + \nabla\phi_i(x) \cdot \nabla\phi_j(x) \right) \, dx}_{=:A_{i,j}} = \underbrace{\int_\Omega f(x)\phi_j(x)}_{=:b_j}$$

to be solved for the unknown coefficients u_i. If we define the matrix $A = (A_{i,j})_{i,j=1}^{N}$ and the vectors $b = (b_j)_{j=1}^{N}$, $u = (u_i)_{i=1}^{N}$, then the above can be written as the linear system $Au = b$ to be solved for u.

In general, the FE discretization of a PDE yields a (usually large) system of linear equations that is commonly solved using iterative methods. There are three main strategies to improve the accuracy of the approximate solution.

(i) The classical approach is to use on each element basis functions of a fixed low polynomial degree, say $p = 1, 2$, and to increase the number of elements in the subdivision. This strategy of local or global refinement of the mesh is called the h-version of the finite element method, where h refers to the diameter of the elements in the subdivision. With this approach, the approximation error decays algebraically (i.e., with polynomial rate) in the number of unknowns.

(ii) An alternative strategy is to keep the mesh fixed and to locally increase the polynomial degree p of the basis functions. This method is called the p-version of the finite element method [28,29] and in the case of a smooth solution leads to exponential convergence with respect to the number of unknowns. But in practical problems the solutions usually are not smooth. In this case the convergence rate of the p-method again degenerates to an algebraic one.

(iii) Exponential convergence can be regained by combining both strategies in the hp-version of the FEM [16],[19],[28]. On parts of the domain where the sought solution is smooth, few coarse elements with basis functions of high polynomial degrees are used, whereas in the presence of singularities (caused, e.g., by re-entrant corners) the polynomial degree is kept low and the mesh is refined locally towards the singularity. The p- and the hp-method are also referred to as high(er) order finite element methods.

FE basis functions are usually defined on some reference element and then transformed to the actual element in the mesh. In the case of one dimension the elements in the subdivision are just intervals and as reference element we choose $\hat{I} = [-1, 1]$.

We give an example for a hierarchic high order finite element basis, i.e., if the polynomial degree is increased, the set of basis functions is incremented keeping the previously basis functions, in contrast to nodal basis functions such as Lagrange polynomials where every time the whole set of basis functions needs to be replaced. The lowest order basis functions are the hat functions $\hat{\phi}_0(x) = \frac{1-x}{2}, \hat{\phi}_1(x) = \frac{1+x}{2}$ that are also referred to as vertex basis functions.

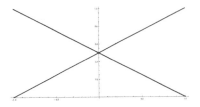 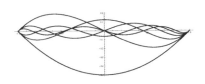

Fig. 2. FE basis functions in 1D, lowest order and higher order for degrees $i = 2, \ldots, 7$

They are the piecewise linear functions that are 1 on the defining vertex and vanish in all other vertices. The vertex basis functions are supported on the two neighbouring elements sharing the defining vertex. For the higher order basis functions of degree $i \geq 2$ a common choice for a conforming basis for our model problem are integrated Legendre polynomials. Different PDEs require different continuity of the given basis and for the present problem the basis functions are requried to be continuous across element interfaces. This property is guaranteed by construction. Let $P_n(x)$ denote the nth Legendre polynomial [1]. Then, for $n \geq 1$ the nth integrated Legendre polynomial $\hat{p}_n^0(x)$ is defined as

$$\hat{p}_n^0(x) := \int_{-1}^{x} P_{n-1}(s) \, ds.$$

Note that for $n \geq 2$, $\hat{p}_n^0(-1) = 0$ (for obvious reasons) and that $\hat{p}_n^0(1) = 0$ (by the Legendre $L^2(-1,1)$-orthogonality). The basis functions $\hat{\phi}_i(x) = \hat{p}_i^0(x)$ for $i \geq 2$ are supported on a single element and are also referred to as cell based basis functions.

In the definition of the hierarchic basis in two and three dimensions besides vertex and cell based basis functions in addtion edge and face based basis functions are constructed. Edge based basis functions are nonzero on the defining edge and vanish on all others and are supported on only those elements sharing the defining edge. Face based basis functions are nonzero on the defining face and vanish on all other faces and are supported on the two elements sharing the defining face. All basis functions are collected in the vector of basis functions $\phi = (\phi_1, \ldots, \phi_N)$. Then the system matrix A is already sparse because only some of the basis functions forming its entries share common support.

One reason, why integrate Legendre polynomials are used as basis functions is that they are orthogonal w.r.t. the inner product $\langle f, g \rangle = \int_{-1}^{1} f'(x)g'(x) \, dx$ and the L^2 inner product $\int_{-1}^{1} \hat{p}_i^0(x)\hat{p}_j^0(x) \, dx$ vanishes if $|i - j| \neq 0, 2$. Hence, the one-dimensional basis introduced above is constructed to yield an even sparser system matrix. For quadrilateral or hexahedral elements the basis functions are typically defined as tensor products of 1D basis functions and thus many of the sparsity properties can be inherited.

But for approximating complicated domains often triangular or tetrahedral elements are preferred. These elements are usually considered as collapsed quadrilateral or hexahedron. The construction of basis functions that yield sparse system matrices for these elements and that satisfy continuity requirements imposed by the underlying PDE (i.e., the underlying function spaces) is not obvious.

3   Sparsity Optimized Basis Functions

Beuchler and Schöberl [8] introduced triangular basis functions that yield sparse system matrices of the type (2.1). These basis functions are defined using integrated Jacobi polynomials [1]. Let $P_n^{(\alpha,\beta)}(x)$ denote the nth Jacobi polynomial. Then, we define the nth integrated Jacobi polynomial as

$$\hat{p}_n^{\alpha}(x) := \int_{-1}^{x} P_{n-1}^{(\alpha,0)}(s)\, ds.$$

Integrated Legendre polynomials are special cases of these polynomials as $P_n(x) = P_n^{(0,0)}(x)$. The sparsity of the system matrix has been proven by explicitly computing the matrix entries by exploiting relations between Jacobi polynomials of different parameters. It has been shown that the rows of the inner block of the system matrix have a constant number of nonzero matrix entries independent of the maximal polynomial degree.

An extension of this construction to tetrahedral elements was presented in [3]. Again the proof of sparsity of the system matrix was accomplished by explicitly computing the matrix entries. In three dimensions this was no longer feasible to be carried out by hand. Hence the computations were handed over to a computer program that evaluated the integrals with a rewriting procedure. In this process relations between Jacobi polynomials of different parameters were again exploited. Let us note that these relations can be discovered and proven using automatic tools for symbolic summation. In our work, we applied the packages Holonomic-Functions [22] and SumCracker [20], both implemented in Mathematica.

Subsequently, the construction of basis functions was extended to cover further function spaces (i.e., other types of PDEs) [6,7]. In all these cases the proof of sparsity was carried out using symbolic computation. The nonzero matrix entries of the different system matrices are rational functions in the polynomial degrees of the basis functions. The sparsity already eases handling the system matrices. Furthermore in [5] an algorithm was presented that computes the matrix entries in optimal complexity based on sum factorization and exploiting the nonzero pattern of the system matrix. In the remainder of this section we want to point out how the techniques of [23] can also be applied to the sparsity optimized basis functions and how recurrences can be used to quickly set up the system matrices.

As a specific example, we consider the following cell based basis functions defined on the triangular reference triangle $\hat{T}$ with vertices $(-1,-1),(1,-1),(0,1)$

$$\hat{\phi}_{i,j}(x,y) = \hat{p}_i^0 \left(\frac{2x}{1-y} \right) \left(\frac{1-y}{2} \right)^i \hat{p}_j^{2i}(y),$$

that are optimal for the example PDE stated in the previous section. In [23] cell based basis functions that are orthogonal on the reference triangle w.r.t. the L^2 inner product were used. In the algorithm to evaluate the gradients of the basis functions, relations between shifts of Jacobi polynomials and their derivatives (that are Jacobi polynomials with different parameters) were used. Furthermore, these relations are required to have coefficients that are independent of the variables x and y as they are to be turned into relations between integrals over products of these basis functions. Such relations can be computed automatically using Koutschan's package HolonomicFunctions. To facilitate computations, we use a different representation of the integrated Jacobi polynomials that relates them to classical Jacobi polynomials,

$$\hat{p}_n^{\alpha}(x) = \frac{1+x}{n} P_{n-1}^{(\alpha-1,1)}(x), \quad \hat{p}_m^0(x) = \frac{x^2-1}{2(m-1)} P_{m-2}^{(1,1)}(x),$$

valid for $n \geq 1$ and $m \geq 2$. Then, in order to obtain the corresponding mixed relations for $\hat{\phi}_{i,j}(x,y)$ as defined above, one proceeds as follows:

$_{\text{In[1]=}}$ **ann = Annihilator$[\hat{\phi}_{i,j}[x,y], \{S[i], S[j], Der[x], Der[y]\}]$**

Here $S[n]$ (and below S_n) denotes the forward shift in n and $\text{Der}[x]$ (and below D_x) the derivative w.r.t. x. This command quickly delivers the annihilating ideal for the given input by performing the necessary closure properties. To determine a relation of the type we need in this annihilating ideal we use the command "FindRelation":

$_{\text{In[2]=}}$ **FindRelation[ann, Eliminate $\rightarrow \{x,y\}$, Pattern $\rightarrow \{\text{-},\text{-},0 \mid 1,0\}]$**

$_{\text{Out[2]=}}$ $\{(2i+j+3)(2i+j+4)(2i+2j+3)S_i^2 S_j^2 D_x - (j+3)(j+4)(2i+2j+3)S_j^4 D_x - 4(j+1)(i+j+2)(2i+j+3)S_i^2 S_j \, D_x + 4(j+3)(i+j+2)(2i+j+1)S_j^3 D_x + j(j+1)(2i+2j+5)S_i^2 D_x - 2(2i+1)(i+j+2)(2i+2j+3)(2i+2j+5)S_i S_j^2 - (2i+j)(2i+j+1)(2i+2j+5) \, S_j^2 D_x\}$

$_{\text{In[3]=}}$ **rel = FindRelation[ann, Eliminate $\rightarrow \{x,y\}$, Pattern $\rightarrow \{\text{-},\text{-},0,0 \mid 1\}]$;**

$_{\text{In[4]=}}$ **Support[rel]**

$_{\text{Out[4]=}}$ $\{\{S_i^2 S_j^4 D_y, S_j^6 D_y, S_i^2 S_j^3 D_y, S_j^5 D_y, S_i^2 S_j^3, S_i^2 S_j^2 D_y, S_j^5, S_j^4 D_y, S_i^2 S_j^2, S_i^2 S_j D_y, S_j^4, S_j^3 D_y, S_i^2 S_j, S_i^2 D_y, S_j^3, S_j^2 D_y\}\}$

The option "Eliminate" specifies which variables are not to occur in the coefficients and the option "Pattern" defines the admissible exponents for the operators, i.e., in the first call above any exponent is allowed for the shift operators, whereas D_x may occur with power at most one and D_y must not be in the result altogether. For the second example we only display the support of the resulting relation for sake of space. For this example, the computation times are still neglible. The same procedure can be carried out for the tetrahedral basis functions; however the comuputational effort increases significantly.

Another way to speed up the computations is to use recurrence relations for the nonzero parts. We continue with the example from the previous section and consider the two integrals constituting the system matrix A separately. If

$$M_{ij;kl} = \int\int_{\hat{T}} \hat{\phi}_{i,j}(x,y)\hat{\phi}_{k,l}(x,y)\,d(x,y),$$

then the matrix built from these entries is nonzero only if $|i-k| \in \{0,2\}$ and $|i-k+j-l| \le 4$. In terms of the offset we have for $f(d,i,j,l) = M_{i,j;i+d,l}$ for $d = 0,2$ that

$$f(d,i,j,l) = \frac{2i+j+l-8+d}{2i+j+l+2+d}f(d,i,j-1,l-1)$$
$$+ \frac{j-l-5-d}{2i+j+l+2+d}f(d,i,j-1,l)$$
$$- \frac{j-l+5-d}{2i+j+l+2+d}f(d,i,j,l-1),$$

with different initial values. This recurrence was determined using multivariate guessing with Kauers' package Guess [21]. Since all matrix entries have been computed explicitly, the correctness of the recurrence can easily be verified.

For the second integral defining the system matrix A, the following integrals need to be computed:

$$K_{ij;kl}^{xx} = \int\int_{\hat{T}} \frac{d}{dx}\hat{\phi}_{i,j}(x,y)\frac{d}{dx}\hat{\phi}_{k,l}(x,y)d(x,y),$$

$$K_{ij;kl}^{yy} = \int\int_{\hat{T}} \frac{d}{dy}\hat{\phi}_{i,j}(x,y)\frac{d}{dy}\hat{\phi}_{k,l}(x,y)d(x,y).$$

The sparsity result for these matrices yields that $K_{ij;kl}^{xx}$ is nonzero only if $k = i$ and $|j-l| \le 2$, and, $K_{ij;kl}^{yy}$ is nonzero only if $|i-k| \in \{0,2\}$ and $|i-k+j-l| \le 2$ (i.e., in both cases we have nonzero only if $|i-k+j-l| \le 2$). With $g(d,i,j,l) = K_{i,j;i+d,l}^{\xi\zeta}$ either of the matrix entries above in terms of the corresponding offsets satisfies the recurrence

$$g(d,i,j,l) = \frac{2i+j+l-6+d}{2i+j+l+d}g(d,i,j-1,l-1) + \frac{j-l-3-d}{2i+j+l+d}g(d,i,j-1,l)$$
$$- \frac{j-l+3-d}{2i+j+l+d}g(d,i,j,l-1),$$

with different initial values. Let us also note that the integrals over the mixed products $\frac{d}{dx}\hat{\phi}_{i,j}(x,y)\frac{d}{dy}\hat{\phi}_{k,l}(x,y)$ (that enter in a more general PDE allowing for a coefficient matrix) satisfy the same recurrence.

4   Symbolic Local Fourier Analysis

The typically large linear systems that arise in FEM are usually solved approximately using some iterative scheme. The multigrid method is such an iterative solver and it operates on (at least) two grids in the finite element discretization. It has two main features: smoothing on the finer grid and error correction on the coarser grid. Intuitively speaking, the smoother is applied to dampen out the oscillatory part of the error. After applying a few such steps, the smooth part of the defect is dominant and the coarse-grid correction takes care of the low-frequency modes of the overall error.

In multigrid theory commonly convergence is proven, but neither sharp nor realistic bounds for convergence rates are given [17]. Local Fourier analysis [10] is a technique to analyze multigrid methods (and also various other numerical methods) that gives quantitative estimates, i.e., it leads to the determination of sharp convergence rates. Although it can be justified rigorously only in special cases such as rectangular domains with uniform grids and periodic boundary conditions, still results obtained with local Fourier analysis can be carried over rigorously to more general classes of problems [11].

In order to obtain a bound for the convergence rate, the supremum of some rational function needs to be computed. Usually this is merely resolved by numerical interpolation. With algorithms carrying out quantifier elimination such as CAD it is possible to determine an exact bound. This combination of classical local Fourier analysis and symbolic computation we refer to as symbolic local Fourier analysis (sLFA) [26,27]. The proposed approach is certainly applicable to different kinds of problems and different types of solvers. Next, we give an overview of the main results of [27].

The particular model problem that we consider is a PDE-constrained optimization problem which has to be discretized so that numerical methods can be applied. It is an optimal control problem of tracking type: given a desired state y_D and a regularization (or cost) parameter $\alpha > 0$, find a state y and a control u s.t. they minimize

$$J(y, u) := \frac{1}{2} \int_\Omega (y(x) - y_D(x))^2 \, dx + \frac{\alpha}{2} \int_\Omega u^2(x) \, dx,$$

subject to the elliptic boundary value problem

$$-\Delta y = u \quad \text{in} \Omega \quad \text{and} \quad y = 0 \quad \text{on} \, \partial\Omega,$$

where $\Omega \subseteq \mathbb{R}^2$ is a given domain with sufficiently smooth boundary. For the discretization we use FEM with rectangular elements and bilinear basis functions (i.e., low order FEM). As a solution method we apply a two-grid algorithm. The steps in the iteration consist of some ν_1 (pre-)smoothings steps, then the coarse grid correction, followed by some ν_2 (post-)smoothing steps. The iteration matrix for this can be written as

$$TG_k^{k-1} := S_k^{\nu_2} \underbrace{\left(I - P_{k-1}^k A_{k-1}^{-1} P_k^{k-1} A_k \right)}_{CG_k^{k-1} :=} S_k^{\nu_1}, \qquad (4.2)$$

where A_k and A_{k-1} denote the finite element system matrix on the finer and coarser level, respectively, P_{k-1}^k and $P_k^{k-1} = \left(P_{k-1}^k \right)^T$ are the intergrid transfer operators prolongation and restriction, I denotes the identity matrix, and S_k is the matrix of the smoothing step

$$S_k = I - \tau \hat{A}_k^{-1} A_k$$

with preconditioner $\hat{A}_k^{-1}$. The matrix CG_k^{k-1} above is the iteration matrix of the coarse grid correction, i.e., at the coarse grid an exact solve is performed. Summarizing, if $\bar{x}_k$ denotes the exact solution on the fine grid, and $x_k^{(n)}, x_k^{(n+1)}$ the nth and $n+1$st iterate, then

$$x_k^{(n+1)} - \bar{x}_k = TG_k^{k-1}\left(x_k^{(n)} - \bar{x}_k\right).$$

The convergence rate of the two-grid method can be bounded from above by the matrix norm of the iteration matrix, i.e.,

$$q \leq q_{TG} = \|TG_k^{k-1}\|.$$

This estimate is sharp if we consider the supremum over all possible starting values or, equivalently, all possible right-hand sides. If $q_{TG} < 1$ then the method converges for all starting values. If furthermore q_{TG} is independent of the mesh size and regularization parameter (or any other parameter) then the convergence is robust and optimal, i.e., the number of iterations does not depend on the parameters. In local Fourier analysis the iteration matrix is replaced by its symbol, where for our purpose it is sufficient to know that it is a matrix of fixed finite dimension. For the model problem at hand with the given discretization it is an 8×8 matrix. Instead of bounding the norm of the iteration matrix TG_k^{k-1} itself, the norm of the symbol is bounded, which can be expressed as the spectral radius of a certain matrix. For the full definitions of these matrices we refer (for sake of space) to [27] and its accompanying notebooks[1]. Up to this point we followed the steps of classical local Fourier analysis and ususally the bound for the spectral radius is approximated using numerical interpolation. After some suitable substitutions the entries of the symbol are rational functions and the supremum of the spectral radius can be computed explictely (at least in theory) using CAD.

As it turned out, only for the one-dimensional case, the spectral radius of the 4×4 two-grid iteration matrix could be estimated fully in an all-at-once approach. In two dimensions, the size of the entries of the involved matrices became prohibitively large. Hence, we proceed in two steps and treat the smoothing rate and the coarse grid correction separately and obtain this way an upper bound for the convergence rate. This task is computationally rather ambitious. Let $\mathcal{N}$ denote the matrix under discussion. It is a symmetric matrix that depends on four parameters $(c_1, c_2, \eta, q) \in [0,1]^2 \times (0,\infty)^2$ (the mesh parameter h and the regularization parameter α are hidden by substitution in the parameter η). After pulling out the common denominator

$$256 \left(16c_2^4 c_1^4 \eta + 16c_2^2 c_1^4 \eta + 4c_1^4 \eta + 16c_2^4 c_1^2 \eta + 16c_2^2 c_1^2 \eta + 4c_1^2 \eta + 4c_2^4 \eta + 4c_2^2 \eta \right.$$
$$\left. + 144c_2^4 c_1^4 - 72c_2^2 c_1^4 + 9c_1^4 - 72c_2^4 c_1^2 - 126c_2^2 c_1^2 + 36c_1^2 + 9c_2^4 + 36c_2^2 + \eta + 36 \right),$$

---

[1] Available for download at http://www.risc.jku.at/people/vpillwei/sLFA/

the matrix has polynomial entries of degrees up to 6 in c_1, c_2 and degrees up to 4 in q. Already the symbolic computation of the eigenvalues is not possible using the built-in functions of a computer algebra system. Since the matrix depends polynomially on q, it is possible to determine the eigenvalues using exact interpolation in q. This way we find that the eigenvalues are given by

$$0, \quad q^4, \quad \left(e(q) + \sqrt{d(q)}\right) \quad \text{and} \quad \left(e(q) - \sqrt{d(q)}\right)$$

with multiplicity two each, where e and d are rational functions in the unknowns c_1, c_2 and η that are too large to be displayed here. With the eigenvalues at hand it is readily verified that the largest eigenvalue in absolute value is $e(q) + \sqrt{d(q)}$. It remains to determine a bound for this eigenvalue uniformly in the parameters c_1, c_2, and η depending on the remaining variable q (the smoothing rate).

In theory, this bound can be determined entirely automatically using CAD. In practice, however, these computations become too complicated to be carried out in reasonable time. By considering the boundary of the parametric domain for (c_1, c_2, η) the following simple guess for this bound could be found:

$$q_{GUESS}(q) := \begin{cases} \frac{q^2 + 3}{4}, & 0 < q \leq Q, \\ q\sqrt{q^2 + 1}, & Q < q < 1. \end{cases}$$

The proof that this guess actually gives the true bound can be carried out with the aid of CAD, if the calculations are broken down to smaller pieces. This bound of the convergence rate is not sharp – in fact numerical experiments indicate much faster convergence. This is because the analysis had to be carried out in two steps. Nonetheless, the result gives quantitative statements on both, the choice of the parameter τ used in the smoothing iteration, and the number of smoothing steps ν that have to be applied. Such results cannot be obtained using a classical analysis based on smoothing and approximation property, even though the choice of the parameters is a key issue in the implementation of a numerical method.

It would be desirable to carry out sLFA all at once on the full iteration matrix. Even though this is currently out of scope of our computational capabilities, the two-step analysis carried out above hints on a heuristic procedure to determine the convergence rate of the full method depending on the damping parameter τ of the smoothing iteration.

Let $\sigma(c_1, c_2, \eta, \tau)$ be the spectral radius of the symbol (an 8×8 matrix) of the two-grid iteration matrix TG_k^{k-1} as defined in (4.2). Then in the all-at-once approach we need to bound $\sigma(c_1, c_2, \eta, \tau)$ uniformly in $(c_1, c_2, \eta) \in [0, 1]^2 \times (0, \infty)$, i.e., we need to compute

$$q_{TG}^2(\tau) = \sup_{(c_1, c_2) \in [0,1]^2} \sup_{\eta > 0} \sigma(c_1, c_2, \eta, \tau).$$

As the supremum is the least upper bound, $q_{TG}^2(\tau)$ is the smallest $\lambda(\tau)$ satisfying

$$\forall (c_1, c_2) \in [0, 1]^2 \, \forall \eta > 0 : \sigma(c_1, c_2, \eta, \tau) < \lambda.$$

Computing $\lambda(\tau)$ could be done by quantifier elimination using CAD entirely automatically, provided that the spectral radius is given (and an algebraic function). The symbol of the iteration matrix is non-symmetric with only few zero entries and again everything can be brought to a common denominator

$$16384(\eta + 36)^2 \left(16c_2^4c_1^4\eta + 16c_2^2c_1^4\eta + 4c_1^4\eta + 16c_2^4c_1^2\eta + 16c_2^2c_1^2\eta \right.$$
$$+ 4c_1^2\eta + 4c_2^4\eta + 4c_2^2\eta + 144c_2^4c_1^4 - 72c_2^2c_1^4 + 9c_1^4 - 72c_2^4c_1^2 - 126c_2^2c_1^2$$
$$\left. +36c_1^2 + 9c_2^4 + 36c_2^2 + \eta + 36\right).$$

The matrix entries are now polynomials of degrees either 9 or 10 in c_1, c_2, degrees either 2 or 3 in η, and degree 4 in τ. Plugging in specific values quickly shows that the general formula for the eigenvalues does not have a simple closed form – other than as the roots of the characteristic polynomial – as the symbol of the coarse grid correction above.

However, both the upper bound computed in the two step analysis as well as the sharp bound in an all-at-once approach for the one-dimensional case were obtained by considering limiting cases at the boundary of the parameter domain [27]. Hence, as heuristics, we propose to consider limiting cases of the symbol TG, compute the eigenvalues there using a computer algebra system, and determine their supremum using CAD. This procedure is backed up by the rigorous analysis carried out in the other cases.

First, if we consider the corner $(c_1, c_2) = (1, 1)$, the symbol degenerates to the singular diagonal matrix $\mathrm{diag}(0, 0, \xi_0, \xi_0, \xi_0, \xi_0, \xi_1, \xi_1)$, where

$$\xi_0 = \frac{\left(9\eta\tau^2 - 24\eta\tau + 16\eta + 1296\tau^2 - 1728\tau + 576\right)^2}{256(\eta + 36)^2},$$

$$\xi_1 = \frac{\left(\eta\tau^2 - 8\eta\tau + 16\eta + 576\tau^2 - 1152\tau + 576\right)^2}{256(\eta + 36)^2},$$

are the nonzero eigenvalues of the matrix. The upper bound for these eigenvalues can be determined easily by considering the limits $\eta \to 0, \infty$.

Secondly, we consider the values $(c_2, \eta) = (1, 0)$, for which the symbol is a singular diagonal matrix with the eigenvalues

$$0, \tfrac{1}{256}\left(c_1\tau - 5\tau + 4\right)^4, \tfrac{1}{256}\left(c_1\tau + 5\tau - 4\right)^4, \tfrac{1}{256}\left(9c_1^2\tau^2 + 9\tau^2 - 24\tau + 16\right)\left(9c_1^4\tau^2 \right.$$
$$\left. +54c_1^2\tau^2 - 72c_1^2\tau + 16c_1^2 + 9\tau^2 - 24\tau + 16\right).$$

The supremum of these eigenvalues for all $c_1 \in [0, 1]$ can then be computed exactly using CAD. Restricting the matrix to values of the remaining parts of the parametric domain does not yield any further information. So it turns out that the spectral radius can be bounded by taking the supremum of these two special cases. This way we obtain the following guess for the supremum of the norm:

$$q_{TG}(\tau) = \max\left\{\frac{1}{16}(\tau - 4)^2, \frac{1}{4}|3\tau - 2|\sqrt{9\tau^2 - 12\tau + 8}\right\}.$$

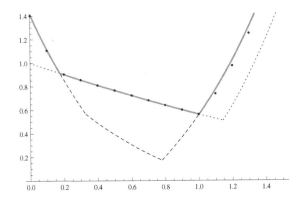

Fig. 3. Guessed bound (gray line) vs brute force interpolation Maxima (black points)

This guess was, as in the rigorous cases, obtained rather quickly. Compared to the large entries in the symbol, the final bound is again rather simple, but not quite of the form that would easily be used as an ansatz for numerical interpolation. Furthermore, since the symbolic bound was found at a point where a singularity occurs, these values cannot be recovered by brute force interpolation. This also becomes apparent when comparing the guessed bound to values obtained using a gridding approach for τ close to zero or τ much larger than one, see Figure 3. We propose to use this heuristic approach also for other PDEs in order to obtain a more precise insight on the dependence of the parameters.

References

1. Andrews, G.E., Askey, R., Roy, R.: Special Functions. Encyclopedia of Mathematics and its Applications, vol. 71. Cambridge University Press (2000)
2. Becirovic, A., Paule, P., Pillwein, V., Riese, A., Schneider, C., Schöberl, J.: Hypergeometric Summation Algorithms for High Order Finite Elements. Computing **78**(3), 235–249 (2006)
3. Beuchler, S., Pillwein, V.: Sparse shape functions for tetrahedral p-FEM using integrated Jacobi polynomials. Computing **80**(4), 345–375 (2007)
4. Beuchler, S., Pillwein, V., Schöberl, J., Zaglmayr, S.: Sparsity optimized high order finite element functions on simplices. In: Langer, U., Paule, P. (eds.) Numerical and Symbolic Scientific Computing: Progress and Prospects. Texts and Monographs in Symbolic Computation, pp. 21–44. Springer, Wien (2012)
5. Beuchler, S., Pillwein, V., Zaglmay, S.: Fast summation techniques for sparse shape functions in tetrahedral hp-FEM. In: Domain Decomposition Methods in Science and Engineering XX, pp. 537–544 (2012)
6. Beuchler, S., Pillwein, V., Zaglmayr, S.: Sparsity optimized high order finite element functions for H(div) on simplices. Numerische Mathematik **122**(2), 197–225 (2012)

7. Beuchler, S., Pillwein, V., Zaglmayr, S.: Sparsity optimized high order finite element functions for H(curl) on tetrahedra. Advances in Applied Mathematics **50**, 749–769 (2013)
8. Beuchler, S., Schöberl, J.: New shape functions for triangular p-fem using integrated Jacobi polynomials. Numer. Math. **103**, 339–366 (2006)
9. Braess, D.: Finite Elements: Theory, Fast Solvers and Applications in Solid Mechanics. Cambridge University Press, Cambridge (2007)
10. Brandt, A.: Multi-level adaptive solutions to boundary-value problems. Math. Comp. **31**, 333–390 (1977)
11. Brandt, A.: Rigorous Quantitative Analysis of Multigrid, I: Constant Coefficients Two-Level Cycle with L_2-Norm. SIAM J. on Numerical Analysis **31**(6), 1695–1730 (1994)
12. Brenner, S.C., Scott, L.R.: The Mathematical Theory of Finite Element Methods. Texts in Applied Mathematics, 2nd edn., vol. 15. Springer, New York (2002)
13. Ciarlet, P.G.: The Finite Element Method for Elliptic Problems. North Holland, Amsterdam (1978)
14. Cluzeau, T., Dolean, V., Nataf, F., Quadrat, A.: Symbolic preconditioning techniques for linear systems of partial differential equations. In: Proceedings of 20th International Conference on Domain Decomposition Methods. accepted for publ. (2012)
15. Cluzeau, T., Dolean, V., Nataf, F., Quadrat, A.: Symbolic methods for developing new domain decomposition algorithms. Rapport de recherche RR-7953, INRIA, May 2012
16. Demkowicz, L.: Computing with hp Finite Elements I. One- and Two-Dimensional Elliptic and Maxwell Problems. CRC Press, Taylor and Francis (2006)
17. Hackbusch, W.: Multi-Grid Methods and Applications. Springer, Berlin (1985)
18. Hong, H., Liska, R., Steinberg, S., Hong, H., Liska, R., Steinberg, S.: Testing stability by quantifier elimination. J. Symbolic Comput. **24**(2), 161–187 (1997). Applications of quantifier elimination, Albuquerque, NM (1995)
19. Karniadakis, G.E., Sherwin, S.J.: Spectral/hp Element Methods for CFD. Numerical Mathematics and Scientific Computation. Oxford University Press, New York, Oxford (1999)
20. Kauers, M.: SumCracker - A Package for Manipulating Symbolic Sums and Related Objects. J. Symbolic Comput. **41**(9), 1039–1057 (2006)
21. Kauers, M.: Guessing Handbook. RISC Report Series 09–07, Research Institute for Symbolic Computation (RISC), Johannes Kepler University Linz, Schloss Hagenberg, 4232 Hagenberg, Austria (2009)
22. Koutschan, Ch.: HolonomicFunctions (User's Guide). Technical Report 10–01, RISC Report Series, University of Linz, Austria, January 2010
23. Koutschan, C., Lehrenfeld, C., Schöberl, J.: Computer algebra meets finite elements: an efficient implementation for maxwell's equations. In: Numerical and Symbolic Scientific Computing: Progress and Prospects. Texts and Monographs in Symbolic Computation, vol. 1, pp. 105–121. Springer, Wien (2012)
24. Langer, U., Reitzinger, S., Schicho, J.: Symbolic methods for the element preconditioning technique. In: Winkler, F., Langer, U. (eds.) SNSC 2001. LNCS, vol. 2630, pp. 293–308. Springer, Heidelberg (2003)
25. Levandovskyy, V., Martin, B.: A symbolic approach to generation and analysis of finite difference schemes of partial differential equations. In: Langer, U., Paule, P. (eds.) Numerical and Symbolic Scientific Computing: Progress and Prospects. Texts & Monographs in Symbolic Computation, pp. 123–156. Springer, Wien (2011)

26. Pillwein, V., Takacs, S.: Smoothing analysis of an all-at-once multigrid approach for optimal control problems using symbolic computation. In: Langer, U., Paule, P. (eds.) Numerical and Symbolic Scientific Computing: Progress and Prospects. Springer, Wien (2011)
27. Pillwein, V., Takacs, S.: A local Fourier convergence analysis of a multigrid method using symbolic computation. J. Symbolic Comput. **63**, 1–20 (2014)
28. Schwab, C.: p- and hp-Finite Element Methods: Theory and Applications in Solid and Fluid Mechanics. Numerical Mathematics and Scientific Computation. Oxford University Press, Oxford (1998)
29. Szabó, B.A., Babuska, I: Finite Element Analysis. John Wiley & Sons (1991)

Approximate Quantum Fourier Transform and Quantum Algorithm for Phase Estimation

Alexander N. Prokopenya[1,2]

[1] Warsaw University of Life Sciences – SGGW
Nowoursynowska st. 159, 02-776 Warsaw, Poland
alexander_prokopenya@sggw.pl
[2] Collegium Mazovia Innovative Higher School,
Sokolowska st. 161, 08-110 Siedlce, Poland

Abstract. A quantum Fourier transform and its application to a quantum algorithm for phase estimation is discussed. It has been shown that the approximate quantum Fourier transform can be successfully used for the phase estimation instead of the full one. The lower bound for the probability to get a correct result in a single run of the algorithm has been obtained. The validity of the results is demonstrated by simulation of the algorithm in case of the phase shift operator using the QuantumCircuit package written in the Wolfram Mathematica language. All relevant calculations and visualizations are done with the Wolfram Mathematica system.

1 Introduction

It is known that some computational tasks being difficult for classical computer may be solved efficiently by a quantum computer, and quantum Fourier transform (QFT) plays an important role in the corresponding quantum algorithms (see [14,16,10,11]). An essential point is that QFT may be carried out efficiently by a quantum circuit constructed entirely out of the Hadamard gates and controlled phase shift gates R_k, $k = 2, 3, ..., n$, where n is a number of qubits in the circuit. A matrix representation of the R_k gate in the computational basis is given by

$$R_k = \begin{pmatrix} 1 & 0 \\ 0 & \exp(2\pi i/2^k) \end{pmatrix} , \qquad (1)$$

where i is the imaginary unit ($i^2 = -1$). Note that a value of the phase shift $2\pi/2^k$ in (1) becomes exponentially small when the number of qubits in the circuit grows up. As a result, high precision controlled phase shift gates are required for a practical implementation of QFT, although creation of such gates is a very difficult technical problem. So it would be very useful if the full Fourier transform could be replaced by the approximate QFT, where only finite degree phase shift gates R_k are involved ($k \leq m < n$). Obviously, in the last case, one can expect that an accuracy of computation decreases, and probability of getting a correct result reduces as well. Nevertheless, analysis of different quantum

© Springer International Publishing Switzerland 2015
V.P. Gerdt et al. (Eds.): CASC 2015, LNCS 9301, pp. 391–405, 2015.
DOI: 10.1007/978-3-319-24021-3_29

algorithms has shown that applying the approximate QFT can yield even better results than the full Fourier transform [3,5].

In the present paper, we consider a quantum algorithm for phase estimation based on QFT that was first proposed in [1] and is an essential part of many quantum algorithms [12,13]. Analysis of this algorithm shows that using the full Fourier transform enables one to estimate the phase accurate to 2^{-n} with a probability greater than $8/\pi^2$, where n is a number of qubits in the quantum memory register used for storing the approximate value of the phase. We have shown that decreasing the accuracy of phase estimation results in increasing the probability of the successful problem solution, and the lower bound on this probability was calculated [15]. The results obtained provide much better estimates than those given in [14].

We have analyzed also the quantum algorithm for phase estimation based on the approximate QFT and have shown that the probability to obtain a correct value of the phase accurate to 2^{-n} may be only a little bit smaller than in case of the full QFT but the number of the phase shift gates involved becomes much smaller. It shows that the approximate QFT may be used successfully instead of the full one if the size of the quantum circuit becomes large, similar results were obtained earlier in [4].

The validity of the results obtained is demonstrated using concrete computations for the case of the quantum phase shift operator. All calculations are done with the Wolfram Mathematica package "QuantumCircuit" that was designed for simulation of quantum computation (see [6,7,8,9]). The package provides a user-friendly interface to specify a quantum circuit, to draw it, and to construct the corresponding unitary matrix for quantum computation defined by the circuit. Using this matrix, one can find the final state of the quantum memory register by its given initial state and to check the operation of the algorithm determined by the quantum circuit.

The paper is organized as follows. In section 2, we describe briefly quantum circuits implementing the full and approximate quantum Fourier transform and define a function generating symbolic matrix that encodes information about the circuit, this function is used later for simulation of the approximate QFT. Section 3 is devoted to analysis of the quantum algorithm for phase estimation based of the full and approximate QFT. In section 4, we simulate the quantum algorithm for phase estimation and compare the results obtained in cases of the full and approximate QFT. And we conclude in section 5.

2   Quantum Fourier Transform

The quantum Fourier transform is defined as (see [14])

$$U_{FT}|x\rangle_n \rightarrow \frac{1}{2^{n/2}} \sum_{y=0}^{2^n-1} \exp\left(\frac{2\pi i}{2^n}xy\right)|y\rangle_n , \tag{2}$$

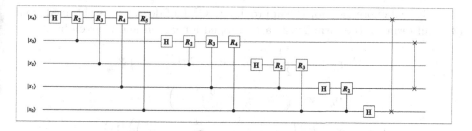

Fig. 1. Quantum circuit implementing the QFT ($n = 5$)

where $|x\rangle_n = |x_{n-1}\ldots x_1 x_0\rangle$ is the basis state of the n-qubit memory register, and the set of numbers $x_j = 0, 1$ ($j = 0, 1, \ldots, n - 1$) provides the binary representation of the integer x ($x = 0, 1, \ldots, 2^{n-1}$).

One can readily check that the right-hand side of (2) can be represented in the form

$$\frac{1}{2^{n/2}} \left(|0\rangle + \exp\left(2\pi i \frac{x_0}{2}\right)|1\rangle\right) \otimes \left(|0\rangle + \exp\left(2\pi i \left(\frac{x_1}{2} + \frac{x_0}{2^2}\right)\right)|1\rangle\right) \otimes \ldots \otimes$$

$$\left(|0\rangle + \exp\left(2\pi i \left(\frac{x_{n-1}}{2} + \ldots + \frac{x_1}{2^{n-1}} + \frac{x_0}{2^n}\right)\right)|1\rangle\right), \qquad (3)$$

where vectors $|0\rangle, |1\rangle$ denote the basis states of the qubits $|x_j\rangle$, and symbol $\otimes$ denotes a tensor product of vectors. This representation enables one to construct a quantum circuit implementing the QFT which can be performed efficiently. Actually, the corresponding quantum circuit contains only n single-qubit gates (the Hadamard gates **H**), $n(n-1)/2$ two-qubit controlled phase shift gates R_k, and $\lfloor n/2 \rfloor$ Swap gates (see Fig. 1, where the standard notations for quantum gates [14] are used).

Quantum circuit implementing the approximate QFT of degree m is obtained from the circuit implementing the full QFT if all controlled phase shift gates of degree greater than m ($m < n$) are dropped (see Fig. 2). Such a circuit contains only $(2n - m)(m - 1)/2$ controlled phase shift gates and so the execution time of the approximate QFT grows as nm instead of n^2 in case of the full QFT.

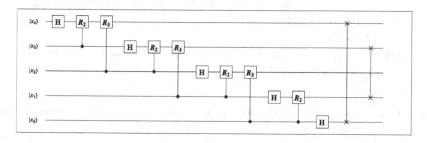

Fig. 2. Quantum circuit implementing the approximate QFT ($n = 5, m = 3$)

Analytical expression for the approximate QFT of degree m is obtained from (3) if all terms of the form $2\pi/2^k$ are dropped for $k > m$ and is given by

$$\tilde{U}_{FT}|x\rangle_n \rightarrow \frac{1}{2^{n/2}} \left(|0\rangle + \exp\left(2\pi i \frac{x_0}{2}\right)|1\rangle\right) \otimes$$

$$\left(|0\rangle + \exp\left(2\pi i \left(\frac{x_1}{2} + \frac{x_0}{2^2}\right)\right)|1\rangle\right) \otimes \ldots \otimes$$

$$\left(|0\rangle + \exp\left(2\pi i \left(\frac{x_{m-1}}{2} + \ldots + \frac{x_1}{2^{m-1}} + \frac{x_0}{2^m}\right)\right)|1\rangle\right) \otimes$$

$$\left(|0\rangle + \exp\left(2\pi i \left(\frac{x_m}{2} + \ldots + \frac{x_2}{2^{m-1}} + \frac{x_1}{2^m}\right)\right)|1\rangle\right) \otimes \ldots \otimes$$

$$\left(|0\rangle + \exp\left(2\pi i \left(\frac{x_{n-1}}{2} + \ldots + \frac{x_{n-m}}{2^m}\right)\right)|1\rangle\right) . \tag{4}$$

In the package "QuantumCircuit", all information about the quantum algorithm is stored in a symbolic matrix whose size is determined by the number of qubits and quantum gates in the circuit (see [6,7]). As a structure of the quantum circuit implementing the approximate QFT is known (see Fig. 2), we can easily define a function generating such a symbolic matrix which will be used later for simulation. The corresponding function **modelFourierApp[n,m]** written in the Wolfram Mathematica language is depicted in Fig. 3. Running a command **mat1 = modelFourierApp[5,3]**, we obtain the symbolic matrix which encodes the quantum circuit shown in Fig. 2. Using the built-in function **matrixU[mat1]**, we can easily compute the unitary matrix corresponding to the quantum circuit encoded by the matrix **mat1**, algorithms for computing such matrices are described in [8].

```
modelFourierApp[n_, m_] := Module[{model, mm, n1, m1},
    m1 = 0;
    Do[m1 = m1 + If[k ≤ m, k, m], {k, n}];
    model = Array[mm, {n, m1 + Floor[n / 2]}] /. mm[i_, j_] → 1;
        n1 = 0;
    Do[
        Do[ If[k == 1, (model[[j, n1 + 1]] = H; n1 = n1 + 1),
            If[k ≤ m, (model[[j, n1 + 1]] = R_k;
                    model[[j + k - 1, n1 + 1]] = C; n1 = n1 + 1)] ],
        {k, n - j + 1}],
        {j, n}];
    Do[ model[[j, n1 + j]] = SW; model[[n - j + 1, n1 + j]] = SW ,
        {j, Floor[n / 2]}];
    model ]
```

Fig. 3. Function generating a symbolic matrix which encodes the quantum circuit implementing the approximate QFT

3   Quantum Algorithm for Phase Estimation

Let us consider a unitary operator U that has the eigenvector $|u\rangle$ and the eigenvalue $\exp(2\pi i\varphi)$, where φ is an unknown real number, $0 \le \varphi < 1$. It is assumed that the quantum circuit implementing the operator U is given, and the quantum memory register can be set to the state $|u\rangle$. Obviously, action of the operator U on the state $|u\rangle$ results only in appearance of the phase factor $\exp(2\pi i\varphi)$ with the unit modulus

$$U|u\rangle = \exp(2\pi i\varphi)|u\rangle , \tag{5}$$

which cannot be observed if we measure the state $|u\rangle$. So the problem is to construct such a quantum circuit which enables us to obtain an information about the unknown phase φ.

3.1   Case of the Full QFT

The quantum algorithm for phase estimation based on the QFT [1] requires the second memory register that is used for storing an approximate value of the phase. At first, each of the n qubits of this memory register $|x_{n-1} \ldots x_0\rangle$ initially set to the state $|0\rangle$ is acted on by the Hadamard gate and is transferred to the state $\frac{1}{\sqrt{2}}(|0\rangle + |1\rangle)$. Using such a qubit as a control one for the controlled-U operator acting on the state $|u\rangle$, one can transfer the phase factor $\exp(2\pi i\varphi)$ to the control qubit while the state $|u\rangle$ doesn't change (see a quantum circuit in the left-hand side of Fig. 4). Applying a sequence of the controlled-U^{2^k} ($k = 0, 1, \ldots, n-1$) operators to the state $|u\rangle$ with the control qubits $|x_k\rangle$, respectively, we transform the memory register $|x_{n-1} \ldots x_0\rangle$ to the state $|\psi\rangle$ given by

$$|\psi\rangle = \frac{1}{2^{n/2}} \left(|0\rangle + \exp\left(2\pi i\varphi 2^{n-1}\right)|1\rangle\right) \otimes \left(|0\rangle + \exp\left(2\pi i\varphi 2^{n-2}\right)|1\rangle\right) \otimes \ldots \otimes$$

$$\left(|0\rangle + \exp\left(2\pi i\varphi 2^1\right)|1\rangle\right) \otimes \left(|0\rangle + \exp\left(2\pi i\varphi 2^0\right)|1\rangle\right) =$$

$$= \sum_{x_{n-1}=0}^{1} \ldots \sum_{x_0=0}^{1} \exp\left(2\pi i\varphi \left(x_{n-1}2^{n-1} + \ldots + x_0 2^0\right)\right)|x_{n-1} \ldots x_0\rangle . \tag{6}$$

The corresponding quantum circuit is depicted in the right-hand side of Fig. 4, where the sequence of the controlled-U^{2^k} operators is denoted as the controlled U_f gate and the n-qubit memory register is used for control.

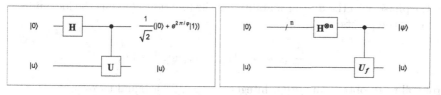

Fig. 4. Quantum circuits for transferring the phase factor to the control qubit (left) and preparing a state $|\psi\rangle$ for the phase estimation (right)

Further, the inverse quantum Fourier transform is applied to the state $|\psi\rangle$, it is given by (2), where the imaginary unit i is replaced by $-i$. Taking into account (3), one can write the final state of the second memory register in the form

$$U_{FT}^{-1}|\psi\rangle = \frac{1}{2^n} \sum_{y_{n-1}=0}^{1} \cdots \sum_{y_0=0}^{1} \left(1 + \exp\left(2\pi i(2^0\varphi - 0.y_{n-1}\ldots y_0)\right)\right) \times$$

$$\times \left(1 + \exp\left(2\pi i(2^1\varphi - 0.y_{n-2}\ldots y_0)\right)\right) \times \ldots \times \left(1 + \exp\left(2\pi i(2^{n-2}\varphi - 0.y_1 y_0)\right)\right) \times$$

$$\times \left(1 + \exp\left(2\pi i(2^{n-1}\varphi - 0.y_0)\right)\right) |y_{n-1}\ldots y_0\rangle , \tag{7}$$

where $0.y_{n-1}\ldots y_0 = y_{n-1}2^{-1} + \ldots + y_0 2^{-n}$, $\ldots$, $0.y_1 y_0 = y_1 2^{-1} + y_0 2^{-2}$, $0.y_0 = y_0 2^{-1}$ denote the corresponding binary fractions.

The final state (7) is a superposition of all vectors $|y_{n-1}\ldots y_0\rangle$ of the computational basis. Measuring this state, we obtain the number

$$y = y_{n-1}2^{n-1} + \ldots + y_1 2^1 + y_0 2^0 \tag{8}$$

with probability

$$p(y) = \cos^2\left(\pi(2^0\varphi - 0.y_{n-1}\ldots y_0)\right) \times \cos^2\left(\pi(2^1\varphi - 0.y_{n-2}\ldots y_0)\right) \times \ldots \times$$

$$\times \cos^2\left(\pi(2^{n-2}\varphi - 0.y_1 y_0)\right) \times \cos^2\left(\pi(2^{n-1}\varphi - 0.y_0)\right) =$$

$$= \frac{1}{2^{2n}} \frac{\sin^2\left(\pi(2^n\varphi - y)\right)}{\sin^2\left(\pi(\varphi - y/2^n)\right)} , \tag{9}$$

that is a square of an absolute value of the coefficient of $|y_{n-1}\ldots y_0\rangle$ in (7).

To estimate the probability of obtaining different values of y, consider the function

$$f(x) = \frac{1}{2^{2n}} \frac{\sin^2\left(\pi 2^n x\right)}{\sin^2\left(\pi x\right)} . \tag{10}$$

It takes the maximum value of 1 if x is an integer. In the case under consideration, the variable $x = \varphi - y/2^n$ takes values in the interval $-1 < x < 1$; therefore, $f(x)$ attains its principal maximum 1 at the unique point $x = 0$. If the phase φ can be represented by the n-bit binary fraction

$$\tilde{\varphi} = 0.\varphi_{n-1}\ldots\varphi_0 = \frac{\varphi_{n-1}}{2} + \frac{\varphi_{n-2}}{2^2} + \ldots + \frac{\varphi_1}{2^{n-1}} + \frac{\varphi_0}{2^n} , \tag{11}$$

then $x = 0$ for $y = 2^n\tilde{\varphi}$ and measuring the memory register gives correct value for the phase with the unit probability. As the function $f(x)$ has the local minima $f = 0$ at the points

$$x = \frac{k}{2^n} , \quad k = \pm 1, \pm 2, \ldots, \pm(2^n - 1) , \tag{12}$$

probability to obtain any other integer $0 \leq y < 2^n$ is equal to zero.

Between the points (12) of local minima, there are points of local maxima of the function $f(x)$ but its corresponding values are very small. Therefore, measuring the memory register in case of an arbitrary phase $\varphi = \tilde{\varphi} + \delta$, $0 \leq \delta < 2^{-n}$, we can obtain one of the two integers $y = 2^n\tilde{\varphi}$ or $y = 2^n\tilde{\varphi} + 1$ with high probability because only two points $x = \delta$ and $x = \delta - 1/2^n$ belong to the domain of the principal maximum of the function $f(x)$ in the neighborhood of the point $x = 0$. The lower bound for this probability is given by

$$p(2^n\tilde{\varphi}) + p(2^n\tilde{\varphi} + 1) = \frac{1}{2^{2n}}\left(\frac{\sin^2(\pi 2^n\delta)}{\sin^2(\pi\delta)} + \frac{\sin^2(\pi 2^n\delta)}{\sin^2(\pi(\delta - 1/2^n))}\right) \geq$$

$$\frac{1}{2^{2n}}\frac{2}{\sin^2(\pi/2^{n+1})} > \frac{8}{\pi^2} . \tag{13}$$

Thus, measuring the state (7) of the memory register, we can determine the phase accurate to 2^{-n} with a probability higher than $8/\pi^2$.

One can consider that if measuring the memory register gives one of the integers in the interval $(2^n\tilde{\varphi} - r + 1) \leq y \leq (2^n\tilde{\varphi} + r)$, where $r \geq 1$, then the problem is solved successfully but the accuracy of the measurements is reduced and is $|\varphi - \tilde{\varphi}| \leq r/2^n$. In this case, the probability of successful phase estimation improves, and it is determined by the relation [15]

$$p_r = \sum_{y=2^n\tilde{\varphi}-r+1}^{2^n\tilde{\varphi}+r} p(y) = \frac{\sin^2(\pi 2^n\delta)}{2^{2n}}\sum_{k=-r}^{r-1}\frac{1}{\sin^2(\pi(\delta + k/2^n))} \geq$$

$$\frac{8}{\pi^2}\left(1 + \frac{r-1}{3(2r+1)}\right) . \tag{14}$$

Thus, we have obtained the following result.

Theorem 1. *Let $\tilde{\varphi}$ be the best n-bit approximation to phase φ, $\tilde{\varphi} \leq \varphi$. The phase estimation algorithm based on the QFT returns one of the integers in the interval $(2^n\tilde{\varphi} - r + 1) \leq y \leq (2^n\tilde{\varphi} + r)$ with probability at least $\frac{8}{\pi^2}\left(1 + \frac{r-1}{3(2r+1)}\right)$, where r ia a positive integer characterizing an accuracy of the measurement.*

Note that the results (13), (14) for probability of successful solution of the problem provide better estimates than those given in [14]. Actually, in case of $r = 8$, for example, probability of success given by (14) is at least $p_r = 0.922$. The estimate obtained in [14] gives larger value of r

$$r = 2^{\lceil \log_2(2+1/(2(1-p_r)))\rceil} - 1 = 15 ,$$

what means a reduction in the phase estimation accuracy.

3.2   Case of the Approximate QFT

To obtain a final state of the memory register in case of application of the approximate QFT it is sufficient to apply the transformation (4), where the imaginary unit i is replaced by $-i$, to the state $|\psi\rangle$ defined by (6). Then the final state takes the form

$$\tilde{U}_{FT}^{-1}|\psi\rangle = \frac{1}{2^n} \sum_{y_{n-1}=0}^{1} \cdots \sum_{y_0=0}^{1} \left(1 + \exp\left(2\pi i(2^0\varphi - 0.y_{n-1}\cdots y_{n-m})\right)\right) \times$$

$$\times \left(1 + \exp\left(2\pi i(2^1\varphi - 0.y_{n-2}\cdots y_{n-m-1})\right)\right) \times \cdots \times$$
$$\left(1 + \exp\left(2\pi i(2^{n-m-1}\varphi - 0.y_m\cdots y_1)\right)\right) \times$$
$$\left(1 + \exp\left(2\pi i(2^{n-m}\varphi - 0.y_{m-1}\cdots y_0)\right)\right) \times \cdots \times$$
$$\times \left(1 + \exp\left(2\pi i(2^{n-1}\varphi - 0.y_0)\right)\right) |y_{n-1}\cdots y_0\rangle , \tag{15}$$

where $0.y_{n-k}\cdots y_{n-k-m+1} = y_{n-k}/2^1 + \ldots + y_{n-k-m+1}/2^m$, $k = 1, 2, \ldots, n - m + 1$ is the m-bit binary fraction.

Again the probability $p(y)$ of obtaining the integer y as a result of measuring the final state of the memory register is determined by the square of the absolute value of the coefficient of the corresponding vector $|y_{n-1}\cdots y_0\rangle$ in (15) and is given by

$$p(y) = \cos^2\left(\pi(2^0\varphi - 0.y_{n-1}\cdots y_{n-m})\right) \times$$
$$\cos^2\left(\pi(2^1\varphi - 0.y_{n-2}\cdots y_{n-m-1})\right) \times \cdots \times$$
$$\times \cos^2\left(\pi(2^{n-m-1}\varphi - 0.y_m\cdots y_1)\right) \times \cos^2\left(\pi(2^{n-m}\varphi - 0.y_{m-1}\cdots y_0)\right) \times \cdots \times$$
$$\cos^2\left(\pi(2^{n-2}\varphi - 0.y_1 y_0)\right) \times \cos^2\left(\pi(2^{n-1}\varphi - 0.y_0)\right) . \tag{16}$$

To analyze the function $p(y)$ it is convenient to rewrite (16) in the form

$$p(y) = \cos^2\left(\pi\left(2^0\left(\varphi - \frac{y}{2^n}\right) + \frac{1}{2^m} 0.y_{n-m-1}\cdots y_0\right)\right) \times$$
$$\cos^2\left(\pi\left(2^1\left(\varphi - \frac{y}{2^n}\right) + \frac{1}{2^m} 0.y_{n-m-2}\cdots y_0\right)\right) \times \cdots \times$$
$$\times \cos^2\left(\pi\left(2^{n-m-1}\left(\varphi - \frac{y}{2^n}\right) + \frac{1}{2^m} 0.y_0\right)\right) \times$$
$$\cos^2\left(\pi 2^{n-m}\left(\varphi - \frac{y}{2^n}\right)\right) \times \cdots \times \cos^2\left(\pi 2^{n-1}\left(\varphi - \frac{y}{2^n}\right)\right) , \tag{17}$$

where (8) has been taken into account.

If the phase can be represented by the n-bit binary fraction (11) then the probability to obtain the number $y = 2^n\tilde{\varphi} + k$, $-2^n\tilde{\varphi} \le k < 2^n(1 - \tilde{\varphi})$ by measuring the memory register is given by

$$p(y) = \cos^2\left(\pi\left(-\frac{k}{2^n} + \frac{1}{2^m} 0.\varphi_{n-m-1}\cdots\varphi_0\right)\right) \times \cdots \times$$
$$\cos^2\left(\pi\left(-\frac{k}{2^{m+1}} + \frac{1}{2^m} 0.\varphi_0\right)\right) \times \cos^2\left(-\frac{\pi k}{2^m}\right) \times \cdots \times \cos^2\left(-\frac{\pi k}{2}\right) =$$
$$\cos^2\left(\pi\left(-\frac{k}{2^n} + \frac{1}{2^m} 0.\varphi_{n-m-1}\cdots\varphi_0\right)\right) \times \cdots \times$$
$$\cos^2\left(\pi\left(-\frac{k}{2^{m+1}} + \frac{1}{2^m} 0.\varphi_0\right)\right) \times \frac{\sin^2(\pi k)}{2^{2m} \sin^2(\pi k/2^m)} . \tag{18}$$

And it is not equal to zero only if $k = 2^m l$, $l = 0, \pm 1, \pm 2, \ldots$. The probability (18) takes a maximum for $l = 0$ or $y = 2^n \tilde{\varphi}$ that is given by

$$p(2^n \tilde{\varphi}) = \cos^2 \left(\frac{\pi}{2^m} 0.\varphi_{n-m-1} \ldots \varphi_0 \right) \times \ldots \times \cos^2 \left(\frac{\pi}{2^m} 0.\varphi_0 \right) \geq$$

$$\left(\cos^2 \left(\frac{\pi}{2^m} \right) \right)^{n-m} . \tag{19}$$

Equation (19) shows clearly that the probability to obtain a correct result $y = 2^n \tilde{\varphi}$ can attain its maximum value of 1 only if either $m = n$ or the all terms φ_{n-k} in (11) are equal to zero for $m < k \leq n$, i.e., $\tilde{\varphi} = 0.\varphi_{n-1} \ldots \varphi_{n-m} 0 \ldots 0$. However, probability (19) remains quite high even in the worst case of $\tilde{\varphi} = 0.\varphi_{n-1} \ldots \varphi_{n-m} 1 \ldots 1$. For example, $p(2^n \tilde{\varphi}) > 0.992$ in case of $n = 20$ and $m \geq 7$ or $p(2^n \tilde{\varphi}) > 0.993$ in case of $n = 50$ and $m \geq 8$.

Note that probability $p(2^n \tilde{\varphi} + 2^m l)$, $(l \neq 0)$, to obtain other values y in case of $\varphi = \tilde{\varphi}$ is easily obtained from (18) and satisfies the following inequality

$$p(2^n \tilde{\varphi} + 2^m l) = \cos^2 \left(\pi \left(-\frac{l}{2^{n-m}} + \frac{1}{2^m} 0.\varphi_{n-m-1} \ldots \varphi_0 \right) \right) \times \ldots \times$$

$$\cos^2 \left(\pi \left(-\frac{l}{2} + \frac{1}{2^m} 0.\varphi_0 \right) \right) < \sin^2 \left(\frac{\pi}{2^m} \right) . \tag{20}$$

In case of $n = 20$, for example, this probability p doesn't exceed 0.0006 for $m \geq 7$ and $p < 0.00015$ for $n = 50$, $m \geq 8$. It should be emphasized that the lower bound (19) for probability to obtain correct result increases if both the number of qubits n and the degree of the approximate QFT m grow up but this increase is much slower in case of m growth in comparison to n.

Now consider a general case when the phase can be represented in the form $\varphi = \tilde{\varphi} + \delta$, $0 \leq \delta < 2^{-n}$ and $y = 2^n \tilde{\varphi} + k$. Then probability (17) takes the form

$$p(y) = \cos^2 \left(\pi \left(2^0 \left(\delta - \frac{k}{2^n} \right) + \frac{1}{2^m} 0.\varphi_{n-m-1} \ldots \varphi_0 \right) \right) \times \ldots \times .$$

$$\cos^2 \left(\pi \left(2^{n-m-1} \left(\delta - \frac{k}{2^n} \right) + \frac{1}{2^m} 0.\varphi_0 \right) \right) \times \frac{\sin^2(\pi(k - 2^n \delta))}{2^{2m} \sin^2(\pi(k - 2^n \delta)/2^m)} . \tag{21}$$

Analysis of (21) shows that probability to obtain one of the two integers $y = 2^n \tilde{\varphi}$ or $y = 2^n \tilde{\varphi} + 1$ by measuring the memory register changes in comparison to the case of the full QFT (see (13)), and the lower bound for this probability becomes

$$p(2^n \tilde{\varphi}) + p(2^n \tilde{\varphi} + 1) > \frac{8}{\pi^2} \left(\cos^2 \left(\frac{\pi}{2^m} \right) \right)^{n-m} . \tag{22}$$

Appearance of the cosine function in the right-hand side of (22) reduces the value of the probability to obtain correct result but not very much. For example, the factor of $8/\pi^2$ in (22) reaches the values 0.992 for $n = 20$, $m = 7$ and 0.993 for $n = 50$, $m = 8$ and increases if m grows up.

Note that the lower bound for the probability of the successful solution of the problem was estimated earlier in [3,4]. But estimation of [3] was quite rough, and it gives only $p \geq 0.119$ for $n = 10$, $m = 5$, for example. The estimate of [4] $p \geq 4/\pi^2 - 1/(4n)$ is much better but it takes into account only a possibility to obtain the value $2^n\tilde{\varphi}$ as a result of measurement while both results $2^n\tilde{\varphi}$ and $2^n\tilde{\varphi} + 1$ which are taken into account in (22) give a correct value of the phase φ accurate to 2^{-n}.

As in case of the full QFT, probability to obtain correct result improves if we reduce the accuracy of measurement and assume that the obtained number y is good if it belongs to the interval $(2^n\tilde{\varphi} - r + 1) \leq y \leq (2^n\tilde{\varphi} + r)$, where $r \geq 1$. Then the accuracy of the measurement is $|\varphi - \tilde{\varphi}| \leq r/2^n$ and the lower bound for the probability of successful phase estimation becomes

$$p_r = \sum_{y=2^n\tilde{\varphi}-r+1}^{2^n\tilde{\varphi}+r} p(y) > \frac{8}{\pi^2}\left(1 + \frac{r-1}{3(2r+1)}\right)\left(\cos^2\left(\frac{\pi}{2^m}\right)\right)^{n-m}. \tag{23}$$

Note that application of the approximate QFT instead of the full one results in reducing the probability of successful phase estimation. However, the factor $\left(\cos^2\left(\pi/2^m\right)\right)^{n-m} \leq 1$ reducing the lower bound of the probability approaches the value close to 1 quite fast if the degree of the approximate QFT m increases. As m grows very slowly in comparison to the number of qubits n in the circuit one can expect that the approximate QFT may be used successfully instead of the full QFT in the quantum algorithm for phase estimation.

4   Simulation

Using the package QuantumCircuit, one can easily simulate the quantum algorithm for phase estimation and demonstrate the validity of the theoretical results obtained above. As a unitary operator U we can consider the phase shift gate $R_z(\varphi)$; its matrix representation in the computational basis is

$$R_z(\varphi) = \begin{pmatrix} 1 & 0 \\ 0 & \exp(2\pi i\varphi) \end{pmatrix}, \tag{24}$$

and the vector $|u\rangle = |1\rangle$ is its eigenvector with the eigenvalue $\exp(2\pi i\varphi)$.

To prepare the state $|\psi\rangle$ (see (6) and Fig. 4) that will be acted on later by the inverse Fourier transform we construct a quantum circuit shown in Fig. 5. Symbolic matrix encoding such a circuit for an arbitrary number of qubits n is generated by the function $mat[n, \varphi]$ depicted in Fig. 6. Note that the circuit in Fig. 5 can be drawn by running the following command written in the Wolfram Mathematica language: $circuit[mat[4, \varphi], \{0, 0, 0, 0, 1\}, \{,,,\}]$.

Initial state $|0\ldots01\rangle$ of the quantum memory register shown in Fig. 5 can be represented by the 2^{n+1}-dimensional vector in which only the second component is 1, and the others are zeros. Such a vector is specified by the function $basisState[1, n + 1]$ depicted in Fig. 7. Then this vector is multiplied by the unitary matrix $matrixU[mat[n, \varphi]]$ corresponding to the quantum circuit shown in Fig. 5, and the result is denoted as $vec1$. Since the target qubit remains in

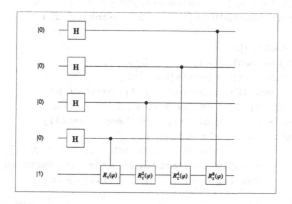

Fig. 5. The quantum circuit preparing the state $|\psi\rangle$ $(n = 4)$

```
mat[n_, φ_] := Block[{mat1},
        mat1 = Table[If[j == 1, H, If[j == n + 2 - i, C, 1] ],
                                {i, n}, {j, n + 1}];
        mat1 = Append[mat1, Table[If[j == 1, 1, R_z^(2^(j-2)) [φ]],
                        {j, n + 1}]];        mat1]
```

Fig. 6. The function generating the symbolic matrix encoding the quantum circuit preparing the state $|\psi\rangle$

the state $|1\rangle$, only the components of the vector $vec1$ with even indexes are not equal to zero, and they are extracted to determine the state $vec2$ of the memory register, where an approximate value of the phase will be written.

The calculations shown in Fig. 7 correspond to the case of $n = 10$, $m = 5$ and the phase $\varphi = 2/7$. Vector $vec2$ is multiplied by the unitary matrix corresponding to the approximate QFT of degree 5 encoded by the symbolic matrix that is generated by the function $modelFourierApp[10, 5]$. Then absolute values of the vector $vec3$ components are squared, and possible results of measurement of the memory register are written in the table $dat1$ together with the corresponding probabilities of their obtaining. Finally the table $dat1$ is visualized with the function $ListPlot$, the corresponding graph is depicted in Fig. 8.

One can readily see that the greatest probabilities are $p_1 = 0.294$ and $p_2 = 0.522$ of obtaining the numbers 292 and 293, respectively. It holds that

$$\frac{292}{2^{10}} < \varphi = \frac{2}{7} < \frac{293}{2^{10}}.$$

```
basisState[x_, n_] := SparseArray[x + 1 -> 1, 2^n]
vec1[n_] := matrixU[mat[n, φ]].basisState[1, n + 1] //
                                                   Normal

vec1a = vec1[10];
vec2 = Table[vec1a[[2 i]], {i, 2^10}];
vec3 = Transpose[Conjugate[matrixU[
            modelFourierApp[10, 5] ]]].vec2 // Normal ;
vec4 = N[Abs[#]^2] & /@ (vec3 /. φ → 2 / 7) // Chop;
dat1 = Table[{i - 1, vec4[[i]]}, {i, Length[vec4]}];
pb = ListPlot[dat1, PlotRange → All,
                        PlotStyle → {Black, PointSize[0.02]},
                    BaseStyle → {FontFamily → "Arial", FontSize → 12},
                    AxesLabel → {"k", "p"}, Filling → Axis]
```

Fig. 7. The Wolfram Mathematica commands for computing the final state (15) and visualizing the result, $n = 10$, $m = 5$, $\varphi = 2/7$

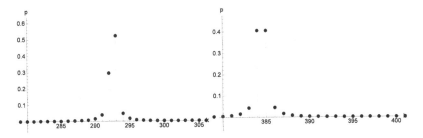

Fig. 8. Probabilities of obtaining different numbers k by measuring the final state for $\varphi = 2/7$ (left) and $\varphi = 769/2048$ (right) in case of $n = 10$, $m = 5$

Thus, the total probability to obtain a correct result accurate to 2^{-10} is $p_1 + p_2 = 0.816$ and is greater than the lower bound for the probability $8/\pi^2$ in case of the full QFT. Obviously, it is greater than the lower bound of the probability given in (22):

$$p_1 + p_2 > \frac{8}{\pi^2} \left(\cos^2\left(\frac{\pi}{2^m}\right)\right)^{n-m} = 0.772 .$$

Note that replacing the phase by $\varphi \to 769/2048$ in calculation of the vector *vec4* (see Fig. 7) and repeating the computation, we obtain the second graph shown in the right-hand side of Fig. 8. In this case, $\tilde{\varphi} = 3/8$, $\delta = 2^{-11}$, and two numbers 384 and 385 are obtained with equal probabilities $p_1 = p_2 = 0.405$. Both results give correct value of φ accurate to 2^{-10} because

$$\frac{384}{2^{10}} < \varphi = \frac{3}{8} + \frac{1}{2^{11}} = \frac{769}{2048} < \frac{385}{2^{10}} ,$$

and inequality (22) is fulfilled as well ($p_1 + p_2 = 0.810 > 0.772$).

The results of computation shown in Fig. 9 demonstrate that decreasing the degree of the approximate QFT to 4 in case of $n = 10$ doesn't reduce much the

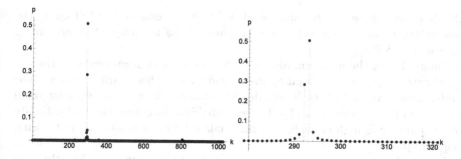

Fig. 9. Probability of obtaining different outputs in case of $\varphi = 2/7$, $n = 10$, $m = 4$

probability to obtain correct value of the phase. Indeed, probabilities to obtain the numbers 292 and 293 are $p_1 = 0.286$ and $p_2 = 0.509$, respectively, and

$$p_1 + p_2 = 0.795 > \frac{8}{\pi^2} \left(\cos^2 \left(\frac{\pi}{2^m} \right) \right)^{n-m} = 0.642 \ .$$

Decreasing the accuracy of the phase estimation to $|\varphi - \tilde{\varphi}| < r/2^{10}$, $r = 10$, we increase the probability to otain correct result to

$$\sum_{k=-r}^{r-1} p(293 + k) = 0.948 > \frac{8}{\pi^2} \left(\cos^2 \left(\frac{\pi}{2^m} \right) \right)^{n-m} \left(1 + \frac{r-1}{3(2r+1)} \right) = 0.714 \ ,$$

and the estimate (23) is satisfied.

Note that if the phase φ is represented by the binary fraction $\varphi = 1/2 + 1/16 + 1/32 = 19/32$ the result is much better because only two components of the vector $vec4$ are not equal to zero, namely, $p(y = 608) = 0.9904$ and $p(y = 96) = 0.0096$. The last result entirely satisfies the estimate (20), where $m = 4$, $l = 32$.

5   Conclusion

In the present paper, we discussed the quantum algorithm for phase estimation based of the quantum Fourier transform proposed first in [1]. This algorithm enables one to obtain the estimate for the phase accurate to 2^{-n}, where n is a number of qubits used for storing an approximate value of the phase. It was shown that the decrease in the accuracy of phase estimation results in increasing the probability of the successful problem solution, and the lower bound (14) for this probability was obtained. Note that this estimate is much better than those in [14].

We have analyzed a possibility to use the approximate QFT instead of the full QFT in the algorithm for phase estimation and established the lower bound for the probability (22) of obtaining a correct result. We have shown that two possible outputs give a correct phase estimate in one run of the algorithm, and this improves the estimate established in [4]. As in case of the full QFT, the probability to obtain correct result increases if the accuracy of measurement decreases (see (23)).

Simulation of the quantum algorithm for phase estimation based on the approximate QFT has shown that approximation doesn't necessarily imply a worse performance. Numerical results obtained by simulation show that even for small degree m of the approximate QFT, the probability to obtain the best estimate of the phase φ is high enough, and this makes it practical alternative to using the full QFT.

It should be noted also that recently a new phase estimation algorithm has been proposed in [2] in which only the third degree approximate QFT is used. But using the only two controlled phase shift gates R_2, R_3 is not sufficient for obtaining a good result in one run of the algorithm. So this algorithm occupies an intermediate position between the original Kitaev's algorithm for phase estimation [12] and the algorithm based on the QFT because several repetitions of the computation are necessary to reach good enough accuracy of measurements. However, the number of repetitions is much lower than in case of Kitaev's algorithm.

Note that the simulation of quantum algorithm for phase estimation was done using the Mathematica package QuantumCircuit, and all relevant computations and visualizations were done with the computer algebra system Mathematica.

References

1. Abrams, D.S., Lloyd, S.: Quantum algorithm providing exponential speed increase for finding eigenvales and eigenvectors. Phys. Rev. Lett. **83**(24), 5162–5165 (1999)
2. Ahmadi, H., Chiang C.-F.: Quantum phase estimation with arbitrary constant-precision phase shift operators (2011). arXiv:quant-ph/1012.4727v4
3. Barenco, A., Ekert, A., Suominen, K.A., Törmä, P.: Approximate quantum Fourier transform and decoherence. Phys. Rev. A **54**(1), 139–146 (1996)
4. Cheung, D.: Improved bounds for the approximate QFT (2004). arXiv:quant-ph/0403071
5. Coppersmith, D.: An approximate Fourier transform useful in quantum factoring (2002). arXiv:quant-ph/0201067
6. Gerdt, V.P., Kragler, R., Prokopenya, A.N.: A Mathematica program for constructing quantum circuits and computing their unitary matrices. Physics of Particles and Nuclei, Letters **6**(7), 526–529 (2009)
7. Gerdt, V.P., Kragler, R., Prokopenya, A.N.: A mathematica package for simulation of quantum computation. In: Gerdt, V.P., Mayr, E.W., Vorozhtsov, E.V. (eds.) CASC 2009. LNCS, vol. 5743, pp. 106–117. Springer, Heidelberg (2009)
8. Gerdt, V.P., Prokopenya, A.N.: Some Algorithms for Calculating Unitary Matrices for Quantum Circuits. Programming and Computer Software **36**(2), 111–116 (2010)
9. Gerdt, V.P., Prokopenya, A.N.: Simulation of quantum error correction my means of QuantumCircuit package. Programming and Computer Software **39**(3), 143–149 (2013)
10. Grover, L.K.: Quantum mechanics helps in searching for a needle in a haystack. Phys. Rev. Lett. **79**, 325–328 (1997)

11. Kitaev, A.Y., Chen, A., Vyalyi, M.: Classical and quantum computation. Graduate Studies in Mathematics, vol. 47. American Mathematical Society (2002)
12. Kitaev, A.Y.: Quantum computations: Algorithms and error correction. Russian Mathematical Surveys **52**(6), 1191–1249 (1997)
13. Mosca, M.: Counting by quantum eigenvalue estimation. Theoretical Computer Science **264**, 139–153 (2001)
14. Nielsen, M., Chuang, I.: Quantum Computation and Quantum Information. Cambridge University Press (2000)
15. Prokopenya, A.N.: Simulation of a quantum algorithm for phase estimation. Programming and Computer Software **41**(2), 98–104 (2015)
16. Shor, P.W.: Polynomial-time algorithms for prime factorization and discrete logarithms on a quantum computer. SIAM J. Comp. **26**(5), 1484–1509 (1997)

Two-Point Boundary Problems with One Mild Singularity and an Application to Graded Kirchhoff Plates

Markus Rosenkranz[1], Jane Liu[3],
Alexander Maletzky[2,*], and Bruno Buchberger[2]

[1] University of Kent, Canterbury, Kent CT2 7NF, United Kingdom
[2] Research Institute for Symbolic Computation (RISC), Hagenberg, Austria
[3] Tennessee Tech University, United States

Abstract. We develop a new theory for treating boundary problems for linear ordinary differential equations whose fundamental system may have a singularity at one of the two endpoints of the given interval. Our treatment follows an algebraic approach, with (partial) implementation in the Theorema software system (which is based on Mathematica). We study an application to graded Kirchhoff plates for illustrating a typical case of such boundary problems.

1 Introduction

The *symbolic treatment of boundary problems* is based on an operator-theoretic formulation of elementary analysis: Integration, differentation, multiplication by smooth functions, and evalation at prescribed points are all viewed as operators on a suitable "function space" (a so-called integro-differential algebra), for example $C^\infty[a, b]$ for two-point boundary value problems on a compact interval $[a, b]$. Various aspects of the resulting algebraic theory of boundary problems have been investigated in [1,2,3].

However, up to now we have always assumed differential equations without singularity or, equivalently, monic differential operators (leading coefficient function being unity). In this paper, we develop for the first time an *algebraic theory* for treating boundary problems with a (mild) *singularity at one endpoint*. For details, we refer to Section 3. Our approach is very different from the traditional analysis setting in terms of the Weyl-Titchmarsh theory of limit points and limit circles [4, §43], [5, §9]. The advantage of our algebraic theory is that it lends itself to a constructive treatment whereas the classical theory is targeted at topological questions (and hence better suited for numerical computation). It would be very interesting to explore their connections; however, this must be left for future work.

As far as we are aware, this is the first genuinely algebraic approach to boundary problems with a (mild) singularity. Hence our emphasis in this paper is to

---

* Supported by the Austrian Science Fund (FWF): Grant W1214-N15, Project DK1.

V.P. Gerdt et al. (Eds.): CASC 2015, LNCS 9301, pp. 406–423, 2015.
DOI: 10.1007/978-3-319-24021-3_30

give an outline of an *algebraic framework* that can cope with such a setting. The purpose of algebraization is to allow for a completely symbolic and exact treatment of such boundary problems. The development of appropriate software is under way, and we provide a short overview in the remainder of this section and in the sample computation at the end of this paper (Section 4).

Regarding the general setup of the *algebraic language for boundary problems*, we refer to the references mentioned above, in particular [3]. At this point, let us just recall some notation. We start from a fixed integro-differential algebra $(\mathcal{F}, \partial, \int)$ over a given field K. The formulation of (local) boundary conditions is in terms of *evaluations*, which are by definition multiplicative linear functionals in $\mathcal{F}^*$. We write $\mathcal{F}_1 \leq \mathcal{F}_2$ if $\mathcal{F}_1$ is a *subspace* of $\mathcal{F}_2$; the same notation is used for subspaces of the dual space $\mathcal{F}^*$. The *orthogonal* of a subspace $\mathcal{B} \leq \mathcal{F}^*$ is defined as

$$\mathcal{B}^\perp = \{u \in \mathcal{F} \mid \beta(u) = 0 \text{ for all } \beta \in \mathcal{B}\}$$

and similarly for the orthogonal of a subspace $\mathcal{A} \leq \mathcal{F}$. We write $[f_1, f_2, \ldots]$ for the *linear K-span* of (possibly infinitely many) elements $f_1, f_2, \cdots \in \mathcal{F}$; the same notation is used for linear spans within $\mathcal{F}^*$. The *zero vector space* (viewed as subspace of any given vector space) is denoted by O.

We write $n^{\underline{k}} = n(n-1) \cdots (n-k+1)$ for the *falling factorial*, where $n \in \mathbb{C}$ could be arbitrary (but will be an integer for our purposes) while k is taken to be a natural number. Note that $n^{\underline{k}} = 0$ for $k > n$. Our main example of an integro-differential algebra will be the space $C^\omega(I)$ of analytic functions on a closed interval $I = [a, b]$. Recall that by definition this is the space of functions that are analytic in some open set containing $[a, b]$.

The development of the new algebraic theory of two-point boundary problems with a mild singularity (whose treatment is really just broached in this paper) was triggered by a collaboration between a *symbolic computation* team (consisting of the first, second and fourth author) and a researcher in *engineering mechancs* (the third author). This underlines the importance and fruitfulness of collaborations between theoretical developments and practical applications. We present the example that had originally led to this research in Section 4.

The connection to *automated theorem proving* is both historical and intrinsic: The first author wrote the original implementation of the `GreenSimplifier`, a software package finding the Green's operator of two-point boundary value problems, in the TH∃OREM∀ system [6,7], during his PhD thesis under the guidance of Bruno Buchberger in cooperation with Heinz Engl. The TH∃OREM∀ system aims to provide computer support to the whole range of mathematical activies—in particular: proving, computing and solving—and to facilitate their *interactions*. This may be seen in the case of boundary problems:

- In order to *solve* a given boundary problem for u in terms of the forcing function f, one determines the Green's operator G that maps f to u.
- For getting G it is sufficient to *compute* a normal form of an operator expression modulo a certain ideal $\mathcal{R}$ of operator relations.

– For ensuring that the normal forms a unique, one must *prove* that the generators of $\mathcal{R}$ form a Gröbner basis (in the original implementation the automated proof was 2000 lines).

For a more comprehensive description of the TH∃OREM∀ project and its underlying philosophy, we refer to [8]. The latest stable release of the software can be found here:

https://www.risc.jku.at/research/theorema/software/

At this juncture, we mention only some of the guiding principles of TH∃OREM∀ project and its *new paradigm* for doing (algorithmic) mathematical research: For the phase of doing research on new theorems and algorithms, it provides a formal language (a version of predicate logic) and an automated reasoning system by which the exploration of the theory is supported. In the phase in which algorithms based on the new theory should be implemented and used in computing examples, TH∃OREM∀ allows to program and execute the algorithms *in the same language*. In the particular case of our approach to solving linear boundary value problems, the fundamental theorem on which the approach is based was proved automatically by checking that the rewrite rules for integro-differential operators forms a Gröbner basis. In a second step, the algorithm for solving linear boundary value problems is expressed again in the TH∃OREM∀ language and can then be called by the users by inputting the linear boundary value problems in a user-friendly notation.

In its current version, the engine for solving boundary problems is bundled in the **GreenGroebner** package of TH∃OREM∀. As an example, consider the boundary problem

$$\begin{array}{l} u'' + \frac{1}{x}u' - \frac{1}{x^2}u = f, \\ u(0) = u(1) = 0, \end{array}$$

which we shall consider in greater detail in Section 2. This can be given to TH∃OREM∀ in the form

$$\text{BPSolve}\left[\left[\begin{array}{l} \text{u}'' + \frac{1}{\text{x}}\text{u}' - \frac{1}{\text{x}^2}\text{u} = \text{f} \\ \text{u}[0] = \text{u}[1] = 0 \end{array}\right], \text{u}, \text{x}, 0, 1\right]$$

leading either to the solution for $u(x)$ as

$$-\frac{1}{2}\frac{1}{x}\left(x^2\int_0^1 \text{f}[\xi]\,\mathrm{d}\xi - x^2\int_0^x \text{f}[\xi]\,\mathrm{d}\xi - x^2\int_0^1 \xi^2\text{f}[\xi]\,\mathrm{d}\xi + \int_0^x \xi^2\text{f}[\xi]\,\mathrm{d}\xi\right)$$

or to the Green's function $g(x, \xi)$ as

$$\begin{cases} \frac{1}{2}\frac{1}{x}\left(-1 + x^2\right)\xi^2 & \Leftarrow & \xi \leq x \\ \frac{1}{2}x\left(-1 + \xi^2\right) & \Leftarrow & x < \xi \end{cases}$$

at the user's request. As one sees from this example, the command BPSolve of the **GreenGroebner** package takes a boundary problem (which may be input in two-dimensional style as above or alternatively in the more traditional form of an ordered pair containing the differential operator and the list of boundary conditions), the dependent variable u, the independent variable x, and the two boundary points 0, 1. The output generated by BPSolve is either the corresponding Green's operator G applied to the generic forcing function f appearing on the right-hand side, or alternatively the Green's function $g(x, \xi)$ associated with G.

In addition to the GreenGroebner package in TH∃OREM∀, a Maple package named IntDiffOp is also available [9]. This package was developed in the frame of Anja Korporal's PhD thesis, supervised by Georg Regensburger and the first author. The Maple package supports also generalized boundary problems (see Section 2 for their relevance to this paper). One advantage of the TH∃OREM∀ system is that both the research phase and the application phase of our method can be formulated and supported within the same logic and software system— which we consider to be quite a novel and promising paradigm for the future.

2   A Simple Example

For illustrating the new ideas, it is illuminating to look at a simple example that exhibits the kind of phenomena that we have to cope with in the Kirchhoff plate boundary problem.

Example 1. Let us start with the *intuitive but mathematically unprecise* statement of the following example: Given a "suitable" forcing function f on the unit interval $I = [0, 1]$, we want to find a "reasonable" solution function u such that

$$\boxed{\begin{aligned} u'' + \tfrac{1}{x}u' - \tfrac{1}{x^2}u = f, \\ u(0) = u(1) = 0. \end{aligned}} \tag{1}$$

But note that the differential operator $T = D^2 + \tfrac{1}{x}D - \tfrac{1}{x^2}$ is singular at the left boundary point $x = 0$ of the interval I under consideration. Hence the first boundary condition $u(0) = 0$ should be looked at with some suspicion. And what function space are we supposed to consider in the first place? If the $\tfrac{1}{x}$ and $\tfrac{1}{x^2}$ are to be taken literally, the space $C^\infty[0, 1]$ will clearly not do. On the other hand, we need functions that are smooth (or at least continuous) at $x = 0$ to make sense of $u(0)$.

In the rest of this section, we shall give one possible solution to the dilemma outlined above. Of course we could resort to using different function space for u and f, and this is in fact the approach one usually takes in Analysis. For our present purposes, however, we prefer to keep the simple paradigm of integro-differential algebras as outlined in Section 1, but we shall modify it to accommodate singularities such as in $\tfrac{1}{x}$ and $\tfrac{1}{x^2}$. Note that these are just poles, so we can take $\mathcal{F}$ to be the subring of the field $\mathcal{M}(I)$ consisting of complex-valued *meromorphic functions* that are regular at all $x \in I$ except possibly $x = 0$. In other words, these are functions that have a Laurent expansion at $x = 0$ with finite principal part, converging in a complex annulus $0 < |z| < \rho$ with $\rho > 1$.

In fact, we will only use the real part $[-1, 1] \setminus \{0\}$ of this annulus. Note that we have of course $\frac{1}{x}, \frac{1}{x^2} \in \mathcal{F}$.

The ring $\mathcal{F}$ is an integro-differential algebra over $\mathbb{C}$ if we use the standard *derivation* $\partial = \frac{d}{dx}$ and the *Rota-Baxter operator*

$$\textstyle\int f := \int_1^x f(\xi) \, d\xi$$

initialized at the regular point $x = 1$.

We have now ensured that the differential operator of (1) has a clear algebraic interpretation $T \in \mathcal{F}[\partial]$. However, the boundary condition $u(0) = 0$ is still dubious. For making it precise, note that the integro-differential algebra $(\mathcal{F}, \partial, \int)$ has only the second of the two boundary evaluations $L, R \colon \mathcal{F} \to \mathbb{C}$ with $L(f) = f(0)$ and $R(f) = f(1)$ in the usual sense of a total function. So while we can interpret the second boundary condition algebraically by $Ru = 0$, the same does not work on the left endpoint. Instead of an evaluation at $x = 0$ we shall introduce the map

$$\mathrm{pp} \colon \sum_{n=N}^{\infty} a_n x^n \mapsto \sum_{n=N}^{-1} a_n x^n$$

that extracts the *principal part* of a function written in terms of its Laurent expansion at $x = 0$. Here and henceforth we assume such expansions of nonzero functions are written with $a_N \neq 0$. If $N \geq 0$ the function is regular at $x = 0$, and the above sum is to be understood as zero. Clearly, $\mathrm{pp} \colon \mathcal{F} \to \mathcal{F}$ is a linear projector, with the complementary projector $\mathrm{reg} := 1_{\mathcal{F}} - \mathrm{pp}$ extracting the *regular part* at $x = 0$. Incidentally, pp and reg are also Rota-Baxter operators of weight -1, which play a crucial role in the renormalization theory of perturbative quantum field theory [10, Ex. 1.1.10].

Finally, we define the functional $C \colon \mathcal{F} \to \mathbb{C}$ that extracts the *constant term* a_0 of a meromorphic function expanded at $x = 0$. Combining C with the monomial multiplication operators, we obtain the coefficient functionals $\langle x^n \rangle := Cx^{-n}$ ($n \in \mathbb{Z}$) that map $\sum_n a_n x^n$ to a_n. In particular, the residue functional is given by $\langle x^{-1} \rangle = Cx$.

Note that for functions regular at $x = 0$, the functional C coincides with the evaluation at the left endpoint, $L \colon \mathcal{F} \to \mathbb{C}, f \mapsto f(0)$. However, for general meromorphic functions, L is undefined and C is *not multiplicative* since for example $C(x \cdot \frac{1}{x}) = 1 \neq 0 = C(x) \, C(\frac{1}{x})$. Hence we refer to C only as a functional but not as an evaluation.

We can now make the boundary condition $u(0) = 0$ precise for our setting over $\mathcal{F}$. What we really mean is that $\lambda(u) = 0$, where $\lambda := \mathrm{pp} + C$ is the projector that extracts the principal part together with the constant term. Extending the algebraic notion of boundary problem to allow for boundary conditions that are not functionals, we may thus view (1) as $(D^2 + \frac{1}{x}D - \frac{1}{x^2}, [\lambda, R]^{\perp})$. In this way we have given a precise meaning to the *formulation* of the boundary problem.

But how are we to go about its *solution* in a systematic manner? Let us first look at the adhoc *standard method* way of doing this—determining the general homogeneous solution, then add the inhomogeneous solution via variation of constants, and finally adapt the integration constants to accommodate the boundary conditions. In our case, one sees that $\mathrm{Ker}(T) = [x, \frac{1}{x}]$ so that the general solution of the homogeneous differential equation is $u(x) = c_1 x + \frac{c_2}{x}$, where $c_1, c_2 \in \mathbb{C}$ are integration constants. Variation of constants [5, p. 74] yields

$$u(x) = c_1 x + \frac{c_2}{x} + \int_1^x \left(\frac{x}{2} - \frac{\xi^2}{2x} \right) f(\xi)\, d\xi \qquad (2)$$

for the inhomogeneous solution. Note that $f \in \mathcal{F}$ may also have singularities as $x = 0$ or any other point $x \in I$ apart from $x = 1$.

Now we need to impose the *boundary conditions*. From $u(1) = 0$ we obtain immediately $c_1 + c_2 = 0$. For the boundary condition at $x = 0$ we have to proceed a bit more cautiously, obtaining

$$u(0) = \lim_{x \to 0} \left(c_1 x + \frac{c_2}{x} + \frac{x}{2} \int_1^x f(\xi)\, d\xi - \frac{1}{2x} \int_1^x \xi^2 f(\xi)\, d\xi \right)$$

$$= \lim_{x \to 0} \left(\frac{c_2}{x} - \frac{1}{2x} \int_1^x \xi^2 f(\xi)\, d\xi \right)$$

where we assume that f is regular at 0. It is clear that the remaining limit can only exist if the integral tends to a finite limit as $x \to 0$, and this is the case by our assumption on f. But then we may apply the boundary condition to obtain $c_2 = \frac{1}{2} \int_1^0 \xi^2 f(\xi)\, d\xi = -\frac{1}{2} \int_0^1 \xi^2 f(\xi)\, d\xi$. This gives the overall solution

$$u(x) = \left(\frac{x}{2} \int_0^1 \xi^2 - \frac{1}{2x} \int_0^1 \xi^2 - \frac{x}{2} \int_x^1 + \frac{1}{2x} \int_x^1 \xi^2 \right) f(\xi)\, d\xi, \qquad (3)$$

which one may write in the standard form $u(x) = \int_0^1 g(x, \xi)\, f(\xi)\, d\xi$ where the *Green's function* is defined as

$$g(x, \xi) = \begin{cases} \frac{x\xi^2}{2} - \frac{\xi^2}{2x} & \text{if } \xi \le x \\ \frac{x\xi^2}{2} - \frac{x}{2} & \text{if } \xi \ge x \end{cases} \qquad (4)$$

in the usual manner. And this is exactly the answer returned by the THEOREM∀ system (Section 1).

How are we to make sense of this in an algebraic way, i.e. without (explicit) use of limits and hence topology? The key to this lies in the projector pp and the functional L, which serve to distill into our algebraic setting what we need from the topology (namely $f(x) = (\mathrm{pp}\, f)(x) + O(1)$ as $x \to 0$). However, there is another complication when compared to boundary problems without singularities as presented in Section 1: We cannot expect a solution $u \in \mathcal{F}$ to (1) for every given forcing function $f \in \mathcal{F}$. In other words, this boundary problem is *not regular* in the sense of [1].

In the past, we have also used the term *singular boundary problem* for such situations (here this seems to be suitable in a double sense but we shall be careful to separate the second sense by sticking to the designation "boundary problems with singularities"). The theory of singular boundary problems was developed in an abstract setting in [11]; applications to boundary problems (without singularities) have been presented in [9]. At this point we shall only recall a few basic facts.

A boundary problem $(T, \mathcal{B})$ is called *semi-regular* if $\mathrm{Ker}(T) \cap \mathcal{B}^\perp = O$. It is easy to see that the boundary problem $(D^2 + \frac{1}{x}D - \frac{1}{x^2}, [\lambda, R]^\perp)$ is in fact semi-regular. Since any $u \in \mathrm{Ker}(T)$ can be written as $u(x) = c_1 x + \frac{c_2}{x}$, the condition $Ru = 0$ implies $c_2 = -c_1$ and hence $u(x) = c_1(x - \frac{1}{x})$. But then $(\lambda u)(x) = -\frac{c_1}{x} = 0$ forces $c_1 = 0$ and hence $u = 0$.

If $(D^2 + \frac{1}{x}D - \frac{1}{x^2}, [\lambda, R]^\perp)$ were a regular boundary problem, we would have $\mathrm{Ker}(T) \dotplus [\lambda, R]^\perp = \mathcal{F}$. However, it is easy to see that there are elements $u \in \mathcal{F}$ that do not belong to $\mathrm{Ker}(T) + [\lambda, R]^\perp$, for example $u(x) = \frac{1}{x^2}$. Hence we conclude that the boundary problem (1) is in fact overdetermined. For such boundary problems $(T, \mathcal{B})$ one can always select a *regular subproblem* $(T, \tilde{\mathcal{B}})$, in the sense that $\tilde{\mathcal{B}} < \mathcal{B}$. In our case, a natural choice is $\tilde{\mathcal{B}} = [\langle x^{-1} \rangle, R]$. This is regular since the evaluation matrix

$$(\langle x^{-1} \rangle, R)(\tfrac{1}{x}, x) = \begin{pmatrix} 1 & 0 \\ 1 & 1 \end{pmatrix}$$

is regular, and the associated kernel projector is $P = \frac{1}{x} \langle x^{-1} \rangle + x \, (R - \langle x^{-1} \rangle)$ by [2, Lem. A.1].

For making the boundary problem (1) well-defined on $\mathcal{F}$ we need one more ingredient: We have to fix a complement $\mathcal{E}$ of $T(\mathcal{B}^\perp)$, the so-called *exceptional space*. Intuitively speaking, this comprises the "exceptional functions" of $\mathcal{F}$ that we decide to discard in order to render (1) solvable. Let us first work out what $T(\mathcal{B}^\perp) \leq \mathcal{F}$ looks like. Since $u(x) = \sum_{n \geq N}^\infty a_n x^n \in \mathcal{B}^\perp$ forces the principal part and constant term to vanish, we may start from $u(x) = \sum_{n>0} a_n x^n$, with the additional proviso that $\sum_{n>0} a_n = 1$. Applying $T = D^2 + \frac{1}{x}D - \frac{1}{x^2}$ to this $u(x)$ yields $\sum_{n>1} a_n (n^2 - 1) x^{n-2}$, which represents an arbitrary element of $C^\omega(I) = [\mathrm{pp}]^\perp$ for a suitable choice of coefficients $(a_n)_{n>1}$ since the additional condition $\sum_{n>0} a_n = 1$ can always be met by choosing $a_1 = 1 - \sum_{n>1} a_n$. But then it is very natural to choose $\mathcal{E} = [\mathrm{reg}]^\perp$ as the required complement. Clearly, the elements of this space $\mathcal{E}$ are the Laurent polynomials of $\mathbb{C}[\frac{1}{x}]$ without constant term.

By Prop. 2 of [9], the *Green's operator* of a generalized boundary problem $(T, \mathcal{B}, \mathcal{E})$ is given by $G = \tilde{G}Q$, where $\tilde{G}$ is the Green's operator of some regular subproblem $(T, \tilde{\mathcal{B}})$ and Q is the projector onto $T(\mathcal{B}^\perp)$ along $\mathcal{E}$. In our case, the latter projector is clearly $Q = \mathrm{reg}$, while the Green's operator $\tilde{G} = (1 - P)T^\diamond$ by [1, Thm. 26], with the kernel projector P as above and the fundamental right inverse $T^\diamond$ given according to [1, Prop. 23] by

$$T^\diamond = \tfrac{1}{2} A_1 x - \tfrac{1}{2x} A_1 x^2,$$

which is essentially just a reformulation of the inhomogeneous part in (2). Following the style of [7] we write here $\int_1^x$ as $A_1 \in \mathcal{F}[\partial, \int]$ to emphasize its role in the integro-differential operator ring. Similarly, we write $F := -LA_1 = \int_0^1$

for the definite integral over the full interval I, which we may regard as a linear functional $C^\omega(I) \to \mathbb{C}$.

Putting things together, it remains to compute

$$G = (1 - P)\, T^\Diamond Q = \left(1 - \tfrac{1}{x}\,\langle x^{-1}\rangle + x\,\langle x^{-1}\rangle - xR\right)\left(\tfrac{1}{2}\,A_1 x - \tfrac{1}{2x}\,A_1 x^2\right) \mathrm{reg}$$
$$= \left(-\tfrac{x}{2}\,A_1 + \tfrac{1}{2x}\,A_1 x^2 - \tfrac{1}{2x}\,Fx^2 + \tfrac{x}{2}\,Fx^2\right)\mathrm{reg},$$

which may be done by the usual *rewrite rules* [1, Tbl. 1] for the operator ring $\mathcal{F}[\partial, \int]$, together with the obvious extra rules that on $\mathrm{Im}(\mathrm{reg}) = C^\omega(I)$ the residual $\langle x^{-1}\rangle$ vanishes and $C = \langle x^0\rangle$ coincides with the evaluation L at zero. Using the standard procedure for extracting the *Green's function*, one obtains exactly (4) when restricting the forcing functions to $f \in C^\omega(I)$. We have thus succeeded in applying the algebraic machinery to regain the solution previoulsy determined by analysis techniques. More than that: The precise form of accessible forcing functions is now fully settled, whereas the regularity assumption in (3) was left somewhat vague (a sufficient condition whose necessity was left open).

3   Two-Point Boundary Problems with One Singularity

Let us now address the general question of specifying and solving boundary problems (as usual: relative to a given fundamental system) that have only *one singularity*; the case of multipliple singularities is left for future investigations. Using a scaling transformation (and possibly a reflection), we may thus assume the same setting as in Section 2, with the *singularity at the origin* and the other boundary point at 1.

For the scope of this paper, we shall also restrict ourselves to a certain subclass of Stieltjes boundary problems $(T, \mathcal{B})$: First of all, we shall allow *only local boundary conditions* in $\mathcal{B}$. This means multi-point conditions and higher-order derivatives (leading to ill-posed boundary problems with distributional Green's functions) are still allowed, but no global parts (integrals); for details we refer to [12, Def. 1]. The second restriction concerns the differential operator $T \in \mathbb{C}(x)[\partial]$, which we require to be *Fuchsian without resonances*. The latter means the differential equation $Tu = 0$ is of Fuchsian type (the singularity is regular), and has fundamental solutions $x^{\lambda_1}\varphi_1(x), \ldots, x^{\lambda_n}\varphi_n(x)$ with each $\varphi_i \in C^\omega(I)$ having order 0, where $\lambda_1, \ldots, \lambda_n$ are the roots of the indicial equation [5, p. 127]. In other words, we do not require logarithms for the solutions. (A sufficient—but not necessary—condition for this is that the roots λ_i are all distinct and do not differ by integers.)

Definition 2. We call $(T, \mathcal{B})$ a *boundary problem with mild singularity* if $T \in \mathbb{C}(x)[\partial]$ is a nonresonant Fuchsian operator and $\mathcal{B}$ a local boundary space.

For a fixed $(T, \mathcal{B})$, we shall then enlarge the *function space* $\mathcal{F}$ of Section 2 by adding $x^{\mu_1}, \ldots, x^{\mu_n}$ as algebra generators, where each μ_i is the fractional part of the corresponding indicial root λ_i. Every element of $\mathcal{F}$ is then a sum of series $x^\mu \sum_{n \geq N} a_n x^n$, with $\mu \in \{\mu_1, \ldots, \mu_n\}$ and $N \in \mathbb{Z}$. The integro-differential structure on $(\mathcal{F}, \partial, \int)$ is determined by setting $\partial x^\mu = \mu\, x^{\mu-1}$, as usual, and by using for $\int$ the integral $\int_1^x$ as we did also in Section 2.

Recall D denotes differentiation and $\mathbf{E}_\xi$ evaluation at $\xi \in \mathbb{R}$. As *boundary functionals* in $\mathcal{B}$ we admit derivatives $\mathbf{E}_\xi D^l$ ($0 < \xi \leq 1, l \geq 0$) and coefficient functionals $\langle x^{k+\mu} \rangle$ ($k \in \mathbb{Z}$) whose action is $x^\mu \sum_{n \geq N} a_n x^n \mapsto a_k$. For functions $f \in C^\omega(I)$ we have of course $\langle x^{k+\mu} \rangle x^\mu f = f^{(k)}(0)/k!$. Since the projectors reg and pp can be expressed in terms of the coefficient functionals, we shall henceforth regard the latter just as convenient abbrevations; boundary spaces are always written in terms of $\mathbf{E}_\xi$ and $\langle x^k \rangle$ only but of course they can be infinite-dimensional. For example, in Section 2 we had the "regularized boundary condition" $\lambda(u) = 0$, which is equivalent to $Lu = 0$ and $\mathrm{pp}(u) = 0$ and hence to $\langle x^k \rangle u = 0$ ($k \leq 0$). Its full boundary space is therefore $\mathcal{B} = [R, \langle x^k \rangle \mid k \leq 0]$.

The first issue that we must now address is the choice of *suitable boundary conditions*: Unlike in the "smooth case" (without singularities), we may not be able to impose n boundary conditions for an n-th order differential equation. The motivating example of Section 2 was chosen to be reasonably similar to the smooth case, so the presence of a singularity was only seen in replacing the boundary evaluation L by its regularized version λ. As explained above, we were effectively adding the extra condition $\mathrm{pp}(u) = 0$ to the standard boundary conditions $u(0) = u(1) = 0$. In other cases, this will not do as the following simple example shows.

Example 3. Consider the nonresonant Fuchsian differential equation $Tu(x) := u'' + \frac{4}{x} u' + \frac{2}{x^2} u = 0$. Note that here the indicial equation has the roots $\lambda_1 = -2$ and $\lambda_2 = -1$, which differ by the integer 1. Nevertheless, we may take $\{\frac{1}{x}, \frac{1}{x^2}\}$ as a fundamental system, so T is indeed nonresonant.

Trying to impose the same (regularized) boundary space $\tilde{\mathcal{B}} = [\mathrm{pp}, L, R]$ as in Example 1, one obtains

$$T(\tilde{\mathcal{B}}^\perp) = \left\{ \sum_{n=-1}^\infty b_n x^n \ \Big| \ \sum_{n=-1}^\infty \frac{b_n}{(n+3)(n+4)} = 0 \right\}$$

after a short calculation. But this means that the forcing functions f in the boundary problem

$$\boxed{\begin{array}{l} u'' + \frac{4}{x} u' + \frac{2}{x^2} u = f \\ \mathrm{pp}(u) = u(0) = u(1) = 0 \end{array}}$$

must satisfy an awkward extra condition (viz. the one on the right-hand side of $T(\tilde{\mathcal{B}}^\perp)$ above). This is not compensated by the slightly enlarged generality of allowing f to have a simple pole at $x = 0$.

In the present case, we could instead impose *initial conditions* at 0 so that $\tilde{\mathcal{B}} = [\mathrm{pp}, L, LD]$. In this case one gets $T(\tilde{\mathcal{B}}^\perp) = C^\omega(I)$, so there is a unique solution for every analytic forcing function.

For a given nonresonant Fuchsian operator $T \in \mathbb{C}(x)[\partial]$, a better approach appears to be the following program (this appears to be a fully algorithmic approach whose steps are made precise below—of course a more comprehensive treatment is needed for analyzing relevant properties and filling in some details):

1. We compute first some boundary functionals $\beta_1, \ldots, \beta_n$ that ensure a *regular subproblem* $(T, \mathcal{B})$ with $\mathcal{B}_n := [\beta_1, \ldots, \beta_n]$. Adding extra conditions (vanishing of all $\langle x^{k+\mu_i} \rangle$ for sufficiently small k) we obtain a boundary space $\mathcal{B} = \mathcal{B}_n + \cdots$ such that $(T, \mathcal{B})$ is semi-regular.
2. If a *particular boundary condition* is desired, it may be "traded" against one of the β_i; if this is not possible, it can be "annexed" to the extra conditions. After these amendments, the subproblem $(T, \mathcal{B}_n)$ is still regular, and $(T, \mathcal{B})$ still semi-regular.
3. Next we compute the corresponding *accessible space* $T(\mathcal{B}^\perp)$. This space might not contain $C^\omega(I)$, as we saw in Example 3 above when we insisted on the conditions $u(0) = u(1)$.
4. We determine a complement $\mathcal{E}$ of $T(\mathcal{B}^\perp)$ as *exceptional space* in $(T, \mathcal{B}, \mathcal{E})$.

Once these steps are completed, we have a regular generalized boundary problem $(T, \mathcal{B}, \mathcal{E})$ whose *Green's operator* G can be computed much in the same way as in Section 2. In detail, we get $G = \tilde{G}Q$, where $\tilde{G}$ is the Green's operator of the regular subproblem $(T, \mathcal{B}_n)$ and Q the projector onto $T(\mathcal{B}^\perp)$ along $\mathcal{E}$. As we shall see, the operators G and Q can be computed as in the usual setting [1]. Let us first address Step 1 of the above program.

Lemma 4. *Let $T \in \mathbb{C}(x)[\partial]$ be a nonresonant Fuchsian differential operator of order n. Then there exists a fundamental system $u_1, \ldots, u_n \in \mathcal{F}$ of T and n coefficient functionals $\beta_1 := \langle x^{\mu_1+k_1} \rangle, \ldots, \beta_n := \langle x^{\mu_n+k_n} \rangle$ ordered as $k_1 + \mu_1 < \cdots < k_n + \mu_n$ so that $\beta(u) \in \mathbb{C}^{n \times n}$ is a lower unitriangular matrix.*

Proof. We start from an arbitrary fundamental system

$$u_1 = x^{\mu_1} \sum_{k \geq k_1} a_{1,k} x^k, \ldots, u_n = x^{\mu_n} \sum_{k \geq k_n} a_{k \geq k_n} x^k$$

of the Fuchsian operator T, where we take $\mu_1, \ldots, \mu_n$ fracational as before and we may assume that $a_{1,k_1}, \ldots, a_{n,k_n} = 1$ so that each fundamental solution u_i has order k_i. (The order of a series $u = x^\mu \sum_{k \geq N} a_k x^k$ is defined as the smallest integer k such that $\langle x^k \rangle u \neq 0$.) We order the fundamental solutions such that $k_1 + \mu_1 \leq \cdots \leq k_n + \mu_n$. We can always achieve strict inequalities as follows. If $i < n$ is the first place where $k_i + \mu_i = k_{i+1} + \mu_{i+1}$ we must also have $\mu_i = \mu_{i+1}$ since $0 \leq \mu_i, \mu_{i+1} < 1$. Therefore we have $k_i = k_{i+1}$, and we can replace u_{i+1} by $u_{i+1} - u_i$ and make it monic so as to ensure $k_i + \mu_i < k_{i+1} + \mu_{i+1}$. Repeating this process at most $n - 1$ times we obtain $k_1 + \mu_1 < \cdots < k_n + \mu_n$. Choosing now the boundary functionals $\beta_1 := \langle x^{\mu_1+k_1} \rangle, \ldots, \beta_n := \langle x^{\mu_n+k_n} \rangle$ as in the statement of the lemma, we have clearly $\beta_i(u_i) = 1$ and $\beta_i(u_j) = 0$ for $j > i$ as claimed. $\square$

In particular we see that $E_n := (\beta_1, \ldots, \beta_n)(u_1, \ldots, u_n)$ has unit determinant, so it is regular. Setting $\mathcal{B}_n := [\beta_1, \ldots, \beta_n]$, we obtain a regular boundary problem $(T, \mathcal{B}_n)$. Note that some of the μ_i may coincide. For each $\mu \in M :=$ $\{\mu_1, \ldots, \mu_n\}$ let k_μ be the smallest of the k_i with $\mu = \mu_i$. We expand the n boundary functionals by suitable curbing constraints to the full boundary space

$$\mathcal{B} := \mathcal{B}_n + [\langle x^{k+\mu} \rangle \mid \mu \in M, \ k < k_\mu] \tag{5}$$

since the inhomogeneous solutions should be at least as smooth (in the sense of pole order) as the homogeneous ones. Note that $(T, \mathcal{B})$ is clearly a semi-regular boundary problem. This achieves Step 1 in our program.

Now for Step 2. Suppose we want to *impose a boundary condition β*, assuming it is of the type discussed above (composed of derivatives $\mathbf{E}_\xi D^l$ and coefficient functionals $\langle x^{k+\mu} \rangle$ for fractional parts μ of indicial roots). We must distinguish two cases:

Trading. If the row vector $r := \beta(u_i)_{i=1,\ldots,n} \in \mathbb{C}^n$ is nonzero, we can express it as a $\mathbb{C}$-linear combination $c_1 r_1 + \cdots + c_n r_n$ of the rows $r_1, \ldots, r_n$ of E_n. Let k be the largest index such that $c_k \neq 0$. Then we may express r_k as a $\mathbb{C}$-linear combination of r and the remaining rows r_i ($i \neq k$), hence we may exchange β_k with β without destroying the regularity of $E_n = (\beta_1, \ldots, \beta_n)(u_1, \ldots, u_n)$.

Annexation. Otherwise, we have $\mathrm{Ker}(T) \leq \beta^\perp$. Together with $\mathrm{Ker}(T) \dotplus \mathcal{B}_n^\perp = \mathcal{F}$ this implies $([\beta] \cap \mathcal{B}_n)^\perp = \beta^\perp + \mathcal{B}_n^\perp = \mathcal{F}$ and hence $\beta \notin \mathcal{B}_n$ by the identities of [2, App. A]. Furthermore, Lemma 4.14 of [11] yields

$$T(\mathcal{B}_n + \beta^\perp) = T(\mathcal{B}_n^\perp \cap \beta^\perp) = T(\mathcal{B}_n^\perp) \cap T(\beta^\perp) = T(\beta^\perp)$$

since we have $T(\mathcal{B}_n^\perp) = \mathcal{F}$ from the regularity of $(T, \mathcal{B})$. But this means that adding β to $\mathcal{B}$ as a new boundary condition necessarily cuts down the space of accessible functions unless β happens to be in the span of the curbing constraints $\langle x^{k+\mu} \rangle \in \mathcal{B}$ added to $\mathcal{B}_n$ in (5).

In the sense of the above discussion (see Example 3), the first case signifies a "natural" choice of boundary condition while the second case means we insist on imposing an extra condition (unless it is a redundant curbing constraint). Repeating these steps as the cases may be, we can successively impose any (finite) number of given boundary conditions. This completes Step 2.

For Step 3 we require the computation of the accessible space $T(\mathcal{B}^\perp)$. We shall now sketch how this can be done algorithmically, starting with a finitary description of the *admissible space* $\mathcal{B}^\perp$ as given in the next proposition. The proof is unfortunately somewhat tedious and long-wided but the basic idea is simple enough: We substitute a series ansatz into the boundary conditions specified in $\mathcal{B}$ to determine a number of lowest-order coefficients; the rest is straightforward if somewhat tedious bureaucracy. For lack of space, we omit the proof of this result and its corollary below; the reader may find the full proofs in the arXiv preprint [13].

Proposition 5. *Let $(T, \mathcal{B})$ be a semi-regular boundary problem of order n with mild singularity such that $(T, \mathcal{B}_n)$ is a regular subproblem. Let $M = \{\mu_1, \ldots, \mu_n\}$*

be the fractional parts of the indicial roots for T. Then we have a direct decomposition of the admissible space $\mathcal{B}^\perp = \bigoplus_{\mu \in M} x^\mu \mathcal{A}_\mu$ with components

$$\mathcal{A}_\mu = [p_{\mu 1}(x), \ldots, p_{\mu l_\mu}(x)] + P_\mu(C^\omega(I)^M) \qquad (\mu \in M). \qquad (6)$$

Here $p_{\mu 1}(x), \ldots, p_{\mu l_\mu}(x) \in \mathbb{C}[x, \frac{1}{x}]$ are linearly independent Laurent polynomials and the linear operators $P_\mu \colon C^\omega(I)^M \to C^\omega(I)$ are defined by

$$P_\mu(b(x)) = x^{j_\mu} b_\mu(x) + \sum_{\nu \in M} \sum_{\xi, j} q_{\mu \nu \xi j}(x)\, \varepsilon_\xi D^j(b_\nu),$$

with $j_\mu \in \mathbb{Z}$ and Laurent polynomials $q_{\mu \nu \xi j}(x) \in \mathbb{C}[x, \frac{1}{x}]$, almost all of which vanish over the summation range $0 < \xi \leq 1$ and $j \geq 0$.

Proof. See [13]. $\qquad\qquad\qquad\qquad\qquad\qquad\qquad\qquad\qquad\qquad\qquad\qquad\qquad$ $\square$

We have established Step 3 of our program since the accessible space $T(\mathcal{B}^\perp)$ can be specified by applying $T \in \mathbb{C}[x, \frac{1}{x}]$ to the generic functions (6) of the admissible space $\mathcal{B}^\perp$. Our next goal is to find a *projector* Q onto $T(\mathcal{B}^\perp)$, which then gives the exceptional space $\mathcal{E} := \mathrm{Ker}(Q) = \mathrm{Im}(1 - Q)$ required for Step 4. This will be easy once we have a corresponding projector onto $\mathcal{B}^\perp$. In fact, the operator P_μ in Proposition 5 is not quite a projector (for one thing, it is not even an endomorphism), but in a sense it is not far away from being one. For seeing this, note that we have

$$\mathcal{F} = \bigoplus_{\mu \in M} x^\mu \, \mathbb{C}((x)), \qquad (7)$$

where $\mathbb{C}((x))$ denotes the field of Laurent series (converging in the punctured unit disk). The direct decomposition (7) just reflects the fact that the x^μ for distinct $\mu \in M$ are linearly independent. Let us write $\langle \mu \rangle \colon \mathcal{F} \to \mathcal{F}$ for the *indicial projector* onto the component $x^\mu \mathbb{C}((x))$ of (7). In other words, $\langle \mu \rangle$ extracts all terms of the form $x^{k+\mu}$ ($k \in \mathbb{Z}$) from a series in $\mathcal{F}$. Note that combinations like $\beta = \varepsilon_\xi D^k \langle \mu \rangle$ provide linear functionals $\beta \in \mathcal{F}^*$ for extracting derivatives of the μ-component of a given series in $\mathcal{F}$.

For writing the projector corresponding to Proposition 5, let us also introduce the *auxiliary operator* $x^\mu \colon \mathbb{C}((x)) \to x^\mu \mathbb{C}((x)) \leq \mathcal{F}$ and its inverse $x^{-\mu}$. Then we shall see that the required projector is essentially a "twisted" version of two kinds of projector: one for splitting off the polynomial part of the occurring series, and one for imposing the derivative terms. For convenience, we shall use *orthogonal projectors* in $\mathbb{C}((x))$, where the underlying inner product is defined by $\langle x^k | x^l \rangle = \delta_{kl}$ for all $k, l \in \mathbb{Z}$. Such projectors are always straightforward to compute (using linear algebra on complex matrices).

Corollary 6. *Using the same notation as in Proposition 5, let $R_\mu, S_\mu \colon \mathbb{C}(x)) \to \mathbb{C}(x)$ be the orthogonal projectors onto $[p_{\mu 1}(x), \ldots, p_{\mu l_\mu}(x)]$ and onto $x^{j_\mu} \mathbb{C}((x))$,* *respectively. Writing* $R'_\mu := x^\mu R_\mu x^{-\mu} \langle \mu \rangle$ *and* $S'_\mu := x^\mu S_\mu x^{-\mu} \langle \mu \rangle$ *for their twisted analogs, there is a linear operator* $P \colon \mathcal{F} \to \mathcal{F}$ *defined by*

$$P = \sum_{\mu \in M} \left(R'_\mu + S'_\mu + \sum_{\nu \in M} \sum_{\xi, j} x^\mu \, q_{\mu\nu\xi j}(x) \, \mathbf{E}_\xi D^j x^{-j\mu - \mu} S'_\nu \right), \tag{8}$$

which is a projector onto $\mathcal{B}^\perp$.

Proof. See [13].   □

For accomplishing Step 4 of our program, it only remains to determine a projector Q onto the accessible space $T(\mathcal{B}^\perp)$ from the projector P onto the admissible space $\mathcal{B}^\perp$ provided in Corollory 6. This can be done easily since Q is *essentially a conjugate* of P except that we use a fundamental right inverse $T^\Diamond$, for want of a proper inverse. (The formula for $friT$ in [1, Prop. 23] and [3, Thm. 20] may be used but recall that in our case $\int = \int_1^x$ so that $\mathbf{E} = \mathbf{E}_1$.)

Proposition 7. *Using the same notation as in Proposition 6, the operator*

$$Q := TPT^\Diamond \colon \mathcal{F} \to \mathcal{F}$$

is a projector onto the accessible space $T(\mathcal{B}^\perp)$.

Proof. Let us first check that Q is a projector. Writing $U := 1 - T^\Diamond T$ for the projector onto $\operatorname{Ker} T$ along $[\mathbf{E}, \mathbf{E}D, \dots, \mathbf{E}D^{n-1}]$ and using $P^2 = P$, we get

$$Q^2 = TP^2 T^\Diamond - TPUPT^\Diamond = TPT^\Diamond = Q$$

provided we can ascertain that $\operatorname{Ker} T \leq \operatorname{Ker} P =: \mathcal{C}$ since then $PU = 0$. We know that $\mathcal{B}^\perp \dotplus \mathcal{C} = \mathcal{F}$ since P is a projector onto $\mathcal{B}^\perp$ along $\mathcal{C}$. On the other hand, we have also $\operatorname{Ker}(T) \dotplus \mathcal{B}_n^\perp = \mathcal{F}$ since $(T, \mathcal{B}_n)$ is regular. Intersecting $\mathcal{C}$ in the former decomposition with the latter yields

$$\mathcal{B}^\perp \dotplus (\mathcal{C} \cap \mathcal{B}_n^\perp) \dotplus (\mathcal{C} \cap \operatorname{Ker} T) = \mathcal{F}. \tag{9}$$

But $\mathcal{B}_n \leq \mathcal{B}$ implies $\mathcal{B}^\perp \leq \mathcal{B}_n^\perp$, so intersecting the decomposition $\mathcal{B}^\perp \dotplus \mathcal{C} = \mathcal{F}$ with $\mathcal{B}_n^\perp$ leads to $\mathcal{B}^\perp \dotplus (\mathcal{C} \cap \mathcal{B}_n^\perp) = \mathcal{B}_n^\perp$. Using the other decompostion $\operatorname{Ker}(T) \dotplus \mathcal{B}_n^\perp = \mathcal{F}$ one more time we obtain

$$\mathcal{B}^\perp \dotplus (\mathcal{C} \cap \mathcal{B}_n^\perp) \dotplus \operatorname{Ker} T = \mathcal{F}. \tag{10}$$

Comparing (9) and (10), we can apply the well-known rule [11, (2.6)] to obtain the identity $\mathcal{C} \cap \operatorname{Ker} T = \operatorname{Ker} T$ and hence the required inclusion $\operatorname{Ker} T \leq \mathcal{C}$.

It remains to prove that $\operatorname{Im}(Q) = T(\mathcal{B}^\perp)$. The inclusion from left to right is obivous, so assume $f = Tu$ with $u \in \mathcal{B}^\perp$. Then $T^\Diamond f = u - Uu$ and hence $PT^\Diamond f = Pu - PUu = u$ because P projects onto $\mathcal{B}^\perp$ and $PU = 0$ from the above. But then we have also $Qf = TPT^\Diamond f = Tu = f$ and in particular $f \in \operatorname{Im}(Q)$.   □

We have now sketched how to carry out the four main steps in our program aimed at the algorithmic treatment of finding/imposing "good" boundary conditions on a Fuchsian differential equation with one (mild) singularity. At the moment we do not have a full implementation of the underlying algorithms in TH∃OREM∀ (or any other system). However, we have implemented a prototype version of some portion of this theory. We shall demonstrate some of its features by with example from engineering mechanics.

4 Application to Functionally Graded Kirchhoff Plates

Circular plates play an important role for many application areas in engineering mechanics and mathematical physics. If the plates are thin (the ratio of thickness to diameter is small enough), one may employ the well-known Kirchhoff-Love plate theory [14], whose mathematical description is essentially two-dimensional (via a linear second-order partial differential equation in two independent variables). We will furthermore restrict ourselves to *circular Kirchhoff plates* so as to have a one-dimensional mathematical model, via a linear ordinary differential equation of second order.

However, we shall not assume homogeneous plates. Indeed, the precise manufacture of *functionally graded* materials is an important branch in engineering mechanics. In the case of Kirchhoff plates, the functional grading is essentially the variable thickness $t = t(r)$ or variable bending rigidity $D = D(r)$ of the plate along its radial profile. (We write r for the radius variable ranging between zero at the center and the outer radius $r = b$.)

Let $w = w(r)$ be the transversal displacement of the plate as a function of its radius. This is the quantity that we try to determine. It induces the *radial and tangential moments* given, respectively by

$$M_r(r) = -D(r)\left(w''(r) + \tfrac{\nu}{r}\,w'(r)\right),$$
$$M_\theta(r) = -D(r)\left(\tfrac{1}{r}\,w'(r) + \nu\,w''(r)\right),$$

where ν is the Poisson's ratio of the plate (which we assume constant). For typical materials, ν may be taken as $\tfrac{1}{3}, \tfrac{1}{4}, \tfrac{1}{5}$ or even 0. A reasonable constitutive law for the bending rigidity is

$$D(r) = \tfrac{E(r)\,t(r)^3}{12(1-\nu^2)}, \tag{11}$$

where $E = E(r)$ is the variable Young's modulus of the plate.

The *equilibrium equation* can then be written as

$$\frac{dM_r}{dr} + \frac{M_r - M_\theta}{r} = Q_r, \tag{12}$$

where $Q_r = Q_r(r)$ is the transverse shear obtained by the free-body equilibrium

$$Q_r(r) = -\frac{1}{2\pi r}\int_0^r q(r)\,2\pi r\,dr = -\frac{1}{r}\int_0^r q(r)\,r\,dr \tag{13}$$

```
BPSolve[ φ''[ρ] + (1/ρ - 3/(1-ρ)) φ'[ρ] - (1/ρ² + 1/((1-ρ)ρ)) φ[ρ] = f[ρ] , φ, ρ, 0, β, Params → {β},
         φ[0] = φ[β] = 0
PostProc → FullSimplify, Output → GreensFunction]
```

$$
\begin{aligned}
&\tfrac{1}{6}\tfrac{1}{(-1+\xi)^2}\xi^2\tfrac{1}{(-1+\rho)^2}\tfrac{1}{\rho}\big(-3+11\xi-15\xi^2-9\xi^3-2\xi^4-9\rho^2-3\xi\rho^2-6\rho^3+2\xi\rho^3+\\
&(3-2\rho)\rho^2\big(-3-3\tfrac{1}{\rho^2}\tfrac{1}{-3+2\beta}\big)+\xi(3-2\rho)\rho^2\big(1+11\tfrac{1}{\rho^2}\tfrac{1}{-3+2\beta}\big)+\\
&3\xi^2\rho^2\big(-15\tfrac{1}{\rho^2}\tfrac{1}{-3+2\beta}\big)-2\xi^2\rho^3\big(-15\tfrac{1}{\rho^2}\tfrac{1}{-3+2\beta}\big)+3\xi^4\rho^2\big(-2\tfrac{1}{\rho^2}\tfrac{1}{-3+2\beta}\big)-\\
&2\xi^4\rho^3\big(-2\tfrac{1}{\rho^2}\tfrac{1}{-3+2\beta}\big)+3\xi^2\rho^2\big(9\tfrac{1}{\rho^2}\tfrac{1}{-3+2\beta}\big)-2\xi^3\rho^3\big(9\tfrac{1}{\rho^2}\tfrac{1}{-3+2\beta}\big)\big)\\[4pt]
&\qquad\qquad\qquad\qquad\qquad\qquad\qquad\qquad\qquad\qquad\qquad\qquad\qquad = \xi \le \rho\\[8pt]
&-\tfrac{1}{6}\tfrac{1}{(-1+\xi)^2}\tfrac{1}{(-1+\rho)^2}\rho(-3+2\rho)\big(-1+3\xi+\xi^2\big(-3-3\tfrac{1}{\rho^2}\tfrac{1}{-3+2\beta}\big)+\\
&\xi^3\big(1+11\tfrac{1}{\rho^2}\tfrac{1}{-3+2\beta}\big)+\xi^4\big(-15\tfrac{1}{\rho^2}\tfrac{1}{-3+2\beta}\big)+\xi^6\big(-2\tfrac{1}{\rho^2}\tfrac{1}{-3+2\beta}\big)+\xi^5\big(9\tfrac{1}{\rho^2}\tfrac{1}{-3+2\beta}\big)\big)\\[4pt]
&\qquad\qquad\qquad\qquad\qquad\qquad\qquad\qquad\qquad\qquad\qquad\qquad\qquad = \rho < \xi
\end{aligned}
$$

Fig. 1. Solving the boundary problem in GreenGroebner

induced by a certain *loading* $q = q(r)$ that may be thought to describe the weight (or other forces) acting in each ring $[r, r + dr]$.

For the calculational treatment of (12) it it useful to introduce the function $\varphi := -w'(r)$, which represents the (negative) slope of the plate profile. In terms of φ, the equilibrium equation is given by

$$
\varphi''(r) + \left(\frac{1}{r} + \frac{D'(r)}{D(r)}\right)\varphi'(r) + \left(\nu\frac{D'(r)}{D(r)} - \frac{1}{r}\right)\frac{\varphi(r)}{r} = \frac{Q_r(r)}{D(r)}.
$$

A typical example of a useful thickness grading is the *linear ansatz* $t = t_0(1 - \frac{r}{b})$, cut off beyond some $a < b$ close to b; this describes a radially symmetric pointed plate with straight edges (like a flat cone). Suppressing the cut-off for the moment and changing the independent variable r to $\rho = r/b$, we have thus thickness $t(\rho) = t_0 \cdot (1 - \rho)$ and from (11) bending rigidity $D(\rho) = D_0 \cdot (1 - \rho)^3$ with $D_0 := Et_0^3/12(1 - \nu^2)$ with E constant. The equilibrium equation becomes

$$
\varphi''(\rho) + \left(\frac{1}{\rho} - \frac{3}{1-\rho}\right)\varphi'(\rho) - \left(\frac{1}{\rho} + \frac{1}{1-\rho}\right)\frac{\varphi(\rho)}{\rho} = \frac{Q_r(\rho)\,b^2}{D_0(1-\rho)^3},
$$

where we have set $\nu = \frac{1}{3}$ for simplicity. Note that the right-hand side of this equation, which we shall designate by $f(\rho)$, is not a fixed function of ρ but depends on our choice of the loading. Hence we consider $f(\rho)$ as a *forcing function*.

For the *boundary conditions* (for once we use this word in its original sense!) we use $w'(0) = w'(a) = 0$, which translates to $\varphi(0) = \varphi(\beta) = 0$ in the $\varphi = \varphi(\rho)$ formulation, with the abbreviation $\beta := a/b < 1$. The condition at $\rho = 0$ is dictated by symmetry, while the one at $\rho = \beta$ describes a clamped end. In summary, we have the boundary problem

$$
\boxed{\begin{aligned}
&\varphi''(\rho) + \left(\frac{1}{\rho} - \frac{3}{1-\rho}\right)\varphi'(\rho) - \left(\frac{1}{\rho} + \frac{1}{1-\rho}\right)\frac{\varphi(\rho)}{\rho} = f(\rho),\\
&\varphi(0) = \varphi(\beta) = 0,
\end{aligned}}
\tag{14}
$$

$$\boxed{\text{BPSolve}\left[\begin{array}{c}\phi''[\rho] + \left(\frac{1}{\rho} - \frac{3}{1-\rho}\right)\phi'[\rho] - \left(\frac{1}{\rho^2} + \frac{1}{(1-\rho)\rho}\right)\phi[\rho] = f[\rho] \\ \phi[0] = \phi[\beta] = 0\end{array}\right], \phi, \rho, 0, \beta, \text{Params} \to \{\beta\},}$$

$$\text{PostProc} \to \text{FullSimplify}\Big]$$

$$\frac{1}{6}\frac{1}{(-1+\rho)^2}\frac{1}{\rho}\Big(\rho^2(-3+2\rho)\int_0^\beta \frac{1}{(-1+\xi)^2}f[\xi]\,d\xi + (3-2\rho)\rho^2\int_0^\rho \frac{1}{(-1+\xi)^2}f[\xi]\,d\xi -$$

$$3(3-2\rho)\rho^2\int_0^\beta \frac{1}{(-1+\xi)^2}\xi f[\xi]\,d\xi + 3\rho^2(-3+2\rho)\int_0^\rho \frac{1}{(-1+\xi)^2}\xi f[\xi]\,d\xi -$$

$$3\int_0^\rho \frac{1}{(-1+\xi)^2}\xi^2 f[\xi]\,d\xi + 3(3-2\rho)\rho^2\int_0^\rho \frac{1}{(-1+\xi)^2}\xi^2 f[\xi]\,d\xi +$$

$$11\int_0^\rho \frac{1}{(-1+\xi)^2}\xi^3 f[\xi]\,d\xi + \rho^2(-3+2\rho)\int_0^\rho \frac{1}{(-1+\xi)^2}\xi^3 f[\xi]\,d\xi -$$

$$15\int_0^\rho \frac{1}{(-1+\xi)^2}\xi^4 f[\xi]\,d\xi + 9\int_0^\rho \frac{1}{(-1+\xi)^2}\xi^5 f[\xi]\,d\xi -$$

$$2\int_0^\rho \frac{1}{(-1+\xi)^2}\xi^6 f[\xi]\,d\xi + (3-2\rho)\rho^2\int_0^\beta \frac{1}{(-1+\xi)^2}\xi^2 f[\xi]\,d\xi\left(-3-3\frac{1}{\beta^2}\frac{1}{-3+2\beta}\right) +$$

$$(3-2\rho)\rho^2\int_0^\beta \frac{1}{(-1+\xi)^2}\xi^3 f[\xi]\,d\xi\left(1+11\frac{1}{\beta^2}\frac{1}{-3+2\beta}\right) +$$

$$(3-2\rho)\rho^2\int_0^\beta \frac{1}{(-1+\xi)^2}\xi^4 f[\xi]\,d\xi\left(-15\frac{1}{\beta^2}\frac{1}{-3+2\beta}\right) +$$

$$(3-2\rho)\rho^2\int_0^\beta \frac{1}{(-1+\xi)^2}\xi^6 f[\xi]\,d\xi\left(-2\frac{1}{\beta^2}\frac{1}{-3+2\beta}\right) +$$

$$(3-2\rho)\rho^2\int_0^\beta \frac{1}{(-1+\xi)^2}\xi^5 f[\xi]\,d\xi\left(9\frac{1}{\beta^2}\frac{1}{-3+2\beta}\right)\Big)$$

Fig. 2. Green's function computed by `GreenGroebner`

which is indeed of the type discussed in Section 3, by a simple scaling from $I = [0, 1]$ to $[0, \beta]$. Its treatment in the `GreenGroebner` package is shown in Figure 1.

The output is the Green's operator G of (14) applied to a generic forcing function $f(\rho)$, giving the solution $\varphi(\rho) = Gf(\rho)$ as an integral

$$\int_0^1 g(\rho, \xi)\, f(\xi)\, d\xi$$

in terms of the Green's function $g(\rho, \xi)$; the latter can also be retrieved explicitly if this is desired (see Figure 2).

Let us now choose a constant loading $q(\rho) = q_0$ so that $Q_r = -q_0 b\rho/2$ by (13). Taking the cut-off into account leads to the forcing function $f(\rho) = C\rho/(1-\rho)^3$ with numerical constant $C := -q_0 b^3/2D_0$, which by suitable scaling can be set to unity. The solution of (14) for a cut-off at $\beta = 0.9$, and the corresponding displacement $w(\rho) = -\int_0^\beta \varphi(\rho)\, d\rho$ comes out as follows (see the graphs in Figure 3):

$$\varphi(\rho) = \Big(3870\rho^3 - 2000\rho^3\log(1-\rho) + 56\rho^3\log(-10(\rho-1)) - 2403\rho^2 - 972\rho$$
$$+ 3000\rho^2\log(1-\rho) - 84\rho^2\log(-10(\rho-1)) - 972\log(1-\rho)\Big)\Big/\Big(5832(\rho-1)^2\rho\Big)$$

$$w(\rho) = \Big(972(\rho-1)\,\text{Li}_2(\rho) + (5814\rho - 6309 + 28\log 10)\rho$$
$$+ (\rho-1)(56\rho\log 10 - (1944\rho - 7281 - 28\log 10)\log(1-\rho))\Big)\Big/\Big(5832(\rho-1)\Big)$$

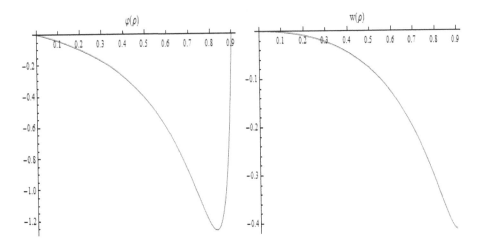

Fig. 3. Displacement and Slope for (14)

Of course one could use different functional gradings and/or loading functions, and the integrals would not always come out in closed form. In this case one can resort to numerical integration (which is also supported by *Mathematica*).

5   Conclusion

Obviously, in this paper we have only scratched the surface of what might become an interesting field of research—the *algebraic* theory of *boundary problems with singularities*. In particular, it would be interesting to see how and to what extent the classical Weyl-Titchmarsh theory is reflected in the algebra.

The case of *singularities at both endpoints* could be the first step, based on a space containing meromorphic functions with poles only at the two endpoints. For the Rota-Baxter operator $\int_c^x$ one must choose an initialization point c distinct from both endpoints. Moreover, the interaction between regularity conditions on the left and on the right endpoint must be studied. This would be an interesting task, where some guidance can be sought from the classical theory.

Acknowledgments. We would like to thank the three anonymous referees, who have helped to improve the presentation of this paper.

References

1. Rosenkranz, M., Regensburger, G.: Solving and factoring boundary problems for linear ordinary differential equations in differential algebras. Journal of Symbolic Computation **43**(8), 515–544 (2008)
2. Regensburger, G., Rosenkranz, M.: An algebraic foundation for factoring linear boundary problems. Ann. Mat. Pura Appl. (4) **188**(1), 123–151 (2009). doi:10.1007/s10231-008-0068-3

3. Rosenkranz, M., Regensburger, G., Tec, L., Buchberger, B.: Symbolic analysis of boundary problems: from rewriting to parametrized Gröbner bases. In: Langer, U., Paule, P. (eds.) Numerical and Symbolic Scientific Computing: Progress and Prospects. Springer, pp. 273–331 (2012)
4. Yosida, K.: Lectures on Differential and Integral Equations. Dover (1991)
5. Coddington, E.A., Levinson, N.: Theory of ordinary differential equations. McGraw-Hill Book Company Inc, New York-Toronto-London (1955)
6. Rosenkranz, M., Buchberger, B., Engl, H.W.: Solving linear boundary value problems via non-commutative Gröbner bases. Appl. Anal. **82**, 655–675 (2003)
7. Rosenkranz, M.: A new symbolic method for solving linear two-point boundary value problems on the level of operators. J. Symbolic Comput. **39**(2), 171–199 (2005)
8. Buchberger, B., Craciun, A., Jebelean, T., Kovacs, L., Kutsia, T., Nakagawa, K., Piroi, F., Popov, N., Robu, J., Rosenkranz, M., Windsteiger, W.: Theorema: Towards computer-aided mathematical theory exploration. Journal of Applied Logic **4**(4), 359–652 (2006). ISSN 1570–8683
9. Korporal, A., Regensburger, G., Rosenkranz, M.: Regular and Singular Boundary Problems in Maple. In: Gerdt, V.P., Koepf, W., Mayr, E.W., Vorozhtsov, E.V. (eds.) CASC 2011. LNCS, vol. 6885, pp. 280–293. Springer, Heidelberg (2011)
10. Guo, L.: An Introduction to Rota-Baxter Algebras. International Press (2012)
11. Korporal, A.: Symbolic Methods for Generalized Green's Operators and Boundary Problems. Ph.D. thesis, Johannes Kepler University, Linz, Austria (November 2012). Abstracted in ACM Communications in Computer Algebra, vol. 46, no. 4, issue 182, December 2012
12. Rosenkranz, M., Serwa, N.: Green's functions for Stieltjes boundary problems. In: ISSAC (to appear, 2015)
13. Rosenkranz, M., Liu, J., Maletzky, A., Buchberger, B.: Two-point boundary problems with one mild singularity and an application to graded kirchhoff plates (May 2015). Preprint on http://arxiv.org/abs/1505.01956
14. Reddy, J.: Theory and analysis of elastic plates and shells. CRC Press, Taylor and Francis (2007)

Analysis of Reaction Network Systems
Using Tropical Geometry

Satya Swarup Samal[1], Dima Grigoriev[2], Holger Fröhlich[3],
and Ovidiu Radulescu[4]

[1] Algorithmic Bioinformatics, Bonn-Aachen International Center for IT,
Dahlmannstraße 2, D-53113, Bonn, Germany
`samal@cs.uni-bonn.de`
[2] CNRS, Mathématiques, Université de Lille, Villeneuve d'Ascq, 59655, France
`dmitry.grigoryev@math.univ-lille1.fr`
[3] Algorithmic Bioinformatics, Bonn-Aachen International Center for IT,
Dahlmannstraße 2, D-53113, Bonn, Germany
`frohlich@bit.uni-bonn.de`
[4] DIMNP UMR CNRS 5235, University of Montpellier, Montpellier, France
`ovidiu.radulescu@univ-montp2.fr`

Abstract. We discuss a novel analysis method for reaction network systems with polynomial or rational rate functions. This method is based on computing tropical equilibrations defined by the equality of at least two dominant monomials of opposite signs in the differential equations of each dynamic variable. In algebraic geometry, the tropical equilibration problem is tantamount to finding tropical prevarieties, that are finite intersections of tropical hypersurfaces. Tropical equilibrations with the same set of dominant monomials define a branch or equivalence class. Minimal branches are particularly interesting as they describe the simplest states of the reaction network. We provide a method to compute the number of minimal branches and to find representative tropical equilibrations for each branch.

1 Introduction

Networks of chemical reactions are widely used in chemistry for modeling catalysis, combustion, chemical reactors, or in biology as models of signaling, metabolism, and gene regulation. Several mathematical methods were developed for analysis of these models such as the study of multiplicity of steady state solutions and detection of bifurcations by stoichiometry analysis, deficiency theorems, reversibility, permanency, etc. [4].

All these methods focus on the number and the stability of the steady states of chemical networks. Beyond steady states, metastable states defined as regions of the phase space where the system has slow dynamics are also important for understanding the behavior of networks. For instance, low dimensional inertial or invariant manifolds gather slow degrees of freedom of the system and are important for model reduction. Invariant manifolds can lose local stability, which

ⓒ Springer International Publishing Switzerland 2015
V.P. Gerdt et al. (Eds.): CASC 2015, LNCS 9301, pp. 424–439, 2015.
DOI: 10.1007/978-3-319-24021-3_31

allow the trajectories to perform large, rapid phase space excursions before slowing down in a different place on the same or on another invariant manifold. [8]. Biological networks have often been modelled as discrete dynamical systems, whose trajectories are sequences of states in a discrete space, see, for instance, the Boolean automata of René Thomas [19]. We think that metastable states of continuous models are good candidates for representing states in a discrete model. It is, therefore, important to know how metastable states are dynamically connected.

We showed elsewhere that tropical geometry methods can be used to approximate such invariant manifolds for systems of polynomial differential equations [13,14,15]. The connection between original dynamics of a specific system [20] to such a method based on invariant manifold is described in [13]. In a nutshell, the slowness of the dynamics on the invariant manifolds follows from the compensation of dominant forces acting on the system, represented as dominant monomials in the differential equations. We have called the equality of dominant monomials tropical equilibration [14,15]. Tropical equilibrations are different from steady states, because in tropically equilibrated systems one has non-compensated weak forces that drive the system slowly, whereas at steady state, the net forces vanish. Furthermore, invariant manifolds can be roughly associated to metastable states because they are regions of phase space where systems dynamics is relatively slower. In this paper, we introduce methods to compute tropical equilibrations and group them into branches that cover the metastable states of the system. These branches of tropical equilibrations form a polyhedral complex. The zeroth homology group (or in other words, the number of connected components) of this complex indicates the possible transitions between the metastable states. Additionally, we explore the structure between the equilibration solutions using directed graphs (to present the inclusion relations among them) and undirected graphs (to present the connectivity among them). Furthermore, one of the biochemical applications of these equilibrations is constructing invariant manifolds and to that extent defining slow variables. We provide a way to identify such slow species and visualise them through heatmaps. Lastly, we benchmark our method against models obtained from the Biomodels database [11] and discuss the findings.

2   Definitions and Settings

We consider biochemical networks described by mass action kinetics

$$\frac{\mathrm{d}x_i}{\mathrm{d}t} = \sum_j k_j S_{ij} x^{\alpha_j},\ 1 \le i \le n, \tag{1}$$

where $k_j > 0$ are kinetic constants, S_{ij} are the entries of the stoichiometric matrix (uniformly bounded integers, $|S_{ij}| < s$, s is small), $\alpha_j = (\alpha_1^j, \ldots, \alpha_n^j)$ are multi-indices, and $x^{\alpha_j} = x_1^{\alpha_1^j} \ldots x_n^{\alpha_n^j}$. We consider that α_i^j are non-negative integers. At this point, we like to mention, there exist approaches to describe such a polynomial system in a graph theoretic way, i.e., by a weighted directed graph and a weighted bipartite graph and to study the number of positive solutions depending on the graph structure [5]. This approach uses decompositions of Newton polytopes to find that parts of the directed graph are related to the existence of positive steady state solutions. In our paper, we do not use these graph theoretic considerations and investigate the different problem of tropical equilibrations.

In the case of slow/fast systems with polynomial dynamics such as (1), the slow invariant manifold is approximated by a system of polynomial equations for the fast species. This crucial point allows us to find a connection with tropical geometry. We introduce now the terminology of tropical geometry needed for the presentation of our results, and refer to [12] for a comprehensive introduction to this field.

Let $f_1, f_2, \ldots, f_k$ be multivariate polynomials, $f_i \in \mathbb{C}[x_1, x_2, \ldots, x_n]$, representing all or a part of the polynomials in the right hand side of (1).

Let us now assume that the variables x_i, $i \in [1, n]$ are written as powers of a small positive parameter ϵ, namely $x_i = \bar{x}_i \epsilon^{a_i}$, where $\bar{x}_i$ has order zero (there are c_i, d_i not depending on ϵ such that $0 < c_i < \bar{x}_i < d_i$). The orders a_i indicate the order of magnitude of x_i. Because ϵ was chosen small, a_i are lower for larger absolute values of x_i. Furthermore, the order of magnitude of monomials $\boldsymbol{x}^{\boldsymbol{\alpha}}$ is given by the dot product $\mu = \langle \boldsymbol{\alpha}, \boldsymbol{a} \rangle$, where $\boldsymbol{a} = (a_1, \ldots, a_n)$. Again, smaller values of μ correspond to monomials with larger absolute values. For each multivariate polynomial f, we define the tropical hypersurface $T(f)$ as the set of vectors $\boldsymbol{a} \in \mathbb{R}^n$ such that the minimum of $\langle \boldsymbol{\alpha}, \boldsymbol{a} \rangle$ over all monomials in f is attained for at least two monomials in f. In other words, f has at least two dominating monomials.

A *tropical prevariety* is defined as the intersection of a finite number of tropical hypersurfaces, namely $T(f_1, f_2, \ldots, f_k) = \cap_{i \in [1,k]} T(f_i)$.

A *tropical variety* is the intersection of all tropical hypersurfaces in the ideal I generated by the polynomials $f_1, f_2, \ldots, f_k$, $T(I) = \cap_{f \in I} T(f)$. The tropical variety is within the tropical prevariety, but the reciprocal property is not always true. There exist algorithms to compute such tropical varieties as in [10].

For our purpose, we slightly modify the classical notion of tropical prevariety. A *tropical equilibration* is defined as a vector $\boldsymbol{a} \in \mathbb{R}^n$ such that $\langle \boldsymbol{\alpha}, \boldsymbol{a} \rangle$ attains its minimum at least twice for monomials of different signs, for each polynomial in the system $f_1, f_2, \ldots, f_k$. Thus, tropical equilibrations are subsets of the tropical prevariety. Our sign condition is needed because we are interested in approximating real strictly positive solutions of polynomial systems (the sum of several dominant monomials of the same sign have no real strictly positive roots).

3   Branches of Tropical Equilibrations

For chemical reaction networks with multiple timescales, it is reasonable to consider that kinetic parameters have different orders of magnitudes.

We, therefore, assume that parameters of the kinetic models (1) can be written as

$$k_j = \bar{k}_j \varepsilon^{\gamma_j}. \tag{2}$$

The exponents γ_j are considered to be integer. For instance, the following approximation produces integer exponents:

$$\gamma_j = \text{round}(\log(k_j)/\log(\varepsilon)), \tag{3}$$

where round stands for the closest integer (with half-integers rounded to even numbers). Without rounding to the closest integer, changing the parameter ε should not introduce variations in the output of our method. Indeed, the tropical prevariety is independent of the choice of ε.

Of course, kinetic parameters are fixed. In contrast, the orders of the species vary in the concentration space and have to be calculated as solutions to the tropical equilibration problem. To this aim, the network dynamics is first described by a rescaled ODE system

$$\frac{d\bar{x}_i}{dt} = \sum_j \varepsilon^{\mu_j - a_i} \bar{k}_j S_{ij} \bar{x}^{\alpha_j}, \tag{4}$$

where

$$\mu_j(a) = \gamma_j + \langle a, \alpha_j \rangle, \tag{5}$$

and $\langle, \rangle$ stands for the dot product.

The r.h.s. of each equation in (4) is a sum of multivariate monomials in the concentrations. The orders μ_j indicate how large are these monomials, in absolute value. A monomial of order μ_j dominates another monomial of order $\mu_{j'}$ if $\mu_j < \mu_{j'}$.

The tropical equilibration problem consists in finding a vector a such that

$$\min_{j, S_{ij} > 0} (\gamma_j + \langle a, \alpha_j \rangle) = \min_{j, S_{ij} < 0} (\gamma_j + \langle a, \alpha_j \rangle) \tag{6}$$

This system can be represented as a set of linear inequalities resulting into a convex polytope. The solutions of this system have a geometrical interpretation. Let us define the extended order vectors $a^e = (1, a) \in \mathbb{R}^{n+1}$ and extended exponent vectors $\alpha_j^e = (\gamma_j, \alpha_j) \in \mathbb{Z}^{n+1}$. Let us consider the equality $\mu_j = \mu_{j'}$. This represents the equation of a n dimensional hyperplane of $\mathbb{R}^{n+1}$, orthogonal to the vector $\alpha_j{}^e - \alpha_{j'}{}^e$:

$$\langle a^e, \alpha_j^e \rangle = \langle a^e, \alpha_{j'}^e \rangle, \tag{7}$$

where $\langle, \rangle$ is the dot product in $\mathbb{R}^{n+1}$. We will see in the next section that the minimality condition on the exponents μ_j implies that the normal vectors $\alpha_j{}^e - \alpha_{j'}{}^e$ are edges of the so-called Newton polytope [9,18].

For each equation i, let us define

$$M_i(a) = \underset{j}{\text{argmin}}(\mu_j(a), S_{ij} > 0) = \underset{j}{\text{argmin}}(\mu_j(a), S_{ij} < 0), \tag{8}$$

in other words M_i denote the set of monomials having the same minimal exponent μ_i.

We call *tropically truncated system*, the system obtained by keeping only the dominating monomials in (4), as follows:

$$\frac{\mathrm{d}\bar{x}_i}{\mathrm{d}t} = \varepsilon^{\mu_i - a_i}\Big(\sum_{j \in M_i(a)} \bar{k}_j S_{ij} \bar{x}^{\alpha_j} \Big). \tag{9}$$

The tropical truncated system is uniquely determined by the index sets $M_i(a)$, therefore, by the tropical equilibration a. Reciprocally, two tropical equilibrations can have the same index sets $M_i(a)$ and truncated systems. We say that two tropical equilibrations a_1, a_2 are equivalent iff $M_i(a_1) = M_i(a_2)$, for all i. Equivalence classes of tropical equilibrations are called *branches*. For each branch there exists a unique convex polytope, cf. (6). The union of branches are subsets of tropical prevariety. It is a subset because we are interested in tropical equilibration of at least two monomials of different signs as expressed in (6). This sign condition is essential as we are interested to approximate positive real solutions of the polynomial system.

Minimal Branches. A branch B with an index set M_i is *minimal* if $M_i' \subset M_i$ for all i where M_i' is the index set of a branch B' implies $B' = B$ or $B' = \emptyset$. In the terminology of convex polytopes, this means a branch B with a convex polytope P_i is *minimal* if $P_i \subset P_i'$ for all i where P_i' is the convex polytope for branch B' implies $B' = B$ or B' is empty. For each index i, relation (7) defines a hyperplane, the tropical equilibration branches are on intersections of k such hyperplanes where k is number of polynomial equations representing the right hand side of (1). Minimal branches are maximal (w.r.t. inclusion) polytopes in the tropical prevariety.

Connected Components of Minimal Branches. Two minimal branches represented by index sets M_i and M_j are connected if there exists a branch with index set M_k such that $M_i \subset M_k$ and $M_j \subset M_k$. In the terminology of convex polytopes, this amounts to checking the intersection between two convex polytopes P_i and P_j (corresponding to minimal branches M_i and M_j) if whether $P_i \cap P_j$ is non void for all $i \neq j$.

4   Algorithm

In this section, we describe an algorithm to compute the tropical equilibrations, test the equilibrations for the equivalence classes (i.e., *branches*) and compute the *minimal branches* (cf. section 3).

4.1   Newton Polytope and Edge Filtering

Given the input polynomial in the form of (4), for each equation and species i, we define a Newton polytope $\mathcal{N}_i$, that is the convex hull of the set of points

α_j^e such that $S_{ij} \neq 0$ and also including together with all the points the half-line emanating from these points in the positive ϵ direction. This is the Newton polytope of the polynomial in right hand side of (4), with the scaling parameter ε considered as a new variable.

As explained in section 2, the tropical equilibrations correspond to vectors $a^e = (1, a) \in \mathbb{R}^{n+1}$ satisfying the optimality condition as per (6). This condition is satisfied automatically on hyperplanes orthogonal to edges of Newton polytope connecting vertices $\alpha_{j'}^e$, $\alpha_{j''}^e$ satisfying the opposite sign condition. Therefore, a subset of edges from the Newton polytope is selected based on the filtering criteria which tells that the vertices belonging to an edge should be from opposite sign monomials as explained in (10).

$$E(P) = \{\{v_1, v_2\} \subseteq \binom{V}{2} \mid \text{conv}(v_1, v_2) \in F_1(P)$$
$$\wedge \, \text{sign}(v_1) \times \text{sign}(v_2) = -1\}, \qquad (10)$$

where v_i is the vertex and V is the vertex set of the Newton polytope, $\text{conv}(v_1, v_2)$ is the convex hull of vertices v_1, v_2 and $F_1(P)$ is the set of 1-dimensional face (edges) of the Newton polytope, $\text{sign}(v_i)$ represents the sign of the monomial which corresponds to vertex v_i. Figure 1 shows an example of Newton polytope construction for a single equation. Further definitions about properties of a polytope and Newton polytope can be found in [9,18].

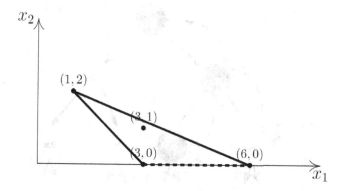

Fig. 1. An example of a Newton polytope for the polynomial $-x_1^6 + x_1^3 x_2 - x_1^3 + x_1 x_2^2$. In this example, all the monomial coefficients have order zero in ϵ and we want to solve the tropical problem $\min(3a_1 + a_2, a_1 + 2a_2) = \min(6a_1, 3a_1)$. The Newton polytope vertices $(6, 0), (3, 0), (1, 2)$ are connected by lines. The point $(3, 1)$ is not a vertex as it lies in the interior of the polytope. This stems to having $\min(3a_1 + a_2, a_1 + 2a_2) = a_1 + 2a_2$ for all tropical solutions, which reduces the number of cases to be tested. The thick edges satisfy the sign condition, whereas the dashed edge does not satisfy this condition. For this example, the solutions of the tropical problem are in infinite number and are carried by the two half-lines $a_1 = a_2 \geq 0$ and $5a_1 = 2a_2 \leq 0$, orthogonal to the thick edges of the Newton polygon

4.2   Computing Branches of Tropical Equilibrations

Using the Newton polytope formulation, one can then solve the tropical equilibration problem in (6) using the edges of Newton polytope (as in (8)). A feasible solution is a vector $(a_1, \dots, a_n)$ satisfying all the equations of system (6) and lies in the intersection of hyperplanes (or convex subsets of these hyperplanes) orthogonal to edges of Newton polytopes obeying the sign conditions. Of course, not all sequences of edges lead to non-void intersections and, thus, feasible solutions. This can be tested by the following linear programming problem resulting from (6):

$$\gamma_j(i) + \langle a, \alpha_j(i) \rangle = \gamma_j'(i) + \langle a, \alpha_j'(i) \rangle \leq \gamma_j'' + \langle a, \alpha_j'' \rangle),$$
$$\text{for all } j'' \neq j, j', \nu_{j''i} \neq 0, \quad i = 1, \dots, n \tag{11}$$

where $j(i), j'(i)$ define the chosen edge of the ith Newton polytope. The set of indices j'' can be restricted to vertices of the Newton polytope, because the inequalities are automatically fulfilled for monomials that are internal to the Newton polytope. From (11), the sequence of edges leading to a feasible solution is actually a set of linear inequalities and hence constitutes a feasible solution system (convex polytope), which was computed using Algorithm 1. Such

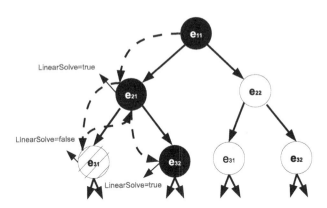

Fig. 2. Pruning strategy. The possible combinations of edges are represented in a tree representation where e_{ij} represents ith edge from jth Newton polytope. An edge set ne_i is the set of edges for ith Newton polytope. The algorithm starts by testing for feasible solution for first pair of edge sets. If a feasible solution is found, the algorithm proceeds further to other edge sets or it backtracks. In the figure, e_{11} and e_{21} are selected from edge sets ne_1, ne_2 and are checked for a feasible solution satisfying (11). If such a solution exists, it moves to e_{31} from the next edge set and again checks for feasible solution, if not then it backtracks to e_{21} and then to e_{32} which results in a feasible solution. Therefore, the sub-tree with root node e_{31} is discarded from future searches and this improves running time. Likewise the branch e_{11} and e_{22} is explored. This approach is similar to the branch and bound algorithm technique. The dashed arrows show the flow of the program

feasible solution systems are actually convex polytopes as defined in (6). The equivalence classes among the feasible solution systems constitute the equivalence classes of tropical equilibrations called *branches*. This was done using the method *equal_polyhedra* implemented in the software package polymake [6]. For instance, in the example of the preceding section, the choice of the edge connecting vertices $(1, 2)$ and $(3, 0)$ leads to the following linear programming problem:

$$a_1 + 2a_2 = 3a_1 \leq 6a_1,$$

whose solution is a half-line orthogonal to the edge of the Newton polygon. The pseudo-code is presented in Algorithm 1. It is clear from the above that the possible choices are exponential. In order to improve the running time of the algorithm, the pruning strategy evaluates (11) in several steps(cf. Algorithm 1 and Fig. 2). It starts with an arbitrary pair of edges and proceeds to add the next edge only when the inequalities (11) restricted to these two pair of edges are satisfied. The pruning method is a heuristic to filter out the infeasible set of edge combinations. A similar approach was undertaken in [3].

Algorithm 1. SolveOrders: Steps of tropical equilibration algorithm

    Input: List of edge sets $ne_1, ne_2, ..., ne_n$ (cf. Fig. 2), and the corresponding
               vertices of Newton polytope

    Output: Set of feasible solution systems (convex polytopes) corresponding to
               orders of the variables $a_1, a_2, ..., a_n$ (tropical equilibration solution
               set)

1 **begin**
2   solutionset $=\{\}$; integer $k=1$; equation $= \{\}$
3   SolveOrders(equation, k, edge-sets, vertices)
4   **if** $k > n$ **then**
5   return
6   **for** $l = 1$ *to number of entries in* ne_k *edge-set* **do**
7   equation(k)* = vertices in l^{th} row
8   inequalities* = all other vertices in ne_1 to ne_k edge-sets
9   **if** *LinearSolve(equation, inequalities) is feasible* **then**
10   **if** $k = n$ **then**
11   add the feasible solution system to solutionset
12   SolveOrders(equation, $k + 1$, $ne_1, .., ne_k$, vertices)
13 *The equations and inequalities are initialised as per (11)

4.3 Computation of Minimal Branches

The *minimal* branches are explained in section 3. Computation was performed using *included_polyhedra* method in polymake. The minimal branches as well as branches contained in minimal branches are represented in Fig. 3 as a directed graph with layers where top most layer are minimal branches.

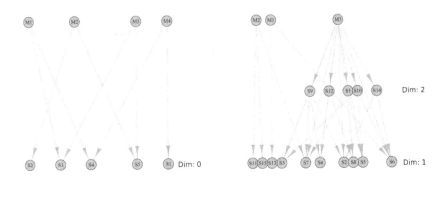

BIOMD0000000035   BIOMD0000000072

Fig. 3. A directed graph in layered form showing the inclusion relations among the different solution branches (for $\varepsilon = 1/11$) for two models namely BIOMD00000000-35,72. Vertices in the graph comprise of polytopes corresponding to solution branches and an directed edge between i and j means j is included in i. The topmost layer contain the minimal solution branches, thereafter the bottom layers are "included" solution branches. The layers of the included solution branches are based on the dimension of the corresponding polytopes (arranged in descending order). Therefore, included solutions in one layer are of same dimensional polytope

Sample Point for Minimal Branches. The polytopes corresponding to minimal solution branches obtained from the previous steps are represented by their facets and affine hull properties which are basically the set of inequalities and equalities. From such a set of inequalities, a sample point $(a_1, \ldots, a_n)$ is computed using Satisfiability Modulo Theories (SMT) solver called Microsoft Z3 software [2] in python programming environment. With Microsoft Z3, one can generate the sample point belonging exclusively to a minimal branch (and not at the intersections of minimal branches) by corresponding Boolean conjugations as shown in the following manner

$$\{T_i \in B_i \wedge T_i \notin N_i, \forall i \in I\} \tag{12}$$

where T_i is a tropical equilibration solution corresponding to B_i where B is set of polytopes corresponding to minimal solutions and N is the set of rest solution branches along with minimal solution branches $B \setminus B_i$. I is an index set denoting the elements of B.

The sample point thus obtained is a feasible solution to (11). For our purpose, the benefit of using Z3 over any existing linear programming software is that it distinguishes strict and non-strict inequality conditions, which allows us to generate the sample point belonging exclusively to a minimal branch.

5 Results

To compute the tropical equilibrations and to demonstrate the running time of our algorithm, 33 models from the r25 version of Biomodels database [11] having polynomial vector field were parsed with PoCaB [16].

5.1 Summary

A summary of the analysis is presented in Table 1. The analysis is performed to compute all possible combinations of Newton polytope edges leading to tropical solutions within a maximal running time of 10000 seconds of CPU time. In practice, we restrict this search space using the tree pruning strategy as explained in Fig. 2. The analysis was repeated with four different choices for ε values. In our framework, the number of variables may not be equal to the number of equations as the conservation laws (that are sums of variables whose total concentration is invariant) are treated as extra linear equations in our framework.

Table 1. Summary of analysis on Biomodels database. Tropical solutions here mean existence of at least one feasible solution from all possible combination of vertices of the Newton polytope. Timed-out means all solutions could not be computed within 10000 secs of computation time. No tropical solution implies that no possible combination of edges could be found resulting in a feasible solution. Model BIOMD0000000289 has solution at ε values 1/5,1/7 and 1/9 but no solutions at 1/23. Model BIOMD0000000108 has no solutions at all values of ε

ε value	Total models considered	Models without tropical solutions	Models with tropical solutions	Average running time (in secs)	Average number of minimal branches
1/5	33	1	32	299.38	3.24
1/7	33	1	32	244.12	3
1/9	33	1	32	309.73	3.75
1/23	33	2	31	3179.32	3.84

5.2 Running Time

A semilog time-plot is presented in Fig. 4(a) which plots the log of running time in milliseconds versus the number of equations in the model. In Fig. 4(b), the pruning ratio, i.e., the efficacy of tree pruning for ε value of 1/5 is plotted. The pruning ratio is the ratio between the number of times the linear programming is invoked with every tree pruning step (cf. Fig. 2) and the possible number of combinations of Newton polytope edges possible without tree pruning. This ratio is, thus, a measure of efficiency achieved due to pruning.

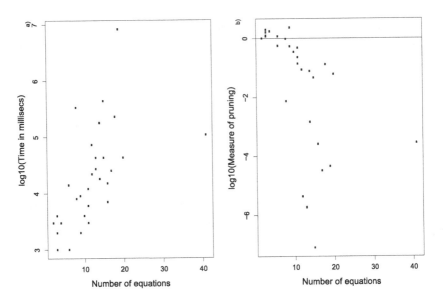

Fig. 4. (a) Plot of running time against number of equations in the model. (b) Pruning ratio for ε value of $1/5$ against number of equations in the model

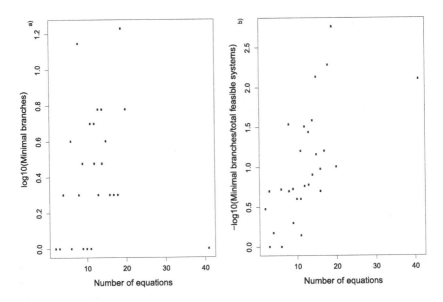

Fig. 5. (a) Minimal branches against number of equations in the model. (b) Ratio of minimal branches to the number of feasible solution systems, i.e., number of feasible edge combinations against number of equations in the model

5.3   Minimal Branches and Role of ε

A semilog plot for minimal solution branches is presented in Fig. 5(a) and a semilog plot in Fig. 5(b) showing the ratio of minimal solution branches to the number of feasible solution systems (obtained from Algorithm 1). It shows that a large proportion of feasible solution systems are either redundant or included in other feasible systems (i.e., inclusion relations).

In order to investigate the effect of different ε values on the number of minimal solutions, a boxplot is presented in Fig. 6(a) for different choices of ε values. In Fig. 6(b), the boxplot shows the ratio of minimal solution branches to the number of feasible solution systems (obtained from Algorithm 1) for different choices of ε values.

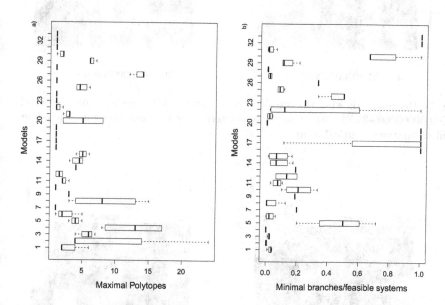

Fig. 6. Boxplots showing (a) Distribution of Minimal branches. (b) Ratio of minimal branches to the number of feasible solution systems. Both distributions are at different ε values: $1/5, 1/7, 1/9, 1/23$

5.4   Slow-Fast Variables

From the tropical equilibrations, we computed $\mu_j - a_i$(cf. (4)) which allow us to order the variables of the model, from the fastest (smallest $\mu_i - a_i$) to the slowest (largest $\mu_i - a_i$). This is a measure to separate the variables into slow and fast which is an important step in constructing the invariant manifold for model reduction [13,14,15,17]. The heatmap in Fig. 7 and Fig. 8 plots this value for minimal solution branch and all solution branches, respectively, for few selected models. For some models, there appears to be a natural clustering (as seen from the dendogram from the hierarchical clustering) which requires further investigation.

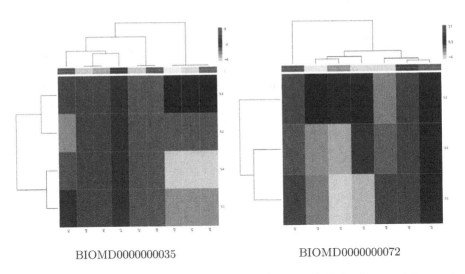

BIOMD0000000035   BIOMD0000000072

Fig. 7. Heatmaps showing the rescaled orders (at $\varepsilon = 1/11$) for four models namely BIOMD00000000-35,72 with hierarchical clustering for variables (horizontal axis) and tropical solutions (vertical axis).

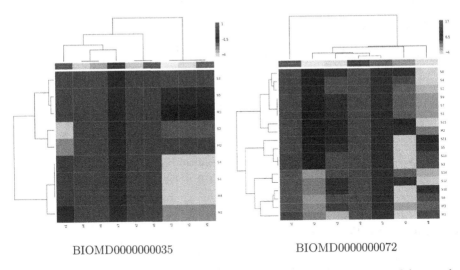

BIOMD0000000035   BIOMD0000000072

Fig. 8. Heatmaps showing the rescaled orders (at $\varepsilon = 1/11$) for four models namely BIOMD00000000-35,72 with hierarchical clustering for variables (horizontal axis) and tropical solutions (vertical axis). The heatmaps include the minimal solution branches (with prefix "M") and other solution branches excluding minimal ones (with prefix "S")

5.5   Connected Components

As described in section 1 and defined in section 3, tropical solutions can be roughly associated with metastable states and these branches (which are convex polytopes) form a polyhedral complex. The number of connected components of this complex indicates the possible number of transitions between the minimal solution branches. Figure 9 depicts such a graph of connected components of minimal solutions branches.

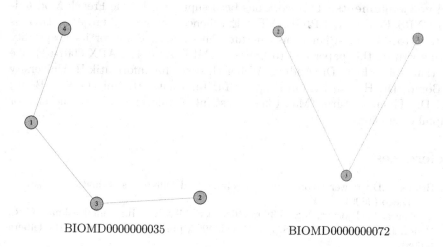

BIOMD0000000035   BIOMD0000000072

Fig. 9. Graph of connected components (at $\varepsilon = 1/11$) for four models namely BIOMD00000000-35,72. All of them have one connected component. The vertices are minimal solution branch and there exists an edge if the intersection between the two vertices is non-void

6   Discussions

We present an algorithm to compute the tropical equilibrations and organise them into *branches* and *minimal branches*. The directed graphs showing the inclusion relations between branches of tropical equilibration solutions reveal a rich structure. In addition, the connectivity of minimal branches are computed which provides an estimate for the possible dynamical transitions between them. One of the applications of tropical equilibration in systems biology is model reduction as demonstrated in [14,17]. The concentration orders depicted in the heatmaps demonstrate the applicability of the algorithm in this direction. Thus, the overall solution structure provides insights into the dynamics of the system. More precisely, for the biochemical models in the Biomodels database, a large number of feasible solution systems were obtained but the number of minimal branches is considerably less. For example, the number of minimal branches at $\varepsilon = 1/5$ ranged from 1 to 17, whereas the total number of solutions ranged from 1 to 9847.

As the dominant terms in the polynomial system are the same for all the tropical solution on branches, it could be that branches correspond to invariant manifolds. In the same spirit, minimal branches could correspond to minimal invariant manifolds. This idea will be pursued in future work. Lastly, we have shown an application of tropical geometry to invariant manifolds defined by polynomial systems but the direct application of tropical geometry to differential equation systems [7,1] is also known.

Acknowledgements. This work has been supported by the French ModRed-Bio CNRS Peps, and EPIGENMED Excellence Laboratory projects. D.G. is grateful to the Max-Planck Institut für Mathematik, Bonn for its hospitality during writing this paper and to Labex CEMPI (ANR-11-LABX-0007-01). We are thankful to Prof. Dr. Andreas Weber (Institut für Informatik II, University of Bonn), Dr. Hassan Errami (Institut für Informatik II, University of Bonn) and Dr. Thomas Sturm (Max-Planck-Institut fürInformatik, Saarbrücken) for helpful discussions.

References

1. Bruno, A.D.: Power Geometry in Algebraic and Differential Equations. Elsevier, San Diego (2000)
2. de Moura, L., Bjørner, N.S.: Z3: an efficient SMT solver. In: Ramakrishnan, C.R., Rehof, J. (eds.) TACAS 2008. LNCS, vol. 4963, pp. 337–340. Springer, Heidelberg (2008)
3. Emiris, I.Z., Canny, J.F.: Efficient incremental algorithms for the sparse resultant and the mixed volume. J. Symb. Comput. **20**(2), 117–149 (1995)
4. Feinberg, M.: Chemical reaction network structure and the stability of complex isothermal reactors-I. the deficiency zero and deficiency one theorems. Chemical Eng. Science **42**(10), 2229–2268 (1987)
5. Gatermann, K., Huber, B.: A family of sparse polynomial systems arising in chemical reaction systems. J. Symb. Comput. **33**(3), 275–305 (2002)
6. Gawrilow, E., Joswig, M.: Polymake: a framework for analyzing convex polytopes. In: Kalai, G., et al. (eds.) Polytopes Combinatorics and Computation, DMV Seminar, vol. 29, pp. 43–73. Springer Basel AG (2000)
7. Grigor'ev, D.Y., Singer, M.F.: Solving ordinary differential equations in terms of series with real exponents. Trans. Amer. Math. Soc. **327**(1), 329–351 (1991)
8. Haller, G., Sapsis, T.: Localized instability and attraction along invariant manifolds. SIAM J. Appl. Dyn. Syst. **9**(2), 611–633 (2010)
9. Henk, M., Richter-Gebert, J., Ziegler, G.M.: 16 basic properties of convex polytopes. Handbook of Discrete and Computational Geometry, pp. 243–270. CRC Press, Boca Raton (2004)
10. Jensen, A.: In: Stillman, M., Verschelde, J., Takayama, N. (eds.) Software for Algebraic Geometry. The IMA Volumes in Mathematics and its Applications, vol. 148 (2008)
11. Le Novere, N., Bornstein, B., Broicher, A., Courtot, M., Donizelli, M., Dharuri, H., Li, L., Sauro, H., Schilstra, M., Shapiro, B., Snoep, J.L., Hucka, M.: BioModels database: a free, centralized database of curated, published, quantitative kinetic models of biochemical and cellular systems **34**(suppl 1), D689–D691 (2006)

12. Maclagan, D., Sturmfels, B.: Introduction to Tropical Geometry. Graduate Studies in Mathematics, vol. 161. Amer. Math. Soc., RI (2015)
13. Noel, V., Grigoriev, D., Vakulenko, S., Radulescu, O.: Tropical geometries and dynamics of biochemical networks application to hybrid cell cycle models. In: Feret, J., Levchenko, A. (eds.) Proc. 2nd International Workshop on Static Analysis and Systems Biology (SASB 2011). Electronic Notes in Theoretical Computer Science, vol. 284, pp. 75–91. Elsevier (2012)
14. Noel, V., Grigoriev, D., Vakulenko, S., Radulescu, O.: Tropicalization and tropical equilibration of chemical reactions. In: Litvinov, G.L. and Sergeev, S.N. (edts.) Tropical and Idempotent Mathematics and Applications, Contemporary Mathematics, vol. 616, pp. 261-275. Amer. Math. Soc. (2014)
15. Radulescu, O., Vakulenko, S., Grigoriev, D.: Model reduction of biochemical reactions networks by tropical analysis methods. Mathematical Model of Natural Phenomena, in press (2015)
16. Samal, S.S., Errami, H., Weber, A.: PoCaB: a software infrastructure to explore algebraic methods for bio-chemical reaction networks. In: Gerdt, V.P., Koepf, W., Mayr, E.W., Vorozhtsov, E.V. (eds.) CASC 2012. LNCS, vol. 7442, pp. 294–307. Springer, Heidelberg (2012)
17. Soliman, S., Fages, F., Radulescu, O.: A constraint solving approach to model reduction by tropical equilibration. Algorithms for Molecular Biology 9(1), 24 (2014)
18. Sturmfels, B.: Solving systems of polynomial equations. CBMS Regional Conference Series in Math., no. 97, pp. 7–8. Amer. Math. Soc., Providence, RI (2002)
19. Thomas, R.: Boolean formalization of genetic control circuits. J. Theoret. Biology 42(3), 563–585 (1973)
20. Tyson, J.J.: Modeling the cell division cycle: cdc2 and cyclin interactions. Proc. National Academy of Sciences 88(16), 7328–7332 (1991)

Safety Verification of Hybrid Systems Using Certified Multiple Lyapunov-Like Functions*

Zhikun She, Dan Song, and Meilun Li

SKLSDE, LMIB and School of Mathematics and Systems Science,
Beihang University, Beijing, China

Abstract. In this paper, we present an efficient, hybrid numeric-symbolic method for safety verification of hybrid systems. To start with, we introduce a function of state, defined as a multiple Lyapunov-like function, whose time derivative along the trajectories is non-negative only outside of the initial set, such that its zero level set separates the unsafe region from all possible trajectories starting from the given initial set. Then, a numerical multiple Lyapunov-like function is computed by using sum of squares decomposition and semi-definite programming. Afterwards, in order to recover the possible unreliability of our numerical solution, we apply a continued fractions based rational recovery technique to this floating result and then obtain a certified one with rational coefficients, such that exact verification can be assured by this certified multiple Lyapunov-like function. Finally, several examples, together with discussions, are provided to illustrate the tractability and advantages of our method.

1 Introduction

A hybrid system [1,4,25] is a formal model that combines discrete and continuous dynamics. Continuous dynamics are associated with each mode, and discrete logic determines how to switch between modes. Hybrid systems play an increasingly important role in transportation networks as part of sophisticated embedded controllers used in the automotive industry and air-traffic management. Especially, as hybrid systems are often part of devices operating in safety-critical situations, the verification of safety properties becomes increasingly important. A hybrid system is considered safe if unsafe states cannot be reached starting from initial safe states.

Verification of discrete systems using temporal logic [5] and verification of continuous systems within the framework of robust control theory [42] are mature. Due to the new features deriving from combining discrete and continuous systems, safety verification of hybrid systems presents more challenges. During the last twenty years, many techniques have been proposed. One method is the exact [3,18] or approximate [7,17,38,37,13] computation of reachable sets. Another important method is the abstraction-based method [2,4,36,27] that first

---

* This work was partly supported by NSFC-11422111, NSFC-11371047, NSFC-11290141 and SKLSDE-2015ZX-06.

V.P. Gerdt et al. (Eds.): CASC 2015, LNCS 9301, pp. 440–456, 2015.
DOI: 10.1007/978-3-319-24021-3_32

abstracts the system and then analyzes the abstraction. In addition, in the past decade, the certificate-based method that directly searches for certificates of correctness (such as inductive invariants and Lyapunov functions) of systems has made significant progress [11,24,25,28,30,10,31,33,32,14,34]. Especially, the sum of squares (SOS) based method was proposed for safety verification [24] by reducing verification problems to SOS problems, which can then be numerically solved by SOS tools [26]. Further, since the sum of squares decomposition with rational coefficients can be computed [22,16], researches on exact safety verification were newly proposed in [40,19,41].

Following the ideas of barrier certificate [24,25] and rational recovery [22], we in this paper present an efficient, hybrid numeric-symbolic method for safety verification of hybrid systems by using multiple Lyapunov-like functions.

To start with, for a given hybrid system, we introduce a function of state, which is defined as a multiple Lyapunov-like function whose time derivative along its trajectories is non-negative only outside both the initial set and the image set of the reset map. Based on such a multiple Lyapunov-like function, the safety of the hybrid system can be assured since its zero level set separates the unsafe region from all possible trajectories starting from the given initial set.

Then, for a polynomial hybrid system, the search for a multiple Lyapunov-like function can be formulated as a sum of squares problem via application of our revised conditions, which is not only a semi-definite programming problem but essentially a linear SOS programming problem which can be solved in polynomial time, instead of a bilinear SOS programming problem which is NP-hard [39].

Afterwards, based on SOSTOOLS [22] and SeDuMi [35], we can obtain a numerical multiple Lyapunov-like function. Moreover, in order to recover the possible unreliability of our numerical solution (i.e., to overcome the numerical errors), we apply a continued fractions based rational recovery technique to this floating multiple Lyapunov-like function and arrive at a certified one with rational coefficients, such that exact safety verification can be assured by this certified multiple Lyapunov-like function.

Finally, several examples are provided to illustrate the tractability and advantages of our method. First, our method can overcome possible unreliability of numerical methods in verification via yielding certified Lyapunov-like functions. Second, our relaxed condition can define larger solution set. Third, we improve the computational efficiency by solving a linear SOS programming problem instead of a bilinear one.

The rest of the paper is organized as follows. Section 2 describes some notions about hybrid systems. In Section 3, multiple Lyapunov-like functions are addressed, associated with the SOS formulation for the safety verification problem. In Section 4, we propose our numeric-symbolic algorithm for obtaining a certified multiple Lyapunov-like function. In Section 5, an illustrating example is presented to show how our method works. In Section 6, four examples are provided to demonstrate the effectiveness of our method with discussions. Finally, we conclude the paper in Section 7.

2 Hybrid Systems Definition

In this section, we introduce some definitions for hybrid systems in the style of [1,24,25]. That is, a hybrid system is a tuple $H = (L, \chi, X_0, X_u, I, F, T)$, where:

1. L is a finite set of locations;
2. $\chi \subseteq \mathbb{R}^n$ is the continuous state space;
3. $X_0 \subseteq L \times \chi$ is the set of initial states;
4. $X_u \subseteq L \times \chi$ is the set of unsafe states, and $X_u \cap X_0 = \emptyset$;
5. $I : L \to 2^\chi$ describes the invariants of the locations;
6. $F : L \times \chi \to \mathbb{R}^n$ is an activity function such that for each $(l, x) \in L \times \chi$, $F(l, x) = f_l(x)$, where $f_l(x)$ is a continuous vector field that governs the flow of the system at location l;
7. $T \subseteq (L \times \chi) \times (L \times \chi)$ is a relation capturing discrete transitions between two locations, that is, a transition $((l, x), (l', x')) \in T$ indicates that from the state (l, x) the system can undergo a discrete jump to the state (l', x').

Let the overall state space of the hybrid system H be $X = L \times \chi$ and denote a state of H as $(l, x) \in X$. In addition, for each tuple $(l, l') \in L \times L$ with $l \neq l'$, we associate a guard set $Guard(l, l') = \{x \in \chi : ((l, x), (l', x')) \in T, x' \in \chi\}$, a continuous reset map $Reset(l, l') : x \in Guard(l, l') \mapsto \{x' \in \chi : ((l, x), (l', x')) \in T\}$, and the image set of the reset map $ImageR(l, l') = \{x' \in \chi : x \in Guard(l, l') \wedge x' \in Reset(l, l')(x)\}$.

Trajectories of the hybrid system H are concatenations of a sequence of continuous flows and discrete transitions. During a continuous flow, the discrete location l is maintained and the continuous state evolves according to the differential equation $\dot{x} = f_l(x)$, as long as x remains inside the invariant set $I(l)$. At a state (l_1, x_1), a discrete transition to (l_2, x_2) can occur if $((l_1, x_1), (l_2, x_2)) \in T$.

Given a hybrid system H and a set of unsafe states $X_u \subseteq X$, the safety verification problem is concerned with proving that all trajectories of the hybrid system H starting from the initial set X_0 cannot enter the unsafe region X_u.

3 Verification Using Multiple Lyapunov-Like Functions

We will first introduce safety verification of continuous systems in Subsection 3.1, extend it to hybrid systems in Subsection 3.2 and then formulate the safety verification problem of polynomial hybrid systems as an SOS problem in Subsection 3.3.

3.1 Safety Verification of Continuous Systems

In this subsection we consider continuous systems of the form:

$$\dot{x} = f(x). \tag{1}$$

Our method for verifying the safety of system (1) relies on the existence of a Lyapunov-like function [28,33], which satisfies conditions not only of the function itself on the initial set and the unsafe set, but also of its time derivative along the flow of the system outside the initial set. These conditions, stated in the theorem below, guarantee the safety of a given system by depicting a 'barrier' between possible system trajectories and the unsafe region. Note that the concept of Lyapunov-like functions is similar to that of barrier certificates [24,25]. However, we use different conditions, making formal verification with SOS tools more feasible.

Theorem 1. *For a given system* (1) *with the sets* χ, χ_0 *and* χ_u *(i.e., the continuous state space, the initial set and the unsafe set, respectively), if there exists a differentiable Lyapunov-like function* $V : \chi \mapsto \mathbb{R}$ *satisfying:*

$$V(x) > 0, \forall x \in \chi_0, \tag{2}$$

$$\frac{\partial V(x)}{\partial x} f(x) > 0, \forall x \in \{x : V(x) \geq 0 \wedge x \notin \chi_0\}, \tag{3}$$

$$V(x) < 0, \forall x \in \chi_u, \tag{4}$$

then the safety of the system (1) *is guaranteed.*

Proof. Suppose that $V(x)$ is a Lyapunov-like function satisfying the conditions in Theorem 1. For any arbitrary trajectory $x(t)$ in χ that starts from some $x_0 \in \chi_0$, we assert that for all t, $V(x(t)) \geq 0$ and thus the safety of the system is guaranteed. Otherwise, there is a time instant t' such that $V(x(t')) < 0$. Let $t_1 = \inf\{t : V(x(t)) < 0\}$. Due to the continuity of $V(x)$, there exists a $\delta_0 > 0$ such that for all $t \in (t_1, t_1 + \delta_0]$, $V(x(t)) < 0$ and $V(x(t_1)) = 0$. By the condition (2), $x(t_1) \notin \chi_0$. Thus, by the condition (3), $\frac{\partial V(x(t_1))}{\partial x} f(x(t_1)) > 0$. Due to the continuity of $\frac{\partial V(x(t))}{\partial x} f(x(t))$, there exists a δ_1 such that for all $t \in [t_1, t_1 + \delta_1]$, $\frac{\partial V(x(t))}{\partial x} f(x(t)) > 0$. So, for all $t \in (t_1, t_1 + \min\{\delta_0, \delta_1\}]$, $V(x(t)) > 0$, contradicting with $V(x(t)) < 0$ when $t \in (t_1, t_1 + \min\{\delta_0, \delta_1\}]$. □

Note that we choose $\frac{\partial V(x)}{\partial x} f(x) > 0$ instead of $\frac{\partial V(x)}{\partial x} f(x) \geq 0$ in the condition (3). The reason for such a choice can be shown from a simple one-dimensional counterexample, where the dynamical system is described by $\dot{x} = -1$, the initial set is defined to be $\chi_0 = \{x | x > 0\}$ and the unsafe set is defined to be $\chi_u = \{x | x < 0\}$. For this example, the system is clearly unsafe, but the function $V(x) = x^3$ satisfies the conditions (2) and (4) and $\frac{\partial V(x)}{\partial x} f(x) \geq 0$ in the set described in condition (3).

Further, note that by the S-Procedure [8], we can formulate the conditions (2), (3) and (4) as a bilinear SOS programming problem, which is an NP-hard problem. Although some iterative methods have been presented in [24,25,6] for efficiently solving bilinear SOS programming, it is still challenging to decrease its computational complexity. In order to obtain a linear SOS programming problem which can then be solved in polynomial time, we would like to derive an alternative condition for condition (3) as follows, whose SOS formulation will be described by the condition (20) in Section 3.3.

Proposition 1. *Let the system* (1) *and the sets* χ, χ_0 *and* χ_u *be given. If there exists a differentiable Lyapunov-like function* $V : \chi \mapsto \mathbb{R}$ *satisfying the conditions* (2), (4) *and*

$$\frac{\partial V(x)}{\partial x} f(x) \geq 0, \forall x \in \{x : x \notin \chi_0\}, \tag{5}$$

then the safety of the system (1) *is guaranteed.*

Proof. Suppose that $V(x)$ is a Lyapunov-like function satisfying the conditions in Proposition 1. For any arbitrary trajectory $x(t)$ in χ that starts from some $x_0 \in \chi_0$, we assert that for all t, $V(x(t)) \geq 0$ and thus the safety of the system is guaranteed. Otherwise, there is a time instant t' such that $V(x(t')) < 0$. Let $t_1 = \inf\{t : V(x(t)) < 0\}$. Due to the continuity of $V(x)$, there exists a $\delta_0 > 0$ such that for all $t \in (t_1, t_1 + \delta_0]$, $V(x(t)) < 0$ and $V(x(t_1)) = 0$. By the condition (2), for all $t \in [t_1, \delta_0]$, $x(t) \notin \chi_0$. Thus, by the condition (3), for all $t \in [t_1, \delta_0]$, $\frac{\partial V(x(t))}{\partial x} f(x(t)) \geq 0$. So, for all $t \in [t_1, t_1 + \delta_0]$, $V(x(t)) \geq 0$, contradicting with $V(x(t)) < 0$ when $t \in (t_1, t_1 + \delta_0]$. □

Remark 1. The solutions of our condition (5) cannot be directly compared to Proposition 3 of [25] since $\forall x \in \{x : V(x) = 0\}$ is weaker than $\forall x \in \{x : x \notin \chi_0\}$ while $\frac{\partial V(x)}{\partial x} f(x) > 0$ is stronger than $\frac{\partial V(x)}{\partial x} f(x) \geq 0$.

3.2   Safety Verification of Hybrid Systems

In this subsection we will extend the safety verification of continuous systems to hybrid systems by using a multiple Lyapunov-like function. Different from a Lyapunov-like function in Subsection 3.1, a multiple Lyapunov-like function depends not only on different single locations, but also on the reset map of the hybrid system. For details, we would like to state the conditions that must be satisfied by a multiple Lyapunov-like function in the following theorem.

Theorem 2. *For a given hybrid system $H = (L, \chi, X_0, X_u, I, F, T)$, assume that there exists a differential multiple Lyapunov-like function, i.e., a set of differential Lyapunov-like functions $\{V_l(x), l \in L\}$ which, for each $l \in L$ and $(l, l') \in L \times L, l \neq l'$, satisfy*

$$V_l(x) > 0, \forall x \in \{x : (l, x) \in X_0\}, \tag{6}$$

$$V_l(x) < 0, \forall x \in \{x : (l, x) \in X_u\}, \tag{7}$$

$$\frac{\partial V_l(x)}{\partial x} f_l(x) > 0, \forall x \in \{x : x \in I(l) \wedge V_l(x) \geq 0 \wedge (l, x) \notin X_0\}, \tag{8}$$

$$V_{l'}(x') > 0, \forall x' \in RS(l, l'), \tag{9}$$

$$\frac{\partial V_{l'}(x)}{\partial x} f_{l'}(x) > 0, \forall x \in \{x : x \in I(l') \wedge V_{l'}(x) \geq 0 \wedge x \notin ImageR(l, l')\}, \tag{10}$$

where

$$RS(l, l') = \{x' : x \in I(l) \wedge x \in Guard(l, l') \wedge V_l(x) \geq 0 \wedge x' \in Reset(l, l')(x)\},$$

then the safety of the hybrid system H is guaranteed.

Proof. Assume that there exists a trajectory $r(t)$ from the initial set to the unsafe set. Then, we try to derive a contradiction.

Let $r(t)$ be a sequence of $r_i(t) = ((l_i, x(t)))$s with $t \in [t_i, t_{i+1})$, where $x(t)$ is a flow in the location l_i satisfying that at each t_{i+1}, $x(t_{i+1}) \in RS(l_i, l_{i+1})$. With the assumption, there exist a location l_k and a time instant t_u with $t_k \leq t_u < t_{k+1}$ such that $r_k(t_u) = (l_k, x(t_u)) \in X_u$. If $k = 0$, then $l_k = l_0, x(t_0) = x(0) \in X_0$. With the conditions (6), (7) and (8), using the conclusion of Theorem 1, it is guaranteed that for all $t \in [t_0, t_1)$, $V_{l_0}(x(t)) \geq 0$, contradicting with $V_{l_0}(x(t_u)) < 0$. If $k > 0$, $x(t_k) \in RS(l_{k-1}, l_k)$ and thus $V_{l_k}(x(t_k)) > 0$ with condition (9). Let $\bar{V} = V_{l_k}, \bar{X}_0 = RS(l_{k-1}, l_k)$ and $\bar{X}_u = X_u$. Then, $\bar{V}$ is a Lyapunov-like function of the system $\dot{x} = f_k(x)$ with initial states $\bar{X}_0$ and unsafe states $\bar{X}_u$. Again from Theorem 1, with conditions (9), (7) and (11), we have that for all $t \in [t_k, t_{k+1})$, $V_{l_k}(x(t)) \geq 0$, contradicting with $V_{l_k}(x(t_u)) < 0$. □

Similarly, the conditions (8), (9) and (10) will lead to a bilinear SOS programming problem which is NP-hard. By the proof of Proposition 1 and Theorem 2, we can similarly derive the following alternative conditions such that linear SOS programming, which can be solved in polynomial time, can be used to search a multiple Lyapunov-like function.

Proposition 2. *For a hybrid system $H = (L, \chi, X_0, X_u, I, F, T)$, suppose that there exists a differential multiple Lyapunov-like function, i.e., a set of differential Lyapunov-like functions $\{V_l(x), l \in L\}$ which, for each $l \in L$ and $(l, l') \in L \times L$, where $l \neq l'$, satisfy the conditions (6)–(7) and*

$$\frac{\partial V_l(x)}{\partial x} f_l(x) \geq 0, \forall x \in \{x : x \in I(l) \wedge (l, x) \notin X_0\}, \tag{11}$$

$$V_{l'}(x') > 0, \forall x' \in ImageR(l, l'), \tag{12}$$

$$\frac{\partial V_{l'}(x')}{\partial x} f_{l'}(x') \geq 0, \forall x' \in \{x' \in I(l') : x' \notin ImageR(l, l')\}, \tag{13}$$

then the safety of the hybrid system H is guaranteed.

3.3   SOS Formulation for Safety Verification

In general, it is difficult to construct multiple Lyapunov-like functions satisfying conditions introduced in Proposition 2. In this subsection, for a class of polynomial hybrid systems, whose vector fields are polynomial and whose set descriptions are semi-algebraic, we will propose a sums of squares (SOS) based method to construct multiple Lyapunov-like functions for safety verification.

Firstly, we introduce the sum of squares decomposition and its corresponding solving techniques. A multivariate polynomial $p(x)$ is a sum of squares if there exist polynomials $p_i(x)$, $i = 1, \ldots, m$, such that $p(x) = \sum_{i=1}^{m} p_i^2(x)$. This is equivalent to the existence of a quadratic form $p(x) = Z^T(x)QZ(x)$ for some positive semi-definite matrix Q and vector of monomials $Z(x)$. And a sum of squares decomposition for $p(x)$ can be computed by using semi-definite programming [22,35,20], which provides a polynomial-time computational relaxation for proving global nonnegativity of multivariate polynomials [21].

Secondly, we would like to extend the above SOS formula for the construction of multiple Lyapunov-like functions. For this, we consider a polynomial hybrid system of form $H = (L, \chi, X_0, X_u, I, F, T)$ such that

$$I(l) = \{x \in \chi : \bigwedge_{k=1}^{w_l} g_{l,k}(x) \geq 0\}, \qquad (14)$$

$$X_0 = \bigcup_{l \in L} X_0(l), \quad X_0(l) = \{x \in \chi : S_l(x) \geq 0\}, \qquad (15)$$

$$X_u = \bigcup_{l \in L} X_u(l), \quad X_u(l) = \{x \in \chi : \bigwedge_{k=1}^{v_l} U_{l,k}(x) \geq 0\}, \qquad (16)$$

$$Guard(l, l') = \{x \in \chi : Gd_{l,l'}(x) \geq 0\}, \qquad (17)$$

$$Reset(l, l')(x) = \{x' \in \chi : x' = R_{l,l'}(x)\}, \qquad (18)$$

where $g_{l,k}(x)$, $S_{l,k}(x)$, $U_{l,k}(x)$, $Gd_{l,l'}(x)$ and $R_{l,l'}(x)$ are all polynomials, and there is a polynomial $IR_{l,l'}(x')$ satisfying

$$ImageR(l, l') = \{x' \in \chi : IR_{l,l'}(x') \geq 0\}. \qquad (19)$$

Note that for some $l \in L$, if $X_0(l)$ is defined to be $\{x \in \chi : \bigvee_{k=1}^{m_l} S_{l,k}(x) \geq 0\}$, where $S_{l,k}(x)$ is a polynomial, we can construct a new hybrid system $H' = (L', \chi, X_0', X_u', I', F', T')$ of forms (14)–(19) such that the safety of H' implies the safety of H as follows:

1. let $L' = (L \setminus \{l\}) \cup \{l_i : i \in \{1, \ldots, m_l\}\}$;
2. for each location $l' \in L'$, if $l' \in L \setminus \{l\}$, let $I'(l') = I(l')$, $F'(l', x) = f_{l'}(x)$, $X_0'(l') = X_0(l')$ and $X_u'(l') = X_u(l')$; if $l' \in \{l_i : i \in \{1, \ldots, m_l\}\}$ and $l' = l_i$, where $i \in \{1, \ldots, m_l\}$, let $I'(l') = I(l)$, $F'(l', x) = f_l(x)$, $X_0'(l') = \{x \in \chi : S_{l,i}(x) \geq 0\}$ and $X_u'(l') = X_u(l)$;
3. for each (l_i', l_j'), where $l_i' \neq l_j'$ and $l_i', l_j' \in L'$,
   (a) if $l_i', l_j' \in L \setminus \{l\}$, let $Guard'(l_i', l_j') = Guard(l_i', l_j')$, $Reset'(l_i', l_j')(x) = Reset(l_i', l_j')(x)$, $ImageR'(l_i', l_j') = ImageR(l_i', l_j')$;
   (b) if $l_i' \in L \setminus \{l\}$ and $l_j' \in \{l_i : i \in \{1, \ldots, m_l\}\}$, let $Guard'(l_i', l_j') = Guard(l_i', l)$, $Reset'(l_i', l_j')(x) = Reset(l_i', l)(x)$, $ImageR'(l_i', l_j') = ImageR(l_i', l)$;
   (c) if $l_i' \in \{l_i : i \in \{1, \ldots, m_l\}\}$ and $l_j' \in L \setminus \{l\}$, let $Guard'(l_i', l_j') = Guard(l, l_j')$, $Reset'(l_i', l_j')(x) = Reset(l, l_j')(x)$, $ImageR'(l_i', l_j') = ImageR(l, l_j')$;
   (d) if $l_i' \in \{l_i : i \in \{1, \ldots, m_l\}\}$ and $l_j' \in \{l_i : i \in \{1, \ldots, m_l\}\}$, there are no associated guard set.

In addition, similar construction can be carried out when $I(l)$, $X_u(l)$, $Guard(l, l')$ and $Reset(l, l')(x)$ are defined to be a corresponding union of forms (14), (16), (17), (18) and (19), respectively.

Now, given a hybrid system $H = (L, \chi, X_0, X_u, I, F, T)$ of forms (14)–(19), by using the conditions stated in Proposition 2, the safety verification problem can be formulated as a sum of squares problem as follows.

Proposition 3. *Given a hybrid system* $H = (L, \chi, X_0, X_u, I, F, T)$ *of forms* (14)–(19), *suppose that there exist a differential multiple Lyapunov-like function* $\{V_l(x), l \in L\}$, *positive numbers* $\epsilon_{l,1}, \epsilon_{l,2}, \epsilon_{l',1}$, *and sums of squares* $\theta_l(x)$, $\xi_{l,i}(x)$, $\varsigma_l(x)$, $\phi_{l,j}(x)$, $\upsilon_l(x)$, $\psi_{l,k}(x)$, $\varphi_l(x')$ *and* $\rho_{l',m}(x')$, *where* $l, l' \in L$ *and* $l \neq l'$, *such that the following expressions:*

$$V_l(x) - \theta_l(x)S_l(x) - \epsilon_{l,1}, \tag{20}$$

$$-V_l(x) - \sum_{i=1}^{\upsilon_l} \xi_{l,i}(x)U_{l,i}(x) - \epsilon_{l,2}, \tag{21}$$

$$\frac{\partial V_l(x)}{\partial x} f_l(x) + \varsigma_l(x)S_l(x) - \sum_{j=1}^{\omega_l} \phi_{l,j}(x)g_{l,j}(x), \tag{22}$$

$$V_{l'}(R_{l,l'}(x)) - \upsilon_l(x)Gd_{l,l'}(x) - \sum_{k=1}^{\omega_l} \psi_{l,k}(x)g_{l,k}(x) - \epsilon_{l',1}, \tag{23}$$

$$\frac{\partial V_{l'}(x')}{\partial x} f_{l'}(x') + \varphi_l(x')IR_{l,l'}(x') - \sum_{m=1}^{\omega_{l'}} \rho_{l',m}(x')g_{l',m}(x'), \tag{24}$$

are sums of squares. Then, the multiple Lyapunov-like function $\{V_l(x), l \in L\}$ *satisfies the conditions in Proposition 2, implying that the safety of the system* H *is guaranteed.*

Proof. Since the expressions (20)–(24) are sums of squares, they are all nonnegative. In addition, $\epsilon_{l,1}$ and $\epsilon_{l,2}$ are positive, and since $\theta_l(x)$, $\xi_{l,i}(x)$, $\varsigma_l(x)$, $\phi_{l,j}$, $\upsilon_l(x)$, $\psi_{l,k}(x)$, $\varphi_l(x')$, $\rho_{l',m}(x')$ are sum of squares, they are all nonnegative.

Taking any state $(l, x) \in X_0$, we have $V_l(x) \geq \theta_l(x)S_l(x) + \epsilon_{l,1} \geq \epsilon_{l,1} > 0$ since $S_l(x) \geq 0$ and $\theta_l(x)S_l(x) \geq 0$, implying that the condition (6) is satisfied. Similarly, the conditions (7), (11)–(13) are satisfied by $\{V_l(x), l \in L\}$.

Thus, by Proposition 2, the safety of the hybrid system is guaranteed.   □

Remark 2. When $R_{l,l'}(x)$ is reversible, the set $ImageR(l, l')$ can be represented by $\{x' \in \chi : Gd_l(R_{l,l'}^{-1}(x')) \geq 0\}$ and the condition (24) can be written as

$$\frac{\partial V_{l'}(x')}{\partial x} f_{l'}(x') + \varphi_l(x')Gd_{l,l'}(R_{l,l'}^{-1}(x')) - \sum_{k=1}^{\omega_{l'}} \rho_{l',k}(x')g_{l',k}(x'). \tag{25}$$

When $R_{l,l'}(x)$ is irreversible, we can replace the image set $ImageR(l, l')$ with an under-approximation for the condition (24).

4   Computation of Certified Multiple Lyapunov-Like Functions

Given a hybrid system $H = (L, \chi, X_0, X_u, I, F, T)$ of forms (14)–(19), for each $l \in L$, we first parameterize the corresponding Lyapunov-like function $V_l(x)$ as

$$V_l(x) = b_{0,l}(x) + \sum_{i=1}^{w} c_{i,l}b_{i,l}(x),$$

where $b_{i,l}(x)$s are monomials in x and $c_{i,l}$s are real parametric coefficients. Note that we usually pre-fixed an upper bound m on the degree of a Lyapunov-like function and let all monomials be of degrees less than or equal to the bound.

Then, using SOS toolboxes such as SOSTOOLS [26], we can solve the sum of squares problem stated in Proposition 3 and obtain a numerical multiple Lyapunov-like function $\{\overline{V_l}(x), l \in L\}$. Due to the numerical errors, such a numerical solution may lead to dissatisfaction of the conditions in Proposition 3. Thus, in this section, we would like to rationalize the coefficients of $\{\overline{V_l}(x), l \in L\}$ and obtain a certified multiple Lyapunov-like function $\{V_l(x), l \in L\}$.

In general, rationalizing is challenging because computers are only good at expressing limited decimal. Recently, a continued fractions and projection based method for exact sum of squares decomposition has been presented in [22] and its corresponding algorithm has been implemented in [26,9]. However, this method currently fits problems which contain only one SOS constraint. Since our SOS problems involve at least three SOS constraints, which can be easily seen from Proposition 3, it is pressingly required to adapt the method for efficiently solving our problems. For this, we first briefly recall the continued fractions as follows. A continued fraction represents a real number r with the following expression:

$$r = a_0 + \cfrac{1}{a_1 + \cfrac{1}{a_2 + \cfrac{1}{\ddots}}},$$

where a_0 is an integer and a_i $(i \geq 1)$ are positive integers. These a_is can be obtained through an iterative process by first representing the number r as the sum of its integer part a_0 and the reciprocal of another number $\overline{a_1}$, and then writing $\overline{a_1}$ as the sum of its integer part a_1 and the reciprocal of $\overline{a_2}$, and so on. Continued fractions are a sensible choice since truncating the series yields best rational approximations [22] in the sense that a best rational approximation of a real number r is a rational number $\frac{a}{b}$ with $b > 0$ such that there is no rational number with smaller denominator which is closer to r.

Then, based on continued fractions, let us explain our essential idea for rationalizing the numerical computational results obtained by SOSTOOLS for exactly assuring the non-negativeness of the expressions (20)–(24). Firstly, for a given precision, we compute a rational $\{V_l(x)\}$ for the expressions (20)–(24) by using continued fractions [15]; Secondly, for each expression in Proposition 3, we use this rational $\{V_l(x)\}$ to construct a corresponding new SOS problem and then numerically solve this new SOS problem for the other sums of squares and the positive numbers; Thirdly, for each expression, rationalize a positive number or a sum of squares, use this rationalized result to construct another SOS problem and then numerically solve it; Fourthly, we repeat the third step until all the required sums of squares and the positive numbers in Proposition 3 are rationalized; Finally, we use the method introduced in [22] by SOSTOOLS [26] to check whether these expressions are all rational sums of squares.

Clearly, if all expressions with the above rationalized results are all rational sums of squares, the exact safety verification is guaranteed. Thus, based on the

---

Algorithm 1. Computing a rational multiple Lyapunov-like function

---

Input: a hybrid system $H = (L, \chi, X_0, X_u, I, F, T)$ with sets $Guard(l, l')$, $Reset(l, l')(x)$ and $ImageR(l, l')$, the upper bound m for the degree of candidate Lyapunov-like functions, and the precision τ for rationalizing.

Output: a certified rational multiple Lyapunov-like function $\{V_l(x)\}$.

1: for each $l \in L$, construct a polynomial $V_l(x) = b_{0,l}(x) + \sum_{i=1}^{m} c_{i,l} b_{i,l}(x)$ with parametric coefficients $c_{1,l}, \ldots, c_{m,l}$, where $b_{i,l}(x)$ are all monomials with degrees less than or equal to m,

2: solve the sum of squares problem in Proposition 3 without objective functions, and obtain the numerical multiple Lyapunov-like function $\{\overline{V_l}(x)\}$,

3: in the light of the given precision τ, compute a rational approximation $\{V_l(x)\}$ of the numerical multiple Lyapunov-like function $\{\overline{V_l}(x)\}$ by continued fractions (i.e., for each pair of coefficients $(\overline{c_{i,l}}, c_{i,l})$ of $\overline{V_l}(x)$ and $V_l(x)$, $\| \overline{c_{i,l}} - c_{i,l} \| \leq \tau$), and substitute this $\{V_l(x)\}$ into the expressions in Proposition 3,

4: for each expression in Proposition 3, use $\{V_l(x)\}$ to construct a corresponding new SOS problem, numerically solve this new SOS problem for the positive numbers and the other sums of squares, and then rationalize a positive number or a sum of squares; substitute this rationalized result into each expression in Proposition 3 and repeat the above process for numerical solving and rationalizing until all the required sums of squares and the positive numbers in Proposition 3 are rationalized,

5: use the SOSTOOLS [26] to check whether all these expressions with obtained rationalized results are rational sums of squares; if the rationalized expressions (20)–(24) are all sums of squares, return $\{V_l(x)\}$ as a certified multiple Lyapunov-like function.

---

above analysis, we present Algorithm 1 for searching for a certified rational multiple Lyapunov-like function satisfying the conditions in Proposition 3.

Remark 3. Generally, the SOS problem solved with semi-definite programming has an objective function, which is usually requested to be linear. However, linear objective functions always reach their optimal values at the boundary of the feasible region such that the rationalized $\{\overline{V_l}'(x)\}$ will possibly fall into the infeasible region. In order to avoid this case, our method does not use any objective function.

5 An Illustrating Example

Example 1. Consider a two-dimensional system $(\dot{x}_1, \dot{x}_2) = (x_1 x_2, x_2^2 + x_1^4)$, where

$$\chi_0 = \{x : S(x) = 0.25 - x_1^2 - (x_2 - 1)^2 \geq 0\},$$
$$\chi_u = \{x : U(x) = 0.16 - (x_1 + 2)^2 - x_2^2 \geq 0\},$$

and the phase portrait of the system is shown in Figure 1.

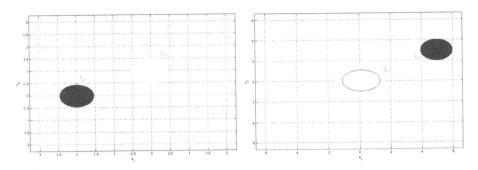

Fig. 1. Phase portrait for Example 1.   **Fig. 2.** Phase portrait for Example 2.

1. Let $V(x) = a + bx_1 + cx_2 + dx_1^2 + ex_1x_2 + fx_2^2$ with parameters a, b, c, d, e and f. We want to find $V(x)$, $\theta(x)$, $\xi(x)$, $\varsigma(x)$, ϵ_1 and ϵ_2 such that

$$V(x) - \theta(x)S(x) - \epsilon_1, \tag{26}$$

$$-V(x) - \xi(x)U(x) - \epsilon_2, \tag{27}$$

$$\frac{\partial V(x)}{\partial x}f(x) + \varsigma(x)S(x) \tag{28}$$

   are sums of squares.
2. After 18 iterations by SOSTOOLS [26], we obtain the below numerical solution:

$$\overline{V}(x) = 0.24448 + 1.1547 \times 10^{-6}x_1 + 2.6072x_2 - 1.1150x_1^2$$
$$-5.4253 \times 10^{-7}x_1x_2 + 0.00034980x_2^2,$$
$$\overline{\theta}(x) = 2.4008, \ \overline{\epsilon_1} = 1.3996, \ \overline{\xi}(x) = 2.3083, \ \overline{\epsilon_2} = 1.4072, \ \overline{\varsigma}(x) =$$
$$2.164 \times 10^{-14}.$$

   Substituting the numerical solution into the expression (28), we obtain that $\frac{\partial \overline{V}(x)}{\partial x}f(x) + \overline{\varsigma}(x)S(x)|_{x=0} = -0.75 \times 2.164 \times 10^{-14} < 0$, implying that the numerical expression $\frac{\partial \overline{V}(x)}{\partial x}f(x) + \overline{\varsigma}(x)S(x)$ is not actually an SOS.
3. In order to avoid the unreliability of the numerical solution, we make use of continued fractions to rationalize $\overline{V}(x)$ with $\tau = 0.01$ and get $V(x) = \frac{11}{45} + \frac{8}{3}x_2 - \frac{10}{9}x_1^2$.
4. Keeping $V(x)$ fixed, let us first consider the expression (26) as follows.
   (a) We first numerically solve the corresponding SOS problem, which only contains the constraint (26), and get numerical solution $\overline{\theta}(x) = 2.4008$ for $\theta(x)$. Then we run the similar rationalization process for $\overline{\theta}(x)$ and get $\theta(x) = \frac{12}{5}$.
   (b) Keeping $V(x)$ and $\theta(x)$ fixed, we run the similar process as above and get a numerical solution $\overline{\epsilon_1} = 1.3996$ and a rationalization $\epsilon_1 = \frac{7}{5}$.
   (c) With all the above rational solutions, we check whether the expression (26) is an SOS by the function *findsos* in SOSTOOLS [26] and get

$$V(x) - \theta(x)S(x) - \epsilon_1 = Z^T(x)QZ(x),$$

$$\text{where } Z(x) = \begin{bmatrix} 1 \\ x_2 \\ x_1 \end{bmatrix} \text{ and } Q = \begin{bmatrix} 29/45 & -16/15 & 0 \\ -16/15 & 12/5 & 0 \\ 0 & 0 & 58/45 \end{bmatrix}. \text{ Clearly, } Q \text{ is}$$

semi-definite and thus the expression (26) is an SOS.

5. We run the similar process for the expressions (27) and (28) and get $\xi(x) = \frac{30}{13}$, $\epsilon_2 = \frac{38}{27}$ and $\varsigma(x) = 0$. It can be certified that the expressions (27) and (28) with the above rational solutions are both sums of squares.

6. Due to Propositions 3 and 2, the safety of this system is guaranteed.

6 More Examples with Discussions

In this section, based on the prototypical implementation, we use four examples continuous/hybrid systems of two/three dementions to illustrate the advantages of our method. All the computations in this subsection were performed on a Hasee notebook of Core II Duo, 2.00 GHz with 2 GB RAM, and each example takes no more than 15 seconds to get the final result.

6.1 More Examples

Example 2. Consider a continuous two-dimensional system:

$$\begin{cases} \dot{x}_1 = -x_2 - x_1 \left(x_1^2 + x_2^2 - 1\right)^3 \\ \dot{x}_2 = x_1 - x_2 \left(x_1^2 + x_2^2 - 1\right)^3 \end{cases},$$

where the initial set is $\chi_0 = \{x \in R^2 : S(x) = 1.5 - x_1^2 - x_2^2 \geq 0\}$, the unsafe set is $\chi_u = \{x \in R^2 : U(x) = 1 - (x_1 - 5)^2 - (x_2 - 3)^2 \geq 0\}$, and the phase portrait of the system is shown in Figure 2, in which the black solid ring is the stable limit cycle of the system.

Letting $V(x) = a + bx_1 + cx_2 + dx_1^2 + ex_1x_2 + fx_2^2$ with parameters a, b, c, d, e and f, we want to solve $V(x), \theta(x), \xi(x), \varsigma(x), \epsilon_1$ and ϵ_2 such that

$$V(x) - \theta(x)S(x) - \epsilon_1,$$
$$-V(x) - \xi(x)U(x) - \epsilon_2,$$
$$\frac{\partial V(x)}{\partial x} f(x) + \varsigma(x)S(x)$$

are sum of squares. By the algorithm in section 4 with $\tau = 0.001$, we get certified solutions below:

$$V(x) = \frac{7}{9} - \frac{x_1^2}{7} - \frac{x_2^2}{7}, \ \theta(x) = \frac{2}{7}, \ \epsilon_1 = \frac{5}{19}, \ \xi(x) = \frac{40}{613}, \ \epsilon_2 = \frac{85}{171}, \ \varsigma(x) = \frac{5}{7}.$$

Example 3. Consider a continuous three-dimensional system:

$$\begin{cases} \dot{x}_1 = -x_1 + x_2 \\ \dot{x}_2 = -x_1 - x_2 - x_1x_3 \\ \dot{x}_3 = -x_3 + x_1x_2 \end{cases},$$

where the initial set is $\chi_0 = \{x \in R^3 : S(x) = 0.0196 - x_1^2 - x_2^2 - (x_3 - 0.2)^2 \geq 0\}$ and the unsafe set is $\chi_u = \{x \in R^3 : U(x) = 0.01 - x_1^2 - x_2^2 - (x_3 + 1)^2 \geq 0\}$.

Similar to Example 2, by Algorithm 1 with $\tau = 0.001$, we can get the following certified solutions:

$$V(x) = \tfrac{3}{13} - \tfrac{3}{5}x_1^2 - \tfrac{5}{7}x_2^2 - \tfrac{5}{7}x_3^2,$$
$$\theta(x) = \tfrac{8}{9}, \ \epsilon_1 = \tfrac{3}{47}, \ \xi(x) = \tfrac{39}{34}, \ \epsilon_2 = \tfrac{3}{34}, \ \varsigma(x) = 0.$$

Example 4. Consider a hybrid system, where the sets $I(l)$, X_0, X_u, $Guard(l, l')$ and $Reset(l, l')(x)$ are defined as follows:

$$I(1) = \left\{x \in R^2 : g_{1,1}(x) = 1 - x_1^2 - x_2^2 \geq 0\right\},$$
$$I(2) = \left\{x \in R^2 : g_{2,1}(x) = x_1^2 + x_2^2 - 2.25 \geq 0\right\};$$
$$X_0 = \left\{(1, x) \in L \times R^2 : S_1(x) = 0.01 - (x_1 - 0.5)^2 - x_2^2 \geq 0\right\};$$
$$X_u = Unsafe(1) \vee Unsafe(2),$$
$$Unsafe(1) = \left\{(1, x) \in L \times R^2 : U_{1,1}(x) = 0.01 - x_1^2 - x_2^2 \geq 0\right\},$$
$$Unsafe(2) = \left\{(2, x) \in L \times R^2 : U_{2,1}(x) = 0.01 - x_1^2 - x_2^2 \geq 0\right\};$$
$$Guard(1, 2) = \left\{x \in R^2 : Gd_{1,2}(x) = x_1^2 + x_2^2 - 1 \geq 0\right\},$$
$$Reset(1, 2)(x) = \left\{x' \in R^2 : x' = R_{1,2}(x) = 2x\right\}.$$

To verify the safety of this hybrid system, we assume that $V_1(x) = a_1 + b_1 x_1 + c_1 x_2 + d_1 x_1^2 + e_1 x_1 x_2 + f_1 x_2^2$ and $V_2(x) = a_2 + b_2 x_1 + c_2 x_2 + d_2 x_1^2 + e_2 x_1 x_2 + f_2 x_2^2$, where a_i, b_i, c_i, d_i, e_i and f_i $(i = 1, 2)$ are parameters. We want to solve $V_1(x)$, $V_2(x)$, $\theta_1(x)$, $\xi_{1,1}(x)$, $\varsigma_1(x)$, $\phi_{1,1}(x)$, $\psi_{1,1}$, $\upsilon_1(x)$, $\xi_{2,1}(x)$, $\varphi_2(x')$, $\rho_{2,1}(x')$, $\epsilon_{1,1}$, $\epsilon_{1,2}$, $\epsilon_{2,1}$ and $\epsilon_{2,2}$ such that

$$V_1(x) - \theta_1(x)S_1(x) - \epsilon_{1,1},$$
$$-V_1(x) - \xi_{1,1}(x)U_{1,1}(x) - \epsilon_{1,2},$$
$$\tfrac{\partial V_1(x)}{\partial x}f_1(x) + \varsigma_1(x)S_1(x) - \phi_{1,1}(x)g_{1,1}(x),$$
$$V_2(R_{1,2}(x)) - \upsilon_1(x)Gd_{1,2}(x) - \epsilon_{2,1},$$
$$-V_2(x) - \xi_{2,1}(x)U_{2,1}(x) - \psi_{1,1}(x)g_{1,1}(x) - \epsilon_{2,2},$$
$$\tfrac{\partial V_2(x')}{\partial x}f_2(x') + \varphi_2(x')Gd_{1,2}(R_{1,2}^{-1}(x')) - \rho_{2,1}(x')g_{2,1}(x')$$

are sums of squares. By the algorithm described in section 4 with $\tau = 0.001$, we get the certified solutions as follows:

$$V_1(x) = -\tfrac{10}{91} + \tfrac{9}{7}x_1^2 + \tfrac{3}{4}x_1 x_2 + \tfrac{41}{33}x_2^2, \ V_2(x) = -\tfrac{341}{57} + \tfrac{7}{4}x_1^2 + \tfrac{7}{4}x_2^2, \ \theta_1(x) = \tfrac{17}{12},$$
$$\epsilon_{1,1} = \tfrac{8}{249}, \ \xi_{1,1}(x) = \tfrac{241}{41}, \ \epsilon_{1,2} = \tfrac{13}{469}, \ \varsigma_1(x) = 0, \ \phi_{1,1}(x) = 0, \ \psi_{1,1}(x) = 0,$$
$$\upsilon_1(x) = \tfrac{72}{11}, \ \epsilon_{2,1} = \tfrac{25}{58}, \ \xi_{2,1}(x) = \tfrac{83}{40}, \ \epsilon_{2,2} = \tfrac{2}{7}, \ \varphi_2(x') = 0, \ \rho_{2,1}(x') = 0.$$

Example 5. Consider a three-dimensional system, where the sets $I(l)$, X_0, X_u, $Guard(l, l')$ and $Reset(l, l')(x)$ are defined as follows:

$$I(1) = \left\{x \in R^2 : g_{1,1}(x) = 0.16 - x_1^2 - x_2^2 - x_3^2 \geq 0\right\},$$
$$I(2) = \left\{x \in R^2 : g_{2,1}(x) = x_1^2 + x_2^2 + x_3^2 - 0.36 \geq 0\right\};$$
$$X_0 = \left\{(1, x) \in L \times R^2 : S_1(x) = 0.01 - (x_1)^2 - x_2^2 - x_3^2 \geq 0\right\};$$
$$X_u = Unsafe(1) \vee Unsafe(2), Unsafe(1) = \emptyset,$$
$$Unsafe(2) = \left\{(2, x) \in L \times R^2 : U_{2,1}(x) = 0.01 - (x_1 - 6)^2 - x_2^2 - x_3^2 \geq 0\right\};$$
$$Guard(1, 2) = \left\{x \in R^2 : Gd_{1,2}(x) = x_1^2 + x_2^2 - 0.09 \geq 0\right\},$$
$$Reset(1, 2)(x) = \left\{x' \in R^2 : x' = R_{1,2}(x) = 2x\right\}.$$

To verify the safety of this hybrid system, we assume that $V_1(x) = a_1 + b_1x_1 + c_1x_2 + d_1x_3 + e_1x_1^2 + f_1x_1x_2 + h_1x_2^2 + r_1x_1x_3 + s_1x_2x_3 + t_1x_3^2$ and $V_2(x) = a_2 + b_2x_1 + c_2x_2 + d_2x_3 + e_2x_1^2 + f_2x_1x_2 + h_2x_2^2 + r_2x_1x_3 + s_2x_2x_3 + t_2x_3^2$, where $a_i, b_i, c_i, d_i, e_i, f_i, h_i, r_i, s_i$ and $t_i(i = 1,2)$ are parameters. We want to solve $V_1(x), V_2(x), \theta_1(x), \varsigma_1(x), \phi_{1,1}(x), \upsilon_1(x), \psi_{1,1}(x), \xi_{2,1}(x), \varphi_2(x'), \rho_{2,1}(x'), \epsilon_{1,1}, \epsilon_{2,1}$ and $\epsilon_{2,2}$ such that

$$V_1(x) - \theta_1(x)S_1(x) - \epsilon_{1,1},$$
$$\tfrac{\partial V_1(x)}{\partial x}f_1(x) + \varsigma_1(x)S_1(x) - \phi_{1,1}(x)g_{1,1}(x),$$
$$V_2(R_{1,2}(x)) - \upsilon_1(x)Gd_{1,2}(x) - \psi_{1,1}(x)g_{1,1}(x) - \epsilon_{2,1},$$
$$-V_2(x) - \xi_{2,1}(x)U_{2,1}(x) - \epsilon_{2,2},$$
$$\tfrac{\partial V_2(x')}{\partial x}f_2(x') + \varphi_2(x')Gd_{1,2}(R_{1,2}^{-1}(x')) - \rho_{2,1}(x')g_{2,1}(x')$$

are sums of squares. By the algorithm described in section 4 with $\tau = 0.1$, we get the certified solutions as follows:

$$V_1(x) = \tfrac{10}{7} + \tfrac{9}{7}x_1^2 + \tfrac{9}{7}x_2^2 + \tfrac{9}{7}x_3^2, \quad V_2(x) = 2 - \tfrac{2}{3}x_1^2 + \tfrac{1}{6}x_1x_2 - \tfrac{7}{10}x_2^2 - \tfrac{3}{5}x_3^2,$$
$$\theta_1(x) = \tfrac{2}{3}, \quad \epsilon_{1,1} = \tfrac{3}{5}, \quad \varsigma_1(x) = 2, \quad \phi_{1,1}(x) = 0, \quad \upsilon_1(x) = 10,$$
$$\psi_{1,1}(x) = \tfrac{41}{3}, \quad \epsilon_{2,1} = \tfrac{1}{3}, \quad \xi_{2,1}(x) = \tfrac{1}{6}, \quad \epsilon_{2,2} = \tfrac{13}{6},$$
$$\varphi_2(x') = \tfrac{2819}{9} - \tfrac{26}{7}x_1' + \tfrac{46}{5}x_2' + \tfrac{843}{2}x_1'^2 + \tfrac{88}{3}x_1'x_2' + \tfrac{470}{9}x_1'x_3' + \tfrac{1067}{3}x_2'^2$$
$$\qquad + \tfrac{45}{2}x_2'x_3' + \tfrac{1082}{3}x_3'^2,$$
$$\rho_{2,1}(x') = \tfrac{793}{10} - \tfrac{6}{7}x_1' + \tfrac{16}{7}x_2' + 102x_1'^2 + \tfrac{27}{4}x_1'x_2' + \tfrac{94}{7}x_1'x_3' + \tfrac{343}{4}x_2'^2$$
$$\qquad + \tfrac{50}{9}x_2'x_3' + \tfrac{600}{7}x_3'^2.$$

6.2 Discussions

In comparison on the algorithmic complexities of those bilinear methods [6], ours keeps higher efficiency since the SOS problem in this paper is linear and then can be solved in polynomial time while the bilinear methods are in general NP-hard. Compared with those linear methods [24,40], since $\tfrac{\partial V(x)}{\partial x}f(x) \geq 0$ is only desired to hold in $\{x \in \mathbb{R}^n : x \notin \chi_0\}$ in this paper while it should hold in the whole state space for those linear methods, our conditions should be able to define a larger solution set such that our method can be applied to handle some systems that those linear methods cannot handle. For example, for systems $\dot{x} = f(x)$ satisfying $\forall V(x)\exists x_0 \in X_0 : \dot{V}(x_0) = 0 \wedge \ddot{V}(x_0) < 0$, those linear methods will fail since for some state x' close enough to the state x_0 along the path, $\dot{V}(x') < 0$ holds. However, our method may find a Lyapunov-like function V for such systems.

In addition, for Examples 3, 4 and 5, similar to Example 1, the numerical solutions are all unreliable. Especially for Example 5, the numerical solution cannot even be rationalized. So different from [24,6], our method can provide a certified rational solution for exact verification, fixing the unreliability in the verification of hybrid systems caused by numerical errors [23]. So even though we get numerical solutions in our method, the final verification results circumvent unreliability and are guaranteed to be precise.

Moreover, the Gauss-Newton iteration-based method in [40] first utilize the classical bilinear method and the , we here use linear programming and continued fractions method

Moreover, [40] first utilizes the classical bilinear method to obtain a numerical solution and then applies the Gauss-Newton iteration to reach a rational solution, while the numerical solution in this paper is computed by solving a linear-programming problem and then rationalized by continued fractions method.

Note that although Proposition 3 in [25] can to some extent define more relaxed conditions than ours, [25] has already pointed out that the set of barrier certificates will no longer be convex and hence a direct computation of a barrier certificate using convex optimization is not possible. Further, note that according to the equation (1) of [16], we can replace the SOS formulation in Proposition 3 by the rational $\frac{SOS}{SOS}$ formulation. Theoretically, the solution space we search will be greater, such that we can verify more systems. However, the corresponding SOS problem will become bilinear or even multi-linear. How to optimize and solve this problem will be our future work.

7   Conclusions

In this paper, we presented an efficient, hybrid symbolic-numeric approach for safety verification of hybrid systems. Firstly, we introduced multiple Lyapunov-like functions satisfying certain constraints different from traditional SOS methods. Then, we proved that the existence of such a multiple Lyapunov-like function guarantees the safety property of the system. Afterwards, making use of the SOS formulation, SOS toolboxes and continued fractions, we obtained a certified rational multiple Lyapunov-like function, which overcomes the numerical errors and thus provides exact safety verification of polynomial hybrid systems. Finally, several examples with comparisons were provided to illustrate the tractability and advantages of our method.

Our next work is to apply an interval based branch-and-relax method [28,29] for safety verification of hybrid systems and then compare the method with our current SOS based method and the VSDP approach [12].

References

1. Alur, R., Courcoubetis, C., Halbwachs, N., Henzinger, T.A., Ho, P.-H., Nicollin, X., Oliviero, A., Sifakis, J., Yovine, S.: The algorithmic analysis of hybrid systems. Theoretical Computer Science **138**, 3–34 (1995)
2. Alur, R., Dang, T., Ivančić, F.: Counterexample-guided predicate abstraction of hybrid systems. Theoretical Computer Science **354**, 250–271 (2006)
3. Anai, H., Weispfenning, V.: Reach set computations using real quantifier elimination. In: Di Benedetto, M.D., Sangiovanni-Vincentelli, A.L. (eds.) HSCC 2001. LNCS, vol. 2034, pp. 63–76. Springer, Heidelberg (2001)
4. Clarke, E.M., Fehnker, A., Han, Z., Krogh, B., Ouaknine, J., Stursberg, O., Theobald, M.: Abstraction and Counterexample-Guided Refinement in Model Checking of Hybrid Systems. International Journal Foundations of Computer Science **14**, 583–604 (2003)
5. Clarke, E.M., Kurshan, R.P.: Computer-aided verification. IEEE Spectrum **33**(6), 61–67 (1996)

6. Christoffer, S., George, J.P., Rafael, W.: Compositional safety analysis using barrier certificates. In: Proceedings of the 15th ACM International Conference on Hybrid Systems: Computation and Control, pp. 15–23 (2012)

7. Chutinan, A., Krogh, B.H.: Computational techniques for hybrid system verification. IEEE Transactions on Automatic Control **48**, 64–75 (2003)

8. Fradkov, A.L., Yakubovich, V.A.: The S-procedure and a duality realations in nonconvex problems of quadratic programming. Vestnik Leningrad Univ. Math **5**, 101–109 (1979)

9. Grayson, D.R., Stillman, M.E.: Macaulay2, a software system for research in algebraic geometry. http://www.math.uiuc.edu/Macaulay2

10. Le Guernic, C., Girard, A.: Reachability analysis of hybrid systems using support functions. In: Bouajjani, A., Maler, O. (eds.) CAV 2009. LNCS, vol. 5643, pp. 540–554. Springer, Heidelberg (2009)

11. Gulwani, S., Tiwari, A.: Constraint-based approach for analysis of hybrid systems. In: Gupta, A., Malik, S. (eds.) CAV 2008. LNCS, vol. 5123, pp. 190–203. Springer, Heidelberg (2008)

12. Härter, V., Jansson, C., Lange, M.: VSDP: a matlab toolbox for verified semidefinte-quadratic-linear programming.
    http://www.ti3.tuhh.de/jansson/vsdp/

13. Huang, Z., Mitra, S.: Proofs from simulations and modular annotations. In: Proc. of the 17th International Conference on Hybrid Systems: Computation and Control, pp. 183–192 (2014)

14. Huang, Z., Fan, C., Mereacre, A., Mitra, S., Kwiatkowska, M.: Invariant verification of nonlinear hybrid automata networks of cardiac cells. In: Biere, A., Bloem, R. (eds.) CAV 2014. LNCS, vol. 8559, pp. 373–390. Springer, Heidelberg (2014)

15. Jones, W., Thron, W.: Continued fractions: analytic theory and applications. In: Encyclopedia of Mathematics and its Applications, vol. 11 (1980)

16. Kaltofen, E., Li, B., Yang, Z., Zhi, L.: Exact certification in global polynomial optimization via sums-of-squares of rational functions with rational coefficients. Journal of Symbolic Computation **47**, 1–15 (2012)

17. Kurzhanski, A.B., Varaiya, P.: Ellipsoidal techniques for reachability analysis. In: Lynch, N.A., Krogh, B.H. (eds.) HSCC 2000. LNCS, vol. 1790, pp. 203–213. Springer, Heidelberg (2000)

18. Lafferriere, G., Pappas, G.J., Yovine, S.: Symbolic reachability computations for families of linear vectorfields. Journal of Symbolic Computation **32**, 231–253 (2001)

19. Lin, W., Wu, M., Yang, Z., Zeng, Z.: Exact safety verification of hybrid systems using sums-of-squares representation. Science China Information Sciences **57**(5), 1–13 (2014)

20. Löfberg, J.: Yalmip: a toolbox for modeling and optimization in MATLAB. In: Proceedings of the CACSD Conference (2004).
    http://control.ee.ethz.ch/joloef/yalmip.php

21. Parrilo, P.A.: Structured Semidefinite Programs and Semialgebraic Geometly Methods in Robustness and Optimization. Ph.D. thesis, California Institute of Technology, Pasadena (2000)

22. Peyrl, H., Parrilo, P.A.: Computing sum of squares decompositions with rational coefficients. Theoretical Computer Science **409**, 269–281 (2008)

23. Platzer, A., Clarke, E.M.: The image computation problem in hybrid systems model checking. In: Bemporad, A., Bicchi, A., Buttazzo, G. (eds.) HSCC 2007. LNCS, vol. 4416, pp. 473–486. Springer, Heidelberg (2007)

24. Prajna, S., Jadbabaie, A.: Safety verification of hybrid systems using barrier certificates. In: Alur, R., Pappas, G.J. (eds.) HSCC 2004. LNCS, vol. 2993, pp. 477–492. Springer, Heidelberg (2004)

25. Prajna, S., Jadbabaie, A., Pappas, G.J.: A framework for worst-case and stochastic safety verification using barrier certificates. IEEE Transactions On Automatic Control **52**, 1415–1428 (2007)

26. Prajna, S., Papachristodoulou, A., Parrilo, P.A.: Introducing SOSTOOLS: a general purpose sum of squares programming solver. In: Proc. IEEE CDC (2002). http://www.cds.caltech.edu/sostools and http://www.aut.ee.ethz.ch/?parrilo/sostools

27. Ratschan, S., She, Z.: Safety Verification of Hybrid Systems by Constraint Propagation-Based Abstraction Refinement. ACM Transactions on Embedded Computing Systems **6**(1), 1–23 (2007). Article No. 8

28. Ratschan, S., She, Z.: Providing a basin of attraction to a target region of polynomial systems by computation of Lyapunov-like functions. SIAM Journal on Control and Optimization **48**, 4377–4394 (2010)

29. Sankaranarayanan, S., Chen, X., Ábrahám, E.: Lyapunov function synthesis using Handelman representations. In: Proceedings of the 9th IFAC Symposium on Nonlinear Control Systems, pp. 576–581 (2013)

30. Sankaranarayanan, S., Sipma, H.B., Manna, Z.: Constructing invariants for hybrid systems. In: Alur, R., Pappas, G.J. (eds.) HSCC 2004. LNCS, vol. 2993, pp. 539–554. Springer, Heidelberg (2004)

31. Amin Ben Sassi, M.: Computation of polytopic invariants for polynomial dynamical systems using linear programming. Automatica **48**, 3114–3121 (2012)

32. She, Z., Li, H., Xue, B., Zheng, Z., Xia, B.: Discovering polynomial Lyapunov functions for continuous dynamical systems. Journal of Symbolic Computation **58**, 41–63 (2013)

33. She, Z., Xue, B.: Computing an invariance kernel with target by computing Lyapunov-like functions. IET Control Theory and Applications **7**, 1932–1940 (2013)

34. She, Z., Xue, B.: Discovering Multiple Lyapunov Functions for Switched Hybrid Systems. SIAM J. Control and Optimization **52**(5), 3312–3340 (2014)

35. Sturm, J.F.: Using SeDuMi 1.02, a MATLAB toolbox for optimization over symmetric cones. Optimization Methods and Software **11**, 625–653 (1999)

36. Tiwari, A., Khanna, G.: Series of abstractions for hybrid automata. In: Tomlin, C.J., Greenstreet, M.R. (eds.) HSCC 2002. LNCS, vol. 2289, pp. 465–478. Springer, Heidelberg (2002)

37. Tiwari, A., Khanna, G.: Nonlinear systems: approximating reach sets. In: Alur, R., Pappas, G.J. (eds.) HSCC 2004. LNCS, vol. 2993, pp. 600–614. Springer, Heidelberg (2004)

38. Tomlin, C.J., Mitchell, I., Bayen, A.M., Oishi, M.: Computational techniques for the verification of hybrid systems. Proc. of the IEEE **91**, 986–1001 (2003)

39. VanAntwerp, J.G., Braatz, R.D.: A tutorial on linear and bilinear matrix inequalities. Journal of Process Control **10**, 363–385 (2000)

40. Wu, M., Yang, Z.: Generating invariants of hybrid systems via sums-of-squares of polynomials with rational coefficients. In: Proc. International Workshop on Symbolic-Numeric Computation, pp. 104–111 (2011)

41. Yang, Z., Lin, W., Wu, M.: Exact Safety Verification of Hybrid Systems Based on Bilinear SOS Representation. ACM Trans. Embedded Comput. Syst **14**(1), 1–19 (2015). Article No. 16

42. Zhou, K., Doyle, J.C., Glover, K.: Robust and Optimal Control. Prentice-Hall Inc., Upper Saddle River (1996)

A New Polynomial Bound and Its Efficiency

Doru Ştefănescu

University of Bucharest, Romania
stef@rms.unibuc.ro

Abstract. We propose a new bound for absolute positiveness of univariate polynomials with real coefficients. We discuss its efficiency with respect to known such bounds and compare it with the threshold of absolute positiveness.

1 Introduction

There are known few efficient bounds for the largest positive root of a univariate polynomial with real coefficients. Besides the classical but forgotten $R+\rho$ bound of J.–L. Lagrange [5], a new insight was given by the results of J.B. Kioustelidis [4] and H. Hong [3].

Theorem 1. (J. B. Kioustelidis [4]) *Let*

$$P(X) = X^d - b_1 X^{d-m_1} - \cdots - b_t X^{d-m_t} + g(X),$$

where $b_1, \ldots, b_t > 0$ and $g \in \mathbb{R}_+[X]$. The number

$$K(P) = 2 \cdot \max\{b_1^{1/m_1}, \ldots, b_t^{1/m_t}\}$$

is an upper bound for the positive roots of P.

Theorem 2. (H. Hong [3]) *Let*

$$P(X) = \sum_{i=1}^{n} a_i X^{d_i},$$

with $d_1 < d_2 < \ldots < d_n = d = \deg(P)$. The number

$$H(P) = 2 \cdot \max_{\substack{i \\ a_i<0}} \min_{\substack{j>i \\ a_j>0}} \left(\frac{|a_i|}{a_j}\right)^{\frac{1}{d_i - d_j}}$$

is an upper bound for the positive roots.

The bounds in Theorems 1 and 2 are bounds of absolute positivity, i.e., they are also bounds for the positive roots of all derivatives.

We note that the bound of Kioustelidis involves only the negative coefficients and the dominant one. On the other hand the bound of Hong involves both

© Springer International Publishing Switzerland 2015
V.P. Gerdt et al. (Eds.): CASC 2015, LNCS 9301, pp. 457–467, 2015.
DOI: 10.1007/978-3-319-24021-3_33

negative and positive coefficients and instead of max it uses min and max. An analysis of the computation of the bound of Hong was done by P. Batra and V. Sharma in [2].

In [6] we obtained new bounds for positive roots, among which the bounds S_1 and S_2 given below.

Theorem 3. (D. Ştefănescu [6]) *Let*

$$P(X) = X^d - b_1 X^{d-m_1} - \cdots - b_t X^{d-m_t} + g(X),$$

with $b_1, \ldots, b_t > 0$ and $g \in \mathbb{R}_+[X]$. The number

$$S_1(P) = max\{(tb_1)^{1/m_1}, \ldots, (tb_t)^{1/m_t}\}$$

is an upper bound for the positive roots of P.

We also obtained the following bound:

Theorem 4. (D. Ştefănescu [6]) *The number*

$$S_2(P) = \max_{1 \le j \le s} \left\{ \left(\frac{b_j}{a_j} \right)^{1/(d_j - e_j)} \right\}$$

is an upper bound for the positive roots of the polynomial

$$P(X) = \sum_{i=1}^{s} \left(a_i X^{d_i} - b_i X^{e_i} \right) + g(X),$$

where $d_1 > d_2 > \ldots > d_1 > 0$, all $a_i, b_i > 0$ and $e_i < d_i$ for all i.

Remark 1. The proofs of theorems 3 and 4 are based on splitting the negative coefficients in convenient sums of coefficients. These bounds were extended by D. Ştefănescu and A. Akritas, see, for example, [1], [7], and [8].

We propose a new bound of absolute positiveness. It is based on the techniques of splitting the negative coefficients of a polynomial in sums of negative coefficients. This allows us to use bounds as in Theorem 4 even for arbitrary polynomials that have signs changes. We also use the max–min device developed by H. Hong in [3]. We discuss the efficiency of our method by comparing it to the bound of Hong and with the threshold of absolute positiveness.

1.1   The Bound

We consider a univariate polynomial P with real coefficients that has at least one positive root. We may suppose that its leading coefficient is positive. We propose a new bound $B(P)$ of absolute positiveness, and we evaluate the ratio between it and the threshold of P.

We remind that the *threshold* of absolute positiveness of an univariate polynomial P with real coefficients is

$$A_P = \inf \{D(P); D(P) \in \{\text{bounds for absolute positiveness of } P\}\}.$$

Theorem 5. *Let* $P(X) = a_1 X^{d_1} + \cdots + a_s X^{d_s} - b_1 X^{e_1} - \cdots - b_t X^{e_t}$ *be a nonconstant univariate polynomial, where* $a_i > 0$, $b_j > 0$ *and* $\deg(P) = d_1 \geq 2$. *The number*

$$B(P) = \max_{1 \leq j \leq t} \min_{\substack{1 \leq i \leq s \\ d_i > e_j}} \left(\frac{b_j}{\beta_j a_i} \right)^{\frac{1}{d_i - e_j}}$$

is an absolute positiveness bound of the polynomial P, *for any* $\beta_1, \beta_2, \ldots, \beta_t > 0$ *such that*

$$\beta_1 + \cdots + \beta_t \leq 1.$$

Proof. We first check that B is a bound for the positive roots of the polynomial P. Let $x > B(P)$. In particular, x is positive. We have

$$P(x) = \sum_{i=1}^{s} a_i x^{d_i} - \sum_{j=1}^{t} b_j x^{e_j}$$

$$\geq \left(\sum_{j=1}^{t} \beta_j \right) \left(\sum_{i=1}^{s} a_i x^{d_i} \right) - \left(\sum_{j=1}^{t} b_j x^{e_j} \right)$$

$$= \sum_{j=1}^{t} \left(\left(\sum_{i=1}^{s} \beta_j a_i x^{d_i} \right) - b_j x^{e_j} \right)$$

$$= \sum_{j=1}^{t} x^{e_j} \left(\left(\sum_{i=1}^{s} \beta_j a_i x^{d_i - e_j} \right) - b_j \right)$$

$$\geq \sum_{j=1}^{t} x^{e_j} \left(\left(\sum_{\substack{i=1 \\ d_i > e_j}}^{s} \beta_j a_i x^{d_i - e_j} \right) - b_j \right)$$

$$= \sum_{j=1}^{t} x^{e_j} \cdot S_j(x),$$

where

$$S_j(x) = -b_j + \beta_j \sum_{\substack{i=1 \\ d_i > e_j}}^{s} a_i x^{d_i - e_j}.$$

Note that such a sum S_j exists for every j, since $d_1 > e_j$.

By the definition of $B(P)$ we have

$$x > B(P) \geq \min_{\substack{1 \leq i \leq s \\ d_i > e_j}} \left(\frac{b_j}{\beta_j a_i} \right)^{\frac{1}{d_i - e_j}} \geq \left(\frac{b_j}{\beta_j a_i} \right)^{\frac{1}{d_i - e_j}}$$

for all i and j such that $d_i > e_j$.

Now we consider $j \in \{1, 2, \ldots, t\}$ and an index i_0 such that $d_{i_0} > e_j$. It follows that

$$S_j(x) > -b_j + \beta_j a_{i_0} \left(\left(\frac{b_j}{\beta_j a_{i_0}} \right)^{\frac{1}{d_{i_0} - e_j}} \right)^{d_{i_0} - e_j} = -b_j + b_j = 0.$$

So all considered sums $S_j(x)$ are positive, we have $P(x) > 0$, so $B(P)$ is a bound for the positive roots.

We shall prove that $B(P)$ is also a bound for the positive roots of the derivative of P. We have

$$P'(X) = \sum_{i=1}^{s} d_i a_i X^{d_i - 1} - \sum_{j=1}^{t} e_j b_j X^{e_j - 1} = \sum_{i=1}^{s} a_i' X^{d_i'} - \sum_{j=1}^{t} b_j' X^{e_j'},$$

where $a_i' = d_i a_i$, respectively, $b_j' = e_j b_j$. For $d_i' > e_j'$ we have $d_i > e_j$, therefore,

$$\frac{b_j'}{a_j'} = \frac{e_j b_j}{d_i a_i} < \frac{b_j}{a_i},$$

so

$$\left(\frac{b_j'}{\beta_j a_j'} \right)^{\frac{1}{d_i' - e_j'}} > \left(\frac{b_j}{\beta_j a_i} \right)^{\frac{1}{d_i - e_j}}.$$

It follows that

$$B(P') = \max_{1 \leq j \leq t} \min_{\substack{1 \leq i \leq s \\ d_i' > e_j'}} \left(\frac{b_j'}{\beta_j a_j'} \right)^{\frac{1}{d_i' - e_j'}} < \max_{1 \leq j \leq t} \min_{\substack{1 \leq i \leq s \\ d_i > e_j}} \left(\frac{b_j}{\beta_j a_i} \right)^{\frac{1}{d_i - e_j}} = B(P).$$

Since $B(P) > B(P')$ it follows that $B(P)$ is also a bound for the positive roots of P'. So it is a bound for the positive roots of all the derivatives of P. □

Remark 2. From the previous Theorem 5 we obtain the bound from Theorem 2 of H. Hong. In fact, we consider

$$\beta_j = \frac{1}{2^{d_1 - e_j}}$$

and we observe that

$$\sum_{j=1}^{t} \beta_j = \frac{1}{2^{d_1-e_1}} + \cdots + \frac{1}{2^{d_1-e_t}}$$

$$= \frac{1}{2^{d_1-e_1}} \left(1 + \frac{1}{2^{e_1-e_2}} + \cdots + \frac{1}{2^{e_1-e_t}} \right)$$

$$< \frac{1}{2^{d_1-e_1-1}}$$

$$\leq 1$$

and

$$\beta_j \leq \frac{1}{2^{d_i-e_j}} \quad \text{for all} \quad i.$$

It follows that

$$\left(\frac{b_j}{\beta_j a_i} \right)^{\frac{1}{d_i - e_j}} \leq \left(\frac{b_j}{a_i} 2^{d_i-e_j} \right)^{\frac{1}{d_i - e_j}} = 2 \left(\frac{b_j}{a_i} \right)^{\frac{1}{d_i - e_j}}.$$

1.2 Auxiliary Results

In order to compare the bound $B(P)$ with the threshold A_P we prove first two auxiliary results.

Lemma 1. *Let*

$$Q(X) = X^d + a_{d-1}X^{d-1} + \cdots + a_{e+1}X^{e+1} - \gamma X^e + b_{e-1}X^{e-1} + \cdots + b_1 X + b_0,$$

where the coefficients a_i, b_j and γ are strictly positive.
The number $\gamma^{1/(d-e)}$ is an absolute positiveness bound for the polynomial P.

Proof. We observe that if $x > \gamma^{1/(d-e)}$, we have $x^{d-e} > \gamma$, so $x^{d-e} - \gamma > 0$.
Therefore,

$$Q(x) = x^e(x^{d-e} - \gamma) + a_{d-1}x^{d-1} + \cdots + a_{d+1}x^{e+1} + b_{e-1}x^{e-1} + \cdots + b_d > 0,$$

We consider now the derivative

$$Q'(X) = dX^{d-1} + \cdots + a_{e+1}(e+1)X^e - \gamma e X^{e-1} + (e-1)b_{e-1}X^{e-2} + \cdots + b_1$$

and we note that we have

$$dx^{d-1} - \gamma e x^{e-1} = dx^{e-1}(x^{d-e} - \frac{e}{d}\gamma) > dx^{e-1}(x^{d-e} - \gamma) > 0.$$

It follows that $P'(x) > 0$. □

Proposition 1. *Let*

$$R(X) = \sum_{k=e+1}^{d} \binom{k}{e} X^{k-e} - \gamma, \quad \text{where} \quad \gamma > 0.$$

The unique positive root α of the polynomial R satisfies the inequality

$$\alpha > (1+\gamma)^{\frac{1}{d}} - 1.$$

Proof. We observe that the polynomial R has real coefficients and a unique change of signs. By Descartes' rule of signs it has a unique positive root, which we denote by α. We observe that

$$\sum_{k=e+1}^{d} \binom{k}{e} X^{k-e} - \gamma = \sum_{k=e}^{d} \binom{k}{e} X^{k-e} - 1 - \gamma$$

$$= -1 - \gamma - \sum_{j=0}^{d-e} \binom{e+j}{e} X^j.$$

On the other hand,

$$\binom{j+e}{j} \leq \binom{d}{j} \quad \text{for all} \quad j = 0, \ldots, d-e,$$

therefore,

$$\sum_{k=e}^{d} \binom{k}{e} \alpha^{k-e} = \sum_{j=0}^{d-e} \binom{e+j}{e} \alpha^j \leq \sum_{j=0}^{d-e} \binom{d}{j} \alpha^j.$$

It follows that

$$\sum_{j=0}^{d-e} \binom{d}{j} \alpha^j \leq \sum_{j=0}^{d-1} \binom{d}{j} \alpha^j = (1+\alpha)^d - \alpha^d$$

Therefore,

$$0 = R(\alpha) = -1 - \gamma + \sum_{k=e}^{d} \binom{k}{e} \alpha^{k-e}$$

$$\leq -1 - \gamma + (1+\alpha)^d - \alpha^d$$

$$< -1 - \gamma + (1+\alpha)^d,$$

Hence α is also a bound for the positive roots of P'. □

Corollary 1. *The positive root of the polynomial*

$$R(X) = \sum_{k=e+1}^{d} \binom{k}{e} X^{k-e} - \gamma$$

lies in the interval $\left((1+\gamma)^{\frac{1}{d}} - 1, \gamma^{1/(d-e)} \right).$

Proof. We observe that $R(X) = S^{(e)}(X)$, where

$$S(X) = X^d + X^{d-1} + \ldots + X^{e+1} - \gamma X^e + X^{e-1} + \ldots + X + 1.$$

Then we apply Lemma 1 and Proposition 1. $\square$

1.3   Comparison with the Threshold

We compare now the bound $B(P)$ with the threshold.

Theorem 6. *Let* $P(X) = a_1 X^{d_1} + \cdots + a_s X^{d_s} - b_1 X^{e_1} - \cdots - b_t X^{e_t}$ *as in Theorem 5. We have*

$$\frac{B(P)}{A_P} < \frac{1}{\left(1 + \frac{1}{\delta}\right)^{1/d} - 1},$$

where

$$\delta = \max\left\{\frac{1}{\beta_1}, \frac{1}{\beta_2}, \ldots, \frac{1}{\beta_t}\right\}.$$

Proof. We observe that

$$B = \max\{B_1, B_2, \ldots, B_t\},$$

where

$$B_j = \min_{d_i > e_j} \left(\frac{b_j}{\beta_j a_i}\right)^{\frac{1}{d_i - e_j}}.$$

Let $j^* \in \{1, 2, \ldots, t\}$ be such that $B = B_{j^*}$. We put

$$b = b_{j^*}, \beta = \beta_{j^*}, e = e_{j^*}.$$

So we have

$$B = \min_{\substack{1 \le i \le s \\ d_i > e}} \left(\frac{b}{\beta\, a_i}\right)^{\frac{1}{d_i - e}}.$$

Therefore,

$$B \le \left(\frac{b}{\beta\, a_i}\right)^{\frac{1}{d_i - e}} \quad \text{for all} \quad i = 1, \ldots, s;\ d_i > e.$$

But $\frac{1}{\beta} \le \delta$. This gives

$$B^{d_i - e} \le \frac{b}{\beta\, a_i} \le \delta \frac{b}{a_i} \quad \text{for all} \quad i = 1, \ldots, s;\ d_i > e,$$

and we obtain

$$a_i \le \delta b \left(\frac{1}{B}\right)^{d_i - e} \quad \text{for all} \quad i = 1, \ldots, s;\ d_i > e.$$

We consider now the derivative of order $e = e_{j*}$ of the polynomial P. Because $B = B(P)$ is a bound for the absolute positiveness of P, we know that B is an upper bound for the positive roots of $P^{(e)}$. On the other hand, we have

$$P^{(e)}(X) = \sum_{i=1}^{s} \frac{d_i!}{(d_i - e)!} a_i X^{d_i - e} - \sum_{j=1}^{t} \frac{e_j!}{(e_j - e)!} b_j X^{e_j - e} .$$

For $x > 0$ we have

$$P^{(e)}(x) = \sum_{i=1}^{s} \frac{d_i!}{(d_i - e)!} a_i x^{d_i - e} - \sum_{j=1}^{t} \frac{e_j!}{(e_j - e)!} b_j x^{e_j - e}$$

$$\le b e! \left(-1 - \delta + \delta \sum_{k=e}^{d} \binom{k}{e} \left(\frac{x}{B} \right)^{k-e} \right) . \tag{1}$$

In fact, we successively have

$$P^{(e)}(x) = \sum_{i=1}^{s} \frac{d_i!}{(d_i - e)!} a_i x^{d_i - e} - \sum_{j=1}^{t} \frac{e_j!}{(e_j - e)!} b_j x^{e_j - e}$$

$$= -b e! - \sum_{\substack{j=1 \\ e_j \ne e}}^{t} \frac{e_j!}{(e_j - e)!} b_j x^{e_j - e} + \sum_{\substack{i=1 \\ d_i > e}}^{s} \frac{d_i!}{(d_i - e)!} a_i x^{d_i - e}$$

$$\le -b e! + \sum_{\substack{i=1 \\ d_i > e}}^{s} \frac{d_i!}{(d_i - e)!} a_i x^{d_i - e}$$

$$\le -b e! + \sum_{\substack{i=1 \\ d_i > e}}^{s} \frac{b}{\beta} \frac{d_i!}{(d_i - e)!} \left(\frac{x}{B} \right)^{d_i - e}$$

$$= b e! \left(-1 + \frac{1}{\beta} \sum_{\substack{i=1 \\ d_i > e}}^{s} \frac{d_i!}{(d_i - e)! e!} \left(\frac{x}{B} \right)^{d_i - e} \right)$$

$$\le b e! \left(-1 + \delta \sum_{\substack{i=1 \\ d_i > e}}^{s} \binom{d_i}{e} \left(\frac{x}{B} \right)^{d_i - e} \right)$$

$$\leq be! \left(-1 + \delta \sum_{k > e}^{d} \binom{k}{e} \left(\frac{x}{B}\right)^{k-e} \right)$$

$$= be! \left(-1 + \delta \left(-1 + \sum_{k=e}^{d} \binom{k}{e} \left(\frac{x}{B}\right)^{k-e} \right) \right)$$

$$= be! \left(-1 - \delta + \delta \sum_{k=e}^{d} \binom{k}{e} \left(\frac{x}{B}\right)^{k-e} \right).$$

We consider the polynomial

$$T(X) = -1 - \frac{1}{\delta} + \sum_{k=e}^{d} \binom{k}{e} X^{k-e}$$

and we observe that

$$T(X) = -\frac{1}{\delta} + \sum_{k=e+1}^{d} \binom{k}{e} X^{k-e}.$$

Let α be the unique positive root of the polynomial T. By proposition 1 we have

$$\alpha > \left(1 + \frac{1}{\delta}\right)^{\frac{1}{d}} - 1. \tag{2}$$

By (1) we have

$$P^{(e)}(\alpha B) \leq \delta \, be! \, T(\alpha) = 0,$$

which proves that

$$\alpha B \leq A_P. \tag{3}$$

In fact, otherwise, for $\alpha B > A_P$, αB would be a bound of absolute positiveness for the polynomial P. In particular, it would follow $P^{(e)}(\alpha B) > 0$, in contradiction with (1).

From (2) and (3), we obtain

$$\frac{B(P)}{A_P} \leq \frac{B(P)}{\alpha} \leq \frac{1}{\left(1 + \frac{1}{\delta}\right)^{1/d} - 1}. \tag{4}$$

$\square$

Remark 3. With the notation in Theorem 6 we also have

$$\frac{B(P)}{A_P} < \frac{d}{\ln\left(1 + \frac{1}{\delta}\right)}.$$

In fact, we search for another estimation of the unique positive root of the polynomial T. We use the method of H. Hoon for the proof of his Quality Theorem 2.3 in [3]. We observe that

$$\binom{k}{e} = \frac{k(k-1)\cdots(k-e+1)}{(k-e)!} \leq \frac{\overbrace{d\cdots d}^{k-e}}{(k-e)!} = \frac{d^{k-e}}{(k-e)!}.$$

This gives, for $x > 0$,

$$\sum_{k=e}^{d}\binom{k}{e}\left(\frac{x}{B}\right)^{k-e} \leq \sum_{k=e}^{d}\frac{d^{k-e}}{(k-e)!}\left(\frac{x}{B}\right)^{k-e} = \sum_{k=0}^{d-e}\frac{1}{(k)!}\left(\frac{dx}{B}\right)^{k}$$

$$\leq \sum_{k=0}^{\infty}\frac{1}{(k)!}\left(\frac{dx}{B}\right)^{k} = e^{\frac{dx}{B}}.$$

(5)

From (1) and (5) it follows that

$$e^{\frac{d\alpha}{B}} \geq 1+\frac{1}{\delta}, \quad \text{which gives} \quad \frac{\alpha}{B} \geq \frac{\ln\left(1+\frac{1}{\delta}\right)}{d}.$$

It follows that

$$\frac{B(P)}{A_P} \leq \frac{B(P)}{\alpha} < \frac{d}{\ln\left(1+\frac{1}{\delta}\right)}.$$

(6)

The estimate (4) is sharp than (6).

Example 1. We consider the polynomial

$$P(X) = X^5 - 2X^4 + 2X^3 - 3X^2 - 2X - 2.$$

We have $d_1 = 4$, $d_2 = 3$, $e_1 = 4$, $e_2 = 2$, $e_3 = 1$, $e_4 = 0$.
Using the notation in Theorem 5 we have

$$B(P) = \max\{B_1, B_2, B_3, B_4\},$$

where

$$B_1 = \left(\frac{b_1}{\beta_1 a_1}\right)^{1/(d_1-e_1)} = \frac{2}{\beta_1},$$

$$B_2 = \min\left\{\left(\frac{b_2}{\beta_2 a_1}\right)^{1/(d_1-e_2)}, \left(\frac{b_2}{\beta_2 a_2}\right)^{1/(d_2-e_2)}\right\} = \min\left\{\left(\frac{3}{\beta_2}\right)^{\frac{1}{3}}, \frac{3}{2\beta_2}\right\},$$

$$B_3 = \min\left\{\left(\frac{b_3}{\beta_3 a_1}\right)^{1/(d_1-e_3)}, \left(\frac{b_3}{\beta_3 a_2}\right)^{1/(d_2-e_3)}\right\} = \min\left\{\left(\frac{2}{\beta_3}\right)^{\frac{1}{4}}, \left(\frac{1}{\beta_3}\right)^{\frac{1}{2}}\right\},$$

$$B_4 = \min\left\{\left(\frac{b_4}{\beta_4 a_1}\right)^{1/(d_1-e_4)}, \left(\frac{b_4}{\beta_4 a_2}\right)^{1/(d_2-e_4)}\right\} = \min\left\{\left(\frac{2}{\beta_4}\right)^{\frac{1}{5}}, \left(\frac{1}{\beta_4}\right)^{\frac{1}{3}}\right\}.$$

$(\beta_1, \beta_2, \beta_3, \beta_4)$	B_1	B_2	B_3	B_4	$B(P)$	A_P	$B(P)/A_P$
$(0.25, 0.25, 0.25, 0.25)$	8.0	6.0	2.0	1.587	8.0	2.069	3.866
$(0.7, 0.1, 0.1, 0.1)$	2.857	2.114	1.821	2.154	2.857	2.069	1.380
$(0.8, 0.1, 0.05, 0.05)$	2.500	2.114	2.514	2.091	2.514	2.069	1.215
$(0.85, 0.09, 0.03, 0.03)$	2.352	2.402	2.857	2.316	2.857	2.069	1.381

The bound of H. Hong [3] gives

$$B_H(P) = 2 \cdot \max\{2, 1.5, 1.259, 1.148\} = 2 \cdot 2.0 = 4.0.$$

We conclude that the choice $(\beta_1, \beta_2, \beta_3, \beta_4) = (0.8, 0.1, 0.05, 0.05)$ in Theorem 5 gives a good estimate for the abosulte positiveness bound of P. The ratio $B(P)/A_P$ is 1.215.

Example 2. We consider the family of polynomials $P_d(X) = X^d - 2X^5 + 3X^4 - 2X^3 - 1$, where $d \geq 6$. We fix $(\beta_1, \beta_2, \beta_3) = (0.4, 0.3, 0.3)$ and we observe that the values of A_P and $B(P)$ decrease when the degree d increases. For example, we have

1. $B(P_6) = \max\{5.0, 2.222, 1.222\} = 5.0$, $A_{P_6} = 1.233$.
2. $B(P_{11}) = \max\{1.307, 3.333, 1.115\} = 3.333$, $A_{P_{11}} = 1.082$.

1.4 Conclusion

We obtained a new bound for absolute positiveness of a univariate polynomial with real coefficients. Its value depends on the choice of a family of positive numbers β_j and this allows the computation of bounds close to the threshold of positiveness.

References

1. Akritas, A.: Linear and quadratic complexity bounds on the values of the positive roots of polynomials. Univ. J. Comput. Sci. **15**, 523–537 (2009)
2. Batra, P., Sharma, V.: Bounds of absolute positiveness of multivariate polynomials. J. Symb. Comput. **45**, 617–628 (2010)
3. Hong, H.: Bounds for absolute positiveness of multivariate polynomials. J. Symb. Comp. **25**, 571–585 (1998)
4. Kioustelidis, J.B.: Bounds for positive roots of polynomials. J. Comput. Appl. Math. **16**, 241–244 (1986)
5. Mignotte, M., Ştefănescu, D.: On an estimation of polynomial roots by Lagrange. IRMA Strasbourg **025/2002**, 1–17 (2002)
6. Ştefănescu, D.: New bounds for positive roots of polynomials. Univ. J. Comput. Sci. **11**, 2125–2131 (2005)
7. Ştefănescu, D.: Bounds for real roots and applications to orthogonal polynomials. In: Ganzha, V.G., Mayr, E.W., Vorozhtsov, E.V. (eds.) CASC 2007. LNCS, vol. 4770, pp. 377–391. Springer, Heidelberg (2007)
8. Ştefănescu, D.: On some absolute positiveness bounds. Bull. Math. Soc. Sci. Math. Roumanie **53**(101), 269–276 (2010)

Distance Evaluation Between an Ellipse and an Ellipsoid

Alexei Yu. Uteshev and Marina V. Yashina

Faculty of Applied Mathematics, St. Petersburg State University,
Universitetskij pr. 35, Petrodvorets, 198504, St. Petersburg, Russia
{alexeiuteshev,marina.yashina}@gmail.com

Abstract. We solve in $\mathbb{R}^n$ the problem of distance evaluation between
a quadric and a manifold obtained as the intersection of another quadric
and a linear manifold. Application of Elimination Theory algorithms for
the system of algebraic equations of the Lagrange multipliers method
results in construction of the *distance equation*, i.e., a univariate algebraic equation one of the zeros of which (generically minimal positive)
coincides with the square of the distance between considered manifolds.
We also deduce the necessary and sufficient algebraic conditions under
which the manifolds intersect and propose an algorithm for finding the
coordinates of their nearest points.

Keywords: Distance between ellipse and ellipsoid, elimination of variables in algebraic system.

1 Introduction

The problem of establishing the closeness of two sets in multidimensional space
is of interest in several branches of computational geometry. For the case of $\mathbb{R}^3$,
evaluation of the relative position of two quadrics or conic sections is a crucial
problem in astronomy and nuclear physics. In the present paper, we treat this
problem in the following statement. Find the Euclidean distance between the
ellipsoid given by the algebraic equation

$$X^\top \mathbf{A}_2 X + 2 B_2^\top X - 1 = 0 \tag{1}$$

and the ellipse given as the intersection of two manifolds

$$X^\top \mathbf{A}_1 X + 2 B_1^\top X - 1 = 0, \tag{2}$$
$$C^\top X = h. \tag{3}$$

Here we assume X, C, B_1, B_2 to be the column vectors from $\mathbb{R}^3$, $h \in \mathbb{R}$ while $\mathbf{A}_1$
and $\mathbf{A}_2$ denote the sign-definite matrices of the order 3.

   We intend to tackle this problem within the approach first outlined in [8] and
then developed in [9] for the problem of distance evaluation between quadric
manifolds in $\mathbb{R}^n$. This approach is based on construction of the so-called *distance*

© Springer International Publishing Switzerland 2015
V.P. Gerdt et al. (Eds.): CASC 2015, LNCS 9301, pp. 468–478, 2015.
DOI: 10.1007/978-3-319-24021-3_34

equation, i.e. a univariate algebraic equation with the set of zeros coinciding with the set of all the critical values of the distance function $(X - Y)^\top (X - Y)$ considered for X satisfying (2)-(3) and Y satisfying (1). Such a construction is carried out with the aid of Elimination Theory methods (briefly sketched in Section 2) and happens to be sufficiently universal in the sense that it gives the result valid for the distance evaluation problem for the manifolds considered in the space of arbitrary dimension. The foregoing theorems of the present paper will be formulated for the general case of $\mathbb{R}^n$ while illuminated with examples from $\mathbb{R}^3$.

Notation. $\mathcal{D}(\cdot)$ (or $\mathcal{D}_x(\cdot)$) denotes the discriminant of a polynomial (subscript denotes the variables w.r.t. which the polynomial is treated); $\mathcal{R}_x(\cdot)$ denotes the resultant of several polynomials; $\mathbf{I}_k$ stands for the kth order identity matrix.

2   Algebraic Preliminaries

We sketch here some results (if not specially indicated, they are borrowed from [2], [5], and [6]) concerning the notion and the properties of the *discriminant* of a univariate or multivariate polynomial. The necessity in this notion will be justified in the foregoing sections: the condition of intersection of the manifolds (1) and (2)–(3) and the distance equation for them are formulated in terms of appropriate discriminants.

For the univariate polynomial

$$g(x) = b_0 x^N + b_1 x^{N-1} + \cdots + b_N \in \mathbb{C}[x], \ b_0 \neq 0, N \geq 2,$$

the discriminant can be formally defined as

$$\mathcal{D}_x(g) \stackrel{def}{=} (-1)^{N(N-1)/2} N^N b_0^{N-1} \prod_{j=1}^{N-1} g(\lambda_j),$$

where $\{\lambda_1, \ldots, \lambda_{N-1}\} \subset \mathbb{C}$ is a set of zeros of $g'(x)$ counted with their multiplicities. Being a symmetric polynomial of the zeros of $g'(x)$, the discriminant can be represented as a polynomial in the coefficients of $g'(x)$.

Theorem 1. *The polynomial $g(x)$ possesses a multiple zero iff $\mathcal{D}_x(g) = 0$. If the multiple zero is unique and has multiplicity 2 then it can be expressed rationally via the coefficients of $g(x)$.*

Example 1. For $N = 3$

$$\mathcal{D}_x(b_0 x^3 + b_1 x^2 + b_2 x + b_3) \equiv D(b_0, b_1, b_2, b_3)$$

where

$$D(b_0, b_1, b_2, b_3) \stackrel{def}{=} b_1^2 b_2^2 - 4 b_1^3 b_3 - 4 b_0 b_2^3 + 18 b_0 b_1 b_2 b_3 - 27 b_0^2 b_3^2.$$

If it vanishes then the cubic polynomial possesses a multiple zero, which can be evaluated via the formula

$$\lambda = \frac{\partial D/\partial b_2}{\partial D/\partial b_3} \equiv \frac{-9\,b_0 b_3 b_1 - b_1^2 b_2 + 6\,b_2^2 b_0}{27\,b_0^2 b_3 - 9\,b_0 b_1 b_2 + 2\,b_1^3}$$

provided that the denominator of the last fraction does not vanish.   △

Theorem 2. *Let the coefficients of g depend polynomially on the parameter h varying within some interval in $\mathbb{R}$. If $a_0 \neq 0$ under this variation then the number of real zeros of $g(x, h)$ alters only when h passes through the value which annihilates the discriminant:*

$$H(h) \overset{def}{=} \mathcal{D}_x(g(x, h)) \,.$$

For the multivariate polynomial $g(X) = g(x_1, \ldots, x_n), \deg g = N \geq 2$, the discriminant is formally defined as

$$\mathcal{D}_X(g) \overset{def}{=} \prod_{j=1}^{\mathfrak{N}} g(\Lambda_j) \,.$$

Here $\Lambda_j = (\lambda_{j_1}, \ldots, \lambda_{j_n}) \in \mathbb{C}^n$ stands for the stationary point of $g(X)$, i.e. a zero of the system $\partial g/\partial x_1 = 0, \ldots, \partial g/\partial x_n = 0$. In a generic case, the latter possesses precisely $\mathfrak{N} = (N - 1)^n$ (Bézout number) zeros in $\mathbb{C}^n$. Being a symmetric polynomial from the zeros of the algebraic equation system, the discriminant can be represented rationally via the coefficients of $g(X)$.

Theorem 3. *The polynomial $g(X)$ possesses a multiple zero (i.e., the zero for which $g = 0, \partial g/\partial x_1 = 0, \ldots, \partial g/\partial x_n = 0$) iff $\mathcal{D}_X(g) = 0$. If the multiple zero is unique and has multiplicity 2 then it can be expressed rationally via the coefficients of $g(X)$.*

Efficient computation of the discriminant can be organized either with the aid of Gröbner basis construction [1], or, alternatively, via an appropriate determinantal representation for the resultant of multivariate polynomials. For the latter approach, further details can be found in [9] where the so-called *Bézout method* is used for representation of the discriminant as the $\mathfrak{N}$th order determinant with the entries rationally depending on the coefficients of the polynomial. In this case, the representation for the potential double zero for g mentioned in the above theorems can be obtained in the form of a ratio of two appropriate minors of the determinant.

Subsequently we will use the following Schur complement formula for the determinant of a block matrix [3]:

$$\det \begin{pmatrix} \mathbf{U} & \mathbf{V} \\ \mathbf{S} & \mathbf{T} \end{pmatrix} = \det \mathbf{U} \det \mathbf{K} \quad \text{with} \quad \mathbf{K} \overset{def}{=} \mathbf{T} - \mathbf{S}\mathbf{U}^{-1}\mathbf{V} \,. \tag{4}$$

Here $\mathbf{U}$ and $\mathbf{T}$ stand for square matrices, and $\mathbf{U}$ is assumed to be non-singular.

3   Intersection Conditions

The problem of finding the intersection conditions for two quadrics of equal dimension in $\mathbb{R}^2$ (like two ellipses) or in $\mathbb{R}^3$ (like two ellipsoids) is sufficiently well investigated, see, for instance, [9], [10], and the references cited therein. As for a similar problem for the two quadrics of <u>distinct</u> dimensions, we have failed to find the relevant sources.

For the case of $\mathbb{R}^3$, the problem of establishing the fact of intersection of the manifolds (1) and (2)–(3) is equivalent to that of finding the conditions for the existence of real solutions for the system of three algebraic equations. Using the fact that one of these equations is linear, the last problem can be reformulated to that of finding the intersection condition for two ellipses lying in the same plane. In [9] such a condition was presented in the form of restrictions on the zero set of some univariate algebraic equation. In the present section, we will modify the suggested approach in order to extend it to the case of the manifolds treated in $\mathbb{R}^n$.

Theorem 4. *The discriminant of the polynomial*

$$\Psi(\lambda,\mu,z) \stackrel{def}{=} \det\left(\begin{bmatrix} \mathbf{A}_2 & B_2 \\ B_2^T & -1-z \end{bmatrix} - \lambda \begin{bmatrix} \mathbf{A}_1 & B_1 \\ B_1^T & -1 \end{bmatrix} - \mu \begin{bmatrix} \mathbb{O} & C \\ C^T & -2\,h \end{bmatrix} \right) \qquad (5)$$

w.r.t. the variables λ, μ is a polynomial in z which can be factored as follows:

$$\mathcal{D}_{\lambda,\mu}\left(\Psi(\lambda,\mu,z)\right) \equiv \Theta(z)Q^2(z) \qquad (6)$$

with the extraneous factor defined with the aid of the bivariate resultant:

$$Q(z) \stackrel{def}{=} \mathcal{R}_{\lambda,\mu}\left(\det(\lambda \mathbf{A}_1 - \mathbf{A}_2), \Psi'_\lambda, \Psi'_\mu\right) .$$

Let all the real zeros of the equation

$$\Theta(z) = 0 \qquad (7)$$

be simple. Manifolds (1) and (2)-(3) intersect iff among the real zeros of this equation there are values of different signs or 0.

Proof. Let us find the extremal values of the function

$$V(X) = X^T \mathbf{A}_2 X + 2\,B_2^T X - 1 \qquad (8)$$

in the manifold (2)-(3). By assumption, the latter is a compact set, therefore, the extremal values are attained by $V(X)$. All these values are of the same sign iff the manifolds (1) and (2)-(3) do not intersect. We now intend to construct a univariate polynomial with the zero set coinciding with that of critical values of $V(X)$ in (2)-(3). For this aim, we apply the Lagrange multipliers method and find the stationary points of the function

$$X^T \mathbf{A}_2 X + 2\,B_2^T X - 1 - \lambda(X^T \mathbf{A}_1 X + 2B_1^T X - 1) - 2\,\mu(C^T X - h) .$$

They satisfy the system

$$\mathbf{A}_2 X + B_2 - \lambda(\mathbf{A}_1 X + B_1) - \mu C = \mathbb{O}$$

from where

$$X = (\mathbf{A}_2 - \lambda\mathbf{A}_1)^{-1}(\lambda B_1 + \mu C - B_2) \qquad (9)$$

provided that matrix $\mathbf{A}_2 - \lambda\mathbf{A}_1$ is nonsingular. Substitute (9) into (2):

$$(\lambda B_1 + \mu C - B_2)^{\top}(\mathbf{A}_2 - \lambda\mathbf{A}_1)^{-1}\mathbf{A}_1(\mathbf{A}_2 - \lambda\mathbf{A}_1)^{-1}(\lambda B_1 + \mu C - B_2)$$

$$+2\, B_1^{\top}(\mathbf{A}_2 - \lambda\mathbf{A}_1)^{-1}(\lambda B_1 + \mu C - B_2) - 1 = 0, \qquad (10)$$

into (3):

$$C^{\top}(\mathbf{A}_2 - \lambda\mathbf{A}_1)^{-1}(\lambda B_1 + \mu C - B_2) - h = 0 \qquad (11)$$

and, finally, into the function (8); for the result of the latter substitution, introduce the new variable z:

$$z = (\lambda B_1 + \mu C - B_2)^{\top}(\mathbf{A}_2 - \lambda\mathbf{A}_1)^{-1}\mathbf{A}_2(\mathbf{A}_2 - \lambda\mathbf{A}_1)^{-1}(\lambda B_1 + \mu C - B_2)$$

$$+2\, B_2^{\top}(\mathbf{A}_2 - \lambda\mathbf{A}_1)^{-1}(\lambda B_1 + \mu C - B_2) - 1\,. \qquad (12)$$

Compose now the linear combination of equalities (10) and (11):

$$0 = (10) \times \lambda + (11) \times 2\,\mu$$

$$= -2\,\mu h - \lambda + 2\,(\lambda B_1 + \mu C)^{\top}(\mathbf{A}_2 - \lambda\mathbf{A}_1)^{-1}(\lambda B_1 + \mu C - B_2)$$

$$+(\lambda B_1 + \mu C - B_2)^{\top}(\mathbf{A}_2 - \lambda\mathbf{A}_1)^{-1}(\lambda\mathbf{A}_1)(\mathbf{A}_2 - \lambda\mathbf{A}_1)^{-1}(\lambda B_1 + \mu C - B_2)$$

$$= -2\,\mu h - \lambda + (\lambda B_1 + \mu C - B_2)^{\top}(\mathbf{A}_2 - \lambda\mathbf{A}_1)^{-1}(\lambda\mathbf{A}_1)(\mathbf{A}_2 - \lambda\mathbf{A}_1)^{-1}(\lambda B_1 + \mu C - B_2)$$

$$+2\, B_2^{\top}(\mathbf{A}_2 - \lambda\mathbf{A}_1)^{-1}(\lambda B_1 + \mu C - B_2)$$

$$+(\lambda B_1 + \mu C - B_2)^{\top}(\mathbf{A}_2 - \lambda\mathbf{A}_1)^{-1}\mathbf{A}_2(\mathbf{A}_2 - \lambda\mathbf{A}_1)^{-1}(\lambda B_1 + \mu C - B_2)$$

Substitution of (12) into the last expression yields the equality

$$\tilde{\Psi}(\lambda, \mu, z) = 0$$

for

$$\tilde{\Psi}(\lambda, \mu, z) \stackrel{def}{=} -2\,\mu h - \lambda + z + 1$$

$$+(\lambda B_1 + \mu C - B_2)^{\top}(\mathbf{A}_2 - \lambda\mathbf{A}_1)^{-1}(\lambda B_1 + \mu C - B_2)\,. \qquad (13)$$

It turns out that partial derivatives of $\tilde{\Psi}(\lambda, \mu, z)$ w.r.t. λ and μ coincide with the left-hand sides of (10) and (11), respectively. Thus, we have reduced the problem of evaluation of the critical values of the function $V(X)$ to that of solution of the system

$$\tilde{\Psi} = 0, \ \tilde{\Psi}'_{\lambda} = 0, \ \tilde{\Psi}'_{\mu} = 0\,.$$

In accordance with the results of Section 2, the elimination of the parameters λ and μ from this system can be performed via the discriminant computation.

With the aid of the Schur formula (4), one can represent $\tilde{\Psi}$ as

$$\tilde{\Psi}(\lambda, \mu, z) \det(\lambda\mathbf{A}_1 - \mathbf{A}_2) \equiv \det \begin{bmatrix} \lambda\mathbf{A}_1 - \mathbf{A}_2 & \lambda B_1 + \mu C - B_2 \\ (\lambda B_1 + \mu C - B_2)^\top & z - \lambda - 2\,\mu h + 1 \end{bmatrix}.$$

The right-hand side of the last identity coincides with (5). This explains the appearance of the discriminant in the expression (6) and also clarifies the meaning of an extraneous factor. □

Elimination of the parameters λ and μ from the system $\Psi = 0$, $\Psi'_\lambda = 0$, $\Psi'_\mu = 0$ is simplified by the fact that Ψ is a quadric polynomial in μ and, therefore, the last equation is linear w.r.t. μ. This results in the reduction of the elimination procedure to computation of the resultant of two bivariate equations.

Example 2. Let

$$\mathbf{A}_1 = \frac{1}{38} \begin{pmatrix} -7 & 2 & 0 \\ 2 & -6 & 2 \\ 0 & 2 & -5 \end{pmatrix}, \quad B_1 = \frac{1}{76} \begin{pmatrix} 23 \\ 8 \\ 17 \end{pmatrix},$$

$$\mathbf{A}_2 = \frac{1}{1828} \begin{pmatrix} 378 & 0 & 2 \\ 0 & 2 & -1 \\ 2 & -1 & 378 \end{pmatrix}, \quad B_2 = \begin{pmatrix} 0 \\ 0 \\ 0 \end{pmatrix}, \quad C = \begin{pmatrix} 1 \\ 1 \\ 1 \end{pmatrix}.$$

Find the values of the parameter h for which the manifolds (1) and (2)–(3) intersect.

Solution. Compute (with the aid of symbolic computation) the polynomial from Theorem 4

$$\Theta(z, h) = {\scriptstyle 150984031237546797599744}\, z^4$$

$$+ {\scriptstyle 24433662208}(-{\scriptstyle 57567207517117}\, h^2 - {\scriptstyle 956726561660206}\, h + {\scriptstyle 4222134951604488})z^3 + \ldots$$

$$+ {\scriptstyle 21793666458122145180}\, h^8 + {\scriptstyle 385456635897336759692}\, h^7 + {\scriptstyle 944734848205614648643}7\, h^6$$

$$- {\scriptstyle 1280563967921745540096}00\, h^5 + {\scriptstyle 199909509333595187064630}\, h^4$$

$$- {\scriptstyle 9375833328721305345566}96\, h^3 + {\scriptstyle 2195820493635262075431736}9\, h^2$$

$$- {\scriptstyle 935320940453794080849030}44\, h + {\scriptstyle 11548285853285742397479362}0.$$

For any specialization of h, it is possible to establish the exact number of real zeros of this polynomial and to separate them (i.e., for instance, to find the exact number of positive or negative ones) by means of purely algebraic procedures applied for the polynomial coefficients. These procedures can be organized either with the aid of Sturm series construction or, alternatively, via application of Hermite's method [4].

In order to find the parameter specializations which separate in the parameter line the domains with distinct combinations for the signs of real zeros of $\Theta(z, h)$, consider two equations. The first one is $\Theta(0, h) = 0$ possessing exactly two real zeros

$$h_1 \approx 3.357841, \quad h_2 \approx 6.231882.$$

When the parameter h, on varying continuously, crosses any of these values, one of the real zeros of the corresponding polynomial $\Theta(z, h)$ changes its sign. The second equation

$$H(h) \overset{def}{=} \mathcal{D}_z(\Theta(z, h)) = 0$$

provides the bifurcation values for the parameter influencing the number of real zeros for the polynomial $\Theta(z, h)$ (in the sense of Theorem 2). For our example,

$$H(h) \equiv 2^{32} 457^{12} (162\, h^2 - 2344\, h + 5377)$$

$$\times (6779563169\, h^2 - 15038696694\, h - 15766015050)^2 \, [H_1(h)]^3 \, ,$$

where $H_1(h)$ stands for the polynomial of degree 6 possessing only imaginary zeros. Real zeros of $H(h)$ are

$$\tilde{h}_1 \approx 2.858769, \quad \tilde{h}_2 \approx 11.610367, \quad \tilde{h}_3 \approx -0.776528, \quad \tilde{h}_4 \approx 2.994767 \, .$$

Polynomial $\Theta(z, h)$ possesses 2 real zeros for $\tilde{h}_1 < h < \tilde{h}_2$; these zeros are of distinct signs iff $h_1 < h < h_2$. As for the values $h > \tilde{h}_2$ or $h < \tilde{h}_1$, the polynomial $\Theta(z, h)$ does not have real zeros. This means: for these parameter values, the plane (3) does not intersect the ellipsoid (2), i.e., conditions (2)–(3) do not define a real curve.

The exceptional cases happen for specializations $h = \tilde{h}_3$ and $h = \tilde{h}_4$. For any of these values, the polynomial $\Theta(z, h)$ possesses a multiple real zero. It turns out that this zero, as a critical value of the objective function (8), corresponds to a pair of imaginary stationary points. $\triangle$

4   Distance Equation

Theorem 5. *Let the manifolds (1) and (2)-(3) do not intersect. The squared distance between them coincides generically with the minimal simple positive zero of the* **distance equation**

$$\mathcal{F}(z) \overset{def}{=} \mathcal{D}_{\lambda_1, \lambda_2, \mu}(\Phi(\lambda_1, \lambda_2, \mu, z)) = 0 \qquad (14)$$

Here

$$\Phi(\lambda_1, \lambda_2, \mu, z) \overset{def}{=} \det \left(\lambda_1 \begin{bmatrix} \mathbf{A}_1 & B_1 \\ B_1^T & -1 \end{bmatrix} + \lambda_2 \begin{bmatrix} \mathbf{A}_2 & B_2 \\ B_2^T & -1 \end{bmatrix} + \mu \begin{bmatrix} \mathbb{O} & C \\ C^T & -2\,h \end{bmatrix} \right.$$

$$\left. - \lambda_1 \lambda_2 \begin{bmatrix} \mathbf{A}_2 \mathbf{A}_1, & \mathbf{A}_2 B_1 \\ B_2^T \mathbf{A}_1, & B_2^T B_1 \end{bmatrix} - \lambda_2 \mu \begin{bmatrix} \mathbb{O} & \mathbf{A}_2 C \\ \mathbb{O} & B_2^T C \end{bmatrix} + \begin{bmatrix} \mathbb{O} & \mathbb{O} \\ \mathbb{O} & z \end{bmatrix} \right). \qquad (15)$$

Proof. We will start with the traditional treatment of the constrained optimization problem via Lagrange multipliers method. Consider the Lagrange function in the form

$$(X - Y)^\top (X - Y) - \lambda_1 (X^\top \mathbf{A}_1 X + 2\, B_1^\top X - 1) - 2\,\mu (C^\top X - h)$$

$$-\lambda_2(Y^\top \mathbf{A}_2 Y + 2B_2^\top Y - 1) \tag{16}$$

Stationary points of this function are given by the system

$$X - Y - \lambda_1(\mathbf{A}_1 X + B_1) - \mu C = \mathbb{O}, \tag{17}$$
$$-X + Y - \lambda_2(\mathbf{A}_2 Y + B_2) = \mathbb{O}, \tag{18}$$
$$Y^\top \mathbf{A}_2 Y + 2B_2^\top Y = 1 \tag{19}$$

accomplished with (2)–(3). Our aim is to eliminate the variables X and Y from this system. Introducing the matrix

$$\mathbf{M} \overset{def}{=} \mathbf{I}_n - \frac{1}{\lambda_1}\mathbf{A}_1^{-1} - \frac{1}{\lambda_2}\mathbf{A}_2^{-1} \tag{20}$$

and the vector

$$P \overset{def}{=} \mathbf{M}^{-1}\left(\mathbf{A}_2^{-1}B_2 - \mathbf{A}_1^{-1}B_1 - \frac{\mu}{\lambda_1}\mathbf{A}_1^{-1}C\right),$$

we resolve the system (17)–(18) w.r.t. X and Y:

$$X = \frac{1}{\lambda_1}\mathbf{A}_1^{-1}(P - \lambda_1 B_1 - \mu C), \tag{21}$$

$$Y = -\frac{1}{\lambda_2}\mathbf{A}_2^{-1}(P + \lambda_2 B_2). \tag{22}$$

Substitute now these expressions into (2),(3), and (19).

$$\frac{1}{\lambda_1^2}(P - \lambda_1 B_1 - \mu C)^\top \mathbf{A}_1^{-1}(P - \lambda_1 B_1 - \mu C)$$

$$+\frac{2}{\lambda_1} B_1^\top \mathbf{A}_1^{-1}(P - \lambda_1 B_1 - \mu C) - 1 = 0, \tag{23}$$

$$\frac{1}{\lambda_2^2}(P + \lambda_2 B_2)^\top \mathbf{A}_2^{-1}(P + \lambda_2 B_2) - \frac{2}{\lambda_2} B_2^\top \mathbf{A}_2^{-1}(P + \lambda_2 B_2) - 1 = 0, \tag{24}$$

$$\frac{1}{\lambda_1} C^\top \mathbf{A}_1^{-1}(P - \lambda_1 B_1 - \mu C) - h = 0. \tag{25}$$

Compose now the linear combination from these equalities:

$$0 = (23) \times \lambda_1 + (24) \times \lambda_2 + (25) \times 2\mu$$

We skip long intermediate transformations of this expression and present only the final result:

$$0 = -\lambda_1 - \lambda_2 - 2\mu h + P^\top P - P^\top \mathbf{M} P$$

$$-\lambda_1\left(B_1 + \frac{\mu}{\lambda_1}C\right)^\top \mathbf{A}_1^{-1}\left(B_1 + \frac{\mu}{\lambda_1}C\right) - \lambda_2 B_2^\top \mathbf{A}_2^{-1}B_2 \tag{26}$$

Consider now the distance function $(X - Y)^\top (X - Y)$ and denote by z its critical values on the pairs of vectors belonging to the corresponding manifolds:

$$z = (X - Y)^\top (X - Y) \overset{(21),(22)}{=} P^\top P.$$

Substitute this into (26):

$$\tilde{\Phi}(\lambda_1, \lambda_2, \mu, z) \overset{def}{=} z - \lambda_1 - \lambda_2 - 2\mu h - P^\top \mathbf{M} P$$

$$-\lambda_1 \left(B_1 + \frac{\mu}{\lambda_1} C \right)^\top \mathbf{A}_1^{-1} \left(B_1 + \frac{\mu}{\lambda_1} C \right) - \lambda_2 B_2^\top \mathbf{A}_2^{-1} B_2 = 0. \tag{27}$$

It turns out that the partial derivatives of $\tilde{\Phi}(\lambda_1, \lambda_2, \mu, z)$ w.r.t. λ_1, λ_2 and μ coincide with the left-hand sides of (23), (24), and (25) correspondingly. Thus, the elimination of the variables X and Y procedure results in the equations

$$\tilde{\Phi} = 0, \ \tilde{\Phi}'_{\lambda_1} = 0, \ \tilde{\Phi}'_{\lambda_2} = 0, \ \tilde{\Phi}'_\mu = 0.$$

Taking into account the results of Section 2, one can perform elimination of the Lagrange multipliers from this system with the aid of discriminant — and that is the reason for its appearance in the statement of the theorem.

To reduce the expression for $\tilde{\Phi}$ to the form (15), one should exploit Schur formula (4):

$$\begin{vmatrix} \lambda_1 \mathbf{A}_1 + \lambda_2 \mathbf{A}_2 - \lambda_1 \lambda_2 \mathbf{A}_2 \mathbf{A}_1 & (\mathbf{I}_n - \lambda_2 \mathbf{A}_2)(\lambda_1 B_1 + \mu C) + \lambda_2 B_2 \\ (\lambda_1 B_1 + \lambda_2 B_2 + \mu C - \lambda_1 \lambda_2 \mathbf{A}_1 B_2)^\top, & z - \lambda_1 - \lambda_2 - 2h\mu - \lambda_2 B_2^\top (\lambda_1 B_1 + \mu C) \end{vmatrix}$$

$$\equiv \lambda_1 \lambda_2 \tilde{\Phi}(\lambda_1, \lambda_2, \mu, z) \det \mathbf{A}_1 \det \mathbf{A}_2 \det \mathbf{M} \equiv \Phi(\lambda_1, \lambda_2, \mu, z).$$

$\square$

Remark 1. The last identity connecting the rational function $\tilde{\Phi}$ with the polynomial Φ gives rise to some extraneous factor in the distance equation (14). This factor is similar to the one appeared in Theorem 4: it equals the square of some polynomial in z. As yet, we failed to establish its structure.

Once the minimal positive zero for the distance equation (14) is evaluated, the corresponding *nearest* points in the manifolds can be found in the following way. Any zero z_* of the distance equation annihilates the discriminant of the polynomial (15) treated as a polynomial in the Lagrange multipliers. This condition guarantees the existence of a multiple zero for $\Phi(\lambda_1, \lambda_2, \mu, z_*)$. According to Theorem 2, this zero $(\lambda_{1*}, \lambda_{2*}, \mu_*)$ can be expressed as a rational function of z_*. Substitution of these parameter values into (21) and (22) yields the coordinates of nearest points in the manifolds.

The algorithm outlined above contains a gap. It might happen that the minimal positive zero for the distance equation corresponds to imaginary values for at least one of $\lambda_{1*}, \lambda_{2*}, \mu_*$, and the algorithm would result in the imaginary points in the manifolds. This problem is avoided by imposing the simplicity restriction for the minimal zero of the distance equation in the statement of the theorem; we refer to [9] for the idea underlying this matter.

$\square$

Elimination of the parameters λ_1, λ_2 and μ from the system $\Phi = 0$, $\Phi'_{\lambda_1} = 0$, $\Phi'_{\lambda_2} = 0$, $\Phi'_\mu = 0$ is simplified by the fact that Φ is a quadric polynomial in μ and, therefore, the last equation is linear w.r.t. μ. This results in the reduction of the elimination procedure to computation of the resultant of three trivariate equations.

Example 3. Find the distance between manifolds (1) and (2)–(3) from Example 2 for the parameter specialization $h = 9$.

Solution. Compute symbolically the discriminant (14). The coefficients of the distance equation[1] are rather huge:

$$\mathcal{F}(z) = \underbrace{1472499059\ldots24000000000000}_{123 \text{ digits}} z^{20} + \cdots + \underbrace{1396257246\ldots166588729088}_{171 \text{ digits}}.$$

It possesses 8 real zeros:

$$z_1 \approx 1.622327, \; z_2 \approx 8.591215, \ldots, z_8 \approx 1193.878245.$$

Thus, the distance between manifolds equals $\sqrt{z_1} \approx 1.273706$.

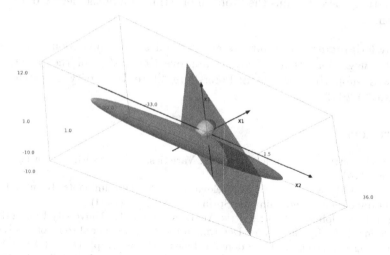

Fig. 1. Ellipse is cut with the plane out of the smaller ellipsoid

To find the coordinates of the nearest points in the manifolds, we first evaluate the multiple zero for the polynomial $\Phi(\lambda_1, \lambda_2, \mu, z_1)$

$$\lambda_{1*} \approx 2.315611, \; \lambda_{2*} \approx -2.819637, \; \mu_* \approx 0.676726,$$

and then substitute these values into (21) and (22):

$$X_* \approx (2.360361, \, 4.117219, \, 2.522420)^\top, \; Y_* \approx (1.487910, \, 4.107009, \, 1.594491)^\top$$

(with the first point lying in the ellipse). $\triangle$

---

[1] On extracting an extraneous factor of the order 8 mentioned in Remark 1.

Remark 2. If one formally sets $C = \mathbb{O}, h = 0$ in expressions (5) and (15), and deletes the subscript μ of the discriminant sign $\mathcal{D}$ in (7) and (14), then the obtained results will be equivalent to those presented in [9] for the condition of intersection of and distance evaluation between quadrics (1) and (2) in $\mathbb{R}^n$.

5   Conclusions

For the problem stated in the present paper, we have exploited the approach suggested in [8]. It is based on the inversion of the traditional approach for the constrained optimization problem which first aimed at finding the stationary point set for the respective distance function and then evaluation of the corresponding critical values. The obtained results look similar to their counterparts for the distance evaluation problem between two quadrics in $\mathbb{R}^n$ [9]. This means: both the condition for the manifold intersection and the distance equation construction is based on an appropriate discriminant as the cornerstone.

The authors are strongly sure that this similarity will also be displayed in the distance evaluation problem between two conic sections in $\mathbb{R}^3$. For further investigation, there remains the problem of estimation of the degree of distance equation.

Acknowledgments. The authors are grateful to the anonymous referees for valuable suggestions that helped to improve the quality of the paper. This work was supported by the St.Petersburg State University research grant # **9.38.674.2013**.

References

1. Cox, D., Little, J., O'Shea, D.: Ideals, Varieties, and Algorithms. Springer, New York (2007)
2. Gelfand, I.M., Kapranov, M.M., Zelevinsky, A.V.: Discriminants, Resultants and Multidimensional Determinants. Springer, New York (1994)
3. Horn, R.A., Johnson, C.R.: Matrix Analysis. Cambridge University Press (1986)
4. Kalinina, E.A., Uteshev, A.Y.: Determination of the number of roots of a polynominal lying in a given algebraic domain. Linear Algebra Appl. **185**, 61–81 (1993)
5. Netto, E.: Rationale Funktionen einer Veränderlichen; ihre Nullstellen. In: Encyklopädie der Mathematischen Wissenschaften, vol. I, Part 1, pp. 227–253. Teubner, Leipzig (1898)
6. Perron, O.: Algebra, vol. 1. De Gruyter, Berlin (1927)
7. Schneider, P.J., Eberly, D.H.: Geometric Tools for Computer Graphics. Elsevier, San Francisco (2003)
8. Uteshev, A.Y., Yashina, M.V.: Distance computation from an ellipsoid to a linear or a quadric surface in IR^n. In: Ganzha, V.G., Mayr, E.W., Vorozhtsov, E.V. (eds.) CASC 2007. LNCS, vol. 4770, pp. 392–401. Springer, Heidelberg (2007)
9. Uteshev, A.Y., Yashina, M.V.: Metric problems for quadrics in multidimensional space. J. Symb. Comput. **68**, 287–315 (2015)
10. Wang, W., Krasauskas, R.: Interference analysis of conics and quadrics. Contemp. Mathematics **334**, 25–36 (2003)

Algebraic General Solutions of First Order Algebraic ODEs

Ngoc Thieu Vo* and Franz Winkler

Research Institute for Symbolic Computation (RISC), Johannes Kepler University,
A-4040 Linz, Austria
tvongoc@risk.jku.at, Franz.Winkler@risc.jku.at

Abstract. In this paper we consider the class of algebraic ordinary differential equations (AODEs), the class of planar rational systems, and discuss their algebraic general solutions. We establish for each parametrizable first order AODE a planar rational system, the associated system, such that one can compute algebraic general solutions of the one from the other and vice versa. For the class of planar rational systems, an algorithm for computing their explicit algebraic general solutions with a given rational first integral is presented. Finally an algorithm for determining an algebraic general solution of degree less than a given positive integer of parametrizable first order AODEs is proposed.

1 Introduction

A first order AODE is a differential equation of the form

$$F(x, y, y') = 0 \tag{1}$$

where F is a polynomial in x, y and y' with coefficients in a field, for instance $\mathbb{Q}$. Solving the differential equation (1) is the problem of finding a function $y = y(x)$ such that $F(x, y(x), y'(x)) = 0$. If furthermore $y(x)$ is a rational (resp. algebraic) function, it is called a rational (resp. algebraic) solution. A solution of (1) may contain an arbitrary constant. Such a solution is called a general solution. For example, $y = x^2 + c$ is a general solution of the differential equation $y' - 2x = 0$.

If F does not explicitly involve the variable of differentiation x, the differential equation (1) is called autonomous. In [5] and [4], Feng and Gao presented an algorithm for determining rational general solutions of autonomous first order AODEs. Later, in [1], Aroca et al. proposed a polynomial time algorithm for computing algebraic general solutions.

The problem of determining rational general solutions of first order AODEs was studied by Ngô and Winkler in [7] and [8]. For using the technique of rational parametrization, the authors add an additional condition to the differential equation (1), that the equation $F(x, y, z) = 0$ defines a rational algebraic surface, i.e.,

---

* Supported by the strategic program "Innovatives OÖ 2010plus" by the Upper Austrian Government.

© Springer International Publishing Switzerland 2015
V.P. Gerdt et al. (Eds.): CASC 2015, LNCS 9301, pp. 479–492, 2015.
DOI: 10.1007/978-3-319-24021-3_35

a surface which is birationally equivalent to the affine plane. A differential equation (1) satisfying such a condition is called a parametrizable first order AODE. The authors associate to each parametrizable first order AODE a certain planar rational system which is named the associated system. The key observation is that rational general solutions of the given differential equation can be obtained from the rational general solutions of its associated system, and vice versa. For the latter system, the authors combine the technique of rational parametrization of algebraic curves and invariant algebraic curves and propose a procedure to find rational general solutions.

This paper aims to extend the results of Ngô and Winkler. In particular, we will study algebraic general solutions of the class of parametrizable (not necessarily autonomous) first order AODEs. It is reasonable for Ngô and Winkler to restrict their work on the class of parametrizable first order AODEs. In fact, a first order AODE having a rational general solution must be parametrizable. Unfortunately, this may no longer be true if we consider algebraic general solutions instead of rational ones. In the other words, a first order AODE which is not parametrizable may have an algebraic general solution.

Section 2 is devoted to recall material needed in subsequent sections. In Section 3 we construct the associated system of a parametrizable first order AODE. The result of this section is an explicit formulation of the correspondence between algebraic general solutions of the given differential equation and its associated system. Then a method for finding explicit algebraic general solutions of planar rational systems will be presented in Section 4. Finally, in Section 5, we study the problem of determining algebraic general solutions of bounded degree for parametrizable first order AODEs.

2   Algebraic General Solutions of AODEs

By k we denote the field of algebraic numbers $\overline{\mathbb{Q}}$. However, theoretically all the discussions are also true for any algebraic closed field of characteristic zero. The derivation is understood as the usual derivation with respect to the variable x.

An algebraic ordinary differential equation (AODE) is a differential equation of the form

$$F(x, y, y', ..., y^{(n)}) = 0 \tag{2}$$

where F, without loss of generality, is an irreducible polynomial in the differential algebra $k(x)\{y\}$. As in [9], for each $\Sigma \subseteq k(x)\{y\}$, by $[\Sigma]$ and $\{\Sigma\}$ we denote the differential ideal and the radical differential ideal, respectively, generated by Σ. Ritt proved in [9] that the radical differential ideal $\{F\}$ generated by F in $k(x)\{y\}$ can be decomposed as $\{F\} = (\{F\} : S_F) \cap \{F, S_F\}$, where S_F is the separant of F, and $(\{F\} : S_F) := \{G \in k(x)\{y\} \mid GS_F \in \{F\}\}$. Furthermore, the first component of the decomposition is a prime differential ideal.

Before coming to the notion of general solutions of AODEs, we recall here the definition of a zero of a differential system. For further details, the reader is referred to [9].

Definition 1. *Let Σ be a subset of $k(x)\{y\}$, and L a differential field extension of $k(x)$ with respect to which the $y_i s$ are indeterminates. An element $\xi \in L$ is called a zero of Σ if all elements of Σ vanish at ξ.*

It is clear that if a differential polynomial F vanishes at ξ, then all derivatives of F vanishes at ξ. Therefore Σ, $[\Sigma]$, and $\{\Sigma\}$ have the same set of zeroes.

Definition 2. *Let I be a prime differential ideal in $k(x)\{y\}$. A zero ξ of I, contained in a differential field extension L of $k(x)$, is called a generic zero of I if every differential polynomial in $k(x)\{y\}$ vanishing at ξ is in I.*

In other words, ξ is a generic zero of the prime differential I if and only if the set $\mathbb{I}(\xi) := \{F \in k(x)\{y\} \mid F(\xi) = 0\}$ is exactly I. It is well-known that every prime differential ideal has a generic zero.

Definition 3. *Let F be as above. A zero of the radical ideal $\{F\}$ is called a solution of the AODE (2). A generic zero of the differential ideal $(\{F\} : S_F)$ is called a general solution of the AODE (2).*

Definition 4. *Let ξ be a solution of the AODE (2) which is contained in a differential field extension L of $k(x)$. Let K be the field of constants of L. ξ is called an algebraic solution of the AODE (2) if there is an irreducible polynomial $G \in K[x, y]$ such that $G(x, \xi) = 0$. In this case, G is called a minimal polynomial of ξ. If furthermore the degree of y in G is 1, then ξ is called a rational solution of the AODE (2). ξ is called an algebraic (resp. rational) general solution if it is a general solution and algebraic (resp. rational).*

Given an algebraic solution ξ of the AODE (2), there is only one minimal polynomial up to multiplication by a non-zero constant in K. If $G(x, y)$ is such a minimal polynomial of ξ, then all roots $y = y(x)$ of the algebraic equation $G(x, y) = 0$ are solutions of the AODE (2), (see [1], Lemma 2.4). Therefore, by abuse of notation, G is sometimes called a solution of the AODE (2) instead of ξ.

The following lemma gives a criterion for deciding whether a solution is general.

Lemma 1. *A solution ξ of the AODE (2) is a general solution if and only if $\forall H \in k(x)\{y\} : H(\xi) = 0 \Rightarrow prem(H, F) = 0$.*

Proof. See [9], chap. II.5 and 6.   □

Proposition 1. *Let ξ be an algebraic solution of the AODE (2) with the minimal polynomial G. If ξ is a general solution, then at least one of the coefficients of G contains a constant which is transcendental over k.*

Proof. Assume that all coefficients of G are in k, then $G = prem(G, F)$. Since $G(x, \xi) = 0$, $G = 0$.   □

A planar rational system is a system of ODEs of the form

$$\begin{cases} s' = M(s,t) \\ t' = N(s,t) \end{cases} \tag{3}$$

where $M = M_1/M_2, N = N_1/N_2 \in k(s,t)$ are reduced rational functions. The definitions of solution, general solution, rational and algebraic solution of a planar rational system are the same as the ones for AODEs. For more details, the reader is referred to [7], [8]. To end this preparatory section, we state here an analogous version of Lemma 1 for planar rational systems.

Lemma 2. *A solution (ξ_1, ξ_2) of the system (3) is a general solution if and only if $\forall H \in k(x)\{s,t\} : H(\xi_1, \xi_2) = 0 \Rightarrow prem(H, \{M_2(s,t)s' - M_1(s,t), N_2(s,t)t' - N_1(s,t), \}) = 0$*

Proof. See [9], chap. II.5 and 6.   □

Proposition 2. *Assume $(\xi_1, \xi_2) \in L^2$ is an algebraic general solution of the system (3), where L is a differential field extension of $k(x)$. Let K be the field of constants of L. Let $G \in K[s,t]$ be an irreducible polynomial such that $G(\xi_1, \xi_2) = 0$. Then at least one of the coefficients of G contains a constant which is transcendental over k.*

Proof. Assume that all coefficients of G are in k, we have

$$G(s,t) = prem(G(s,t), \{M_2(s,t)s' - M_1(s,t), N_2(s,t)t' - N_1(s,t)\})$$

is in $k[s,t]$. Since $G(\xi_1, \xi_2) = 0$, lemma 2 yields $G = 0$. This is a contradiction.   □

3   The Associated System of a Parametrizable First Order AODE

We now restrict our work to the class of parametrizable first order AODEs. A parametrizable first order AODE is an AODE of the form

$$F(x, y, y') = 0 \tag{4}$$

where $F \in k(x)\{y\} - k(x)[y]$ is a differential polynomial of order 1 such that the algebraic surface defined by $F(x, y, z) = 0$ is rational. Observe that only irreducible polynomials can generate rational surfaces.

In this section we will present a procedure for finding an algebraic general solution of the differential equation (4). The method is inherited from the idea of Ngô and Winkler [7]. For self-containedness, the associated system of the differential equation (4) will be briefly reconstructed here. Several facts relating to its algebraic general solutions will be investigated.

Definition 5. *Let F be as above. The algebraic surface $S \subset \mathbb{A}_k^3$ defined by $F(x, y, z) = 0$ is called the solution surface of the differential equation (4).*

As mentioned above, we only consider the class of first order AODEs whose solution surfaces are rational. In other words, there is a rational map $P : \mathbb{A}_k^2 \to S \subset \mathbb{A}_k^3$ defined by $P(s,t) := (\chi_1(s,t), \chi_2(s,t), \chi_3(s,t))$ where $\chi_1, \chi_2, \chi_3 \in k(s,t)$ such that $F(P(s,t)) = 0$, and P is invertible. Such P is called a *proper parametrization* of the surface S. From now on, we always assume that the parametrizable first order AODE (4) is equipped with the proper parametrization P.

Now let us fix an algebraic general solution $\xi = \xi(x)$ of the differential equation (4), i.e. $F(x, \xi(x), \xi'(x)) = 0$. Therefore $\xi \in \overline{K(x)} - \overline{k(x)}$. Denote $(s(x), t(x)) := P^{-1}(x, \xi(x), \xi'(x))$. Since P is proper, $(s(x), t(x))$ is a pair of algebraic functions satisfying $P(s(x), t(x)) = (x, \xi(x), \xi'(x))$. Therefore

$$\begin{cases} \chi_1(s(x), t(x)) = x \\ \chi_2(s(x), t(x)) = \chi_3(s(x), t(x)) \end{cases}$$

Differentiating both sides of the first equation, and expanding the second one gives us a linear system on $s'(x)$ and $t'(x)$.

$$\begin{cases} s'(x)\frac{\partial}{\partial s}\chi_1(s(x),t(x)) + t'(x)\frac{\partial}{\partial t}\chi_1(s(x),t(x)) = 1 \\ s'(x)\frac{\partial}{\partial s}\chi_2(s(x),t(x)) + t'(x)\frac{\partial}{\partial t}\chi_2(s(x),t(x)) = \chi_3(s(x),t(x)) \end{cases}$$

Since P is birational, the Jacobian matrix

$$\begin{bmatrix} \frac{\partial \chi_1}{\partial s} & \frac{\partial \chi_2}{\partial s} & \frac{\partial \chi_3}{\partial s} \\ \frac{\partial \chi_1}{\partial t} & \frac{\partial \chi_2}{\partial t} & \frac{\partial \chi_3}{\partial t} \end{bmatrix}$$

has generic rank 2. Without loss of generality, we can always assume that the determinant

$$g := \begin{vmatrix} \frac{\partial \chi_1}{\partial s} & \frac{\partial \chi_2}{\partial s} \\ \frac{\partial \chi_1}{\partial t} & \frac{\partial \chi_2}{\partial t} \end{vmatrix}$$

is non-zero. Furthermore, we claim that $g(s(x), t(x)) \neq 0$. This is asserted by the following lemma.

Lemma 3. *With notation as above. Then*

$$\forall R \in k(s,t): \quad R(s(x), t(x)) = 0 \Rightarrow R = 0$$

Proof (see also in [8], Theorem 3.14). Let $P(x, y, z) := R(P^{-1}(x, y, z))$, where P^{-1} is the inverse of P. Then $P(x, \xi(x), \xi'(x)) = R(P^{-1}(x, \xi(x), \xi'(x))) = R(s(x), t(x)) = 0$. Since $\xi(x)$ is a general solution of the differential equation (4), the numerator of P, say P_1, must satisfy $prem(P_1(x, y, y'), F) = 0$. Therefore there is $n \in \mathbb{N}$ such that $S_F^n P_1(x, y, y')$ can be written as a linear combination of F and its derivatives with coefficients in $k(x)\{y\}$, where S_F is the separant of F. Since $P_1(x, y, y')$ is of order at most 1, there is $H \in k(x)\{y\}$ such that

$$S_F^n P_1(x, y, y') = H.F$$

Substituting $\mathcal{P}(s,t)$ for (x, y, y') and using the fact that $F(\mathcal{P}(s,t)) = 0$, we obtain

$$S_F(\mathcal{P}(s,t))^n P_1(\mathcal{P}(s,t)) = 0.$$

Thus $P_1(\mathcal{P}(s,t)) = 0$. Therefore $P(\mathcal{P}(s,t)) = 0$. Finally,

$$R(s,t) = R(\mathcal{P}^{-1}(\mathcal{P}(s,t))) = P(\mathcal{P}(s,t)) = 0.$$

□

Now the linear system for $s'(x)$ and $t'(x)$ can be solved by Cramer's rule. Thus, $(s(x), t(x))$ is an algebraic solution of the planar rational system:

$$\begin{cases} s' = \dfrac{\chi_3(s,t)\frac{\partial}{\partial t}\chi_1(s,t) - \frac{\partial}{\partial t}\chi_2(s,t)}{g(s,t)} \\ t' = \dfrac{\frac{\partial}{\partial s}\chi_2(s,t) - \chi_3(s,t)\frac{\partial}{\partial s}\chi_1(s,t)}{g(s,t)} \end{cases} \tag{5}$$

Definition 6. *The system (5) is called the associated system of the differential equation (4) with respect to the proper parametrization $\mathcal{P}$.*

We summarize here the result of the construction of the associated system above.

Theorem 1. *Consider the parametrizable first order AODE (4)*

$$F(x, y, y') = 0$$

and its associated system (5) with respect to a given proper parametrization $\mathcal{P}$. If $y = y(x)$ is an algebraic general solution of the differential equation (4), then

$$(s(x), t(x)) := \mathcal{P}^{-1}(x, y(x), y'(x))$$

is an algebraic general solution of the associated system.

Proof. $(s(x), t(x))$ is an algebraic solution of the associated system as we established. Lemma 3 asserts that it is in fact a general solution. □

Theorem 2. *Consider the parametrizable first order AODE (4)*

$$F(x, y, y') = 0$$

and its associated system (5) with respect to a given proper parametrization $\mathcal{P}$. If $(s(x), t(x))$ is an algebraic general solution of the associated system, then

$$y(x) := \chi_2(s(2x - \chi_1(s(x), t(x))), t(2x - \chi_1(s(x), t(x))))$$

is an algebraic general solution of the differential equation (4).

Proof. As in the construction, $s(x), t(x)$ must satisfies the following system:

$$\begin{cases} \chi_1'(s(x), t(x)) = 1 \\ \chi_2'(s(x), t(x)) = \chi_3(s(x), t(x)) \end{cases}$$

The first relation yields $c := \chi_1(s(x), t(x)) - x$ is an arbitrary constant. Thus we have

$$\begin{cases} \chi_1(s(x-c), t(x-c)) = x \\ \chi_2(s(x-c), t(x-c)) = y(x) \\ \chi_3(s(x-c), t(x-c)) = y'(x) \end{cases}$$

Therefore $y(x)$ is an algebraic general solution of the differential equation (4).

It remains to prove that $y(x)$ is a general solution. To this end, consider an arbitrary $G \in k(x)\{y\}$ such that $G(y(x)) = 0$. Since F is of order 1, $prem(G, F) \in k(x)[y, y']$. Let $R \in k[x, y, y']$ be the numerator of $prem(G, F)$. Observe that $R(x, y(x), y'(x)) = 0$. It implies $R(\mathcal{P}(s(x-c), t(x-c))) = 0$. Since c can be chosen arbitrarily, we have $R(\mathcal{P}(s(x), t(x))) = 0$. Now, by application of Lemma 3 we get $R(\mathcal{P}(s, t)) = 0$. So $R(x, y, z) = R(\mathcal{P}(\mathcal{P}^{-1}(x, y, z))) = 0$, and consequently $prem(G, F) = 0$. Hence $y(x)$ is a general solution. $\qquad\square$

The previous two theorems establish a one-to-one correspondence between algebraic general solutions of a paramatrizable first order AODE and algebraic general solutions of its associated system which is a planar rational system. Furthermore the correspondence is formulated explicitly. Once an algebraic general solution of its associated system is known, the corresponding algebraic general solution of the given parametrizable first order AODE can be determined immediately. The problem of finding an algebraic general solution of paramatrizable first order AODEs can be reduced to the problem of determining an algebraic general solution of a planar rational system.

4   Explicit Algebraic Solutions of Planar Rational System

This section is devoted to the problem of determinating explicitly an algebraic general solution of the planar rational system (3)

$$\begin{cases} s' = M(s, t) \\ t' = N(s, t) \end{cases}$$

where $M = M_1/M_2, N = N_1/N_2$ are rational functions in s and t. Whereas the problem of finding explicit algebraic solutions of planar rational systems has received only little attention in the literature, the problem of finding implicit algebraic solutions, or in other words, finding irreducible invariant algebraic curves and rational first integral, has been heavily studied. Some historical details and recent results which are helpful for our proofs will be recalled. By combining these results and the idea for finding algebraic general solutions of autonomous first order AODEs of Aroca et. al. (see [1]), we will present an algorithm for determining an algebraic general solution of a planar rational system with a given rational first integral.

Definition 7. *An algebraic curve defined by $G(s,t) = 0$ is called an invariant algebraic curve of the system (3) if $M_1 N_2 \frac{\partial G}{\partial s} + M_2 N_1 \frac{\partial G}{\partial t} = GH$ for some $H \in k[s,t]$. In this case, H is called the cofactor of G.*

Definition 8. *A differentiable function $W(s,t)$ in two variables s,t with coefficients over k is a first integral of the system (3) if it is not a constant function and $M \frac{\partial W}{\partial s} + N \frac{\partial W}{\partial t} = 0$. If, furthermore, W is a rational function, it is called a rational first integral.*

It is not hard to see that the set of all first integrals of the system (3) together with constant functions constitutes a field. The intersection of this field and $k(s,t)$ is exactly the set of all rational first integrals with constants in k. If the system has a rational first integral, there is a non-composite reduced rational function, say F, such that every rational first integral has the form $u(F(s,t))$ for some univariate rational function u with coefficients over k (see [2]). Such an F is unique up to a composition with a homography. In particular, instead of finding all rational first integrals, looking for a non-composite one is enough. In [2], Bostan et al. presented an efficient algorithm for computing a non-composite rational first integral with a degree bound of planar polynomial systems.

On the other hand, the set of rational first integrals, and all invariant algebraic curves of the system (3) does not change if we multiply the right hand side of the two differential equations of the system by the same non-zero rational function in $k(s,t)$. Therefore it suffices to consider planar polynomial systems for studying invariant algebraic curves and rational first integrals.

The following theorem is a classical result on the relation between irreducible invariant algebraic curves and rational first integrals of a planar rational system. We recall it here for technical purposes. For more details, the reader is referred to classical literature on rational first integrals, for instance [12].

Theorem 3. *There is a natural number N such that the planar rational system (3) has a rational first integral if and only if the system has more than N irreducible invariant algebraic curves. Furthermore, if $W = \frac{P}{Q}$ is a reduced rational first integral then every irreducible invariant algebraic curves is defined by an irreducible factor of $c_1 P - c_2 Q$, where c_1, c_2 are arbitrary constants.*

Next we will consider the problem of finding an explicit algebraic general solution $(s(x), t(x))$ of a planar rational system with a given irreducible invariant algebraic curve. The following property is a motivation for the idea.

Proposition 3. *If the parametrizable first order AODE $F(x,y,y') = 0$ has an algebraic general solution, then its associated system with respect to a proper parametrization has a rational first integral.*

Proof. If the differential equation (4) has an algebraic general solution, then so does its associated system. By applying Proposition 2, the associated system must have an irreducible invariant algebraic curve $G(s,t) = 0$ such that G is monic and at least one of the coefficients of G contains a constant which is transcendental over k. In the other words, the associated system has infinitely many irreducible invariant algebraic curves. Thus it has a rational first integral. $\square$

Theorem 4. *Assume that $W = \frac{P}{Q}$ is a reduced rational first integral of the system (3), and that $(s(x), t(x))$ is an algebraic solution in which not both $s(x)$ and $t(x)$ are constants. Then $(s(x), t(x))$ is an algebraic general solution if and only if $W(s(x), t(x))$ is a constant which is transcendental over k.*

Proof. Assume that $(s(x), t(x))$ is an algebraic general solution of the system (3), then $W'(s(x), t(x)) = 0$. Therefore $W(s(x), t(x)) = c$ is constant. If $c \in k$, then $P - cQ \in k[s,t]$ has an irreducible factor in $k[s,t]$ vanishing at $(s(x), t(x))$. This cannot happen due to Proposition 2. Hence $c \notin k$. Since k is algebraically closed, c is transcendental over k.

Conversely, assume that $(s(x), t(x))$ is a non-constant algebraic solution of the system (3) and such that $W(s(x), t(x)) = c$, where c is a constant being transcendental over k. Let G be an irreducible polynomial such that $G(s(x), t(x)) = 0$. Since $P - cQ$ also vanishes at $(s(x), t(x))$, G must be an irreducible factor of $P - cQ$. As in ([10], Chapter 3, Theorem 3.6), G has the form $A + \alpha B$ for some $A, B \in k[s,t]$, $B \neq 0$, and $\alpha \in \overline{k(c)}$ which is still transcendental over k.

Now let $H \in k(x)\{s,t\}$ be a differential polynomial such that $H(s(x), t(x)) = 0$. We denote $\tilde{H} := prem(H, \{\tilde{M}, \tilde{N}\})$ where $\tilde{M} := M_2 s' - M_1$ and $\tilde{N} := N_2 t' - N_1$. To finish the proof, we need to show that $\tilde{H} = 0$. It is clear that $\tilde{H} \in k(x)\{s,t\}$ and satisfies $\tilde{H}(s(x), t(x)) = 0$. Consider both $G = A + \alpha B$ and $\tilde{H}$ as polynomials in s, t with coefficients over $k(\alpha, x)$. Then they both vanish at along $(s(x), t(x))$, and G is, again, irreducible. Thus $\tilde{H}$ must be divisible by G. This is only possible in the case $\tilde{H} = 0$, because α is transcendental not only over k but also over $k(x)$. Hence $(s(x), t(x))$ is a general solution. $\square$

The following corollary is an immediately consequence of the theorem. It helps us to split a planar rational system into two autonomous first order AODEs, which lead us to the algorithm for determining explicit algebraic general solutions of the planar rational system.

Corollary 1. *Assume that $W = \frac{P}{Q}$ is a reduced rational first integral of the system (3), and that $(s(x), t(x))$ is an algebraic general solution. Then*

i. *$s(x)$ is an algebraic general solution of the autonomous first order AODE $F_1(s, s') = 0$, where $F_1(s, r) := res_t(P(s,t) - cQ(s,t), rM_2(s,t) - M_1(s,t))$*
ii. *$t(x)$ is an algebraic general solution of the autonomous first order AODE $F_2(t, t') = 0$, where $F_2(t, r) := res_s(P(s,t) - cQ(s,t), rN_2(s,t) - N_1(s,t))$.*

Fortunately, the problem of finding algebraic general solutions of autonomous first order AODEs is investigated. In [1], Aroca et. al. proposed a criterion to decide whether an autonomous first order AODE has an algebraic general solution and gave an algorithm to compute such a solution in the affirmative case. Combining the previous theorem and the corollary, together with the result of Aroca et. al., we arrive at Algorithm 1 for computing explicit algebraic general solutions of planar rational systems with a given rational first integral will be proposed next. For determining a rational first integral, one can use the package "RationalFirstIntegrals" which has been implemented by A. Bostan et. al. [2].

Algorithm 1. Solving planar rational system

Require: The planar rational system (3), and $W = \frac{P}{Q}$ a reduced rational first integral.
Ensure: An algebraic general solution $(s(x), t(x))$.
1: If $M = 0$, then $s(x) = c$ and $t(x)$ is an algebraic general solution of $t' = N(c, t)$
2: If $N = 0$, then $t(x) = c$ and $s(x)$ is an algebraic general solution of $s' = M(s, c_1)$
3: Compute $F_1 := res_t(P - c_1 Q, M_2(s, t)s' - M_1(s, t))$
4: $\mathcal{S} :=$ the set of all irreducible factors of F_1 in $\overline{k(c)}[s, s']$ containing s'
5: **for all** $H \in \mathcal{S}$ **do**
6:     **if** The differential equation $H(s, s') = 0$ has no algebraic solution **then**
7:         Return "No algebraic general solution."
8:     **else**
9:         $s(x) :=$ an algebraic solution of $H(s', s) = 0$
10:        $t(x) :=$ a solution of the equation $W(s(x), t) = c_1$
11:        **if** $s'(x) - M(s(x), t(x)) = t'(x) - N(s(x), t(x)) = 0$ **then**
12:            Return "$(s(x + c_2), t(x + c_2))$"
13:        **end if**
14:    **end if**
15: **end for**
16: Return "No algebraic general solution"

Example 1. Consider the planar rational system

$$\begin{cases} s' = t \\ t' = \frac{t^2}{2s} \end{cases} \tag{6}$$

By multiplying the right hand sides of the two differential equations of the system by $\frac{2s}{t}$, we obtain a new system which shares the same set of rational first integrals and invariant algebraic curves:

$$\begin{cases} s' = 2s \\ t' = t \end{cases}$$

Using the package *RationalFirstIntegrals* of A. Bostan et. al. (see [2]), we can determine a non-composite rational first integral of the last system, for instance $W = \frac{-64s}{100s - t^2}$. W is also a non-composite rational first integral of the system (6). Now we can use Algorithm 1 to find an algebraic general solution of the system (6). First we set $F_1(s, r) := res_t(r - t, -64s - c(100s - t^2)) = (64 + 100c)s - cr$.

Solving the differential equation $F_1(s, s') = 0$ (by using the algorithm of Aroca et al., or just by integrating) yields the algebraic solution $s(x) := \frac{1}{c}(16 + 25c)x^2$. Next, we find $t(x)$ by solving the algebraic equation $W(s(x), t) = c$. It gives two candidates $\pm\frac{1}{c}(16 + 25c)x$. By substituting into the system (6), we conlude that only $(s(x), t(x)) := \left(\frac{1}{c}(16 + 25c)x^2, \frac{1}{c}(16 + 25c)x\right)$ is an algebraic solution.

Example 2. Consider the planar rational system

$$\begin{cases} s' = \frac{t^2}{s} \\ t' = \frac{t^3}{2s^2-1} \end{cases} \tag{7}$$

A rational first integral, for intance $W = \frac{s^2-1}{t^4}$, can be found by a process similiar to the one in the previous example. Let $F_1(s, r) := res_t(sr - t^2, s^2 - 1 - ct^4) = (cs^2r^2 - s^2 + 1)^2$. Solving the differential equation $F_1(s, s') = 0$ yields algebraic solution $s(x) := \pm\sqrt{\frac{x^2}{c} + 1}$. Next, we find $t(x)$ by solving the algebraic equation $W(s(x), t) = c$. Therefore $t(x) = \pm\sqrt{\frac{x}{c}}$. In this example, $\left(\pm\sqrt{\frac{x^2}{c} + 1}, \pm\sqrt{\frac{x}{c}}\right)$ are algebraic solutions.

5   Algebraic General Solutions with Bounded Degree of Parametrizable First Order AODEs

In this section, we will combine these results to study further the problem of finding an algebraic general solution of the parametrizable first order AODE (2). In particular, given a parametrizable first order AODE and a positive integer n, an algorithm for finding an algebraic general solution whose minimal polynomial has total degree less than or equal to n will be presented.

Consider the parametrizable first order AODE (2)

$$F(x, y, y') = 0$$

with a given proper parametrization $\mathcal{P}(s, t) := (\chi_1(s, t), \chi_2(s, t), \chi_3(s, t))$ for some $\chi_1, \chi_2, \chi_3 \in k(s, t)$. Assume that $y = y(x) \in \overline{K(x)}$ is an algebraic general solution of the differential equation (2), where K is a transcendental field extension of k. Let $Y(x, y) \in K[x, y]$ be a minimal polynomial of $y(x)$. As we mentioned in Section 3, we sometimes call $Y(x, y) = 0$ an algebraic general solution of the differential equation (2) instead of $y(x)$.

By $\deg(Y)$ and $\deg_x(Y)$ we denote the total degree of $Y(x, y)$ and the partial degree of Y w.r.t. x, respectively.

Theorem 5. *With notation as above, and assume that*

$$(\sigma_1(x, y, z), \sigma_2(x, y, z)) := \mathcal{P}^{-1}(x, y, z)$$

the inverse map of $\mathcal{P}$. If the differential equation (2) has an algebraic general solution $Y(x, y) = 0$ with $\deg(Y) \leq n$, then the associated system has a rational first integral whose total degree is less than or equal to $m :=$ $n^3\left(\deg_x(\sigma_1) + \deg_y(\sigma_1) + \deg_x(\sigma_2) + \deg_y(\sigma_2)\right) + 2n^2\left(\deg_z(\sigma_1) + \deg_z(\sigma_2)\right)$.

Proof. Denote $s(x) := \sigma_1(x, y(x), y'(x))$ and $t(x) := \sigma_2(x, y(x), y'(x))$, then the pair $(s(x), t(x))$ is an algebraic general solution of the associated system. Let $G(s, t) \in K[s, t]$ be an irreducible polynomial such that $G(s(x), t(x)) = 0$. $G(s, t) = 0$ is in fact an irreducible invariant algebraic curve of the associated system with coefficients in K. We will first claim that $\deg(G) \leq m$.

Denote

$$Q_1(x, y) := \sigma_1\left(x, y, -\frac{\frac{\partial}{\partial x}Y(x, y)}{\frac{\partial}{\partial y}Y(x, y)}\right) \quad \text{and} \quad Q_2(x, y) := \sigma_2\left(x, y, -\frac{\frac{\partial}{\partial x}Y(x, y)}{\frac{\partial}{\partial y}Y(x, y)}\right)$$

which are rational functions in $k(x, y)$. Then we have $s(x) = Q_1(x, y(x))$ and $t(x) = Q_2(x, y(x))$. The degree of x and y in Q_1 and Q_2 can be bounded in terms of σ_1, σ_2 and $n = \deg(Y)$ as follows:

$$\deg_x(Q_1) \leq n.\deg_x(\sigma_1) + \deg_z(\sigma_1) \tag{8}$$

$$\deg_y(Q_1) \leq n.\deg_y(\sigma_1) + \deg_z(\sigma_1) \tag{9}$$

$$\deg_x(Q_2) \leq n.\deg_x(\sigma_2) + \deg_z(\sigma_2) \tag{10}$$

$$\deg_y(Q_2) \leq n.\deg_y(\sigma_2) + \deg_z(\sigma_2) \tag{11}$$

Now in order to get minimal polynomials of $s(x)$ and $t(x)$, using the resultant is a fast way. In particular, minimal polynomials of $s(x)$ and $t(x)$ divide

$$R_1(s, x) := res_y(numer(Q_1) - s.denom(Q_1), Y(x, y))$$

$$R_2(s, x) := res_y(numer(Q_2) - t.denom(Q_2), Y(x, y))$$

respectively, where $numer(Q_1)$ is the numerator of Q_1 and $denom(Q_1)$ the denominator one. Therefore $H(s, t) := res_x(R_1(s, x), R_2(t, x))$ is a polynomial in $K[s, t]$ satisfying $H(s(x), t(x)) = 0$. It implies that G must be divide H. From the definition of the resultant, one can determine immediately an upper bound for the total degree of H, and thus of G. In fact, $\deg_s(H) \leq \deg_s(R_1).\deg_x(R_2) \leq N^2(\deg_x(Q_2) + \deg_y(Q_2))$. Equivalently, we also have $\deg_t(H) \leq N^2(\deg_x(Q_1) + \deg_y(Q_1))$. Combining with (8), (9),(10) and (11) yields $\deg(G) \leq m$.

Moreover, since $(s(x), t(x))$ is an algebraic general solution, $G(s, t) = 0$ can be seen as the class of all irreducible invariant algebraic curves of the associated system. Therefore its degree bound is also a degree bound for the non-composite rational first integral. This concludes the proof.   □

As an immediate consequence, the theorem leads us to Algorithm 2 for finding an algebraic general solution $Y(x, y) = 0$ with $\deg(Y) \leq n$ of the differential equation (2).

Algorithm 2. Finding an algebraic general solution with total degree $\leq n$ of a parametrizable first order AODE

Require: The differential equation (2) with a proper parametrization $\mathcal{P}$; a positive integer n.

Ensure: An algebraic general solution $Y(x,y) = 0$ such that $\deg(Y) \leq n$.

1: $(\sigma_1, \sigma_2) := \mathcal{P}^{-1}$

2: $m := n^3 \left(\deg_x(\sigma_1) + \deg_y(\sigma_1) + \deg_x(\sigma_2) + \deg_y(\sigma_2) \right) + 2n^2 \left(\deg_z(\sigma_1) + \deg_z(\sigma_2) \right)$

3: $M(s,t) := \dfrac{\chi_3(s,t)\frac{\partial}{\partial t}\chi_1(s,t) - \frac{\partial}{\partial t}\chi_2(s,t)}{\frac{\partial}{\partial s}\chi_1(s,t)\frac{\partial}{\partial t}\chi_2 1(s,t) - \frac{\partial}{\partial t}\chi_1(s,t)\frac{\partial}{\partial s}\chi_2(s,t)}$

4: $N(s,t) := \dfrac{\frac{\partial}{\partial s}\chi_2(s,t) - \chi_3(s,t)\frac{\partial}{\partial s}\chi_1(s,t)}{\frac{\partial}{\partial s}\chi_1(s,t)\frac{\partial}{\partial t}\chi_2 1(s,t) - \frac{\partial}{\partial t}\chi_1(s,t)\frac{\partial}{\partial s}\chi_2(s,t)}$

5: If the system $\{s' = M(s,t); t' = N(s,t)\}$ has no rational first integral of total degree at most m, then return "No algebraic general solution of total degree at most n"

6: $W :=$ a rational first integral of degree at most m of the system; the system may be solved by Algorithm 1

7: If the system has no algebraic general solution, then return "No algebraic general solution of total degree at most n"

8: $(s(x), t(x)) :=$ an algebraic general solution of the system

9: $y(x) := \chi_2(s(2x - \chi_1(s(x),t(x))), t(2x - \chi_1(s(x),t(x))))$

10: $Y(x,y) :=$ a minimal polynomial of $y(x)$

11: If $\deg Y > n$, then return "No algebraic general solution of total order at most n"

12: Return "$Y(x,y) = 0$".

Example 3. Consider the differential equation

$$4x(x - y)y'^2 + 2xyy' - 5x^2 + 4xy - y^2 = 0 \tag{12}$$

The solution surface of the differential equation (12) is rational. It is parametrized by the rational map

$$P(s,t) := \left(s, -\frac{t^2 - 5ts + 5s^2}{s}, \frac{t^2 - 4st + 5s^2}{2s(t - 2s)} \right)$$

The associated system of the differential equation (12) with respect to $\mathcal{P}$ is

$$\begin{cases} s' = 1 \\ t' = \frac{t^2 - 3s^2}{2s(t - 2s)} \end{cases}$$

The inverse map of the parametrization is $(\sigma_1(x,y,z), \sigma_2(x,y,z)) := (x, \frac{4xz - y}{2z - 1})$. Therefore if we look for an algebraic general solution $Y(x,y) = 0$ with $\deg(Y) \leq 2$, we need to find a rational first integral of total degree at most 32 of the associated system. In this case, the associated system has the rational first integral $W = \frac{t^2 - 4st + 3s^2}{s}$ which is of total degree 2. (This suggests that our degree bound is not optimal). Thus it has an algebraic solution $(s(x), t(x)) := (x, \bar{t}(x,c))$, where $\bar{t}(x,c)$ is a root of the algebraic equation $t^2 - 4xt + 3x^2 - cx = 0$. By applying Theorem 2,

$$y(x) = \frac{\sqrt{cx(cx + 1)} - 1}{c}$$

is an algebraic general solution of the differential equation (12).

6   Conclusion

We have studied an algorithmic way to solve planar rational systems whenever they are given together with a rational first integral. A one-to-one relation between algebraic general solutions of parametrizable first order AODEs and their associated systems has been established. This has led to an algorithm for finding an algebraic general solution $Y(x, y) = 0$ with $\deg(Y) \leq n$, with a given positive integer n, of a parametrizable first order AODE.

Although Algorithm 2 is complete, the degree bound for the rational first integral derived from n is probably not optimal. Finding a tighter degree bound would be an interesting problem.

References

1. Aroca, J.M., Cano, J., Feng, R., Gao, X.-S.: Algebraic general solutions of algebraic ordinary differential equations. In: Kauers, M. (ed.) ISSAC 2005, pp. 29–36. ACM Press, New York (2005)
2. Bostan, A., Chèze, G., Cluzeau, T., Weil, J.A.: Efficient algorithms for computing rational first integrals and Darboux polynomials of planar vector fields. arXiv (2013)
3. Carnicer, M.M.: The Poincar Problem in the Nondicritical Case. Ann. of Math. **140**(2), 289294 (1994)
4. Feng, R., Gao, X.-S.: Rational general solution of algebraic ordinary differential equations. In: Gutierrez, J. (ed.) ISSAC 2004, pp. 155–162. ACM Press, New York (2004)
5. Feng, R., Gao, X.-S.: A polynomial time algorithm for finding rational general solutions of first order autonomous ODEs. Journal of Symbolic Computation **41**(7), 739–762 (2006)
6. Kolchin, E.R.: Differential Algebra and Algebraic Groups. Pure and Applied Mathematics, vol. 54. Academic Press, New York (1973)
7. Ngô, L.X.C., Winkler, F.: Rational general solutions of first order non-autonomous parametrizable ODEs. Journal of Symbolic Computation **45**(12), 1426–1441 (2010)
8. Ngô, L.X.C., Winkler, F.: Rational general solutions of planar rational systems of autonomous ODEs. Journal of Symbolic Computation **46**(10), 1173–1186 (2011)
9. Ritt, J.F.: Differential Algebra. Dover Publications Inc., New York (1955)
10. Schinzel, A.: Polynomials with Special Regard to Reducibility. Cambridge University Press (2000)
11. Sendra, J.R., Winkler, F., Pérez-Díaz, S.: Rational Algebraic Curves, A Computer Algebra Approach. Algorithms and Computation in Mathematics, vol. 22. Springer-Verlag, Heidelberg (2008)
12. Singer, M.F.: Liouvillian first integrals of differential equations. Transaction of the American Mathematics Society **333**(2), 673–688 (1992)

Author Index

Printed in the United States
By Bookmasters